Selected Titles in This Series

(Continued in the back of this publication)

Mirror Symmetry and Algebraic Geometry

Mathematical
Surveys
and
Monographs

Volume 68

Mirror Symmetry and Algebraic Geometry

David A. Cox

Sheldon Katz

2010 *Mathematics Subject Classification.* Primary 14-02; Secondary 81-02.

ABSTRACT. This monograph is an introduction to the mathematics of mirror symmetry, with a special emphasis on its algebro-geometric aspects. Topics covered include the quintic threefold, toric geometry, Hodge theory, complex and Kähler moduli, Gromov-Witten invariants, quantum cohomology, localization in equivariant cohomology, and the recent work of Lian-Liu-Yau and Givental on the Mirror Theorem. The book is written for algebraic geometers and graduate students who want to learn about mirror symmetry. It is also a reference for specialists in the field and background reading for physicists who want to see the mathematical underpinnings of the subject.

Library of Congress Cataloging-in-Publication Data

Cox, David A.
    Mirror symmetry and algebraic geometry / David A. Cox, Sheldon Katz.
       p. cm. — (Mathematical surveys and monographs, 0076-5376; v. 68)
    Includes bibliographical references and index.
    ISBN 0-8218-1059-6 (alk. paper)
    1. Mirror symmetry. 2. Geometry, Algebraic. 3. Mathematical physics. I. Katz, Sheldon, 1956– II. Title. III. Series: Mathematical surveys and monographs; no. 68.

QC174.17.S9 C69   1999
516.3′62—dc21

98-55564
CIP

AMS softcover ISBN: 978-0-8218-2127-5

*To our families, who have endured with good grace our preoccupation
with this book.*

D. A. C.

S. K.

Contents

Preface

The field of mirror symmetry has exploded onto the mathematical scene in recent years. This is a part of an increasing connection between quantum field theory and many branches of mathematics.

It has sometimes been said that quantum field theory combines 20th century physics with 21st century mathematics. Physicists have gained much experience with mathematical manipulations in situations which have not yet been mathematically justified. They are able to do this in part because experiment can help them differentiate between which manipulations are feasible, and which are clearly wrong. Those manipulations that survive all known tests are presumed to be valid until evidence emerges to the contrary.

Based on this evidence, physicists are confident about the validity of mirror symmetry. One of the tools they use with great virtuosity is the Feynman path integral, which performs integration with complex measures over infinite dimensional spaces, such as the space of C^∞ maps from a Riemann surface to a Calabi-Yau threefold. This is not rigorous mathematics, yet these methods led to the 1991 paper of Candelas, de la Ossa, Green and Parkes [**CdGP**] containing some astonishing predictions about rational curves on the quintic threefold. These predictions went far beyond anything algebraic geometry could prove at the time.

The challenge for mathematicians was to understand what was going on and, more importantly, to prove some of the predictions made by the physicists. In this book, we will see that algebraic geometers have made substantial progress, though there is still a long way to go. The process of creating a mathematical foundation for aspects of mirror symmetry has given impetus to new fields of algebraic geometry. Examples include quantum cohomology, Kontsevich's definition of a stable map, the complexified Kähler moduli space of a Calabi-Yau threefold, Batyrev's duality between certain toric varieties, and Givental's notion of quantum differential equations. Mirror symmetry has also led to advances in deformation theory leading to the theory of the virtual fundamental class, as well as a previously unknown connection between algebraic and symplectic deformation theory. Even though we still don't know what mirror symmetry really "is", the predictions that mirror symmetry makes about Gromov-Witten invariants can now be proved mathematically in many cases.

Goal of the Book

Perhaps the greatest obstacle facing a mathematician who wants to learn about mirror symmetry is knowing where to start. Currently, many references are scattered throughout journals, and many mathematical ideas exist solely in the physics literature, which is difficult for mathematicians to read. Our primary goal is to give an introduction to the algebro-geometric aspects of mirror symmetry. We include sufficient detail so that the reader will have the major ideas and definitions spelled out, and explicit references to the literature when space constraints prohibit more detail. We explain both the rigorous mathematics as well as the intuitions borrowed from physics which are not yet theorems. We do this because we have two primary target audiences in mind: mathematicians wanting to learn about mirror symmetry, and physicists who know about mirror symmetry wanting to learn about the mathematical aspects of the subject.

Mirror symmetry is connected to several branches of mathematics (and there are even broader connections between physical theories in various dimensions and many areas of mathematics). We focus on the connection between algebraic geometry and mirror symmetry, although we discuss closely related areas such as symplectic geometry. By restricting our focus in this way, we hope to give a reasonably self-contained introduction to the subject.

The book begins with a general introduction to the ideas of mirror symmetry in Chapter 1. Then Chapter 2 discusses the quintic threefold and explains how mirror symmetry leads to the enumerative predictions of [**CdGP**]. Chapter 3 reviews toric geometry, and Chapter 4 describes mirror constructions due to Batyrev, Batyrev-Borisov, and Voisin-Borcea. The next four chapters (Chapters 5, 6, 7 and 8) flesh out the mathematics needed to formulate a precise version of mirror symmetry. These chapters cover maximally unipotent monodromy, Yukawa couplings, complex and Kähler moduli, the mirror map, Gromov-Witten invariants, and quantum cohomology. This will enable us to state some Mirror Conjectures at the end of Chapter 8.

The next three chapters (Chapters 9, 10 and 11) are dedicated to proving some instances of mirror symmetry. Equivariant cohomology and localization play a crucial role in the proofs, so that these are reviewed in Chapter 9. These methods also give powerful tools for computing Gromov-Witten invariants. In order to explain Givental's approach to the Mirror Theorem, we need the gravitational correlators and quantum differential equations discussed in Chapter 10. Finally, Chapter 11 describes the work of Lian, Liu and Yau [**LLY**] and Givental [**Givental2, Givental4**] on the Mirror Theorem.

The mathematics discussed in Chapters 1–11 is wonderful but highly nontrivial. Later in the preface we will give some guidance for how to read these chapters.

The book concludes with Chapter 12, which brings together all of the open problems mentioned in earlier chapters and discusses some of the many aspects of mirror symmetry not covered in the text. Finally, there are appendices on singular varieties and physical theories.

We tried to make the bibliography fairly complete, but it has been difficult to keep up with the amazing number of high-quality papers being written on mirror symmetry and related subjects. We apologize to our colleagues for the many recent papers not listed in the bibliography.

Relation to Physics

For mathematicians, one frustration of mirror symmetry is the difficulty of getting insight into the physicist's intuition. There is no question of the power of this intuition, for it is what led to the discovery of the mirror phenomenon. But getting access to it requires a substantial study of quantum field theory. A glance at Appendix B, which discusses some of the physical theories involved, will indicate the magnitude of this task. Understanding the physics literature on mirror symmetry requires an extensive background, more than provided in this book. Appendix B has the more modest goal of introducing the reader to some of the topics in the physics literature which are relevant to mirror symmetry.

While this book was written to address the mathematics of mirror symmetry, we also hope to show how the mathematics reflects the spirit of the physics. With this thought in mind, we begin Chapter 1 with a discussion of the physics which led to mirror symmetry. We use terminology from physics freely, though we don't assume that the reader knows any quantum field theory. The idea is to convey the sense that mirror symmetry is completely natural from the point of view of certain conformal field theories. This is the most "physical" chapter of the book. Subsequent chapters will concentrate on the mathematics, though we will pause occasionally to comment on the relationship between the mathematics and the physics.

An important aspect of the role of physics is that mathematically sophisticated physicists helped discover the mathematical foundation for mirror symmetry. Algebraic geometers can take pride in the wonderful theories they created to explain parts of mirror symmetry, but at the same time we should also recognize that physicists provided more than just predictions—they often suggested the appropriate objects to study, accompanied in some cases by mathematically rigorous descriptions. This will become clear by checking the references given in the text— a surprising number, even in the purely mathematical parts of the book, refer to physics papers. There is no question of the debt we owe to our colleagues in physics.

How to Read the Book

Mirror symmetry is a wonderful story, but its telling requires lots of details in many different areas of algebraic geometry. It is easy to get lost, especially if you try to read the book cover-to-cover. Fortunately, this isn't the only way to read the book.

Our basic suggestion is that you should begin with Chapters 1 and 2. As already mentioned, Chapter 1 explains some of the physics, and it also introduces two key ideas, the A-model of a Calabi-Yau manifold V, which encodes the enumerative information we want, and the B-model of the mirror V°, which we can compute using Hodge theory. Then Chapter 2 shows what this looks like in the case of the quintic threefold $V \subset \mathbb{P}^4$ and in the process derives the enumerative predictions made in [**CdGP**]. This chapter ends with a preview of the proofs of mirror symmetry from Chapter 11.

After reading the first two chapters, there are various ways you can proceed, depending on your mathematical interests and expertise. To help you choose, here is a description of some of the highlights of the remaining chapters:

Chapter 3. Readers familiar with toric geometry can skip most of this chapter. Section 3.5 introduces reflexive polytopes, which are used in the Batyrev mirror construction.

Chapter 4. Section 4.1 describes the Batyrev mirror construction and gives some evidence for the mirror relation. Section 4.2 explains how this applies to the quintic threefold. K3 surfaces are used in Section 4.4 to construct some interesting mirror pairs of Calabi-Yau threefolds.

Chapter 5. Maximally unipotent monodromy is an important part of mirror symmetry and is defined in Section 5.2. Readers interested in computational techniques for projective hypersurfaces, toric hypersurfaces and hypergeometric equations should look at Sections 5.3, 5.4 and 5.5, while those interested in the Hodge theory of Calabi-Yau threefolds should read Section 5.6 very carefully.

Chapter 6. We consider complex moduli in Section 6.1 and Kähler moduli in Section 6.2. The two discussions are interwoven because of the relation between the two predicted by mirror symmetry. The main example we work out concerns toric hypersurfaces, so that the reader will need the Batyrev mirror construction from Chapter 4. Readers interested in moduli of Calabi-Yau manifolds, Kähler cones, and the global aspects of mirror symmetry will want to read these sections carefully. Section 6.3 discusses the mirror map and has more on hypergeometric equations, which are used to construct the mirror map in the toric case.

Chapter 7. With the exception of some examples, Chapter 7 is independent of the earlier chapters. The main objects of study are Gromov-Witten invariants. Sections 7.1.1, 7.1.2 and 7.3.1 are essential reading. Otherwise:

- The discussion of the virtual fundamental class in Sections 7.1.3–7.1.6 is more technical and can be skipped at the first reading. The one exception is Example 7.1.6.1, which gives an important formula for some Gromov-Witten invariants of the quintic threefold. The virtual fundamental class is used in various places in Chapters 9, 10 and 11.
- Readers interested in symplectic geometry should read Sections 7.2 and 7.4.4 carefully.
- Readers interested in enumerative geometry will want to look at Section 7.4. Some of the examples given here will be revisited in Chapter 8.

One surprise in Section 7.4.4 is the subtle relation between the instanton number n_{10} and the number of degree 10 rational curves on the quintic threefold.

Chapter 8. This chapter uses the Gromov-Witten invariants of Chapter 7 to define the two flavors of quantum cohomology, small and big. Everyone should read Section 8.1.1 for the small quantum product and Section 8.1.2 for some examples. Also, some knowledge of the Gromov-Witten potential is also useful. This is covered in Sections 8.2.2, 8.3.1 and 8.3.3. Then:

- Readers interested in enumerative geometry should read Sections 8.1, 8.2 and 8.3 carefully.
- Readers interested in Hodge theory will want to look at Section 8.5, which uses quantum cohomology to construct the A-variation of Hodge structure on the cohomology of a Calabi-Yau manifold.

A highlight of the chapter is Section 8.6, which formulates various Hodge-theoretic versions of the mirror conjecture.

Chapter 9. Sections 9.1 and 9.2.1 are required reading for anyone wanting to understand the proofs of mirror symmetry given in Chapter 11. This especially includes Example 9.2.1.3, which computes the Gromov-Witten invariant $\langle I_{0,0,d} \rangle$ using an equivariant version of the formula given in Example 7.1.6.1. Sections 9.2.2 and 9.2.3 prove some of the assertions about Gromov-Witten invariants of Calabi-Yau threefolds made in Section 7.4.4 and require a detailed knowledge of the virtual fundamental class.

Chapter 10. Readers only interested in the [**LLY**] approach to the Mirror Theorem can skip this chapter. For Givental's approach, however, the reader will need to read about gravitational correlators (Section 10.1.1–10.1.3), flat sections of the Givental connection (Section 10.2.1), and the J-function and quantum differential equations (Section 10.3). For readers with an interest in Hodge theory, the A-model variation of Hodge structure is discussed in Sections 10.2.2 and 10.3.2. This leads to a nice connection between Picard-Fuchs operators and relations in quantum cohomology.

Chapter 11. Here we discuss the recent proofs of the Mirror Theorem. There are two approaches to consider:

- The [**LLY**] approach to the Mirror Theorem for the quintic threefold is covered in Section 11.1. This requires knowledge of the essential sections of Chapters 7, 8 and 9 mentioned earlier.
- For Givental's version of the Mirror Theorem, one needs in addition the sections of Chapter 10 indicated above. In Sections 11.2.1, 11.2.3 and 11.2.4, we discuss the Mirror Theorem for nef complete intersections in $\mathbb{P}^n$, and then in Section 11.2.5 we consider what happens when the ambient space is a smooth toric variety.

In particular, we explain how both of these approaches prove all of the predictions for the quintic threefold made in Chapter 2. Another very interesting case is presented in Example 11.2.5.1, which concerns Calabi-Yau threefolds which are minimal desingularizations of degree 8 hypersurfaces in $\mathbb{P}(1,1,2,2,2)$. This example of a toric hypersurface makes numerous appearances in Chapters 3, 5, 6 and 8, so that the reader will need to look back at these earlier examples in order to fully appreciate what we do in Example 11.2.5.1.

We should also mention that in many cases, our proofs are not complete, for we often refer to the literature for certain details of the argument. The same applies to background material. For some topics (such as equivariant cohomology in Chapter 9) we review the basic facts, while for others (such as algebraic stacks in Chapter 7) we give references to the literature. We hope that this unevenness in the level is not too unsettling to the reader.

Acknowledgements

We are happy to thank numerous people who have assisted us in various ways. The authors would in particular like to express their appreciation to P. Aspinwall, P. Berglund, C. Borcea, P. Candelas, R. Dijkgraaf, B. Greene, T. Hübsch, A. Klemm, P. Mayr, G. Moore, D. R. Morrison, M. R. Plesser, D. van Straten, and C. Vafa, for numerous conversations over the past few years which helped us develop our views of mirror symmetry and understand the relevant mathematics and physics.

Thanks also to the Institut Mittag-Leffler for its hospitality during the spring of 1997. A seminar on mirror symmetry held that spring discussed portions of this book and helped us improve its exposition immensely. It is a pleasure to thank all of the participants. At the risk of inadvertently omitting someone, we in particular want to acknowledge P. Aluffi, V. Batyrov, K. Behrend, P. Belorousski, I. Ciocan Fontanine, A. Elezi, G. Ellingsrud, C. Faber, B. Fantechi, W. Fulton, E. Getzler, L. Göttsche, T. Graber, B. Kim, A. Kresch, D. Laksov, J. Li, A. Libgober,

R. Pandharipande, K. Ranestad, E. Rødland, D. E. Sommervoll, S. A. Strømme, M. Thaddeus, and E. Tjøtta for their help.

During the process of writing the book, we sent draft chapters to various colleagues. In addition to some of the above people who helped with this, we also would like to thank E. Cattani, E. Ionel, B. Lian, K. Liu, A. Mavlyutov, T. Parker, G. Pearlstein, Y. Ruan, C. Taubes, G. Tian, P. Vermeire and S.-T. Yau for their helpful comments. We would also like to thank B. Dwork and J. Kollár for useful conversations. In addition, we would like to thank the reviewers of an earlier draft of the book for their feedback. We are also very grateful to A. Greenspoon for his careful reading of the final manuscript. Any errors which remain are, of course, our responsibility.

We also thank the seminar members at Oklahoma State University and the Valley Geometry Seminar for listening to many talks on mirror symmetry. Their feedback and support are greatly appreciated.

During the period when this book was being written, D.A.C. was partially supported by the NSF, and S.K. was supported by the NSF and NSA. We are grateful to these organizations for their support.

Finally, we would like to thank Sergei Gelfand and Sarah Donnelly of the AMS for their editorial guidance and Barbara Beeton of the AMS technical support staff for her help with $\mathcal{A}_{\mathcal{M}}\mathcal{S}$-LaTeX.

Our Hope

Mirror symmetry is an active area of research in algebraic geometry, with plenty to keep mathematicians busy for many years. This book should be regarded as a preliminary report on the current state of the subject—the definitive text on mirror symmetry has yet to be written. Nevertheless, we hope that the reader will find the connections explored here to be an exciting and continuing story.

November, 1998

David A. Cox
Sheldon Katz

Notation

Here is some of the notation we will use in the book.

General Notation:

M	Complex or symplectic manifold
X and V	Algebraic variety and Calabi-Yau variety
Θ_X and T_X	Tangent sheaf and bundle of X
ω_X and K_X	Canonical bundle and divisor of X
$\widehat{\Omega}^p_X$	Zariski p-forms on an orbifold X
$M(X)$	Mori cone of effective 1-cycles on X
$A_k(X)$	k^{th} Chow group of X
$\langle a, b \rangle$ or $g(a, b)$	Cup product pairing on cohomology
$\mathbb{P}^n$ and $\mathbb{P}(q_0, \dots, q_n)$	Projective space and weighted projective space

Toric Varieties and Polytopes (Chapter 3):

Σ and $\Sigma(1)$	Fan and its 1-dimensional cones
Δ and Δ°	Polytope and its polar (or dual) polytope
$L(\Delta)$	Laurent polynomials with exponents in Δ
X_Σ and $\mathbb{P}_\Delta$	Toric variety of the fan Σ and the polytope Δ
$\mathrm{cpl}(\Sigma)$	Cone of convex support functions on Σ

Hodge Theory and Yukawa Couplings (Chapter 5):

$\nabla = \nabla^{\text{GM}}$	Gauss-Manin connection
$\mathcal{F}^\bullet$ and $\mathcal{W}_\bullet$	Hodge and weight filtrations
$\mathcal{H}$ and $\mathcal{H}_\mathbb{Z}$	Hodge bundle $\mathcal{F}^0$ and its integer subsheaf
$\mathcal{T}_j$ and N_j	Monodromy transformation and its logarithm
$\langle \frac{\partial}{\partial t_i}, \frac{\partial}{\partial t_j}, \frac{\partial}{\partial t_k} \rangle - Y_{ijk}$	Normalized Yukawa coupling or B model correlation function

Complex and Kähler Moduli (Chapter 6):

$\mathcal{M}$ and $\overline{\mathcal{M}}$	Complex moduli and a compactification
$\mathcal{M}_{\mathrm{poly}}$ and $\overline{\mathcal{M}}_{\mathrm{poly}}$	Polynomial moduli and a compactification
$\mathcal{M}_{\mathrm{simp}}$ and $\overline{\mathcal{M}}_{\mathrm{simp}}$	Simplified moduli and a compactification
$K(V)$ and $K(V)_{\mathbb{C}}$	Kähler cone and complexified Kähler space
$\mathcal{KM}$ and $\overline{\mathcal{KM}}$	Kähler moduli and a compactification
$\mathcal{KM}_{\mathrm{toric}}$ and $\overline{\mathcal{KM}}_{\mathrm{toric}}$	Toric Kähler moduli and a compactification

Gromov-Witten Invariants and Quantum Cohomology (Chapters 7 and 8):

$\overline{M}_{g,n}(X,\beta)$ and $\overline{\mathcal{M}}_{g,n}(X,\beta)$	Coarse moduli space and fine moduli stack of n-pointed genus g stable maps of class β
$[\overline{M}_{g,n}(X,\beta)]^{\mathrm{virt}}$	Virtual fundamental class
$I_{g,n,\beta}(\alpha_1,\dots,\alpha_n)$ and $\langle I_{g,n,\beta}\rangle(\alpha_1,\dots,\alpha_n)$	Gromov-Witten class and invariant
n_β and n_d	Instanton number in general and for the quintic threefold
$N_\beta = \langle I_{0,0,\beta}\rangle$ and $N_d = \langle I_{0,0,d}\rangle$	0-pointed genus 0 Gromov-Witten invariant in general and for the quintic threefold
$*_{\mathrm{small}}$ and $*$	Small and big quantum product ($*_{\mathrm{small}}$ is sometimes denoted $*$)
$\langle a,b,c\rangle$	Three-point function or A-model correlation function
$\Phi(\gamma)$	Genus 0 Gromov-Witten potential
$\nabla = \nabla^{\mathrm{GW}}$	A-model connection

Chern Classes and Equivariant Cohomology (Chapter 9):

$c_i(\mathcal{E})$ and $c_i^T(\mathcal{E})$	Chern class and equivariant Chern class of $\mathcal{E}$
$\mathrm{Euler}(\mathcal{E})$ and $\mathrm{Euler}_T(\mathcal{E})$	Euler class and equivariant Euler class of $\mathcal{E}$
$\int_{X_T}$	Equivariant integral
$\lambda_0,\dots,\lambda_n$ and $\hbar$	Ring generators of $H^*(BT)$ and $H^*(B\mathbb{C}^*)$, which generate $H^*(BG)$ for $G = \mathbb{C}^* \times T$, $T = (\mathbb{C}^*)^{n+1}$
H and p	Hyperplane class and equivariant hyperplane class
$q_0,\dots,q_n$	Fixed points of standard T-action on $\mathbb{P}^n$

Gravitational Correlators and the J-Function (Chapter 10):

$\langle \tau_{d_1}\gamma_1, \ldots, \tau_{d_n}\gamma_n \rangle_{g,\beta}$	Gravitational correlator
$\langle \gamma_1, \ldots, \gamma_n \rangle_{g,\beta}$	Alternate notation for $\langle I_{g,n,\beta}\rangle(\gamma_1,\ldots,\gamma_n)$
$\langle\langle \tau_{d_1}\gamma_1, \ldots, \tau_{d_n}\gamma_n \rangle\rangle_g$	Genus g gravitational coupling
$\Phi_g^{\mathrm{grav}}(\gamma)$	Genus g gravitational Gromov-Witten potential
$*_{\mathrm{grav}}$	Gravitational quantum product
∇^g and $\widetilde{\nabla}^g$	Givental connection and dual Givental connection
s_j	Flat section of the Givental connection
$\langle\langle \frac{T_a}{\hbar-c} \rangle\rangle_0$	Symbolic notation for $\sum_{n=0}^{\infty} \hbar^{-(n+1)}\langle\langle \tau_n T_a, T_j \rangle\rangle_0$
$J = J_X$	Givental J-function of the smooth variety X

The Mirror Theorem (Chapter 11):

$\mathcal{V} = \oplus_{i=1}^{\ell}\mathcal{O}_{\mathbb{P}^n}(a_i)$	Vector bundle used in the Mirror Theorem
$\mathcal{V}_d$ and $\mathcal{V}_{d,k}$	Bundle on $\overline{M}_{0,0}(\mathbb{P}^n, d)$ and $\overline{M}_{0,k}(\mathbb{P}^n, d)$ induced by $\mathcal{V}$
M_d and N_d	Compact notation for $\overline{M}_{0,0}(\mathbb{P}^1 \times \mathbb{P}^n, (1,d))$ and $\mathbb{P}(H^0(\mathbb{P}^1, \mathcal{O}_{\mathbb{P}^1}(d))^{d+1})$
$p_{i,r}$	Fixed points of G-action on N_d, $G = \mathbb{C}^* \times T$
κ	Equivariant hyperplane class in $H_G^*(N_d)$
$\mathcal{R}_T$ and $\mathcal{R}_G$	Field of fractions of $H^*(BT)$ and $H^*(BG)$
$\hat{P}$ and $\hat{Q}$	Important Euler data
$HG[B](t)$	Cohomology-valued function of $B = \{B_d\}$
$\Psi(t)$	The mirror map
$I_{\mathcal{V}}$ and $J_{\mathcal{V}}$	Givental's cohomology-valued functions
$\mathcal{V}'_{d,k,1}$	A certain subbundle of $\mathcal{V}_{d,k}$ used in defining $J_{\mathcal{V}}$
$*_X$	Modified quantum product on $H^*(\mathbb{P}^n)$ determined by a complete intersection $X \subset \mathbb{P}^n$
$\overline{I}_{\mathcal{V}}$	Cohomology-valued function used in the Quantum Hyperplane Section Principle
I_T and J_T	Equivariant versions of $I_{\mathcal{V}}$ and $J_{\mathcal{V}}$
$Z_{i,\mathcal{V}}$	A collection of functions for $0 \le i \le n$ which determine J_T uniquely

Our conventions for citations are explained at the beginning of the Bibliography.

CHAPTER 1

Introduction

Mirror symmetry has made some surprising predictions in algebraic geometry, ranging from the number of rational curves on a quintic threefold to the structure of certain moduli spaces. These are wonderful problems to work on and, as indicated in the preface, have led to some very interesting mathematics. Yet to understand where these predictions come from, the algebraic geometer must plunge into the language of physics, which is unfamiliar and sometimes frustratingly nonrigorous. Hence, to begin our survey of the algebraic geometry of mirror symmetry, we will start with the motivations for mirror symmetry and a discussion of what mirror symmetry means in physics. Our treatment will be somewhat incomplete, since it will involve many terms from physics which may be new to the reader. Nevertheless, we hope to convey some of the intuition behind this remarkable phenomenon.

We will then discuss three-point functions (which are crucial to the enumerative predictions of mirror symmetry) and the physical reasons why Calabi-Yau manifolds appear in the theory. Finally, at the end of the chapter, we will return to the more familiar world of mathematics and give the reader a preview of the algebraic geometry to be explored in the remaining chapters of the book.

1.1. The Physics of Mirror Symmetry

The goal of this section is to give the reader a feeling for why mirror symmetry should occur and what it should imply. From the point of view of physics, mirror symmetry arises naturally from standard constructions in supersymmetric string theory, and our discussion will begin with some elementary remarks about strings and supersymmetry. The reader should be assured that no previous knowledge of physics is assumed! Our aim is to convey the flavor of these physical theories and in the process enhance the reader's intuition for the resulting mathematics. A detailed understanding of the physics is not necessary, though later chapters will refer to the physics presented in this chapter. As general references for string theory, the reader can consult [**GSW, Polchinski2**].

In string theory, physical processes are described by the propagation of a string in spacetime. A propagating string traces out a surface, called the *world sheet* Σ of the string. Classical fields can be described as functions, sections of bundles, etc. on the world sheet, and quantizing leads to a two-dimensional quantum field theory. This theory has a generalized Hilbert space of states, together with a collection of observables, which become self-adjoint operators on the space of states. The other key ingredients of the theory are the *action* S obtained by integrating a Lagrangian over the world sheet Σ, and the *correlation functions*

$$(1.1) \qquad \langle \phi(x_1), \ldots, \phi(x_n) \rangle = \int [D\phi] \phi(x_1) \ldots \phi(x_n) e^{iS(\phi)},$$

1

where ϕ is an observable. This Feynman integral is over all possible world sheets and is mathematically undefined at present. We discuss such theories in more detail in Appendix B. For now, let's concentrate on their general features.

String theories are nice because they eliminate some of the problems which occur when a particle splits into two particles. As the following picture shows, representing the particle by a point leads to a singularity, while the string representation is a smooth 2-manifold with boundary:

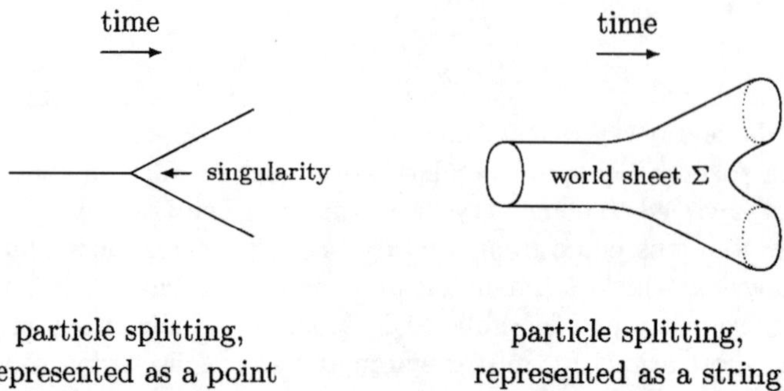

particle splitting,   particle splitting,

represented as a point   represented as a string

However, string theories still have some undesirable features, including many infinities which require renormalization. A remarkable discovery in recent times is that *supersymmetry* can eliminate many of these difficulties. Supersymmetry transforms bosons (particles with integer spins and symmetric wavefunctions) into fermions (particles with half-integer spins and antisymmetric wavefunctions) and vice-versa. Although supersymmetry has not been experimentally verified to date, supersymmetric theories have become very important in theoretical physics because of their nice behavior.

Another ingredient we need is that the world sheet Σ has a conformal structure, and our supersymmetric string theory needs to be equivalent under conformal equivalence. Hence this theory is a *superconformal field theory* (SCFT for short). The Lie algebra of the symmetry group of such a theory is a superconformal algebra. This algebra contains the conformal algebra (the Lie algebra of the group of conformal transformations of the world sheet) as a subalgebra, and it also contains the supersymmetry transformations.

The superstring theories come in four basic types: type I, type IIA, type IIB, and, of greatest interest to us, *heterotic*. Heterotic string theory is an $N = 2$ SCFT because there are two supersymmetries. In such a theory, the equations of motion for the fermions decouple into left- and right-moving solutions, which means that there are actually four supersymmetries, two left-moving and two right-moving. For this reason, heterotic string theories are more properly called $(2, 2)$ theories, as there are two independent supersymmetries in each of the left- and right-moving sectors of the theory.

The $N = 2$ superconformal algebra contains two copies of the usual superconformal algebra, and hence has two $u(1)$ subalgebras, one in the right-moving sector of the theory which infinitesimally rotates the two supersymmetries, and the other in the left-moving sector which acts similarly. A noncanonical choice of generator for each $u(1)$ can be made by ordering the two supersymmetry transformations. If the order of the supersymmetries is reversed, the result is to change the sign of the generator of the $u(1)$. The respective generators of these subalgebras are denoted by $(Q, \bar{Q})$ and are only well-defined up to sign. We regard $(Q, \bar{Q})$ as operators on the Hilbert space of states, which decomposes into eigenspaces under the action of $u(1) \times u(1)$. As we will see, these eigenspaces can be very interesting.

So far, we've discussed heterotic string theories in the abstract. The next step is to actually construct a such a theory, which is where algebraic geometry enters the picture. There are several ways this can be done, but for us, the most important is the *nonlinear sigma model* (sigma model for short) determined by a *Calabi-Yau threefold*[1] V and a *complexified Kähler class* $\omega = B + iJ$ on V. Here, B and J are elements of $H^2(V, \mathbb{R})$, with J a Kähler class.

From the input data (V, ω), there is a geometric construction of an $N = 2$ SCFT which explicitly gives distinct roles for the various supersymmetries; hence in this context there is a canonical choice for $(Q, \bar{Q})$. This gives an explicit choice of the $u(1) \times u(1)$ representation on our Hilbert space, and one can compute that for $p, q \geq 0$, $(Q, \bar{Q})$ has eigenspaces:

$$\begin{aligned}
(p, q) \text{ eigenspace} &\simeq H^q(V, \wedge^p T_V) \\
(-p, q) \text{ eigenspace} &\simeq H^q(V, \Omega_V^p).
\end{aligned}$$

(1.2)

Appendix B.2 describes more fully what it means to be a nonlinear sigma model and Section 1.3 explains how the Calabi-Yau condition arises from the physics.

The most important fields in a heterotic string theory are associated to elements of $H^1(V, T_V)$ and $H^1(V, \Omega_V^1)$, corresponding to eigenvalues $(1, 1)$ and $(-1, 1)$ respectively. The operators corresponding to elements in these spaces are called *marginal operators*. These are important partly because they are closely related to the moduli of sigma models coming from (V, ω). Intuitively, SCFT moduli are obtained by simultaneously varying the complex structure on V and the complexified Kähler class $\omega = B + iJ$, although there are extra discrete SCFT identifications on the moduli spaces which we will ignore for the moment. While readers should be familiar with the complex moduli, the idea of "complexified Kähler moduli" may be new. We will describe this in more detail in Section 1.4.

It follows from this description that the SCFT moduli space has two foliations, one of whose leaves can be described as "V constant", while the leaves of the other are "ω constant." This leads to the following picture:

(1.3)

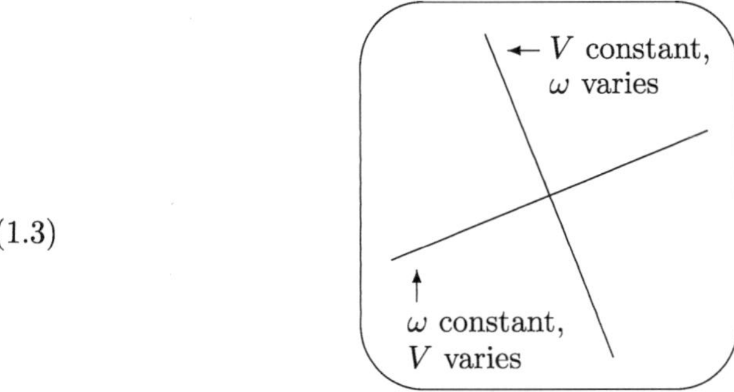

SCFT moduli near (V, ω)

[1]For now, a Calabi-Yau threefold is a smooth compact connected threefold with vanishing first Betti number and trivial canonical class. Later, we will allow certain singularities.

In spite of this picture, we should emphasize that the SCFT moduli space is *not* a product of the complex structure and Kähler structure moduli spaces, not even locally. In fact, the Kähler moduli space of ω can depend on the complex structure of V [**Wilson2**]. In general, this is only well-defined if we have in mind a fixed complex structure on V. On the other hand, it follows from [**Wilson2**] that for a sufficiently generic Calabi-Yau threefold V, the Kähler moduli of ω is independent of the complex structure of V. These issues will be discussed in more detail in Chapter 6. Hence, although the above picture is useful at a conceptual level, it does not reflect the subtleties of the SCFT moduli space.

Now that we have a better idea of how Calabi-Yau threefolds and complexified Kähler classes gives interesting physical theories, it is time to explain where mirror symmetry comes from. The basic starting point lies in the sign indeterminacy of $(Q, \bar{Q})$. We mentioned above that $(Q, \bar{Q})$ are only well-defined up to sign, yet the sigma model coming from (V, ω) makes a very specific choice. If we changed Q to $-Q$ and left $\bar{Q}$ as is, we would interchange the (p, q) and $(-p, q)$ eigenspaces, which by (1.2) would interchange $H^q(V, \wedge^p T_V)$ and $H^q(V, \Omega_V^p)$. This is not possible since these are vector spaces of different dimensions in general. Yet from the physical point of view, such a sign change is reasonable. This asymmetry suggests that maybe the sign change corresponds to the sigma model arising from a *different* pair (V°, ω°). If such a pair (V°, ω°) exists, we say that (V, ω) and (V°, ω°) are a *mirror pair*. More formally, we have the following definition from physics.

PHYSICS DEFINITION 1.1.1. *(V, ω) and (V°, ω°) form a* mirror pair *if their sigma models induce isomorphic superconformal field theories whose $N = 2$ superconformal representations are the same up to the above sign change.*

Note that this is not a mathematical definition since the SCFT associated to (V, ω) is not rigorously defined. However, in Chapter 8, we will give a careful definition of a *mathematical mirror pair*. This definition will incorporate many of the properties predicted by mirror symmetry.

It is these properties to which we now turn our attention. If (V, ω) and (V°, ω°) are a mirror pair, then we get isomorphic SCFT's. But what does this mean about the mathematics? One of the major goals of this book is to understand the mathematical consequences of mirror symmetry.

To see what mirror symmetry tells us about V and V°, first note that if we combine (1.2) with the eigenvalue change $(p, q) \leftrightarrow (-p, q)$, we get isomorphisms

$$(1.4) \qquad \begin{aligned} H^q(V, \wedge^p T_V) &\simeq H^q(V^\circ, \Omega_{V^\circ}^p) \\ H^q(V, \Omega_V^p) &\simeq H^q(V^\circ, \wedge^p T_{V^\circ}). \end{aligned}$$

Since V is Calabi-Yau, it has a nonvanishing holomorphic 3-form Ω, and cup product with Ω gives a (noncanonical) isomorphism $H^q(V, \wedge^p T_V) \simeq H^q(V, \Omega_V^{3-p})$. The same is true for V°, so that (1.4) can be written as

$$(1.5) \qquad \begin{aligned} H^q(V, \Omega_V^{3-p}) &\simeq H^q(V^\circ, \Omega_{V^\circ}^p) \\ H^q(V, \Omega_V^p) &\simeq H^q(V^\circ, \Omega_{V^\circ}^{3-p}). \end{aligned}$$

These isomorphisms have a nice interpretation in terms of the Hodge diamond. Since V is a smooth threefold, its Hodge numbers $h^{p,q}(V) = \dim H^q(V, \Omega_V^p)$ have the symmetries $h^{p,q}(V) = h^{q,p}(V) = h^{3-p,3-q}(V) = h^{3-q,3-p}(V)$, and since V is Calabi-Yau, we also have $b_1(V) = 0$ and $\Omega_V^3 \simeq \mathcal{O}_V$. This implies that $h^{1,0}(V) = 0$ and $h^{3,0}(V) = 1$. Furthermore, $h^{0,2}(V) = \dim H^2(V, \mathcal{O}_V) = \dim H^1(V, \mathcal{O}_V) =$

$h^{0,1}(V) = 0$, where the second equality follows from Serre duality and $\Omega_V^3 \simeq \mathcal{O}_V$. Thus the Hodge diamond of V is as follows:

$$
\begin{array}{ccccccc}
 & & & 1 & & & \\
 & & 0 & & 0 & & \\
 & 0 & & h^{1,1}(V) & & 0 & \\
1 & & h^{2,1}(V) & & h^{2,1}(V) & & 1 \\
 & 0 & & h^{1,1}(V) & & 0 & \\
 & & 0 & & 0 & & \\
 & & & 1 & & &
\end{array}
$$

If we now compare the Hodge diamonds of a mirror pair V and V°, (1.5) implies that $h^{p,q}(V) = h^{3-p,q}(V^\circ)$, which shows that the Hodge diamond of V° is the reflection (or mirror image) of the Hodge diamond of V about a $45°$ line. This is where the name "mirror symmetry" comes from.

The isomorphisms (1.4) and (1.5) are actually the first of a series of increasingly impressive consequences of mirror symmetry. The next interesting implication of being a mirror pair concerns moduli spaces. To see where moduli enter the picture, note that (1.4) gives isomorphisms $H^1(V, T_V) \simeq H^1(V^\circ, \Omega^1_{V^\circ})$ and $H^1(V, \Omega^1_V) \simeq H^1(V^\circ, T_{V^\circ})$. This naturally identifies the tangent space to the complex moduli of V with the tangent space to the Kähler moduli of ω° (see Section 1.4 for a definition), and similarly identifies the tangent space to the Kähler moduli of ω with the tangent space to the complex moduli of V°. Thus the complex moduli space of V is locally isomorphic to the Kähler moduli space of ω°, and similarly the complex moduli space of V° is locally isomorphic to the Kähler moduli space of ω. These local isomorphisms are collectively called *the mirror map*. In Chapter 6, we will study complex and Kähler moduli in more detail and give a careful definition of the mirror map.

Recall from our (slightly inaccurate) picture (1.3) that the SCFT moduli space of (V, ω) has two foliations, one where the leaves are "V constant" and the other with leaves "ω constant." This means that if (V, ω) and (V°, ω°) are a mirror pair, then we get the following local isomorphism of SCFT moduli spaces:

(1.6)

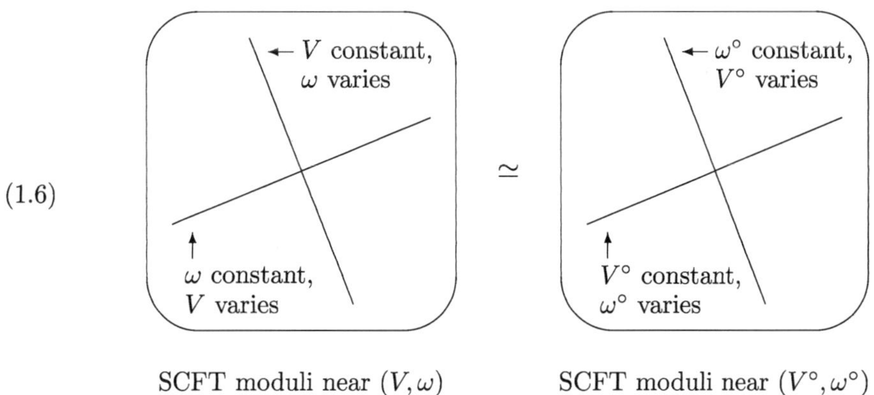

SCFT moduli near (V, ω) SCFT moduli near (V°, ω°)

This picture also clarifies why it makes no sense to speak of "the" mirror manifold of a Calabi-Yau manifold V: given V, we can vary ω freely in one leaf of the foliation of the SCFT moduli space, which on the mirror side would cause ω° to be fixed and V° to vary freely. So if anything, "the mirror of V" should be the class ω°, together with the moduli space of those deformations of V° on which ω° makes sense as a complexified Kähler class.

In addition to what we've discussed so far, the existence of a mirror pair has further consequences, not all of which are understood yet. The basic idea is that *any* quantity that can be defined in terms of the SCFT can in principle be computed using two different constructions for the SCFT. In the best cases, these quantities can be computed in terms of the geometry of V and V°. Note that due to the sign change in Q and the fact that $H^q(V, \Omega_V^p)$ corresponds to $H^q(V^\circ, \wedge^p T_{V^\circ})$, it is expected that the geometric calculations will be different for the different models. The best example is given by the correlation functions of the SCFT, already mentioned in (1.1). These will be discussed in more detail in Section 1.2 and are the key to the enumerative predictions made by mirror symmetry. We will see that a computation on the mirror family can yield amazing results about the original Calabi-Yau manifold.

In the physics literature, mirror symmetry is a rich phenomenon. In addition to the mirror symmetry for nonlinear sigma models discussed so far, mirror symmetry has also been observed for some non-geometric types of SCFT's, including *minimal models* and *Landau-Ginzburg orbifolds*. To explain how mirror symmetry works in these cases, one needs to take the "orbifold" of a SCFT by a finite group. This begins with the subtheory consisting of invariant fields, but since the resulting subtheory is not stable under the flow of time (i.e., is not unitary), extra fields are added to get a unitary theory which is again a SCFT (actually, the physics is a bit more subtle—see Appendix B.4 for the details). If we quotient out by a carefully chosen group action, we get the same physical theory we started with, but with a change in the sign of the Q eigenvalues. This version of "mirror symmetry" predates the discovery of mirror symmetry for nonlinear sigma models.

Early evidence for mirror symmetry of Calabi-Yau threefolds was given by lists of Calabi-Yau hypersurfaces in weighted projective spaces (or their quotients by finite groups). The Hodge numbers of these hypersurfaces exhibited a striking (but far from perfect) symmetry. For some of these hypersurfaces, mirror symmetry was demonstrated in [**GP1**] by first showing mirror symmetry for certain Landau-Ginzburg theories (as mentioned above) and then relating these theories to the sigma models of the hypersurfaces. As we will see in Chapter 4, all of these weighted projective hypersurfaces are a subclass of those that arise from Batyrev's reflexive polytope construction [**Batyrev4**], as observed in [**CdK**]. It is conjectured (and widely believed) that Calabi-Yau threefolds coming from reflexive polytopes are mirror symmetric, and more generally, that the larger class of toric complete intersections [**Borisov1**] is mirror symmetric. Much evidence has been given in the last few years [**CdGP, Font, Morrison1, CdFKM, HKTY1, CFKM, HKTY2, BK1, AGM1, ES1, BKK, Kontsevich2, MP1, Givental2, Givental4, LLY**]. In Chapter 11, we will outline two related approaches to the *Mirror Theorem*, which establishes the equality of certain correlation functions of Calabi-Yau toric complete intersections and their conjectured mirrors.

1.2. Three-Point Functions

The correlation functions defined in (1.1) are objects of intrinsic interest in a SCFT. In physics, they arise naturally in the study of successive generations of particles. The most common correlation function is the *three-point function*, which describes interactions between particles from three generations, not necessarily distinct. In the Standard Model of elementary particle physics, a *generation* of particles is a collection of particles with particular types of interactions under the electric, weak nuclear, and strong nuclear forces. Experiments indicate that there are 3 generations of particles. One of these generations includes the most familiar particles, namely the electron and its accompanying neutrino, and the up and down quarks (which are the constituents of the proton and the neutron). The other known generations contain the more exotic quarks and leptons.

We will consider three-point functions for the nonlinear sigma model coming from a Calabi-Yau threefold V and a complexified Kähler class ω. The most interesting three-point functions are the *Yukawa couplings*, which come from the marginal operators discussed in the previous section. These correspond to $H^1(V, \Omega_V^1)$ and $H^1(V, T_V)$, which gives two types of Yukawa coupling to consider.

We begin with the Yukawa coupling coming from $H^1(V, \Omega_V^1)$. To each element of $H^1(V, \Omega_V^1)$, the sigma model associates a 27-dimensional vector space of fields, which form an irreducible representation of E_6. In order to connect string theory to the physical world, this vector space is presumed to contain a generation of elementary particles. The Yukawa coupling between three generations corresponding to elements ω_i of $H^1(V, \Omega_V^1)$, $i = 1, 2, 3$, is a physically important coupling between three particles, one from each of the respective generations of particles.

In this situation, the Yukawa coupling is calculated by Feynman path integral techniques to be

$$
(1.7) \qquad
\begin{aligned}
\langle \omega_1, \omega_2, \omega_3 \rangle &= \int_V \omega_1 \wedge \omega_2 \wedge \omega_3 \; + \\
&\quad \sum_{\beta \neq 0} n_\beta \textstyle\int_\beta \omega_1 \int_\beta \omega_2 \int_\beta \omega_3 \; \frac{e^{2\pi i \int_\beta \omega}}{1 - e^{2\pi i \int_\beta \omega}},
\end{aligned}
$$

where the sum is over homology classes $\beta \in H_2(V, \mathbb{Z})$ and n_β is naively the number of rational curves in the homology class β. A careful definition of n_β requires an understanding of *Gromov-Witten invariants*, which will be discussed in detail in Chapter 7. The theory predicts that the n_β don't change if we deform V, so that this Yukawa coupling depends on ω but not on V. In Chapter 8, we will see that the coupling (1.7) is closely related to the quantum cohomology ring of V.

The Yukawa coupling just described is sometimes called the *A-model correlation function*. The latter terminology arises from the *A-model* described in Appendix B.2. This is a "twisted" version of the SCFT, which means that certain fields are locally the same as in the sigma model, but globally are sections of certain twists of the bundles that they were originally sections of. These couplings are identified with the corresponding three-point functions in the A-model.

Let us also say a few words about where (1.7) comes from. Notice that the first term of this formula is just cup product:

$$
(1.8) \qquad \int_V \omega_1 \wedge \omega_2 \wedge \omega_3.
$$

From the physics point of view, two things happen in going from (1.8) to the A-model correlation function (1.7). The first is a *non-renormalization theorem*, which says that from a perturbative point of view, there are no quantum corrections needed. As mentioned in the previous section, this is one of the nice consequences of supersymmetry. However, there are also world sheet non-perturbative corrections to be considered, which in this case are the *holomorphic instantons*. These are nonconstant holomorphic maps $\Sigma \to V$, where Σ is a compact Riemann surface. When we treat the A-model correlation function more carefully in Chapters 7 and 8, we will see that Σ can have nodal singularities and more than one component. In the A-model correlation function, the only instantons needed are those where Σ has genus 0. Naively, these are what the n_β count in formula (1.7). In the terminology of Chapter 7, we call n_β an *instanton number*.

The second Yukawa coupling to consider comes from $H^1(V, T_V)$. Here, elements of $H^1(V, T_V)$ correspond to the conjugate 27-dimensional representation of E_6, and we get a Yukawa coupling associated to three generations corresponding to elements θ_i of $H^1(V, T_V)$, $i = 1, 2, 3$. The Yukawa coupling in this case is given by

$$(1.9) \qquad \langle \theta_1, \theta_2, \theta_3 \rangle = \int_V \Omega \wedge (\theta_1 \cdot \theta_2 \cdot \theta_3 \cdot \Omega),$$

where Ω is a holomorphic 3-form on V. The expression $\theta_1 \cdot \theta_2 \cdot \theta_3 \cdot \Omega$ is defined by the composition

$$S^3 H^1(V, T_V) \otimes H^0(V, \Omega_V^3) \mapsto H^3(V, \wedge^3 T_V \otimes \Omega_V^3) \simeq H^3(V, \mathcal{O}_V) \simeq H^{0,3}(V).$$

Alternatively, one can think of this as

$$\int_V \Omega \wedge (\nabla_{\theta_1} \nabla_{\theta_2} \nabla_{\theta_3} \Omega),$$

where ∇ is the Gauss-Manin connection.[2] Note that $\langle \theta_1, \theta_2, \theta_3 \rangle$ is not yet well-defined, since Ω can be multiplied by any constant. There is however a natural normalization which we will describe later. This Yukawa coupling is clearly independent of the complexified Kähler class ω but depends on the complex structure of V (since Ω is a holomorphic 3-form).

The Yukawa coupling (1.9) is sometimes called the *B-model correlation function*, since it is identical with the corresponding three-point function in a different twisted theory, the *B-model* described in Appendix B.2.

To explain where the formula (1.9) comes from, we proceed as in the A-model case. Beginning with the "cup product"

$$(1.10) \qquad \int_V \Omega \wedge (\theta_1 \cdot \theta_2 \cdot \theta_3 \cdot \Omega),$$

the same non-renormalization theorem applies, and we have the same holomorphic instantons as before. The crucial observation is that these instantons enter via cup product. In the A-model case, ω_i appears in (1.7) via $\int_\beta \omega_i$, which can be thought of as the cup product $g_\beta \cup \omega_i \in H^6(V, \mathbb{C}) \simeq \mathbb{C}$, where $g_\beta \in H^4(V, \mathbb{C}) = H^{2,2}(V)$ is the Poincaré dual of β. But in the B-model case, we have $\theta_i \in H^1(V, T_V) \simeq H^1(V, \Omega_V^2) = H^{2,1}(V)$, and since $g_\beta \in H^{2,2}(V)$, their cup product $g_\beta \cup \theta_i$ lies in $H^{4,3}(V) = 0$. Hence the instantons don't interact with (1.10), which is the crude reason why (1.10) equals the B-model correlation function (1.9).

---

[2]This is how the Yukawa coupling appears in [**CdGP**]. As we will see in Proposition 5.6.1, Hodge theory leads us to introduce an additional minus sign into the definition of the Yukawa coupling. There are also physical reasons for adding a minus sign, as discussed in [**Cd**]. Since Ω can be multiplied by an arbitrary constant, the sign is of little concern for now. On the other hand, in Section 5.6.4, we will fix a choice of Ω so will have to be careful about the sign. Our definition of the *normalized Yukawa coupling* will include this sign.

The absence of instanton corrections in (1.9) is extremely important. It means that we can compute the B-model Yukawa coupling *exactly* using standard methods of algebraic geometry. This procedure will be explained in Chapter 5.

Suppose now that (V, ω) and (V°, ω°) are a mirror pair, i.e., the sigma model associated to (V, ω) gives the same SCFT as the sigma model associated to (V°, ω°) with the appropriate sign change. This gives a natural isomorphism $H^1(V, \Omega_V^1) \simeq H^1(V^\circ, T_{V^\circ})$, and since three-point functions are intrinsic to the SCFT, it should follow that the A-model correlation function arising from the sigma model on (V, ω) coincides with the (appropriately normalized) B-model correlation function arising from the sigma model on (V°, ω°). This is one of the major mathematical consequences of mirror symmetry. The actual details of this identification are a bit more complicated because of the role of the mirror map, but this equality of A-model and B-model correlation functions is certainly plausible.

Note that the properties of the correlation functions are consistent with mirror symmetry. We have already observed that the A-model correlation function associated to (V, ω) depends on ω but not on V. It follows from mirror symmetry and the local identification of moduli spaces discussed above that the B-model correlation function associated to (V°, ω°) should depend on V° but not on ω°. As noted above, the B-model does indeed have this property.

As we will see with the example of the quintic threefold, being able to identify the three-point function of the A-model with the three-point function of the B-model of its mirror was used in [**CdGP**] to make some remarkable predictions for numbers of rational curves on the quintic threefold. This example is discussed in detail in the next chapter. In general, some of the most important consequences of mirror symmetry arise from the combination of the two following facts:

- The equality of the A-model and B-model correlation functions.
- The ability to compute the B-model function exactly.

Together, these allow one to compute Gromov-Witten invariants on Calabi-Yau threefolds, which in turn give a wealth of enumerative information. All of this, of course, depends on our ability to prove mathematically that these physical consequences are indeed correct.

What we have just described can be thought of as the "classical" approach to the consequences of mirror symmetry, where the A-model and B-model correlation functions are the primary objects of interest. However, in the years since the discovery of mirror symmetry, the focus has shifted a bit. In particular, the relation between the B-model correlation function and the Gauss-Manin connection is much better understood, which on the A-model side has led to the development of quantum cohomology and the A-variation of Hodge structure. Also, the work of Givental and of Lian, Liu, and Yau on the Mirror Theorem has introduced other new objects of interest—we will see that equivariant cohomology and localization play an important role in the proofs of the Mirror Theorem. We will address all of these ideas in subsequent chapters.

We close this section with a final observation about the formulas (1.7) and (1.9). They seem asymmetric since the first is much more complicated than the second. Fortunately, mirror symmetry easily accounts for this discrepancy. We noted above that the A-model correlation function should vary with complexified Kähler class ω, while the B-model function should vary with complex structure on V. Since the cup product in (1.8) is a purely topological invariant, it is clear that other terms are needed if we are to have a nontrivial dependence on ω. On the other hand, the cup product in (1.10) already depends on the complex structure, since Ω is a holomorphic 3-form on V. So it is conceivable that no further terms are needed.

1.3. Why Calabi-Yau Manifolds?

Our next task is to explain why mirror symmetry applies only to Calabi-Yau manifolds, especially Calabi-Yau threefolds. A full explanation of why nonlinear sigma models need Calabi-Yau manifolds would require a considerable detour into physics. We give a partial treatment of this topic in Appendices B.2 and B.3. For now, we will content ourselves with sketching some of the ideas. The starting point is the assumption that space-time should be a 10-dimensional manifold with a semi-Riemannian metric of signature $(9, 1)$. This manifold should locally be a product $M_{3,1} \times V$, where $M_{3,1}$ is the usual space-time of special relativity, and V is a compact 6-dimensional Riemannian manifold. The basic intuition is that V is too small to be seen at macroscopic scales but is essential for the quantum aspects of the theory. In any dimension other than 10, the conformal symmetry discussed earlier in the section does not survive the process of quantization.

As usual, the physics is nontrivial and involves some unfamiliar terms. The idea is to approach a nonlinear sigma model by first considering other theories which are more elementary from the physics point of view. In particular, one starts with an $N = 1$ supergravity theory in the low energy limit. This gets rid of the fermionic fields, and then supersymmetry and other consistency requirements impose strong conditions on the metric on Riemannian manifold V. In particular, one finds that the holonomy group of the metric equals $SU(3)$. This has some nice consequences:

- (differential geometry) The metric on V is Ricci flat, i.e., its Ricci curvature tensor vanishes identically. This implies $b_1(V) = 0$.
- (algebraic geometry) V has a complex structure such that $c_1(V) = 0$, and the metric is Kähler for this complex structure.

Hence we see that V is Calabi-Yau, as desired.

So far, we only have $N = 1$ supersymmetry in spacetime. After some further work, this theory can be reinterpreted as an $N = 2$ SCFT on the world sheet, although to preserve superconformal invariance, we need to deform the above Ricci flat metric. A precise description of this deformation is not known (this is still an open question in physics), but one can show that the new metric lies in the same Kähler class as the old one, so that we still have a Calabi-Yau threefold.

For a more complete description of how the Ricci curvature arises (from the physics point of view) and references to the literature, the reader should consult Chapter 0 of [**Hübsch**].

1.4. The Mathematics of Mirror Symmetry

From a mathematical point of view, the formulation of mirror symmetry given in Section 1.1 poses serious problems. For example, the definition of an $N = 2$ SCFT involves integrals over the space of *all* maps $\Sigma \to V$. Such integrals have yet to be defined rigorously. So a mathematical proof of mirror symmetry would involve an isomorphism between objects (the sigma models of (V, ω) and (V°, ω°)) which aren't well-defined mathematically. Even the $N = 2$ SCFT moduli spaces pictured in Section 1.1 are not well-defined!

What are mathematicians to do in this situation? One approach would be to avoid SCFT's altogether by concentrating on careful definitions of Kähler moduli spaces, Gromov-Witten invariants, etc. and then trying to prove that these objects behave as predicted by mirror symmetry. Another approach would be to embrace the physics and use its intuitions to see more deeply into the algebraic geometry, notably in the predictions mirror symmetry makes concerning the Gromov-Witten invariants of Calabi-Yau toric complete intersections. A third would be to formulate a purely mathematical version of mirror symmetry. For example, [**Kontsevich3**] proposes that the mirror of a complex manifold V is a certain symplectic manifold V° and that mirror symmetry might be formulated as an equivalence of derived categories relating coherent sheaves on V to a category built from Lagrangian submanifolds of V°. Another fascinating although still somewhat speculative approach is to attempt to geometrically *construct* the mirror manifold as a moduli space of special Lagrangian submanifolds [**SYZ**].[3] In practice, all of these approaches have been used, which is why mirror symmetry is such an exciting field.

In this book, we will follow mainly the first approach, with occasional comments on the physics of the situation. Thus, our goal is to discuss the algebraic geometry involved in understanding the mathematical aspects of Sections 1.1 and 1.2. One difference is that, unlike the physical theories, we will work with Calabi-Yau manifolds of arbitrary dimension, not just dimension three. However, when we try to formulate a mathematical definition of mirror pair in Chapter 8, we will be most successful in the case of Calabi-Yau threefolds.

We begin with a careful definition of Calabi-Yau. Since the mirror symmetry constructions to be studied in Chapter 4 sometimes produce singular varieties, we need a definition which allows certain types of singularities.

DEFINITION 1.4.1. *A* Calabi-Yau variety *is a d-dimensional normal compact variety V which satisfies the following conditions:*

(i) V has at most Gorenstein canonical singularities.

(ii) The dualizing sheaf of V is trivial, i.e., $\widehat{\Omega}_V^d \simeq \mathcal{O}_V$.

(iii) $H^1(V, \mathcal{O}_V) = \cdots = H^{d-1}(V, \mathcal{O}_V) = \{0\}$.

If in addition V has at most Gorenstein $\mathbb{Q}$-factorial terminal singularities, we say that V is a minimal *Calabi-Yau variety.*

In Appendix A, we review the dualizing sheaf $\widehat{\Omega}_V^d$ and the definitions of Gorenstein, canonical and terminal singularities. The Calabi-Yau threefolds considered in Section 1.1 certainly satisfy this definition.

Some other terms used in Section 1.1 deal with moduli of various sorts. The space of all complex structures on a given manifold V is a well known object in algebraic geometry, but the idea of *Kähler moduli* may be unfamiliar. Recall from Section 1.1 that we had a Calabi-Yau threefold V with a complexified Kähler class $\omega = B + iJ \in H^2(V, \mathbb{C})$ such that J was Kähler. However, if we change ω by an integral class, we don't change the physical theory, since the definition of nonlinear sigma model only uses $\exp(2\pi i \int_\Sigma \phi^*(\omega))$ for maps $\phi : \Sigma \to V$.[4] This quantity is unchanged if we change ω by an element of $H^2(V, \mathbb{Z})$. Thus, in defining Kähler moduli, we should mod out by the image of $H^2(V, \mathbb{Z})$. This leads to the following definition (which as before allows some singularities).

---

[3]In general, symplectic geometry plays an important role in mirror symmetry, though we will concentrate more on the algebro-geometric aspects.

[4]You can see this in the A-model correlation function (1.7). The full details can be found in Appendix B.2.

DEFINITION 1.4.2. *Let V be a projective orbifold with $h^{2,0}(V) = 0$. Then:*

(*i*) *The* Kähler cone *is the subset $K(V) \subset H^2(V, \mathbb{R}) = H^{1,1}(V, \mathbb{R})$ consisting of all Kähler classes.*

(*ii*) *The* complexified Kähler space *is the quotient*

$$K_{\mathbb{C}}(V) = \{\omega \in H^2(V, \mathbb{C}) : \mathrm{Im}(\omega) \in K(V)\}/\mathrm{im}\, H^2(V, \mathbb{Z}),$$

where $\mathrm{im}\, H^2(V, \mathbb{Z})$ is the image of the natural map $H^2(V, \mathbb{Z}) \to H^2(V, \mathbb{C})$.

(*iii*) *The* complexified Kähler moduli space *is the quotient $K_{\mathbb{C}}(V)/\mathrm{Aut}(V)$.*

Actually, the Kähler moduli space[5] as it arises in SCFT differs from this slightly. The Kähler moduli space as we have defined it receives *quantum corrections* which will modify its properties slightly. In particular, the theory may not converge for all ω, and so we will be forced to restrict our attention to complexified Kähler classes $\omega = B + iJ$ where J is sufficiently positive. Nevertheless, the larger space that we defined here is mathematically interesting and will be one our primary objects of study. We will consider some of the related subtleties in Chapter 6.

In Appendix A, we will review the definitions of orbifold and Kähler class on an orbifold. Since $h^{2,0} = 0$, the Kähler cone $K(V)$ is an open convex cone in $H^2(V, \mathbb{R})$. This tells us that the complexified Kähler space $K_{\mathbb{C}}(V)$ is a well-behaved object. However, in order to determine the structure of the Kähler moduli space $K_{\mathbb{C}}(V)/\mathrm{Aut}(V)$, we need to know how the automorphism group $\mathrm{Aut}(V)$ acts on the Kähler cone. We will return to this subject in Chapter 6.

Notice that Definitions 1.4.1 and 1.4.2 seem to involve slightly different types of singularities. Fortunately, by a result of [**Reid1**], a Gorenstein orbifold has canonical singularities. This means that any orbifold satisfying the second and third parts of Definition 1.4.1 is automatically Calabi-Yau. Hence, for the purposes of mirror symmetry, *Calabi-Yau orbifolds* are a natural class to work with.

1.5. What's Next?

We now describe how the next ten chapters will take us from here to a proof of the Mirror Theorem. We begin in Chapter 2 with a careful description of the mirror of a smooth quintic threefold $V \subset \mathbb{P}^4$. By carrying out the strategy outlined in Section 1.2, we will get some explicit predictions for the number of rational curves on V of given degree. We will revisit this example several times during subsequent chapters as we develop more mathematical background and eventually justify all of the computations which appear in Chapter 2.

In trying to generalize the example of the quintic, it was soon realized that toric geometry had an important role to play in mirror symmetry. Hence Chapter 3 will explore various ways of describing toric varieties, and then Chapter 4 will describe the known mirror symmetry constructions, many of which use toric geometry. This will give us a large supply of examples which should satisfy mirror symmetry and provide a good testing ground for the mathematics. We should also mention that there are even physical theories (the *gauged linear sigma models* of [**Witten5**] to be described in Appendix B.5) which explicitly use toric varieties. Support for mirror symmetry in the context of the gauged linear sigma model is given in [**MP1**].

Our next task is to describe and compute the B-model Yukawa coupling. This will be done in Chapter 5. It turns out that finding the correct coordinates for

---

[5]In discussing the Kähler moduli space, we frequently drop the adjective "complexified" when the meaning is clear from context.

calculating this Yukawa coupling requires a good understanding of the moduli space at certain boundary points. Hence, in Chapter 6 we will study the structure and compactifications of these moduli spaces. As mentioned above, we will also consider Kähler moduli. In fact, we will see that certain basic facts about Kähler moduli give insight into what the compactification of the usual moduli space should look like. In Chapter 7, we will discuss Gromov-Witten invariants. Definitions have been given in both the symplectic and algebraic categories, and either can be used to give a mathematical definition of the A-model Yukawa coupling.

By the end of Chapter 7, we will have everything needed to give a precise formulation of "classical" mirror symmetry, which asserts that certain correlation functions are compatible via the mirror map. But starting in Chapter 8, we will explore a deeper understanding of the subject. Two new ingredients are *quantum cohomology*, which can be thought of as working out the algebraic and enumerative implications of Gromov-Witten invariants, and the *A-model variation of Hodge structure*, which is a natural consequence of quantum cohomology. The basic idea is that mirror symmetry really involves an isomorphism, via the mirror map, of two variations of Hodge structure: one on the B-model (the usual VHS coming from complex moduli), and the other on the A-model (the A-variation of Hodge structure). This version of the Mirror Theorem will be formulated at the end of Chapter 8. It turns out that the desired equality of correlation functions follows immediately from this isomorphism of variations of Hodge structure.

The Mirror Theorem, when formulated using variations of Hodge structure, is still an open question, although recent work of Givental [**Givental2, Givental4**] and Lian, Liu and Yau [**LLY**] represents substantial progress toward proving this form of the theorem. We will discuss the ideas of these papers in Chapter 11. As we will see, this will require the introduction of new techniques and new objects of study. In particular, *equivariant cohomology* and *localization* will play an important role in the proof. In Chapter 9, we will review some of the basic definitions and theorems, and we will use these methods to prove some interesting results about Gromov-Witten invariants. Then Chapter 10 discusses an extension of Gromov-Witten invariants called *gravitational correlators*. These invariants will enable us to describe the flat sections of the A-model connection. We will also define the Givental *J*-function and explain its relation to *quantum differential equations*. Finally, in Chapter 11, we will discuss the Mirror Theorems stated in [**Givental2, Givental4**] and [**LLY**]. A brief preview of what is involved will be given at the end of Chapter 2, and then the full details for the quintic threefold will be presented in Chapter 11.

These chapters cover a lot of mathematics, and reading them straight through would be a somewhat daunting task. We suggest that the reader start with Chapter 2. From here, there are several ways to proceed, depending on the reader's interests. The preface offers some guidance for what to read, and a glance at the table of contents may also be useful. Our hope is that once you start reading, the intrinsic interest of the subject will draw you in. Mirror symmetry is a fascinating story, and it is fun to see how the various pieces fit together.

CHAPTER 2

The Quintic Threefold

The quintic threefold was the first example for which mirror symmetry was used to make enumerative predictions [**CdGP**]. In this chapter, we will review this example, following the approach of [**CdGP**], with some modifications based on the more mathematical exposition of [**Morrison2**]. In Section 2.6 we will give two alternative methods for calculating the same enumerative data based on [**Givental2**] and [**LLY**]. We will also revisit the quintic threefold several times in later chapters to illustrate the various methods used to study mirror symmetry.

2.1. The A-Model Correlation Function of the Quintic Threefold

It is well known that a smooth quintic hypersurface $V \subset \mathbb{P}^4$ is Calabi-Yau (this also follows from the theory to be developed in Chapter 4). We consider the nonlinear sigma model associated to (V, ω), where $\omega = B + iJ$ is a complexified Kähler class on V. If H denotes the hyperplane class, then the A-model correlation function (1.7) from Chapter 1 simplifies to give the formula

$$(2.1) \qquad \langle H, H, H \rangle = 5 + \sum_{d-1}^{\infty} n_d d^3 \frac{q^d}{1 - q^d}.$$

Naively, n_d is the number of rational curves of degree d contained in V, and $q = e^{2\pi i \int_\ell \omega}$, where ℓ is a line in V. The formula for q can be written $q = e^{2\pi i t}$ if we put $\omega = tH$. We will see in the next section that q is actually a local coordinate on the compactified Kähler moduli space. Furthermore, as explained in Chapter 1, this formula depends on ω but not on the complex structure of V, since it is an A-model correlation function.

When algebraic geometers first learned of (2.1), it seemed a bit strange. For one thing, we don't even know a general quintic threefold has finitely many rational curves of a given degree d. According to the Clemens conjecture, this should be true, but so far the conjecture is known only for $d \leq 9$ [**JK1**]. Yet (2.1) seems to assume the conjecture for all d. This unsettling beginning is then compounded by mirror symmetry, which claims to give an explicit formula for the number n_d of such curves, and it does this for *all $d \geq 1$*.

We will give a firm foundation for (2.1) in Chapter 7. The key player is the *Gromov-Witten invariant* $\langle I_{0,3,d} \rangle (H, H, H)$. This invariant can be defined in a variety of ways. From the symplectic point of view, one considers C^∞ maps $f : \mathbb{P}^1 \to V$ which obey an appropriately deformed Cauchy-Riemann equation. Imposing appropriate conditions on the image of 3 points in $\mathbb{P}^1$, it can be shown that there are only finitely many such curves in each homology class. Then $\langle I_{0,3,d} \rangle (H, H, H)$ is defined to be the total number of such curves, counted with sign according to orientation (there is no longer a preferred orientation coming from a complex structure).

15

There are actually several ways to make this precise, all of which will be reviewed in Section 7.2.

A different approach to defining n_d has been proposed in [**Kontsevich2**] using the moduli space of stable maps $f : C \to V$, where C has genus 0 (but need not be irreducible). The resulting definition of Gromov-Witten invariant is rather sophisticated, involving *algebraic stacks* and *virtual fundamental classes*. These topics will be discussed in Section 7.1.

Once we have a rigorous definition of the Gromov-Witten invariant, we can then define the A-model correlation function to be the formal sum

$$\langle H, H, H \rangle = \sum_{d=1}^{\infty} \langle I_{0,3,d} \rangle (H, H, H)\, q^d.$$

To see how this relates to (2.1), fix d and suppose that $k|d$. If $C \subset V$ is a smooth rational curve of degree k with normal bundle $\mathcal{O}_V(-1) \oplus \mathcal{O}_V(-1)$, then a map $f : \mathbb{P}^1 \to C$ of degree d/k gives a stable map of the sort counted by $\langle I_{0,3,d} \rangle (H, H, H)$. We will see in Chapter 7 that the family of such f's contributes a factor of k^3 to $\langle I_{0,3,d} \rangle (H, H, H)$. Since this happens for each C of degree $k|d$, it makes sense that

$$(2.2) \qquad\qquad \langle I_{0,3,d} \rangle (H, H, H) = \sum_{k|d} n_k k^3$$

when $d \geq 1$. Given this equation and the above definition of $\langle H, H, H \rangle$, (2.1) follows by easy power series manipulations.

Unfortunately, this argument is far from complete and seems to assume the Clemens conjecture. So to make everything rigorous, we will *define* the n_d using equation (2.2). Then the question becomes, how do the n_d relate to rational curves in V? In Section 7.4.4, we will prove that for $d \leq 9$, n_d is precisely the number of rational curves of degree d on the quintic threefold. This is less obvious than it seems, for although the Clemens Conjecture is true for $d \leq 9$, some rational curves in V are singular. In particular, V contains 17,601,000 6-nodal rational curves of degree 5 [**Vainsencher**], and some care is required to see how they enter into n_5.

For a long time, people expected that once the Clemens Conjecture was proved, it would follow immediately that in (2.2), n_d should be the number of rational curves of degree d in V. However, it was recently observed by R. Pandharipande that even if the Clemens Conjecture is true for $d = 10$, the number n_{10} defined by (2.2) does *not* give the number of degree 10 rational curves on V. The reason is that double covers of the 6-nodal rational curves of degree 5 contribute more than the expected factor of 5^3. We will prove this carefully in Chapters 7 and 11. As a consequence, we will no longer refer to n_d as "the number of rational curves of degree d". Rather, in the terminology of Chapter 7, we adopt the more neutral name *instanton number*.

Our discussion of n_{10} reflects one of the themes of Chapter 7, which is the subtle interplay between Gromov-Witten invariants and enumerative geometry. But in spite of the difficulties which can arise, we will see in Chapter 7 that Gromov-Witten invariants contain a lot of enumerative information and that mirror symmetry has stimulated a flurry of activity in enumerative algebraic geometry.

In this chapter, we will use mirror symmetry to compute the A-model correlation function (2.1) of the quintic threefold V, guided by the strategy outlined in Chapter 1. The following sections contain some wonderful computations which will culminate in explicit predictions for the n_d. Our discussion will be somewhat unsatisfying since we use nonrigorous physical theories to compute numbers whose enumerative meaning is unclear. But we will see in subsequent chapters that every step of the computation can be made rigorous. In particular, we will prove in Chapter 11 that if we define n_d using (2.2) and the symplectic or algebraic definition of $\langle I_{0,3,d} \rangle (H, H, H)$, then the predictions for n_d made by mirror symmetry are straightforward corollaries of the Mirror Theorem.

2.2. The Quintic Mirror

Since $h^{1,1}(V) = 1$ holds for the quintic threefold V, its mirror should satisfy $\dim H^1(V^\circ, T_{V^\circ}) = 1$. The mirror of the quintic is therefore a one-parameter family of Calabi-Yau manifolds. Greene and Plesser [**GP1**] describe the mirror as a resolution of singularities of a family of hypersurfaces in $\mathbb{P}^4/G$. Here, G is the group

$$(2.3) \qquad G = \{\, (a_1, \dots, a_5) \in \mathbb{Z}_5^5 : \textstyle\sum_i a_i \equiv 0 \bmod 5 \,\}/\mathbb{Z}_5,$$

where the $\mathbb{Z}_5$ is embedded diagonally, and $g = (a_1, \dots, a_5) \in G$ acts on $\mathbb{P}^4$ as

$$g \cdot (x_1, \dots, x_5) = (\mu^{a_1} x_1, \dots, \mu^{a_5} x_5),$$

where $\mu = e^{2\pi i/5}$ is a primitive fifth root of unity.

Using ψ as a parameter, the hypersurfaces are defined by the equation

$$(2.4) \qquad x_1^5 + x_2^5 + x_3^5 + x_4^5 + x_5^5 + \psi\, x_1 x_2 x_3 x_4 x_5 = 0,$$

which is invariant under the action of G and hence defines a family of hypersurfaces in $\mathbb{P}^4/G$. (In [**CdGP**], this equation is written with -5ψ in place of ψ.) The quotient $\mathbb{P}^4/G$ is singular, and the hypersurfaces (2.4) inherit these singularities. However, as long as $\psi \neq -5\mu^i$, $0 \le i \le 4$, the hypersurface is a Calabi-Yau orbifold in $\mathbb{P}^4/G$. Even better, we can resolve singularities (simultaneously for all ψ) without destroying the Calabi-Yau property—this is straightforward to establish locally. Thus, we define V° to be the family of resolved hypersurfaces coming from (2.4). To emphasize that we have a one-parameter family of varieties, we will sometimes denote the mirror family by V_ψ°.

This construction of V_ψ° is due to Greene and Plesser and involves the superconformal field theories discussed in Chapter 1. However, we will see in Chapter 4 that Batyrev gave a purely mathematical description of V_ψ° which generalizes to Calabi-Yau hypersurfaces in toric varieties.

As noted above, the equation (2.4) defines a singular variety. Since we are dealing with a threefold, there is no canonical resolution of singularities for (2.4). An explicit choice of resolution is given in [**Morrison2**], but there are *many* others. Fortunately, different choices still have the same complex moduli and in addition have the same B-model correlation function. We will let V_ψ° refer to any one of these choices. (In Chapter 6 we will learn that the different choices reflect the Kähler moduli of the mirror. In terms of physics, they gives different "phases" in the same physical theory.)

Our next task is to flesh out the strategy discussed in Section 1.2 for computing (2.1) in terms of the mirror V_ψ°. The first step is to understand the moduli spaces involved. For the complex moduli of the mirror, notice that the map $(x_1, \dots, x_5) \mapsto (\mu^{-1} x_1, x_2, \dots, x_5)$ induces an isomorphism $V_\psi^\circ \simeq V_{\mu\psi}^\circ$.[1] Hence ψ^5 is well-defined

---

[1] This is obvious for the hypersurfaces in $\mathbb{P}^4/G$, and one can show that the isomorphism extends to the resolution—see [**Morrison2**].

on the complex structure moduli space of V_ψ°, so that

$$x = \psi^{-5}$$

is a local coordinate for the complex moduli (we will soon see the reason for this choice of x). Furthermore, one can check that singularities of V_ψ° occur for $\psi = -5\mu^i$, $0 \leq i \leq 4$ (noted earlier) and also for $\psi = \infty$. In terms of the moduli coordinate x, this means that the complex moduli space has boundary points $x = -5^{-5}$ and $x = 0$.

According to mirror symmetry, the complex moduli of V° should correspond to the Kähler moduli of the quintic threefold V. Since H generates $H^2(V, \mathbb{Z}) \simeq \mathbb{Z}$, we can write $\omega = tH$, and since H is Kähler, $\omega = tH$ is a complexified Kähler class precisely when t is in the upper half plane. According to Definition 1.4.2, the Kähler space $K_{\mathbb{C}}(V)$ is the quotient of the complexified Kähler classes modulo $H^2(V, \mathbb{Z})$. It follows that $q = e^{2\pi i t}$ induces an isomorphism

$$q : K_{\mathbb{C}}(V) \simeq \Delta^*.$$

Thus q is a local parameter for Kähler moduli space, with $q = 0$ as a boundary point. An important but easy observation is that if ℓ is a line in V, then q can be written

$$(2.5) \qquad\qquad q = e^{2\pi i t} = e^{2\pi i \int_\ell \omega}.$$

Thus the A-model correlation function (2.1) is naturally a function in the local parameter q for Kähler moduli.

Assuming mirror symmetry, we get a local isomorphism between the Kähler moduli of V and the complex moduli of V°. In terms of the illustration (1.6), the "V constant, ω varies" slice maps locally to the "ω° constant, V° varies" slice on the mirror side. However, the key word is "local": we need appropriate *local* coordinates for each moduli space. For the Kähler moduli of V, this coordinate is clearly q since the three-point function (2.1) is a power series in q. Since q is a local coordinate at a *boundary* point, the same should be true on the mirror side.

For the complex moduli of V°, we have two boundary points to choose from, $x = 0$ and $x = -5^{-5}$. We claim that $x = 0$ corresponds to $q = 0$, so that x is the desired local coordinate. The reason is as follows. Back on the quintic threefold V, cup product with H gives an endomorphism $\cup H$ of $\oplus_{p=0}^3 H^{p,p}(V)$ which is maximally nilpotent (meaning $(\cup H)^3 \neq 0$ but $(\cup H)^4 = 0$). We will see in Chapters 5 and 6 that under mirror symmetry, we expect $\cup H$ to correspond to the logarithm of the monodromy about the point of the complex moduli space of V° corresponding to $q = 0$. Hence, mirror symmetry tells us to look for *maximally unipotent monodromy*, and of the above boundary points, this occurs only at $x = 0$. Chapter 5 will give a careful definition of maximally unipotent monodromy.

In order to compute the A-model correlation function (2.1), we have two remaining tasks:

- The local coordinates q and x describe corresponding boundary points of the Kähler and complex moduli spaces, but the local isomorphism given by mirror symmetry is *not* $q = x$. From the physics point of view, there are *quantum corrections* to take into account. Hence we need to compute x as a function of q, and vice versa. Once this is done, we will have an explicit local mapping between the complex moduli space of V° and the Kähler moduli space of V. This is usually called the *mirror map*.

- We have to compute and correctly normalize the B-model correlation function, and then use the mirror map to write it as a function of q. This will give an explicit formula for (2.1).

We now turn our attention to the first of these tasks.

2.3. The Mirror Map

One way to think of the mirror map is that q is a canonically chosen coordinate on the Kähler moduli space of V. Thus we want to find a "canonical" coordinate on the moduli space of V°. As will be explained in Chapter 6, we can do this because the monodromy at $x = 0$ is maximally unipotent. Following [**Morrison2**], there is a minimal integral vanishing cycle γ_0 near $x = 0$, so that γ_0 is invariant under monodromy. Furthermore, there is a minimal integral cycle γ_1 transforming under monodromy about $x = 0$ as $\gamma_1 \mapsto \gamma_1 + m\gamma_0$ for some $m \in \mathbb{Z}$. In the case of the quintic, we have $m = 1$ [**Morrison2**]. It follows that for a holomorphic 3-form Ω on V°, the quantity $\int_{\gamma_1} \Omega / \int_{\gamma_0} \Omega$ transforms under monodromy into $(\int_{\gamma_1} \Omega / \int_{\gamma_0} \Omega) + 1$, so that

$$\exp\left(2\pi i \int_{\gamma_1} \Omega / \int_{\gamma_0} \Omega\right)$$

is a well-defined function in a neighborhood of $x = 0$. Furthermore, γ_0 is unique up to sign, and γ_1 is unique up to the same sign and an integer multiple of γ_0. Hence the above quantity is canonically determined on the moduli space of V°, and we also see that it is independent of Ω since V° is Calabi-Yau. It is then natural to assert that the mirror map is given by $t = \int_{\gamma_1} \Omega / \int_{\gamma_0} \Omega$ and

$$(2.6) \qquad q = \exp\left(2\pi i \int_{\gamma_1} \Omega / \int_{\gamma_0} \Omega\right).$$

A physics argument for this assertion is given in [**BCOV2**]. This mirror map can be mathematically proven to be the correct one (Theorem 11.1.1 or Theorem 11.2.2).

To express $\int_{\gamma_1} \Omega$ and $\int_{\gamma_0} \Omega$ explicitly as functions of x, we will follow the approach of [**Morrison1**]. For any 3-cycle γ on V°, the integral $y = \int_\gamma \Omega$ is a *period* of the family. It is well known that the periods satisfy the Picard-Fuchs equation, which for a threefold with $h^3(V^\circ) = 4$ is a differential equation of the form

$$y'''' + f_1 y''' + f_2 y'' + f_3 y' + f_4 y = 0,$$

where the f_i depend on the moduli coordinate x and differentiation is with respect to x. In general, Picard-Fuchs equations can be calculated by several methods: explicit calculation of the periods [**CdGP**], the Griffiths-Dwork method of reduction of pole order [**Morrison1, Font, KT1**], generalized hypergeometric equations as in [**BvS, HKTY1**], or finally, by an explicit formula for toric complete intersections according to an assertion in [**Givental4**]. The last three of these methods will be discussed in Chapter 5. Alternative ways of organizing the data together with proofs of different versions of the mirror theorem are given in Chapter 11. We will give a sneak preview of these approaches in Section 2.6.

If we choose for Ω the 3-form

$$(2.7) \qquad \begin{aligned} \Omega = \mathrm{Res}\Big(&\frac{\psi}{x_1^5 + \ldots + x_5^5 + \psi x_1 \cdots x_5}(x_1 dx_2 \wedge dx_3 \wedge dx_4 \wedge dx_5 - \\ &x_2 dx_1 \wedge dx_3 \wedge dx_4 \wedge dx_5 + x_3 dx_1 \wedge dx_2 \wedge dx_4 \wedge dx_5 - \\ &x_4 dx_1 \wedge dx_2 \wedge dx_3 \wedge dx_5 + x_5 dx_1 \wedge dx_2 \wedge dx_3 \wedge dx_4)\Big), \end{aligned}$$

then the methods of Chapter 5 give the Picard-Fuchs equation

$$\text{(2.8)} \qquad \begin{aligned} 0 = &\left(x\frac{d}{dx}\right)^4 y + \frac{2\cdot5^5 x}{1+5^5 x}\left(x\frac{d}{dx}\right)^3 y + \frac{7\cdot5^4 x}{1+5^5 x}\left(x\frac{d}{dx}\right)^2 y \\ &+ \frac{2\cdot5^4 x}{1+5^5 x}\left(x\frac{d}{dx}\right)y + \frac{24\cdot5x}{1+5^5 x}y. \end{aligned}$$

This equation has a regular singular point at $x = 0$. Note also that there is a singularity at the other boundary point $x = -5^{-5}$.

The periods $y_0 = \int_{\gamma_0}\Omega$ and $y_1 = \int_{\gamma_0}\Omega$ satisfy this equation, and the monodromy properties of γ_0 and γ_1 imply that y_0 and $2\pi i\, y_1 - (\log x)y_0$ are single-valued at $x = 0$. Then, using standard methods for solving (2.8), we find that y_0 and y_1 are determined uniquely up to

$$\begin{aligned} y_0 &\mapsto b_1\, y_0 \\ y_1 &\mapsto b_1\, y_1 + b_2\, y_0, \end{aligned}$$

where $b_1, b_2 \in \mathbb{C}$ and $b_1 \neq 0$ (this is similar to the uniqueness for γ_0, γ_1 except that we have lost the integer lattice). It follows that we can compute

$$q = e^{2\pi i\, y_1/y_0}$$

up to a constant $c_1 = \exp(2\pi i\, b_2/b_1)$. The result is that the mirror map is

$$\text{(2.9)} \qquad q = c_1(x - 770x^2 + \cdots),$$

with inverse

$$\text{(2.10)} \qquad x = \frac{q}{c_1} + 770\left(\frac{q}{c_1}\right)^2 + \cdots.$$

In Section 2.5, we will use mirror symmetry to show that $c_1 = -1$. There are other ways to arrive at this normalization, for example this is a natural choice in implementing the method of Frobenius in the solution of generalized hypergeometric functions. This idea appears in [BvS], and we will return to this point in Chapter 6 when we discuss the mirror map in general.

Using methods for solving generalized hypergeometric equations, it is possible to get closed forms for many of these formulas. For example, up to a constant, y_0 is given by

$$\text{(2.11)} \qquad y_0 = \sum_{n=0}^{\infty} \frac{(5n)!}{(n!)^5}(-1)^n x^n = 1 - 120x + \cdots,$$

and the mirror map can be shown to be (assuming $c_1 = -1$)

$$\text{(2.12)} \qquad q = -x\exp\left(\frac{5}{y_0(x)}\sum_{n=1}^{\infty}\frac{(5n)!}{(n!)^5}\left[\sum_{j=n+1}^{5n}\frac{1}{j}\right](-1)^n x^n\right).$$

The details of this derivation will appear in Section 6.3. We will see another approach in Section 2.6. Using (2.12) and Dwork's theory of p-adic hypergeometric series, one can prove that q is a power series in x with integer coefficients [LY2].

2.4. The B-Model Correlation Function

We now turn to the problem of computing the Yukawa coupling on V°. The crucial but elementary observation is that the Yukawa coupling also satisfies a linear differential equation, which can easily be derived from the Picard-Fuchs equations. By formula (1.9) from Section 1.2, we have

$$(2.13) \qquad \langle x\tfrac{d}{dx}, x\tfrac{d}{dx}, x\tfrac{d}{dx} \rangle = \int_{V^\circ} \Omega \wedge \Omega''',$$

where the prime denotes $x\tfrac{d}{dx}$. Here, we assume that Ω is given by (2.7). We could replace Ω with its multiple by any nonvanishing function of x. Which multiple to choose is the *normalization problem*, which we will postpone until later in the section. As noted in the footnote in Section 1.2, normalization will also insert a minus sign on the right hand side of (2.13). This will be explained in Section 5.6.4.

We claim that the Yukawa coupling $Y = \int_{V^\circ} \Omega \wedge \Omega'''$ satisfies the differential equation

$$(2.14) \qquad \left(x\frac{d}{dx}\right)Y = \frac{-5^5 x}{1 + 5^5 x} Y.$$

To see this, first observe that by Griffiths transversality, Ω' and Ω'' have no $(0,3)$ component. Hence

$$(2.15) \qquad \int_{V^\circ} \Omega \wedge \Omega = \int_{V^\circ} \Omega \wedge \Omega' = \int_{V^\circ} \Omega \wedge \Omega'' = 0.$$

Differentiating the last equality twice gives

$$(2.16) \qquad \int_{V^\circ} \Omega \wedge \Omega'''' + 2 \int_{V^\circ} \Omega' \wedge \Omega''' = 0.$$

On the other hand, $Y = \int_{V^\circ} \Omega \wedge \Omega'''$ gives $Y' = \int_{V^\circ} \Omega' \wedge \Omega''' + \int_{V^\circ} \Omega \wedge \Omega''''$. Using (2.16), we conclude that $Y' = \tfrac{1}{2} \int_{V^\circ} \Omega \wedge \Omega''''$. However, the Picard-Fuchs equation (2.8) implies that

$$(2.17) \qquad 0 = \Omega'''' + \frac{2 \cdot 5^5 x}{1 + 5^5 x}\Omega''' + \frac{7 \cdot 5^4 x}{1 + 5^5 x}\Omega'' + \frac{2 \cdot 5^4 x}{1 + 5^5 x}\Omega' + \frac{24 \cdot 5x}{1 + 5^5 x}\Omega$$

in cohomology. The result (2.14) now follows by wedging Ω with the right hand side of (2.17) and using (2.15).

The solution of (2.14) is

$$(2.18) \qquad Y = \frac{c_2}{1 + 5^5 x},$$

where c_2 is a constant which will be determined in the next section. Thus we know the Yukawa coupling up to a constant. Notice that it is singular at the boundary point $x = -5^{-5}$, as might be expected.

However, we are not done, for we still need to normalize the Yukawa coupling, since the formula (2.7) for Ω can be multiplied by any nonvanishing function of x. A natural way to normalize Ω is by demanding that $\int_{\gamma_0} \Omega \equiv 1$ after normalization. With the choice of Ω from (2.7), we use the period

$$y_0(x) = \int_{\gamma_0} \Omega$$

and replace Ω by $\Omega/y_0(x)$. This achieves the desired normalization. Because of (2.15), one easily shows that the effect of this normalization on the Yukawa coupling

Y is to replace Y by $Y/y_0(x)^2$. Hence, for this choice of Ω, we get the *normalized Yukawa coupling*

$$(2.19) \qquad \langle x\tfrac{d}{dx}, x\tfrac{d}{dx}, x\tfrac{d}{dx}\rangle = \frac{c_2}{(1+5^5 x)y_0(x)^2}.$$

Note that this expression is essentially unaffected by the introduction of the minus sign mentioned above, since the sign can be incorporated into the choice of c_2. This may be the reason why this sign has not to our knowledge appeared in the mathematics literature before now. Chapter 5 will give general methods for computing the normalized Yukawa coupling and discuss its surprisingly strong connection to the Gauss-Manin connection.

2.5. Putting It All Together

We can now use the formulas from the previous two sections to compute the A-model correlation function of the quintic threefold. Under mirror symmetry, we get an isomorphism

$$H^1(V,\Omega^1_V) \simeq H^1(V^\circ, T_{V^\circ})$$

which is compatible with the A-model and B-model correlation functions. Thus, if $\theta \in H^1(V^\circ, T_{V^\circ})$ corresponds to $H \in H^1(V, \Omega^1_V)$ via mirror symmetry, then as explained in Chapter 1, we should have

$$\langle H, H, H\rangle = \langle \theta, \theta, \theta\rangle.$$

We find θ as follows. Thinking of $H^1(V,\Omega^1_V)$ as the tangent space to the Kähler moduli, H is a vector field on the Kähler moduli space of V. Writing a general complexified Kähler class as tH, this vector field is identified with d/dt. In terms of the local coordinate $q = e^{2\pi i t}$ for Kähler moduli, we have $\frac{d}{dt} = 2\pi i\, q\frac{d}{dq}$. Under the mirror map $q = q(x)$, $\frac{d}{dq}$ maps to $\frac{dx}{dq}\frac{d}{dx}$, and it follows that

$$H = 2\pi i\, q\frac{d}{dq} \;\longmapsto\; \theta = 2\pi i\, q\frac{dx}{dq}\frac{d}{dx}.$$

Then $\langle H, H, H\rangle = \langle \theta, \theta, \theta\rangle$ gives the equation

$$(2.20) \quad \langle H, H, H\rangle = \Big(2\pi i q\frac{dx}{dq}\Big)^3 \Big\langle \frac{d}{dx}, \frac{d}{dx}, \frac{d}{dx}\Big\rangle = \Big(2\pi i \frac{q}{x}\frac{dx}{dq}\Big)^3 \Big\langle x\frac{d}{dx}, x\frac{d}{dx}, x\frac{d}{dx}\Big\rangle.$$

This assertion, which is the equality of the A-model and B-model correlation functions via the mirror map, is mirror symmetry for the quintic threefold.

To make this equation more explicit, note that we calculated $\langle x\tfrac{d}{dx}, x\tfrac{d}{dx}, x\tfrac{d}{dx}\rangle$ in (2.19). Hence, using (2.19), the mirror map (2.9) and (2.10), and the formula for $y_0(x)$ given in (2.11), we obtain

$$\langle H, H, H\rangle = \frac{c_2}{(1+5^5 x)y_0(x)^2}\Big(2\pi i \frac{q}{x}\frac{dx}{dq}\Big)^3 =$$

$$\frac{c_2}{(1+5^5 x)(1-120x+\cdots)^2}\Big(2\pi i\, c_1(1-770x+\cdots)(1+1540\frac{q}{c_1^2}+\cdots)\Big)^3.$$

Converting wholly to q coordinates via (2.10), this simplifies to

$$(2.21) \qquad \frac{(2\pi i)^3 c_2}{(1+5^5(\frac{q}{c_1})+\cdots)(1-240(\frac{q}{c_1})+\cdots)}\Big(1+770(\frac{q}{c_1})+\cdots\Big)^3.$$

We next determine the constants c_1 and c_2. The basic idea is to choose them so that the above power series agrees with what we know about the formula

$$(2.22) \qquad \langle H, H, H \rangle = 5 + \sum_{d=1}^{\infty} n_d d^3 \frac{q^d}{1 - q^d} = 5 + 2875q + \cdots$$

given in (2.1). First, since the constant term of this series is 5, comparison with (2.21) shows that

$$c_2 = \frac{5}{(2\pi i)^3}.$$

Second, we claim that $c_1 = -1$. This is because with the above choice of c_2, (2.21) gives a power series expansion

$$\langle H, H, H \rangle = 5 - 2875\frac{q}{c_1} + \cdots .$$

It is known classically that the generic quintic threefold contains 2875 lines, which is why $n_1 = 2875$ in (2.22). Hence, in order for these series to be equal, we must have $c_1 = -1$. This assertion is not mere wishful thinking, for the desired value $c_1 = -1$ follows from the original calculation in [**CdGP**]. As mentioned earlier, we will give a systematic method for determining c_1 when we study the mirror map in Chapter 6.

Given these choices for c_1 and c_2, we can summarize mirror symmetry for the quintic threefold V as follows. First, by (2.11),

$$(2.23) \qquad y_0(x) = \sum_{n=0}^{\infty} \frac{(5n)!}{(n!)^5} (-1)^n x^n$$

is the unique (up to a constant) holomorphic solution of the Picard-Fuchs equation (2.8) of the mirror V°. There is also the solution

$$(2.24) \qquad y_1(x) = y_0(x) \log(-x) + 5 \sum_{n=1}^{\infty} \frac{(5n)!}{(n!)^5} \left[\sum_{j=n+1}^{5n} \frac{1}{j} \right] (-1)^n x^n$$

with a logarithmic singularity, and by (2.12), the mirror map is

$$(2.25) \qquad q = \exp(y_1/y_0) = -x \exp\left(\frac{5}{y_0(x)} \sum_{n=1}^{\infty} \frac{(5n)!}{(n!)^5} \left[\sum_{j=n+1}^{5n} \frac{1}{j} \right] (-1)^n x^n \right).$$

Then mirror symmetry for the quintic threefold, as stated in (2.20), asserts that

$$(2.26) \qquad 5 + \sum_{d=1}^{\infty} n_d d^3 \frac{q^d}{1 - q^d} = \frac{5}{(1 + 5^5 x) y_0(x)^2} \left(\frac{q\, dx}{x\, dq} \right)^3.$$

Recall the left hand side is the A-model correlation function of V and the right hand side is the B-model correlation function of V°, given by the normalized Yukawa coupling regarded as a function of q.

If we expand the right hand side of this equation, we obtain

$$(2.27) \qquad 5 + \sum_{d=1}^{\infty} n_d d^3 \frac{q^d}{1 - q^d} - 5 + 2875\frac{q}{1 - q} + 609250 \cdot 2^3 \frac{q^2}{1 - q^2} + \cdots ,$$

so that mirror symmetry implies that $n_2 = 609250$. Computing additional terms of this series, we get the following predictions for the instanton numbers n_d of the quintic threefold:

d	n_d
1	2875
2	609250
3	317206375
4	242467530000
5	229305888887625

The first four entries agree with known enumerative results: $n_1 = 2875$ is classical (as noted earlier), $n_2 = 609250$ has been shown in [**Katz2**], $n_3 = 317206375$ has been shown in [**ES1**], and $n_4 = 24246753000$ has been shown in [**Kontsevich2**]. These numbers are computed for all $d \leq 10$ in [**CdGP**].

For general d, equation (2.26) asserts that the numbers 2875, 609250, ... appearing in the series on the right hand side of (2.27) are precisely the instanton numbers n_d, which are defined in terms of the Gromov-Witten invariants of the quintic threefold by (2.2). We will prove in Section 7.4.4 that for $d \leq 9$, the numbers n_d are indeed equal to the number of rational curves of degree d on a generic quintic threefold. For $d \geq 10$, we still have the numbers n_d, but as already indicated, their relation to the rational curves of degree d is more subtle.

The numbers in the above table suggest that the n_d should have some remarkable divisibility properties. It is easy to see that the n_d are rational, but it is unknown if they are integers (though this is true for all that have been computed). However, if we assume that the n_d are integral, then one can prove that $5^3 | n_d$ for all d not divisible by 5 [**LY1**]. It is conjectured that $5^3 | n_d$ is true for *all* d.

So far, we have said nothing about how to prove (2.26). This is the essential content of the *Mirror Theorem*, which is our next topic of discussion.

2.6. The Mirror Theorems

The equation (2.26) given in the preceding sections is based on the "classical" formulation of mirror symmetry, where the primary goal is to prove that certain correlation functions can be identified via the mirror map. In order to justify these calculations, one needs to prove a *Mirror Theorem*. In this section, we will introduce three different versions of the Mirror Theorem, all of which imply "classical" mirror symmetry for the quintic threefold. Rather than state the theorems in general, we will instead discuss what they look like for the quintic threefold V and its mirror V°. Complete statements will appear in Chapters 8 and 11.

2.6.1. Hodge Theory. The mirror map constructed in Section 2.3 expresses the Kähler parameter q of the quintic threefold V as an explicit function of the moduli parameter x of the mirror V°. This allows us to use q as a new local parameter for the moduli of V°. Then, in Section 2.4, we determined the normalized Yukawa coupling (2.19). In Chapter 5, this normalized coupling will be denoted Y. The enumerative predictions of the previous section came by expressing Y in terms of q and equating it with the A-model correlation function (2.22).

However, we will learn in Chapter 5 that once we express Y in terms of q, we also completely determine the variation of Hodge structure on $H^3(V^\circ, \mathbb{C})$. More precisely, in the proof of Proposition 5.6.1, we will see that in an appropriate basis, the Gauss-Manin connection ∇ of V° has connection matrix

$$(2.28) \qquad \begin{pmatrix} 0 & 0 & 0 & 0 \\ 1 & 0 & 0 & 0 \\ 0 & Y & 0 & 0 \\ 0 & 0 & 1 & 0 \end{pmatrix}$$

provided we use q as moduli parameter.

This shows that the normalized B-model correlation function Y is not just some strange function suggested by physics, but rather is integral to the Hodge structure of the mirror family. Furthermore, since Y is supposed to correspond to the A-model correlation function, this suggests that there should be a variation of Hodge structure on the A-model side. This is our first hint that mirror symmetry can be formulated in terms of variations of Hodge structure. We will develop this idea more fully in Chapter 8 when we define the *A-variation of Hodge structure*. The associated connection, called the *A-model connection*, is constructed by means of quantum cohomology. In particular, for the quintic threefold V, we will see in Example 8.5.4.1 that the A-model connection has (2.28) as its connection matrix, provided Y is the A-model correlation function (2.22).

Chapter 8 will give a careful definition of a *mathematical mirror pair*. For the quintic threefold and its mirror, this means that the A-variation of Hodge structure of V should be isomorphic, via the mirror map, to the variation of Hodge structure coming from the mirror family V°. In Chapter 8, we will show that this version of mirror symmetry is equivalent to the equality of A-model and B-model correlation functions given in (2.26).

Proving mirror symmetry, either in its classical form or the more sophisticated Hodge-theoretic version just discussed, is not easy. In particular, the proof requires the introduction of some new objects of study. As we will now explain, these new objects lead in turn to new versions of the Mirror Theorem.

2.6.2. Givental's I and J Functions. In Givental's approach to the Mirror Theorem, a key role is played by two cohomology-valued formal functions denoted I_V and J_V. Let's first discuss J_V. This function is defined in terms of *gravitational correlators*, which can be thought of as generalizations of Gromov-Witten invariants. We actually begin with the Givental J-function. In Section 10.3.2, we will see that for the quintic threefold, J is given by the formula

$$(2.29) \qquad J = e^{(t_0 + t_1 H)/\hbar}\left(1 + \hbar^{-2}\sum_{d\geq 0} N_d\, q^d\, d\ell - 2\hbar^{-3}\sum_{d\geq 0} N_d\, q^d\, pt\right),$$

where $q = e^{t_1}$, ℓ is the class of a line in V, pt is the class of a point, and

$$(2.30) \qquad N_d = \langle I_{0,0,d}\rangle = \sum_{k\mid d} n_{\frac{d}{k}} k^{-3}.$$

Here, the $n_{d/k}$ are the instanton numbers we want to compute. Earlier in the chapter, we mentioned the Gromov-Witten invariant $\langle I_{0,3,d}\rangle(H,H,H)$. Properties of Gromov-Witten invariants from Chapter 7 imply $\langle I_{0,3,d}\rangle(H,H,H) = d^3\langle I_{0,0,d}\rangle$, so that the above formula for N_d is equivalent to (2.2).

The J-function plays a central role in Givental's theory of *quantum differential operators* from Section 10.3.1. For our immediate purposes, the key feature of (2.29) is that J contains all of the enumerative information we're interested in.

The function J takes values in the cohomology ring of V. However, in order to compare this to the function I_V described below, we need something that takes values in $H^*(\mathbb{P}^4)$. We can do this by defining $J_V = i_!(J)$, where $i : V \to \mathbb{P}^4$ is the inclusion map and $i_!$ is the Gysin map. In terms of the formula (2.29) for J, one easily obtains

$$(2.31) \qquad J_V = e^{(t_0+t_1 H)/\hbar} 5H \left(1 + \hbar^{-2} \sum_{d \geq 0} N_d q^d d \tfrac{1}{5} H^2 - 2\hbar^{-3} \sum_{d \geq 0} N_d q^d \tfrac{1}{5} H^3 \right).$$

(In Chapter 11, we will take a slightly different approach to defining J_V.) Here, the important fact is that the n_d are uniquely determined by J_V.

The second of Givental's functions is I_V, which is also a cohomology-valued function. For the quintic threefold V, the formula for I_V from Example 11.2.1.3 is

$$I_V = e^{(t_0+t_1 H)/\hbar} 5H \sum_{d=0}^{\infty} e^{dt_1} \frac{(5H+\hbar)\cdots(5H+5d\hbar)}{((H+\hbar)\cdots(H+d\hbar))^5}.$$

Although I_V takes values in the cohomology of $\mathbb{P}^4$, one of Givental's observations is that I_V is determined by the periods of the mirror V°. Thus I_V contains the moduli information for the mirror, and Givental's version of the Mirror Theorem states that I_V and J_V are related in an especially nice way. Our explanation of the proof uses the *Quantum Hyperplane Section Principle*, which involves localization formulas for equivariant cohomology on moduli spaces of stable maps. All of this will be discussed in Chapter 11.

In order to see how this relates to the computations done in earlier sections, let's simplify the above formula for I_V by taking $t_0 = 0$ and $\hbar = 1$. This gives

$$(2.32) \qquad I_V = e^{t_1 H} 5H \sum_{d=0}^{\infty} e^{dt_1} \frac{\prod_{m=1}^{5d}(5H+m)}{\prod_{m=1}^{d}(H+m)^5}.$$

We can expand I_V, writing

$$(2.33) \qquad I_V = 5H \left(y_0(t_1) + y_1(t_1)H + y_2(t_1)H^2 + y_3(t_1)H^3 \right).$$

After the change of variables $x = -e^{t_1}$, we will show in Chapter 5 that I_V satisfies the Picard-Fuchs equation of the mirror V°, as given in (2.8). In fact, using the Frobenius method, we will show in Chapter 6 that y_0, y_1, y_2, y_3 form a basis of solutions. Furthermore, y_0 is the series from (2.23), and similarly y_1 is (2.24). Hence the mirror map (2.25) arises naturally in Givental's formulation as $q = \exp(y_1/y_0)$.

Givental's version of the Mirror Theorem is equivalent to the assertion that $J_V = I_V/y_0$ after changing coordinates via the mirror map (see Theorem 11.2.2). Since J_V involves the n_d by (2.30) and (2.31) and I_V is given by the explicit formula (2.32), it follows that we can compute as many instanton numbers n_d as desired.

However, to complete the circle of ideas, we should explain how this relates to correlation functions. We first note that the normalized Yukawa coupling Y of V° from Section 2.5 arises naturally from I_V. To see why, consider I_V/y_0. If we expand up to terms containing e^{t_1} times a polynomial in t_1, we obtain

$$I_V/y_0 = 5H \left(1 + (t_1 + 770e^{t_1} + \cdots)H + (\frac{t_1^2}{2} + 770te^{t_1} + 575e^{t_1} + \cdots)H^2 \right.$$
$$\left. + (\frac{t_1^3}{6} + 385t_1^2 e^{t_1} + 575t_1 e^{t_1} - 1150e^{t_1} + \cdots)H^3 \right).$$

If we write this in terms of q, we get

$$I_\mathcal{V}/y_0 = 5H\left(1 + sH + \left(\frac{s^2}{2} + 575q + \cdots\right)H^2 + \left(\frac{s^3}{6} + 575sq - 1150q + \cdots\right)H^3\right),$$

where $s = \log q$. Thus, if $'$ denotes $\frac{d}{ds} = q\frac{d}{dq}$, one computes that

$$
\begin{aligned}
(2.34) \qquad (I_\mathcal{V}/y_0)'' &= 5(1 + 575q + \cdots)H^3 + 5(s + 575sq + \cdots)H^4 \\
&= (5 + 2875q + \cdots)(H^3 + sH^4).
\end{aligned}
$$

Notice that the first factor in the bottom line begins $5 + 2875q + \cdots$, which agrees with the normalized Yukawa coupling Y, at least up to terms of degree one in q. In fact, this factor is precisely the normalized coupling Y—this is Givental's observation that $I_\mathcal{V}$ is determined by the periods of the mirror. To prove this, we use the connection matrix (2.28), which implies that the normalized Picard-Fuchs equation can be written in the form

$$\frac{d^2}{ds^2}\left(\frac{d^2 y/ds^2}{Y}\right) = 0.$$

One easily sees that this equation has solutions f_0, f_1, f_2, f_3 such that $f_0 = 1$, $f_1 = s$, $f_2'' = Y$ and $f_3'' = Ys$. In (2.33), we wrote down solutions y_0, y_1, y_2, y_3 of the unnormalized Picard-Fuchs equation (2.8). Normalizing means dividing by y_0, so that we have solutions $y_0/y_0 = 1$, $y_1/y_0 = s$, y_2/y_0 and y_3/y_0. The wonderful fact is that y_2/y_0 and y_3/y_0 are essentially f_2 and f_3. More formally, we have the identities

$$(2.35) \qquad 5\frac{d^2}{ds^2}\frac{y_2}{y_0} = Y, \quad 5\frac{d^2}{ds^2}\frac{y_3}{y_0} = Ys.$$

To prove the first equation, note that $y_2/y_0 = a + bs + cf_2 + df_3$ for some constants a, b, c, d, so that $(y_2/y_0)'' = cf_2'' + df_3'' = cY + dYs$. By (2.34), we have $(y_2/y_0)'' = 1 + 575q + \cdots$, and since $Y = 5 + 2875q + \cdots$, it follows that $c = 1/5$ and $d = 0$. This proves the first equation, and argument for the second is similar.

Using (2.35), we see that $(I_\mathcal{V}/y_0)'' = Y(H^3 + sH^4)$, and we will prove similarly in Proposition 10.3.4 that $J_\mathcal{V}'' = \langle H, H, H\rangle(H^3 + sH^4)$. Hence Theorem 11.2.2, as described above, implies $\langle H, H, H\rangle = Y$, which proves the desired equality (2.26) of A-model and B-model correlation functions. This gives "classical" mirror symmetry for the quintic threefold.

2.6.3. Euler Data and the Gromov-Witten Potential.

A different approach to the Mirror Theorem is presented in [**LLY**]. This paper, which contains a complete proof of mirror symmetry for the quintic threefold, will be discussed in Chapter 11. Here is a small preview of what's involved.

The paper [**LLY**] introduces several objects of interest for us to consider. The first is *Euler data*. This is a compatible collection of elements in the equivariant cohomology of certain projective spaces which "linearize" the moduli spaces of stable maps. There are two special sets of Euler data, denoted $\hat{Q}$ and $\hat{P}$, which roughly speaking code the information for the A-model and B-model correlation functions respectively.

From Euler data, one can construct certain cohomology-valued functions, where the values are now in equivariant cohomology. For example, the Euler data $\hat{P}$ for the quintic threefold gives the function

$$HG[\mathcal{I}(\hat{P})](t_1) = e^{-pt_1/\hbar} \sum_{d \geq 0} \frac{\prod_{m=0}^{5d}(5p - m\hbar)}{\prod_{m=1}^{d}\prod_{k=0}^{4}(p - \lambda_k - m\hbar)} e^{dt_1}$$

(see the discussion following Definition 11.1.8). Here, p is the equivariant hyperplane section and the λ_k are certain equivariant cohomology classes. If we let $\lambda_k \to 0$ (in Chapter 9, we will call this the "nonequivariant limit"), then we get ordinary cohomology. Thus p becomes H, and then setting $\hbar = -1$ gives the formula for I_V given in (2.32). In particular, (2.33) becomes

$$(2.36) \qquad \lim_{\lambda \to 0} HG[\mathcal{I}(\hat{P})](t_1) = I_V = 5H\left(y_0(t_1) + y_1(t_1)H + y_2(t_1)H^2 + y_3(t_1)H^3\right).$$

As before, this allows us to write the mirror map as $q = \exp(y_1/y_0)$. We will use $\Psi(t_1)$ to denote the quotient y_1/y_0, so that the mirror map is $q = \exp(\Psi(t_1))$. If we let $q = e^s$, then the mirror map is $s = \Psi(t_1) = \frac{y_1(t_1)}{y_0(t_1)}$.

So far, we've been dealing with the Euler data $\hat{P}$ which describes the complex moduli of V°. There is also the Euler data $\hat{Q}$ which is determined by the Gromov-Witten invariants of V. The easiest way to understand this is via the *Gromov-Witten potential*, which is defined by

$$(2.37) \qquad \Phi(s) = \frac{5}{6}s^3 + \sum_{d=1}^{\infty} N_d\, q^d = \frac{5}{6}s^3 + \sum_{d=1}^{\infty} \langle I_{0,0,d} \rangle\, q^d,$$

where N_d is given by (2.30) (and $q = e^s$). The function $\Phi(s)$ is a close cousin of the A-model correlation function, and in fact one sees easily that (2.22) is $\frac{d^3}{ds^3}\Phi(s)$. We will study the Gromov-Witten potential in detail in Chapter 8.

Now consider the Euler data $\hat{Q}$. This gives the equivariant cohomology-valued function $HG[\mathcal{I}(\hat{Q})](s)$, and in Section 11.1, we will show that

$$(2.38) \qquad \lim_{\lambda \to 0} HG[\mathcal{I}(\hat{Q})](s) = 5H\left(1 + sH + \frac{\Phi'}{5}H^2 + \frac{s\Phi' - 2\Phi}{5}H^3\right),$$

where $'$ is $\frac{d}{ds}$. (The version of this formula given in Chapter 11 has an $\hbar$ which as above we have set to -1.) This formula reflects the above comment that $\hat{Q}$ is determined by Gromov-Witten invariants.

Then the key step in the proof is to show that after the substitution $s = \Psi(t_1)$ coming from the mirror map, the limit (2.38) coincides with the limit (2.36), up to a factor of y_0. More precisely, Lemma 11.1.15 will prove that

$$\lim_{\lambda \to 0} HG[\mathcal{I}(\hat{Q})](\Psi(t_1)) = \frac{1}{y_0} \lim_{\lambda \to 0} HG[\mathcal{I}(\hat{P})](t_1).$$

Combining this with the formulas (2.38) and (2.36) and comparing coefficients, one easily obtains the identity

$$(2.39) \qquad \Phi(\Psi(t_1)) = \frac{5}{2}\left(\frac{y_1}{y_0}\frac{y_2}{y_0} - \frac{y_3}{y_0}\right).$$

This is the Mirror Theorem for the quintic threefold as presented in [**LLY**].

Since $\Phi(s)$ is the Gromov-Witten potential (2.37), formula (2.39) enables us to compute n_d for arbitrary d. But how does this relate to three-point functions? We've already noted that $\frac{d^3}{ds^3}\Phi$ is the A-model correlation function. Mirror symmetry for the quintic threefold says that under the mirror map, this equals the

B-model correlation function Y. Hence the formulation of mirror symmetry given in (2.26) follows from (2.39), provided we can show that

$$(2.40) \qquad \frac{d^3}{ds^3} \frac{5}{2} \left(\frac{y_1}{y_0} \frac{y_2}{y_0} - \frac{y_3}{y_0} \right) = Y$$

(where as usual the y_i are functions of s via the mirror map).

To prove this, one first uses (2.35) to show that

$$\frac{d^2}{ds^2} \frac{5}{2} \left(\frac{y_1}{y_0} \frac{y_2}{y_0} - \frac{y_3}{y_0} \right) = \frac{d^2}{ds^2} \frac{5}{2} \left(s \frac{y_2}{y_0} - \frac{y_3}{y_0} \right) = 5 \frac{d}{ds} \frac{y_2}{y_0}.$$

We leave the straightforward proof to the reader. Then, differentiating a third time and using (2.35) again, (2.40) follows immediately. This completes the proof of mirror symmetry for the quintic threefold, modulo some substantial work to be done between here and Chapter 11.

In this approach to the Mirror Theorem, notice that the normalized Yukawa coupling Y appears only at the end of the computation—all of the intermediate steps involve new objects to study. The same thing happens in Givental's approach, though the intermediate objects are slightly different. Hence, as we develop the mathematics needed to formulate mirror symmetry in Chapters 3 through 8, the reader should keep in mind that some fundamentally new ideas will be needed in Chapter 11 in order to prove mirror symmetry. This will involve a lot of work, but the wonder of the final result will more than justify the effort.

CHAPTER 3

Toric Geometry

This chapter will explore the theory of toric varieties over the complex numbers. We will assume some familiarity with the classical way of describing toric varieties in terms of cones and fans, as in [**Danilov, Fulton3, Oda**], though we will also describe more recent constructions of toric varieties which use polytopes and homogeneous coordinates. We will also discuss those aspects of toric geometry most relevant to mirror symmetry, including Kähler cones, symplectic reduction, the GKZ decomposition, Fano toric varieties and reflexive polytopes, and automorphism groups. The chapter will end with some examples.

3.1. Cones and Fans

We begin with a summary of toric varieties, mostly to fix notation and terminology. Proofs can be found in [**Danilov, Fulton3, Oda**].

Let $M \simeq \mathbb{Z}^n$ be a free Abelian group of rank n, and let $N = \text{Hom}(M, \mathbb{Z})$ be its dual. The pairing between $m \in M$ and $v \in N$ is given by $\langle m, v \rangle \in \mathbb{Z}$. A *rational polyhedral cone* $\sigma \subset N_\mathbb{R} = N \otimes \mathbb{R}$ is a subset of the form

$$\sigma = \left\{ \sum_{i=1}^s \lambda_i u_i : \lambda_i \geq 0 \right\}$$

where $u_1, \ldots, u_s \in N$. We say that σ is *strongly convex* if $\sigma \cap (-\sigma) - \{0\}$. The *dimension* of σ is the dimension of the subspace of $N_\mathbb{R}$ spanned by σ. Every cone σ has a *dual cone* $\check{\sigma}$ defined by

$$\check{\sigma} = \{ m \in M_\mathbb{R} : \langle m, v \rangle \geq 0 \text{ for all } v \in \sigma \}.$$

Standard theory shows that $\check{\sigma}$ is also a rational polyhedral cone. Given $m \in M \cap \check{\sigma}$, the subset

$$\tau = \{ v \in \sigma : \langle m, v \rangle = 0 \} \subset \sigma$$

is a *face* of σ. Every face of σ is again a rational polyhedral cone and a face of a face is a face.

A *fan* Σ in $N_\mathbb{R}$ consists of a finite collection of strongly convex rational polyhedral cones in $N_\mathbb{R}$ satisfying:

- If $\sigma \in \Sigma$, then every face of σ is also in Σ.
- If $\sigma, \tau \in \Sigma$, then $\sigma \cap \tau$ is a face of each.

The set $|\Sigma| = \cup_{\sigma \in \Sigma} \sigma \subset N_\mathbb{R}$ is the *support* of Σ. For each d, $\Sigma(d)$ denotes the set of d-dimensional cones of Σ.

The 1-dimensional cones, sometimes called the *1-skeleton* of the fan, are especially important. We reserve the letter ρ to stand for elements of $\Sigma(1)$. For each ρ, let v_ρ be the unique generator of the semigroup $\rho \cap N$. Using these generators, a cone $\sigma \in \Sigma$ can be written

$$\sigma = \left\{ \sum_{\rho \subset \sigma} \lambda_\rho v_\rho : \lambda_\rho \geq 0 \right\}.$$

31

The $v_\rho \in \sigma$ are the *generators* of σ. A standard abuse of notation is to identify ρ with v_ρ. Thus we may write $v_\rho \in \Sigma(1)$. If $r = |\Sigma(1)|$ is the number of 1-dimensional cones, we sometimes denote the v_ρ's by $v_1, \ldots, v_r$.

We next recall the classical definition of toric variety. Given a fan Σ, each cone $\sigma \in \Sigma$ gives the affine toric variety

$$X_\sigma = \mathrm{Spec}(\mathbb{C}[M \cap \check{\sigma}]),$$

where $\mathbb{C}[M \cap \check{\sigma}]$ is the $\mathbb{C}$-algebra with generators χ^m for each $m \in M \cap \check{\sigma}$ and relations $\chi^m \chi^{m'} = \chi^{m+m'}$. The toric variety X_Σ is obtained from these affine pieces by gluing together X_σ and X_τ along $X_{\sigma \cap \tau}$.

If Σ is a fan in $N_\mathbb{R} \simeq \mathbb{R}^n$, then the toric variety X_Σ is a Cohen-Macaulay algebraic variety of dimension n. Since the affine toric variety corresponding to the trivial cone $\{0\}$ is the torus $T_N = N \otimes \mathbb{C}^* = \mathrm{Spec}(\mathbb{C}[M])$, we see that T_N is an affine open subset of X_Σ. The action of the torus on itself extends to an algebraic action of T_N on X_Σ. This is where the name *toric variety* comes from. When there is no danger of confusion, we will write X instead of X_Σ.

The torus action on X has only finitely many orbits, and there is an inclusion-reversing one-to-one correspondence between orbit closures and cones of Σ. In particular, each $\rho \in \Sigma(1)$ corresponds to an irreducible T_N-invariant divisor $D_\rho \subset X$. Furthermore, $m \in M$ gives a character $\chi^m : T_N \to \mathbb{C}^*$, and regarding χ^m as a rational function on X, we have

$$(3.1) \qquad \mathrm{div}(\chi^m) = \sum_\rho \langle m, v_\rho \rangle D_\rho.$$

The properties of the fan Σ strongly affect the geometry of the toric variety X. For example:

- X is complete (i.e., compact) if and only if $|\Sigma| = N_\mathbb{R}$. Such a fan is called *complete*.
- X is smooth (i.e., nonsingular) if and only if for every cone in Σ, its generators are part of a $\mathbb{Z}$-basis of N. Such a fan is called *smooth*.
- X is an orbifold (i.e., a V-manifold) if and only if the generators of every cone in Σ are linearly independent over $\mathbb{R}$. We say that Σ and X are *simplicial*.

A general toric variety has only mild singularities (being Cohen-Macaulay), though we will sometimes require that X be simplicial. As explained in Appendix A, a simplicial toric variety behaves like a manifold in many ways.

The Chow group $A_{n-1}(X)$ of Weil divisors modulo linear equivalence can be computed directly from the fan. We will always assume that the 1-dimensional cones span $N_\mathbb{R}$. Then, using (3.1), we get an exact sequence

$$(3.2) \qquad 0 \longrightarrow M \longrightarrow \mathbb{Z}^{\Sigma(1)} \longrightarrow A_{n-1}(X) \longrightarrow 0$$

where $m \in M$ maps to $(\langle m, v_\rho \rangle) \in \mathbb{Z}^{\Sigma(1)}$ and $(a_\rho) \in \mathbb{Z}^{\Sigma(1)}$ maps to the divisor class of $\sum_\rho a_\rho D_\rho$. Thus $A_{n-1}(X)$ has rank $r - n$.

Sitting inside the Chow group $A_{n-1}(X)$ is the Picard group $\mathrm{Pic}(X)$, which consists of Cartier divisors modulo linear equivalence. A Weil divisor $D = \sum_\rho a_\rho D_\rho$ is Cartier if and only if for each $\sigma \in \Sigma$, there is $m_\sigma \in M$ such that $\langle m_\sigma, v_\rho \rangle = -a_\rho$ whenever $\rho \subset \sigma$. The function $\phi_D : |\Sigma| \to \mathbb{R}$, defined by $\phi_D(v) = \langle m_\sigma, v \rangle$ for $v \in \sigma$, is the *support function* of D.

If X is smooth, then $\mathrm{Pic}(X) = A_{n-1}(X)$. In the simplicial case, the Picard group has finite index in the Chow group. The Picard group is always torsion free when X is complete, while $A_{n-1}(X)$ can have torsion, even if X is simplicial.

If X is complete and the divisor $D = \sum_\rho a_\rho D_\rho$ is Cartier, we can determine whether D is ample or generated by global sections as follows:

- D is generated by global sections $\Leftrightarrow \langle m_\sigma, v_\rho \rangle \geq -a_\rho$ whenever $\rho \not\subset \sigma$.
- D is ample $\Leftrightarrow \langle m_\sigma, v_\rho \rangle > -a_\rho$ whenever $\rho \not\subset \sigma$ and σ is n-dimensional.

We say that ϕ_D is *upper convex* in the first case, and *strictly upper convex* in the second. In the next section, we will interpret these conditions in terms of certain polytopes.

3.2. Polytopes and Homogeneous Coordinates

The classical definition of toric variety, as presented in the previous section, involves gluing together affine toric varieties. Recently, other ways of creating toric varieties have been discovered, and this section will explore two of these alternate constructions, one coming from polytopes and the other generalizing the usual homogeneous coordinates of $\mathbb{P}^n$. A third construction, involving symplectic reduction, will be discussed in Section 3.3.3.

3.2.1. Polytopes. We begin by setting terminology. A *polytope* $\Delta \subset M_\mathbb{R}$ is the convex hull of a finite set of points. The *dimension* of Δ is the dimension of the subspace spanned by the differences $\{m_1 - m_2 : m_1, m_2 \in \Delta\}$. We say that Δ is *integral* if the vertices of Δ lie in M (equivalently, Δ is the convex hull of a finite subset of M). Finally, a *facet* is a codimension one face of Δ.

Given polytopes $\Delta_1, \ldots, \Delta_k$ in $M_\mathbb{R}$, we can create new polytopes in several ways. For example, the convex hull of $\Delta_1 \cup \cdots \cup \Delta_k$ is denoted

$$\mathrm{Conv}(\Delta_1, \ldots, \Delta_k).$$

We can also define their *Minkowski sum*, which is the set

$$\Delta_1 + \cdots + \Delta_k = \{m_1 + \cdots + m_k : m_i \in \Delta_i\}.$$

If $k\Delta = \Delta + \cdots + \Delta$ (k times), note that $k\Delta = \{km : m \in \Delta\}$ by convexity. The mirror symmetry constructions discussed in Section 4.3 make essential use of convex hulls and Minkowski sums.

Polytopes arise naturally when dealing with toric varieties. For example, when X is complete and $D = \sum_\rho a_\rho D_\rho$ is Cartier, then

$$(3.3) \qquad \begin{aligned} \Delta_D &= \{m \in M_\mathbb{R} : \langle m, v \rangle \geq \phi_D(v) \text{ for all } v \in N_\mathbb{R}\} \\ &= \{m \in M_\mathbb{R} : \langle m, v_\rho \rangle \geq -a_\rho \text{ for all } \rho\} \end{aligned}$$

is a polytope.[1] One also has $\Delta_{kD} = k\Delta_D$, $\Delta_{D+\mathrm{div}(\chi^m)} = \Delta_D - m$, and

- D is generated by global sections $\Leftrightarrow \Delta_D$ is the convex hull of the set $\{m_\sigma : \sigma \in \Sigma(n)\}$.
- D is ample $\Leftrightarrow m_\sigma \neq m_\tau$ for $\sigma \neq \tau$ in $\Sigma(n)$ and Δ_D is an n-dimensional polytope with vertices $\{m_\sigma : \sigma \in \Sigma(n)\}$.

In either case, Δ_D is an integral polytope. Furthermore, when D is ample, there is a bijective correspondence between nonempty faces of Δ_D and cones of Σ. In particular, facets of Δ correspond to elements of $\Sigma(1)$. Proofs of these assertions can be found in [**Danilov, Fulton3, Oda**].

---

[1] The second equality can still be used to define Δ_D even if D is not Cartier.

Another useful observation is that $\Delta_D \cap M$ is naturally identified with the T_N eigenvectors of $H^0(X, \mathcal{O}(D))$. In other words, there is a T_N-equivariant map

$$(3.4) \qquad H^0(X, \mathcal{O}(D)) \simeq \bigoplus_{m \in \Delta_D \cap M} \mathbb{C}\chi^m.$$

To see why this is true, think of sections of $\mathcal{O}(D)$ as rational functions f on X such that $(f) + D \geq 0$. By (3.1), $\operatorname{div}(\chi^m) + D \geq 0$ is equivalent to $\langle m, v_\rho \rangle \geq -a_\rho$ for all ρ, and the isomorphism now follows easily (see [**Fulton3**, Section 3.4] for details).

3.2.2. Toric Varieties via Polytopes. We now describe how to start with an n-dimensional integral polytope and construct a toric variety. We will follow the approach of [**Batyrev4**]. Given Δ, consider "monomials" $t_0^k \chi^m$ where $m \in k\Delta$. These monomials multiply by the rule $t_0^k \chi^m \cdot t_0^l \chi^{m'} = t_0^{k+l} \chi^{m+m'}$ since $m \in k\Delta$ and $m' \in l\Delta$ imply $m + m' \in (k+l)\Delta$. The $\mathbb{C}$-algebra generated by the $t_0^k \chi^m$ is denoted S_Δ, and this ring is graded by declaring that

$$\deg(t_0^k \chi^m) = k.$$

We call S_Δ the *polytope ring* of Δ. Then let

$$\mathbb{P}_\Delta = \operatorname{Proj}(S_\Delta)$$

be the corresponding projective variety. We will write $\mathbb{P}$ instead of $\mathbb{P}_\Delta$ if the context is clear.

To describe $\mathbb{P} = \mathbb{P}_\Delta$ as a toric variety, we need to exhibit its fan in $N_\mathbb{R}$. Given a nonempty face $F \subset \Delta$, consider the cone

$$\check{\sigma}_F = \{\lambda(m - m') : m \in \Delta, \ m' \in F, \ \lambda \geq 0\} \subset M_\mathbb{R}.$$

Its dual is a cone $\sigma_F \subset N_\mathbb{R}$. Putting these cones together gives the fan $\{\sigma_F : F \text{ is a nonempty face of } \Delta\}$, which we call the *normal fan* of the polytope Δ. This is a complete fan in $N_\mathbb{R}$, and one can prove that $\mathbb{P}$ is the toric variety determined by the normal fan [**Batyrev4**].

When the origin is an interior point of Δ, the normal fan is especially easy to visualize. First define the *polar polytope*

$$\Delta^\circ = \{v \in N_\mathbb{R} : \langle m, v \rangle \geq -1 \text{ for all } m \in \Delta\} \subset N_\mathbb{R}.$$

The polytope Δ° contains the origin as an interior point and $(\Delta^\circ)^\circ = \Delta$. Furthermore, each i-dimensional face F of Δ corresponds to a $(n - 1 - i)$-dimensional face F° of Δ°, determined by the condition that $\langle m', v \rangle = -1$ when $m' \in F$ and $v \in F^\circ$. Using the polytope Δ°, we can describe $\mathbb{P}$ as follows.

LEMMA 3.2.1. *The fan in $N_\mathbb{R}$ obtained from the cones over the proper faces of Δ° is the normal fan of Δ and hence gives the toric variety $\mathbb{P} = \mathbb{P}_\Delta$.*

PROOF. It suffices to show that $\sigma_F = \{\lambda v : v \in F^\circ, \ \lambda \geq 0\}$ for every proper face F of Δ. First observe that $v \in \sigma_F$ if and only if $\langle m, v \rangle \geq \langle m', v \rangle$ for all $m \in \Delta$ and $m' \in F$. Also note that $v \neq 0$ implies $\langle m', v \rangle < 0$, since otherwise, we would have $\langle m, v \rangle \geq 0$ for all $m \in \Delta$. This is impossible since 0 is an interior point of Δ.

If $v \in F^\circ$, then $\langle m, v \rangle \geq \langle m', v \rangle = -1$, so that $\lambda v \in \sigma_F$ follows easily. Going the other way, suppose $\langle m, v \rangle \geq \langle m', v \rangle$ for all $m \in \Delta$ and $m' \in F$. We can assume $v \neq 0$. For a fixed m' in the relative interior of F, $\langle m', v \rangle$ is negative (as noted above), so that we can write $v = \lambda v'$ where $\lambda > 0$ and $\langle m, v' \rangle \geq \langle m', v' \rangle = -1$ for all $m \in \Delta$. Hence $v' \in \Delta^\circ$, and then $\langle m', v' \rangle = -1$ implies $v' \in F^\circ$ since m' is in the relative interior of F. $\qquad\square$

It follows from Lemma 3.2.1 that in $\mathbb{P}_\Delta$, the T_N orbit closures are in one-to-one correspondence with the proper faces of Δ°, hence by the polar duality are in one-to-one correspondence with the nonempty faces F of Δ. This correspondence can in fact be seen directly in terms of the polytope rings as follows.

Let $F \subset \Delta$ be a face. There is a canonical surjection of polytope rings $\pi_F : S_\Delta \to S_F$ defined by

$$\pi_F(t_0^k \chi^m) = \begin{cases} t_0^k \chi^m & m \in kF \\ 0 & m \in k\Delta - kF. \end{cases}$$

This surjection induces a natural inclusion of toric varieties $\mathbb{P}_F \hookrightarrow \mathbb{P}_\Delta$. There is a particularly simple description of this inclusion if S_Δ is generated by $\Delta \cap M$. In that case, we can think of $\mathbb{P}_\Delta$ as embedded in a projective space whose homogeneous coordinates are in one-to-one correspondence with the points of $\Delta \cap M$. Then $\mathbb{P}_F$ is obtained from $\mathbb{P}_\Delta$ by setting to zero all the coordinates corresponding to points of $(\Delta - F) \cap M$.

If we start from an integral polytope Δ containing the origin in its interior, the polar Δ° need not be integral. In Section 3.5, we will study a particular class of polytopes, the *reflexive polytopes* of Batyrev, where the polar is integral. These polytopes play an important role in many mirror symmetry constructions.

Returning to the case of an arbitrary n-dimensional integral polytope Δ, we observe that the toric variety $\mathbb{P} = \mathbb{P}_\Delta$ comes equipped with an ample line bundle. This is because $\mathbb{P} = \mathrm{Proj}(S_\Delta)$ has the ample bundle $\mathcal{O}_\mathbb{P}(1)$. From the toric point of view, we describe this line bundle as follows. Consider the function $\phi_\Delta : N_\mathbb{R} \to \mathbb{R}$ defined by

$$\phi_\Delta(v) = \min_{m \in \Delta} \langle m, v \rangle.$$

One can prove that ϕ_Δ is the support function of a divisor D_Δ on $\mathbb{P}$. Furthermore, ϕ_Δ is strictly upper convex, $\mathcal{O}_\mathbb{P}(D_\Delta) \simeq \mathcal{O}_\mathbb{P}(1)$, and the polytope Δ_{D_Δ} associated to D_Δ is exactly Δ (see [**Batyrev4, Oda**]). Hence, when a toric variety comes from a polytope, it is a projective variety with a specific choice of ample divisor.

One way to see $\mathbb{P}_\Delta$ and its projective embedding more concretely is to choose a basis for M. This corresponds to picking coordinates $t_1, \ldots, t_n$ for the torus T_N. Then, if $m \in M$ is written $m = (a_1, \ldots, a_n)$, we have $\chi^m = \Pi_{i=1}^n t_i^{a_i}$, so we can write t^m instead of χ^m. Now, given a polytope Δ, pick k so that kD_Δ is very ample. The lattice points $k\Delta \cap M = \{m_1, \ldots, m_s\}$ give monomials t^{m_i}, which define a map $T_N \to \mathbb{P}^{s-1}$ by sending $t \in T_N$ to $(t^{m_1}, \ldots, t^{m_s})$. Then $\mathbb{P}_\Delta$ is the closure of the image of this map in $\mathbb{P}^{s-1}$.

Also, for any $k \geq 0$, we have the vector space of Laurent polynomials

$$(3.5) \qquad L(k\Delta) = \left\{ f : f = \textstyle\sum_{m \in k\Delta \cap M} a_m t^m, a_m \in \mathbb{C} \right\}.$$

Each $f \in L(k\Delta)$ gives the affine hypersurface $Z_f \subset T_N$ defined by $f = 0$, and its closure $\overline{Z}_f \subset \mathbb{P}_\Delta$ is the corresponding divisor. Thus, by thinking of elements of M as exponent vectors of Laurent monomials, we can work with $\mathbb{P}_\Delta$ using coordinates. In some situations, one starts with a Laurent polynomial f and defines Δ to be the convex hull of its exponent vectors (this is the *Newton polytope* of f). We will use the following notation:

$$(3.6) \qquad \begin{aligned} l(k\Delta) &= |k\Delta \cap M| = \dim(L(k\Delta)) \\ l^*(k\Delta) &= |\{m \in k\Delta \cap M : m \text{ is not in any facet of } k\Delta \cap M\}|. \end{aligned}$$

The numbers $l(k\Delta)$ and $l^*(k\Delta)$ have many interesting properties and appear in various contexts in mirror symmetry.

We conclude with some remarks about the polytope ring S_Δ defined above. First observe that the polynomials $t_0^k f$, for $f \in L(k\Delta)$, are precisely the elements of degree k in S_Δ. This gives isomorphisms

$$(3.7) \qquad (S_\Delta)_k \simeq L(k\Delta) \simeq H^0(\mathbb{P}, \mathcal{O}_\mathbb{P}(kD_\Delta)),$$

where the second isomorphism follows from (3.4).

We can also interpret S_Δ as an affine coordinate ring. Namely, let $\check{\sigma}_\Delta \subset \mathbb{R} \times M_\mathbb{R}$ be the cone over $\{1\} \times \Delta$. Note that $(k, m) \in \check{\sigma}_\Delta \cap (\mathbb{Z} \times M)$ if and only if $m \in k\Delta$. Thus, if $\sigma_\Delta \subset \mathbb{R} \times N_\mathbb{R}$ is the dual of $\check{\sigma}_\Delta$, it follows that S_Δ is the coordinate ring of the affine toric variety given by σ_Δ. By [**Danilov**], this implies that S_Δ is Cohen-Macaulay.

3.2.3. Homogeneous Coordinate Rings. In the discussion immediately before (3.5), we introduced coordinates $t_1, \ldots, t_n$ on the torus of a toric variety. These coordinates are very useful, but it would also be nice to have global coordinates, similar to homogeneous coordinates on projective space. With this goal in mind, we define the *homogeneous coordinate ring* of a toric variety.

If $X = X_\Sigma$ is given by a fan Σ in $N_\mathbb{R}$, introduce a variable x_ρ for each $\rho \in \Sigma(1)$, and consider the polynomial ring

$$S = \mathbb{C}[x_\rho : \rho \in \Sigma(1)].$$

A monomial in S is written $x^D = \Pi_\rho x_\rho^{a_\rho}$, where $D = \sum_\rho a_\rho D_\rho$ is an effective torus-invariant divisor on X (this uses the correspondence $\rho \leftrightarrow D_\rho$), and we say that x^D has degree

$$\deg(x^D) = [D] \in A_{n-1}(X).$$

Thus, S is graded by the Chow group $A_{n-1}(X)$. Given a divisor class $\alpha \in A_{n-1}(X)$, S_α denotes the graded piece of S of degree α. We often write the variables as $x_1, \ldots, x_r$, where x_i corresponds to the cone generator v_i and $r = |\Sigma(1)|$. Then $S = \mathbb{C}[x_1, \ldots, x_r]$.

DEFINITION 3.2.2. *The ring S, together with the grading defined above, is the* homogeneous coordinate ring *of X.*

For $\mathbb{P}^n$, we will see in Section 3.7 that the homogeneous coordinate ring is $\mathbb{C}[x_0, \ldots, x_n]$ with the usual grading. Also, for $\mathbb{P}^n \times \mathbb{P}^m$, the coordinate ring is the appropriate graded ring of bihomogeneous polynomials. In the physical theories associated with mirror symmetry, the variables $x_\rho \in S$ correspond to certain observables, though the observables by themselves have no physical meaning. Rather, one uses certain monomials in the observables, which are taken modulo an ideal of S—this is the *chiral ring*.

Note that the graded ring S depends only on the 1-skeleton of Σ. Thus there are potentially many fans with the same 1-skeleton, and hence many toric varieties with the same coordinate ring. This will be amplified in Section 3.4.

To represent the full fan Σ in S, we use the ideal $B(\Sigma)$ defined as follows. Given a cone $\sigma \in \Sigma$, let

$$\hat{x}_\sigma = \Pi_{\rho \not\subset \sigma} x_\rho$$

be the product of the variables "not in" σ, and then define the ideal

$$B(\Sigma) = \langle \hat{x}_\sigma : \sigma \in \Sigma \rangle \subset S.$$

For $\mathbb{P}^n$, this ideal is just the "irrelevant" ideal $\langle x_0, \dots, x_n \rangle$. In general, $B(\Sigma)$ determines Σ uniquely.

The graded pieces of S have a nice cohomological interpretation. If a divisor $D = \sum_\rho a_\rho D_\rho$ is not Cartier, it won't determine a line bundle, but we still get a reflexive sheaf $\mathcal{O}_X(D)$. This sheaf is the subsheaf of the sheaf of rational functions on X whose sections over an open set $U \subset X$ are rational functions with poles on U of order at most the order specified by D. (For more details on the sheaves determined by Weil divisors, see [**Reid1**, Appendix to §1].) As we saw in (3.7), there is an isomorphism

$$H^0(X, \mathcal{O}_X(D)) \simeq L(\Delta_D),$$

where Δ_D is defined by the second line of (3.3). From here, it is easy to see that the map sending the Laurent monomial t^m to $\Pi_\rho x_\rho^{\langle m, v_\rho \rangle + a_\rho}$ induces an isomorphism

$$(3.8) \qquad H^0(X, \mathcal{O}_X(D)) \simeq S_\alpha,$$

where $\alpha = [D] \in A_{n-1}(X)$.

In the projective case, we can combine this with the isomorphism (3.7), and it follows easily that if D is ample and $\alpha = [D]$, then we get a ring isomorphism

$$S_{\Delta_D} \simeq \bigoplus_{k=0}^\infty S_{k\alpha},$$

where S_{Δ_D} the polytope ring from Section 3.2.1. Thus S contains the polytope rings S_{Δ_D} coming from all possible ample divisors D on X.

We can also describe differential forms on X in terms of the homogeneous coordinates. This will be done in Chapter 5 when we compute the Yukawa coupling for ample hypersurfaces in X.

3.2.4. Toric Varieties via Homogeneous Coordinates. We next show how the usual construction $\mathbb{P}^n = (\mathbb{C}^{n+1} - \{0\})/\mathbb{C}^*$ of projective space generalizes to an arbitrary toric variety. As in Section 3.1, assume that the 1-dimensional cones of Σ span $N_\mathbb{R}$ (this is automatically true if Σ is complete). Define the group

$$G = \mathrm{Hom}_\mathbb{Z}(A_{n-1}(X), \mathbb{C}^*).$$

If we apply $\mathrm{Hom}_\mathbb{Z}(-, \mathbb{C}^*)$ to the exact sequence (3.2), we get an exact sequence

$$(3.9) \qquad 1 \longrightarrow G \longrightarrow (\mathbb{C}^*)^{\Sigma(1)} \longrightarrow T_N \longrightarrow 1.$$

We have used the fact that $\mathrm{Hom}_\mathbb{Z}(M, \mathbb{C}^*) = N \otimes \mathbb{C}^* = T_N$. Thus G is isomorphic to $(\mathbb{C}^*)^{r-n}$ times a finite cyclic group (this finite group is present precisely when $A_{n-1}(X)$ has torsion). Then $G \to (\mathbb{C}^*)^{\Sigma(1)}$ gives an action of G on $\mathbb{C}^{\Sigma(1)} = \mathrm{Spec}(S)$, where $a = (a_\rho) \in \mathbb{C}^{\Sigma(1)}$ and $g \in G$ map to

$$g \cdot a = (g([D_\rho]) \, a_\rho)$$

(remember $g : A_{n-1}(X) \to \mathbb{C}^*$ by the definition of G). Also, the ideal $B(\Sigma) \subset S$ defines a variety $Z(\Sigma) \subset \mathbb{C}^{\Sigma(1)}$. This data determines X as follows.

THEOREM 3.2.3. *If the 1-dimensional cones of Σ span $N_\mathbb{R}$, then:*

(i) *X is the categorical quotient of $\mathbb{C}^{\Sigma(1)} - Z(\Sigma)$ by G.*

(ii) *X is the geometric quotient of $\mathbb{C}^{\Sigma(1)} - Z(\Sigma)$ by G if and only if X is simplicial.*

This result was first proved in [**Audin**] in the simplicial case, though the idea of this construction arose independently several times (see [**Cox**] for a proof and list of references).

The description of X given in Theorem 3.2.3 yields a simple description of the inclusion-reversing correspondence between orbit closures and cones of Σ that was mentioned in Section 3.1. Namely, to each $\sigma \in \Sigma$ we associate the linear subspace

$$\{x \in \mathbb{C}^{\Sigma(1)} \mid x_\rho = 0 \text{ for all } \rho \in \sigma\}.$$

This subspace descends to a subvariety of X_Σ which is easily checked to be the orbit closure corresponding to σ. In particular, the toric divisor D_i associated to the edge v_i is simply the hypersurface defined by $x_i = 0$.

The variety $Z(\Sigma)$ in Theorem 3.2.3 has a simple structure. A subset $\mathcal{S} \subset \Sigma(1)$ is a *primitive collection* if $\mathcal{S}$ is not the set of 1-dimensional cones of some cone $\sigma \in \Sigma$ but every proper subset of $\mathcal{S}$ is contained in some cone in the fan. Then

$$(3.10) \qquad\qquad\qquad Z(\Sigma) = \bigcup_{\mathcal{S}} \mathbf{V}(\mathcal{S}),$$

where the union is over all primitive collections $\mathcal{S}$ and $\mathbf{V}(\mathcal{S})$ is the subspace defined by $x_\rho = 0$ for $\rho \in \mathcal{S}$ (see [**BC**] for a proof).

If X is simplicial, Theorem 3.2.3 tells us that we really have homogeneous coordinates: a point of X has coordinates $a = (a_\rho) \in \mathbb{C}^{\Sigma(1)} - Z(\Sigma)$, and $b \in \mathbb{C}^{\Sigma(1)} - Z(\Sigma)$ gives the same point of X if and only if $b = g \cdot a$ for some $g \in G$. In particular, if $f \in S$ is homogeneous (meaning $f \in S_\alpha$ for some α), then the equation $f = 0$ defines a hypersurface in X. More generally, any homogeneous ideal $I \subset S$ defines a subvariety $\mathbf{V}(I) \subset X$. The toric ideal-variety correspondence is explored in [**Cox**].

In the non-simplicial case, the relation between $\mathbb{C}^{\Sigma(1)} - Z(\Sigma)$ and X is not as direct, though homogeneous ideals of S still determine subvarieties of X. One interesting thing that can happen (even in the simplicial case) is that a hypersurface $Y \subset X$ given by a homogeneous equation $f = 0$ need not be Cartier. This means that although Y is defined globally by $f = 0$, it need not be defined by a single equation locally on X.

3.3. Kähler Cones and Symplectic Geometry

This section will describe the Kähler cone of a simplicial toric variety and explain how symplectic geometry can be used to construct toric varieties.

3.3.1. The Kähler Cone of a Toric Variety. For a smooth projective toric variety X, we have $A_{n-1}(X) \simeq H^2(X, \mathbb{Z})$, which implies $H^2(X, \mathbb{R}) = H^{1,1}(X, \mathbb{R})$. More generally, if X is simplicial and projective, we have $A_{n-1}(X) \otimes \mathbb{R} \simeq H^2(X, \mathbb{R})$, so that $H^2(X, \mathbb{R}) = H^{1,1}(X, \mathbb{R})$ still holds. Recall from Chapter 1 that the *Kähler cone* in $H^{1,1}(X, \mathbb{R})$ consists of all Kähler classes on X (see Appendix A for the definition of Kähler in the orbifold case). Under the above isomorphism, we get a cone in $A_{n-1}(X) \otimes \mathbb{R}$, which we also call the Kähler cone.

Let $A_{n-1}^+(X) \otimes \mathbb{R}$ be the cone generated by the divisor classes $[D_i]$, where $v_i \leftrightarrow D_i$ for $i = 1, \ldots, r$. The Kähler cone sits inside $A_{n-1}^+(X) \otimes \mathbb{R}$ and is described as follows. Let $a = \sum_{i=1}^{r} a_i[D_i] \in A_{n-1}^+(X) \otimes \mathbb{R}$. Since X is simplicial, for each $\sigma \in \Sigma$, we can find $m_\sigma \in M_{\mathbb{R}}$ with $\langle m_\sigma, v_i \rangle = -a_i$ when $v_i \in \sigma$. We say that a is *convex* if $\langle m_\sigma, v_i \rangle \geq -a_i$ for $\sigma \in \Sigma$ and $1 \leq i \leq r$. Then define

$$(3.11) \qquad \mathrm{cpl}(\Sigma) = \{a \in A_{n-1}^+(X) \otimes \mathbb{R} : a \text{ is convex}\}.$$

Since X is projective and simplicial, $\mathrm{cpl}(\Sigma)$ is an $(r-n)$-dimensional convex cone. Furthermore, a is in the interior of $\mathrm{cpl}(\Sigma)$ if and only if a is *strictly convex*, which means $\langle m_\sigma, v_i \rangle > -a_i$ whenever σ is n-dimensional and $v_i \notin \sigma$ (see [**OdP**]). We can now describe the Kähler cone of X as follows:

PROPOSITION 3.3.1. *If* $X = X_\Sigma$ *is projective and simplicial, then the Kähler cone of* X *equals the interior of the cone* $\mathrm{cpl}(\Sigma)$ *defined in* (3.11).

PROOF. The Kähler cone is contained in the interior of $\mathrm{cpl}(\Sigma)$ by [**Baily2**], and the opposite inclusion is proved in [**AGM1**, Section 4]. $\qquad\square$

Since $\mathrm{cpl}(\Sigma)$ is a strongly convex polyhedral cone, this proposition shows that the Kähler cone of a projective simplicial toric variety has an especially nice structure.

3.3.2. The Mori Cone. If X_Σ is simplicial and complete, its *Mori cone* $M(X_\Sigma)$ is the cone of effective 1-cycles in $A_1(X_\Sigma) \otimes \mathbb{R} \simeq H_2(X_\Sigma, \mathbb{R})$. This cone is sometimes denoted $NE(X_\Sigma)_\mathbb{R}$ or $\overline{NE}(X_\Sigma)_\mathbb{R}$. For our purposes, we are interested in $M(X_\Sigma)$ because it is dual to the Kähler cone, so that by Proposition 3.3.1, $\mathrm{cpl}(\Sigma)$ and $M(X_\Sigma)$ are dual.

As might be expected, we can describe $M(X_\Sigma)$ explicitly in terms of the toric data of the fan Σ. To see how this can be done, let $\sigma \in \Sigma(n-1)$ be a cone of dimension $n-1$, so that the corresponding orbit closure $C_\sigma \subset X_\Sigma$ is a curve. By an observation of Reid [**Reid3**, Prop. 1.6], $M(X_\Sigma)$ is the cone generated by the C_σ.

We can turn this into something more explicit as follows. Let $v_1, \ldots, v_{n-1}$ be the integral generators of σ, and note that σ is contained in precisely two n-dimensional cones. The first of these cones is the span of σ and one more primitive integral generator v_n, and similarly the second cone is spanned by σ and some other generator v_{n+1}. It follows that we get a linear relation

$$(3.12) \qquad \sum_{i=1}^{n+1} \lambda_i v_i = 0.$$

In this relation, we may assume that $\lambda_1, \ldots, \lambda_{n+1}$ are relatively prime integers with $\lambda_n, \lambda_{n+1} > 0$ since v_n and v_{n+1} lie on opposite sides of σ. Hence the λ_i give a relation λ_σ in

$$(3.13) \qquad \Lambda_\mathbb{Q} = \{(\lambda_\rho)_{\rho \in \Sigma(1)} : \lambda_\rho \in \mathbb{Q}, \ \textstyle\sum_\rho \lambda_\rho v_\rho = 0\}.$$

Taking the dual of (3.2) and tensoring with $\mathbb{Q}$ gives a natural isomorphism

$$A_1(X_\Sigma) \otimes \mathbb{Q} \simeq \Lambda_\mathbb{Q}.$$

Under this isomorphism, we can relate the curve $C_\sigma \in A_1(X_\Sigma) \otimes \mathbb{Q}$ to the relation $\lambda_\sigma \in \Lambda_\mathbb{Q}$ as follows.

LEMMA 3.3.2. *Let* $\sigma \in \Sigma(n-1)$. *Then there is a positive constant* c_σ *such that under the above isomorphism,* C_σ *corresponds to* $c_\sigma \lambda_\sigma$, *where* λ_σ *is the relation determined by* (3.12).

PROOF. First observe that $A_1(X_\Sigma)\otimes\mathbb{Q}$ is dual to $A_{n-1}(X_\Sigma)\otimes\mathbb{Q}$ via intersection product, while $\Lambda_\mathbb{Q}$ is dual to $A_{n-1}(X_\Sigma)\otimes\mathbb{Q}$ via dot product on $\mathbb{Q}^{\Sigma(1)}$ (using (3.2) and (3.13)). Since the isomorphism $A_1(X_\Sigma)\otimes\mathbb{Q}\simeq\Lambda_\mathbb{Q}$ is compatible with these dualities, it suffices to find $c_\sigma > 0$ such that

$$D_\rho \cdot C_\sigma = c_\sigma D_\rho \cdot \lambda_\sigma \quad \text{for all } \rho \in \Sigma(1).$$

If ρ is not one of the v_i's, then one easily sees that both sides of this equation are 0. On the other hand, if ρ is one of the v_i (as usual, we identify ρ with the unique generator of $\rho \cap N$), then $D_{v_i} \cdot \lambda_\sigma = \lambda_i$. But [**Reid3**, Prop. 2.7] shows that

$$(3.14) \qquad\qquad \lambda_j\, D_{v_i} \cdot C_\sigma = \lambda_i\, D_{v_j} \cdot C_\sigma, \quad 1 \le i, j \le n+1.$$

We've already seen that $\lambda_{n+1} > 0$, and since σ and v_{n+1} determine an n-dimensional cone of Σ, it follows that $D_{v_{n+1}} \cap C_\sigma$ consists of a single point ($=$ the fixed point corresponding to the n-dimensional cone). Thus $D_{v_{n+1}} \cdot C_\sigma > 0$, so that $c_\sigma = \lambda_{n+1}/(D_{v_{n+1}} \cdot C_\sigma)$ is positive. Using (3.14), one sees easily that c_σ has the desired property. $\qquad\square$

Note that if X_Σ is smooth in a neighborhood of C_σ, then $c_\sigma = 1$ in Lemma 3.3.2 by [**Fulton3**, p. 99].

One nice consequence of Lemma 3.3.2 is the following ampleness criterion.

COROLLARY 3.3.3. *A divisor* $D = \sum_{\rho\in\Sigma(1)} a_\rho D_\rho$ *is ample if and only if it is Cartier and satisfies* $(a_\rho)_{\rho\in\Sigma(1)} \cdot \lambda_\sigma > 0$ *for all* $\sigma \in \Sigma(n-1)$.

PROOF. Suppose that D is Cartier. Since the C_σ generate the Mori cone, it follows that D is ample if and only if $D \cdot C_\sigma > 0$ for all $\sigma \in \Sigma(n-1)$. The corollary now follows immediately from Lemma 3.3.2. $\qquad\square$

When X_Σ is smooth, there is a second description of the Mori cone $M(X_\Sigma)$ which involves the primitive collections defined in Section 3.2.4. The details can be found in [**Batyrev1**], and the corresponding ampleness criterion appears in [**Batyrev2**].

3.3.3. Symplectic Geometry. Symplectic geometry plays an important role in mirror symmetry. Here, we will review some of the basic definitions. A good reference for symplectic geometry is [**Audin**].

A *symplectic manifold* is a real manifold endowed with a closed, nondegenerate 2-form ω. The symplectic structure converts functions into vector fields as follows: if f is a C^∞ function on the manifold, then there is a unique vector field X_f on the manifold with the property that $\omega(X, X_f) = X(f)$ for any vector field X. We call X_f the *Hamiltonian* of f.

For us, the most important example is $\mathbb{C}^r$ endowed with the symplectic form $\omega = \sum_{i=1}^r dx_i \wedge dy_i$, where $z_i = x_i + iy_i$. A special feature of $\mathbb{C}^r$ is that the natural action of $U(1)^r = (S^1)^r$ on $\mathbb{C}^r$ is *symplectic*, which means that $g^*\omega = \omega$ for all $g \in U(1)^r$.

In fact, the action of $U(1)^r$ is *Hamiltonian*, which means the following. The Lie algebra of $U(1)^r$ is $\mathbb{R}^r$, where $\lambda = (\lambda_1, \dots, \lambda_r) \in \mathbb{R}^r$ maps under the exponential map to $\exp(i\lambda) = (\exp(i\lambda_1), \dots, \exp(i\lambda_r)) \in U(1)^r$. Given λ, we get the flow $v \to \exp(it\lambda) \cdot v$ on $\mathbb{C}^r$ with vector field

$$X_\lambda = \sum_{i=1}^r \lambda_i\left(-y_i \frac{\partial}{\partial x_i} + x_i \frac{\partial}{\partial y_i}\right).$$

Then being Hamiltonian means that for all λ in the Lie algebra, the vector field X_λ is the Hamiltonian of some function on $\mathbb{C}^r$.

We will prove this using the *moment map*

$$\mu : \mathbb{C}^r \longrightarrow (\mathbb{R}^r)^*$$

defined by $\mu(z_1, \ldots, z_r) = (1/2)(|z_1|^2, \ldots, |z_r|^2)$ (this uses the basis of $(\mathbb{R}^r)^*$ dual to the standard basis of $\mathbb{R}^r$). The key point is that $\lambda = (\lambda_1, \ldots, \lambda_r) \in \mathbb{R}^r$ can be regarded as a map $\lambda : (\mathbb{R}^r)^* \to \mathbb{R}$, and composing this with the moment map gives $\lambda \circ \mu : \mathbb{C}^r \to \mathbb{R}$. An easy calculation shows that

$$X_\lambda = X_{\lambda \circ \mu},$$

which proves that we have a Hamiltonian action.

In general, moment maps and Hamiltonian actions play an important role in symplectic geometry. In Chapter 7, we will discuss other aspects of symplectic geometry when we give the symplectic definition of Gromov-Witten invariant.

3.3.4. Toric Varieties via Symplectic Reduction. We next explain how toric varieties are related to symplectic geometry by the process known as *symplectic reduction*. Let $X = X_\Sigma$ be determined by the fan Σ in $N_\mathbb{R} \simeq \mathbb{R}^n$. Then, as in Section 3.2.3, we have the group $G = \mathrm{Hom}_\mathbb{Z}(A_{n-1}(X), \mathbb{C}^*)$. The maximal compact subgroup of G is

$$G_\mathbb{R} = \mathrm{Hom}_\mathbb{Z}(A_{n-1}(X), U(1)),$$

with Lie algebra $\mathfrak{g}_\mathbb{R} = \mathrm{Hom}_\mathbb{Z}(A_{n-1}(X), \mathbb{R})$. Thus $\mathfrak{g}_\mathbb{R}^* = A_{n-1}(X) \otimes \mathbb{R}$.

If we let $r = |\Sigma(1)|$, the inclusion $G \subset (\mathbb{C}^*)^r$ induces $G_\mathbb{R} \subset U(1)^r$. This gives a Hamiltonian action of $G_\mathbb{R}$ on $\mathbb{C}^r$ whose moment map is easily seen to be the composition

$$\mu_\Sigma : \mathbb{C}^r \xrightarrow{\mu} (\mathbb{R}^r)^* \xrightarrow{p} \mathfrak{g}_\mathbb{R}^* = A_{n-1}(X) \otimes \mathbb{R} \simeq \mathbb{R}^{r-n},$$

where μ is the moment map for $U(1)^r$ and p comes from the exact sequence

$$(3.15) \qquad 0 \longrightarrow M_\mathbb{R} \longrightarrow (\mathbb{R}^r)^* \xrightarrow{p} A_{n-1}(X) \otimes \mathbb{R} \longrightarrow 0$$

obtained from (3.2) by tensoring with $\mathbb{R}$. The map μ_Σ is easy to describe explicitly. First note that in (3.15), $(b_1, \ldots, b_r) \in \mathbb{R}^r$ annihilates the image of $M_\mathbb{R}$ if and only if $\sum_{i=1}^r b_i v_i = 0$ in $N_\mathbb{R}$. All such $(b_1, \ldots, b_r)$ form a subspace $\Lambda_\mathbb{R} \subset \mathbb{R}^r$ of dimension $r - n$ (note $\Lambda_\mathbb{R} = \Lambda_\mathbb{Q} \otimes \mathbb{R}$, where $\Lambda_\mathbb{Q}$ is from (3.13)). If we pick a basis $(b_{1,j}, \ldots, b_{r,j})$, $1 \le j \le r - n$, of $\Lambda_\mathbb{R}$, then we get an isomorphism $A_{n-1}(X) \otimes \mathbb{R} \simeq \mathbb{R}^{r-n}$ such that p becomes

$$p(x_1, \ldots, x_r) = \big(\textstyle\sum_{i=1}^r b_{i,1} x_i, \ldots, \sum_{i=1}^r b_{i,r-n} x_i\big).$$

It follows that μ_Σ is given by the formula

$$(3.16) \qquad \mu_\Sigma(z_1, \ldots, z_r) = (1/2)\big(\textstyle\sum_{i=1}^r b_{i,1}|z_i|^2, \ldots, \sum_{i=1}^r b_{i,r-n}|z_i|^2\big).$$

We also see that μ_Σ is constant on $G_\mathbb{R}$-orbits.

When X is simplicial, we saw above that $A_{n-1}(X) \otimes \mathbb{R} \simeq H^{1,1}(X, \mathbb{R})$, and by Proposition 3.3.1, the Kähler cone of X can be identified with the interior of the cone $\mathrm{cpl}(\Sigma) \subset A_{n-1}(X) \otimes \mathbb{R}$. We can then construct X as follows.

THEOREM 3.3.4. *If the toric variety $X = X_\Sigma$ is projective and simplicial and $a \in A_{n-1}(X) \otimes \mathbb{R}$ is Kähler, then $\mu_\Sigma^{-1}(a) \subset \mathbb{C}^r - Z(\Sigma)$, and the natural map*

$$\mu_\Sigma^{-1}(a)/G_\mathbb{R} \longrightarrow (\mathbb{C}^r - Z(\Sigma))/G = X$$

is an orbifold diffeomorphism. Furthermore, the symplectic form ω on $\mathbb{C}^r$, when restricted to $\mu_\Sigma^{-1}(a)$, descends to a symplectic form on $\mu_\Sigma^{-1}(a)/G_{\mathbb{R}}$ whose cohomology class is identified with $a \in H^2(X, \mathbb{R})$ via the above diffeomorphism.

PROOF. In the statement of the theorem, note that ω is *not* symplectic when restricted to $\mu_\Sigma^{-1}(a)$. Going to the quotient is exactly what is needed to make it nondegenerate. This process is called *symplectic reduction*. Since $G_{\mathbb{R}}$ can have finite stabilizers, the quotient $\mu_\Sigma^{-1}(a)/G_{\mathbb{R}}$ has the natural structure of a symplectic orbifold (see Appendix A).

We will use the notation of (3.11), where $a \in A_{n-1}(X) \otimes \mathbb{R}$ is written in the form $a = \sum_{i=1}^r a_i[D_i]$. Since a is Kähler, Proposition 3.3.1 tells us a is in the interior of $\text{cpl}(\Sigma)$, i.e., a is strictly convex. Then consider the convex set $\Delta \subset M_{\mathbb{R}}$ defined by $\langle m, v_i \rangle \geq -a_i$. Although Δ need not be integral (or even rational), the key point is that the strict convexity of a implies that the polytope Δ is *combinatorially dual* to Σ, which means that the proper faces of Δ are given by $\{m \in \Delta : \langle m, v_i \rangle = -a_i$ for $v_i \in \sigma\}$ as σ ranges over the cones of Σ. When the a_i are integral, this is one of the properties of ampleness discussed in Section 3.2.1. The case when the a_i are rational follows easily, and then the general case is proved using a continuity argument.

Once we know that the fan of X is combinatorially dual to Δ, we can use Theorem 1.4 of [**Guillemin**, Appendix 1] to conclude that $\mu_\Sigma^{-1}(a)/G_{\mathbb{R}} \simeq X$. Furthermore, the final assertion of the theorem follows from equation (1.6) of [**Guillemin**, Appendix 2] (the λ_i Guillemin uses are $-a_i$ in our notation). Guillemin's arguments are given in the smooth case, but can be modified to work when X is simplicial. $\square$

The construction of toric varieties by means of symplectic reduction has some interesting consequences for physics. In particular, there are certain physical theories, called *gauged linear sigma models*, which take as their starting point a toric variety as described by symplectic reduction. Details can be found in Appendix B.5.

3.4. The GKZ Decomposition

When we study Kähler moduli in Chapter 6, we will see that we need to enlarge the Kähler moduli space defined in Chapter 1. The rough idea is that two Calabi-Yau manifolds related by a flop have "adjacent" Kähler cones, which will enable us to glue together the corresponding Kähler moduli spaces.

In the toric context, the idea of "adjacent" Kähler cones arises naturally in the *GKZ decomposition*. To see where this comes from, note that by (3.2), the Chow group $A_{n-1}(X)$ of a toric variety $X = X_\Sigma$ depends only on the set $\Sigma(1)$ of 1-dimensional cones of Σ. Thus two toric varieties with the same 1-dimensional cones have canonically isomorphic Chow groups of divisors. We can then ask how their respective Kähler cones sit inside the Chow group tensored with $\mathbb{R}$.

3.4.1. The Decomposition. To make this idea precise, fix a finite set of strongly convex rational 1-dimensional cones Ξ in $N_{\mathbb{R}}$, and let Σ be a fan with $\Sigma(1) = \Xi$. Since $A_{n-1}(X) \otimes \mathbb{R}$ depends only on Ξ, we will write this vector space as $A(\Xi)$. Similarly, the effective divisor classes form the cone $A_{n-1}^+(X) \otimes \mathbb{R}$ which also depends only on Ξ. This will be denoted $A^+(\Xi)$. It follows that $A^+(\Xi)$ is a cone in $A(\Xi) \simeq \mathbb{R}^{r-n}$ (where $r = |\Xi|$).

If Σ is a projective simplicial fan with $\Sigma(1) = \Xi$, then by Proposition 3.3.1, the closure of its Kähler cone is $\mathrm{cpl}(\Sigma)$. Using the above notation, we get the $(r-n)$-dimensional cone $\mathrm{cpl}(\Sigma) \subset A^+(\Xi)$.

Now suppose we vary over all possible such Σ's. This gives a collection of cones lying in $A^+(\Xi)$. The naive hope would be that these cones fill out all of $A^+(\Xi)$. In practice, this doesn't happen, and to get what remains, we must consider projective simplicial fans with $\Sigma(1) \subset \Xi$. Here, $A_{n-1}(X_\Sigma) \otimes \mathbb{R}$ may differ from $A(\Xi)$, but we can still define a cone $\mathrm{cpl}(\Sigma) \subset A^+(\Xi)$ as follows. Given $a = \sum_{i=1}^r a_i[D_i] \in A^+(\Xi)$, for $\sigma \in \Sigma$, we can find $m_\sigma \in M_\mathbb{R}$ such that $\langle m_\sigma, v_i \rangle = -a_i$ whenever $v_i \in \sigma$ and v_i generates an element of $\Sigma(1)$. We say that a is Σ-*convex* if $\langle m_\sigma, v_i \rangle \geq -a_i$ for all $\sigma \in \Sigma$ and $1 \leq i \leq r$. Then, as in (3.11), we define

$$\mathrm{cpl}(\Sigma) = \{a \in A^+(\Xi) : a \text{ is } \Sigma\text{-convex}\}.$$

This cone again has dimension $r - n$ by [**OdP**]. The remarkable fact is that these cones fill up $A^+(\Xi)$ in a very nice manner, giving what [**OdP**] call the *Gelfand-Kapranov-Zelevinsky (or GKZ) decomposition*. A precise description is as follows.

THEOREM 3.4.1. *Let Ξ be a finite set of strongly convex rational 1-dimensional cones in $N_\mathbb{R}$. As Σ ranges over all projective simplicial fans with $\Sigma(1) \subset \Xi$, the cones $\mathrm{cpl}(\Sigma)$ and their faces form a fan in $A(\Xi)$ whose support is $A^+(\Xi)$.*

In Section 3.7, we will give an example of a GKZ decomposition consisting of two cones $\mathrm{cpl}(\Sigma)$ and $\mathrm{cpl}(\Sigma')$ such that $\Sigma(1) = \Xi$ and $\Sigma'(1) \subsetneq \Xi$.

In general, one can ask when two cones in the GKZ decomposition have a common face of codimension one. The answer is very nice: this happens when the corresponding fans in $N_\mathbb{R}$ are related by subdividing (adding or subtracting a single 1-dimensional cone) or by a "flop"—see [**OdP**] for the details. We will discuss flops in Section 6.2.3. We should also mention that non-simplicial fans correspond to certain cones of dimension $< r - n$ in the GKZ decomposition.

3.4.2. Secondary Fans and Gale Transforms. The GKZ decomposition can be enlarged to give a complete fan whose support is all of $A(\Xi)$. This can be done as follows. First identify Ξ with a subset of $N_\mathbb{R}$ by associating to each 1-dimensional cone in Ξ its primitive integral generator. Then define the subset $\Xi^+ \subset N_\mathbb{R} \times \mathbb{R}$ to be

$$(3.17) \qquad \Xi^+ = (\Xi \cup \{0\}) \times \{1\}.$$

The fan of interest is the *secondary fan* of Ξ^+. Its definition is analogous to that of the GKZ decomposition. We replace the projective simplicial fans occurring in the GKZ decomposition by *regular triangulations* $\mathcal{T}$ of the convex hull $\mathrm{Conv}(\Xi^+)$. The vertices of the simplices in $\mathcal{T}$ must be elements of Ξ^+, and regularity means that $\mathcal{T}$ can be defined by a strictly convex function in a manner analogous to the way that a projective simplicial fan Σ can be defined by a strictly convex support function $a \in \mathrm{cpl}(\Sigma)$. To $\mathcal{T}$ is associated the cone $\mathcal{C}(\mathcal{T}) \subset A(\Xi)$ of strictly convex support functions which define the same triangulation $\mathcal{T}$. The cones $\mathcal{C}(\mathcal{T})$ are the maximal cones of the secondary fan, which is a complete fan in $A(\Xi)$.

Given a triangulation $\mathcal{T}$, the cone $\mathcal{C}(\mathcal{T})$ has an alternate description as follows. Using (3.2), one sees that $\Xi^+ \subset N_\mathbb{R} \times \mathbb{R}$ gives

$$(3.18) \qquad 0 \longrightarrow M_\mathbb{R} \oplus \mathbb{R} \longrightarrow \mathbb{R}^{\Xi^+} \longrightarrow A(\Xi) \longrightarrow 0.$$

To each $v \in \Xi^+$ we associate the element $e_v^* \in A(\Xi)$ which is the image under the surjection in (3.18) of the standard basis element of $\mathbb{R}^{\Xi^+}$ corresponding to v. The set of elements $\{e_v^* : v \in \Xi^+\}$ is called the *Gale transform* of Ξ^+. The reader can check that if $\Sigma(1) = \Xi$ and $v = (\rho, 1) \in \Xi^+ - \{(0, 1)\}$, then $e_v^* = [D_\rho]$ is the class of the toric divisor corresponding to $\rho \in \Sigma(1)$ in $A(\Xi) = A_{n-1}(X_\Sigma) \otimes \mathbb{R}$.

Now suppose $\mathcal{T}$ is a regular triangulation of $\mathrm{Conv}(\Xi^+)$ as described above. Then each simplex $\sigma \in \mathcal{T}$ gives a cone $\mathcal{C}(\sigma) \subset A(\Xi)$ whose generators are e_v^* for those v which are not vertices of σ, and one can show that

$$\mathcal{C}(\mathcal{T}) = \bigcap_{\sigma \in \mathcal{T}} \mathcal{C}(\sigma).$$

If $(0, 1)$ is one of the vertices of the triangulation $\mathcal{T}$, then since $(0, 1) \in \mathrm{Conv}(\Xi^+) \subset N_\mathbb{R} \times \{1\} \simeq N_\mathbb{R}$ is an interior point, we get a complete fan Σ whose cones are equal to the cones formed from the vertex $(0, 1)$ over the faces of the simplices $\sigma \in \mathcal{T}$. The resulting fan Σ satisfies $\Sigma(1) \subset \Xi$ and $\mathcal{C}(\mathcal{T}) = \mathrm{cpl}(\Sigma)$. This process can be reversed, and it follows that the secondary fan contains the GKZ fan as a subfan. A more complete description of the secondary fan can be found in [**GKZ2**], and its relation to the GKZ decomposition is given in [**OdP**].

Using (3.18), we see that the dual space $A(\Xi)^*$ is

$$A(\Xi)^* = \left\{ \lambda = (\lambda_v) \in \mathbb{R}^{\Xi^+} : \sum_{v \in \Xi^+} \lambda_v \cdot v = 0 \right\}.$$

This is the vector space of linear relations among elements of Ξ^+ (and is analogous to $\Lambda_\mathbb{Q}$ defined in (3.13)). Note that $A(\Xi)^*$ has an obvious integer lattice consisting of those relations with integer coefficients. Furthermore, each $e_v^* \in A(\Xi)$ in the Gale transform is a linear functional on $A(\Xi)^*$, and one easily calculates that it is given by $e_v^*(\lambda) = \lambda_v$. In Chapter 6, we will use this as follows. Fix a basis of $A(\Xi)^*$. Then, regarding each basis vector as a row $\lambda = (\lambda_v)$, the basis gives a matrix, and the *columns* of this matrix give the e_v^*, provided we use the dual basis of $A(\Xi)$. This description of the e_v^* makes the Gale transform easy to work with.

An especially nice case is when $|\Xi^+| = n + 3$ (n is the rank of the lattices N and M). Here, $A(\Xi)$ has dimension 2, and the Gale transform $\{e_v^*\}$ is a set of vectors in the plane. The secondary fan in this case is easily seen to be the complete 2-dimensional fan whose 1-dimensional cones are spanned by the vectors of the Gale transform, as noted explicitly in [**DHSS**].

3.4.3. Relation to Symplectic Reduction.

We conclude this section by examining the relation between the moment map and the GKZ decomposition. First observe that the moment map μ_Σ depends only on $\Sigma(1)$, since the exact sequence (3.15) is determined by $\Sigma(1)$. Thus we can write the moment map as $\mu_\Xi : \mathbb{C}^\Xi \to A(\Xi)$. Its image is $A^+(\Xi)$, so that

$$\mu_\Xi : \mathbb{C}^\Xi \longrightarrow A^+(\Xi).$$

Since the group G acting on $\mathbb{C}^\Xi$ depends only on Ξ, we write this group as $G(\Xi)$.

Now suppose we have an $(r - n)$-dimensional cone $\mathrm{cpl}(\Sigma) \subset A^+(\Xi)$ in the GKZ decomposition. We can take a in its interior and form the quotient

$$\mu_\Xi^{-1}(a)/G(\Xi)_\mathbb{R}.$$

If $\Sigma(1) = \Xi$ and Σ is projective and simplicial, this is X_Σ by Theorem 3.3.4. When $\Sigma(1)$ is a proper subset of Ξ, the theorem no longer applies, but the following generalization tells us that we still get X_Σ.

THEOREM 3.4.2. *If Σ is a projective simplicial fan with $\Sigma(1) \subset \Xi$ and $a \in A^+(\Xi)$ is in the interior of $\mathrm{cpl}(\Sigma)$, then there is a natural orbifold diffeomorphism*

$$\mu_\Xi^{-1}(a)/G(\Xi)_\mathbb{R} \simeq X_\Sigma.$$

PROOF. Besides the group $G(\Xi)$ and the moment map μ_Ξ, we have $G(\Sigma)$ and μ_Σ coming from Σ. To relate these, let $C = \Xi - \Sigma(1)$, so that $\mathbb{Z}^\Xi = \mathbb{Z}^{\Sigma(1)} \oplus \mathbb{Z}^C$. The projection $\mathbb{Z}^\Xi \to \mathbb{Z}^{\Sigma(1)}$ induces an exact sequence

$$(3.19) \qquad 1 \longrightarrow G(\Sigma) \longrightarrow G(\Xi) \longrightarrow (\mathbb{C}^*)^C \longrightarrow 1$$

such that relative to $\mathbb{C}^\Xi = \mathbb{C}^{\Sigma(1)} \oplus \mathbb{C}^C$, $G(\Sigma)$ acts trivially on the second factor. Furthermore, we can relate the moment maps through the following commutative diagram

$$\begin{array}{ccccc}
\mu_\Xi : \mathbb{C}^\Xi & \longrightarrow & \mathbb{R}^\Xi & \longrightarrow & A(\Xi) \\
\downarrow & & \downarrow & & \downarrow \\
\mu_\Sigma : \mathbb{C}^{\Sigma(1)} & \longrightarrow & \mathbb{R}^{\Sigma(1)} & \longrightarrow & A_{n-1}(X_\Sigma) \otimes \mathbb{R},
\end{array}$$

where the vertical arrows are the natural projections.

If $a = \sum_{i=1}^r a_i[D_i]$ is in the interior of $\mathrm{cpl}(\Sigma)$, then the proof of [**OdP**, Proposition 3.1] shows that a is strictly convex, which means $\langle m_\sigma, v_i \rangle > -a_i$ whenever $\sigma \in \Sigma$ is n-dimensional and v_i is not a generator of σ. Let a_0 be the image of a in $A_{n-1}(X_\Sigma) \otimes \mathbb{R}$. It follows that a_0 is also strictly convex. By Theorem 3.3.4, it thus suffices to show that

$$\mu_\Sigma^{-1}(a_0)/G(\Sigma)_\mathbb{R} \simeq \mu_\Xi^{-1}(a)/G(\Xi)_\mathbb{R}.$$

Since Σ is complete, each $v_i \in C$ (as usual, we blur the distinction between 1-dimensional cones and their generators) lies in some n-dimensional cone $\sigma \in \Sigma$, so that $v_i = \sum_j c_{ij} v_j$, where $c_{ij} \geq 0$ and the summation is over those j for which v_j is a generator of σ (this convention will apply for the rest of this proof). Then the strict convexity condition $\langle m_\sigma, v_i \rangle > -a_i$ gives the inequality

$$a_i > \sum_j c_{ij} a_j \quad \text{for } v_i \in C.$$

Let the coordinates of $\mathbb{C}^\Xi$ be $z_1, \ldots, z_r$, and choose indices so that $v_1, \ldots, v_k \in \Sigma(1)$ and $v_{k+1}, \ldots, v_r \in C$. Then one can show without difficulty that $(z_1, \ldots, z_r) \in \mu_\Xi^{-1}(a)$ if and only if $(z_1, \ldots, z_k) \in \mu_\Sigma^{-1}(a_0)$ and

$$(3.20) \qquad \tfrac{1}{2}|z_i|^2 = \tfrac{1}{2}\sum_j c_{ij}|z_j|^2 + a_i - \sum_j c_{ij} a_j \quad \text{for } k+1 \leq i \leq r.$$

Now define a map $\phi : \mu_\Sigma^{-1}(a_0) \to \mu_\Xi^{-1}(a)$ by sending $(z_1, \ldots, z_k) \in \mu_\Sigma^{-1}(a_0)$ to $(z_1, \ldots, z_r) \in \mu_\Xi^{-1}(a)$, where $z_{k+1}, \ldots, z_r$ are the positive real numbers satisfying (3.20) (the right hand side is positive by the above inequality). Then ϕ is equivariant with respect to the actions of $G(\Sigma)_\mathbb{R}$ and $G(\Xi)_\mathbb{R}$, and every $G(\Xi)_\mathbb{R}$ orbit meets the image of ϕ by (3.19). Finally, suppose $g\,\phi(z) = \phi(w)$ for $z, w \in \mu_\Sigma^{-1}(a_0)$ and $g \in G(\Xi)_\mathbb{R}$. Since g doesn't affect absolute values, (3.20) shows that g acts trivially on $z_{k+1}, \ldots, z_r$. Furthermore, these numbers are nonzero, which implies $g \in G(\Sigma)_\mathbb{R}$ by (3.19). We conclude that ϕ induces a bijection on orbits, which is easily seen to be an orbifold diffeomorphism. $\qquad\square$

Theorem 3.4.2 shows that the moment map μ_Ξ can be used to construct not just one toric variety but *all* projective simplicial toric varieties X_Σ with $\Sigma(1) \subset \Xi$. This is related to the phenomenon of *multiple mirrors*, which will be explored in Chapter 4. The *gauged linear sigma models* of [**Witten5**] are also related to this. In these models, an element $a \in A(\Xi)$ is a parameter, and one gets different physical theories depending on where a lies. If a is in the interior of a cone $\mathrm{cpl}(\Sigma)$, then the theory involves the corresponding toric variety X_Σ, while if it lies outside $A^+(\Xi)$, one gets quite different theories (e.g., Landau-Ginzburg theories). This phenomenon is called *phases* in the physics literature—see, for example, [**MP1**].

3.5. Fano Varieties and Reflexive Polytopes

The anticanonical class of projective space is ample, and more generally, any smooth complete variety with this property is said to be *Fano*. For the purposes of mirror symmetry, we need to consider *singular* Fano varieties as well. To define this, recall from Appendix A that a Cohen-Macaulay variety V has a dualizing sheaf $\widehat{\Omega}_V^n$, where $n = \dim(V)$, and that $\widehat{\Omega}_V^n$ is a line bundle if and only if V is Gorenstein.

Being Fano means that the dual of the dualizing sheaf is ample. In particular, this indicates that we might want $\widehat{\Omega}_V^n$ to be a line bundle, so that such a variety is Gorenstein. Hence we have the following definition.

DEFINITION 3.5.1. *A complete n-dimensional Gorenstein variety V is* Fano *if the dual of the line bundle $\widehat{\Omega}_V^n$ is ample.*

For a toric variety, we can characterize these ideas as follows. First recall that the dualizing sheaf on an arbitrary toric variety X has the simple description

$$\widehat{\Omega}_X^n = \mathcal{O}_X\big(-\textstyle\sum_\rho D_\rho\big)$$

(see [**Danilov, Fulton3, Oda**]). In terms of the canonical divisor K_X, this means that $K_X = -\sum_\rho D_\rho$. It follows that X is Gorenstein if and only if $\sum_\rho D_\rho$ is Cartier. Then Fano toric varieties can be characterized as follows.

LEMMA 3.5.2. *A complete toric variety X of dimension n is Fano if and only if $\sum_\rho D_\rho$ is Cartier and ample.*

Since every ample divisor comes from a polytope, the ample divisor $-K_X$ on a Fano toric variety determines a polytope Δ, which has some very special properties. This is where we encounter Batyrev's notion of a *reflexive polytope* [**Batyrev4**]. Here is the precise definition.

DEFINITION 3.5.3. *A n-dimensional integral polytope $\Delta \subset M_{\mathbb{R}} \simeq \mathbb{R}^n$ is* reflexive *if the following two conditions hold:*
  (*i*) *All facets Γ of Δ are supported by an affine hyperplane of the form $\{m \in M_{\mathbb{R}} : \langle m, v_\Gamma \rangle = -1\}$ for some $v_\Gamma \in N$.*
  (*ii*) $\mathrm{Int}(\Delta) \cap M = \{0\}$.

Reflexive polytopes have a very pretty combinatorial duality. Let Δ be an integral polytope, and let Δ° be the *polar* polytope defined in Section 3.2.1. Besides $(\Delta^\circ)^\circ = \Delta$, [**Batyrev4**] shows that the basic duality between Δ and Δ° is as follows.

LEMMA 3.5.4. Δ *is reflexive if and only if* Δ° *is reflexive.*

Reflexive polytopes are interesting in this context because of the following result, which characterizes when $\mathbb{P}_\Delta$ is Fano.

PROPOSITION 3.5.5. Δ *is reflexive if and only if* $\mathbb{P}_\Delta$ *is Fano.*

PROOF. Let Δ be reflexive. Then the v_Γ in Definition 3.5.3 are precisely the cone generators of the fan of $\mathbb{P}_\Delta$. It follows easily that the support function ϕ_Δ takes the value -1 on each v_Γ. The corresponding ample divisor on $\mathbb{P}_\Delta$ is thus $\sum_\rho D_\rho$, which means that $\mathbb{P}_\Delta$ is Fano. The converse is equally easy to prove. $\square$

The simplest example of a Fano toric variety is $\mathbb{P}^n$. The next case to consider is weighted projective space, where the answer is slightly more interesting.

LEMMA 3.5.6. *Let $X = \mathbb{P}(q_0, \ldots, q_n)$ be a weighted projective space, and let $q = \sum_{i=0}^n q_i$. Then X is Fano if and only if $q_i | q$ for all i.*

PROOF. The fan of $X = \mathbb{P}(q_0, \ldots, q_n)$ has cone generators $v_0, \ldots, v_n$ which satisfy $\sum_{i=0}^n q_i v_i = 0$ in the n-dimensional lattice N, and the maximal cones of the fan are generated by the n-element subsets of $\{v_0, \ldots, v_n\}$. We first determine when the divisor $D = \sum_{i=0}^n D_i$ is Cartier.

For each i, one easily sees that there is a unique $m_i \in M \otimes \mathbb{Q}$ such that $\langle m_i, v_j \rangle = -1$ for all $j \neq i$. Then the divisor D is Cartier if and only if $m_i \in M$ for all i. However, the relation $\sum_{i=0}^n q_i v_i = 0$ implies that

$$\langle m_i, v_i \rangle = \frac{\sum_{j \neq i} q_j}{q_i} = \frac{q}{q_i} - 1.$$

Since $v_0, \ldots, v_n$ generate N, we see that $m_i \in M$ if and only if $q_i | q$ for all i. Once we know that D is Cartier, the ampleness criterion of Section 3.1 reduces to the inequality $\langle m_i, v_i \rangle > -1$, which is certainly true in this case. $\square$

Section 3.7 will give further examples of toric Fano varieties. In any given dimension, there are only finitely many reflexive polytopes up to unimodular transformation, which means that there are only finitely many toric Fano varieties of dimension n up to isomorphism (see [**Batyrev4**]). Smooth toric Fano varieties have been classified in low dimensions (see [**Batyrev6, Oda**]), and attempts are underway to classify *all* 4-dimensional reflexive polytopes (see [**KS3, Skarke**]).

Since reflexive polytopes come in pairs Δ, Δ°, we get toric Fano varieties $\mathbb{P}_\Delta$, $\mathbb{P}_{\Delta^\circ}$ which are in some sense dual. In Chapter 4, we will use these toric varieties to create "dual" families of Calabi-Yau hypersurfaces which are important in mirror symmetry.

3.6. Automorphisms of Toric Varieties

This section will describe the automorphism group of a complete simplicial toric variety X. These results are due to Demazure [**Demazure**], who computed $\mathrm{Aut}(X)$ when X was smooth. In [**Cox**], it was shown that Demazure's description of $\mathrm{Aut}(X)$ also holds in the simplicial case.

If X is given by the fan Σ, Theorem 3.2.3 gives the quotient representation $X = (\mathbb{C}^{\Sigma(1)} - Z(\Sigma))/G$, where $G = \mathrm{Hom}_{\mathbb{Z}}(A_{n-1}(X), \mathbb{C}^*)$. Also recall the homogeneous coordinate ring $S = \mathbb{C}[x_\rho : \rho \in \Sigma(1)]$, which is graded by $A_{n-1}(X)$.

We will construct automorphisms of $X = (\mathbb{C}^{\Sigma(1)} - Z(\Sigma))/G$ by finding automorphisms of $\mathbb{C}^{\Sigma(1)} - Z(\Sigma)$ which commute with the action of G. One obvious class of such automorphisms is given by elements of the "big torus" $(\mathbb{C}^*)^{\Sigma(1)}$. The exact sequence (3.9)

$$1 \longrightarrow G \longrightarrow (\mathbb{C}^*)^{\Sigma(1)} \longrightarrow T_N \longrightarrow 1$$

shows that this class gives the automorphisms of X coming from the torus, i.e., $T_N \subset \mathrm{Aut}(X)$.

The next class of automorphisms comes from roots. A *root* of X consists of a pair (x_ρ, x^D) where x_ρ is one of our variables and $x^D \in S$ is a monomial distinct from x_ρ of the same degree, i.e., $\deg(x^D) = \deg(x_\rho)$ in $A_{n-1}(X)$. Note that x_ρ cannot divide x^D, since otherwise x^D/x_ρ would be a nontrivial monomial of degree 0, which can't exist since X is complete.

The root (x_ρ, x^D) determines an automorphism of $\mathbb{C}^{\Sigma(1)}$ as follows. Write a point of $\mathbb{C}^{\Sigma(1)}$ as $(x_\rho, \mathbf{x})$, where $\mathbf{x}$ is the vector indexed by $\Sigma(1) - \{\rho\}$. Then we get the 1-parameter family of automorphisms

$$(3.21) \qquad\qquad y_\lambda(x_\rho, \mathbf{x}) = (x_\rho + \lambda \mathbf{x}^D, \mathbf{x}),$$

where $\lambda \in \mathbb{C}$ and $\mathbf{x}^D$ is the monomial x^D evaluated at $\mathbf{x}$, which makes sense since x_ρ doesn't divide x^D. One can prove that y_λ is an automorphism of $\mathbb{C}^{\Sigma(1)} - Z(\Sigma)$ which commutes with the action of G and hence descends to give an element of $\mathrm{Aut}(X)$.

The final class of automorphisms comes from symmetries of the fan Σ, i.e., automorphisms of N which permute the cones of Σ. In particular, such automorphisms permute the 1-dimensional cones and hence give automorphisms of $\mathbb{C}^{\Sigma(1)} - Z(\Sigma)$ which commute with the action of G.

THEOREM 3.6.1. *If X is a simplicial complete toric variety, then the three classes of automorphisms coming from the torus, roots, and fan symmetries generate* $\mathrm{Aut}(X)$. *Furthermore, the first two classes coming from the torus and roots generate the connected component of the identity of* $\mathrm{Aut}(X)$, *and*

$$\dim(\mathrm{Aut}(X)) = n + \text{number of roots} = n + \sum_\rho(\dim(S_{\deg(x_\rho)}) - 1).$$

For example, the homogeneous coordinate ring of $\mathbb{P}^n$ is the usual graded ring $\mathbb{C}[x_0, \dots, x_n]$. This implies that the roots are all pairs (x_i, x_j) for $i \neq j$. There are $(n+1)^2 - (n+1)$ such pairs, so $\dim(\mathrm{Aut}(\mathbb{P}^n)) = (n+1)^2 - 1$ by Theorem 3.6.1. This, of course, is the dimension of $\mathrm{PGL}(n + 1, \mathbb{C})$. The references [**Cox, Demazure, Oda**] give more details about the structure of $\mathrm{Aut}(X)$ as an algebraic group.

We next describe a geometric method, due to [**AGM1**], for computing the dimension of $\mathrm{Aut}(X)$. Suppose that the simplicial toric variety X is Gorenstein. Hence the anticanonical class $-K_X = \sum_\rho D_\rho$ is Cartier, which gives the polytope

$$\Delta = \Delta_{-K_X} = \{m \in M_\mathbb{R} : \langle m, v_\rho \rangle \geq -1\}.$$

Note that 0 is in the interior of Δ, so that $\dim(\Delta) = n$ (but Δ need not be integral).

PROPOSITION 3.6.2. *If X is simplicial and Gorenstein, and Δ is as above, then*

$$\dim(\mathrm{Aut}(X)) = n + \sum_\Gamma l^*(\Gamma),$$

where the sum is over all facets $\Gamma \subset \Delta$.

PROOF. Recall from (3.6) that $l^*(\Gamma)$ is the number of lattice points in the relative interior of the facet Γ. To prove the proposition, let (x_ρ, x^D) be a root, and write $x^D = \Pi_{\rho' \neq \rho} x_{\rho'}^{a_{\rho'}}$. By the exact sequence (3.2), we know that $\deg(x_\rho) = \deg(x^D)$ if and only if there is $\widetilde{m} \in M$ such that

$$(3.22) \qquad \begin{aligned} \langle \widetilde{m}, v_\rho \rangle &= -1 \\ \langle \widetilde{m}, v_{\rho'} \rangle &= a_{\rho'} \quad \text{for } \rho' \neq \rho. \end{aligned}$$

Since Δ is defined by the inequalities $\langle m, v_\rho \rangle \geq -1$ and $\langle m, v_{\rho'} \rangle \geq -1$ for $\rho' \neq \rho$, the inequalities (3.22) tell us that $\widetilde{m} \in \Delta$ since $a_{\rho'} \geq 0 > -1$ for $\rho' \neq \rho$. In fact, the latter inequalities and $\langle \widetilde{m}, v_\rho \rangle = -1$ imply that $\widetilde{m}$ is in the interior of the facet of Δ defined by $\langle m, v_\rho \rangle = -1$. Conversely, given $\widetilde{m}$ in the interior of this facet, one easily checks that (x_ρ, x^D) is a root when $x^D = \Pi_{\rho' \neq \rho} x_{\rho'}^{\langle \widetilde{m}, v_{\rho'} \rangle + 1}$. $\qquad\square$

For an example of how this works, note that $\mathbb{P}^4$ is Fano, so that $\mathbb{P}^4 = \mathbb{P}_\Delta$, where $\Delta = \Delta_{-K_{\mathbb{P}^4}}$ is reflexive. We know the automorphism group $\mathrm{Aut}(\mathbb{P}^4)$, but what about the automorphisms of $X^\circ = \mathbb{P}_{\Delta^\circ}$, where Δ° is the dual polytope of Δ? In Section 4.2, we will see that the vertices of the simplex Δ° are given by

$$(-1, -1, -1, -1)$$
$$(1, 0, 0, 0)$$
$$(0, 1, 0, 0)$$
$$(0, 0, 1, 0)$$
$$(0, 0, 0, 1).$$

Observe that the facets of this polytope have *no* interior points. Since Δ° is the polytope coming from the anticanonical class on X°, we see that X° has *no* roots by Proposition 3.6.2. It follows that the connected component of the identity of $\mathrm{Aut}(X^\circ)$ is precisely the torus of X°. In particular, although the original toric variety $X = \mathbb{P}^4$ has a large automorphism group, the automorphism group of its dual X° is relatively small.

3.7. Examples

This section will present three examples which illustrate various aspects of the theory of toric varieties discussed so far. Working out the details of the examples is a good exercise for readers less familiar with toric geometry.

Example 3.7.1. The most basic example of a toric variety is $\mathbb{P}^n$. Here, $N = \mathbb{Z}^n$, and the 1-dimensional cones are the standard basis $e_1, \ldots, e_n$ and $e_0 = -\sum_{i=1}^n e_i$ (as usual, we make no distinction between 1-dimensional cones and their generators). This gives a smooth fan with cones generated by all proper subsets of $\{e_0, \ldots, e_n\}$. The exact sequence (3.2) becomes

$$0 \longrightarrow \mathbb{Z}^n \longrightarrow \mathbb{Z}^{n+1} \longrightarrow \mathbb{Z} \longrightarrow 0,$$

where $(a_0, \ldots, a_n) \in \mathbb{Z}^{n+1}$ maps to $a_0 + \cdots + a_n$. Thus, the homogeneous coordinate ring is $\mathbb{C}[x_0, \ldots, x_n]$ with the usual grading, and the construction of $\mathbb{P}^n$ from Theorem 3.2.3 is the usual $\mathbb{P}^n = (\mathbb{C}^{n+1} - \{0\})/\mathbb{C}^*$.

The moment map is given by $\mu(z_0, \ldots, z_n) = (1/2)\sum_{i=0}^n |z_i|^2$, so that $\mu^{-1}(r)$ is a sphere, and the construction of Theorem 3.3.4 is the Hopf fibration $S^{2n+1} \to \mathbb{P}^n$ with fiber S^1.

Finally, the divisor D_0 corresponding to e_0 is ample, and the associated polytope is the simplex

$$\Delta_{D_0} = \{(c_1, \ldots, c_n) : c_i \geq 0, \; c_1 + \cdots + c_n \leq 1\}.$$

Example 3.7.2. Our next example involves a weighted projective space and its toric resolution. In $N = \mathbb{Z}^4$, consider the vectors

$$v_0 = (-1, -2, -2, -2), \quad v_1 = e_1, \quad v_2 = e_2, \quad v_3 = e_3, \quad v_4 = e_4.$$

These vectors satisfy the relation $v_0 + v_1 + 2v_2 + 2v_3 + 2v_4 = 0$. Using the fan formed by proper subsets of $\{v_0, v_1, v_2, v_3, v_4\}$, we get the weighted projective space $\mathbb{P}(1, 1, 2, 2, 2)$, which is Fano by Lemma 3.5.6. The exact sequence (3.2) is

$$0 \longrightarrow \mathbb{Z}^4 \longrightarrow \mathbb{Z}^5 \longrightarrow \mathbb{Z} \longrightarrow 0,$$

where $(a_0, a_1, a_2, a_3, a_4) \in \mathbb{Z}^5$ maps to $a_0 + a_1 + 2a_2 + 2a_3 + 2a_4$. Thus, the homogeneous coordinate ring is $\mathbb{C}[x_0, x_1, x_2, x_3, x_4]$, where x_0, x_1 have degree 1 and x_2, x_3, x_4 have degree 2.

To resolve the singularities of $\mathbb{P}(1, 1, 2, 2, 2)$, we take the singular cone generated by v_0, v_1 and add the generator

$$v_5 = (1/2)(v_0 + v_1) = (0, -1, -1, -1).$$

This gives a new fan Σ where each 4-dimensional cone of the original fan containing v_0, v_1 gets split into two cones of Σ. Then Σ is smooth, so that the corresponding toric variety $X = X_\Sigma$ is a resolution of singularities of $\mathbb{P}(1, 1, 2, 2, 2)$.

For X, one sees that the group $G = (\mathbb{C}^*)^2$ acts on $\mathbb{C}^{\Sigma(1)} = \mathbb{C}^6$ via

$$(\lambda, \mu)(x_0, x_1, x_2, x_3, x_4, x_5) = (\lambda\, x_0, \lambda\, x_1, \mu\, x_2, \mu\, x_3, \mu\, x_4, \lambda^{-2}\mu\, x_5).$$

Furthermore, the only two primitive collections are $\{v_0, v_1\}$ and $\{v_2, v_3, v_4, v_5\}$. By (3.10), $Z(\Sigma) = \{x_0 = x_1 = 0\} \cup \{x_2 = x_3 = x_4 = x_5 = 0\}$. Thus

$$X = (\mathbb{C}^6 - Z(\Sigma))/(\mathbb{C}^*)^2.$$

We can also describe the moment map and GKZ decomposition. Since the relations among $v_0, v_1, v_2, v_3, v_4, v_5$ are generated by $v_2 + v_3 + v_4 + v_5 = 0$ and $v_0 + v_1 - 2v_5 = 0$, (3.16) shows that the moment map is given by

$$\mu_\Sigma(z_0, z_1, z_2, z_3, z_4, z_5) = \tfrac{1}{2}(|z_2|^2 + |z_3|^2 + |z_4|^2 + |z_5|^2, |z_0|^2 + |z_1|^2 - 2|z_5|^2).$$

The GKZ decomposition of $A_3^+(X_\Sigma) \otimes \mathbb{R} \subset A_3(X_\Sigma) \otimes \mathbb{R} = \mathbb{R}^2$ is especially simple since there are only two projective simplicial fans with generators contained in $\Sigma(1)$: the fan Σ of X, and the original fan Σ' of the weighted projective space $\mathbb{P}(1, 1, 2, 2, 2)$. We leave it as an exercise for the reader to verify that the GKZ decomposition of $A_3^+(X_\Sigma) \otimes \mathbb{R}$ is as follows:

(3.23)

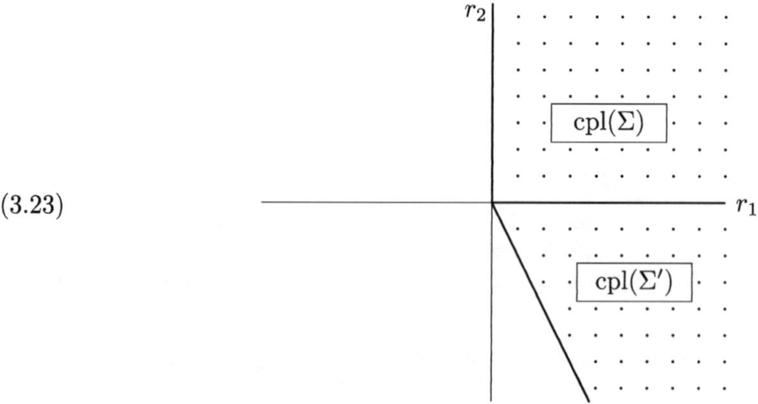

Turning to the secondary fan, note that $\Xi = \{v_0, \ldots, v_5\}$, so that by (3.17), Ξ^+ consists of the seven points $(v_0, 1), \ldots, (v_5, 1), (0, 1)$. A basis of all relations among these is given by the rows of the matrix

$$\begin{pmatrix} 0 & 0 & 1 & 1 & 1 & 1 & -4 \\ 1 & 1 & 0 & 0 & 0 & -2 & 0 \end{pmatrix}.$$

It follows from Section 3.4.2 that in terms of the dual basis of $A(\Xi) \simeq \mathbb{R}^2$, the columns of this matrix give the generators of the edges of the secondary fan. In fact, we see from (3.23) that the first six columns of the above matrix give the GKZ decomposition, and the last column enlarges it to a complete fan by inserting an edge along the negative r_1-axis.

Example 3.7.3. Let $M = \mathbb{Z}^3$, and consider the cube $\Delta \subset M_\mathbb{R}$ centered at the origin with vertices $(\pm 1, \pm 1, \pm 1)$. This gives the toric variety $X = \mathbb{P}_\Delta$. To describe the fan of X, note that the polar $\Delta^\circ \subset N_\mathbb{R}$ is the octahedron with vertices $\pm e_1, \pm e_2, \pm e_3$. Thus the normal fan is formed from the faces of the octahedron, giving a fan Σ whose 3-dimensional cones are the octants of $\mathbb{R}^3$. It follows easily that $X = \mathbb{P}^1 \times \mathbb{P}^1 \times \mathbb{P}^1$. Furthermore, the divisor $\mathcal{O}_X(1)$ coming from Δ is the anticanonical divisor $-K_X$. In particular, X is Fano.

We have the following pictures of Δ and Δ°:

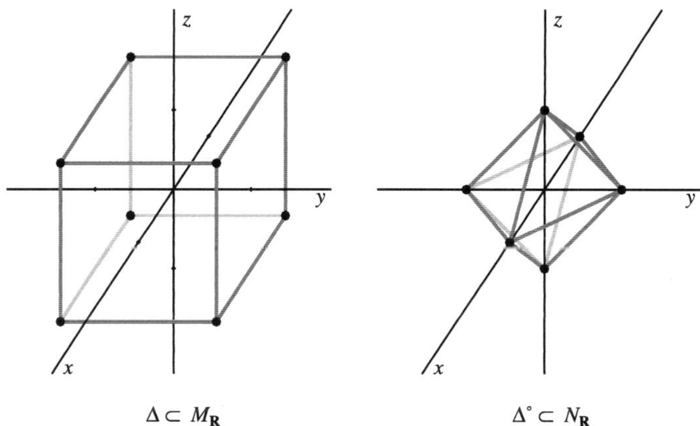

$$\Delta \subset M_\mathbf{R} \qquad\qquad \Delta^\circ \subset N_\mathbf{R}$$

It is easy to check that the cube $\Delta \subset M_\mathbb{R}$ and the octahedron $\Delta^\circ \subset N_\mathbb{R}$ are dual reflexive polytopes in the sense of Section 3.5. In particular, Δ° gives a "dual" toric variety $X^\circ = \mathbb{P}_{\Delta^\circ}$, which is determined by the normal fan of Δ° (= the fan in $M_\mathbb{R}$ formed by cones over the faces of the cube Δ). Hence we have a pair of "dual" toric varieties, X and X°. It is interesting to observe that X is smooth while X° is rather singular. In fact, the 3-dimensional cones of Σ° are not even simplicial—they're all infinite pyramids. However, since Δ° is reflexive, we know that X° is Fano and that $\Delta^\circ \subset N_\mathbb{R}$ is the polytope associated to the anticanonical divisor $-K_{X^\circ}$.

Note that Δ and Δ° also have quite different numbers and geometries of lattice points. For $\Delta^\circ \subset N_\mathbb{R}$, the only lattice points in N are the origin and vertices, while $\Delta \subset M_\mathbb{R}$ has many more since the midpoints of the edges and the centers of the faces lie in M.

We can also describe the GKZ decomposition of X°. The maximal cones of the decomposition correspond to projective simplicial fans whose cone generators lie in $\Sigma^\circ(1) = \{\text{vertices of the cube}\}$. It is clear that the only way to do this is to subdivide each face of the cube into two triangles along a diagonal. For each face, this can be done in two ways, so that the GKZ decomposition has $2^6 = 64$ cones. However, in Chapter 4, we will see that for the purposes of mirror symmetry, we need to form the GKZ decomposition using projective simplicial fans whose cone generators lie in $\Delta \cap M$, which is much larger than just the vertices of the cube. Hence, the resulting GKZ decomposition has *many* more than 2^6 cones.

This example shows how we can start with a smooth toric variety (such as $\mathbb{P}^1 \times \mathbb{P}^1 \times \mathbb{P}^1$) and by a process of "duality" wind up with something singular. It is no accident that the same sort of thing happened in Chapter 2 with the quintic threefold and its mirror. Chapter 4 will explore this phenomenon in detail.

CHAPTER 4

Mirror Symmetry Constructions

This chapter will describe various constructions of mirror manifolds. Most (but not all) of the constructions involve toric varieties, and in particular we will give a toric interpretation of mirror symmetry for the quintic hypersurface.

4.1. The Batyrev Mirror Construction

In Chapter 1, we discussed why mirror symmetry is reasonable from the point of view of superconformal field theory, but we didn't explain how to find the mirror of a given Calabi-Yau threefold. The first explicit examples of mirror manifolds in physics were given by Greene and Plesser [**GP1**], who obtained the mirror as a quotient of a Fermat hypersurface in a weighted projective space. An example of this construction is the mirror of the quintic threefold given in Chapter 2. Roan observed that toric geometry provides a natural framework within which to describe this class of mirrors [**Roan1**], and later, Batyrev found a beautiful toric description which greatly generalizes the Greene-Plesser construction.

Batyrev's construction makes essential use of reflexive polytopes, which were introduced in Section 3.5. There, we saw that reflexive polytopes Δ correspond to toric Fano varieties and that Δ has a dual Δ° which is also reflexive. In this section, we will explain how the duality between Δ and Δ° results in a duality between families of Calabi-Yau hypersurfaces in certain toric varieties closely related to $\mathbb{P}_\Delta$ and $\mathbb{P}_{\Delta^\circ}$.

4.1.1. Calabi-Yau Toric Hypersurfaces. The first step in Batyrev's construction is to observe that Calabi-Yau hypersurfaces in toric varieties arise naturally from reflexive polytopes. A reflexive polytope Δ gives the toric variety $\mathbb{P}_\Delta$. For the purposes of mirror symmetry, however, $\mathbb{P}_\Delta$ may be too singular. We need something where we can do Hodge theory, and for this purpose we will use simplicial fans which refine the fan of $\mathbb{P}_\Delta$. Recall that the fan of $\mathbb{P}_\Delta$ is the normal fan of Δ, which consists of the cones over proper faces of $\Delta^\circ \subset N_\mathbb{R}$. Furthermore, since Δ is reflexive, it follows that the cone generators of the normal fan are the vertices of Δ°, which lie in $\Delta^\circ \cap N - \{0\}$. We will consider the following fans in $N_\mathbb{R}$.

DEFINITION 4.1.1. *Given a reflexive polytope $\Delta \subset M_\mathbb{R}$, a fan Σ in $N_\mathbb{R}$ is a* projective subdivision *if it has the following properties:*

- *Σ refines the normal fan of Δ.*
- *$\Sigma(1) \subset \Delta^\circ \cap N - \{0\}$.*
- *X_Σ is projective and simplicial.*

Furthermore, we say that Σ is maximal *if $\Sigma(1) = \Delta^\circ \cap N - \{0\}$.*

Maximal projective subdivisions Σ can be shown to exist [**OdP**]. There are several choices for such a Σ, and as indicated in Section 3.4, each choice gives a

53

different cone in the GKZ decomposition (and a distinct phase in a related gauged linear sigma model—see Appendix B). This will be discussed in more detail in Section 4.1.4.

Given a projective subdivision Σ, we get a toric variety X_Σ and a birational morphism $f : X_\Sigma \to \mathbb{P}_\Delta$ which have the following nice properties.

LEMMA 4.1.2. *If Σ is a projective subdivision, then:*

(i) X_Σ *is a Gorenstein orbifold.*

(ii) Δ *is the polytope associated to* $-K_{X_\Sigma}$.

(iii) $-K_{X_\Sigma}$ *is semi-ample, meaning that* $-K_{X_\Sigma}$ *is generated by global sections and* $(-K_{X_\Sigma})^n > 0$.

(iv) *The map* $f : X_\Sigma \to \mathbb{P}_\Delta$ *is* crepant, *meaning that* $f^*(K_{\mathbb{P}_\Delta}) = K_{X_\Sigma}$.

Furthermore, if Σ is a maximal projective subdivision, then X_Σ has terminal singularities.

PROOF. Since Δ° is reflexive, each facet Γ° of Δ° is defined by $\langle m, v \rangle = -1$ for some $m = m_{\Gamma^\circ} \in M$. Then $\langle m, v_\rho \rangle = -1$ when $v_\rho \in \Gamma^\circ$, and it easily follows that $-K_{X_\Sigma} = \sum_\rho D_\rho$ is Cartier by the criterion given in Section 3.1. Hence $X = X_\Sigma$ is Gorenstein, and it is an orbifold since Σ is simplicial. Also, the equation $\langle m, v_\rho \rangle = -1$ shows that the polytope associated to the divisor $-K_X$ is exactly Δ. By duality, the m's are the vertices of Δ, and it follows from Section 3.2.1 that $-K_X$ is generated by global sections. Finally, by [**Fulton3**, Sect. 5.3], the intersection number $(-K_{X_\Sigma})^n$ is $n!\mathrm{vol}(\Delta)$, which is positive since Δ is n-dimensional.

To prove the remaining parts of the lemma, we will study how adding new cone generators to Σ affects the canonical class of X. Recall from the discussion following Definition 3.5.1 that for any toric variety, K_X is the Weil divisor $-\sum_{\rho \in \Sigma(1)} D_\rho$. Let $\sigma \in \Sigma$ be a cone, and let X' be the toric variety obtained by adding a new cone generator $v' \in N$ which lies in the interior of σ. We know that the generators v_ρ of σ lie on some facet Γ° of Δ°, so that $\langle m, v_\rho \rangle = -1$ (where $m = m_{\Gamma^\circ}$ is as above). The key calculation is that if $g : X' \to X$ is the natural map, then

$$(4.1) \qquad K_{X'} = g^*(K_X) - (\langle m, v' \rangle + 1)D'.$$

where D' is the divisor on X' corresponding to v'. This equality is to be interpreted as a linear equivalence of Weil divisors. Equation (4.1) can be proved, for example, using the techniques of [**Reid4**, Section 4]. We just have to examine the coefficient of D' in (4.1). For this, note that on the affine open $X_\sigma \subset X$, K_X coincides with $\mathrm{div}(\chi^m)$, and therefore the coefficient of D' in g^*K_X is $\langle m, v' \rangle$.

If $v' \in \Delta^\circ \cap N$, then v' lies in the facet Γ°, which implies $\langle m, v' \rangle = -1$, and $K_{X'} = g^*(K_X)$ follows immediately from (4.1). Since $\Sigma(1) \subset \Delta^\circ \cap N$, X_Σ is obtained from $\mathbb{P}_\Delta$ by adding a succession of new vertices in $\Delta^\circ \cap N$, and $K_{X_\Sigma} = f^*(K_{\mathbb{P}_\Delta})$ follows easily.

On the other hand, if Σ is maximal, then $\Sigma(1) = \Delta^\circ \cap N - \{0\}$. Hence any new cone generator v' we add must lie *outside* of Δ° (since 0 is the only interior point by the definition of reflexive), which implies $\langle m, v' \rangle + 1 < 0$ in (4.1). By the discussion in Appendix A, we see that X has terminal singularities. $\square$

We now show how to create Calabi-Yau varieties using the anticanonical linear system on X_Σ and $\mathbb{P}_\Delta$.

PROPOSITION 4.1.3. *If Δ is a reflexive polytope of dimension n, then the general member $\bar{V} \in |-K_{\mathbb{P}_\Delta}|$ is a Calabi-Yau variety of dimension $n-1$. Furthermore, if Σ is a projective subdivision and $X = X_\Sigma$, then:*

(i) The general member $V \in |-K_X|$ is a Calabi-Yau orbifold.

(ii) If Σ is maximal, then the general member $V \in |-K_X|$ is a minimal Calabi-Yau orbifold.

PROOF. We first consider $\bar{V}$. By the definition of Calabi-Yau variety given in Section 1.4, we must show that $\bar{V}$ has canonical singularities, a trivial canonical sheaf, and vanishing cohomology $H^k(\bar{V}, \mathcal{O}_{\bar{V}}) = 0$ for $0 < k < n-1$.

Since the Fano toric variety $\mathbb{P}_\Delta$ is Gorenstein, it has at most canonical singularities (see [**Batyrev4**]). Then a version of the Bertini theorem guarantees that the general member $\bar{V} \in |-K_{\mathbb{P}_\Delta}|$ also has at most canonical singularities [**Reid1**, Theorem 1.13]. Also, note that

$$\widehat{\Omega}_{\bar{V}}^{n-1} \simeq \widehat{\Omega}_{\mathbb{P}_\Delta}^{n}(-K_{\mathbb{P}_\Delta}) \otimes \mathcal{O}_{\bar{V}} \simeq \mathcal{O}_{\bar{V}},$$

where the first isomorphism is the adjunction formula (which holds since $\mathbb{P}_\Delta$ is Cohen-Macaulay and $-K_{\mathbb{P}_\Delta}$ is Cartier). The final step is to show that $H^k(\bar{V}, \mathcal{O}_{\bar{V}}) = 0$ for $0 < k < n-1$. Since $\mathcal{O}_{\mathbb{P}_\Delta}(-\bar{V}) \simeq \mathcal{O}_{\mathbb{P}_\Delta}(K_{\mathbb{P}_\Delta}) = \widehat{\Omega}_{\mathbb{P}_\Delta}^n$, we get an exact sequence

$$0 \longrightarrow \widehat{\Omega}_{\mathbb{P}_\Delta}^n \longrightarrow \mathcal{O}_{\mathbb{P}_\Delta} \longrightarrow \mathcal{O}_{\bar{V}} \longrightarrow 0,$$

which gives the long exact sequence

$$\cdots \longrightarrow H^k(\mathbb{P}_\Delta, \mathcal{O}_{\mathbb{P}_\Delta}) \longrightarrow H^k(\bar{V}, \mathcal{O}_{\bar{V}}) \longrightarrow H^{k+1}(\mathbb{P}_\Delta, \widehat{\Omega}_{\mathbb{P}_\Delta}^n) \longrightarrow \cdots$$

However, $H^k(\mathbb{P}_\Delta, \mathcal{O}_{\mathbb{P}_\Delta}) = 0$ for $k > 0$ and, by Serre-Grothendieck duality, we have $H^{k+1}(\mathbb{P}_\Delta, \widehat{\Omega}_{\mathbb{P}_\Delta}^n) \simeq H^{n-k-1}(\mathbb{P}_\Delta, \mathcal{O}_{\mathbb{P}_\Delta})^* = 0$ for $k < n-1$. This implies the desired vanishing of $H^k(\bar{V}, \mathcal{O}_{\bar{V}})$, and we conclude that $\bar{V}$ is Calabi-Yau.

Now let Σ be a projective subdivision and set $X = X_\Sigma$. Since $-K_X$ is semi-ample by Lemma 4.1.2, the linear system $|-K_X|$ has no basepoints. Furthermore, X is an orbifold since Σ is simplicial, and then the Bertini theorem (applied to the fixed loci of the local quotients defining X as an orbifold) guarantees that the general member $V \in |-K_X|$ is a suborbifold of X. Everything we did above remains true, and it follows that V is a Calabi-Yau orbifold.

Finally, suppose that Σ is maximal. According to Definition 1.4.1, V is a minimal Calabi-Yau provided it has Gorenstein $\mathbb{Q}$-factorial terminal singularities. Since V is already a Gorenstein orbifold, it automatically has Gorenstein $\mathbb{Q}$-factorial singularities. Hence, we need only show that V has terminal singularities. However, the ambient space X is terminal by Lemma 4.1.2, so we are done by using Bertini as in the proof of [**Reid1**, Theorem 1.13]. $\square$

Since $f : X_\Sigma \to \mathbb{P}_\Delta$ is crepant, it follows that the Calabi-Yau hypersurfaces $V \subset X_\Sigma$ constructed above are the proper transforms via f of general anticanonical hypersurfaces $\bar{V} \subset \mathbb{P}_\Delta$. One can also show that the induced map $V \to \bar{V}$ is crepant.

When Σ is a maximal projective subdivision, the variety $V \subset X_\Sigma$ is a minimal model (in the sense of Mori theory) of $\bar{V} \subset \mathbb{P}_\Delta$. Batyrev calls V a *maximal projective crepant partial desingularization* of $\bar{V}$, or MPCP-desingularization for short.

The situation is especially nice when V is a threefold. Since 3-dimensional Gorenstein orbifold terminal singularities are smooth by Appendix A, it follows that for a maximal projective subdivision, the minimal Calabi-Yau threefolds $V \subset X_\Sigma$ constructed in Proposition 4.1.3 are smooth.

4.1.2. The Mirror of a Calabi-Yau Toric Hypersurface. We can now describe the Batyrev mirror construction. Fix a reflexive polytope Δ, and let Σ be a maximal projective subdivision of its normal fan, as in Definition 4.1.1. This gives the family of Calabi-Yau hypersurfaces $V \subset X_\Sigma$ for $V \in |-K_{X_\Sigma}|$, and as explained in Chapter 1, each member of this family comes equipped with a complexified Kähler cone, so that we obtain a family $\{(V,\omega)\}$ consisting of all possible pairs of a Calabi-Yau hypersurface and a complexified Kähler class.

However, Δ reflexive implies that Δ° is reflexive, so that we can repeat the above construction using Δ°. Thus, for a maximal projective subdivision Σ° of the normal fan of Δ° in $M_{\mathbb{R}}$, $|-K_{X_{\Sigma^\circ}}|$ yields a family of Calabi-Yau hypersurfaces $V^\circ \subset X_{\Sigma^\circ}$. When we add in the complexified Kähler cones, we get a family $\{(V^\circ,\omega^\circ)\}$ called the *Batyrev mirror* of the original family $\{(V,\omega)\}$.

In physics, it is believed that these families are mirror pairs in the sense of Physics Definition 1.1.1. More precisely, we have the following conjecture:

PHYSICS CONJECTURE 4.1.4. *The families $\{(V,\omega)\}$ and $\{(V^\circ,\omega^\circ)\}$ induce isomorphic superconformal field theories whose $N = 2$ superconformal representations are the same, up to the sign change discussed in Section 1.1.*

Another way to say this is that the families $\{(V,\omega)\}$ and $\{(V^\circ,\omega^\circ)\}$ each represent open subsets of the *same* component of the moduli space of $N = 2$ superconformal field theories (SCFTs). Strictly speaking, the conjecture only applies to the case when V and V° are threefolds (so Δ has dimension four), although the Batyrev mirror construction applies to reflexive polytopes of arbitrary dimension.

This mirror symmetry conjecture has not been proved in physics (but see [**MP2**] for an attempt). The Greene-Plesser [**GP1**] construction gives a physics proof that $\{(V,\omega)\}$ and $\{(V^\circ,\omega^\circ)\}$ form a mirror pair in the special case of Fermat hypersurfaces of the appropriate degree in certain weighted projective spaces. The general case is still open.

Given the lack of a rigorous foundation for SCFTs, Physics Conjecture 4.1.4 is not a conjecture in the standard mathematical sense. But it is possible to formulate precise mathematical consequences of this equality of SCFTs. In particular, we saw in Chapter 1 that mirror symmetry implies that the following should be true:

- The cohomology group $H^{1,1}(V)$ (resp. $H^{1,1}(V^\circ)$) should be isomorphic to $H^{n-2,1}(V^\circ)$ (resp. $H^{n-2,1}(V)$). (Recall that V and V° have dimension $n-1$, where $n = \dim(\Delta)$.)
- According to the picture (1.6) of mirror symmetry, the Kähler moduli of V (resp V°) should be locally isomorphic (via the mirror map) to the complex moduli of V° (resp V).
- The A-model correlation function of V (resp V°) should correspond (via the mirror map) to the suitably normalized B-model correlation function of V° (resp V).

The challenge for mathematicians is to prove these consequences. Substantial progress has been made when V° is the Batyrev mirror of V. In Section 4.1.3, we will dispose of the first bullet, and the second will be addressed in Chapter 6 when we study the complex and Kähler moduli of a Calabi-Yau toric hypersurface. The third bullet will have to wait until Chapter 8, where we will define mathematical mirror pairs and formulate the Hodge-Theoretic Toric Mirror Conjecture. Chapter 11 will discuss the proofs of special cases.

4.1.3. Hodge Numbers of Batyrev Mirrors. In order to show that two finite dimensional vector spaces are isomorphic, it suffices to show that they have the same dimension. Hence the first bullet above asks us to compare the Hodge numbers of a Calabi-Yau toric hypersurface V and its Batyrev mirror V°.

As above, we will assume that Δ is an n-dimensional reflexive polytope and that V is a general anticanonical hypersurface in X_Σ, where Σ is a maximal projective subdivision of the normal fan of Δ. This process, applied to Δ°, gives the Batyrev mirror family V°. As proved in [**Batyrev4, AGM1**], the Hodge numbers of V and V° are related as follows:

THEOREM 4.1.5. *If V° is the Batyrev mirror of V, then*

$$h^{1,1}(V) = h^{n-2,1}(V^\circ) \quad and \quad h^{n-2,1}(V) = h^{1,1}(V^\circ).$$

PROOF. We will begin by defining some interesting subspaces of $H^{1,1}(V)$ and $H^{n-2,1}(V)$. For each $v_i \in \Sigma(1)$, let $D_i \subset X_\Sigma$ be the corresponding T_N-invariant divisor. These divisors restrict to divisors D_i' on the hypersurface $V \subset X_\Sigma$ whose cohomology classes generate a subspace (over $\mathbb{C}$) denoted

$$(4.2) \qquad\qquad H^{1,1}_{\text{toric}}(V) \subset H^{1,1}(V).$$

Turning our attention to $H^{n-2,1}(V) \simeq H^1(V, T_V)$, first note that although V may be singular, it is a Gorenstein orbifold with at worst terminal singularities, so that $H^1(V, T_V)$ classifies the infinitesimal deformations of V by Proposition A.4.2. Then the infinitesimal deformations determined by the hypersurfaces $V \in |-K_{X_\Sigma}|$ give a subspace denoted

$$(4.3) \qquad\qquad H^{n-2,1}_{\text{poly}}(V) \subset H^{n-2,1}(V).$$

We claim that the dimensions of these subspaces can be computed in terms of Δ and Δ° as follows:

$$(4.4) \qquad\qquad h^{1,1}_{\text{toric}}(V) = l(\Delta^\circ) - n - 1 - \sum_{\Gamma^\circ} l^*(\Gamma^\circ)$$

$$(4.5) \qquad\qquad h^{n-2,1}_{\text{poly}}(V) = l(\Delta) - n - 1 - \sum_{\Gamma} l^*(\Gamma),$$

where the sums are over all facets Γ° and Γ of Δ° and Δ respectively. Here, l and l^* are as defined in (3.6). Once this claim is proved, the equations

$$h^{1,1}_{\text{toric}}(V) = h^{n-2,1}_{\text{poly}}(V^\circ) \quad and \quad h^{n-2,1}_{\text{toric}}(V) = h^{1,1}_{\text{poly}}(V^\circ)$$

follow immediately since changing from V to V° interchanges the roles of Δ and Δ° in (4.4) and (4.5). We should also mention that it is possible to construct a natural isomorphism $H^{1,1}_{\text{toric}}(V) \simeq H^{n-2,1}_{\text{poly}}(V^\circ)$ called the *monomial-divisor mirror map* (see [**AGM1**]).

We first prove (4.4). Suppose that $v_i \in \Sigma(1)$ is an interior point of a facet of Δ°. Then the standard description of toric blowups shows that $f(D_i)$ is a point, where $f : X_\Sigma \to \mathbb{P}_\Delta$ is the natural map. It follows immediately that in X_Σ, D_i is disjoint from a general $V \in |-K_{X_\Sigma}|$. Thus, these elements don't contribute to $H^{1,1}_{\text{toric}}(V)$, so that if we set

$$\Sigma(1)' - \{v_i \in \Sigma(1) : v_i \text{ is } not \text{ in the interior of a facet of } \Delta^\circ\},$$

then $H^{1,1}_{\text{toric}}(V)$ is generated by D_i' for $v_i \in \Sigma(1)'$. To see what the relations between the D_i' are, note that $m \in M$ gives a character χ^m of T_N, hence a rational function on X_Σ. By (3.1), we know the divisor of χ^m, and if we restrict this formula to V, we obtain a linear equivalence

$$(4.6) \qquad \sum_{v_i \in \Sigma(1)'} \langle m, v_i \rangle D_i' \sim 0.$$

In [**AGM1**], it is shown that these are the only relations between the D_i', which implies

$$H^{1,1}_{\text{toric}}(V) \simeq (\mathbb{Z}^{\Sigma(1)'}/M) \otimes \mathbb{C},$$

where the inclusion of $M \to \mathbb{Z}^{\Sigma(1)'}$ is given by $m \mapsto (\langle m, v_i \rangle)$ as required by (4.6). Hence

$$h^{1,1}_{\text{toric}}(V) = |\Sigma(1)'| - n.$$

In the notation of (3.6), we see that $|\Sigma(1)| = l(\Delta^\circ)-1$ since $\Sigma(1) = \Delta^\circ \cap N - \{0\}$ (this is where we use the maximality of the projective subdivision Σ). Then, removing the interior points of the facets gives $|\Sigma(1)'| = l(\Delta^\circ) - 1 - \sum_{\Gamma^\circ} l^*(\Gamma^\circ)$, and (4.4) follows easily.

Now consider $h^{n-2,1}_{\text{poly}}(V)$. We will describe the moduli coming from the family $V \in |-K_{X_\Sigma}|$. Our discussion will be informal, with a careful treatment to follow in Chapter 6. Lemma 4.1.2 showed that Δ is the polytope associated to $-K_{X_\Sigma}$. Thus, a Laurent polynomial $f \in L(\Delta)$ gives the affine hypersurface $Z_f \subset T_N$, and we can assume that $V = \overline{Z_f} \subset X_\Sigma$. Since multiplying f by a scalar doesn't affect V, varying f gives a parameter space of polynomial deformations of dimension $l(\Delta) - 1$. But we must take into account the automorphisms of X_Σ. Since Δ is associated to $-K_{X_\Sigma}$, Proposition 3.6.2 implies that the automorphism group has dimension

$$\dim(\text{Aut}(X_\Sigma)) = n + \sum_\Gamma l^*(\Gamma).$$

It now follows easily that the polynomial moduli have dimension given in (4.5).

To complete the proof of the theorem, we need to understand the "non-toric" divisor classes and the "non-polynomial" deformations of V. Batyrev [**Batyrev4**] showed that in each case, the "correction term" is determined by the codimension 2 faces of Δ and Δ°. The formulas are:

$$(4.7) \qquad h^{1,1}(V) = l(\Delta^\circ) - n - 1 - \sum_{\Gamma^\circ} l^*(\Gamma^\circ) + \sum_{\Theta^\circ} l^*(\Theta^\circ) l^*(\widehat{\Theta}^\circ)$$

$$(4.8) \qquad h^{n-2,1}(V) = l(\Delta) - n - 1 - \sum_\Gamma l^*(\Gamma) + \sum_\Theta l^*(\Theta) l^*(\widehat{\Theta}),$$

where in the first equation, Γ° refers to codimension 1 faces of Δ°, Θ° refers to codimension 2 faces of Δ°, and $\widehat{\Theta}^\circ$ refers to the face of Δ dual to Θ°. The meanings for the second equation are similar. As above, applying these formulas to V and V° interchanges the roles of Δ and Δ°, so that the theorem follows immediately from (4.7) and (4.8).

The correction term in (4.7) arises from divisors D_i whose intersection with a general V is reducible. We've already seen that we can ignore those D_i corresponding to v_i lying in the interior of a facet Γ°. In general, suppose that v_i lies in the interior of some face $\Theta^\circ \subset \Delta^\circ$. If $\widehat{\Theta}^\circ \subset \Delta$ denotes the dual face, then $\mathbb{P}_{\widehat{\Theta}^\circ} \subset \mathbb{P}_\Delta$ is an orbit closure, and $f(D_i) = \mathbb{P}_{\widehat{\Theta}^\circ}$ under the blowup map $f : X_\Sigma \to \mathbb{P}_\Delta$. If $\bar{V} = f(V)$ is the corresponding anticanonical hypersurface of $\mathbb{P}_\Delta$, it follows that $D_i' = V \cap D_i$ and $\bar{V} \cap \mathbb{P}_{\widehat{\Theta}^\circ}$ have the same number of irreducible components.

Recall that the face Θ° and its dual $\widehat{\Theta}^\circ$ satisfy $\dim \Theta^\circ + \dim \widehat{\Theta}^\circ = n - 1$. It follows that if Θ° has codimension ≥ 3, then $\widehat{\Theta}^\circ$ has dimension ≥ 2. Thus $\mathbb{P}_{\widehat{\Theta}^\circ}$ has dimension ≥ 2, which implies that $\bar{V} \cap \mathbb{P}_{\widehat{\Theta}^\circ}$ is irreducible by Bertini. We conclude that D_i' is irreducible when v_i lies in a face of Δ° of codimension ≥ 3.

On the other hand, if Θ° has codimension 2, then $\widehat{\Theta}^\circ$ has dimension 1 and $\mathbb{P}_{\widehat{\Theta}^\circ}$ is a curve. In this situation, [**Fulton3**, Sect. 5.3] implies that

$$\bar{V} \cdot \mathbb{P}_{\widehat{\Theta}^\circ} = \mathrm{vol}(\widehat{\Theta}^\circ) = l^*(\widehat{\Theta}^\circ) + 1,$$

where vol means normalized volume, and the second equality holds since $\widehat{\Theta}^\circ$ has dimension 1. This tells us that $\bar{V} \cap \mathbb{P}_{\widehat{\Theta}^\circ}$ consists of $l^*(\widehat{\Theta}^\circ) + 1$ points, so that D_i' has $l^*(\widehat{\Theta}^\circ) + 1$ connected components. One can show that these components give independent classes in $H^{1,1}(V)$, and since their sum D_i' is already in $H^{1,1}_{\mathrm{toric}}(V)$, we see that each interior point of Θ° adds $l^*(\widehat{\Theta}^\circ)$ new classes. Since Σ is maximal, we use *all* interior points of Θ°, which explains the correction term in (4.7).

As for the correction term in (4.8), we will explain why the formula is reasonable and refer to [**Batyrev4, Voisin3**] for a complete proof. If we let $Z_f = V \cap T_N \subset T_N$ be the affine hypersurface determined by V, then the complement $V - Z_f$ equals $\sum_i V \cap D_i$, which is a divisor with normal crossings (in the orbifold sense). Hence the Gysin spectral sequence gives an exact sequence

$$\longrightarrow \bigoplus_i H^{n-3}(V \cap D_i, \mathbb{C}) \longrightarrow H^{n-1}(V, \mathbb{C}) \longrightarrow Gr^W_{n-1} H^{n-1}(Z_f, \mathbb{C}) \longrightarrow 0.$$

All of these groups have pure Hodge structures, from which one can show that

$$(4.9) \qquad h^{n-2,1}(V) = h^{n-2,1}(Gr^W_{n-1} H^{n-1}(Z_f, \mathbb{C})) + \sum_i h^{n-3,0}(V \cap D_i).$$

By the work of Danilov and Khovanskii [**DK**, 5.9], the first term on the right hand side can be computed to be

$$h^{n-2,1}(Gr^W_{n-1} H^{n-1}(Z_f, \mathbb{C})) = l^*(2\Delta) - (n+1)l^*(\Delta) - \sum_\Gamma l^*(\Gamma).$$

However, being reflexive implies $l^*(\Delta) = 1$, and $l^*(2\Delta) = l(\Delta)$ also follows from being reflexive. Hence the above formula simplifies to

$$h^{n-2,1}(Gr^W_{n-1} H^{n-1}(Z_f, \mathbb{C})) = l(\Delta) - n - 1 - \sum_\Gamma l^*(\Gamma).$$

It remains to show that the second term on the right hand side of (4.9) is equal to the correction term in (4.8).

One way to see this is as follows. Suppose that v_i lies in the interior of a face $\widehat{\Theta} \subset \Delta^\circ$, and let $\Theta \subset \Delta$ denote the dual face. As above, this gives an orbit closure $\mathbb{P}_\Theta \subset \mathbb{P}_\Delta$ such that $f(D_i) = \mathbb{P}_\Theta$. Then we get a map $V \cap D_i \to \bar{V} \cap \mathbb{P}_\Theta$, which in turn induces a natural map

$$H^{n-3,0}(\bar{V} \cap \mathbb{P}_\Theta) \to H^{n-3,0}(V \cap D_i).$$

Most of the time, $h^{n-3,0}(\bar{V} \cap \mathbb{P}_\Theta) = 0$, but in the special case when Θ is simplicial and $\bar{V} \cap \mathbb{P}_\Theta$ has dimension $n-3$, we have the formula

$$(4.10) \qquad\qquad h^{n-3,0}(\bar{V} \cap \mathbb{P}_\Theta) = l^*(\Theta)$$

from [**DK**, 5.5]. Note that $\dim \bar{V} \cap \mathbb{P}_\Theta = n - 3$ is equivalent to Θ being a face of codimension 2. It follows that if $\Theta \subset \Delta$ is simplicial of codimension 2, then every one of the $l^*(\widehat{\Theta})$ interior lattice points $v_i \in \widehat{\Theta}$ contributes $l^*(\Theta)$ holomorphic $(n-3)$-forms. In this way, the correction term $\sum_\Theta l^*(\Theta)l^*(\widehat{\Theta})$ arises naturally. Unfortunately, this argument does not work in general. To actually compute the sum on the right hand side of (4.9), one needs to decompose D_i into a disjoint union of orbits O and compute the appropriate $h^{p,q}$ numbers for each $V \cap O$ using the techniques of [**DK**]. As mentioned above, the details of this argument can be found in [**Batyrev4, Voisin3**]. $\square$

Historically, large numbers of families of Calabi-Yau varieties were first constructed as quasi-smooth hypersurfaces in weighted projective space—eventually, 7555 examples were found [**KS, KS2**]. When the Hodge numbers of these examples were plotted, the resulting graph exhibited a striking *partial* symmetry, which was some of the earliest evidence for mirror symmetry. After the publication of [**Batyrev4**], it was checked in [**CdK**] that for each of the 7555 varieties on the list, the translate to the origin of the Newton polytope of the general Calabi-Yau hypersurface is in fact reflexive. Thus these examples are all special cases of the Batyrev mirror construction. Section 4.4.3 will say more about the techniques involved.

The equality of Hodge numbers given in Theorem 4.1.5 says, for example, that $H^{1,1}(V)$ is isomorphic to $H^1(V^\circ, T_{V^\circ}) \simeq H^{n-2,1}(V^\circ)$, but doesn't specify the isomorphism. According to mirror symmetry, the mirror map relating the Kähler moduli of V to the complex moduli of V° should be defined over at least an open subset of the Kähler moduli space of V, and its derivative should give a canonical isomorphism $H^{1,1}(V) \simeq H^{n-2,1}(V^\circ)$. Also, in the course of the above proof, we mentioned the *monomial-divisor mirror map*, which is a natural isomorphism between $H^{1,1}_{\mathrm{toric}}(V) \simeq H^{n-2,1}_{\mathrm{poly}}(V^\circ)$. In Chapter 6, we will discuss this isomorphism in more detail and explain its relation to the mirror map.

We should also mention that according to (1.5), mirror symmetry suggests that more generally, we should have equalities of Hodge numbers

$$(4.11) \qquad\qquad h^{p,q}(V) = h^{n-1-p,q}(V^\circ)$$

when V° is the Batyrev dual of V (remember that V and V° have dimension $d = n - 1$ in this case). Theorem 4.1.5 covers the cases $(p, q) = (1, 1)$ and $(n-2, 1)$, and the case when $q = 0$ is trivial. For a threefold, this is all that is needed, but for dimension 4 and greater, it is an open question as to whether (4.11) holds.

In [**BB2, BD**], the *string theoretic Hodge numbers* $h^{p,q}_{\mathrm{st}}(V)$ are defined when $V \subset X_\Sigma$ is a Calabi-Yau toric hypersurface of the type we've been considering. These numbers agree with the usual Hodge numbers $h^{p,q}(V)$ when V is smooth or when $q = 1$. More importantly, these numbers satisfy

$$h^{p,q}_{\mathrm{st}}(V) = h^{n-1-p,q}_{\mathrm{st}}(V^\circ)$$

for all $p + q = \dim(V)$ when V° is the Batyrev dual of V. It follows that (4.11) holds in the smooth case when $p + q = \dim(V)$, but as already mentioned, this is still open if either V or V° is a singular minimal Calabi-Yau variety.

4.1.4. Multiple Mirrors. In the Batyrev mirror construction, there may be more than one choice of maximal projective subdivision for Σ and Σ°. This seems to lead to a multiplicity of mirror families. For example, if Σ and Σ' both satisfy $\Sigma(1) = \Sigma'(1) = \Delta^\circ \cap N - \{0\}$, then we get two families $V \subset X_\Sigma$ and $V' \subset X_{\Sigma'}$. Is it possible for $V^\circ \subset X_{\Sigma^\circ}$ to be the mirror of both families? The answer is yes, which is part of the "multiple mirror" phenomenon discussed in [**AGM2**]. Chapter 6 will discuss multiple mirrors in detail. However, if we restrict to toric Kähler moduli as in (4.2), then the rough idea is that Σ and Σ' correspond to distinct cones in the GKZ decomposition, and the Kähler moduli of the families V and V' give disjoint open sets of the *same* global Kähler moduli space.

One can also make different choices on the other side. Here, maximal projective subdivisions Σ° and Σ'° give families V° and V'°. We want the complex moduli of these families to correspond to the Kähler moduli of the family V. Hence V° and V'° should have the *same* complex moduli. If we restrict to polynomial deformations as in (4.3), this follows from the proof of Theorem 4.1.5. Namely, the polynomial moduli of V° correspond to an open subset of $L(\Delta^\circ)$ modulo the action of the automorphism group of X_{Σ°. But V'° is defined by polynomials in the *same* space $L(\Delta^\circ)$, and using results of [**Cox**], one sees that $\mathrm{Aut}(X_{\Sigma^\circ})$ and $\mathrm{Aut}(X_{\Sigma'^\circ})$ have the *same* connected component of the identity. It follows that the polynomial moduli spaces look the same, at least locally. All of this will be explained carefully in Chapter 6.

4.2. The Quintic Threefold, Revisited

We now return to the quintic threefold $V \subset \mathbb{P}^4$, which was discussed in Chapter 2. Since V is an anticanonical hypersurface in the Fano toric variety $\mathbb{P}^4$, we can describe its mirror family using Batyrev's construction.

The anticanonical class $D_0 + \cdots + D_4$ on $\mathbb{P}^4$ corresponds to the reflexive polytope $\Delta \subset M_\mathbb{R} \simeq \mathbb{R}^5$ whose vertices are

$$
\begin{aligned}
&(-1,-1,-1,-1)\\
&(4,-1,-1,-1)\\
&(-1,4,-1,-1)\\
&(-1,-1,4,-1)\\
&(-1,-1,-1,4).
\end{aligned}
$$

(4.12)

We can think of Δ as the Newton polytope of quintics in $\mathbb{P}^4$. To see this, consider a Laurent polynomial determined by Δ as in (3.5), say

$$
f = \sum_{i=1}^s a_i t^{m_i}, \quad \Delta \cap M = \{m_1, \ldots, m_s\},
$$

and then observe that its homogenization

$$
F(x_0, x_1, x_2, x_3, x_4) = x_0 x_1 x_2 x_3 x_4 f(x_1/x_0, x_2/x_0, x_3/x_0, x_4/x_0),
$$

from (3.8) is an arbitrary quintic in x_0, x_1, x_2, x_3, x_4.

To find the mirror family of V, we use the dual polytope $\Delta^\circ \subset N_\mathbb{R}$, which is easily seen to have vertices

$$
\begin{aligned}
&(-1,-1,-1,-1)\\
&(1,0,0,0)\\
&(0,1,0,0)\\
&(0,0,1,0)\\
&(0,0,0,1).
\end{aligned}
$$

(4.13)

The toric variety $\mathbb{P}_{\Delta^\circ}$ is determined by the normal fan of Δ° in $M_\mathbb{R}$, which is given by the cones over the faces of Δ. The cone generators are thus (4.12) and generate a lattice $M_1 \subset M$ of index 125. The quotient M/M_1 is

$$
G = \left\{ (a_0, a_1, a_2, a_3, a_4) \in (\mathbb{Z}_5)^5 : \sum_{i=0}^4 a_i \equiv 0 \bmod 5 \right\} / \mathbb{Z}_5
$$

where $\mathbb{Z}_5 \subset (\mathbb{Z}_5)^5$ is the diagonal subgroup. Since $M_\mathbb{R} = (M_1)_\mathbb{R}$, we can view the normal fan of Δ° as a fan in $(M_1)_\mathbb{R}$. This fan has generators consisting of a $\mathbb{Z}$-basis for M_1 together with the negative of their sum. This shows that using the lattice M_1, the normal fan is the standard fan for $\mathbb{P}^4$. If we then switch to the larger lattice M, standard results in toric geometry imply that $\mathbb{P}_{\Delta^\circ}$ is the quotient of $\mathbb{P}^4$ by $M/M_1 \simeq G$, where G acts on $\mathbb{P}^4$ exactly as described in Chapter 2.

The homogeneous coordinate ring S of $\mathbb{P}_{\Delta^\circ}$ can be described as follows. It is the polynomial ring $S = \mathbb{C}[x_0, x_1, x_2, x_3, x_4]$, which is graded by the Chow group $A_3(\mathbb{P}_{\Delta^\circ})$. From (3.2), we get the exact sequence

$$0 \longrightarrow \mathbb{Z}^4 \longrightarrow \mathbb{Z}^5 \longrightarrow \mathbb{Z} \oplus G \longrightarrow 0,$$

where the first map has the matrix whose rows are the rows of (4.12), and the second sends $(a_0, a_1, a_2, a_3, a_4) \in \mathbb{Z}^5$ to

$$[a_0, a_1, a_2, a_3, a_4] = \left(\textstyle\sum_{i=0}^4 a_i, (-a_1 - a_2 - a_3 - a_4, a_1, a_2, a_3, a_4)\right) \in \mathbb{Z} \oplus G.$$

Thus $A_3(\mathbb{P}_{\Delta^\circ}) \simeq \mathbb{Z} \oplus G$ and the grading on S is obtained by letting a monomial $x_0^{a_0} \cdots x_4^{a_4}$ have degree $[a_0, a_1, a_2, a_3, a_4] \in \mathbb{Z} \oplus G$.

It follows that the anticanonical class of $\mathbb{P}_{\Delta^\circ}$ has degree $[1, 1, 1, 1, 1] = (5, 0) \in \mathbb{Z} \oplus G$, and the only monomials in S of this degree are x_i^5 and $x_0 x_1 x_2 x_3 x_4$. Hence, we get hypersurfaces $\bar{V}^\circ \subset \mathbb{P}_{\Delta^\circ}$ defined by

$$(4.14) \qquad a_0\, x_0^5 + a_1\, x_1^5 + a_2\, x_2^5 + a_3\, x_3^5 + a_4\, x_4^5 + a_5\, x_0 x_1 x_2 x_3 x_4 = 0.$$

In concrete terms, $\bar{V}^\circ$ is the quotient by G of the hypersurface in $\mathbb{P}^4$ defined by the above equation. Notice that the monomials in (4.14) correspond to the integral points in Δ, which are the origin and the points (4.13). Also, the relation

$$(x_0^5)(x_1^5)(x_2^5)(x_3^5)(x_4^5) = (x_0 x_1 x_2 x_3 x_4)^5$$

reflects the fact that the five vectors in (4.13) add up to five times the origin.

The equation (4.14) differs from the 1-parameter family of varieties we discussed in Chapter 2. But recall that we are interested in the moduli of these hypersurfaces. The automorphism group of $\mathbb{P}_{\Delta^\circ}$ is the torus T_M, and using these automorphisms, we can reduce to the 1-parameter family

$$x_0^5 + x_1^5 + x_2^5 + x_3^5 + x_4^5 + \psi\, x_0 x_1 x_2 x_3 x_4 = 0$$

considered in Chapter 2. Chapter 6 will give a systematic method for doing this.

Finally, to get the mirror of the quintic hypersurface V, we pick a maximal projective subdivision Σ° in $M_\mathbb{R}$. Such a fan refines the normal fan of Δ° and satisfies $\Sigma^\circ(1) = \Delta \cap M - \{0\}$. Then the mirror family is given by the hypersurfaces $V^\circ \subset X_{\Sigma^\circ}$ which are the proper transforms of $\bar{V}^\circ \subset \mathbb{P}_{\Delta^\circ}$.

Note that $M \cap \Delta - \{0\}$ has *lots* of points besides the vertices, and hence there are *many* choices for Σ°. The computational problem of identifying *all* such Σ° greatly exceeds current computer limitations. A specific choice for Σ° is described in the appendix to [**Morrison2**].

4.3. Toric Complete Intersections

Generalizing the toric hypersurfaces studied in Section 4.1, one can also ask if mirror symmetry holds for Calabi-Yau complete intersections in toric varieties. In this section, we will answer this question by describing two mirror symmetry constructions due to Batyrev and Borisov [**Borisov1, BB1**] which involve *nef-partitions* and *reflexive Gorenstein cones*.

4.3.1. Nef-Partitions. Suppose that $X = \mathbb{P}_\Delta$ is an n-dimensional toric variety corresponding to a reflexive polytope Δ. Let D_ρ denote the toric divisor associated to $\rho \in \Sigma(1)$, where Σ is the normal fan of Δ. Given a partition $\Sigma(1) = I_1 \cup \cdots \cup I_k$ into k disjoint subsets, we get the divisors $E_j = \sum_{\rho \in I_j} D_\rho$ such that $-K_X = E_1 + \cdots + E_k$.

DEFINITION 4.3.1. *The decomposition* $\Sigma(1) = I_1 \cup \cdots \cup I_k$ *is called a nef-partition if for each j, E_j is a Cartier divisor spanned by its global sections.*

If Δ_j is the polytope corresponding to E_j, then being a nef-partition implies that

$$\Delta = \Delta_1 + \cdots + \Delta_k.$$

By abuse of language, the decompositions $-K_X = E_1 + \cdots + E_k$ and $\Delta = \Delta_1 + \cdots + \Delta_k$ are also called nef-partitions.

General sections of $\mathcal{O}(E_1), \ldots, \mathcal{O}(E_k)$ define a complete intersection $V \subset X$ of dimension $d = n - k$. Using $-K_X = E_1 + \cdots + E_k$ and the adjunction formula, we see that $\bar{V} \subset \mathbb{P}_\Delta$ is a Calabi-Yau variety. As in Section 4.1, a maximal projective subdivision Σ' of the normal fan of Δ gives a map $X_{\Sigma'} \to \mathbb{P}_\Delta$, and the proper transform $V \subset X_\Sigma$ of $\bar{V} \subset \mathbb{P}_\Delta$ is again a complete intersection which is now a *minimal* Calabi-Yau [**BB1**, Prop. 4.15].

To describe the mirror family of V, we as usual regard elements of $\Sigma(1)$ as cone generators, which in the reflexive case are precisely the vertices of the polar polytope $\Delta^\circ \subset N_\mathbb{R}$. Then the nef-partition $\Sigma(1) = I_1 \cup \cdots \cup I_k$ gives polytopes

$$\nabla_j = \mathrm{Conv}(\{0\} \cup I_j) \subset N_\mathbb{R}.$$

This immediately implies that

$$\Delta^\circ = \mathrm{Conv}(\nabla_1 \cup \cdots \cup \nabla_k).$$

The basic result of [**Borisov1**] is that the Minkowski sum

$$\nabla = \nabla_1 + \cdots + \nabla_k$$

is again a reflexive polytope and, furthermore, $\nabla = \nabla_1 + \cdots + \nabla_k$ is a nef-partition. In the terminology of [**BB1**], this is called the *dual nef-partition*.

Using the dual nef-partition, we get Calabi-Yau complete intersections $\bar{V}^\circ \subset \mathbb{P}_\nabla$, and choosing a maximal projective subdivision Σ° of the normal fan of ∇, we get a family of minimal Calabi-Yau complete intersections $V^\circ \subset X_{\Sigma^\circ}$. The conjecture is that

$$V^\circ \text{ is the mirror family of } V.$$

This conjecture is consistent with a construction of mirror symmetry for complete intersections in projective space proposed in [**LTe**].

As evidence for this conjecture, Batyrev and Borisov prove that

$$(4.15) \qquad\qquad h^{p,1}(V) = h^{d-p,1}(V^\circ),$$

where $d = n - k$ is the dimension of V and V°. More generally, using the string theoretic Hodge numbers $h^{p,q}_{\mathrm{st}}$ discussed in Section 4.1.3, one can prove

$$(4.16) \qquad\qquad h^{p,q}_{\mathrm{st}}(V) = h^{d-p,q}_{\mathrm{st}}(V^\circ)$$

(see [**BB2**]). Since $h^{p,1}(V) = h^{p,1}_{\mathrm{st}}(V)$, we see that (4.15) is a special case of (4.16).

4.3.2. Gorenstein Cones. An generalization of the combinatorial duality of nef-partitions comes from the study of *Gorenstein cones*, which are defined as follows. Suppose we have a lattice $\bar{M}$ with dual $\bar{N}$.

DEFINITION 4.3.2. *Let $\sigma \subset \bar{M}_{\mathbb{R}}$ be a rational polyhedral cone.*

(i) σ *is a* Gorenstein cone *if σ is strongly convex and there is $n_\sigma \in \bar{N}$ such that $\langle v, n_\sigma \rangle = 1$ for every generator v of σ.*

(ii) σ *is a* reflexive Gorenstein cone *if both σ and its dual $\check{\sigma}$ are Gorenstein cones.*

It is clear that σ is a reflexive Gorenstein cone if and only if $\check{\sigma}$ is. Furthermore, if σ is reflexive, it must have maximal dimension, which implies that $n_\sigma \in \bar{N}$ is uniquely determined. Then $\check{\sigma}$ has a similarly defined $m_{\check{\sigma}} \in \bar{M}$, and the integer

$$r = \langle m_{\check{\sigma}}, n_\sigma \rangle$$

is called the *index* of σ (or $\check{\sigma}$).

We first explain how reflexive Gorenstein cones relate to the duality of reflexive polytopes. Given a polytope $\Delta \subset M_{\mathbb{R}}$, we extend the lattice M to $\bar{M} = M \times \mathbb{Z}$, and similarly we have $\bar{N} = N \times \mathbb{Z}$. We let $\sigma \subset \bar{M}$ be the cone spanned by the elements $\{ (v, 1) \mid v \in \Delta \}$. In other words,

$$(4.17) \qquad \sigma = \{(\lambda v, \lambda) : v \in \Delta, \ \lambda \geq 0\}.$$

Note that σ is a Gorenstein cone. Then an easy argument shows that σ is reflexive if and only if Δ is. Furthermore, when Δ is reflexive, then the index is $r = 1$ and $\check{\sigma}$ is constructed from Δ° in the same way σ was constructed from Δ.

We can also describe the duality of nef-partitions using reflexive Gorenstein cones. Given a reflexive polytope $\Delta \subset M_{\mathbb{R}}$ and a nef-partition $\Delta = \Delta_1 + \cdots + \Delta_k$, we define the lattice $\bar{M} = M \times \mathbb{Z}^k$ with dual lattice $\bar{N} = N \times \mathbb{Z}^k$. Then let σ be the cone

$$(4.18) \qquad \sigma = \big\{ (\textstyle\sum_{i=1}^k \lambda_i v_i, \lambda_1, \ldots, \lambda_k) : v_i \in \Delta_i, \ \lambda_i \geq 0 \big\}.$$

Notice how this generalizes (4.17).

PROPOSITION 4.3.3. σ *is a reflexive Gorenstein cone of index k and its dual cone $\check{\sigma}$ is constructed by applying* (4.18) *to the dual nef-partition $\nabla = \nabla_1 + \cdots + \nabla_k$.*

PROOF. The proof is given in [**BB1**]. $\square$

There is an interesting geometric significance to the dual cone $\check{\sigma}$, where σ is as in (4.18). This cone is the support of a fan $\bar{\Sigma}$ whose associated toric variety is the total space of the vector bundle

$$(4.19) \qquad \mathcal{O}(E_1) \oplus \cdots \oplus \mathcal{O}(E_k)$$

on $X = \mathbb{P}_\Delta$. The toric variety X may be thought of as embedded in $X_{\bar{\Sigma}}$ as the zero section, so that V can be constructed as a subvariety of $X_{\bar{\Sigma}}$. Also, $\bar{\Sigma}$ can be extended to a complete fan whose associated toric variety is the projective bundle

$$\mathbb{P}(\mathcal{O}(E_1) \oplus \cdots \oplus \mathcal{O}(E_k))$$

over X. The sections of $\mathcal{O}(E_1), \ldots, \mathcal{O}(E_k)$ giving the complete intersection V also determine a global section of (4.19). The corresponding hypersurface

$$(4.20) \qquad Y' \subset \mathbb{P}(\mathcal{O}(E_1) \oplus \cdots \oplus \mathcal{O}(E_k))$$

is intimately related to V. This is the *Cayley trick* described in [**BB3, GKZ2**].

The hypersurface Y' of (4.20) is not a Calabi-Yau variety, but it is a *generalized Calabi-Yau variety* in the sense of Schimmrigk [**BB3, Schimmrigk**]. We can create a generalized Calabi-Yau variety using *any* reflexive Gorenstein cone σ as follows. From σ, we get the semigroup $\sigma \cap \bar{M}$, and we define the *degree* of $m \in \sigma \cap \bar{M}$ to be $\langle m, n_\sigma \rangle$. This makes $\mathbb{C}[\sigma \cap \bar{M}]$ into a graded ring, which gives the toric variety

$$\mathbb{P}_\sigma = \mathrm{Proj}(\mathbb{C}[\sigma \cap \bar{M}]).$$

If $\mathcal{O}_{\mathbb{P}_\sigma}(1)$ is the tautological sheaf on $\mathbb{P}_\sigma$, then a global section of $\mathcal{O}_{\mathbb{P}_\sigma}(1)$ defines a subvariety

$$Y \subset \mathbb{P}_\sigma$$

which Batyrev and Borisov call the *generalized Calabi-Yau variety associated to* σ. Of course, since the dual cone $\check{\sigma}$ is also a reflexive Gorenstein cone, we get the dual family

$$Y^\circ \subset \mathbb{P}_{\check{\sigma}}$$

which is conjectured to be the mirror of $Y \subset \mathbb{P}_\sigma$ in the appropriate sense (see [**BB3**, Conjecture 2.17]).

This construction includes the Calabi-Yau hypersurfaces and complete intersections considered so far. More precisely,

- When σ is given by (4.17), then $\mathbb{P}_\sigma$ is precisely the toric variety $\mathbb{P}_\Delta$ (and $\mathbb{C}[\sigma \cap \bar{M}]$ is the ring S_Δ of Section 3.2.1), and the hypersurfaces $Y \subset \mathbb{P}_\sigma$ are precisely the Calabi-Yau hypersurfaces $V \subset \mathbb{P}_\Delta$ studied in Section 4.1.
- When σ is given by (4.18), then there is a birational crepant morphism

$$\mathbb{P}(\mathcal{O}(E_1) \oplus \cdots \oplus \mathcal{O}(E_k)) \longrightarrow \mathbb{P}_\sigma$$

  such that $Y \subset \mathbb{P}_\sigma$ is the image of $Y' \subset \mathbb{P}(\mathcal{O}(E_1) \oplus \cdots \oplus \mathcal{O}(E_k))$.

Furthermore, in each case, mirror symmetry reduces to the duality of Gorenstein cones.

Besides these examples we already know, the Gorenstein cone construction helps explain some cases where the usual formulation of mirror symmetry breaks down. This is most evident in the case of *rigid* Calabi-Yau threefolds. These have no nontrivial deformations, so that $H^{2,1}(V) = 0$. A mirror V° in the usual sense would have $H^{1,1}(V^\circ) = 0$, which is impossible for a Kähler manifold. But "mirrors" have still been found in some cases. For example, let E be the unique elliptic curve with an automorphism of order 3, and let $V = E^3/G$, where G is the group generated by the diagonal action of the automorphism. One can show that V is a (singular) rigid Calabi-Yau whose mirror, as calculated in [**CDP**], is a finite quotient of the 7-dimensional Fermat hypersurface in $\mathbb{P}^8$. In [**BB3**], this is shown to be consistent with the duality of Gorenstein cones.

4.4. The Voisin-Borcea Construction

In this section, we describe a different construction of mirror families due independently to Voisin and Borcea. The inspiration comes from the work of Nikulin on K3 surfaces with involution. He found 75 families of such surfaces whose invariants exhibited a striking partial symmetry [**Nikulin**]. Using these surfaces, Voisin and Borcea construct families of Calabi-Yau threefolds which are conjectured to be mirror symmetric. We will sketch some of the proofs, though the reader should refer to the original papers [**Voisin1, Borcea2**] for full details.

4.4.1. Threefolds from K3 Surfaces and Elliptic Curves. We begin by constructing Calabi-Yau threefolds using certain K3 surfaces and elliptic curves with involution. More precisely, consider the following data:

- A K3 surface S and an involution $i : S \to S$ acting by -1 on $H^{2,0}(S)$. The fixed locus of i consists of a disjoint union of smooth curves $C_1, \ldots, C_N$. We explicitly assume that $N > 0$.
- An elliptic curve E and an involution $j : E \to E$ with quotient $\mathbb{P}^1$. We denote the four fixed points of j by p_1, p_2, p_3, p_4.

The involutions i of S and j of E induce an involution k of $E \times S$ defined by $k(e, s) = (j(e), i(s))$ for $(e, s) \in E \times S$. The fixed point locus of k consists of the $4N$ curves $C_{r,s} = \{p_r\} \times C_s$. If we blow up $E \times S$ along these $4N$ curves to obtain the threefold $\widetilde{E \times S}$ with smooth exceptional divisor $D = \cup D_{r,s}$, then the involution k naturally lifts to an involution $\tilde{k}$ of $\widetilde{E \times S}$. We put

$$(4.21) \qquad V = \widetilde{E \times S}/\tilde{k}.$$

We can regard $D = \cup D_{r,s}$ as a divisor on V with $4N$ irreducible components. Then one can prove the following.

LEMMA 4.4.1. *V is a smooth Calabi-Yau threefold.*

Let g_s denote the genus of the curve $C_s \subset S$ and put $N' = \sum_{s=1}^{N} g_s$. We compute the Hodge components of V as follows.

THEOREM 4.4.2. $h^{1,1}(V) = 11 + 5N - N'$ *and* $h^{2,1}(V) = 11 + 5N' - N$.

REMARK. From these formulas, observe that a similarly constructed V° which reversed the roles of N and N' would be a candidate for a mirror manifold. This is precisely what Voisin and Borcea do.

PROOF. We begin by giving an explicit decomposition of $H^{1,1}(V)$. First, the components of the divisor $D = \cup D_{r,s}$ give $4N$ classes in $H^{1,1}(V)$. Next, we get some classes from the smooth surface $T = S/i$. We compute $h^{1,1}(T)$ by identifying each curve $C_s \subset S$ with its image in T and then calculating the Euler characteristic:

$$(4.22) \qquad \begin{aligned} e(T) &= e(T - \cup_s C_s) + \sum_s e(C_s) \\ &= \frac{1}{2}e(S - \cup_s C_s) + \sum_s e(C_s) \\ &= \frac{1}{2}\left(24 - \sum_s(2 - 2g_s)\right) + \sum_s(2 - 2g_s) \\ &= 12 + N - N'. \end{aligned}$$

Denote by i^* the action on the cohomology of S induced by i. By looking at i^*-invariants of the cohomology of S, it is immediate that the cohomology of T satisfies $h^0(T) = h^4(T) = 1$ and $h^1(T) = h^3(T) = h^{2,0}(T) = h^{0,2}(T) = 0$. Combining these invariants with (4.22) gives

$$(4.23) \qquad h^{1,1}(T) = 10 + N - N'.$$

The cohomology classes in $H^{1,1}(T)$ can be pulled back to V via the natural projection $\pi_1 : V \to T$, and the resulting elements of $H^{1,1}(V)$ will be identified with $H^{1,1}(T)$.

Finally, the elements of $H^{1,1}(E)$ pull back to $\tilde{k}$-invariant cohomology classes on $H^{1,1}(\widetilde{E \times S})$ via the natural projection $\widetilde{E \times S} \to E$, and hence can be identified with cohomology classes on V.

It turns out that the three types of classes just described are independent over $\mathbb{C}$ and span $H^{1,1}(V)$. Thus we get a direct sum decomposition

$$(4.24) \qquad H^{1,1}(V) = \oplus_{r,s}\mathbb{C} \cdot [D_{r,s}] \oplus H^{1,1}(T) \oplus H^{1,1}(E)$$

This decomposition implies $h^{1,1}(V) = 4N + (10 + N - N') + 1 = 11 + 5N - N'$, which proves the desired formula for $h^{1,1}(V)$.

For the second assertion of the theorem, we identify $H^{2,1}(V)$ with the $\tilde{k}$-invariants of $H^{2,1}(\widetilde{E \times S})$ and again list three types of classes. First, note that $D_{r,s}$ is by construction a projective bundle over $C_{r,s}$ and thus gives a family of curves in $\widetilde{E \times S}$ parameterized by $C_{r,s}$. There is an associated Abel-Jacobi mapping $H^{1,0}(C_{r,s}) \to H^{2,1}(\widetilde{E \times S})$ (see [**CG**]) whose image we identify with $H^{1,0}(C_{r,s})$. Next, let $H^{1,1}(S)^-$ denote the -1 eigenspace for the action of i on $H^{1,1}(S)$. Since j induces -1 on $H^1(E)$ and i acts as -1 on $H^{2,0}(S)$, we also get classes from $H^{1,1}(S)^- \otimes H^{1,0}(E)$ and $H^{2,0}(S) \otimes H^{0,1}(E)$ via the Künneth decomposition. Again, the three types of classes just described are independent over $\mathbb{C}$ and span $H^{2,1}(V)$. Noting that $H^{1,0}(E)$ and $H^{0,1}(E)$ are each 1-dimensional, we can write this (non-canonically) as

$$(4.25) \qquad H^{2,1}(V) \simeq \oplus_{r,s}H^{1,0}(C_{r,s}) \oplus H^{1,1}(S)^- \oplus H^{2,0}(S).$$

This shows that $h^{2,1}(V) = 4N' + h^{1,1}(S)^- + 1$. However, if $H^{1,1}(S)^+$ denotes the $+1$ eigenspace for the action of i^* on $H^{1,1}(S)$, then equation (4.23) implies

$$(4.26) \qquad h^{1,1}(S)^+ = 10 + N - N', \qquad h^{1,1}(S)^- = 10 + N' - N.$$

The desired formula for $h^{2,1}(V)$ now follows immediately. $\square$

For later purposes, we also need the following invariant of the involution i. The intersection form $\langle \cdot, \cdot \rangle$ is unimodular on $H^2(S, \mathbb{Z})$, but not necessarily on $H^2(S, \mathbb{Z})^+$ or $H^2(S, \mathbb{Z})^-$. However, using results of Nikulin on involutions of K3 lattices [**AN, Dolgachev2, Nikulin**], one can show that there is an integer a such that

$$(4.27) \qquad \begin{aligned} 2^a &= \text{discriminant of } \langle \cdot, \cdot \rangle \text{ on } H^2(S, \mathbb{Z})^+ \\ &= \text{discriminant of } \langle \cdot, \cdot \rangle \text{ on } H^2(S, \mathbb{Z})^-. \end{aligned}$$

We have the following explicit formula for a.

LEMMA 4.4.3. $a = 12 - N - N'$.

PROOF. The key step is the construction of an exact sequence

$$0 \longrightarrow H^2(T, \mathbb{Z}) \longrightarrow H^2(S, \mathbb{Z})^+ \longrightarrow (\mathbb{Z}_2)^{N-1} \longrightarrow 0,$$

the inclusion being induced by pullback. We omit the construction here. Since the form $\langle \cdot, \cdot \rangle$ on $H^2(S, \mathbb{Z})^+$ restricts to twice the unimodular form on $H^2(T, \mathbb{Z})$, the exact sequence implies that the discriminant is $2^{b_2(T) - 2(N-1)}$. The lemma now follows from (4.23). $\square$

4.4.2. The Voisin-Borcea Mirror. We now describe the mirror family V° of the Calabi-Yau threefold $V = \widetilde{E \times S}/\tilde{k}$. As remarked earlier in Section 4.4.1, Theorem 4.4.2 shows that if there were a mirror family of the same type, then the roles of N and N' would necessarily be reversed. Construction of the family V° requires the following technical lemma.

LEMMA 4.4.4. *If $(N, N') \neq (5, 1)$, then $H^2(S, \mathbb{Z})^-$ contains a hyperbolic plane.*

The statement in [**Voisin1**] is in fact slightly stronger. The proof uses the results mentioned above of Nikulin on involutions of K3 lattices.

Assuming that $(N, N') \neq (5, 1)$, we choose a hyperbolic plane $P \subset H^2(S, \mathbb{Z})^-$ and let r_P denote the reflection in P. This means that P is the $+1$ eigenspace of r_P and $P^\perp$ is the -1 eigenspace. We put $i_1^* = r_P \cdot i^*$. It can be shown (again using Nikulin's results on involutions of K3 lattices) that i_1^* is an involution arising from an involution i_1 on some K3 surface S_1. Furthermore, an easy calculation shows that

$$(4.28) \qquad H^2(S_1, \mathbb{Z})^- = P \oplus H^2(S, \mathbb{Z})^+.$$

Now construct the Calabi-Yau threefolds V° from (S_1, i_1) and an elliptic curve E_1 with involution j_1 by the same construction leading to V. As explained in [**Voisin1**], there is a natural choice for (E_1, j_1). Let the invariants for V° be (N_1, N_1'). Then we have the following result.

PROPOSITION 4.4.5. $N_1 = N'$ *and* $N_1' = N$.

PROOF. Let 2^{a_1} be the discriminant of the intersection form on $H^2(S_1, \mathbb{Z})^-$. Since $P \subset H^2(S, \mathbb{Z})$ is unimodular, (4.27) and (4.28) imply $a_1 = a$, so that

$$12 - N_1 - N_1' = 12 - N - N'$$

by Lemma 4.4.3. Equation (4.28) also implies $h^{1,1}(S_1)^- = h^{1,1}(S)^+$. When combined with (4.26), we obtain

$$10 + N_1' - N_1 = 10 + N - N'.$$

The desired equalities $N_1 = N'$ and $N_1' = N$ follow immediately. $\qquad\square$

The resulting family of Calabi-Yau threefolds V° is conjectured to be the mirror family of V. In support of the mirror relation between V and V°, we get $h^{1,1}(V) = h^{2,1}(V^\circ)$ and $h^{2,1}(V) = h^{1,1}(V^\circ)$ by Theorem 4.4.2 and Proposition 4.4.5. In addition, Voisin and Borcea offer other evidence for the mirror relation.

The first evidence, due to Voisin, concerns the relation between complex and Kähler moduli of V and V°. The complex moduli of V can be described as follows. We start by considering the moduli coming from the K3 surface S. Consider the domain

$$D = \{\omega \in \mathbb{P}(H^2(S, \mathbb{C})^-) : \omega \cdot \omega = 0,\ \omega \cdot \bar{\omega} > 0\},$$

which is almost a period domain for K3 surfaces with involution i' such that $i'^* = i^*$. More precisely, a generic point $\omega \in D$ corresponds to a K3 surface S' with involution i' satisfying $i'^* = i^*$ and $\mathbb{C} \cdot \omega = H^{2,0}(S')$, and the moduli space of pairs (S, i) is a discrete quotient of an explicit second category subset $U \subset D$. By fixing the elliptic curve (E, j), we get an embedding of the moduli space of (S, i) into the complex moduli space of V. So we think of D as essentially being slightly larger than a leaf of a foliation of the complex moduli space of V.

Turning to the Kähler moduli of V, define the domain

$$D' = \{\alpha \in H^2(S,\mathbb{C})^+ : (\operatorname{Re}\alpha)^2 > 0\}.$$

Note that D' can be identified with a subset of $H^2(T,\mathbb{C})$ which roughly speaking is slightly larger the the Kähler moduli space of T. Furthermore, via pullback by $\pi_1 : V \to T$, elements of the Kähler moduli of T can be added to complexified Kähler classes of V to obtain new complexified Kähler classes. Therefore the Kähler moduli space of T can be identified with the leaves of a foliation of the Kähler moduli space of V, and the domain D' is slightly larger than this leaf in the Kähler moduli of V. Notice how this is similar to what we just did for complex moduli.

The K3 surface S gave the domains D and D' related to complex and Kähler moduli of V. On the mirror side, we have the surface S_1 constructed above. Let D_1 and D_1' denote the corresponding domains for S_1. If V and V° are indeed mirrors, then it would be consistent for there to be a relationship between D and D_1' and between D' and D_1. Voisin proves the following result.

THEOREM 4.4.6. $D \simeq D_1'$ and $D' \simeq D_1$.

Another type of evidence provided by Voisin is the asymptotic behavior of the Yukawa coupling. On $H^{1,1}(V)$, we have the coupling given by cup product, and as defined in Chapter 1, $H^1(V,T_V)$ has the Yukawa coupling

$$(4.29) \qquad \langle \theta_1, \theta_2, \theta_3 \rangle = \int_V \Omega \wedge (\nabla_{\theta_1}\nabla_{\theta_2}\nabla_{\theta_3}\Omega),$$

where Ω is a holomorphic 3-form on V (see (1.9)). The isomorphism $H^1(V,T_V) \simeq H^{2,1}(V)$ gives a coupling on $H^{2,1}(V)$ also called the Yukawa coupling.

Using the decompositions given in (4.24) and (4.25), Voisin gives a careful description of the the couplings on $H^{1,1}(V)$ and $H^{2,1}(V^\circ)$ and proves that cup product on $H^{1,1}(V)$ can be identified asymptotically with the Yukawa coupling on $H^{2,1}(V^\circ)$.

To explain this last statement, recall from Chapter 1 that the B-model correlation function on $H^{2,1}(V^\circ)$ is precisely the Yukawa coupling described above, while the A-model correlation function on $H^{1,1}(V)$ is given by the formula

$$\langle \omega_1, \omega_2, \omega_3 \rangle = \int_V \omega_1 \wedge \omega_2 \wedge \omega_3 \ +$$

$$\sum_{\beta \neq 0} n_\beta \int_\beta \omega_1 \int_\beta \omega_2 \int_\beta \omega_3 \ \frac{e^{2\pi i \int_\beta \omega}}{1 - e^{2\pi i \int_\beta \omega}}$$

(see (1.7)). Now suppose the complexified Kähler class of V approaches infinity, meaning $\int_\beta \operatorname{Im}(\omega) \to \infty$ for all effective curves β (this is called the "large radius limit" in the physics literature). It follows that

$$\langle \omega_1, \omega_2, \omega_3 \rangle \to \int_V \omega_1 \wedge \omega_2 \wedge \omega_3,$$

which is cup product. In this situation, we say that asymptotically, the A-model correlation function becomes cup product.

Since the B-model correlation function of a mirror manifold V° coincides with the A-model correlation function of the original manifold V, we expect that the B-model correlation function should have similar behavior. This is what Voisin proves, namely that under suitable hypotheses, the Yukawa coupling on V° has a limit which coincides with cup product on V [**Voisin1**, Section 3].

4.4.3. Relation to the Batyrev Construction. A different kind of evidence for mirror symmetry between V and its Voisin-Borcea mirror V° is given by Borcea, who shows that some instances of this construction are special cases of Batyrev's mirror duality. For instance, consider the anticanonical hypersurface

$$(4.30) \qquad\qquad u^4 + v^4 = y^6 + z^6 + w^6$$

in the weighted projective space $\mathbb{P}(3,3,2,2,2)$ (the variables u,v,y,z,w have respective weights $3,3,2,2,2$). To relate this to the threefolds (4.21), observe that $x^2 = u^4 + v^4$ defines an elliptic curve E in $\mathbb{P}(2,1,1)$ and $x^2 = y^6 + z^6 + w^6$ defines a (singular) K3 surface in $\mathbb{P}(3,1,1,1)$. In each case, the map sending x to $-x$ is an involution. Borcea shows that the Calabi-Yau threefold $V = \widetilde{E \times S}/\tilde{k}$ is a desingularization of above hypersurface in $\mathbb{P}(3,3,2,2,2)$. Notice also that in S, the fixed locus of the involution is given by $y^6 + z^6 + w^6 = 0$, which is a smooth curve of genus 10. Thus $(N, N') = (1, 10)$ in this case.

Using the Voisin-Borcea construction, the family (4.30) has a mirror family V°_{VB} with invariants $(N, N') = (10, 1)$. In [**Borcea2**], Borcea shows that this family contains the anticanonical hypersurface

$$(4.31) \qquad\qquad u^4 + v^4 = y^5 + yz^5 + zw^6$$

in the weighted projective space $\mathbb{P}(25, 25, 20, 16, 14)$. In this case, E is again $x^2 = u^4 + v^4$ in $\mathbb{P}(2,1,1)$, but S_1 is the singular $K3$ surface defined by $x^2 = y^5 + yz^5 + zw^6$ in $\mathbb{P}(25, 10, 8, 7)$. As above, the desingularization of (4.31) is $V^\circ_{\mathrm{VB}} = \widetilde{E \times S_1}/\tilde{k}$. The fixed locus of the involution is $y^5 + yz^5 + zw^6 = 0$ in S_1. When S_1 is desingularized, the fixed locus becomes an elliptic curve plus nine $\mathbb{P}^1$'s, which explains why $(N, N') = (10, 1)$.

The basic claim is that V°_{VB} coincides with the Batyrev mirror of (4.30), which we will denote V°_{Bat}. Borcea proves this in [**Borcea2**] using his bipyramid construction, but we will give a more elementary proof using the *generalized Berglund-Hübsch transposition rule* from [**CdK**]. The duality between (4.30) and (4.31) is not obvious because $\mathbb{P}(3,3,2,2,2)$ is a Fano toric variety but $\mathbb{P}(25,25,20,16,14)$ is not (see Lemma 3.5.6). So how do we apply Batyrev duality?

The answer involves a technique used in the physics literature whereby a Batyrev mirror is represented by a hypersurface in a non-Fano weighted projective space. Rather than describe the general case, we will explain how it works for $\mathbb{P}(q)$, $q = (3,3,2,2,2)$ (see [**CdK**, Sect. 3.1] for more details). Thinking of $\mathbb{P}(q)$ in toric terms, we have

$$M = \{a \in \mathbb{Z}^5 : q \cdot a = 0\} \quad \text{and} \quad N = \mathbb{Z}^5/\mathbb{Z}q,$$

and the cone generators of the fan of $\mathbb{P}(q)$ are $w_i = [e_i] \in N$, where $[e_i]$ denotes the coset of the ith standard basis vector $e_i \in \mathbb{Z}^5$. Also, the polytope of the anticanonical divisor is

$$\Delta = \{a \in M_{\mathbb{R}} : \langle a, w_i \rangle \geq -1\} = \{a \in M_{\mathbb{R}} : a \cdot e_i \geq -1\}.$$

Note that Δ is reflexive since $\mathbb{P}(q)$ is Fano.

Now consider the following points in $\Delta \cap M$:

$$
\begin{aligned}
a_0 &= (\ \ 3, -1, -1, -1, -1) \\
a_1 &= (-1, \ \ 3, -1, -1, -1) \\
a_2 &= (-1, -1, \ \ 4, \ \ 0, -1) \\
a_3 &= (-1, -1, -1, \ \ 4, \ \ 0) \\
a_4 &= (-1, -1, -1, -1, \ \ 5).
\end{aligned}
$$
(4.32)

These points span a simplex with 0 in its interior. Hence some positive linear combination of the a_i's equals zero, and one can check that

$$
(4.33) \qquad\qquad 25\, a_0 + 25\, a_1 + 20\, a_2 + 16\, a_3 + 14\, a_4 = 0.
$$

Furthermore, all other relations are multiples of this one. It follows that $\hat{q} = (25, 25, 20, 16, 14)$ is the unique positive integral relatively prime relation among the a_i's. This gives the weighted projective space $\mathbb{P}(\hat{q}) = \mathbb{P}(25, 25, 20, 16, 14)$, which has

$$
\widehat{M} = \{a \in \mathbb{Z}^5 : \hat{q} \cdot a = 0\} \quad \text{and} \quad \widehat{N} = \mathbb{Z}^5/\mathbb{Z}\hat{q}.
$$

The fan of $\mathbb{P}(\hat{q})$ has the cone generators $\hat{w}_i = [e_i] \in \widehat{N}$.

Since Δ is reflexive, $\mathbb{P}(q) = \mathbb{P}_\Delta$ is Fano and the dual polytope Δ° gives the Fano toric variety $\mathbb{P}_{\Delta^\circ}$, which we denote $\mathbb{P}(q)^\circ$. The fan of $\mathbb{P}(q)^\circ$ lies in $M_\mathbb{R}$ and has cones determined by the faces of Δ. To relate $\mathbb{P}(q)^\circ$ to the non-Fano weighted projective space $\mathbb{P}(\hat{q})$, we will use the map $\phi : \widehat{N} \to M$ which sends $\hat{w}_i \in \widehat{N}$ to $a_i \in M$. Using (4.33), it is easy to see that ϕ is well-defined and induces an isomorphism $\widehat{N}_\mathbb{R} \simeq M_\mathbb{R}$. Since ϕ takes the cone generator $\hat{w}_i$ to $a_i \in \Delta \cap M$, which is also a cone generator, we can find a maximal projective subdivision Σ° compatible with the fan of $\mathbb{P}(\hat{q})$. This means that the fan in $\widehat{N}_\mathbb{R}$ obtained by pulling back Σ° via ϕ refines the fan of $\mathbb{P}(\hat{q})$. This gives a toric blowup $\widehat{X} \to \mathbb{P}(\hat{q})$, and it follows that ϕ induces a morphism

$$
\phi_* : \widehat{X} \longrightarrow X_{\Sigma^\circ}.
$$

By standard theory, ϕ_* is a finite morphism whose degree is the index of $\phi(\widehat{N}) \subset M$. If we let $d = q \cdot (1,1,1,1,1) = 12$ and $\hat{d} = \hat{q} \cdot (1,1,1,1,1) = 100$, then the index is computed by the formula

$$
[M : \phi(\widehat{N})] = \frac{\det(A + \mathbf{1})}{d\hat{d}} = \frac{2400}{12 \cdot 100} = 2,
$$

where A is the 5×5 matrix whose rows are given by (4.32) and $\mathbf{1}$ is the 5×5 matrix with all entries equal to 1. This formula is stated in [**CdK**, Sect. 3.1] without proof, but the argument is not difficult. Thus $\phi_* : \widehat{X} \to X_{\Sigma^\circ}$ is a $2 : 1$ cover.

The next step is to identify $V_{\mathrm{VB}}^\circ \subset \widehat{X}$ with the pullback of an anticanonical hypersurface $V_{\mathrm{Bat}}^\circ \subset X_{\Sigma^\circ}$. The vertices of the reflexive simplex $\Delta^\circ \subset N_\mathbb{R}$ are the w_i, which give the Laurent polynomial $f = \sum_{i=0}^{4} c_i t^{w_i} \in L(\Delta^\circ)$ (this is not the most general Laurent polynomial in $L(\Delta^\circ)$, but it does have the correct Newton polytope). Thus we can assume that V_{Bat}° is defined by $f = 0$. To see what this gives in $\widehat{X}$, we need to study the map $\phi^t : N \to \widehat{M}$ which is dual to ϕ. If we think of $\mathbb{Z}^5$ in terms of column vectors and use the natural basis and coordinates for $\widehat{N}$ and M, then the composition

$$
\mathbb{Z}^5 \longrightarrow \widehat{N} \xrightarrow{\ \phi\ } M \longrightarrow \mathbb{Z}^5
$$

is multiplication by A^t, where (as above) A is the matrix whose rows are given by
(4.32). Taking duals, it follows that the composition

$$\mathbb{Z}^5 \longrightarrow N \xrightarrow{\phi^t} \widehat{M} \longrightarrow \mathbb{Z}^5$$

is multiplication by A. This means that $\phi^t(w_i) = \hat{a}_i$, where $\hat{a}_i$ is the ith column
of A. Thus $f = 0$ on X_{Σ° pulls back to $\hat{f} = 0$, where $\hat{f} = \sum_{i=0}^{4} c_i t^{\hat{a}_i}$. The final
step is to homogenize with respect to the variables u, v, y, z, w. We leave it to the
reader to show that this is done by adding 1 to every entry of $\hat{a}_i$. It follows that
the exponent vectors we get are the columns of the matrix

$$A + \mathbf{1} = \begin{pmatrix} 4 & 0 & 0 & 0 & 0 \\ 0 & 4 & 0 & 0 & 0 \\ 0 & 0 & 5 & 1 & 0 \\ 0 & 0 & 0 & 5 & 1 \\ 0 & 0 & 0 & 0 & 6 \end{pmatrix}.$$

Thus, in terms of u, v, y, z, w, we can write $\hat{f} = 0$ as

$$\hat{f} = c_0\, u^4 + c_1\, v^4 + c_2\, y^5 + c_3\, yz^5 + c_4\, zw^6 = 0.$$

Since (4.31) is an equation of this form, we can assume that V_{VB}° is defined by $\hat{f} = 0$.

All of this proves that ϕ_* induces a $2:1$ map $\phi_* : V_{\mathrm{VB}}^\circ \to V_{\mathrm{Bat}}^\circ$. This doesn't
seem quite right, since we want the two to be the same. Fortunately, there is a
simple explanation. In working with $V_{\mathrm{VB}}^\circ = \widetilde{E \times S_1}/\tilde{k}$, dividing E by a point of
order 2 gives a degree 2 isogeny $E \to E'$ which commutes with the involution. It
follows that we get a $2:1$ cover $V_{\mathrm{VB}}^\circ \to V_{\mathrm{VB}}'^\circ = \widetilde{E' \times S_1}/\tilde{k}$. One can prove that
$\phi_* : V_{\mathrm{VB}}^\circ \to V_{\mathrm{Bat}}^\circ$ is precisely this sort of map, so that we can identify V_{Bat}° with $V_{\mathrm{VB}}'^\circ$
(the proof uses Borcea's bipyramid construction—see the remarks following (29) in
[**Borcea2**]). Now comes the crucial observation: since E and E' are deformations
of each other, so are V_{VB}° and $V_{\mathrm{VB}}'^\circ$. Thus, in terms of mirror *families*, the Batyrev
mirror of (4.30) is (4.31). This shows that, at least in the case $(N, N') = (1, 10)$,
the Voisin-Borcea construction is consistent with Batyrev mirror duality.

The paper [**Borcea2**] lists other pairs (N, N') for which the Voisin-Borcea
construction can be interpreted in terms of Batyrev mirror duality. However, since
not every Calabi-Yau is a toric hypersurface, it is possible that not all (N, N') can
be treated in this way. Still, the fact that two vastly different mirror constructions
(Batyrev and Voisin-Borcea) coincide in a case like (4.30) is a strong indication of
the reality of mirror symmetry. On the other hand, the differences between these
constructions also illustrate how far we are from having a truly general method for
finding mirrors.

A final remark concerns the list of 7555 weighted projective spaces mentioned
in Section 4.1.3. Like $\mathbb{P}(25, 25, 20, 16, 14)$, most of the weighted projective spaces
on the list were *not* Fano. The list exhibited only a partial symmetry, and people worried about the "missing mirrors". The resolution of this difficulty involved
Batyrev mirror duality together with a variant of the above technique for interpreting Batyrev duals using non-Fano weighted projective spaces. From this point of
view, the "missing mirrors" correspond to those Batyrev duals which couldn't be
represented by a weighted projective space. See [**CdK**] for details and examples.

CHAPTER 5

Hodge Theory and Yukawa Couplings

In this chapter, we recall and develop some ideas from Hodge theory, and apply this to the calculation of the Yukawa couplings of a Calabi-Yau manifold V. To do this, we must first understand the Picard-Fuchs differential equations satisfied by the periods of integrals on V. We illustrate two methods for doing this.

In Section 5.1 we review variations of Hodge structures, Picard-Fuchs equations, and boundary behavior, and in Section 5.2, we define *maximally unipotent boundary points*, which will play an important role in determining the mirror map. In Section 5.3 we explain the method developed by Griffiths and Dwork, whereby cohomology classes on V are represented by residues of rational differentials on projective space. Two examples are given in Section 5.4. In Section 5.5, we study an alternate approach to computing Picard-Fuchs equations which involves the *generalized hypergeometric equations* of Gelfand, Kapranov, and Zelevinsky. We then apply all these to the calculation of Yukawa couplings in Section 5.6.

Two unexpected aspects of this chapter are the cohomology-valued solution $\tilde{I}$ of the GKZ system from Section 5.5.3, which will appear in our treatment of the mirror theorem in Chapter 11, and the relation between normalized Yukawa couplings and the Gauss-Manin connection from Section 5.6.3, which will play a prominent role in the A-variation of Hodge structure discussed in Chapter 8.

5.1. Hodge Theory

In this section, we review the Hodge theory used in mirror symmetry. We follow the spirit of [**Morrison2**] while generalizing to families with higher dimensional parameter spaces. Some general references to variations of Hodge structure can be found in [**Griffiths1, Griffiths2, Schmid, CaK**]. Particulars about Calabi-Yau variations can be found in [**BG**].

5.1.1. Hodge Structures and Their Variations. If X is an n-dimensional smooth complex projective variety, then $H^k(X, \mathbb{C})$ has a Hodge decomposition

$$H^k(X, \mathbb{C}) \simeq \bigoplus_{p+q=k} H^{p,q}(X).$$

where $\overline{H^{p,q}(X)} = H^{q,p}(X)$ relative to the real structure determined by $H^k(X, \mathbb{R})$. When we add the integer lattice coming from $H^k(X, \mathbb{Z})$, we get a *Hodge structure of weight k*.

We will often formulate the Hodge structure on $H^k(X, \mathbb{C})$ in terms of the *Hodge filtration* $F^\bullet(X) - \{F^p(X)\}_{p=0}^n$, which is defined by

$$F^p(X) = \bigoplus_{a \geq p} H^{a,k-a}(X).$$

73

Note that $F^p(X) \oplus \overline{F^{k-p+1}(X)} = H^k(X, \mathbb{C})$ and that

$$H^{p,q}(X) = F^p(X) \cap \overline{F^q(X)}.$$

We also have $H^{p,q}(X) \simeq F^p(X)/F^{p+1}(X)$. Thus the Hodge structure on $H^k(X, \mathbb{C})$ is uniquely determined by the integer structure and the Hodge filtration.

Finally, a Kähler class ω on X determines the *primitive cohomology group* $H_0^k(X, \mathbb{C})$ or $PH^k(X)$ in $H^k(X, \mathbb{C})$, and the pairing

$$Q(\alpha, \beta) = (-1)^{k(k-1)/2} \int_X \omega^{n-k} \wedge \alpha \wedge \beta, \quad \alpha, \beta \in H_0^k(X, \mathbb{C})$$

makes primitive cohomology into a *polarized Hodge structure*. This means that

$$(5.1) \qquad Q(i^{p-q}\alpha, \overline{\alpha}) > 0, \quad \text{for all } \alpha \neq 0 \text{ in } H_0^{p,q}(X, \mathbb{C}), \ p + q = k.$$

See [**CaK, Griffiths2**] for more about polarized Hodge structures.

The subspaces $H^{p,q}(X)$ do not vary holomorphically with families. However, the relative version of $F^p(X)$ is very nicely behaved. Suppose that we have a smooth morphism $\pi : \mathcal{X} \to S$ of relative dimension n, where S is smooth and quasi-projective. Put $X_t = \pi^{-1}(t)$ for any $t \in S$. Then the cohomology groups $H^k(X_t, \mathbb{C})$ fit together to form the locally free sheaf $\mathcal{F}^0 = R^k\pi_*\mathbb{C} \otimes \mathcal{O}_S$ on S. The spaces $F^p(X_t)$ may be seen to fit together to form a locally free subsheaf $\mathcal{F}^p$.

By construction, $\mathcal{F}^0$ contains a local system, the locally constant sheaf $R^k\pi_*\mathbb{C}$. This uniquely determines a flat connection ∇ on $\mathcal{F}^0$, the *Gauss-Manin connection* ∇, whose flat (or horizontal) sections coincide with the local system $R^k\pi_*\mathbb{C}$. Concretely, $\nabla : \mathcal{F}^0 \to \mathcal{F}^0 \otimes \Omega_S^1$ is defined by

$$\nabla(s \otimes f) = s \otimes df,$$

where s is a section of $R^k\pi_*\mathbb{C}$ and f is a function on S. The Gauss-Manin connection satisfies *Griffiths transversality* $\nabla(\mathcal{F}^p) \subset \mathcal{F}^{p-1} \otimes \Omega_S^1$.

If we set $\mathcal{H} = \mathcal{F}^0$, then $\mathcal{H}$ has the locally constant subsheaf $\mathcal{H}_{\mathbb{C}} = R^k\pi_*\mathbb{C}$, and this in turn has the subsheaf $\mathcal{H}_{\mathbb{Z}}$ of integer sections (the image of $R^k\pi_*\mathbb{Z} \to R^k\pi_*\mathbb{C}$). We call $(\mathcal{H}, \nabla, \mathcal{H}_{\mathbb{Z}}, \mathcal{F}^{\bullet})$ a *variation of Hodge structure*.

5.1.2. Picard-Fuchs Equations. For the rest of this section, we will study the variation of Hodge structure on the middle-dimensional cohomology group $H^n(X_t, \mathbb{C})$, $n = \dim(X_t)$. Let $\Omega(t)$ be a fixed local section of $\mathcal{F}^n$ at a point of $p \in S$, and let $\mathcal{D}$ be the sheaf of linear differential operators on S. In local analytic coordinates $z_1, \dots, z_r$ at p, we have the ring of differential operators

$$\mathcal{D} = \mathbb{C}\{z_1, \dots, z_r\}\left[\frac{\partial}{\partial z_1}, \dots, \frac{\partial}{\partial z_r}\right],$$

where $\mathbb{C}\{z_1, \dots, z_r\}$ is the ring of convergent power series in $z_1, \dots, z_r$. Then the Gauss-Manin connection ∇ determines an $\mathcal{O}_S$-homomorphism $\phi : \mathcal{D} \to \mathcal{F}^0$ determined by the rule

$$\phi(X_1 \cdots X_\ell) = \nabla_{X_1} \cdots \nabla_{X_\ell} \Omega(t)$$

for vector fields X_i on S. This gives $\mathcal{F}^0$ the structure of a $\mathcal{D}$-module.

The ideal $I = \ker(\phi)$ consists of the differential operators annihilating $\Omega(s)$. We call I the *Picard-Fuchs ideal*. This ideal plays an important role in the theory. For example, we will see later that knowing I allows us to calculate the Yukawa couplings. If the above map ϕ is surjective, which happens for Calabi-Yau threefolds by [**BG**], then the $\mathcal{D}$-module structure on $\mathcal{F}^0$ is completely determined by the Picard-Fuchs ideal I.

The equations $D \cdot y = 0$ for $D \in I$ are the *Picard-Fuchs equations*, and a local section y of $\mathcal{O}_S$ is a *solution* if $D \cdot y = 0$ for all $D \in I$. Since

$$(5.2) \qquad \frac{\partial}{\partial z_i} \int_{\gamma_t} \Omega(t) = \int_{\gamma_t} \nabla_{\partial/\partial z_i} \Omega(t)$$

for a (locally) constant homology n-cycle γ_t, it follows that the periods $y = \int_{\gamma_t} \Omega(t)$ of Ω are solutions of the Picard-Fuchs equations. We will often write periods as $y = \langle g(t), \Omega(t) \rangle$, where $g(t) \in H^n(X_t, \mathbb{C})$ is the Poincaré dual of γ_t (thus $g(t)$ is a flat section of ∇). Note that a differential operator D is in the Picard-Fuchs ideal if and only if $D \cdot y = 0$ for *all* periods $y = \langle g, \Omega \rangle$.

Although we have only constructed the Picard-Fuchs equations locally, by analytic continuation they make sense as algebraic differential equations over all of S if we restrict our attention to the periods of algebraic differentials.

5.1.3. Degenerations of Hodge Structures. Suppose that the smooth family $\pi : \mathcal{X} \to S$ can be completed to a flat family $\bar\pi : \bar{\mathcal{X}} \to \bar{S}$, where $\bar{S}$ is a smooth compactification of S with normal crossings boundary divisor $D = \cup_i D_i = \bar{S} - S$.

The bundle $\mathcal{F}^0$ on S has a *canonical extension* $\bar{\mathcal{F}}^0$ on $\bar{S}$ [**Deligne1**]. The Gauss-Manin connection does *not* necessarily extend to a connection on $\bar{\mathcal{F}}^0$ because it can acquire singularities. The singularities are very mild (regular singular points), which means that ∇ extends to a map $\bar\nabla : \bar{\mathcal{F}}^0 \to \bar{\mathcal{F}}^0 \otimes \Omega^1_{\bar{S}}(\log D)$. The sheaf $\Omega^1_{\bar{S}}(\log D)$ is a subsheaf of the sheaf of rational 1-forms on $\bar{S}$. In local coordinates $z_1, \ldots, z_r$ for $\bar{S}$ such that D is defined by $z_1 \cdots z_l = 0$, the sheaf $\Omega^1_{\bar{S}}(\log D)$ is generated by

$$\frac{dz_1}{z_1}, \ldots, \frac{dz_l}{z_l}, dz_{l+1}, \ldots, dz_r.$$

There is also monodromy to consider. Let $\gamma_j = \gamma_j(u)$ be a small loop going around the boundary divisor D_j based at $\gamma_j(0) = \gamma_j(1) = t \in S$. A cohomology class $\eta \in H^n(X_t, \mathbb{C})$ can be uniquely lifted to a ∇-flat section $\eta(u) \in H^n(X_{\gamma_j(u)}, \mathbb{C})$ over $[0, 1]$ such that $\eta(0) = \eta$. The *monodromy transformation* (or *Picard-Lefschetz transformation*)

$$\mathcal{T}_j : H^n(X_t, \mathbb{C}) \longrightarrow H^n(X_t, \mathbb{C})$$

is defined by $\mathcal{T}_j(\eta) = \eta(1)$. Up to conjugation, it only depends on D_j and not on the choice of γ_j. The monodromy theorem [**Landman**] says that for some integer $m \geq 0$,

$$(\mathcal{T}_j^m - I)^{n+1} = 0.$$

Thus $\mathcal{T}_j$ is quasi-unipotent, with index of unipotency at most $n + 1$. In particular, the eigenvalues of $\mathcal{T}_j$ are roots of unity.

5.1.4. The Limiting Mixed Hodge Structure. Assume that the monodromy is unipotent and that we are at a point $p \in \bar{S}$ such that $D = \bar{S} - S$ is defined by $z_1 \cdots z_r = 0$, where $z_1, \ldots, z_r$ are local coordinates at p as before. Hence we may assume $S = (\Delta^*)^r$, $\bar{S} = \Delta^r$ and $p = 0$. Let $N_j = \log(\mathcal{T}_j)$, where $\mathcal{T}_j$ is the monodromy given by going counterclockwise around the jth factor of $(\Delta^*)^r$. In this situation, the canonical extension of $\mathcal{F}^0$ can be described as follows. Let s be a flat multi-valued section of $\mathcal{H} = \mathcal{F}^0$ over $(\Delta^*)^r$. Then $\exp(-\frac{1}{2\pi i} \sum_j \log(z_j) N_j) s$ is single-valued and extends to a section of the canonical extension $\bar{\mathcal{F}}^0$. This property is easily seen to characterize $\bar{\mathcal{F}}^0$.

Furthermore, the nilpotent orbit theorem [**Schmid**] implies that the bundles $\mathcal{F}^p$ extend to $\bar{\mathcal{F}}^p$, which are subbundles of the canonical extension. At $0 \in \Delta^r$, $\bar{\mathcal{F}}^p$ gives the *limiting Hodge filtration* $\bar{\mathcal{F}}_0^\bullet = F_{\lim}^\bullet$. One can show that $N_j(F_{\lim}^p) \subset F_{\lim}^{p-1}$. Thus we get a linear map

$$N_j : F_{\lim}^p / F_{\lim}^{p+1} \longrightarrow F_{\lim}^{p-1} / F_{\lim}^p.$$

If $\delta_j = z_j\, \partial/\partial z_j$, then $\bar{\nabla}_{\delta_j}(\bar{\mathcal{F}}^p) \subset \bar{\mathcal{F}}^{p-1}$ by Griffiths transversality. It follows that $\bar{\nabla}_{\delta_j}$ induces a linear map

$$\bar{\nabla}_{\delta_j} : F_{\lim}^p / F_{\lim}^{p+1} \longrightarrow F_{\lim}^{p-1} / F_{\lim}^p.$$

Using the nilpotent orbit theorem, one sees that these maps are related by

$$(5.3) \qquad\qquad \bar{\nabla}_{\delta_j} = \frac{-1}{2\pi i} N_j.$$

We also have a natural integer structure over $0 \in \Delta^r$. Let s be a (flat) multi-valued section of $\mathcal{H}_{\mathbb{Z}}$ over $(\Delta^*)^r$. Then, as above, $\tilde{s} = \exp(-\frac{1}{2\pi i} \sum_j \log(z_j) N_j)\, s$ is a section of the canonical extension $\bar{\mathcal{F}}^0$, and $\tilde{s}(0)$ is an integral element over 0. If we transport this back to $H^n(X_t, \mathbb{C})$ using $\exp(\frac{1}{2\pi i} \sum_j \log(z_j) N_j)$, then we get the usual integer structure determined by $H^n(X_t, \mathbb{Z})$.

Another important ingredient is the *monodromy weight filtration*

$$W_\bullet : W_0 \subset W_1 \subset \cdots \subset W_{2n-1} \subset W_{2n} = H^n(X_t, \mathbb{C}),$$

which is defined in terms of the action of $N = \sum_j a_j N_j$, $a_j > 0$, on $H^n(X_t, \mathbb{C})$. For instance,

$$
\begin{aligned}
W_0 &= \mathrm{im}(N^n) \\
W_1 &= \mathrm{im}(N^{n-1}) \cap \ker(N) \\
W_2 &= \mathrm{im}(N^{n-2}) \cap \ker(N) + \mathrm{im}(N^{n-1}) \cap \ker(N^2) \\
&\ \ \vdots \\
W_{2n-1} &= \ker(N^n).
\end{aligned}
$$

(5.4)

Formulas for the other W_k can be found in [**Griffiths2**]. The main properties of the monodromy weight filtration are:

- $N(W_k) \subset W_{k-2}$.
- N^k induces an isomorphism $N^k : W_{n+k}/W_{n+k-1} \simeq W_{n-k}/W_{n-k-1}$.
- $F_{\lim}^\bullet$ induces a Hodge structure of weight k on W_k/W_{k-1}.

The last item says that $(W_\bullet, F_{\lim}^\bullet)$ is a *mixed Hodge structure*. A more complete discussion can be found in [**Griffiths2**]. The monodromy weight filtration will play an important role in Section 5.2.

5.1.5. Monodromy and the Picard-Fuchs Equation. For simplicity, assume that $\dim S = 1$, and let z be a coordinate in a disk $\Delta \subset \bar{S}$ centered at $p \in \bar{S} - S$. Choosing a basis $\omega_1, \ldots, \omega_r$ of $\bar{\mathcal{F}}^0$ over Δ, the connection ∇ is completely determined by its *connection matrix* (Γ_{ij}), defined by $\nabla_{d/dz}\omega_i = \sum_j \Gamma_{ij}\omega_j$.

Γ_{ij} has at worst a simple pole at $z = 0$ since ∇ has regular singular points. Then we get the *residue matrix*

$$(5.5) \qquad \mathrm{Res}(\nabla) = \mathrm{Res}_{z=0}(\Gamma_{ij}),$$

which by [**Deligne1**], has the following properties:

- The eigenvalues λ of $\mathrm{Res}(\nabla)$ are rational and satisfy $0 \le \lambda < 1$.
- $\exp(-2\pi i \mathrm{Res}(\nabla))$ is conjugate to the monodromy $\mathcal{T}$.
- $\mathcal{T}$ is unipotent if and only if $\mathrm{Res}(\nabla)$ is nilpotent.

In practice, we might not know a basis of $\bar{\mathcal{F}}^0$. Assume instead that $\omega_1, \dots, \omega_r$ is a basis over the punctured disk Δ^*. If the connection matrix has at most simple poles at $z = 0$, then the residue matrix $\mathrm{Res}(\nabla)$ still gives useful information:

- The eigenvalues of the monodromy $\mathcal{T}$ are $\exp(2\pi i\lambda)$ as λ ranges over the eigenvalues of $\mathrm{Res}(\nabla)$.
- $\mathcal{T}$ is unipotent if and only if $\mathrm{Res}(\nabla)$ has integer eigenvalues.
- If no two distinct eigenvalues of $\mathrm{Res}(\nabla)$ differ by an integer, then $\mathcal{T}$ is conjugate to $\exp(-2\pi i \mathrm{Res}(\nabla))$.

To see what the Picard-Fuchs equations look like in this case, let $\delta = z\, d/dz$. For a fixed section $\Omega(z)$ of $\bar{\mathcal{F}}^n$, suppose that we have a relation in $\bar{\mathcal{F}}^0$ of the form

$$\nabla_\delta^m \Omega + f_1(z)\nabla_\delta^{m-1}\Omega + \cdots + f_m(z)\Omega = 0,$$

where the $f_i(z)$ are all analytic at $z = 0$. As explained above, this implies that $\delta^m + f_1(z)\delta^{m-1} + \cdots + f_m(z)$ is in the Picard-Fuchs ideal, and the periods $y = \int_{\gamma_z} \Omega(z)$ are solutions of the Picard-Fuchs equation

$$(5.6) \qquad \delta^m y + f_1(z)\delta^{m-1}y + \cdots + f_m(z)y = 0,$$

possibly multiple-valued because of the effect of monodromy on γ_z.

Let's now connect this to ordinary differential equations, as in [**CL**]. The equation (5.6) has regular singular points as defined in standard texts in differential equations. The solutions are governed by the *indicial equation*, which is obtained from (5.6) by replacing δ with an auxiliary variable λ:

$$\lambda^m + f_1(0)\lambda^{m-1} + \cdots + f_m(0) = 0.$$

Suppose that λ is a root of the indicial equation with the property that there is no root λ' such that $\lambda' - \lambda$ is a positive integer. Then the Picard-Fuchs equation (5.6) has a solution of the form $y_0 = z^\lambda f_0(z)$ near $z = 0$, with f analytic and nonvanishing at 0. Thus analytically continuing y_0 around $z = 0$ gives $\exp(2\pi i\lambda)\, y_0$. Furthermore, if λ is a root of multiplicity $\mu > 1$, then the classical method of Frobenius gives in addition to $y_0 = z^\lambda f(z)$ other solutions $y_1, \dots, y_{\mu-1}$ such that analytically continuing y_i around $z = 0$ gives $\exp(2\pi i\lambda)\,(y_i + y_{i-1})$ for $1 \le i \le \mu - 1$. This implies the asymptotic behavior

$$y_i \sim c\, z^\lambda \left(\frac{\log z}{2\pi i}\right)^i, \quad 0 \le i \le \mu - 1$$

as $z \to 0$, where c is a nonzero constant. The number μ depends on the Jordan canonical form of the companion matrix of the indicial equation.

The above Picard-Fuchs equation came from a section Ω of $\bar{\mathcal{F}}^n$. If we switch to $\Omega' = g(z)\Omega$, where $g(z)$ is meromorphic on S, we get a new Picard-Fuchs equation with solutions $\tilde{y} = g(z)y$, where y is a solution of (5.6). In this situation, one can show that the new equation has regular singular points with indicial equation

$$(\lambda - l)^m + f_1(0)(\lambda - l)^{m-1} + \cdots + f_m(0) = 0,$$

where the Laurent expansion of $g(z)$ at $z = 0$ is $g(z) = a_l z^l + \cdots$, $a_l \neq 0$.

In the next section, we will see that if we are at the right kind of boundary point, then the relation between the Picard-Fuchs equation and the monodromy is especially close.

5.2. Maximally Unipotent Monodromy

In this section, we will discover that the mirror of the quintic threefold has some special monodromy behavior which is crucial to mirror symmetry.

5.2.1. 1-Dimensional Moduli. We begin our discussion of maximally unipotent monodromy with the case of a Calabi-Yau threefold V whose complex moduli space S is 1-dimensional. An example is the mirror of the quintic threefold described in Chapter 2. We need not have a universal family over S, but we can construct families locally over finite covers of S. The resulting Picard-Fuchs equations make sense as algebraic differential equations provided that we start with algebraic sections of $\mathcal{F}^3$, and so can be analytically continued over all of S. Let $\bar{S}$ as above be a compactification of S, and suppose we have a boundary point $p \in \bar{S} - S$ which has *maximally unipotent monodromy*. This means that the monodromy $\mathcal{T} : H^3(V, \mathbb{C}) \to H^3(V, \mathbb{C})$ is unipotent, and since $(\mathcal{T} - I)^4 = 0$ by the monodromy theorem, maximally unipotent implies $(\mathcal{T} - I)^3 \neq 0$. It follows that $N = \log(\mathcal{T})$ also satisfies $N^3 \neq 0$.

Our first goal is to explain how the theory of the last section applies to this situation. We will see that maximally unipotent monodromy has some strong consequences concerning periods and the Picard-Fuchs equation. A first observation is that $h^{2,1}(V) = 1$ since V has 1-dimensional complex moduli. It follows that $h^3(V) = 4$, and since $\mathcal{T}$ is maximally unipotent, we can find a basis g_0, g_1, g_2, g_3 of $H^3(V, \mathbb{C})$ such that $\mathcal{T}$ is given by the matrix

$$(5.7) \qquad \mathcal{T} = \begin{pmatrix} 1 & 1 & 0 & 0 \\ 0 & 1 & 1 & 0 \\ 0 & 0 & 1 & 1 \\ 0 & 0 & 0 & 1 \end{pmatrix}.$$

Each g_i is dual to a homology class $\gamma_i \in H_3(V, \mathbb{C})$. Given a fixed local basis Ω of $\bar{\mathcal{F}}^3$ near $z = 0$, we want to describe the behavior of the periods $y_j = \int_{\gamma_j} \Omega$ near 0.

We first follow [**Morrison2**]. Any $g \in H^3(V, \mathbb{C})$ extends to a multiple-valued flat section $g(z)$ of $\bar{\mathcal{F}}^0$, and $\exp(-\frac{\log z}{2\pi i} N)\, g(z)$ is single-valued and analytic at $z = 0$ by the nilpotent orbit theorem. When $g = g_0$, we have $\exp(-\frac{\log z}{2\pi i} N)\, g_0(z) = g_0(z)$ since $\mathcal{T}(g_0) = g_0$ implies $N(g_0) = 0$. Thus $g_0(z)$ is single-valued at $z = 0$. Since

$$y_0 = \int_{\gamma_0} \Omega = \langle g_0(z), \Omega(z) \rangle,$$

it follows that y_0 is single-valued and analytic at $z = 0$. We will also see below that y_0 is nonvanishing at $z = 0$. Turning our attending to y_1, y_2, y_3, note that $\mathcal{T}(g_i) = g_i + g_{i-1}$ for $i \geq 1$, and since $y_i = \langle g_i(z), \Omega(z) \rangle$, it follows that analytically continuing y_i around $z = 0$ gives $y_i + y_{i-1}$. Furthermore, since $\exp(-\frac{\log z}{2\pi i} N)\, g_i(z)$ is a single-valued function, one easily gets an asymptotic formula of the form

$$y_i \sim c \left(\frac{\log z}{2\pi i}\right)^i, \qquad 0 \le i \le 3$$

as $z \to 0$, where c is a nonzero constant.

This is reminiscent of the differential equation theory discussed in Section 5.1.5. So we next explain what the Picard-Fuchs equation looks like in this case. Here is the key result.

PROPOSITION 5.2.1. *Fix a Calabi-Yau threefold V with a 1-dimensional complex moduli space S and a local coordinate z at a boundary point of S. Also assume Ω is a local section of $\mathcal{F}^0$ giving a Picard-Fuchs equation*

$$\delta^4 y + f_1(z)\delta^3 y + f_2(z)\delta^2 y + f_3(z)\delta y + f_4(z)y = 0,$$

where $\delta = z\, d/dz$. Then:

(i) *The monodromy $\mathcal{T}$ is unipotent if and only if the roots of the indicial equation are all integers.*

(ii) *The monodromy $\mathcal{T}$ is maximally unipotent if and only if the indicial equation is of the form $(\lambda - l)^4 = 0$ for some integer l.*

PROOF. The above Picard-Fuchs equation implies that in $\mathcal{F}^0$, we have

$$(5.8) \qquad \nabla_\delta^4 \Omega + f_1(z)\nabla_\delta^3 \Omega + f_2(z)\nabla_\delta^2 \Omega + f_3(z)\nabla_\delta \Omega + f_4(z)\Omega = 0.$$

By [**BG**], we also know that $\Omega, \nabla_\delta \Omega, \nabla_\delta^2 \Omega, \nabla_\delta^3 \Omega$ must be generically linearly independent. In particular, we can assume they form a basis of $\mathcal{F}^0$ in a punctured neighborhood Δ^* of $z = 0$ (since Ω might *not* be a section of $\bar{\mathcal{F}}^3$, we might not have a basis at $z = 0$).

The basis $\Omega, \nabla_\delta \Omega, \nabla_\delta^2 \Omega, \nabla_\delta^3 \Omega$ enables us to compute the connection matrix of ∇. Applying $\nabla_{d/dz}$ to the basis and using (5.8) yields

$$\nabla_{d/dz}(\Omega) = \frac{1}{z}\nabla_\delta \Omega$$

$$\nabla_{d/dz}(\nabla_\delta \Omega) = \frac{1}{z}\nabla_\delta^2 \Omega$$

$$\nabla_{d/dz}(\nabla_\delta^2 \Omega) = \frac{1}{z}\nabla_\delta^3 \Omega$$

$$\nabla_{d/dz}(\nabla_\delta^3 \Omega) = -\frac{f_4(z)}{z}\Omega - \frac{f_3(z)}{z}\nabla_\delta \Omega - \frac{f_2(z)}{z}\nabla_\delta^2 \Omega - \frac{f_1(z)}{z}\nabla_\delta^3 \Omega.$$

The connection matrix has at worst simple poles at $z = 0$, so we can use the theory developed in Section 5.1.5. The residue matrix $\mathrm{Res}(\nabla)$ in this basis is

$$(5.9) \qquad \mathrm{Res}(\nabla) = \begin{pmatrix} 0 & 1 & 0 & 0 \\ 0 & 0 & 1 & 0 \\ 0 & 0 & 0 & 1 \\ -f_4(0) & -f_3(0) & -f_2(0) & -f_1(0) \end{pmatrix},$$

which is the companion matrix of the indicial equation.

Then $\mathcal{T}$ is unipotent $\Leftrightarrow$ the eigenvalues of $\mathrm{Res}(\nabla)$ are integers $\Leftrightarrow$ the roots of the indicial equation are integers, which proves the first part of the proposition.

As for the second part, suppose that the indicial equation is $(\lambda - l)^4 = 0$ for some integer l. If we replace Ω by $z^{-l}\Omega$, then the discussion at the end of Section 5.1 shows that we can assume that the indicial equation is $\lambda^4 = 0$. Then (5.9) becomes

$$\mathrm{Res}(\nabla) = \begin{pmatrix} 0 & 1 & 0 & 0 \\ 0 & 0 & 1 & 0 \\ 0 & 0 & 0 & 1 \\ 0 & 0 & 0 & 0 \end{pmatrix}.$$

Since no two distinct roots differ by an integer, the properties of $\mathrm{Res}(\nabla)$ imply that $\mathcal{T}$ is conjugate to $\exp(2\pi i \mathrm{Res}(\nabla))$, which implies that $\mathcal{T}$ is maximally unipotent.

Conversely, suppose that $\mathcal{T}$ is maximally unipotent. It suffices to find a Picard-Fuchs equation whose indicial equation is $\lambda^4 = 0$. Let Ω be a local basis of $\bar{\mathcal{F}}^3$. We claim that Ω, $\nabla_\delta(\Omega)$, $\nabla_\delta^2(\Omega)$ and $\nabla_\delta^3(\Omega)$ form a basis of $\bar{\mathcal{F}}^0$ near $z = 0$.

We will prove this using the monodromy weight filtration described in Section 5.1.4. By (5.4), $W_6 = H^3(V, \mathbb{C})$ and $W_5 = \ker(N^3)$, where $N = \log(\mathcal{T})$. Then $\dim(W_6/W_5) = 1$ follows from (5.7). Since $F_{\mathrm{lim}}^\bullet$ induces a Hodge structure of weight 6 on W_6/W_5 and $F_{\mathrm{lim}}^6 = F_{\mathrm{lim}}^5 = F_{\mathrm{lim}}^4 = 0$, we must have $W_6/W_5 = H^{3,3}$. But W_6/W_5 and F_{lim}^3 are both 1-dimensional, and it follows that $F_{\mathrm{lim}}^3 \cap W_5 = \{0\}$. This gives the direct sum decomposition

$$(5.10) \qquad\qquad F_{\mathrm{lim}}^3 \oplus W_5 = H^3(V, \mathbb{C}).$$

The local basis Ω of $\bar{\mathcal{F}}^3$ gives a nonzero element of $\Omega(0) \in F_{\mathrm{lim}}^3$. Since $W_5 = \ker(N^3)$, (5.10) implies $N^3(\Omega(0)) \neq 0$, and it follows from (5.7) that $\Omega(0)$, $N(\Omega(0))$, $N^2(\Omega(0))$, $N^3(\Omega(0))$ are a basis of $H^3(V, \mathbb{C})$. Iterating (5.3) implies that

$$\nabla_\delta^j(\Omega)(0) \equiv \left(\tfrac{-1}{2\pi i}\right)^j N^j(\Omega(0)) \bmod F_{\mathrm{lim}}^{4-j},$$

and then, using $N(F_{\mathrm{lim}}^p) \subset F_{\mathrm{lim}}^{p-1}$ and $\dim(F_{\mathrm{lim}}^p) = 4 - p$, it follows easily that Ω, $\nabla_\delta(\Omega)$, $\nabla_\delta^2(\Omega)$, $\nabla_\delta^3(\Omega)$ are linearly independent at $z = 0$. Hence they form a basis of $\bar{\mathcal{F}}^0$ in a neighborhood of $z = 0$.

In particular, we can express $\nabla_\delta^4(\Omega)$ in terms of this basis, which gives a Picard-Fuchs equation of the form (5.8). It remains to prove that $f_1(0) = f_2(0) = f_3(0) = f_4(0) = 0$. This is now easy: using our basis to compute the connection matrix and residue matrix of ∇ as above, we get the formula (5.9) for $\mathrm{Res}(\nabla)$. Since we formed this matrix using a basis of $\bar{\mathcal{F}}^0$, we know from Section 5.1 that $\mathrm{Res}(\nabla)$ is nilpotent since $\mathcal{T}$ is unipotent. Thus the indicial equation is $\lambda^4 = 0$, and the proposition is proved. $\qquad\square$

There are two comments to make concerning the proof of Proposition 5.2.1. First, one can prove $W_{2k} = W_{2k+1}$ and

$$(5.11) \qquad\qquad F_{\mathrm{lim}}^p \oplus W_{2p-1} = F_{\mathrm{lim}}^p \oplus W_{2p-2} = H^3(V, \mathbb{C})$$

for $0 \leq p \leq 3$, generalizing (5.10). It follows that the mixed Hodge structure $(W_\bullet, F_{\mathrm{lim}}^\bullet)$ is a *Hodge-Tate structure* in this case. In general, the relevance of Hodge-Tate structures to mirror symmetry is discussed in [**Deligne2**].

A second comment concerns the discussion preceding Proposition 5.2.1, where we claimed that the solution $y_0 = \langle g_0, \Omega \rangle$ was nonvanishing at $z = 0$. To see why, first note that $\langle W_0, W_5 \rangle = 0$ follows easily from (5.4) and

$$\langle N(a), b \rangle + \langle a, N(b) \rangle = 0, \qquad a, b \in H^3(V, \mathbb{C}).$$

Since $g_0 \in W_0$ and $\Omega(0)$ spans F^3_{lim}, we must have $\langle g_0, \Omega(0) \rangle \neq 0$ by (5.10) and the nondegeneracy of cup product. Thus $y_0(0) = \langle g_0, \Omega(0) \rangle$ is nonzero.

For an example of how to use Proposition 5.2.1, consider the quintic mirror family from Chapter 2. The Picard-Fuchs equation is

$$(5.12) \qquad \begin{aligned} 0 = {}& \left(z \frac{d}{dz} \right)^4 y + \frac{2 \cdot 5^5 z}{1 + 5^5 z} \left(z \frac{d}{dz} \right)^3 y + \frac{7 \cdot 5^4 z}{1 + 5^5 z} \left(z \frac{d}{dz} \right)^2 y \\ & + \frac{2 \cdot 5^4 z}{1 + 5^5 z} \left(z \frac{d}{dz} \right) y + \frac{24 \cdot 5 z}{1 + 5^5 z} y \end{aligned}$$

by (2.8) (we will derive this in Section 5.4). In this equation, the variable z (called x in Chapter 2) is $z = \psi^{-5}$, where ψ is the parameter in the defining equation of the quintic mirror (2.4). Since the indicial equation is $\lambda^4 = 0$, we have maximally unipotent monodromy at $z = 0$ by Proposition 5.2.1.

On the other hand, the quintic mirror family has two other boundary points at $z = -5^{-5}, \infty$. It is easy to see that the monodromy is not maximally unipotent at $z = \infty$. This follows from (5.12) by a coordinate change $w = 1/z$ and calculating the roots of the indicial equation to be $\{1/5, 2/5, 3/5, 4/5\}$. By Proposition 5.2.1, ∞ is not even unipotent. Adapting coordinates to $z = -5^{-5}$, we get integral roots of the indicial equation, hence unipotent monodromy. But the roots are not all equal, so that the monodromy cannot be maximally unipotent by the proposition. In fact, one can show that $N^2 = 0$ in this case.

5.2.2. r-Dimensional Moduli. We next generalize the definition of maximally unipotent monodromy. For a Calabi-Yau manifold V of dimension d (the case when V is singular will be considered in Chapter 6), let S be the full complex moduli space of V. Then S is a smooth manifold of dimension $r = h^{d-1,1}(V)$ **[Bogomolov, Tian, Todorov]**. Let $\bar{S}$ as in Section 5.1 be a compactification of S with normal crossing boundary divisor $D = \bar{S} - S$. We focus attention on boundary points $p \in \bar{S} - S$ which are the intersection of $r = \dim(S)$ boundary divisors $D_1, \ldots, D_r$ (we will study natural compactifications of S in more detail in Chapter 6.) We will be interested in boundary points of the following type.

DEFINITION 5.2.2. **[Morrison3]** *The point $p = D_1 \cap \cdots \cap D_r$ is a maximally unipotent boundary point if the following conditions hold.*

   (i) The monodromy transformations T_i are all unipotent.
   (ii) Put $N_i = \log(T_i)$ and let $N = \sum_i a_i N_i$ be a linear combination with all $a_i > 0$. If $W_{\bullet}$ is the monodromy weight filtration (5.4) determined by N, then $\dim W_0 = \dim W_1 = 1$ and $\dim W_2 = 1 + r$.
   (iii) Let $g_0, \ldots, g_r$ be a basis of W_2 such that g_0 spans W_0 and define m_{ij} via $N_i(g_j) = m_{ij} g_0$ for $1 \leq i, j \leq r$. Then the matrix (m_{ij}) is invertible.

We can explain this definition as follows. Each $g_0 \in W_0$ gives a solution $y_0 = \langle g_0, \Omega \rangle$ of the Picard-Fuchs equations which is analytic at the maximally unipotent boundary point p. Then $\dim W_0 = 1$ says that up to a constant multiple, the Picard-Fuchs equations have a *unique* solution which is holomorphic at p. Also note that each $g_i \in W_2$ gives a solution $\langle g_i, \Omega \rangle$ of the Picard-Fuchs equations which has at worst logarithmic growth along the divisors D_j. Hence $\dim W_2 = 1 + r$ means that we have r independent solutions with logarithmic growth along the D_j. Finally, the invertibility of the matrix (m_{ij}) in condition *(iii)* allows us to change to a new basis

$$(5.13) \qquad g'_k = \sum_{j=1}^{r} g_j m^{jk}, \quad (m^{jk}) = (m_{jk})^{-1}$$

with the property $N_j(g'_k) = \delta_{jk} g_0$. This implies that $y_k = \langle g'_k, \Omega \rangle$ is a solution of the Picard-Fuchs equations which is holomorphic at a general point of D_j for $j \neq k$ and has logarithmic growth along D_k. More precisely, locally near p, y_k can be written in the form

$$y_k = y_0 \frac{\log z_k}{2\pi i} + \text{holomorphic}.$$

In practice, this is what one looks for when dealing with a maximally unipotent boundary point. We will see in Chapter 6 that the mirror map is built from the functions $y_0, y_1, \ldots, y_r$.

Another way to understand condition (iii) comes from mirror symmetry. Suppose that V and V° are mirrors. As we will see in Section 6.3, the mirror map identifies the Kähler moduli space of V° with a neighborhood of a maximally unipotent boundary point of the complex moduli space of V. On the mirror side, we will use quantum cohomology in Chapter 8 to construct a flat connection on the Kähler moduli space of V° called the *A-model connection*. Locally, this connection lives on a punctured polydisc $(\Delta^*)^r$ whose coordinate axes correspond to a set T_i, $1 \leq i \leq r$, of generators of the Kähler cone of V°. Each axis gives rise to a monodromy transformation $\widetilde{T}_i$ for this flat connection, and the logarithm $\widetilde{N}_i = \log(\widetilde{T}_i)$ is the endomorphism given by cup product with $-T_i$ on $\oplus_p H^{p,p}(V^\circ)$. Conjecturally, the mirror map identifies the A-model connection with the Gauss-Manin connection in such a way that:

- N_i is identified with $\widetilde{N}_i = \cdot \cup (-T_i)$, and thus $N = \sum_i a_i N_i$ is identified with cup product with the negative of an ample class $D = \sum_i a_i T_i$.
- g_0 is identified with the cohomology class of a point.
- g_i is identified with the Poincaré dual of a curve C_i, where C_i is a suitable basis of $H_2(V^\circ, \mathbb{Z})$.

Then, under the mirror map, the matrix (m_{ij}) is the matrix of intersection products $(D_i \cdot C_j)$, which is necessarily invertible by Poincaré duality. Hence the third condition of Definition 5.2.2 is completely natural from the point of view of mirror symmetry.

At a point of maximally unipotent monodromy, $N^d : W_{2d}/W_{2d-1} \simeq W_0$ implies that $N^d \neq 0$, so that N is unipotent of maximal index (since $N^{d+1} = 0$ by the monodromy theorem). In particular, when dealing with threefolds with 1-dimensional complex moduli, it follows easily that Definition 5.2.2 is equivalent to $N^3 \neq 0$. More generally, for a Calabi-Yau of dimension d with 1-dimensional complex moduli, Definition 5.2.2 is equivalent to $N^d \neq 0$ [**Morrison2**] .

Finally, we should mention how the integer structure of the weight filtration interacts with Definition 5.2.2. Suppose that we are at a maximally unipotent boundary point, and let $g_0, g_1, \ldots, g_r$ be a $\mathbb{Z}$-basis of W_2. Then one easily shows that the numbers m_{jk} defined by $N_j(g_k) = m_{jk} g_0$ are integers. In [**Morrison6**], Morrison conjectures that the matrix (m_{jk}) is invertible over $\mathbb{Z}$, or equivalently, that the m^{jk} in (5.13) are integers. This *Integrality Conjecture* has been verified for the mirror of the quintic threefold [**Morrison2**], and other cases of the conjecture are discussed in [**Morrison6**].

5.3. The Griffiths-Dwork Method

We now turn to a method for calculating the Picard-Fuchs ideal due to Griffiths and Dwork. This technique was introduced for projective hypersurfaces and more recently has been extended to hypersurfaces in weighted projective spaces or more generally projective toric varieties.

5.3.1. Projective Hypersurfaces. We begin with a hypersurface $V \subset \mathbb{P}^n$ of degree ℓ. We have the residue map

$$\mathrm{Res}: \ H^n(\mathbb{P}^n - V) \to H^{n-1}(V)$$

which we will describe presently. We can represent elements of $H^n(\mathbb{P}^n - V)$ by holomorphic n-forms on $\mathbb{P}^n$ with poles along V. To do this, we define

$$\Omega_0 = \sum_j (-1)^j x_j dx_0 \wedge \ldots \wedge \widehat{dx_j} \wedge \ldots \wedge dx_n.$$

Note that Ω_0 is a section of $\Omega^n_{\mathbb{P}^n}(n+1)$, which is a trivial sheaf. This means that Ω_0 is up to scalars the unique holomorphic n-form on $\mathbb{P}^n$ which is homogeneous of degree $n+1$.

If V is defined by a degree ℓ homogeneous equation $f = 0$, then by [**Griffiths1**], elements of $H^n(\mathbb{P}^n - V)$ can be represented by forms

$$\frac{P\Omega_0}{f^k}, \qquad \deg(P) = k\ell - (n+1),$$

where P is also homogeneous. For any topological $(n-1)$-cycle γ in V, let $T(\gamma)$ be the tube over γ (a circle bundle over γ contained in $\mathbb{P}^n - V$). Then the residue is defined by

$$(5.14) \qquad \int_\gamma \mathrm{Res} \frac{P\Omega_0}{f^k} = \int_{T(\gamma)} \frac{P\Omega_0}{f^k}.$$

Addition of an exact form to $P\Omega_0/f^k$ does not alter the right hand side of (5.14), so that Res is well-defined on cohomology classes. Since $\ell H \sim V$, where H is the hyperplane class, it follows that $\mathrm{Res}(P\Omega_0/f^k) \cdot H = 0$. Define the *primitive cohomology*

$$PH^{n-1}(V) = \{\eta \in H^{n-1}(V) \mid \eta \cdot H = 0\}.$$

The primitive cohomology is equal to the full cohomology if the dimension of V is odd. In general, the image of Res is contained in $PH^{n-1}(V)$, and the map $\mathrm{Res}: H^n(\mathbb{P}^n - V) \to PH^{n-1}(V)$ is surjective by [**Griffiths1**]. In the sequel, we will follow common practice and frequently drop the "Res" from our notation and view $P\Omega_0/f^k$ as a cohomology class or form on V.

It turns out that $P\Omega_0/f^k$ actually lies in $F^{n-k}PH^{n-1}(V)$ by [**Griffiths1**]. If a family $\mathcal{V}$ is obtained by letting the defining polynomial f and numerator coefficient P depend on a parameter s, we calculate

$$\nabla_{d/ds}\Big(\frac{P\Omega_0}{f^k}\Big) = \frac{(-kPf' + fP')\Omega_0}{f^{k+1}} \in \mathcal{F}^{n-k-1},$$

where the primes denote differentiation with respect to s. This verifies Griffiths transversality for projective hypersurfaces.

An effective procedure was developed in [**Griffiths1**] for calculating modulo exact forms. For each i, let G_i be a homogeneous polynomial of degree $(k-1)\ell - n$, and put

$$\eta = \frac{1}{f^{k-1}} \sum_{i<j} (x_i G_j - x_j G_i) dx_0 \wedge \cdots \wedge \widehat{dx_i} \wedge \cdots \wedge \widehat{dx_j} \wedge \cdots \wedge dx_n.$$

Then

$$(5.15) \qquad d\eta = \frac{((k-1)\sum_j G_j \frac{\partial f}{\partial x_j})\Omega_0}{f^k} - \frac{(\sum_j \frac{\partial G_j}{\partial x_j})\Omega_0}{f^{k-1}}.$$

The point is that using (5.15), any form $P\Omega_0/f^k$ with P lying in the Jacobian ideal

$$J(f) = \langle \partial f/\partial x_0, \ldots, \partial f/\partial x_n \rangle$$

can be reduced modulo an exact form to a form having a lower order pole. The converse is also true: if $P\Omega_0/f^k$ reduces modulo an exact form to an element of F^{n-k-1}, then P lies in the Jacobian ideal. In other words, the map $P \mapsto P\Omega_0/f^k$ defines an isomorphism

$$(5.16) \qquad (S/J(f))_{k\ell-(n+1)} \simeq PH^{n-k,k-1}(V),$$

where $S = \mathbb{C}[x_0, \ldots, x_n]$ and the subscript denotes the graded piece in degree $k\ell - (n+1)$. Thus, by choosing a basis of $(S/J(f))_{k\ell-(n+1)}$, any form $(P\Omega_0)/f^k$ with P of degree $k\ell - (n+1)$ can be reduced modulo exact forms and forms with lower order poles to a linear combination of the forms coming from the basis of $(S/J(f))_{k\ell-(n+1)}$

We next compute the Picard-Fuchs equations. For simplicity, we will explain the method when V is a Calabi-Yau hypersurface in $\mathbb{P}^n$, which means $\ell = n + 1$. In this case, $\omega = \mathrm{Res}(\Omega_0/f)$ is a holomorphic $(n-1)$-form on V. Then the *Griffiths-Dwork method* proceeds as follows:

- Suppose V is defined by a single equation f depending on a parameter $s \in S$, where S is 1-dimensional. Then $\omega = \mathrm{Res}(\Omega_0/f)$ is a holomorphic $(n-1)$-form on the family of hypersurfaces $\mathcal{V}$ defined by f as s varies. We will assume that f depends on s in a polynomial fashion.
- Choose a basis for the primitive cohomology of $\mathcal{V}$ represented by a collection of conveniently chosen forms $\omega_i = P_i\Omega/f^{k_i}$ for $1 \le i \le r$ (so that $r = h^{n-1}(V) - 1$).
- Repeatedly differentiate ω with respect to s to get sections

$$\omega, \nabla_{d/ds}(\omega), \ldots, \nabla^r_{d/ds}(\omega)$$

of $\mathcal{F}^0$. Each of these can be expressed in terms of the basis modulo exact forms. This is done as follows. Working from the highest order pole downward, we find relations between the polynomials in the numerator of $\nabla^i_{d/ds}(\omega)$ and the numerators of the chosen basis modulo $J(f)$, most conveniently using Gröbner basis techniques in the ring $\mathbb{C}(s)[x_0, \ldots, x_n]$. Then we use (5.15) to express the differentiated form modulo an exact form as a linear combination of the basis elements and a form η with a lower order pole. The coefficients in the linear combination will be rational functions in the parameter s. Then the process is repeated, this time applied to η.

- Since we have $r + 1$ sections and $\mathcal{F}^0$ has rank r, then there is necessarily a linear relation between $\omega, \nabla_{d/ds}(\omega), \ldots, \nabla^r_{d/ds}(\omega)$ with coefficients in $\mathbb{C}(s)$. This gives an element of the Picard-Fuchs ideal, and the resulting ordinary differential equation is the Picard-Fuchs equation.

In the next section, we will apply the Griffiths-Dwork method to the quintic mirror. Also, when the defining equation f of the hypersurface depends on more than one parameter, it is easy to modify the method to obtain elements of the Picard-Fuchs ideal.

Furthermore, it is easy to extend the Griffiths-Dwork method to hypersurfaces in weighted projective spaces, following [**Dolgachev2**]. If the weighted projective space is $\mathbb{P}(q_0, \ldots, q_n)$, then we put

$$\Omega_0 = \sum_{j=0}^{n} (-1)^{j-1} q_j x_j dx_0 \wedge \cdots \wedge \widehat{dx_j} \wedge \cdots \wedge dx_n$$

The other modifications are straightforward. An example of how this works will be given in the next section.

5.3.2. Toric Hypersurfaces. We will next describe how the Griffiths-Dwork method applies to ample hypersurfaces in toric varieties. We follow the treatment in [**BC**]. Let $X = X_\Sigma$ be an n-dimensional projective simplicial toric variety with $\Sigma(1) = \{\rho_1, \ldots, \rho_r\}$, and let $S = \mathbb{C}[x_1, \ldots, x_r]$ be its homogeneous coordinate ring, as defined in Section 3.2.3. Let $\beta \in A_{n-1}(X)$ be the first Chern class of an ample line bundle on X, and let S_β be the set of polynomials in S which are homogeneous of degree β in the sense of Section 3.2.3 (recall that S is graded by $A_{n-1}(X)$). Then, for a general $f \in S_\beta$, the hypersurface $V \subset X$ defined by $f = 0$ is quasi-smooth with β as its divisor class.

The ideas of the Griffiths-Dwork method go over in a natural way. We again need an analogue of the form Ω_0 in this context. This is to be a section of $\widehat{\Omega}_X^n$ which is homogeneous of degree $\beta_0 = \sum_{i=1}^{r} [D_i]$, since $\widehat{\Omega}_X^n \simeq \mathcal{O}_X(-\sum_{i=1}^{r} D_i)$. Note also that if $\beta_i = \deg(x_i)$, then $\beta_0 = \sum_{i=1}^{r} \beta_i$.

Fix a basis $e_1, \ldots, e_n$ for M, and for each subset $I = \{i_i, \ldots, i_n\} \subset \{1, \ldots, r\}$ with n elements, put

$$\det(e_I) = \det(\langle e_j, \rho_{i_k} \rangle_{1 \leq j,k \leq n}),$$

where we abuse notation by letting ρ_i also denote the primitive integral generator of the edge $\rho_i \in \Sigma(1)$. Also let

$$dx_I = dx_{i_1} \wedge \cdots \wedge dx_{i_n}, \qquad \widehat{x_I} = \prod_{i \notin I} x_i.$$

We now can now define

$$(5.17) \qquad \Omega_0 = \sum_{|I|=n} \det(e_I) \widehat{x_I} dx_I.$$

Note that Ω_0 has degree β_0. Furthermore, Ω_0 is actually a form on X (not just on $\mathbb{C}^r$), as explained in [**BC**]. We can again write n-forms on X with poles along V in the form

$$\frac{P\Omega_0}{f^k}, \qquad \deg(P) = k\beta - \beta_0.$$

To reduce the pole order, we need for each $1 \leq i \leq r$ an $(n-1)$-form Ω_i of degree $\beta_0 - \beta_i$ whose exact definition will not concern us. What will be important is the formula

$$(5.18) \qquad d\left(\frac{P\Omega_i}{f^k}\right) = \frac{(f\frac{\partial P}{\partial x_i} - kP\frac{\partial f}{\partial x_i})\Omega_0}{f^{k+1}}.$$

We define the Jacobian ideal here to be

$$J(f) = \langle \partial f/\partial x_1, \dots, \partial f/\partial x_r \rangle.$$

We can again use Gröbner basis techniques and the Jacobian ideal to express arbitrary forms in terms of a given basis modulo exact forms by reducing the pole order.

In the toric case, we use a slightly different notion of primitive cohomology:

$$PH^{n-1}(V) = H^{n-1}(V)/(\text{im } H^{n-1}(X)).$$

This again coincides with the usual cohomology if the dimension of V is odd. We also have a residue map

$$\text{Res} : H^n(X - V) \to PH^{n-1}(V).$$

In the classical case, Res is an isomorphism, but here, we instead have an exact sequence

$$0 \longrightarrow H^{n-2}(X) \xrightarrow{\cup[V]} H^n(X) \longrightarrow H^n(X - V) \xrightarrow{\text{Res}} PH^{n-1}(V) \longrightarrow 0.$$

Since $H^n(X) = H^{n/2,n/2}(X)$, we see that (5.16) gets replaced with the isomorphism

$$(S/J(f))_{k\beta-\beta_0} \simeq PH^{n-k,k-1}(V), \qquad k \neq (n/2)+1,$$

and, for $k = (n/2)+1$, the exact sequence

$$0 \to H^{n-2}(X) \to H^n(X) \to (S/J(f))_{((n/2)+1)\beta-\beta_0} \to PH^{(n/2)-1,n/2}(V) \to 0$$

(see [**BC**]). Note that we only have to worry about the exact sequence when n is even, which occurs when the hypersurface V is odd-dimensional (e.g., a Calabi-Yau threefold).

It is also possible to avoid the exact sequence by changing $J(f)$ slightly. Instead of the Jacobian ideal, consider the ideal quotient

$$J_1(f) = \langle x_1 \partial f/\partial x_1, \dots, x_r \partial f/\partial x_r \rangle : x_1 \cdots x_r.$$

Then [**BC**] shows that if $V \subset X$ is *nondegenerate* (meaning that V is quasi-smooth and meets all torus orbits in X quasi-transversely), then we have isomorphisms

$$(S/J_1(f))_{k\beta-\beta_0} \simeq PH^{n-k,k-1}(V)$$

for all k. To explain the appearance of the ideal $J_1(f)$, note that the operators $x_i\partial/\partial x_i$ are invariant under the torus action, and taking the ideal quotient is necessary in order that $J(f) \subset J_1(f)$. Also, $J_1(f)$ appears naturally in studying gauged linear sigma models (see [**MP1**]).

Hence, using either $J(f)$ (when n is odd) or $J_1(f)$ (for any n), the Griffiths-Dwork method for computing the Picard-Fuchs equation of an ample hypersurface can be applied to any projective simplicial toric variety.

A drawback of this method is the insistence that the hypersurface be ample and that the ambient space be simplicial. For instance, in the situation of the Batyrev mirror construction, recall from Section 4.1 that we start with a reflexive polytope Δ and an ample anticanonical hypersurface $\bar{V} \subset \mathbb{P}_\Delta$. Then, taking a maximal projective subdivision Σ of the normal fan of Δ, we get the minimal Calabi-Yau $V \subset X_\Sigma$ which is the proper transform of $\bar{V}$. However, $\mathbb{P}_\Delta$ may fail to be simplicial, although $\bar{V}$ is ample, and V may fail to be ample, although X_Σ is simplicial. Thus it can happen that the Griffiths-Dwork method applies to neither $\bar{V}$ nor V.

5.4. Examples

We will present two examples which illustrate how the methods of the previous section can be used to compute Picard-Fuchs equations.

Example 5.4.1. We first consider the quintic mirror family, following the spirit of [**Morrison1**]. By (2.4), the quintic mirror is defined by the equation

$$(5.19) \qquad x_1^5 + x_2^5 + x_3^5 + x_4^5 + x_5^5 + \psi\, x_1 x_2 x_3 x_4 x_5 = 0,$$

modulo the action of the group G from (2.3). More precisely, as we saw in Section 4.2, if $W \subset \mathbb{P}^4$ is defined by the above equation, then the mirror V° is a resolution of the quotient $W/G \subset \mathbb{P}^4/G = \mathbb{P}_\Delta^\circ$, where Δ° has the vertices (4.13). Furthermore, by equation (4.8) from Chapter 4, we have

$$(5.20) \qquad \begin{aligned} h^{2,1}(V^\circ) &= l(\Delta^\circ) - 5 - \sum_{\Gamma^\circ} l^*(\Gamma^\circ) + \sum_{\Theta^\circ} l^*(\Theta^\circ) l^*(\widehat{\Theta}^\circ) \\ &= h^{2,1}_{\text{poly}}(V^\circ) + \sum_{\Theta^\circ} l^*(\Theta^\circ) l^*(\widehat{\Theta}^\circ), \end{aligned}$$

where Γ° refers to codimension 1 faces of Δ°, Θ° refers to codimension 2 faces of Δ°, and $\widehat{\Theta}^\circ$ refers to the 1-dimensional face of Δ dual to Θ°. An easy computation shows that the only lattice points of Δ° are the five vertices listed in (4.13), together with $(0,0,0,0)$. It follows that

$$h^{2,1}(V^\circ) = h^{2,1}_{\text{poly}}(V^\circ) = 6 - 5 - 0 + 0 = 1,$$

so that V° has 1-dimensional complex moduli which comes from varying the ψ in (5.19). Recall from Chapter 2 that $z = \psi^{-5}$ is a local parameter in the moduli space (the variable $x = \psi^{-5}$ from Chapter 2 is now called z).

Since V° is a non-ample divisor in a toric blowup of $\mathbb{P}_{\Delta^\circ}$, we can't apply the Griffiths-Dwork method to V°. Fortunately, W/G is ample and $\mathbb{P}_{\Delta^\circ}$ is simplicial, so that we will instead let $V^\circ = W/G$ since the Picard-Fuchs equations do not change under resolution. It follows that we can apply the methods of Section 5.3, provided we use G-invariant forms $P\Omega_0/f^k$ on $\mathbb{P}^4 - W$ to represent cohomology classes on V°.

We begin with the G-invariant 3-form $\omega_1 = \psi\Omega_0/f$, which appeared in (2.7) in Chapter 2. Our goal is to find a relation between $\omega_1, \delta\omega_1, \ldots, \delta^4\omega_1$, where $\delta = z\, d/dz = -\frac{1}{5}\psi\, d/d\psi$. For this purpose, we need a basis for cohomology. Consider the G-invariant forms

$$\omega_j = \frac{(-1)^{j-1}\psi^j (\prod_i x_i)^{j-1}\Omega_0}{f^j}, \qquad j = 1,\ldots,4.$$

The factor of ψ^j has been inserted so that ω is invariant under the substitution $x_1 \mapsto \mu^{-1}x_1$, $\psi \mapsto \mu\psi$ which induces the isomorphism $V_\psi^\circ \simeq V_{\mu\psi}^\circ$ (as discussed in Section 2.2). In particular, this implies that the Picard-Fuchs equations for ω_1 can be rewritten in terms of $z = \psi^{-5}$.

Then $\mathrm{Res}(\omega_j) \in F^{4-j}H^3(V^\circ, \mathbb{C})$, and an easy Gröbner basis calculation shows that this class has nonzero projection into $H^{4-j,j-1}(V^\circ)$. Since $h^{3,0} = h^{2,1} = h^{1,2} = h^{0,3} = 1$, it follows that the residues of the ω_j, $j = 1,2,3,4$, are a basis for

$H^3(V^\circ)$. Thus any G-invariant form $P\Omega_0/f^k$ is equivalent modulo exact forms and forms with lower order poles to a multiple of ω_k for $k = 1, 2, 3, 4$, and any form with poles of any order is equivalent to a linear combination of these four forms modulo exact forms. The pole order reduction formula (5.15) can be invoked using Gröbner basis techniques to perform these calculations.

We will apply this process to $\omega_1, \delta\omega_1, \ldots, \delta^4\omega_1$. Since

$$(5.21) \qquad \delta\omega_j = -\frac{j}{5}(\omega_j + \omega_{j+1})$$

for any j, it follows easily that

$$(5.22) \qquad \begin{pmatrix} \omega_1 \\ \delta\omega_1 \\ \delta^2\omega_1 \\ \delta^3\omega_1 \end{pmatrix} = \begin{pmatrix} 1 & 0 & 0 & 0 \\ -\frac{1}{5} & -\frac{1}{5} & 0 & 0 \\ \frac{1}{25} & \frac{3}{25} & \frac{2}{25} & 0 \\ -\frac{1}{125} & -\frac{7}{125} & -\frac{12}{125} & -\frac{6}{125} \end{pmatrix} \begin{pmatrix} \omega_1 \\ \omega_2 \\ \omega_3 \\ \omega_4 \end{pmatrix}.$$

This takes care of $\omega_1, \delta\omega_1, \delta^2\omega_1, \delta^3\omega_1$, and as for $\delta^4\omega_1$, repeatedly applying the Gröbner basis calculation and reduction of pole order to $\delta^4\omega_1$ gives

$$\delta^4\omega_1 = \frac{5z}{1 + 5^5 z}\omega_1 + \frac{75z}{1 + 5^5 z}\omega_2 + \frac{250z}{1 + 5^5 z}\omega_3 + \frac{300z}{1 + 5^5 z}\omega_4$$

modulo exact forms. If we write this as

$$\delta^4\omega_1 = \left(\frac{5z}{1 + 5^5 z} \quad \frac{75z}{1 + 5^5 z} \quad \frac{250z}{1 + 5^5 z} \quad \frac{300z}{1 + 5^5 z} \right) \begin{pmatrix} \omega_1 \\ \omega_2 \\ \omega_3 \\ \omega_4 \end{pmatrix},$$

then (5.22) easily gives the Picard-Fuchs equation (5.12).

Example 5.4.2. We next explain how the Griffiths-Dwork method works for the mirror family of a Calabi-Yau hypersurface in a blowup of $\mathbb{P}(1, 1, 2, 2, 2)$. As we will soon see, the mirror has two-dimensional complex moduli, which means that the Picard-Fuchs equations will be partial differential equations.

For the standard fan of $\mathbb{P}(1, 1, 2, 2, 2)$, the generators of the 1-dimensional cones are

$$(5.23) \qquad \begin{aligned} \{(-1, -2, -2, -2), (1, 0, 0, 0), (0, 1, 0, 0), \\ (0, 0, 1, 0), (0, 0, 0, 1)\}. \end{aligned}$$

These are the vertices of a reflexive polytope $\Delta^\circ \subset N_{\mathbb{R}}$. Using the methods discussed in Section 4.2, one easily computes that $\mathbb{P}_{\Delta^\circ}$ is the quotient of $\mathbb{P}(1, 1, 2, 2, 2)$ by a finite group H.

An easy computation shows that the only lattice points of Δ° are the five listed above, together with $(0, -1, -1, -1) = \frac{1}{2}(-1, -2, -2, -2) + \frac{1}{2}(1, 0, 0, 0)$ and $(0, 0, 0, 0)$. It follows from equation (5.20) that

$$h^{2,1}(V^\circ) = h^{2,1}_{\text{poly}}(V^\circ) = 2,$$

so that V° has two-dimensional complex moduli. Furthermore, they are polynomial moduli, which means that we can detect them by varying the defining equation of V°.

As in the previous example, we can assume that V° is an anticanonical hypersurface in $\mathbb{P}_{\Delta^\circ}$. Then V° is ample and $\mathbb{P}_{\Delta^\circ}$ is simplicial, so that we can compute Picard-Fuchs equations using the Griffiths-Dwork method.

Since Δ° is the convex hull of the points (5.23), we compute the polar polytope Δ to be the convex hull of the points

$$(5.24) \qquad \begin{aligned} \{&(-1,-1,-1,-1), (7,-1,-1,-1), (-1,3,-1,-1), \\ &(-1,-1,3,-1), (-1,-1,-1,3)\}. \end{aligned}$$

We denote the vertices of Δ by v_i in the order given. The v_i generate the 1-dimensional cones of the fan of $\mathbb{P}_{\Delta^\circ}$. Note that $\mathbb{P}_\Delta = \mathbb{P}(1,1,2,2,2)$, and if Σ is the unique maximal projective subdivision of the normal fan of Δ, we see that X_Σ is a single blowup of $\mathbb{P}(1,1,2,2,2)$ obtained by inserting the edge $(0,-1,-1,-1)$ into the fan determined by (5.23). The family V° is now seen to be the family described in the first sentence of this example.

The homogeneous coordinate ring of $\mathbb{P}_{\Delta^\circ}$ is just $\mathbb{C}[x_1,\ldots,x_5]$ and is graded by $A_3(\mathbb{P}_{\Delta^\circ})$, which in this case contains torsion. We need to determine the monomials of degree $\beta_0 = \sum_{i=1}^5 \deg(x_i) \in A_3(\mathbb{P}_{\Delta^\circ})$, which in concrete terms are the H-invariant monomials of degree 8 (where $\mathbb{P}_{\Delta^\circ} = \mathbb{P}(1,1,2,2,2)/H$). The easiest way to do this is using the isomorphism (3.8), which implies that the desired monomials are given by

$$\prod_i x_i^{\langle v,v_i\rangle + 1}, \qquad v \in \Delta^\circ \cap N.$$

Since we already know the points of $\Delta^\circ \cap N$, we get the monomials

$$x_1^8, \ x_2^8, \ x_3^4, \ x_4^4, \ x_5^4, \ x_1^4 x_2^4, \ x_1 x_2 x_3 x_4 x_5.$$

These are just the H-invariant monomials of weight 8 in the weighted projective space. From the point of view of weighted projective spaces, this is precisely the Greene-Plesser orbifold construction. In general, one can show that the Batyrev mirror of a weighted projective space which admits a Fermat hypersurface always coincides with the Greene-Plesser construction (see [**BK2**]).

So the general anticanonical hypersurface in $\mathbb{P}_{\Delta^\circ}$ is given by linear combinations of these H-invariant monomials. Using the torus action to rescale the variables x_i, we can put the equation of an anticanonical hypersurface in the form

$$(5.25) \qquad f = z_2\, x_1^8 + x_2^8 + z_1\, x_3^4 + x_4^4 + x_5^4 + x_1 x_2 x_3 x_4 x_5 + x_1^4 x_2^4 = 0,$$

where z_1, z_2 are parameters. This choice has been made so that $z_1 = z_2 = 0$ is a maximally unipotent boundary point. We will let δ_i denote $z_i\, d/dz_i$. The Hodge numbers are $h^{3,0} = h^{0,3} = 1$ and (as computed above) $h^{2,1} = h^{1,2} = 2$. We choose the basis

$$\begin{aligned}
\omega_1 &= & &\frac{\Omega_0}{f} \\[4pt]
\omega_2 &= & \delta_1 \omega_1 &= -\frac{z_1 x_3^4 \Omega_0}{f^2} \\[4pt]
\omega_3 &= & \delta_2 \omega_1 &= -\frac{z_2 x_1^8 \Omega_0}{f^2} \\[4pt]
\omega_4 &= & \delta_1^2 \omega_1 &= \omega_2 + \frac{2z_1^2 x_3^8 \Omega_0}{f^3} \\[4pt]
\omega_5 &= & \delta_1 \delta_2 \omega_1 &= \frac{2z_1 z_2 x_1^8 x_3^4 \Omega_0}{f^3} \\[4pt]
\omega_6 &= & \delta_1^2 \delta_2 \omega_1 &= \omega_5 - \frac{6z_1^2 z_2 x_1^8 x_3^8 \Omega_0}{f^4}
\end{aligned}$$

We can now generate Picard-Fuchs equations by expressing various $\delta_i^j \omega_1$ in terms of this basis.

For example, we compute

$$\delta_2^2 \omega_1 = \omega_3 + \frac{2z_2^2 x_1^{16} \Omega_0}{f^3}$$

and accordingly attempt to reduce x_1^{16} to an expression involving only x_3^8 and $x_1^8 x_3^4$ using the Jacobian ideal. Using Gröbner basis techniques, we compute the relation

$$z_2(1 - 4z_2)x_1^{16} - z_1^2 x_3^8 + 4z_1 z_2 x_1^8 x_3^4 \in J(f).$$

In fact, it is easy to verify that

$$z_2(1 - 4z_2)x_1^{16} - z_1^2 x_3^8 + 4z_1 z_2 x_1^8 x_3^4 = -\frac{1}{16}\left(\sum_{i=i}^{3} A_i x_i \frac{\partial f}{\partial x_i}\right),$$

where we have put

$$A_1 = 2(4z_2 - 1)x_1^8 + x_1 x_2 x_3 x_4 x_5 - 4x_1^4 x_2^4$$
$$A_2 = 2x_1^8$$
$$A_3 = 4z_1 x_3^4 - 16z_2 x_1^8 - x_1 x_2 x_3 x_4 x_5$$

and combine this with (5.18) to reduce the pole order of $\delta_2^2 \omega_1$. Repeating this process to reduce the pole order further, we arrive at the Picard-Fuchs equation

(5.26) $$(1 - 4z_2)\delta_2^2 \omega_1 - z_2(\delta_1^2 - 4\delta_1\delta_2 - \delta_1 + 2\delta_2)\omega_1 = 0.$$

This example (using a slightly different form of equation (5.26)) was worked out in detail in [**CdFKM**] by different methods (see also [**HKTY1**]). One can prove that the point $z_1 = z_2 = 0$ is a maximally unipotent boundary point, as has been remarked earlier. One needs another Picard-Fuchs equation to verify this in detail. For now, we content ourselves with the observation that the form of equation (5.26) leads quickly to $N_2^2 = 0$, so N_2 by itself cannot define the monodromy weight filtration of a maximally unipotent boundary point. But any positive linear combination $a_1 N_1 + a_2 N_2$ satisfies $(a_1 N_1 + a_2 N_2)^3 \neq 0$, and in fact defines the right kind of monodromy weight filtration.

We will return to this example in Example 5.5.2.1, and again in the next chapter. Note that a smooth compactification of the moduli space with normal crossings boundary necessarily has many components. The one constructed in [**CdFKM**] nonetheless has only one maximally unipotent boundary point. The other intersection points of pairs of boundary divisors are not maximally unipotent.

5.5. Hypergeometric Equations

We now describe the hypergeometric systems of Gelfand, Kapranov and Zelevinsky and discuss some examples which arise naturally in mirror symmetry.

5.5.1. The $\mathcal{A}$-System. Let $\mathcal{A} = \{v_1, \ldots, v_k\} \subset \mathbb{Z}^{n+1}$ be a collection of $k > n + 1$ points lying in an integral affine hyperplane. Denote the coordinates of v_i by v_{ij}. Introduce variables λ_i for each $v_i \in \mathcal{A}$, and fix a vector $\hat{\beta} = (\hat{\beta}_1, \ldots, \hat{\beta}_{n+1}) \in \mathbb{C}^{n+1}$. To this data, we can associate the $\mathcal{A}$-*system of hypergeometric equations*, a system of differential equations in the λ_i. We let ∂_i denote the differential operator $\partial/\partial\lambda_i$, and we use δ_i for the logarithmic derivative $\lambda_i \partial_i = \lambda_i \partial/\partial\lambda_i$.

The $\mathcal{A}$-system is comprised of two types of differential operators, denoted Z_j and $\square_\ell$. The operators Z_j are defined by

(5.27) $$Z_j = \left(\sum_i v_{ij}\delta_i\right) - \hat{\beta}_j, \qquad j = 1, \ldots, n+1.$$

To describe the $\square_\ell$, we need some more notation. We let Λ be the lattice of relations among the elements of $\mathcal{A}$:

$$(5.28) \qquad \Lambda = \left\{ \ell = (\ell_i) \in \mathbb{Z}^k : \sum_i \ell_i v_i = 0 \right\}.$$

By our assumption on $\mathcal{A}$, we have $\Lambda \neq \emptyset$. Then define

$$(5.29) \qquad \square_\ell = \prod_{\ell_i > 0} \partial_i^{\ell_i} - \prod_{\ell_i < 0} \partial_i^{-\ell_i}, \qquad \ell \in \Lambda.$$

DEFINITION 5.5.1. *The $\mathcal{A}$-system of hypergeometric equations is the system of differential equations $Z_j \Phi = 0$, $1 \leq j \leq n+1$ and $\square_\ell \Phi = 0$, $\ell \in \Lambda$, where Z_j and $\square_\ell$ are respectively defined in (5.27) and (5.29).*

This system is frequently referred to as the GKZ system, after the work of Gelfand, Kapranov, and Zelevinsky [**GKZ1**].

5.5.2. Calabi-Yau Toric Hypersurfaces. Now let $\Delta \subset M_\mathbb{R}$ be a reflexive polytope of dimension n, and let X_Σ be the toric variety associated to a maximal projective subdivision Σ of the normal fan of Δ. We consider the family of anticanonical hypersurfaces in X_Σ, which are minimal Calabi-Yau subvarieties by Proposition 4.1.3. Our task is to find Picard-Fuchs equations for this family. As we've already noted, these equations are independent of the choice of subdivision, and coincide with those for the family $\mathcal{V}$ of anticanonical hypersurfaces in $\mathbb{P}_\Delta$, which is a family of Calabi-Yau varieties by Proposition 4.1.3. So we need only study this family of Calabi-Yau hypersurfaces in $\mathbb{P}_\Delta$.

Pick an ordering of the points m_i of $\Delta \cap M$ such that $m_0 = 0$. Since Δ is reflexive, we know that the sections of the anticanonical bundle can be identified with the Laurent polynomials

$$(5.30) \qquad f = \sum_i \lambda_i t^{m_i} = \lambda_0 + \sum_{i>0} \lambda_i t^{m_i} \in L(\Delta),$$

where the λ_i are parameters. Then the family of Calabi-Yau varieties $V \subset \mathbb{P}_\Delta$ is defined by the equation $f = 0$ for $f \in L(\Delta)$.

To make contact with hypergeometric equations, let

$$(5.31) \qquad \mathcal{A} = (\Delta \cap M) \times \{1\} \subset M \times \mathbb{Z} \simeq \mathbb{Z}^{n+1},$$

and put $\hat{\beta} = (0, \dots, 0, -1)$. Note that $\ell \in \Lambda$ implies both $\sum_i \ell_i m_i = 0$ and $\sum_i \ell_i = 0$. The latter implies that the $\square_\ell$ from (5.29) are homogeneous differential operators.

We claim that equations of the $\mathcal{A}$-system are Picard-Fuchs equations, that is, the periods of certain differentials on V satisfy the $\mathcal{A}$-system of hypergeometric equations.

Following [**Batyrev3**], we use the n-form

$$(5.32) \qquad \omega = \frac{1}{f}\,\eta, \qquad \eta = \frac{dt_1}{t_1} \wedge \dots \wedge \frac{dt_n}{t_n},$$

where $(t_1, \dots, t_n)$ are coordinates for the torus $T \subset \mathbb{P}_\Delta$. For $\ell = (\ell_i) \in \Lambda$ and I a subset of indices, we put $\ell_I = \sum_{i \in I} \ell_i$ and compute

$$(5.33) \qquad \left(\prod_{i \in I} \partial_i^{\ell_i} \right) \omega = \frac{(-1)^{\ell_I} (\ell_I)! \prod_{i \in I} t^{\ell_i m_i}}{f^{\ell_I + 1}}\,\eta.$$

Since $\ell \in \Lambda$ implies $\sum_{\ell_i > 0} \ell_i m_i = \sum_{\ell_i < 0} (-\ell_i) m_i$, we have

$$\prod_{\ell_i > 0} t^{\ell_i m_i} = \prod_{\ell_i < 0} t^{-\ell_i m_i},$$

and then $\square_\ell \omega = 0$ easily follows from (5.33) and $\sum_{\ell_i > 0} \ell_i = \sum_{\ell_i < 0}(-\ell_i)$. If Φ is any period of ω, it follows immediately that $\square_\ell \Phi = 0$.

It is equally easy to show that $Z_j \Phi = 0$ for $1 \le j \le n+1$. Consider the $(n-1)$-form

$$\eta_j = \frac{dt_1}{t_1} \wedge \cdots \wedge \frac{\widehat{dt_j}}{t_j} \wedge \cdots \wedge \frac{dt_n}{t_n}, \qquad 1 \le j \le n.$$

One computes that

$$d\left(\frac{1}{f}\,\eta_j\right) = \frac{(-1)^j}{f^2}\left(\sum_i \lambda_i m_{ij} t^{m_i}\right)\eta,$$

where m_{ij} are the coordinates of m_i. Since $\hat{\beta}_j = 0$ for $1 \le j \le n$, this easily implies that $Z_j \omega$ is exact for $j < n+1$, and therefore $Z_j \Phi = 0$ for these j. As for $j = n+1$, we have $\hat{\beta}_{n+1} = -1$, and since the last coordinate of $(m_i, 1) \in \mathcal{A}$ is always 1, we obtain

$$Z_{n+1}\omega = \left(\sum_i \lambda_i \partial_i\right)\omega + \omega = -\frac{\sum_i \lambda_i t^{m_i}}{f^2}\,\eta + \omega = 0,$$

as desired.

The form ω in (5.32) is defined on $T - Z_f$, where $T \subset \mathbb{P}_\Delta$ is the torus and $Z_f = V \cap T$ is the affine hypersurface determined by $f \in L(\Delta)$. Then the residue of ω is an $(n-1)$-form on Z_f. The affine point of view is explained in more detail in [**Batyrev3**]. However, we want Picard-Fuchs equations for periods on V and not just the affine part Z_f. This by done by using the homogeneous coordinate ring $\mathbb{C}[x_1, \ldots, x_r]$ of $\mathbb{P}_\Delta$. The x_i relate to the t_j via the formula

$$t_j = \prod_{i=1}^r x_i^{\langle e_j, \rho_i \rangle},$$

and under this substitution, we have

$$\Omega_0 = x_1 \cdots x_r\,\eta,$$

where Ω_0 is defined in (5.17). Proofs of these facts can be found in [**BC**]. The formula for Ω_0 is easy to understand, since the form η has simple poles on the components $D_1, \ldots, D_r$ of $\mathbb{P}_\Delta - T$. On the other hand, the section $x_1 \cdots x_r$ of $\mathcal{O}_{\mathbb{P}_\Delta}\left(\sum_{j=1}^r D_j\right)$ corresponding to $0 \in \Delta$ vanishes simply on each toric divisor D_j, and $\Omega_0 = x_1 \cdots x_r\,\eta$ follows since $\widehat{\Omega}^n_{\mathbb{P}_\Delta} \simeq \mathcal{O}_{\mathbb{P}_\Delta}\left(-\sum_{j=1}^r D_j\right)$. Then, under the above change of variables, we have

$$\omega = \frac{\eta}{f(t_1, \ldots, t_n)} = \frac{x_1 \cdots x_r\,\eta}{x_1 \cdots x_r\,f(t_1, \ldots, t_n)} = \frac{\Omega_0}{f(x_1, \ldots, x_r)},$$

where $f(x_1, \ldots, x_r)$ is the homogenization of $f(t_1, \ldots, t_n)$ as defined in (3.8). Thus ω is one of the forms considered in Section 5.3, so that its residue is a holomorphic n-form on V. We conclude that the $\mathcal{A}$-hypergeometric equations are Picard-Fuchs equations. One technicality is that $\mathbb{P}_\Delta$ might not be simplicial, but once we pull back to a resolution, there is no difficulty.

The $\mathcal{A}$-hypergeometric equations are defined on the space $L(\Delta)$, which has the λ_i as coordinates. However, we are interested in differential equations on the moduli space, which is a quotient of $L(\Delta)$ (more precisely, the quotient gives the polynomial part of the moduli). This topic will be discussed in detail in Chapter 6, so for now we will content ourselves with two of the more obvious aspects of moduli.

First, we have the natural action of the torus T on $L(\Delta)$ given by

$$\nu \cdot f = \nu \cdot \sum_i \lambda_i t^{m_i} = \sum_i \nu^{m_i} \lambda_i t^{m_i}, \quad \nu \in T.$$

Since this action is induced by the natural action of T on itself, it follows that for each $\nu \in T$, $f \in L(\Delta)$ and $\nu \cdot f$ define isomorphic hypersurfaces in $\mathbb{P}_\Delta$, so that f and $\nu \cdot f$ give the same point in moduli.

The second action we have is the $\mathbb{C}^*$ action given by $f \mapsto c \cdot f$ for $c \in \mathbb{C}^*$. Since rescaling gives the same hypersurface, f and $c \cdot f$ also give the same moduli point. Putting these all together, we get an action of $T \times \mathbb{C}^*$ on $L(\Delta)$ such that the quotient maps to the moduli space.

Since $\omega = \eta/f$ is determined by $f \in L(\Delta)$, $T \times \mathbb{C}^*$ acts on the period integrals of ω. One can show that the equations $Z_j \Phi = 0$, $1 \leq j \leq n$, are the infinitesimal form of the invariance of the periods under the action of T, and similarly, $Z_{n+1}\omega = 0$ follows from invariance under rescaling. See [**HLY1**] for the details.

If we work with the subset $(\mathbb{C}^*)^{\Delta \cap M} \subset L(\Delta)$ where $\lambda_i \neq 0$ for all i, the quotient $(\mathbb{C}^*)^{\Delta \cap M}/(T \times \mathbb{C}^*)$ is easy to describe since it is a quotient of tori. This is where the set Λ from (5.28) comes into play. Choose a basis $\ell_j = (\ell_{ji})$ of Λ, and define functions

$$z_j = \prod_i \lambda_i^{\ell_{ji}}.$$

It is easy to check that the z_j are invariant under the action of $T \times \mathbb{C}^*$ and give coordinates for the quotient. One of our goals will be to find a basis for Λ such that the z_j will be local coordinates on the complex moduli space near the maximally unipotent boundary point under study. Then the periods which are analytic near the maximally unipotent boundary point will be analytic in the z_j. We will return to this point in Chapter 6.

We will want to write the $\mathcal{A}$-hypergeometric equations in terms of the variables z_j. But first, let us put them in a more useful form for dealing with moduli. The problem is that ω is not invariant under our $\mathbb{C}^*$ action $\lambda_i \mapsto c\lambda_i$. For this reason, we will choose the form

$$(5.34) \qquad \widetilde{\omega} = \lambda_0 \omega = \frac{\lambda_0}{f}\,\eta,$$

which now is homogeneous of degree 0 in the λ_i. Since the periods of ω satisfy the $\mathcal{A}$-system, we see that the periods $\widetilde{\Phi}$ of $\widetilde{\omega}$ satisfy the equations

$$Z_j \lambda_0^{-1} \widetilde{\Phi} = 0, \quad \Box_\ell \lambda_0^{-1} \widetilde{\Phi} = 0$$

for $1 \leq j \leq n+1$ and $\ell \in \Lambda$.

The equations $Z_j \lambda_0^{-1}\widetilde{\Phi} = 0$ are the infinitesimal form of the invariance of $\widetilde{\Phi}$ under the action of $T \times \mathbb{C}^*$, so that once we take the quotient by $T \times \mathbb{C}^*$, our remaining task is to change the equations $\Box_\ell \lambda_0^{-1} = 0$ into z_j coordinates. We do this by repeated use of the identity of differential operators

$$(5.35) \qquad \delta_i \lambda_i^p = \lambda_i^p (\delta_i + p).$$

We illustrate the method for the quintic mirror. Here, the relevant generator for Λ is $\ell = (1,1,1,1,1,-5)$ (recall that Λ consists of relations among the elements of $\mathcal{A} - (\Delta \cap M) \times \{1\}$), yielding the coordinate

$$(5.36) \qquad z = \lambda_1 \cdots \lambda_5/\lambda_0^5.$$

If we homogenize the equation (5.30) in this case, we get

$$\lambda_1\, x_1^5 + \lambda_2\, x_2^5 + \lambda_3\, x_3^5 + \lambda_4\, x_4^5 + \lambda_5\, x_5^5 + \lambda_0\, x_1 x_2 x_3 x_4 x_5 = 0.$$

This gives equation (2.4) by putting $\lambda_1 = \cdots = \lambda_5 = 1$ and $\lambda_0 = \psi$, and then the above moduli coordinate is $z = \psi^{-5}$, agreeing with the coordinate z chosen in Chapter 2 (recall that we now use z instead of x for the moduli coordinate). We must convert the $\mathcal{A}$-hypergeometric equation

$$\Box_\ell \lambda_0^{-1} = (\partial_1 \cdots \partial_5 - \partial_0^5)\lambda_0^{-1} = 0$$

to an equation in z. Multiplying $\Box_\ell \lambda_0^{-1}$ on the left by $\lambda_0 \cdots \lambda_5 = z\lambda_0^6$ gives

$$\delta_1 \cdots \delta_5 - z\lambda_0^6 \partial_0^5 \lambda_0^{-1} = \delta_1 \cdots \delta_5 - z\lambda_0^5 \delta_0 (\lambda_0^{-1}\delta_0)^4 \lambda_0^{-1},$$

and repeated application of (5.35) from right to left yields the equation

$$\delta_1 \cdots \delta_5 - z(\delta_0 - 5)(\delta_0 - 4) \cdots (\delta_0 - 1) = 0.$$

Now let δ denote $z\, d/dz$. From (5.36) we obtain the identities

$$\delta_1 = \cdots = \delta_5 = \delta, \ \delta_0 = -5\delta$$

when we restrict attention to functions of z alone. We thus finally arrive at

$$(5.37) \qquad\qquad \delta^5 + z(5\delta + 5)(5\delta + 4) \cdots (5\delta + 1) = 0.$$

We immediately see the limitation of this method—the fifth order operator (5.37) does not generate the Picard-Fuchs ideal, since we are missing the fourth order operator (5.12).

In general, the system of differential equations in the coordinates z_i arising from the $\mathcal{A}$-system contains the periods among its solutions, but there will typically be other solutions. The problem is to identify a larger system of differential equations which vanish only on the periods. There are several strategies for doing this.

One method for finding the full set of Picard-Fuchs equations is to begin with the hypergeometric equations from the $\mathcal{A}$-system and augment them with other Picard-Fuchs equations. Typically, this is done by adding in lower order equations found using the Griffiths-Dwork method. See [**BKK**] for some examples.

Another common method is to factor a Picard-Fuchs equation arising from the $\mathcal{A}$-system. As explained in Section 5.1.2, the resulting lower order operator D is Picard-Fuchs provided it annihilates all periods. Fortunately, there are many cases where it suffices to check that D annihilates just *one* period y. Suppose, for example, that $D \cdot y = 0$ and that analytically continuing y over the whole moduli space gives a basis y_i of solutions of the Picard-Fuchs equations. Then $D \cdot y = 0$ implies $D \cdot y_i = 0$ for all i. It follows that D annihilates all periods and hence is a Picard-Fuchs operator.

For an example of how this works, consider the quintic mirror. We have the fifth order equation (5.37), and commuting z to the right, we obtain

$$\delta^5 + (5\delta)(5\delta - 1) \cdots (5\delta - 4)z = 0,$$

and so factoring out a δ on the left gives the fourth order equation

$$(5.38) \qquad\qquad \delta^4 + 5(5\delta - 1)(5\delta - 2)(5\delta - 3)(5\delta - 4)z = 0,$$

which is equivalent to (5.12). A systematic method for this sort of factoring is described in [**HKTY1**]. To prove that this is a Picard-Fuchs equation, consider

$$y_0(z) = \sum_{n=0}^{\infty} \frac{(5n)!}{(n!)^5}(-1)^n z^n$$

from (2.11). In (6.48) we will see that y_0 really is a period, and one easily checks that y_0 is a solution of of (5.38). Furthermore, [**CdGP**] shows that under analytic continuation, y_0 gives a basis of periods. It follows that (5.38) is a Picard-Fuchs equation for the quintic mirror.

In order to implement this strategy in general, one needs to know an explicit period satisfying the $\mathcal{A}$-system. One such period, taken from [**Batyrev3**, Prop. 14.6], will be given in equation (6.48) in Chapter 6. In fact, the period y from (6.48) leads to a slightly different method for generating Picard-Fuchs equations. Here, the idea is to deduce the differential equations from recurrence relations among the coefficients of y. This technique is used for several one parameter families in [**BvS**]. One drawback of this method is that it is not known in general if analytic continuation of (6.48) leads to a basis of solutions, though one expects this to be the case for Calabi-Yau threefolds. See [**BvS**] for more details.

Unfortunately, factorization might not give all of the missing Picard-Fuchs operators, because the $\mathcal{A}$-system has a second source of incompleteness due to automorphisms. The GKZ equations $Z_j\Phi = 0$ reflect invariance under the torus of $\mathbb{P}_\Delta$, but to get the moduli of $f = 0$, we need to mod out by the action of $\mathrm{Aut}(\mathbb{P}_\Delta)$. We saw in Section 3.6 that $\mathrm{Aut}(\mathbb{P}_\Delta)$ includes automorphisms arising from roots. These automorphisms render some of the coordinates λ_i ineffective (the ones associated to interior points of facets of Δ). Instead of adding more equations, we can simply omit these λ_i, remembering that there may be residual discrete identifications of the moduli space (for an example, see [**CFKM**]). We will adopt this approach in Chapter 6 when we define the *simplified moduli space* $\overline{\mathcal{M}}_{\mathrm{simp}}$ of the toric hypersurface V. A closely related method is to extend the $\mathcal{A}$-system by the infinitesimal form of the extra automorphisms, as described in [**HKTY1, HLY1**]. These complement the equations $Z_j\Phi = 0$, which (as just mentioned) are the infinitesimal form of the automorphisms induced by the torus.

The solution (6.48) of the GKZ-system mentioned above is one of the many wonderful formulas for solutions of hypergeometric equations. These are generalizations of the classical hypergeometric power series. This is discussed in [**GKZ1, HLY1**], and we will give some explicit examples and solution methods when we discuss the mirror map in Section 6.3.

We illustrate our discussion by returning to an example already considered.

Example 5.5.2.1. In Section 5.4, we discussed the mirror V° of the degree 8 hypersurfaces in $\mathbb{P}(1,1,2,2,2)$. In this case, homogenizing (5.30) leads to an equation of the form

$$\lambda_1\,x_1^8 + \lambda_2\,x_2^8 + \lambda_3\,x_3^4 + \lambda_4\,x_4^4 + \lambda_5\,x_5^4 + \lambda_6\,x_1^4x_2^4 + \lambda_0\,x_1x_2x_3x_4x_5 = 0.$$

These monomials were computed in Example 5.4.2. Using the reflexive polytope Δ° determined by (5.23), one computes that a basis of Λ is given by $\ell_1 = (0,0,1,1,1,1,-4)$ and $\ell_2 = (1,1,0,0,0,-2,0)$. Thus we get moduli coordinates

$$z_1 = \frac{\lambda_3\lambda_4\lambda_5\lambda_6}{\lambda_0^4}, \quad z_2 = \frac{\lambda_1\lambda_2}{\lambda_6^2}.$$

Note that setting $\lambda_0 = \lambda_2 = \lambda_4 = \lambda_5 = \lambda_6 = 1$ leads to $z_1 = \lambda_3$ and $z_2 = \lambda_1$, which agrees with (5.25). Then from

$$\lambda_0\lambda_1\lambda_2 \square_{\ell_2} \lambda_0^{-1} = \lambda_0\lambda_1\lambda_2(\partial_1\partial_2 - \partial_6^2)\lambda_0^{-1} = 0,$$

we obtain by the methods discussed above

$$(5.39) \qquad \delta_{z_2}^2 - z_2(\delta_{z_1} - 2\delta_{z_2})(\delta_{z_1} - 2\delta_{z_2} - 1) = 0,$$

where we have put $\delta_{z_1} = z_1\,\partial/\partial z_1$, $\delta_{z_2} = z_2\,\partial/\partial z_2$ and used the identities

$$(5.40) \qquad \delta_1 = \delta_2 = \delta_{z_2}, \; \delta_6 = \delta_{z_1} - 2\delta_{z_2},$$

which hold on functions of z_1 and z_2. The reader should note that some easy algebra transforms (5.39) into the Picard-Fuchs equation (5.26) found earlier using the Griffiths-Dwork method (the δ_i appearing in (5.26) are the operators denoted here by δ_{z_i}).

A similar computation using (5.40) and the further identities $\delta_3 = \delta_4 = \delta_5 = \delta_{z_1}$, $\delta_0 = -4\delta_{z_1}$ shows that the GKZ equation

$$\lambda_0\lambda_3\lambda_4\lambda_5\lambda_6 \square_{\ell_1} \lambda_0^{-1} = \lambda_0\lambda_3\lambda_4\lambda_5\lambda_6(\partial_3\partial_4\partial_5\partial_6 - \partial_0^4)\lambda_0^{-1} = 0$$

leads to the Picard-Fuchs equation

$$\delta_{z_1}^3(\delta_{z_1} - 2\delta_{z_2}) - (4\delta_{z_1})(4\delta_{z_1} - 1)(4\delta_{z_1} - 2)(4\delta_{z_1} - 3)z_1 = 0.$$

Hence by factorization, we get the third order equation

$$\delta_{z_1}^2(\delta_{z_1} - 2\delta_{z_2}) - 4(4\delta_{z_1} - 1)(4\delta_{z_1} - 2)(4\delta_{z_1} - 3)z_1 = 0,$$

which is equivalent to

$$(5.41) \qquad \delta_{z_1}^2(\delta_{z_1} - 2\delta_{z_2}) - 4z_1(4\delta_{z_1} + 1)(4\delta_{z_1} + 2)(4\delta_{z_1} + 3) = 0.$$

This equation is satisfied when applied to the analytic period (6.50), and as shown in [**CdFKM**], this period analytically continues to a basis of solutions. By the discussion preceding this example, we conclude that (5.41) is in fact a Picard-Fuchs equation.

The Picard-Fuchs equations (5.39) and (5.41) have ranks 3 and 2 respectively and hence have a solution space of dimension at most 6. Since $h^3(V^\circ) = 6$, we have found generators for all of the Picard-Fuchs equations.

5.5.3. Nef Complete Intersections. So far, we've discussed Picard-Fuchs equations and the GKZ system, but we haven't said much about solutions. This will have to wait until Section 6.3, where we will construct the mirror map using solutions of the Picard-Fuchs equations. But we thought it worthwhile to describe here one interesting solution to the GKZ system which arises from a toric complete intersection.

Let $X = \mathbb{P}_\Delta$ be an n-dimensional toric variety corresponding to a reflexive polytope $\Delta \subset M_{\mathbb{R}}$ with normal fan Σ. For ease of exposition, we assume that X is smooth, although this assumption can be weakened somewhat. We consider a generalization of the complete intersections studied in Section 4.3.1.

DEFINITION 5.5.2. *A nef complete intersection on X is a complete intersection $Y = \cap_{i=1}^k Y_i$ such that each Y_i is the zero locus of a general section of a line bundle $\mathcal{L}_i$ generated by global sections, and $-(K_X + \sum_{i=1}^k \mathcal{L}_i)$ is nef.*

Note that the last condition is slightly stronger than saying that $-K_Y$ is nef. We are requiring that

$$\left(K_X + \sum_{i=1}^k \mathcal{L}_i\right) \cdot C \leq 0$$

for all curves $C \subset X$. By the adjunction formula, saying that $-K_Y$ is nef merely requires this condition only for curves $C \subset Y$. The cone of effective curves of Y and X can differ, as we will discuss in Section 6.2.3.

Note that a nef-partition $-K_X = \sum_{i=1}^k E_i$ as defined in Section 4.3.1 automatically defines a nef complete intersection. In this case, $\mathcal{L}_i = \mathcal{O}(E_i)$, so that $-(K_X + \sum_{i=1}^k \mathcal{L}_i)$ is trivial and hence clearly nef.

We now put $\overline{N} = N \oplus \mathbb{Z}^k$ and for $1 \leq i \leq k$ choose $a_{i\rho}$ such that

$$\mathcal{L}_i = \mathcal{O}\big(\textstyle\sum_{\rho \in \Sigma(1)} a_{i\rho} D_\rho\big).$$

Then consider the set

$$(5.42) \qquad \mathcal{A} = \{(v_\rho, a_{1\rho}, \ldots, a_{k\rho}) : \rho \in \Sigma(1)\} \cup \{(0, e_1), \ldots, (0, e_k)\} \subset \overline{N},$$

where e_i is the ith standard basis vector in $\mathbb{C}^k$. Note how this generalizes (5.31), which is the case $k = 1$ and $\mathcal{L}_1 = \mathcal{O}(\sum_\rho D_\rho)$.

Let $r = |\Sigma(1)|$ as usual. For each $\rho \in \Sigma(1)$, we put $w_\rho = (v_\rho, a_{1\rho}, \ldots, a_{k\rho})$, and for $1 \leq i \leq k$ we put $u_i = (0, e_i)$. Then (5.42) can be written as $\mathcal{A} = \{w_\rho,\ \rho \in \Sigma(1);\ u_i,\ 1 \leq i \leq k\}$.

The set $\mathcal{A}$ is related to the smooth toric variety X as follows.

LEMMA 5.5.3. *Let Λ be the lattice of relations among elements of $\mathcal{A}$. Then*

(i) For each $\beta \in H_2(X, \mathbb{Z})$, we have a relation

$$\textstyle\sum_\rho (D_\rho \cdot \beta) w_\rho - \sum_i (c_1(\mathcal{L}_i) \cdot \beta) u_i = 0$$

in $\overline{N}$. Hence β gives an element of the lattice Λ.

(ii) The induced map $H_2(X, \mathbb{Z}) \to \Lambda$ is an isomorphism.

PROOF. First note that the exact sequence (3.2) implies that for all $m \in M$, $\sum_\rho \langle m, v_\rho \rangle [D_\rho] = 0$ in $H^2(X, \mathbb{Z})$. Thus $\sum_\rho \langle m, v_\rho \rangle (D_\rho \cdot \beta) = 0$ when $\beta \in H_2(X, \mathbb{Z})$, and since this holds for all m, we must have $\sum_\rho (D_\rho \cdot \beta) v_\rho = 0$ in N. Using this together with $c_1(\mathcal{L}_i) = \sum_\rho a_{i\rho} D_\rho$ and the definitions of w_ρ and u_i, the desired relation follows easily.

To prove the second part of the lemma, consider the commutative diagram

$$
\begin{array}{ccccccc}
0 & \to & \Lambda & \to & \mathbb{Z}^{\mathcal{A}} & \to & \overline{N} \\
  &   & \downarrow & & \downarrow & & \downarrow \\
0 & \to & H_2(X, \mathbb{Z}) & \to & \mathbb{Z}^{\Sigma(1)} & \to & N
\end{array}
$$

The top row is exact by the definition of Λ, and the bottom row is exact, arising from (3.2) by dualizing. The middle vertical map sends w_ρ to v_ρ and u_i to 0, while the rightmost vertical map is the natural projection. The leftmost vertical map is defined by the commutative diagram. If N' is the image of $\mathbb{Z}^{\Sigma(1)} \to N$, then the image of $\mathbb{Z}^{\mathcal{A}} \to \overline{N}$ is easily seen to be $N' \oplus \mathbb{Z}^k$. From here, a diagram chase shows that $\Lambda \to H_2(X, \mathbb{Z})$ is an isomorphism. One also can check that the inverse of this map is the map $H_2(X, \mathbb{Z}) \to \Lambda$ described in the statement of the lemma. $\qquad\square$

For simplicity, we will now assume that the Kähler cone $K(X)$ is simplicial. We will return to a discussion of issues surrounding this assumption in Section 6.2.3. This implies that we can choose a basis $\beta_1, \ldots, \beta_{r-n}$ of $H_2(X, \mathbb{Z})$ such that the β_i

generate the integral classes $M(X)_\mathbb{Z}$ of the Mori cone $M(X) \subset H^2(X, \mathbb{R})$ generated by homology classes of effective curves on X. For the rest of this section, we fix an ordering of the elements of $\Sigma(1)$, although it will frequently be convenient to continue denoting generators of $\Sigma(1)$ by v_ρ. Fixing such an ordering, we associate variables $\lambda_1, \ldots, \lambda_{r+k}$ to the generators of $\mathcal{A}$ in the order given in (5.42). We will also often denote the variable associated to $w_\rho \in \mathcal{A}$ as λ_ρ.

We put $\overline{T} = \overline{N} \otimes \mathbb{C}^*$, and introduce the $\overline{T}$-invariant variables

$$(5.43) \qquad q_j = \left(\prod_\rho \lambda_\rho^{D_\rho \cdot \beta_j} \right) \Big/ \left(\prod_{i=1}^k \lambda_{r+i}^{c_1(\mathcal{L}_i) \cdot \beta_j} \right), \qquad 1 \le j \le r - n.$$

Let $T_1, \ldots, T_{r-n}$ be the generators of $K(X)$ dual to $\beta_1, \ldots, \beta_{r-n}$ and define

$$q^\beta = \prod_j q_j^{T_j \cdot \beta} \quad \text{for } \beta \in M(X)_\mathbb{Z} \subset H_2(X, \mathbb{Z}).$$

We will see much more of this notation when we discuss quantum cohomology in Chapter 8. We also put $t_j = \log q_j$, so that $q_j = e^{t_j}$.

Inspired by [**Givental4**], we will consider the cohomology-valued expression

$$\tilde{I} = e^{\sum_j t_j T_j} \sum_{\beta \in M(X)_\mathbb{Z}} q^\beta \frac{\left(\prod_{i=1}^k \prod_{m=1}^{c_1(\mathcal{L}_i) \cdot \beta} (c_1(\mathcal{L}_i) + m) \right) \left(\prod_\rho \prod_{m=-\infty}^0 (D_\rho + m) \right)}{\prod_\rho \prod_{m=-\infty}^{D_\rho \cdot \beta} (D_\rho + m)}.$$

The exponential $e^{\sum_j t_j T_j}$ is a polynomial in $H^*(X)[t_1, \ldots, t_{r-n}]$, while the sum can be viewed as an element of $H^*(X)[[q_1, \ldots, q_{r-n}]]$. Note that $c_1(\mathcal{L}_i) \cdot \beta \ge 0$ (since $\mathcal{L}_i$ is generated by global sections) and all but finitely many factors of $D_\rho + m$ cancel from the numerator and denominator. This slightly cumbersome way of defining $\tilde{I}$ is needed to handle the possibility that $D_\rho \cdot \beta < 0$. Since $-(K_X + \sum_{i=1}^k \mathcal{L}_i)$ is nef, it is conjectured that the series for $\tilde{I}$ converges for q_j sufficiently small.

Let $\mathcal{V}$ be the vector bundle $\oplus_{i=1}^k \mathcal{L}_i$, which has Euler class given by $\text{Euler}(\mathcal{V}) = \prod_{i=1}^k c_1(\mathcal{L}_i)$. In Chapter 11, we will see that the cohomology-valued formal function

$$(5.44) \qquad\qquad\qquad I_\mathcal{V} = \text{Euler}(\mathcal{V})\, \tilde{I}$$

plays an important role in Givental's version of the Mirror Theorem for toric complete intersections (see (11.73) and Theorem 11.2.16).

In the special case when $X = \mathbb{P}^4$, $k = 1$ and $\mathcal{V} = \mathcal{L}_1 = \mathcal{O}_{\mathbb{P}^4}(5H)$, the complete intersection is the quintic threefold. Then $\tilde{I}$ is given by

$$(5.45) \qquad\qquad \tilde{I} = e^{t_1 H} \sum_{d=0}^\infty e^{dt_1} \frac{\prod_{m=1}^{5d} (5H + m)}{\prod_{m=1}^d (H + m)^5},$$

and $I_\mathcal{V} = 5H\, \tilde{I}$ becomes

$$(5.46) \qquad\qquad I_\mathcal{V} = e^{t_1 H} 5H \sum_{d=0}^\infty e^{dt_1} \frac{\prod_{m=1}^{5d} (5H + m)}{\prod_{m=1}^d (H + m)^5}.$$

This last formula appeared in Chapter 2 as equation (2.32).

The function $\tilde{I}$ is related to hypergeometric equations as follows.

PROPOSITION 5.5.4. *The formal function $\tilde{I}$ satisfies the $\mathcal{A}$-system associated to* (5.42) *with $\hat{\beta} = \vec{0}$.*

PROOF. In the statement of this proposition, $\hat{\beta}$ of course refers to the vector needed to define the operators Z_i in (5.27). As we discussed in Section 5.5.2, the assertion that $Z_i(\tilde{I}) = 0$ for all i amounts to the $\overline{T}$-invariance of the formal function $\tilde{I}$. This follows immediately, since the q_j were constructed to be $\overline{T}$-invariant, and the λ_i appear in $\tilde{I}$ only through the q_j.

We now consider the operators $\square_\beta$ for $\beta \in H_2(X, \mathbb{Z}) \simeq \Lambda$. By Lemma 5.5.3, β gives the relation $\sum_\rho (D_\rho \cdot \beta) w_\rho - \sum_i (c_1(\mathcal{L}_i) \cdot \beta) u_i = 0$. It follows that $\square_\beta$ can be written as

$$(5.47) \qquad \square_\beta = \prod_{D_\rho \cdot \beta > 0} \partial_\rho^{D_\rho \cdot \beta} \prod_{c_1(\mathcal{L}_i) \cdot \beta < 0} \partial_{r+i}^{-c_1(\mathcal{L}_i) \cdot \beta} - \prod_{D_\rho \cdot \beta < 0} \partial_\rho^{-D_\rho \cdot \beta} \prod_{c_1(\mathcal{L}_i) \cdot \beta > 0} \partial_{r+i}^{c_1(\mathcal{L}_i) \cdot \beta},$$

where $\partial_\rho = \partial/\partial\lambda_\rho$ and $\partial_{r+i} = \partial/\partial\lambda_{r+i}$. We need to show that $\square_\beta \tilde{I} = 0$,

As usual, we put $\delta_\rho = \lambda_\rho \partial_\rho$ and $\delta_{r+i} = \lambda_{r+i} \partial_{r+i}$. We also put $\delta_{q_j} = q_j \partial/\partial q_j = \partial/\partial t_j$. Our methods applied to (5.43) give the identities

$$(5.48) \qquad \begin{aligned} \delta_\rho &= \sum_j (D_\rho \cdot \beta_j) \delta_{q_j} \\ \delta_{r+i} &= -\sum_j (c_1(\mathcal{L}_i) \cdot \beta_j) \delta_{q_j} \end{aligned}$$

when applied to $\overline{T}$-invariant functions. We also compute that

$$(5.49) \qquad \delta_{q_j} \left(e^{\sum_i t_i T_i} q^\beta \right) = (T_j + (T_j \cdot \beta)) \left(e^{\sum_i t_i T_i} q^\beta \right).$$

Combining (5.48) and (5.49), we obtain

$$(5.50) \qquad \begin{aligned} \delta_\rho \left(e^{\sum_i t_i T_i} q^\beta \right) &= \left(D_\rho + (D_\rho \cdot \beta) \right) \left(e^{\sum_i t_i T_i} q^\beta \right) \\ \delta_{r+i} \left(e^{\sum_i t_i T_i} q^\beta \right) &= -\left(c_1(\mathcal{L}_i) + (c_1(\mathcal{L}_i) \cdot \beta) \right) \left(e^{\sum_i t_i T_i} q^\beta \right) \end{aligned}$$

In (5.50), we have used the identity of divisors $D = \sum_j (D \cdot \beta_j) T_j$ for $D = D_\rho$ and $D = c_1(\mathcal{L}_i)$. This identity follows because T_j is the dual basis to β_j.

The operator $\square_\beta$ is defined in terms of ∂_ρ and ∂_{r+i}. In order to write this in terms of $\delta_\rho = \lambda_\rho \partial_\rho$ and $\delta_{r+i} = \lambda_{r+i} \partial_{r+i}$, we will consider the operator

$$\square_\beta' = \prod_{D_\rho \cdot \beta > 0} \lambda_\rho^{D_\rho \cdot \beta} \prod_{c_1(\mathcal{L}_i) \cdot \beta < 0} \lambda_{r+i}^{-c_1(\mathcal{L}_i) \cdot \beta} \square_\beta$$

Then we compute that

$$(5.51) \qquad \begin{aligned} \square_\beta' =\ & \prod_{D_\rho \cdot \beta > 0} \delta_\rho (\delta_\rho - 1) \cdots (\delta_\rho - D_\rho \cdot \beta + 1) \times \\ & \prod_{c_1(\mathcal{L}_i) \cdot \beta < 0} \delta_{r+i} (\delta_{r+i} - 1) \cdots (\delta_{r+i} + c_1(\mathcal{L}_i) \cdot \beta + 1) \\ & - q^\beta \prod_{D_\rho \cdot \beta < 0} \delta_\rho (\delta_\rho - 1) \cdots (\delta_\rho + D_\rho \cdot \beta + 1) \times \\ & \prod_{c_1(\mathcal{L}_i) \cdot \beta > 0} \delta_{r+i} (\delta_{r+i} - 1) \cdots (\delta_{r+i} - c_1(\mathcal{L}_i) \cdot \beta + 1) \end{aligned}$$

Using (5.51) and (5.50), we can now compute $\square_\beta' \tilde{I}$ to be

$$\square_\beta' \tilde{I} = e^{(\sum t_j T_j)} \times$$

$$(5.52) \qquad \left(\sum_{\beta' \in M(X)_{\mathbb{Z}}} q^{\beta'} \frac{\left(\prod_{i=1}^k \prod_{m=1}^{c_1(\mathcal{L}_i) \cdot \beta'} (c_1(\mathcal{L}_i) + m) \right) \left(\prod_\rho \prod_{m=-\infty}^0 (D_\rho + m) \right)}{\prod_\rho \prod_{m=-\infty}^{D_\rho \cdot (\beta' - \beta)} (D_\rho + m)} - \right.$$

$$\left. \sum_{\beta'' \in M(X)_{\mathbb{Z}}} q^{\beta'' + \beta} \frac{\left(\prod_{i=1}^k \prod_{m=1}^{c_1(\mathcal{L}_i) \cdot (\beta'' + \beta)} (c_1(\mathcal{L}_i) + m) \right) \left(\prod_\rho \prod_{m=-\infty}^0 (D_\rho + m) \right)}{\prod_\rho \prod_{m=-\infty}^{D_\rho \cdot (\beta'')} (D_\rho + m)} \right).$$

In this expression for $\square'_\beta \tilde{I}$, note that the term in the first sum for $\beta' \in M(X)_{\mathbb{Z}}$ cancels the term in the second sum containing $\beta'' \in M(X)_{\mathbb{Z}}$ provided $\beta' = \beta'' + \beta$. After this cancelation, we are left with two classes of terms: those in the first sum corresponding to $\beta' \in M(X)_{\mathbb{Z}}$ such that $\beta' - \beta \notin M(X)_{\mathbb{Z}}$, and those in the second sum corresponding to $\beta'' \in M(X)_{\mathbb{Z}}$ such that $\beta'' + \beta \notin M(X)_{\mathbb{Z}}$.

We begin with the first class of these terms. Thus, suppose $\beta' \in M(X)_{\mathbb{Z}}$ and $\beta' - \beta \notin M(X)_{\mathbb{Z}}$. It follows that $e^{(\sum t_j T_j)} q^{\beta'}$ appears in the first sum of (5.52) but not the second. Our goal is to prove that the coefficient of $e^{(\sum t_j T_j)} q^{\beta'}$ is zero. Our assumption $\beta' - \beta \notin M(X)_{\mathbb{Z}}$ implies there is an ample divisor D such that $D \cdot (\beta' - \beta) < 0$. We may without loss of generality assume that D is T-invariant, and write $D = \sum_\rho a_\rho D_\rho$ with $a_\rho \geq 0$. Let

$$S = \{\rho \in \Sigma(1) : D_\rho \cdot (\beta' - \beta) < 0\}.$$

Note that $S \neq \emptyset$ since $D \cdot (\beta' - \beta) < 0$.

We claim that the $\Pi_{\rho \in S} D_\rho = 0$ in $H^*(X)$. To see this, it suffices to prove that $\cap_{\rho \in S} D_\rho = \emptyset$. So suppose that this intersection is nonempty. Then $\cap_{\rho \in S} D_\rho$ contains at least one fixed point of the torus action. If σ is the corresponding n-dimensional cone of Σ, then $\rho \subset \sigma$ for all $\sigma \in S$. Since $D = \sum a_\rho D_\rho$ is Cartier, we can find $m_\sigma \in M$ with $\langle m_\sigma, v_\rho \rangle = -a_\rho$ for all edges ρ of σ (and in particular, for all elements of S). Here, $v_\rho \in N$ is the generator of the edge ρ. But D is also ample, so the ampleness criterion given at the end of Section 3.1 implies that $\langle m_\sigma, v_\rho \rangle \geq -a_\rho$ for all $\rho \in \Sigma(1)$. Then D is linearly equivalent to the divisor $D' = D + \mathrm{div}(\chi^{m_\sigma})$. Replacing D by D', we may therefore assume that

$$(5.53) \qquad\qquad D = \sum_{\rho \notin S} a_\rho D_\rho, \quad a_\rho \geq 0.$$

Intersecting both sides of (5.53) with $\beta' - \beta$, we see that the left hand side is negative and the right hand side is nonnegative, a contradiction. This proves our claim.

Now consider the coefficient of $e^{(\sum t_j T_j)} q^{\beta'}$ in the first sum of (5.52), and fix $\rho \in S$. The definition of S implies $D_\rho \cdot (\beta' - \beta) < 0$, and then it follows from (5.52) that D_ρ is a factor of this coefficient. This holds for all $\rho \in S$, so that the coefficient is a multiple of $\Pi_{\rho \in S} D_\rho = 0$. Thus the coefficient is zero.

It remains to study the second class of terms, which correspond to $\beta'' \in M(X)_{\mathbb{Z}}$ such that $\beta'' + \beta \notin M(X)_{\mathbb{Z}}$. The argument is entirely analogous and is omitted. $\square$

An immediate corollary of Proposition 5.5.4 is that the formal function $I_{\mathcal{V}} = \mathrm{Euler}(\mathcal{V})\,\tilde{I}$ defined in (5.44) is also a solution of the $\mathcal{A}$-system.

Proposition 5.5.4 gives interesting solutions to the GKZ system for $\mathcal{A}$ as in (5.42). This means that if we expand $\tilde{I}$ in any basis for $H^*(X)$, the coefficients satisfy the GKZ equations. But how does this relate to Picard-Fuchs equations? In the case of a Calabi-Yau complete intersection $V \subset X$ associated to a nef-partition, we have the mirror family V° constructed in Section 4.3.1, and one can prove that the $\mathcal{A}$-system of differential equations are among the Picard-Fuchs equations of the mirror family V°, generalizing the case of hypersurfaces considered in Section 5.5.2. (The reader may worry that we had $\hat{\beta} = (0, \ldots, 0, -1)$ for the hypersurface case but $\hat{\beta} = (0, \ldots, 0)$ for a complete intersection. This difference is insignificant, arising from the choice of the holomorphic form ω in (5.34).)

As an example of what this means, let $X = \mathbb{P}^4$ and $\mathcal{V} = \mathcal{L}_1 = \mathcal{O}_{\mathbb{P}^4}(5H)$. Then expanding $\tilde{I}$ from (5.45) using the usual basis of $H^*(\mathbb{P}^4)$ gives

$$(5.54) \qquad \tilde{I} = y_0 + y_1\,H + y_2\,H^2 + y_3\,H^3 + y_4\,H^4.$$

It follows that $y_0, \dots, y_4$ are solutions of the fifth order GKZ equation (5.37).

However, we have seen that the GKZ system is smaller than the full set of Picard-Fuchs equations contain more than just the GKZ system. Thus, while $\tilde{I}$ satisfies the GKZ system, it can happen that $\tilde{I}$ does not satisfy *all* of the Picard-Fuchs equations. For example, we saw in Section 5.5.2 that (5.37) factors to give a fourth order equation, which is the Picard-Fuchs equation of the quintic mirror. Section 6.3.4 will show that in (5.54), only $y_0, \dots, y_3$ give solutions of the Picard-Fuchs equation—y_4 is spurious. This is why we consider $I_{\mathcal{V}} = \mathrm{Euler}(\mathcal{V})\,\tilde{I}$, which transforms (5.54) into

$$(5.55) \qquad I_{\mathcal{V}} = 5H(y_0 + y_1\,H + y_2\,H^2 + y_3\,H^3)$$

since $H^5 = 0$. The theory developed in Section 6.3.4 will imply that $y_0, \dots, y_3$ are a basis of solutions of the Picard-Fuchs equations of the quintic mirror. We will also see how the formula for $\tilde{I}$ given in (5.45) arises naturally from the Frobenius method applied to the Picard-Fuchs equation.

More generally, when the complete intersection determined by the $\mathcal{L}_i$ is Calabi-Yau, the function $I_{\mathcal{V}} = \prod_i c_1(\mathcal{L}_i)\,\tilde{I}$ defined in (5.44) should satisfy the Picard-Fuchs equations for the mirror. This is asserted without proof in [**Givental4**]. In explicit examples, it is often straightforward to check that the coefficients of $I_{\mathcal{V}}$ satisfy the Picard-Fuchs equations of V°. For example, this is true for the mirror of the quintic threefold. In these examples, justification typically proceeds by considering the dimension of various spaces of cohomology classes and using the irreducibility of the $\mathcal{D}$-module associated to the Picard-Fuchs system.

In Chapter 11, we will first encounter (5.46) and (5.55) in the more general form used in [**LLY**]. Specifically, for $\mathcal{O}_{\mathbb{P}^4}(5)$, equation (11.25) reduces to

$$HG[\mathcal{I}(\hat{P})](t_1) = e^{-pt_1/\hbar} \sum_{d \geq 0} \frac{\prod_{m=0}^{5d}(5p - m\hbar)}{\prod_{m=1}^{d}\prod_{k=0}^{4}(p - \lambda_k - m\hbar)}e^{dt_1}.$$

This formula takes values in equivariant cohomology, and p is the equivariant hyperplane class. If we set $\hbar = -1$ and reduce to ordinary cohomology by letting $\lambda_k \to 0$, then we get

$$\lim_{\lambda \to 0} HG[\mathcal{I}(\hat{P})](t_1) = e^{t_1 H}5H \sum_{d \geq 0} \frac{\prod_{m=1}^{5d}(5H + m)}{\prod_{m=1}^{d}(H + m)^5}e^{dt_1},$$

which is precisely $I_{\mathcal{V}}$ in this case. The authors of [**LLY**] chose the "HG" notation to emphasize the hypergeometric interpretation of these functions.

Some of the items introduced here will play an important role in subsequent chapters. For example, in Section 6.2.3, we will see how the q_j from (5.43) give natural coordinates on the Kähler moduli space of a Calabi-Yau hypersurface $V \subset X$. Also, as already mentioned, the function $I_{\mathcal{V}}$ will appear prominently in Chapter 11 when we discuss the mirror theorem. In particular, Example 11.2.5.2 will show that Proposition 5.5.4 implies some nontrivial relations in the quantum cohomology ring of toric complete intersections.

5.6. Yukawa Couplings

In this section, we will focus on the special case of Calabi-Yau threefolds. As in Chapter 1, the *Yukawa coupling* is defined by the formula

$$\langle\theta_1,\theta_2,\theta_3\rangle = \int_V \Omega \wedge (\nabla_{\theta_1}\nabla_{\theta_2}\nabla_{\theta_3}\Omega)$$

for $\theta_i \in H^1(V,T_V)$ (see (1.9)). If we think of $H^1(V,T_V)$ as the tangent space to the moduli space at a maximally unipotent boundary point p, the moduli coordinates $z_1,\dots,z_r$ give a basis $\partial/\partial z_i$ of the tangent space. However, since the Gauss-Manin connection ∇ has a regular singular point at p, it is more convenient to work with $\delta_i = z_i\partial/\partial z_i$. Hence we will be interested in computing

$$\langle\delta_i,\delta_j,\delta_k\rangle = \int_V \Omega \wedge (\nabla_{\delta_i}\nabla_{\delta_j}\nabla_{\delta_k}\Omega).$$

5.6.1. The 1-Dimensional Case. We start by considering a Calabi-Yau threefold V with 1-dimensional complex moduli. Suppose we are at a maximally unipotent boundary point with coordinate z. For $\delta = z\,d/dz$, we can write the Picard-Fuchs equation as

$$\nabla_\delta^4\Omega + f_1(z)\nabla_\delta^3\Omega + f_2(z)\nabla_\delta^2\Omega + f_3(z)\nabla_\delta\Omega + f_4(z)\Omega = 0.$$

This equation allows us to determine the Yukawa couplings up to a constant. We have already seen this done for the quintic mirror in Chapter 2, and the process generalizes easily. Griffiths transversality implies $\int_V \Omega \wedge \nabla_\delta^2\Omega = 0$, and differentiating twice leads to the equation

$$\int_V \nabla_\delta\Omega \wedge \nabla_\delta^3\Omega = -\frac{1}{2}\int_V \Omega \wedge \nabla_\delta^4\Omega,$$

just as in (2.16). Letting $Y = \int_V \Omega \wedge \nabla_\delta^3\Omega$ denote the Yukawa coupling, we now compute

$$\begin{aligned}
\delta Y &= \int_V \Omega \wedge \nabla_\delta^4\Omega + \int_V \nabla_\delta\Omega \wedge \nabla_\delta^3\Omega \\
&= \frac{1}{2}\int_V \Omega \wedge \nabla_\delta^4\Omega \\
&= -\frac{1}{2}f_1(z)\int_V \Omega \wedge \nabla_\delta^3\Omega.
\end{aligned}$$

The last equality arises from the Picard-Fuchs equation together with the vanishing of $\int_V \Omega \wedge \nabla_\delta^i\Omega$ for $i \le 2$ by Griffiths transversality. Hence we get

$$\delta Y = -\frac{1}{2}f_1(z)\,Y,$$

which can be solved for Y up to a constant factor, as was done for the quintic mirror.

5.6.2. The General Case. Essentially the same method works in the general setting, though the algebra gets messier. For simplicity, suppose that V is a Calabi-Yau threefold with 2-dimensional complex moduli. Given a maximally unipotent boundary point with coordinates z_1, z_2 such that $z_i = 0$ define the boundary divisors, we set $\delta_i = z_i\,d/dz_i$ and define

$$K^{ab} = \int_V \Omega \wedge \nabla_{\delta_1}^a \nabla_{\delta_2}^b\Omega, \quad a+b=3.$$

Thus K^{ab} is one of the Yukawa couplings

$$\langle \delta_1, \delta_1, \delta_1 \rangle, \quad \langle \delta_1, \delta_1, \delta_2 \rangle, \quad \langle \delta_1, \delta_2, \delta_2 \rangle, \quad \langle \delta_2, \delta_2, \delta_2 \rangle.$$

However, we also have Griffiths transversality, which here becomes

$$(5.56) \qquad \int_V \Omega \wedge \nabla^a_{\delta_1} \nabla^b_{\delta_2} \Omega = 0, \quad a + b \le 2.$$

Repeated differentiation yields identities relating the quantities

$$(5.57) \qquad W^{ab} = \int_V \Omega \wedge \nabla^a_{\delta_1} \nabla^b_{\delta_2} \Omega, \quad a + b = 4$$

to the partial derivatives of the Yukawa couplings K^{ab}. Combining these with the Picard-Fuchs equations, one gets a solvable system of first order differential equations in the Yukawa couplings. See [**HKTY1**] for more details.

To see what this means in practice, we return to a familiar example.

Example 5.6.2.1. The mirror V° of the degree 8 hypersurface in $\mathbb{P}(1,1,1,2,2)$ has 2-dimensional complex moduli with coordinates z_1, z_2 from Example 5.5.2.1. There, we worked out two Picard-Fuchs equations(5.39) and (5.41), which after some algebra can be written as

$$(5.58) \qquad 0 = (1 - 4z_2)\delta_2^2 - z_2(\delta_1^2 - 4\delta_1\delta_2 - \delta_1 + 2\delta_2)$$

$$(5.59) \qquad 0 = (1 - 256z_1)\delta_1^3 - 2\delta_1^2\delta_2 - 8z_1(48\delta_1^2 + 22\delta_1 + 3).$$

These equations give some immediate relations among the K^{ab}. For example, the Picard-Fuchs equation (5.59) and Griffiths transversality (5.56) easily imply

$$0 = (1 - 256z_1)K^{30} - 2K^{21},$$

which tells us that

$$K^{21} = \frac{1}{2}(1 - 256z_1)K^{30}.$$

Similarly, multiplying (5.58) on the left by δ_1 and δ_2 and then solving in terms of K^{30} gives

$$K^{12} = \frac{z_2(-1 + 512z_1)}{1 - 4z_2} K^{30}$$

$$K^{03} = \frac{z_2(1 - 256z_1 + 4z_2 - 3072z_1z_2)}{2(1 - 4z_2)^2} K^{30}.$$

Hence, it suffices to determine the single Yukawa coupling K^{30}. We do this by finding the first order partial differential equation it satisfies. To begin, first multiply (5.59) on the left by δ_1. Proceeding as above and using $\delta_1 z_1 = z_1(\delta_1 + 1)$, one easily obtains

$$(5.60) \qquad (1 - 256z_1)W^{40} - 2W^{31} - 640z_1 K^{30} = 0$$

where W^{ab} is from (5.57). However, repeated differentiation of (5.56) gives the identities

$$W^{40} = 2\delta_1 K^{30}$$

$$W^{31} = \tfrac{3}{2}\delta_1 K^{21} + \tfrac{1}{2}\delta_2 K^{30}$$

$$W^{22} = \delta_1 K^{12} + \delta_2 K^{21}.$$

A systematic method for writing down all such identities can be found in [**HKTY1**]. If we combine equation (5.60) with these identities and the above formula for K^{21}, then we get the equation

$$(1 - 256z_1)\,\delta_1 K^{30} - 2\delta_2 K^{30} - 512z_1 K^{30} = 0.$$

We can derive a second equation by multiplying (5.58) on the left by δ_1^2. This yields

$$(1 - 4z_2)W^{22} - z_2(W^{40} - 4W^{31} - K^{30} + 2K^{21}) = 0,$$

and proceeding as above leads to the equation

$$-256z_1 z_2\,\delta_1 K^{30} + \tfrac{1}{2}(1 - 256z_1 + 1024z_1 z_2)\,\delta_2 K^{30} = 0.$$

(There is a lot of cancelation!)

We thus have two first order linear partial differential equations for K^{30}. Using some algebra and $\delta_i = z_i \partial/\partial z_i$, we can rewrite these equations as

$$\frac{\partial K^{30}}{\partial z_1} = \frac{512(1 - 256z_1 + 1024z_1 z_2)}{D(z_1, z_2)}$$

$$\frac{\partial K^{30}}{\partial z_2} = \frac{512^2 z_1^2}{D(z_1, z_2)},$$

where

$$D(z_1, z_2) = (1 - 256z_1)^2 - 512^2 z_1^2 z_2.$$

It follows easily that the Yukawa coupling K^{30} is given by the formula

$$K^{30} = \frac{c}{D(z_1, z_2)}, \quad c \neq 0.$$

From here, we immediately get formulas for K^{21}, K^{12} and K^{03} as well. These answers agree with those given in the appendix to [**HKTY1**].

The denominators of the Yukawa couplings K^{ab} all involve $D(z_1, z_2)$ and $1-4z_2$. In Section 6.1.4, we will show that in the complex moduli space of V°, the equations $D(z_1, z_2) = 0$ and $1 - 4z_2 = 0$ define the two components of the discriminant locus (= all points (z_1, z_2) where the defining equation (5.25) becomes singular). Hence the appearance of these polynomials in the denominators of the Yukawa couplings is no surprise.

5.6.3. Normalized Yukawa Couplings and the Gauss-Manin Connection.

Our next task is to discuss the *normalized Yukawa coupling*. As we will see, this requires a careful choice of both the 3-form Ω and the moduli coordinates.

We begin by noting that Ω is far from unique—we can multiply it by any nonvanishing function on the moduli space. As in Chapter 2, we normalize Ω as follows. At a maximally unipotent boundary point, $\dim(W_0) = 1$. By the definition of the monodromy weight filtration $W_\bullet$, W_0 is invariant under monodromy. Let g_0 be a generator of $W_0 \cap H^3(V, \mathbb{Z})$, so that g_0 is well-defined up to sign. Then up to a constant factor, $y_0 = \langle g_0, \Omega \rangle$ is the unique period which is analytic at the boundary point. As in Section 2.4, we then normalize the 3-form by replacing Ω by Ω/y_0. The argument of Chapter 2 shows that this replaces a Yukawa coupling Y with Y/y_0^2, in perfect analogy with the discussion prior to (2.19). As in [**Morrison2**], we note that Y/y_0^2 is independent of the choice of g_0.

The next step in normalizing the Yukawa coupling is to choose a canonical moduli variable. Here, we will discover a very interesting relation between Yukawa couplings and the Gauss-Manin connection ∇. For simplicity, we begin with the case where the complex moduli of the Calabi-Yau threefold is 1-dimensional, and assume that we are at a maximally unipotent boundary point. As above, we have the 3-form Ω and the period $y_0 = \langle g_0, \Omega \rangle$. Then normalize Ω by replacing it with Ω/y_0, which as above takes Y to Y/y_0^2. With this replacement, we have $\langle g_0, \Omega \rangle = 1$. The following proposition shows that this has a dramatic effect on the form of the Picard-Fuchs equation, provided we pick the correct local coordinate.

PROPOSITION 5.6.1. *Let V be a Calabi-Yau threefold with 1-dimensional complex moduli, and suppose that we are at a maximally unipotent boundary point which satisfies the integrality conjecture of Section 5.2.2. Then there is a local coordinate q such that the Picard-Fuchs equation of the normalized 3-form Ω is given by*

$$\nabla_\delta^2 \left(\frac{\nabla_\delta^2 \Omega}{Y} \right) = 0,$$

where δ denotes $2\pi i q\, d/dq$ and Y is the Yukawa coupling defined by

$$Y = - \int_V \Omega \wedge \nabla_\delta^3 \Omega.$$

PROOF. Our proof is based on ideas from [**Deligne2**]. We begin with a local coordinate z at the maximally unipotent boundary point. In a neighborhood of $z = 0$, we have the Hodge filtration $\mathcal{F}^\bullet$ of $\mathcal{H} = \mathcal{F}^0$, and the weight filtration $W_\bullet$ gives a filtration $W_\bullet$ of $\mathcal{H}$. Above $z = 0$, we know that $(W_\bullet, F_{\lim}^\bullet)$ is Hodge-Tate by (5.11). As observed in [**Deligne2**, Sect. 6], this implies that

$$\mathcal{F}^p \oplus W_{2p-1} = \mathcal{F}^p \oplus W_{2p-2} = \mathcal{H}$$

in a neighborhood of 0. It follows that $(W_\bullet, \mathcal{F}^\bullet)$ is a variation of mixed Hodge structure of Hodge-Tate type, so that for each p, $\mathcal{F}^p$ induces on Gr_{2p}^W a 1-dimensional pure Hodge structure of type (p, p). If we set $I^{p,p} = W_{2p} \cap \mathcal{F}^p$, then we have a natural isomorphism

$$I^{p,p} \simeq Gr_{2p}^W.$$

Since $W_\bullet$ is flat, ∇ induces a connection ∇_{2p}^W on Gr_{2p}^W. We also have $N(W_{2p}) \subset W_{2p-2} = W_{2p-1}$ (where N is the monodromy logarithm), so that N acts trivially on Gr_{2p}^W. Hence ∇_{2p}^W has trivial monodromy on Gr_{2p}^W, which thus extends naturally to $z = 0$. Since Gr_{2p}^W is 1-dimensional and has a natural structure over $\mathbb{Z}$, it follows that in a neighborhood of 0, we can find an integer valued ∇_{2p}^W-flat section which spans Gr_{2p}^W and is unique up to ± 1. Under the above isomorphism, this gives a section of $I^{p,p}$ which we will denote e_{3-p}.

The sections e_0, e_1, e_2, e_3 have some very nice properties. First, N is an infinitesimal automorphism of $\langle \alpha, \beta \rangle = \int_V \alpha \wedge \beta$, which in combination with (5.4) and $W_4 = W_5$ implies $\langle W_0, W_4 \rangle = \langle W_2, W_2 \rangle = \{0\}$. Hence the only nonzero pairings are $\langle e_0, e_3 \rangle$ and $\langle e_1, e_2 \rangle$, and we can compute these in Gr^W. It follows that $\langle e_0, e_3 \rangle$ and $\langle e_1, e_2 \rangle$ must be ± 1 since $\langle \cdot, \cdot \rangle$ is a perfect pairing over $\mathbb{Z}$. Thus we can pick e_0, e_1, e_2, e_3 such that

$$(5.61) \qquad \langle e_0, e_3 \rangle = -1 \quad \text{and} \quad \langle e_1, e_2 \rangle = 1.$$

The reason for this sign choice will soon become clear.

Now let $\delta = 2\pi i z\, d/dz$. Since e_{3-p} maps to a ∇^W_{2p}-flat section of Gr^W_{2p}, $\nabla(e_{3-p})$ lies in $W_{2p-1} = W_{2p-2}$. But e_{3-p} is also a section of $\mathcal{F}^p$, so that $\nabla_\delta(e_{3-p})$ lies in $\mathcal{F}^{p-1}$. This shows that $\nabla_\delta(e_{3-p})$ is a section of $I^{p-1,p-1}$, and it follows that

$$\nabla_\delta(e_0) = Y_1\, e_1, \quad \nabla_\delta(e_1) = Y_2\, e_2, \quad \nabla_\delta(e_2) = Y_3\, e_3, \quad \nabla_\delta(e_3) = 0$$

for some functions Y_1, Y_2, Y_3. However, since $\langle e_0, e_2 \rangle$ is identically zero, we have

$$0 = \delta\langle e_0, e_2 \rangle = \langle \nabla_\delta(e_0), e_2 \rangle + \langle e_0, \nabla_\delta(e_2) \rangle = \langle Y_1\, e_1, e_2 \rangle + \langle e_0, Y_3\, e_3 \rangle = Y_1 - Y_3,$$

where the last equality uses (5.61). It follows that $Y_1 = Y_3$.

Since ∇ has regular singular points and $\delta = 2\pi i z\, d/dz$, Y_3 is holomorphic at 0. Let us prove that $Y_3(0) = \pm 1$. By (5.3), we know that over 0,

$$\bar{\nabla}_\delta = -N$$

when regarded as a map $F^1_{\lim}/F^2_{\lim} \to F^0_{\lim}/F^1_{\lim}$. This shows that $\bar{\nabla}_\delta(e_2) = -N(e_2)$. However, $\nabla_\delta(e_2) = Y_3\, e_3$ clearly implies $\bar{\nabla}_\delta(e_2) = Y_3(0)e_3$. It follows that $N(e_2) = -Y_3(0)e_3$. But over 0, e_3 is a $\mathbb{Z}$-basis of W_0 and e_2, e_3 is a $\mathbb{Z}$-basis of W_2. Then $N(e_2) = \pm e_3$ is a consequence of the integrality conjecture described at the end of Section 5.2.2. We conclude that $Y_3(0) = \pm 1$, as desired.

We can then assume that $N(e_2) = -e_3$ by replacing e_1, e_2 with $-e_1, -e_2$ if necessary (note that (5.61) remains true). This implies $Y_3(0) = 1$, and it follows that the function

$$q = \exp\left(\int Y_3(z)\frac{dz}{z}\right)$$

gives a local coordinate at 0 such that

$$Y_3(z)\, 2\pi i q\, \frac{d}{dq} = 2\pi i z\, \frac{d}{dz}.$$

Redefining δ to be $2\pi i q\, d/dq$ yields $\nabla_\delta(e_2) = e_3$, and since $Y_1 = Y_3$, we also have $\nabla_\delta(e_0) = e_1$. Thus, relative to the basis e_0, e_1, e_2, e_3 and using q as local coordinate, ∇_δ has the connection matrix

$$(5.62) \qquad \begin{pmatrix} 0 & 0 & 0 & 0 \\ 1 & 0 & 0 & 0 \\ 0 & Y & 0 & 0 \\ 0 & 0 & 1 & 0 \end{pmatrix}$$

for $Y = Y_2$. We will eventually identify Y with the normalized Yukawa coupling.

The next step is to show that e_0 is the normalized 3-form Ω. We know that e_0 is a nonvanishing section of $\mathcal{F}^3$, so that $\Omega = a(q)\, e_0$ for some holomorphic function $a(q)$. The above matrix shows

$$\nabla_\delta^j \Omega \equiv \delta^j(a)\, e_0 \bmod e_1, e_2, e_3,$$

which implies that $a(q)$ satisfies the same Picard-Fuchs equation Ω does. Since we are at a maximally unipotent boundary point, there is only one holomorphic solution (up to a constant), and since Ω is normalized, one solution is $\langle g_0, \Omega \rangle = 1$. This shows that $a(q)$ is a constant a, hence $\Omega = a\, e_0$. However, recall that g_0 is an integral basis of W_0, so we can assume $e_3 = g_0$. Then, since $\langle \cdot, \cdot \rangle$ is odd, we have

$$1 = \langle g_0, \Omega \rangle = \langle e_3, a\, e_0 \rangle = -a\,\langle e_0, e_3 \rangle = a,$$

where the last equality is by (5.61). This proves that $e_0 = \Omega$. From here, one easily obtains the desired Picard-Fuchs equation

$$(5.63) \qquad \nabla_\delta^2\left(\frac{\nabla_\delta^2\Omega}{Y}\right) = 0.$$

The final step in the proof is to show that Y is the Yukawa coupling. This is easy, for we've already seen that $\Omega = e_0$, and then (5.62) shows that $\nabla_\delta^3\Omega = \delta(Y)\,e_2 + Ye_3$, and then (5.61) implies

$$\int_V \Omega \wedge \nabla_\delta^3\Omega = \int_V e_0 \cup (\delta(Y)\,e_2 + Ye_3) = \langle e_0, \delta(Y)\,e_2 + Ye_3\rangle = -Y.$$

This completes the proof of the proposition. $\qquad\qquad\square$

After having all of the signs work out so carefully during the proof, it may seem odd to have the minus sign appear in the final formula for Y. However, this is completely natural, for in the equation

$$(5.64) \qquad Y = -\int_V \Omega \wedge \nabla_\delta^3\Omega,$$

the right hand side is the pairing $Q(\alpha, \beta) = (-1)^{3(3-1)/2}\int_V \alpha \cup \beta$ which polarizes the Hodge structure on $H^3(V, \mathbb{C})$ (see (5.1)). According to mirror symmetry, Y should correspond to a certain A-model correlation function on the mirror V°, which can similarly be defined using a polarization on the *A-variation of Hodge structure*, which we will discuss in Section 8.5. The need for the sign will become clear when we compare these two polarized variations of Hodge structure. In the language of Section 5.6.4, (5.64) is the *normalized Yukawa coupling* in the case of 1-dimensional moduli.

The proof of Proposition 5.6.1 given in [**Deligne2**] is more sophisticated than what we did above. Starting with our original local coordinate z, Deligne fixes $1 \le i \le 3$ and considers the extension of Hodge structures

$$(5.65) \qquad 0 \longrightarrow Gr_{2i-2}^W \longrightarrow \mathcal{W}_{2i}/\mathcal{W}_{2i-4} \longrightarrow Gr_{2i}^W \longrightarrow 0.$$

Over each point, the extension we get is classified by a number in $\mathbb{C}^*$, so that as $(\mathcal{W}_\bullet, \mathcal{F}^\bullet)$ varies in a neighborhood of $z = 0$, the extension is classified by a function $q_i(z)$. Then the functions we denoted Y_1, Y_2, Y_3 can be expressed in terms of the q_i as $Y_{4-i} = \frac{z}{q_i}\frac{dq_i}{dz}$, $1 \le i \le 3$. (A minus sign occurs in [**Deligne2**, Sect. 7] since he is working in the dual situation.) Thus the special coordinate q of Proposition 5.6.1 is $q_1 = q_3$, which classifies (5.65) for $i = 1, 3$.

Proposition 5.6.1 generalizes nicely to the case when the Calabi-Yau threefold has r-dimensional moduli. If we assume the integrality conjecture from Section 5.2.2, then we get the following description of the Gauss-Manin connection near a maximally unipotent boundary point, which we think of as $0 \in \Delta^r$.

As before, we have $I^{p,p} = W_{2p} \cap \mathcal{F}^p \simeq Gr_{2p}^W$, though $I^{1,1}$ and $I^{2,2}$ now have rank r. The basis e_0, e_1, e_2, e_3 used in the proof of the proposition generalizes to bases

$$e_0 \text{ of } I^{3,3}, \quad e_j, \; 1 \le j \le r, \text{ of } I^{2,2}, \quad e^k, \; 1 \le k \le r, \text{ of } I^{1,1}, \quad e^0 \text{ of } I^{0,0}$$

such that

$$(5.66) \qquad \langle e_0, e^0\rangle = -1 \quad \text{and} \quad \langle e_j, e^k\rangle = \delta_{jk}, \; 1 \le j, k \le r,$$

and all other pairings vanish. Note how (5.61) is the special case $r = 1$. Furthermore, as in the proof of Proposition 5.6.1, we can assume that these basis elements extend to 0 and are integral there.

The methods used to prove Proposition 5.6.1 (or the more powerful methods of [**Deligne2**]) then imply that we can find special local coordinates $q_1, \dots, q_r$ of $0 \in \Delta^r$ such that for $\delta_j = 2\pi i q_j \, \partial/\partial q_j$, we have

$$
\begin{aligned}
\nabla_{\delta_i} e^0 &= 0 \\
\nabla_{\delta_i} e^k &= \delta_{ik}\, e^0, \quad 1 \le k \le r \\
\nabla_{\delta_i} e_j &= \sum_{k=1}^{r} Y_{ijk}\, e^k, \quad 1 \le j \le r \\
\nabla_{\delta_i} e_0 &= e_i
\end{aligned}
$$

(5.67)

for some functions Y_{ijk}. This shows how the matrix (5.62) generalizes to the case of r-dimensional moduli.

By the argument of Proposition 5.6.1, we know that e_0 is the normalized 3-form Ω, and then (5.67) implies that the Y_{ijk} are given by

$$
(5.68) \qquad Y_{ijk} = -\int_V \Omega \wedge \nabla_{\delta_k} \nabla_{\delta_i} \nabla_{\delta_j} \Omega.
$$

In Section 5.6.4, we will define the Y_{ijk} to be the *normalized Yukawa couplings*. Then the above discussion shows that the Gauss-Manin connection is uniquely determined by the normalized Yukawa couplings.

We next show that the functions Y_{ijk} all come from a single "potential function". Introduce variables t_j such that $q_j = e^{2\pi i t_j}$, so that $\partial/\partial t_j = 2\pi i q_j \, \partial/\partial q_j = \delta_j$. Then we have the following lemma.

LEMMA 5.6.2. *There is a holomorphic function* $\Phi(t_1, \dots, t_r)$ *such that*

$$
Y_{ijk} = \frac{1}{(2\pi i)^3} \frac{\partial^3 \Phi}{\partial t_i \partial t_j \partial t_k} \quad \text{for all} \quad i, j, k.
$$

PROOF. The key point is that ∇ is flat, so that the composition

$$
\mathcal{H} \xrightarrow{\nabla} \mathcal{H} \otimes \Omega_S^1 \xrightarrow{\nabla} \mathcal{H} \otimes \Omega_S^2
$$

is zero (where $S = (\Delta^*)^r$). Using (5.67), one computes that

$$
0 = \nabla^2(e_0) = \sum_{i,j,k=1}^{r} Y_{ijk} e^k \otimes dt_i \wedge dt_j.
$$

This implies that $Y_{ijk} = Y_{jik}$ for all i, j, k. Furthermore, one also computes that for $1 \le j \le r$, we have

$$
0 = \nabla^2(e_j) = \sum_{i,k=1}^{r} Y_{ijk} e^0 \otimes dt_k \wedge dt_i + \sum_{i,k,\ell=1}^{r} \frac{\partial Y_{ijk}}{\partial t_\ell} e^k \otimes dt_\ell \wedge dt_i.
$$

This implies first, that $Y_{ijk} = Y_{kji}$ for all i, j, k, and second, that

$$
\frac{\partial Y_{ijk}}{\partial t_\ell} = \frac{\partial Y_{\ell jk}}{\partial t_i}, \quad \text{for all} \quad i, j, k, \ell.
$$

Hence Y_{ijk} is completely symmetric with respect to i, j, k. Then, using the above condition on the partial derivatives, we easily conclude the existence of Φ. $\square$

Notice that if we define ∇ by (5.67), then the existence of the potential function Φ is *equivalent* to the flatness of ∇.

We have thus proved that near a maximally unipotent boundary point $0 \in \Delta^r$ satisfying the integrality conjecture, the Gauss-Manin connection of a family of Calabi-Yau threefolds is determined by the single function Φ. Since Φ is a function of $t_j = \frac{1}{2\pi i} \log q_j$, it is multi-valued on $(\Delta^*)^r$, but its third derivatives are the normalized Yukawa couplings, which are single-valued and extend holomorphically to Δ^r. Furthermore, using Φ, one can write down explicitly lots of interesting things about the variation of Hodge structure on $H^3(V, \mathbb{C})$. For example, we will see in (8.53) that one can write down a basis of multi-valued flat sections in terms of the potential function Φ.

One of the real surprises is that a natural analog of Φ exists on the mirror side, where it is called the *Gromov-Witten potential*. We will explore this analogy in detail in Sections 8.5 and 8.6.

5.6.4. Normalized Yukawa Couplings and the Mirror Map. Proposition 5.6.1 shows that the Gauss-Manin connection is determined by the Yukawa coupling, provided we use the normalized 3-form and the correct local coordinate. Now a miracle occurs: *the mysterious coordinate q is precisely the mirror map!* We won't define the mirror map in general until Section 6.3, but we can understand the case of 1-dimensional moduli if we follow Chapter 2. There, recall formula (2.6), which gave the mirror map as

$$(5.69) \qquad \tilde{q} = \exp\left(2\pi i \int_{\gamma_1} \Omega / \int_{\gamma_0} \Omega\right),$$

where $\int_{\gamma_0} \Omega$ is holomorphic and $\int_{\gamma_1} \Omega / \int_{\gamma_0} \Omega$ increases by 1 when we go around $z = 0$. If we let g_0, g_1 be the Poincaré duals of γ_0, γ_1, then this translates into $g_0 \in W_0$, $g_1 \in W_2$, and $N(g_1) = g_0$, where N is the logarithm of the monodromy. Hence g_0, g_1 satisfy the definition of maximally unipotent monodromy, and the formula for $\tilde{q}$ becomes

$$(5.70) \qquad \tilde{q} = \exp\left(2\pi i \langle g_1, \Omega \rangle\right)$$

since Ω normalized implies $\langle g_0, \Omega \rangle = 1$.

We can prove that this is the coordinate q from Proposition 5.6.1 as follows. We have the basis e_0, e_1, e_2, e_3 constructed in the proof of the proposition, and the proof also shows that $g_0 = e_3$, $\Omega = e_0$, and $\nabla_\delta(e_2) = e_3$. It follows easily that

$$g_1 = -e_2 + \frac{\log q}{2\pi i}\, e_3$$

is flat, where q is now the coordinate from Proposition 5.6.1. Going once clockwise around the origin increases $\frac{\log q}{2\pi i}$ by 1, so that $\mathcal{T}(g_1) = g_1 + g_0$, which in turn implies $N(g_1) = g_0$. By definition, e_2 and e_3 lie in W_2 and W_0 respectively, and we conclude that g_1 is a flat section in W_2 satisfying $N(g_1) = g_0$. Hence this is the g_1 we want, from which we obtain

$$\langle g_1, \Omega \rangle = \frac{1}{2\pi i} \log q.$$

It follows that the coordinate q is the mirror map $\tilde{q}$ defined in (5.70).

More generally, once we define the mirror map in general in Chapter 6, it will follow that in the case of a Calabi-Yau threefold with r dimensional moduli, the mirror map is given by the special coordinates $q_1, \dots, q_r$ appearing in our discussion (5.66) and (5.67) of the r-dimensional generalization of Proposition 5.6.1.

Assuming this for the time being, we will call $q_1, \ldots, q_r$ the *mirror coordinates*, and using them, we can finally give a formal definition of the normalized Yukawa coupling.

DEFINITION 5.6.3. *Suppose we have a family of Calabi-Yau threefolds which satisfies the integrality conjecture at a maximally unipotent boundary point. If $q_1, \ldots, q_r$ are the mirror coordinates defined above and Ω is the normalized 3-form, then the* normalized Yukawa couplings *are defined by*

$$Y_{ijk} = -\int_V \Omega \wedge \nabla_{\delta_k} \nabla_{\delta_i} \nabla_{\delta_j} \Omega,$$

where $\delta_j = 2\pi i q_j\, \partial/\partial q_j$.

The Y_{ijk} are also often called the *B-model correlation functions*. There is one case where we can explicitly compute the normalized Yukawa coupling.

Example 5.6.4.1. Let V° be the quintic mirror. By [**Morrison2**], we know that V° satisfies the integrality conjecture, so that its normalized Yukawa coupling Y is defined. We claim that Y is given by the formula

$$(5.71) \qquad Y = \frac{5}{(1 + 5^5 z)y_0(z)^2} \left(\frac{q\, dz}{z\, dq} \right)^3,$$

where z is the usual moduli variable and

$$y_0(z) = \sum_{n=0}^{\infty} \frac{(5n)!}{(n!)^5} (-1)^n z^n$$

$$y_1(z) = y_0(z) \log(-z) + 5 \sum_{n=1}^{\infty} \frac{(5n)!}{(n!)^5} \left[\sum_{j=n+1}^{5n} \frac{1}{j} \right] (-1)^n z^n$$

$$q = \exp(y_1/y_0).$$

Note that (5.71) is precisely the B-model correlation function which appears in the statement of mirror symmetry for the quintic threefold (2.26) given in Chapter 2.

Section 6.3.4 will prove that y_0 and y_1 are solutions of the Picard-Fuchs equation of the quintic mirror. The formula $q = \exp(y_1/y_0)$ is how the mirror map is defined in [**CdGP**], and then [**Morrison2**, App. C] shows that this agrees with the definition of mirror map given in (5.69). Hence q is the mirror coordinate.

If Ω is the 3-form defined in (2.7), then the methods of Section 5.6.1 applied to the Picard-Fuchs equation (5.12) from Example 5.4.1 show that

$$\int_{V^\circ} \Omega \wedge \nabla_{z\frac{d}{dz}} \nabla_{z\frac{d}{dz}} \nabla_{z\frac{d}{dz}} \Omega = \frac{1}{1 + 5^5 z} \times \text{some constant,}$$

just as in Chapter 2. Normalizing replaces Ω with $\widetilde{\Omega} = \Omega/\langle g_0, \Omega \rangle$. Since y_0 and $\langle g_0, \Omega \rangle$ are holomorphic solutions of the Picard-Fuchs equation, y_0 is a nonzero constant multiple of $\langle g_0, \Omega \rangle$, so that as explained in Section 5.6.3,

$$\int_{V^\circ} \widetilde{\Omega} \wedge \nabla_{z\frac{d}{dz}} \nabla_{z\frac{d}{dz}} \nabla_{z\frac{d}{dz}} \widetilde{\Omega} = \frac{1}{(1 + 5^5 z)y_0(z)^2} \times \text{another constant.}$$

Finally, when we switch from z to the mirror coordinate q, we have

$$\delta = 2\pi i q \frac{d}{dq} = 2\pi i \frac{q\, dz}{z\, dq} z \frac{d}{dz},$$

so that by Griffiths transversality, the normalized Yukawa coupling is

$$Y = -\int_{V^\circ} \widetilde{\Omega} \wedge \nabla_\delta \nabla_\delta \nabla_\delta \widetilde{\Omega}$$

$$(5.72) \qquad = \frac{c}{(1 + 5^5 z) y_0(z)^2} \left(\frac{q\,dz}{z\,dq}\right)^3$$

for some nonzero constant c. Then (5.71) will follow once we show $c = 5$. Since evaluating the above equation at $q = 0$ gives $Y(0) = c$, it suffices to show $Y(0) = 5$.

In terms of Chapter 2, the constant c was written $(2\pi i)^3 c_2$, and we set $c_2 = 5/(2\pi i)^3$ in the discussion following (2.22) to get the correct leading term as predicted by mirror symmetry. But now, in this chapter, we can predict the constant *in advance* by computing the normalized Yukawa coupling. More precisely, we will compute $Y(0)$ by combining the Hodge theory developed earlier in this chapter with explicit computations from [**CdGP, Morrison2**].

We will need the notation of the proof of Proposition 5.6.1. Recall that $e_0 = \widetilde{\Omega}$ is part of a basis e_0, e_1, e_2, e_3 of the canonical extension $\overline{\mathcal{H}}$ which over $q = 0$ is a basis of the integral structure.

Now consider $\overline{\nabla}_\delta^3(e_0)$ at $q = 0$. First, by (5.3), we know that over $q = 0$,

$$\overline{\nabla}_\delta = -N$$

on the graded pieces of the limiting Hodge filtration, so that modulo $F^1_{\lim}$, we have $\overline{\nabla}_\delta^3(e_0)(0) = (-N)^3(e_0(0))$. However, we also have $\overline{\nabla}_\delta^3(e_0) = Y e_3 + \delta(Y) e_2$ from the connection matrix (5.62). Putting these together, we obtain

$$N^3(e_0(0)) = -Y(0) e_3(0).$$

Now let s_i be the flat section of $\mathcal{H}$ such that the value of $\exp(-t_1 N) s_i$ at $q = 0$ is $e_i(0)$. Since the e_i are an integral basis at $q = 0$, the s_i are an integral basis of $\mathcal{H}_{\mathbb{Z}}$. Then it follows from the previous equation that

$$(5.73) \qquad\qquad N^3(s_0) = -Y(0) s_3.$$

Finally, we use the explicit computation of the monodromy of the quintic mirror family given in [**CdGP**]. As explained in [**Morrison2**, App. C], there are integral flat sections β^2, α_2 of $\mathcal{H}$ which are part of a $\mathbb{Z}$-basis and satisfy $\langle \alpha_2, \beta^2 \rangle = 1$ and $N^3(\beta^2) = -5\alpha_2$. (In [**Morrison2**], this is written with a $+$ sign because there, the boundary point is ∞, while here it is 0, and the inversion map taking ∞ to 0 reverses the orientation of the generator of the monodromy group.)

Comparing (5.73) and $N^3(\beta^2) = -5\alpha_2$, we see that s_3 and α_2 are integral generators of W_0, so that multiplying β^2, α_2 by -1 if necessary, we can assume $\alpha_2 = s_3$. Since $N^3 : W_6/W_5 \simeq W_0$ and $N^3(Y(0)\beta^2 - 5s_0) = 0$, we infer that $Y(0)\beta^2 - 5s_0 \in W_5$. Then $\langle W_5, W_0 \rangle = 0$ implies $\langle Y(0)\beta^2 - 5s_0, s_3 \rangle = 0$. But we also know that $\langle \alpha_2, \beta^2 \rangle = 1$, and (5.61) implies $\langle s_0, s_3 \rangle = -1$. From here, we immediately conclude that $Y(0) = 5$, as desired.

There are two further comments we wish to make about the example just discussed. The first is that $Y(0) = 5$ and (5.73) imply that

$$N^3(s_0) = -5 s_3.$$

It turns out that this equation makes perfect sense from the point of view of mirror symmetry. When we study the A-model connection of the quintic threefold in Section 8.5, we will see that the monodromy N of this connection is naturally isomorphic to cup product with $-H$, where H is the hyperplane section of the quintic threefold. Then the above monodromy equation is consistent with $H^3 = 5$.

The second observation to make is that proving (5.71) required some *extremely* detailed knowledge about the quintic mirror family. In addition to the types of computations discussed in this chapter, we also needed explicit facts about the monodromy so that we could satisfy the integrality conjecture, compute the mirror map, and determine $Y(0)$.

The moral of this discussion is that although the definition of normalized Yukawa coupling has wonderful theoretical properties, it is very hard to compute in practice. What happens more often is that one only partially normalizes the Yukawa coupling, which means normalizing the 3-form Ω and switching to mirror coordinates, but not worrying about the value at 0. In such cases, we only know the Y_{ijk} up to a constant, though because of Lemma 5.6.2, there is only *one* constant to worry about. We will take this approach when we formulate a toric version of the mirror theorem.

This section began with the Yukawa coupling, then discussed Picard-Fuchs equations and connection matrices, and wound up with the mirror map and the notion of a normalized Yukawa coupling. This is a remarkable turn of events, but the story will get even better. In Chapter 8, we will study the *A-model connection*, which is built from quantum cohomology. In particular, Example 8.5.4.1 will show that if a Calabi-Yau threefold V° has $h^2(V^\circ) = 1$, then the middle cohomology $\oplus_p H^{p,p}(V^\circ)$ has a basis such that matrix of the A-model connection is *precisely* (5.62), except that Y is now the A-model correlation function discussed in Section 1.2. The situation gets richer in Chapter 10, where Example 10.3.2.1 discusses V° from the point of view of *quantum differential equations* and derives the operator

$$\left(\hbar \frac{d}{dt_1}\right)^2 \left(\frac{(\hbar\frac{d}{dt_1})^2}{Y}\right),$$

where Y is again the A-model correlation function of V°. This is strikingly similar to the Picard-Fuchs equation (5.63) (especially if we let $q = e^{2\pi i t_1}$, so that $\delta = d/dt_1$). Finally, when the Calabi-Yau threefold V is a toric hypersurface and V° is its Batyrev mirror, the Hodge-Theoretic Toric Mirror Conjectures discussed in Section 8.6.4 assert (roughly speaking) that the Gauss-Manin connection on $H^3(V, \mathbb{C})$ coincides with the A-model connection on $\oplus_p H^{p,p}(V^\circ)$ under the coordinate change given by the mirror map. An immediate corollary will be the equality of the two Yukawa couplings.

But there is serious work to be done before this can happen. First of all, we need to give a careful definition of the mirror map. This requires a detailed study of complex and Kähler moduli. There are many compactifications of the complex moduli space, giving potentially many maximally unipotent boundary points to choose from, and the picture on the Kähler side is equally complicated. These issues will be explored in Chapter 6.

The other missing ingredient is a careful definition of the A-model correlation function or three-point function mentioned above. For this, we will need a careful study of Gromov-Witten invariants. These invariants and their relation to enumerative geometry will be discussed in Chapter 7. Then, when we combine this with quantum cohomology from Chapter 8, we will have all the ingredients needed to state a mathematical version of mirror symmetry. Proving some special cases of this will be the main topic of Chapter 11.

CHAPTER 6

Moduli Spaces

The goal of this chapter is to describe the complex and Kähler moduli spaces and to explain how to construct the mirror map between the Kähler moduli space of a Calabi-Yau manifold V and the complex moduli space of its mirror V°.

Our discussion will reveal that the complex and Kähler moduli have a surprising richness. We will devote special attention to Calabi-Yau hypersurfaces in toric varieties, for here one can describe explicitly (though somewhat conjecturally) what the complex and Kähler moduli look like. Actually, we will only examine the polynomial and toric parts of these moduli spaces, but these will be plenty interesting.

The intimate connection between complex and Kähler moduli is mediated by the mirror map, which will be defined at the end of the chapter. The difficulty is that one can't fully understand complex moduli without knowing about Kähler moduli, and vice versa. Furthermore, both require knowing the mirror map, yet a full understanding of the mirror map will have to wait until we study quantum cohomology and quantum differential equations in Chapters 8 and 11. Hence our discussion, while ostensibly proceeding from complex moduli to Kähler moduli to the mirror map, will in fact weave back and forth between these topics.

6.1. Complex Moduli

In this section, we will discuss the complex moduli of Calabi-Yau manifolds. It is shown in [**Viehweg**] that *polarized* Calabi-Yau manifolds $(V, \mathcal{L})$ are parameterized by a smooth quasi-projective variety. We will focus our attention on a connected component $\mathcal{M}$ of this moduli space.

Our goal is to find a convenient compactification $\overline{\mathcal{M}}$ of $\mathcal{M}$. We will adopt a utilitarian approach, so that we will be flexible in interpreting the meaning of "convenient". At a minimum, we will want the compactification divisor to have normal crossings near the maximally unipotent boundary points that are relevant for mirror symmetry. By resolution of singularities, a compactification with normal crossings always exists but is far from unique. However, it is not clear why a such a compactification should have maximally unipotent monodromy.

After a discussion of maximally unipotent monodromy, our primary focus will be on the moduli space of Calabi-Yau hypersurfaces arising from Batyrev's construction described in Section 4.1. We will then describe some examples in detail.

6.1.1. Maximally Unipotent Boundary Points. Suppose we have a family of Calabi-Yau manifolds over the punctured polydisc $(\Delta^*)^r \subset \Delta^r$ such that the monodromy is maximally unipotent at the origin. This is what we call a *maximally unipotent boundary point*. Let $z_1, \ldots, z_r$ be coordinates of Δ^r, and let $g_0, \ldots, g_r$ be as in the definition of maximally unipotent monodromy (Definition 5.2.2). We then define holomorphic functions $q_1, \ldots, q_r$ by the formula [**Morrison6**]

$$(6.1) \qquad \frac{1}{2\pi i} \log q_k = \frac{1}{\langle g_0, \Omega \rangle} \sum_{j=1}^{r} \langle g_j, \Omega \rangle m^{jk},$$

where m^{jk} is the inverse matrix to the matrix m_{ij} from Definition 5.2.2.

Let's compute the behavior of q_k near the origin. From $N_i g_j = m_{ij} g_0$, we note that $\langle g_j, \Omega \rangle \sim m_{ij} \langle g_0, \Omega \rangle \log(z_i)/(2\pi i)$ along $z_i = 0$. Substituting these asymptotics for all z_i into (6.1), we get $q_k \sim c \, z_k$ near the origin for some constant $c \neq 0$. Thus $(q_1, \ldots, q_r)$ define local coordinates near our maximally unipotent boundary point. Furthermore, these coordinates are almost unique. If we change $g_0, \ldots, g_r$ to $g_j' = \sum_{k=0}^{r} c_{jk} g_k$, then we get new local coordinates $(q_1', \ldots, q_r')$, and one can show that

$$(6.2) \qquad q_k' = e^{2\pi i (c^{k0}/c_{00})} q_k, \qquad c^{k0} = \sum_{j=1}^{r} c_{j0} m'^{jk},$$

where (m'^{jk}) is inverse to the matrix determined by $N_j(g_k') = m'_{jk} g_0'$. In Section 6.3, we will discuss how to determine the q_k uniquely.

Unfortunately, the "uniqueness" just mentioned assumes that we're working on a fixed normal crossings compactification of $(\Delta^*)^r$. Hence we need to determine how the q_k change when we pass to a different compactification. This need only be done locally, so consider a holomorphic map $\phi : \Delta^r \to \Delta^r$ which is an isomorphism when the source and target are restricted to $(\Delta^*)^r$, and which maps coordinate hyperplanes to coordinate hyperplanes. Shrinking the disks if necessary, it is easy to see that coordinates $w = (w_1, \ldots, w_r)$ on the source and $z = (z_1, \ldots, z_r)$ on the target can be chosen so that the map is given by

$$(6.3) \qquad z_j = \phi(w) = \prod_i w_i^{a_{ij}},$$

where the a_{ij} are nonnegative integers such that the matrix (a_{ij}) is invertible over $\mathbb{Z}$.

We claim that $w = 0$ is a maximally unipotent boundary point in the source if and only if $z = 0$ is maximally unipotent in the target. To prove this, we first compare the monodromies about the divisors $z_i = 0$ and the divisors $w_i = 0$. Let T_i^z and N_i^z respectively denote the monodromy and the logarithm of monodromy about the divisor $z_i = 0$, and similarly let T_i^w and N_i^w denote the monodromy and the logarithm of monodromy about the divisor $w_i = 0$. If we follow a path circling around $w_i = 0$ once, say

$$\theta \mapsto (p_1, \ldots, p_{i-1}, p_i e^{2\pi i \theta}, p_{i+1}, \ldots, p_r),$$

then substituting into (6.3) shows that the composition with ϕ gives a path winding a_{ij} times around $z_j = 0$. Thus, for each i, we obtain the formula

$$T_i^w = \prod_j (T_j^z)^{a_{ij}}.$$

Now suppose that the T_j^z are unipotent. Since they commute, the T_i^w are also unipotent, and the converse follows by the invertibility of (a_{ij}). This allows us to take logarithms, which gives

$$(6.4) \qquad N_i^w = \sum_j a_{ij} N_j^z.$$

For later purposes, note that this can be written as

$$(6.5) \qquad \begin{pmatrix} N_1^w \\ \vdots \\ N_r^w \end{pmatrix} = A \begin{pmatrix} N_1^z \\ \vdots \\ N_r^z \end{pmatrix},$$

where $A = (a_{ij})$.

We next compare the monodromy weight filtrations. On the source, this is determined by $N = \sum_i a_i N_i^w$ with $a_i > 0$. However, by (6.4), we have

$$N = \sum_i a_i N_i^w = \sum_j \Big(\sum_i a_i a_{ij} \Big) N_j^z.$$

We are assuming $a_{ij} \geq 0$ for all i, j, and it follows that $\sum_i a_i a_{ij} > 0$ since (a_{ij}) is invertible. This shows that we can use N on the target as well, so that source and target have the same monodromy weight filtration. From here, it is easy to see that Definition 5.2.2 is satisfied at $w = 0$ if and only if it holds at $z = 0$, and our claim is proved.

In this situation, we also use q_i^z and q_i^w to denote the coordinates determined in the differing local compactifications by (6.1). Then we compute

$$(6.6) \qquad q_j^z = \prod_i (q_i^w)^{a_{ij}} .$$

Note the similarity to (6.3).

It follows that one maximally unipotent boundary point gives rise to *lots* of others by means of (6.3), which in turn gives *lots* of local coordinates $(q_1, \dots, q_r)$, all related by (6.6). In Section 6.2.1, we will see that on the Kähler side, the large radius limit points have local coordinates with *exactly* the same indeterminacy.

Returning to the global situation, suppose that $\overline{\mathcal{M}}$ is a smooth normal crossings compactification of the complex moduli space $\mathcal{M}$. If $\overline{\mathcal{M}}$ has a maximally unipotent boundary point, then the above discussion shows that blowing-up gives other compactifications with several maximally unipotent boundary points. And it could also happen that $\overline{\mathcal{M}}$ is itself a blow-up. In fact, $\overline{\mathcal{M}}$ could have several maximally unipotent boundary points which are identified under a blow-down to a singular compactification.

One way to sort out these possibilities is to introduce an equivalence relation among maximally unipotent boundary points. Following [**Morrison6**], we make the following definition.

DEFINITION 6.1.1. *Let $\overline{\mathcal{M}}$ be a smooth normal crossings compactification of the moduli space $\mathcal{M}$. We say that two maximally unipotent boundary points $p, q \in \overline{\mathcal{M}}$ are equivalent if there is a connected set $\Xi \subset \overline{\mathcal{M}} - \mathcal{M}$ containing p and q and a local system $\mathcal{L}$ on the cotangent bundle of $\overline{\mathcal{M}}$ defined on a neighborhood of Ξ having logarithmic poles on the boundary such that the local system is spanned by the 1-forms $d \log q_k$ for q_k as in (6.1) near each of the points p and q.*

It is easy to see that the multiple maximally unipotent boundary points created by a blow-up are equivalent in this sense.

One of the basic ideas of mirror symmetry is that each equivalence class of maximally unipotent boundary points of the complex moduli space of the Calabi-Yau manifold V should correspond to a suitably defined equivalence class of large radius limit points (to be discussed in Section 6.2.1) on the Kähler moduli space of its mirror V°.

In practice, starting from a particular choice of $\overline{\mathcal{M}}$, one usually blows down as much as possible until we get a smooth normal crossings compactification which can't be blown down any further. This makes the equivalence classes as small as possible. We will see an example of this in Section 6.1.4.

We also note that the equivalence relation of Definition 6.1.1 can be defined on more general compactifications of $\mathcal{M}$. This is done in [**Morrison3**].

Finally, notice that we've said nothing about the *existence* of maximally unipotent boundary points. This is still conjectural, though they are present in every example computed to date, and as we will soon see, the moduli of toric hypersurfaces have naturally occurring compactifications with distinguished boundary points conjectured to be maximally unipotent (see [**HLY2**] for a nice partial result). In general, mirror symmetry predicts the existence of very special compactifications of $\mathcal{M}$ which not only have maximally unipotent boundary points but also reflect the structure of the Kähler moduli of the mirror. We will have more to say about this later in the chapter.

6.1.2. Complex Moduli of Toric Hypersurfaces.

We will study the complex moduli of the Calabi-Yau toric hypersurfaces. Recall from Section 4.1 that for a reflexive polytope Δ, the anticanonical hypersurfaces $\bar{V} \subset \mathbb{P}_\Delta$ are Calabi-Yau, though possibly singular. If Σ is a maximal projective subdivision of the normal fan of Δ, then we proved that the proper transform of $\bar{V}$ is again anticanonical, giving a family of minimal Calabi-Yau hypersurfaces $V \subset X = X_\Sigma$. Thus V is a Gorenstein orbifold with at worst terminal singularities and is smooth when $\dim(V) = 3$.

Rather than deal with the whole complex moduli space $\mathcal{M}$ of V, we will concentrate on the "polynomial" part $\mathcal{M}_{\mathrm{poly}} \subset \mathcal{M}$ consisting of those complex structures which can be realized as hypersurfaces in X. We will see later that if no points of Δ are interior points of codimension two faces, then $\mathcal{M}_{\mathrm{poly}}$ is the full complex moduli space.

To construct the polynomial moduli space, we begin with the vector space $L(\Delta \cap M)$ of Laurent polynomials of the form $\sum_{m \in \Delta \cap M} \lambda_m t^m$. As in Section 3.2.1, t^m is the character corresponding to m. (In Chapter 3, $L(\Delta \cap M)$ was denoted $L(\Delta)$, but here we want to make the dependence on the lattice more explicit.) By (3.7), we have a natural isomorphism

$$H^0(X, \mathcal{O}_X(-K_X)) \simeq L(\Delta \cap M).$$

We can think of a polynomial in $L(\Delta \cap M)$ as defining the affine part of an anticanonical hypersurface in X. Also, $-K_X = \sum_\rho D_\rho$ implies that

$$\Delta \cap M = \{ m \in M : \langle m, v_\rho \rangle \geq -1 \text{ for all } \rho \}.$$

Since the pullback of a canonical differential is again a canonical differential, the automorphism group $\mathrm{Aut}(X)$ acts naturally on the projective space $\mathbb{P}(L(\Delta \cap M))$, so that the polynomial moduli space is the quotient

$$\mathcal{M}_{\mathrm{poly}} = \mathbb{P}(L(\Delta \cap M))/\mathrm{Aut}(X).$$

To make this precise, we need to replace $\mathbb{P}(L(\Delta \cap M))$ with the subset corresponding to quasi-smooth hypersurfaces, and we also need to worry about the existence of the quotient. The latter concern is serious since $\mathrm{Aut}(X)$ need not be reductive. However, as pointed out in [**BC**, Sect. 13], one can find a nonempty invariant open set $U \subset \mathbb{P}(L(\Delta \cap M))$ such that a geometric quotient $\mathcal{M}_{\mathrm{poly}}$ exists. Hence we can construct the polynomial moduli, at least generically.

It would be nice to get a clearer picture of this moduli space, and for this reason, we next discuss the *simplified polynomial moduli space* $\mathcal{M}_{\mathrm{simp}}$ introduced in [**AGM1**]. Let $(\Delta \cap M)_0$ be the set of points of $\Delta \cap M$ which do not lie in the interior of any facet of Δ, so that $L((\Delta \cap M)_0)$ is the vector space of Laurent polynomials of the form

$$\sum_{m \in (\Delta \cap M)_0} \lambda_m t^m.$$

Since the torus $T = T_N$ acts on t^m via $\nu \cdot t^m = \nu^m t^m$ for $\nu \in T$, we get the induced action on $L((\Delta \cap M)_0)$, where

$$(6.7) \qquad \nu \cdot \lambda_m = \nu^m \lambda_m, \quad \nu \in T.$$

It follows that T acts on $\mathbb{P}(L((\Delta \cap M)_0))$, and one can check that under the inclusion $T \subset \mathrm{Aut}(X)$, this is compatible with the action of $\mathrm{Aut}(X)$ on $\mathbb{P}(L(\Delta \cap M))$.

We now define the simplified polynomial moduli space $\mathcal{M}_{\mathrm{simp}}$ to be

$$\mathcal{M}_{\mathrm{simp}} = \mathbb{P}(L((\Delta \cap M)_0))/T.$$

As before, we actually need to restrict to the subset of $\mathbb{P}(L((\Delta \cap M)_0))$ corresponding to quasi-smooth hypersurfaces, and we also need to worry about the existence of the quotient. Fortunately, the latter is not so bad since T is reductive and (6.7) linearizes the action. Hence we can not only construct $\mathcal{M}_{\mathrm{simp}}$ but also compactify it using the GIT quotient [**Mumford1**]

$$\overline{\mathcal{M}}_{\mathrm{simp}} = \mathbb{P}(L((\Delta \cap M)_0))//T.$$

For our purposes, however, the most interesting compactification is provided by the *Chow quotient* from [**KSZ1**]. We will see below that this compactification has a natural toric structure given by the secondary fan.

Our basic claim is that the obvious map $\phi : \mathcal{M}_{\mathrm{simp}} \to \mathcal{M}_{\mathrm{poly}}$ is a local isomorphism, at least generically. This is the *dominance conjecture* from [**AGM1**], which we now prove.

PROPOSITION 6.1.2. *The map* $\phi : \mathcal{M}_{\mathrm{simp}} \to \mathcal{M}_{\mathrm{poly}}$ *is generically étale.*

PROOF. We will let $\mathrm{Aut}_0(X)$ denote the connected component of the identity of $\mathrm{Aut}(X)$. As noted in [**Cox**], $\mathrm{Aut}_0(X)$ has finite index in $\mathrm{Aut}(X)$, so that it suffices to show that the natural map

$$\mathbb{P}(L((\Delta \cap M)_0))/T \longrightarrow \mathbb{P}(L(\Delta \cap M))/\mathrm{Aut}_0(X)$$

is dominant.

Let $I = (\Delta \cap M) - (\Delta \cap M)_0$ be the set of integral points of Δ which are in the interior of a facet. By the proof of Proposition 3.6.2, each $m \in I$ determines a 1-parameter family of automorphisms $y_{m,u} : X \to X$, where $u \in \mathbb{C}$ is the parameter. To describe $y_{m,u}$, we use the homogeneous coordinate ring $S = \mathbb{C}[x_\rho]$ from Section 3.2.3. Namely, since m is the interior of a facet, there is ρ such that $\langle m, v_\rho \rangle = 1$ and $\langle m, v_{\rho'} \rangle > 1$ for all $\rho' \neq \rho$. If we let $\mathbf{x}^D = \Pi_{\rho' \neq \rho} x_{\rho'}^{\langle m, v_{\rho'} \rangle}$, then the corresponding 1-parameter family of automorphisms is given by

$$y_{m,u}(x_\rho, \mathbf{x}) = (x_\rho + u\, \mathbf{x}^D, \mathbf{x}).$$

Here, as in (3.21), we have represented points of $\mathbb{C}^{\Sigma(1)}$ by $(x_\rho, \mathbf{x})$, where $\mathbf{x}$ is a vector indexed by $\Sigma(1) - \{\rho\}$.

Consider the $|I|$ parameter family of automorphisms

$$(6.8) \qquad y_\mathbf{u} = \prod_{m_i \in I} y_{m_i, u_i}.$$

An ordering of I needs to be chosen to define the right hand side of (6.8), since the y_{m_i, u_i} need not commute. If we combine the inclusion $T \subset \mathrm{Aut}_0(X)$ with the map $\mathbf{u} \mapsto y_\mathbf{u}$, we get a map

$$T \times \mathbb{C}^I \longrightarrow \mathrm{Aut}_0(X)$$

which is easily seen to be dominant. Hence it suffices to consider the action of $T \times \mathbb{C}^I$ on $\mathbb{P}(L(\Delta \cap M))$, and the proposition will follow provided we can show that generically, the orbits of $\mathbb{C}^I$ have dimension $|I|$ and meet $\mathbb{P}(L((\Delta \cap M)_0))$ transversely.

We begin by computing the action of $y_{m,u}$ on a Laurent monomial $t^\alpha \in L(\Delta \cap M)$. To do this, one needs to map t^α to the monomial $\Pi_\rho x_\rho^{\langle \alpha, v_\rho \rangle + 1}$ (see Section 3.2.3), apply $y_{m,u}$, and then map back to $L(\Delta \cap M)$. If $\langle m, v_\rho \rangle = -1$ determines the facet m lies in, then one can compute that

$$y_{m,u}(t^\alpha) = \begin{cases} t^\alpha & \text{if } \langle \alpha, v_\rho \rangle = -1 \\ t^\alpha + u(\langle \alpha, v_\rho \rangle + 1)\, t^{\alpha+m} + \text{h.o.t.} & \text{if } \langle \alpha, v_\rho \rangle > -1. \end{cases}$$

One can check that $\langle \alpha, v_\rho \rangle > -1$ implies that $\alpha + m \in \Delta \cap M$. See Example 6.1.4.2 for a sample computation, including the higher order terms omitted above.

Since Δ is reflexive, the origin 0 is contained in $(\Delta \cap M)_0$, so that $1 = t^0 \in L((\Delta \cap M)_0)$. Then one easily obtains

$$y_\mathbf{u}(1) = 1 + \sum_{m_i \in I} u_i\, t^{m_i} + \text{h.o.t.}$$

This shows that the orbit of $\mathbb{C}^I$ acting on 1 has the correct dimension, and its tangent space at 1 is clearly complementary to $L((\Delta \cap M)_0)$. Because the constant term of $y_\mathbf{u}(1)$ is always 1, the same is true once we projectivize, and the proposition follows easily. $\qquad\square$

This proposition implies that $\mathcal{M}_{\mathrm{simp}}$ is a finite cover of $\mathcal{M}_{\mathrm{poly}}$. From knowledge of the automorphism group of X, we can calculate the finite group acting on $\mathcal{M}_{\mathrm{simp}}$ explicitly. This has been done in an example in [**CFKM**], and Example 6.1.4.2 below will illustrate the technique in a slightly simpler situation.

We next compute the dimensions of $\mathcal{M}_{\mathrm{poly}}$ and the full complex moduli space $\mathcal{M}$ and determine when the two are equal.

PROPOSITION 6.1.3. *Let $V \subset X = X_\Sigma$ be a minimal Calabi-Yau toric hypersurface, where Σ is a maximal projective subdivision of the normal fan of the n-dimensional reflexive polytope Δ. Then:*

(*i*) *The polynomial moduli space $\mathcal{M}_{\mathrm{poly}}$ of V has dimension*

$$\dim(\mathcal{M}_{\mathrm{poly}}) = l(\Delta) - n - 1 - \sum_\Gamma l^*(\Gamma),$$

where $l(\Delta) = |\Delta \cap M|$, the sum is over all facets Γ of Δ, and $l^*(\Gamma)$ is the number of lattice points in the relative interior of Γ.

(*ii*) *The complex moduli space $\mathcal{M}$ of V has dimension*

$$\dim(\mathcal{M}) = l(\Delta) - n - 1 - \sum_{\Gamma} l^*(\Gamma) + \sum_{\Theta} l^*(\Theta) l^*(\widehat{\Theta}^\circ),$$

where the second sum is over all codimension two faces of Δ, and $\widehat{\Theta}^\circ$ is the 2-dimensional face of Δ° dual to Θ.

(*iii*) *If no points of $\Delta \cap M$ are interior points of codimension two faces, then $\mathcal{M} = \mathcal{M}_{\text{poly}}$.*

PROOF. Computing the dimension of $\mathcal{M}_{\text{simp}}$ is easy, for generically, the action of T on $\mathbb{P}(L((\Delta \cap M)_0))$ has finite stabilizers since Δ has dimension n. Thus

$$\dim(\mathcal{M}_{\text{simp}}) = |(\Delta \cap M)_0| - n - 1,$$

which is also the dimension of $\mathcal{M}_{\text{poly}}$. Since $l(\Delta) = |(\Delta \cap M)_0| + \sum_{\Gamma} l^*(\Gamma)$, the first part of the proposition follows.

For the second part, we note that since V is a minimal Calabi-Yau orbifold, its infinitesimal deformations are classified by $H^1(V, \Theta_V) \simeq H^1(V, \widehat{\Omega}_V^{n-2})$ by Proposition A.4.3 (remember that V has dimension $n - 1$). Furthermore, by a result of Ran [**Ran2**], these deformations are unobstructed since the singularities of V have codimension at least 4 by Proposition A.2.2. It follows that the dimension of $\mathcal{M}$ is $h^{n-2,1}(V)$. This Hodge number is computed in (4.8), which proves the second part of the proposition. The final part follows immediately from the other two, so we are done. $\qquad\square$

This proposition relates nicely to Proposition 4.1.5 in Chapter 4. The tangent space to $\mathcal{M}$ at a generic point V is $H^1(V, \Theta_V) \simeq H^{n-2,1}(V)$, and what we called $H_{\text{poly}}^{n-2,1}(V) \subset H^{n-2,1}(V)$ in Section 4.1 is the tangent space to $\mathcal{M}_{\text{poly}}$ at V. Thus the first part of Proposition 6.1.3 gives a rigorous proof of (4.5).

Now that we know more about $\mathcal{M}_{\text{poly}}$, our next task is to compactify it and see if we can find some maximally unipotent boundary points. Actually, we will work with $\mathcal{M}_{\text{simp}}$ because it has an especially nice compactification. As mentioned earlier, we will use the Chow quotient

$$(6.9) \qquad \overline{\mathcal{M}}_{\text{simp}} = \mathbb{P}(L((\Delta \cap M)_0)) /\!/ T$$

from [**KSZ1**]. This quotient is a toric variety, and we begin by describing the torus involved. First note that $L((\Delta \cap M)_0)$ contains the torus $(\mathbb{C}^*)^{(\Delta \cap M)_0}$. Homotheties give a natural map $\mathbb{C}^* \to (\mathbb{C}^*)^{(\Delta \cap M)_0}$ (we need this to projectivize), and (6.7) shows that the action of T comes from the map $T \to (\mathbb{C}^*)^{(\Delta \cap M)_0}$ given by the characters t^m for $m \in (\Delta \cap M)_0$. Thus we have the torus

$$(6.10) \qquad T_0 = (\mathbb{C}^*)^{(\Delta \cap M)_0} / \text{im}(T \times \mathbb{C}^*)$$

which is contained in the above quotient. We can get rid of the $\mathbb{C}^*$ by using the distinguished point $0 \in (\Delta \cap M)_0$. More precisely, if we let $\Xi = (\Delta \cap M)_0 - \{0\}$, then

$$T_0 = (\mathbb{C}^*)^{\Xi} / \text{im}(T).$$

The 1-parameter subgroups of T_0 form the lattice

$$(6.11) \qquad N_0 = \mathbb{Z}^{\Xi} / N \text{ modulo torsion},$$

where the inclusion $N \to \mathbb{Z}^{\Xi}$ is given by $u \mapsto (\langle u, m_1 \rangle, \dots, \langle u, m_s \rangle)$ for $\Xi = \{m_1, \dots, m_s\}$.

The lattice N_0 is important, for the GKZ decomposition of $\Xi = (\Delta \cap M)_0 - \{0\}$, as defined in Section 3.4, lives naturally in the vector space $N_0 \otimes \mathbb{R}$. Furthermore, by Theorem 3.4.1, the maximal cones in this fan are the cones $\mathrm{cpl}(\Sigma^\circ)$ for all fans Σ° in $M_\mathbb{R}$ which are projective, simplicial, and satisfy $\Sigma^\circ(1) \subset \Xi$.

This is closely related to the Batyrev mirror of V. Recall how the construction works: if $\Delta^\circ \subset N_\mathbb{R}$ is the polar polytope and Σ° is a maximal projective subdivision of the normal fan of Δ°, then the Batyrev mirror is the family of anticanonical hypersurfaces $V^\circ \subset X_{\Sigma^\circ}$. According to Definition 4.1.1, Σ° is projective and simplicial, refines the normal fan of Δ°, and satisfies $\Sigma^\circ(1) = \Delta \cap M - \{0\}$. However, we will see in Section 6.2.3 that we can replace the last condition with $\Sigma^\circ(1) = (\Delta \cap M)_0 - \{0\} = \Xi$ without changing V°. Once we do this, $V^\circ \subset X_{\Sigma^\circ}$ gives one of the cones $\mathrm{cpl}(\Sigma^\circ)$ in the GKZ decomposition of Ξ.

It follows that for each choice of a Batyrev mirror $V^\circ \subset X_{\Sigma^\circ}$, we get a distinguished maximal cone in the GKZ decomposition. In the toric variety corresponding to the GKZ decomposition, these cones give distinguished points on the toric boundary. They may be singular points, but as we will soon see, their resolutions are conjectured to give the desired maximally unipotent boundary points. Also, as a preview of Section 6.2.3, we should mention that the interior of $\mathrm{cpl}(\Sigma^\circ)$ is the Kähler cone of X_{Σ° and is naturally isomorphic to the toric part of the Kähler cone of V°. This is the beginning of the isomorphism between the complex moduli of V and the Kähler moduli of its mirror V°.

We can now give a careful description of the Chow quotient (6.9). The GKZ decomposition of Ξ from Theorem 3.4.1 is not complete, but as described in Section 3.4, can be naturally enlarged to a complete fan called the *secondary fan* of Ξ. For later purposes, we note that the secondary fan is constructed from the set

$$(6.12) \qquad \Xi^+ = (\Xi \cup \{0\}) \times \{1\} = (\Delta \cap M)_0 \times \{1\} \subset M \times \mathbb{Z}.$$

Then [**KSZ2**] shows that the Chow quotient $\overline{\mathcal{M}}_{\mathrm{simp}}$ is the toric variety determined by the secondary fan of Ξ. The relation between the Chow quotient and the various GIT quotients is explored in [**KSZ1**]. We should also mention that $\overline{\mathcal{M}}_{\mathrm{simp}}$ has a natural projective embedding given by the *secondary polytope* of [**GKZ2**].

The toric compactification $\overline{\mathcal{M}}_{\mathrm{simp}}$ contains distinguished toric boundary points corresponding to the different choices of the mirror family. However, $\overline{\mathcal{M}}_{\mathrm{simp}}$ isn't always smooth, so that we need to resolve singularities before talking about maximally unipotent boundary points. Hence we pick a refinement of the secondary fan which gives a smooth toric variety. The resulting smooth compactification of $\mathcal{M}_{\mathrm{simp}}$ will be denoted $\widetilde{\mathcal{M}}_{\mathrm{simp}}$.

This smooth compactification has the property that each Batyrev mirror $V^\circ \subset X_{\Sigma^\circ}$ gives finitely many distinguished boundary points of $\widetilde{\mathcal{M}}_{\mathrm{simp}}$ corresponding to the cones subdividing $\mathrm{cpl}(\Sigma^\circ)$. Conjecturally, these are the maximally unipotent boundary points we seek.

CONJECTURE 6.1.4. *Let $\widetilde{\mathcal{M}}_{\mathrm{simp}}$ be a smooth toric resolution of the Chow quotient $\overline{\mathcal{M}}_{\mathrm{simp}}$, and let $V^\circ \subset X_{\Sigma^\circ}$ be a Batyrev mirror of V. Then the points of $\widetilde{\mathcal{M}}_{\mathrm{simp}}$ corresponding to the cones subdividing $\mathrm{cpl}(\Sigma^\circ)$ are all maximally unipotent boundary points. Furthermore, these points are all equivalent under the equivalence relation given in Definition 6.1.1.*

An incomplete proof of this conjecture appeared in [**BK1**]. If $H^{1,1}(V)$ generates $\oplus_p H^{p,p}(V)$ under cup product, then the conjecture would follow easily from a proof that the coefficients of the formal function I_V defined in (11.73) are indeed the periods of the mirror family V°, as asserted in [**Givental4**] (see the discussion in Section 5.5.3). In particular, this would apply when V is a threefold. It can be shown that Conjecture 6.1.4 holds for Examples 6.1.4.1 and 6.1.4.2 below. The proof for Example 6.1.4.1 follows for instance from the computations given in the paper [**CdFKM**].

One of the motivations for Conjecture 6.1.4 is that under mirror symmetry, the logarithm of monodromy around a boundary divisor is identified with cup product with the negative of the corresponding generator of the cone subdividing $\mathrm{cpl}(\Sigma^\circ)$, as mentioned in the discussion following the definition of maximally unipotent boundary points (Definition 5.2.2). In Chapter 8, we will see this from a more sophisticated point of view when we discuss the A-variation of Hodge structure.

For the rest of this chapter, we will assume that Conjecture 6.1.4 is true.

The maximally unipotent boundary points described in Conjecture 6.1.4 come from very special cones in the secondary fan. There are other cones in the GKZ decomposition that correspond to Σ°'s which don't use all of the points of Ξ or fail to refine the normal fan of Δ°, and there are also the cones we get when we enlarge the GKZ decomposition to the secondary fan. Conjecture 6.1.4 says nothing about the boundary points of $\overline{\mathcal{M}}_{\mathrm{simp}}$ corresponding to such cones. However, in Section 6.2.3, we will see that *all* of these cones have interesting physical interpretations—each one will give a different physical theory.

We should also mention that there is a canonical refinement of the secondary fan given by the *Gröbner fan* [**MR, Sturmfels1**]. This gives a compactification of $\mathcal{M}_{\mathrm{simp}}$ which is a blow-up of $\overline{\mathcal{M}}_{\mathrm{simp}}$ and hence potentially less singular. This compactification is used in [**HLY1, HLY2**], where it is shown that there exist boundary points of *maximal degeneracy*, which means that locally there exists a unique (up to a multiple) holomorphic period integral. This is a necessary condition for maximally unipotent and hence would follow from Conjecture 6.1.4.

Our discussion so far has concentrated on blowing up $\overline{\mathcal{M}}_{\mathrm{simp}}$, as if $\overline{\mathcal{M}}_{\mathrm{simp}}$ were the "minimal" compactification of $\mathcal{M}_{\mathrm{simp}}$. But there are situations where combining cones of the secondary fan into a larger cone is completely natural. This gives a compactification $\overline{\mathcal{M}}'_{\mathrm{simp}}$ such that $\overline{\mathcal{M}}_{\mathrm{simp}}$ is a blowup of $\overline{\mathcal{M}}'_{\mathrm{simp}}$. We will see an example of this behavior in Example 6.1.4.2, and a careful study of how and when this should be done can be found in Section 6.2.3.

Another point to note is that besides having multiple mirrors, there are also multiple versions of our original family $V \subset X_\Sigma$, because of the potentially many choices for the maximal projective subdivision Σ. Our description of $\overline{\mathcal{M}}_{\mathrm{simp}}$ makes no mention of Σ, for the good reason that these families all have the same polynomial moduli. What changes is the Kähler cone of V, since each Σ corresponds to a different part of the enlarged Kähler moduli space of V to be discussed in Section 6.2.3. In Chapter 4, the multiple choices for Σ and Σ° seemed to be a nuisance because they prevented us from making a canonical choice for the mirror. But as we are now learning, the "multiple mirror" phenomenon is essential to a global understanding of complex and Kähler moduli.

The compactification $\overline{\mathcal{M}}_{\mathrm{simp}}$ has a toric boundary which contains the maximally unipotent boundary points just discussed. But there are points in $\overline{\mathcal{M}}_{\mathrm{simp}} - \mathcal{M}_{\mathrm{simp}}$ not in the toric boundary, for the torus $T_0 \subset \overline{\mathcal{M}}_{\mathrm{simp}}$ has many points which correspond to singular hypersurfaces $V \subset X_\Sigma$. Actually, since V and X_Σ may be singular, it is usually easier to speak of *nondegenerate* hypersurfaces. For $V \subset X_\Sigma$, this means that for each torus orbit $\mathcal{O} \subset X_\Sigma$, the intersection $V \cap \mathcal{O}$ is either empty or smooth of codimension one. Since X_Σ is an orbifold, it follows easily that every nondegenerate hypersurface of X_Σ is a suborbifold.

In $L((\Delta \cap M)_0)$, the *discriminant locus* $\mathcal{D}$ is the set of all Laurent polynomials f such that $f = 0$ fails to be nondegenerate. By [**GKZ2**, Chap. 10], the defining equation of $\mathcal{D}$ is given by the *principal A-determinant*. If we let $A = (\Delta \cap M)_0 = \Xi \cup \{0\}$, this is a polynomial in the coefficients of f and is defined by

$$E_A(f) = \mathrm{Res}_A(t_1 \partial f/\partial t_1, \dots, t_n \partial f/\partial t_n, f),$$

where Res_A is the sparse resultant defined in [**GKZ2**, Chap. 8] and $t_1, \dots, t_n$ are coordinates of the torus T. This polynomial is actually reducible, and one has a factorization

$$E_A(f) = \prod_{\Gamma \subset \Delta} D_{A \cap \Gamma}(f_\Gamma)^{m(\Gamma)}.$$

We can understand the right hand side as follows. The *A-discriminant* D_A defines the Zariski-closure of the set of polynomials in $L((\Delta \cap M)_0)$ which have a singular point in the torus T (so $D_A = 1$ if this set has codimension > 1). Then, for each face $\Gamma \subset \Delta$, one defines f_Γ to be the sum of those terms of f corresponding to elements of $\Gamma \cap A$. Finally, in the above equation, the exponent $m(\Gamma)$ is a positive integer equal to 1 in nice cases.

It follows from this description that the discriminant locus has potentially many connected components. Fortunately, it often happens that many of the $D_{A \cap \Gamma}$'s are 1. For instance, the discriminant locus in Example 6.1.4.1 below has exactly two components. We should also mention that the discriminant locus $\mathcal{D}$ is precisely the discriminant locus of the GKZ hypergeometric system (to be described below).

When we descend from $L((\Delta \cap M)_0)$ to the Chow quotient, the image of the discriminant locus is again denoted $\mathcal{D}$. The *principal component* of the discriminant is the hypersurface defined by $D_A = 0$. When V is a threefold, its generic point parametrizes hypersurfaces V with a node (three dimensional nodes are often called *conifolds* in the physics literature). The other nonempty components of the discriminant parametrize singular hypersurfaces with singularities that may be more complicated.

In addition to the discriminant locus just described, the toric boundary of $\overline{\mathcal{M}}_{\mathrm{simp}}$ also parametrizes singular hypersurfaces, although the parameter values λ_i may not be well-defined. If we think of this in terms of a GIT quotient, the problem is that these parameter values might correspond to non-stable hypersurfaces. Hence the hypersurfaces themselves may be ill-defined when thought of in $\overline{\mathcal{M}}_{\mathrm{simp}}$. For example, the limit $\psi \to \infty$ of the quintic mirror family (2.4) can be taken to be $x_1 x_2 x_3 x_4 x_5 = 0$, yet after the coordinate change $x_5 \mapsto x_5/\psi$, the limit is $x_1^4 + x_2^4 + x_3^4 + x_4^4 + x_5^4 + x_1 x_2 x_3 x_4 x_5 = 0$.

6.1.3. Hypergeometric Equations and Moduli.

In Section 5.5, we defined the GKZ hypergeometric equations and discussed (rather naively) how they behave on the moduli of V. Now that we know more about complex moduli, we can get a better idea of what the GKZ system looks like.

The hypergeometric equations constructed in Section 5.5 used $\Delta \cap M$. Here, because we want to work with the moduli space $\overline{\mathcal{M}}_{\mathrm{simp}}$, we will use $(\Delta \cap M)_0$

instead. Following (5.31), this means that we will use the GKZ system built from the set

$$(\Delta \cap M)_0 \times \{1\} = \Xi^+$$

identified in (6.12). As in Chapter 5, we label the points of $(\Delta \cap M)_0$ as m_i with $m_0 = 0$. Then the module of relations among the points of Ξ^+ is

$$(6.13) \qquad M_0 = \big\{\ell = (\ell_i) \in \mathbb{Z}^{\Xi^+} : \textstyle\sum_i \ell_i m_i = 0, \ \sum_i \ell_i = 0\big\}.$$

From (6.10), we see that M_0 can be identified as the character group of the torus T_0, so that N_0 from (6.11) is the dual of M_0. The GKZ hypergeometric system consists of operators $\Box_\ell$ for $\ell \in M_0$ and Z_j for $j = 1, \ldots, n+1$. Precise definitions of $\Box_\ell$ and Z_j can be found in Section 5.5.

In Chapter 5, we used $(\mathbb{C}^*)^{\Delta \cap M}/\mathrm{im}(T \times \mathbb{C}^*)$ as a crude approximation of the moduli space. The subspace $(\mathbb{C}^*)^{(\Delta \cap M)_0}/\mathrm{im}(T \times \mathbb{C}^*)$ is precisely the torus $T_0 \subset \overline{\mathcal{M}}_{\mathrm{simp}}$. Then $f \in L((\Delta \cap M)_0)$ gives the n-form

$$\widetilde{\omega} = \frac{\lambda_0}{f} \frac{dt_1}{t_1} \wedge \cdots \wedge \frac{dt_n}{t_n}$$

on T_0, and we showed that $\widetilde{\omega}$ satisfies the equations $\Box_\ell \lambda_0^{-1}\widetilde{\omega} = 0$ and $Z_j \lambda_0^{-1}\widetilde{\omega} = 0$. The latter simply express the invariance of $\widetilde{\omega}$ under the actions of T and $\mathbb{C}^*$ (because of the extra factor of λ_0 in $\widetilde{\omega}$), so that $\widetilde{\omega}$ descends to a rational n-form on $\overline{\mathcal{M}}_{\mathrm{simp}}$. Hence the GKZ system on $\overline{\mathcal{M}}_{\mathrm{simp}}$ is given by

$$\Box_\ell \lambda_0^{-1}\widetilde{\omega} = 0, \quad \ell \in M_0.$$

To exploit these equations, we need to express them in local coordinates on $\overline{\mathcal{M}}_{\mathrm{simp}}$. Each $\ell = (\ell_i) \in M_0$ gives

$$(6.14) \qquad z = \prod_i \lambda_i^{\ell_i}$$

which is invariant under $T \times \mathbb{C}^*$, so picking a basis of M_0 gives coordinates $z_1, \ldots, z_r$ of the torus T_0. The methods of Section 5.5 then enable us to write the equations $\Box_\ell \lambda_0^{-1}\widetilde{\omega} = 0$ in terms of the z_i, though Section 5.5 didn't specify which basis of M_0 to use.

Now, given Conjecture 6.1.4, we know exactly how to pick the coordinates. Fix a smooth toric resolution $\widetilde{\mathcal{M}}_{\mathrm{simp}}$ of $\overline{\mathcal{M}}_{\mathrm{simp}}$ and let σ be a maximal cone in the fan of the resolution which subdivides a cone $\mathrm{cpl}(\Sigma^\circ)$ coming from a Batyrev mirror $V^\circ \subset X_{\Sigma^\circ}$ of V. Since $\widetilde{\mathcal{M}}_{\mathrm{simp}}$ is smooth, the generators of σ form a basis of N_0, and the dual basis of M_0 gives coordinates $z_1, \ldots, z_r$ which are local coordinates for the point of $\widetilde{\mathcal{M}}_{\mathrm{simp}}$ corresponding to σ.

We will see a nice example of this in Section 6.1.4 below. We should also mention that the special coordinates $z_1, \ldots, z_r$ constructed above are part of the mirror map, to be discussed in more detail in Section 6.3.

6.1.4. Examples. We will now give two examples of the complex moduli of Calabi-Yau toric hypersurfaces.

Example 6.1.4.1. We first consider the mirror family of anticanonical hypersurfaces in the toric blowup of $\mathbb{P}(1, 1, 2, 2, 2)$ which was considered previously in

Sections 5.4, 5.5 and 5.6. In the notation of Example 5.4.2, we have the reflexive polytope Δ° with vertices

$$(-1,-2,-2,-2),\ (1,0,0,0), (0,1,0,0),\ (0,0,1,0)\ ,(0,0,0,1).$$

We want to study the complex moduli of $V^\circ \subset X_{\Sigma^\circ}$, where Σ° is a maximal projective subdivision refining the normal fan of Δ°.

Let's first show that $\overline{\mathcal{M}} = \overline{\mathcal{M}}_{\mathrm{poly}} = \overline{\mathcal{M}}_{\mathrm{simp}}$. In Example 5.4.2, we noted that $\Delta^\circ \cap N$ consists of the 5 vertices of Δ° together with $(0,-1,-1,-1) = \frac{1}{2}(-1,-2,-2,-2) + \frac{1}{2}(1,0,0,0)$ and $(0,0,0,0)$. This gives a codimension 2 face of Δ° with an interior lattice point, but the dual dimension 2 face of Δ has no interior lattice points, so that $\overline{\mathcal{M}} = \overline{\mathcal{M}}_{\mathrm{poly}}$ by Proposition 6.1.3.

Note also that the facets of Δ° have no interior lattice points. Hence $\mathrm{Aut}(X_{\Sigma^\circ})$ has no roots by Proposition 3.6.2, which means that the torus T° of X_{Σ° is the connected component of the identity of the automorphism group. However, we have $\mathrm{Aut}(X_{\Sigma^\circ})/T^\circ \simeq \mathbb{Z}/2\mathbb{Z} \times S_3$ because of the automorphisms of the normal fan of Δ° (we get $\mathbb{Z}/2\mathbb{Z}$ by switching the first two vertices of Δ° and S_3 by permuting the remaining three). We need to show that these automorphisms act trivially on $\overline{\mathcal{M}}_{\mathrm{simp}}$. To prove this, note that $f \in L((\Delta \cap M)_0)$ can be written

$$f = \lambda_1 t_1^{-1} t_2^{-2} t_3^{-2} t_4^{-2} + \lambda_2 t_1 + \lambda_3 t_2 + \lambda_4 t_3 + \lambda_5 t_4 + \lambda_6 t_2^{-1} t_3^{-1} t_4^{-1} + \lambda_0.$$

The automorphism switching t_2 and t_3 (one of the S_3 ones) induces an automorphism of $L((\Delta \cap M)_0)$ which switches λ_3 and λ_4. However, one can check that $(1, \lambda_4/\lambda_3, \lambda_3/\lambda_4, 1) \in T^\circ$ has the same effect on f, so that this automorphism acts trivially on $(T^\circ \times \mathbb{C}^*)$-orbits. The other automorphisms are handled similarly, and we conclude that $\overline{\mathcal{M}}_{\mathrm{poly}} = \overline{\mathcal{M}}_{\mathrm{simp}}$.

The next step is to construct the secondary fan, which is built from $\Xi^+ = (\Delta \cap M)_0 \times \{1\}$. Thus

$$\Xi^+ = \big\{(-1,-2,-2,-2,1),\ (1,0,0,0,1),\ (0,1,0,0,1),$$
$$(0,0,1,0,1), (0,0,0,1,1),\ (0,-1,-1,-1,1),\ (0,0,0,0,1)\big\}.$$

By (6.13), M_0 is the lattice of linear relations among the points of Ξ^+. A basis is given by the rows of the matrix

$$(6.15) \qquad \begin{pmatrix} 0 & 0 & 1 & 1 & 1 & 1 & -4 \\ 1 & 1 & 0 & 0 & 0 & -2 & 0 \end{pmatrix}.$$

We also get the dual basis of N_0. To describe the secondary fan, we use Section 3.4, which implies that in the rank two case, the secondary fan is the complete fan in $N_0 \otimes \mathbb{R}$ which has the columns of the matrix (6.15) as 1-dimensional cone generators. Thus, the secondary fan is obtained by enlarging the GKZ decomposition (3.23) by inserting the cone generated by $(-1,0)$ and completing.

Looking back at (3.23), we see that V° has a unique Batyrev mirror, namely the toric blowup of $\mathbb{P}(1,1,2,2,2)$ described in Example 3.7.2. The corresponding cone in the GKZ decomposition is the first quadrant, which means that the local coordinates z_1, z_2 we want are given by rows of the matrix (6.15). Using (6.14), the rows give

$$z_1 = \frac{\lambda_3 \lambda_4 \lambda_5 \lambda_6}{\lambda_0^4}, \qquad z_2 = \frac{\lambda_1 \lambda_2}{\lambda_6^2}.$$

As noted in Section 5.5, we can transform the general $f \in L((\Delta \cap M)_0)$ into the form

$$(6.16) \qquad f = z_2\, t_1^{-1} t_2^{-2} t_3^{-2} t_4^{-2} + t_1 + z_1\, t_2 + t_3 + t_4 + t_2^{-1} t_3^{-1} t_4^{-1} + 1$$

using $T^\circ \times \mathbb{C}^*$. Hence this way of writing f, first used in (5.25), was no accident—it's precisely what's needed to give a maximally unipotent boundary point by Conjecture 6.1.4.

Outside of the toric boundary, the discriminant locus $\mathcal{D} \subset \overline{\mathcal{M}}_{\mathrm{simp}}$ has two components in this case: the principal component defined by $D_A = 0$, for $A = (\Delta \cap M)_0$, and the component $D_{\Gamma \cap A} = 0$ associated to the edge Γ of Δ with

$$\Gamma \cap M = \{(-1,-2,-2,-2), (1,0,0,0), (0,-1,-1,-1)\} = \{m_1, m_2, m_6\}.$$

To determine D_A, one writes f in the form (6.16) and determines the values of z_1, z_2 for which the equations $f = t_i \partial f / \partial t_i = 0$ have a solution in T°. This can be done by hand, and one finds that

$$D_A(f) = 512^2\, z_1^2 z_2 - (1 - 256\, z_2)^2.$$

Turning to the edge Γ, we have $f_\Gamma = z_2\, t^{m_1} + t^{m_2} + t^{m_6}$. Since $m_1 + m_2 = 2\, m_6$, we see that $t^{m_1} t^{m_2} = (t^{m_6})^2$, and one finds that $D_{A \cap \Gamma}(f_\Gamma) = 1 - 4 z_2$. The Picard-Fuchs equation (5.26) degenerates when $1 - 4 z_2 = 0$, so that this equation picks out one component of the discriminant locus. As for the other component $D_A = 0$, the Yukawa couplings computed in Example 5.6.1 show that the GKZ system also degenerates on this component.

Example 6.1.4.2. We consider the reflexive polytope Δ which is the convex hull of the points

$$(6.17) \qquad \begin{matrix} (-1,-2,-3,-7), (1,0,0,0), (0,1,0,0), \\ (0,0,1,0), (0,0,0,1), (0,0,\;\;1,\;\;2) \end{matrix}$$

in $\mathbb{Z}^4 = M$. The Batyrev construction gives mirror families $V \subset X_\Sigma$ and $V^\circ \subset X_{\Sigma^\circ}$. Our goal here is to describe the complex moduli of V. We will consider the Kähler moduli of V° in Section 6.2.3.

A first observation is that the lattice points of Δ are the six vertices from (6.17) together with the origin $(0,0,0,0)$ and the point $(0,0,0,-1)$, which is interior to the facet spanned by all vertices except the fifth. It follows that $\overline{\mathcal{M}} = \overline{\mathcal{M}}_{\mathrm{poly}}$ since the codimension 2 faces have no interior points (Proposition 6.1.3). Since we have exactly one interior point in a facet, the same proposition implies that we have two-dimensional moduli.

Having an interior point in a facet also tells that $\mathrm{Aut}(X)$ is larger than the torus T. However, we still have $\overline{\mathcal{M}}_{\mathrm{poly}} = \overline{\mathcal{M}}_{\mathrm{simp}}$, as we will now show. We begin by describing the action of the extra automorphism arising from $(0,0,0,-1)$ on the space $L(\Delta \cap M)$. Although this can be done directly, it is instructive to do this using homogeneous coordinates. We recall that the fan Σ for X_Σ satisfies $\Sigma(1) = \Delta^\circ \cap N - \{0\}$. One checks that Δ° is the convex hull of the points

$$\begin{matrix} (13,-1,-1,-1), (1,-1,3,-1), (-1,6,-1,-1), (-1,-1,3,-1), \\ (-1,-1,-1,-1), (-1,-1,-1,1), (-1,0,3,-1). \end{matrix}$$

There are numerous other lattice points of Δ°. To simplify matters, recall that the hypersurfaces V are all pullbacks of anticanonical hypersurfaces in $\mathbb{P}_\Delta$. We can therefore perform the calculation using homogeneous coordinates associated to the normal fan itself. We will use the above order for the homogeneous coordinates $x_1, \dots, x_7$.

We know from Section 3.6 that the only root of $\mathrm{Aut}(X_\Sigma)$ comes from $(0,0,0,-1)$ since this is the only lattice point interior to a facet of Δ. Since the facet containing this point is dual to the vertex $(-1,-1,-1,1)$ of Δ°, one easily sees that the corresponding root is $(x_6, x_1 x_2 x_3 x_4 x_5 x_7)$. The corresponding 1-parameter family of automorphisms is given by $y_u(x_6) = x_6 + u\, x_1 x_2 x_3 x_4 x_5 x_7$ and $y_u(x_i) = x_i$ for $i \neq 6$.

Recall from the discussion preceding Proposition 3.6.2 that the Laurent monomial $t^m \in L(\Delta \cap M)$ gives the monomial $t^m = \Pi_i x_i^{\langle m, v_i \rangle + 1}$ in the homogeneous coordinate ring. In this formula, the v_i are the vertices of Δ° listed above. The points of $\Delta \cap M$ will be written as follows: the points of (6.17) are $m_1, \dots, m_6$ in the order written, while $m_7 = (0,0,0,-1)$ and $m_0 = (0,0,0,0)$. Then we obtain

$$
\begin{aligned}
t^{m_0} &= x_1 x_2 x_3 x_4 x_5 x_6 x_7 \\
t^{m_1} &= x_4^2 x_5^{14} \\
t^{m_2} &= x_1^{14} x_2^2 \\
t^{m_3} &= x_3^7 x_7 \\
t^{m_4} &= x_2^4 x_4^4 x_7^4 \\
t^{m_5} &= x_6^2 \\
t^{m_6} &= x_1^4 x_3^4 x_5^4 \\
t^{m_7} &= x_1^2 x_2^2 x_3^2 x_4^2 x_5^2 x_7^2
\end{aligned}
$$

The 1-parameter family of automorphisms y_u acts on these monomials via

$$
\begin{aligned}
y_u\big(t^{m_0}\big) &= t^{m_0} + u\, t^{m_7} \\
y_u\big(t^{m_5}\big) &= t^{m_5} + 2u\, t^{m_0} + u^2\, t^{m_7} \\
y_u\big(t^{m_i}\big) &= t^{m_i}, \quad i \neq 0, 5.
\end{aligned}
$$

We now prove that the natural map $\overline{\mathcal{M}}_{\mathrm{simp}} \to \overline{\mathcal{M}}_{\mathrm{poly}}$ is generically one-to-one. The first step is to show that the automorphisms of the lattice which preserve Δ act trivially on $\overline{\mathcal{M}}_{\mathrm{simp}}$—this is done using the methods of Example 6.1.4.1. Hence we need only consider the connected component of the identity of $\mathrm{Aut}(X_\Sigma)$, which is generated by T and the automorphisms y_u. Take a point of $\overline{\mathcal{M}}_{\mathrm{simp}}$ represented by $f = \sum_{i=0}^{6} \lambda_i t^{m_i}$ and apply y_u. The coefficient of t^{m_7} in $y_u(f)$ is $u\,\lambda_0 + u^2 \lambda_5$. We set this coefficient to 0 in order to find all points of $\overline{\mathcal{M}}_{\mathrm{simp}}$ which have the same image as f in $\overline{\mathcal{M}}_{\mathrm{poly}}$. There are two solutions: $u = 0$ and $u = -\lambda_0/\lambda_5$. The nontrivial solution $u = -\lambda_0/\lambda_5$ gives $y_u(f) = -\lambda_0 t^{m_0} + \sum_{i=1}^{6} \lambda_i t^{m_i}$. But this represents the same point of $\overline{\mathcal{M}}_{\mathrm{simp}}$ as f, since it can be obtained from f by the action of $(-1,-1,-1,-1,-1) \in T \times \mathbb{C}^*$. We conclude that $\overline{\mathcal{M}}_{\mathrm{simp}} = \overline{\mathcal{M}}_{\mathrm{poly}}$.

It follows that we can compactify the complex moduli of V using the secondary fan, which here is built from $\Xi^+ = (\Delta \cap M)_0 \times \{1\}$. We choose a basis for the linear relations M_0, which here is given by the rows of

$$
\begin{pmatrix}
-2 & -2 & -4 & 1 & 0 & 7 & 0 \\
1 & 1 & 2 & 0 & 1 & -3 & -2
\end{pmatrix}.
$$

As in the previous example, the edges of the secondary fan can be read off from the columns of the above matrix. So $\overline{\mathcal{M}}_{\mathrm{simp}}$ is the toric variety associated to the fan given by the following picture:

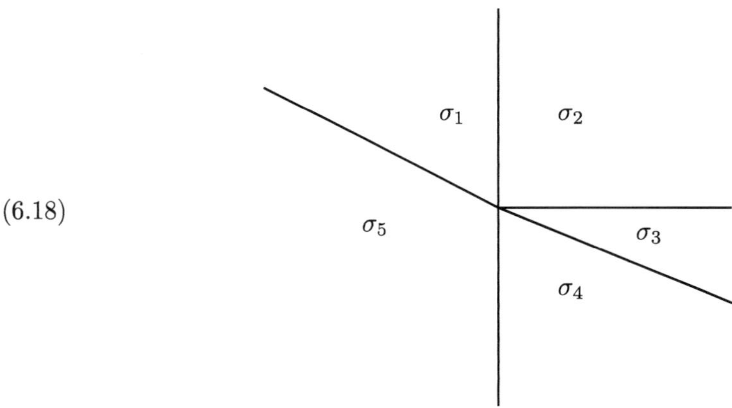

(6.18)

When we compute the Kähler moduli of the mirror V° in Section 6.2.3, we will see that the GKZ decomposition consists of the cones $\sigma_1, \sigma_2, \sigma_3$ (remember that the GKZ decomposition is a subfan of the secondary fan). However, only σ_1 and σ_2 are of the form $\mathrm{cpl}(\Sigma^\circ)$ for an appropriate projective subdivision of the normal fan of Δ°. Thus only σ_1 and σ_2 give maximally unipotent boundary points of $\overline{\mathcal{M}}_{\mathrm{simp}}$.

Of these two cones, only σ_2 is smooth, so to apply our theory to σ_1, we must subdivide. This will give several maximally unipotent boundary points of a resolution $\widetilde{\mathcal{M}}_{\mathrm{simp}}$. However, there is another way to approach this problem, for we can *combine* σ_1 and σ_2 into a single cone $\sigma = \sigma_1 \cup \sigma_2$. If we use σ with the other cones $\sigma_3, \sigma_4, \sigma_5$ of (6.18), we get a blowdown $\overline{\mathcal{M}}_{\mathrm{simp}} \to \overline{\mathcal{M}}'_{\mathrm{simp}}$, and the point $p \in \overline{\mathcal{M}}'_{\mathrm{simp}}$ corresponding to σ is smooth. One can prove that p is a maximally unipotent boundary point and that under the blowup $\widetilde{\mathcal{M}}_{\mathrm{simp}} \to \overline{\mathcal{M}}'_{\mathrm{simp}}$, all maximally unipotent boundary points of $\widetilde{\mathcal{M}}_{\mathrm{simp}}$ map to p and are equivalent in the sense of Definition 6.1.1. Hence, from the point of view of maximally unipotent monodromy, $\overline{\mathcal{M}}'_{\mathrm{simp}}$ is the best compactification to use.

In Section 6.2.3, we will return to this example from the perspective of the Kähler moduli space of the mirror manifold V°. In particular, we will learn the systematic reason for why we should combine σ_1 and σ_2 into the single cone σ.

6.2. Kähler Moduli

We now turn to Kähler moduli. A Calabi-Yau manifold V of dimension ≥ 2 has a Kähler cone $K(V)$ and, as in Definition 1.4.2, a complexified Kähler space

$$K_{\mathbb{C}}(V) = \{\omega \in H^2(V, \mathbb{C}) : \mathrm{Im}(\omega) \in K(V)\}/\mathrm{im}\, H^2(V, \mathbb{Z}).$$

Then the *complexified Kähler moduli space* is the quotient

$$\mathcal{KM}(V) = K_{\mathbb{C}}(V)/\mathrm{Aut}(V).$$

When there is no danger of confusion, we will write $\mathcal{KM}$ instead of $\mathcal{KM}(V)$. The goal of this section is to understand the structure of $\mathcal{KM}$ and in particular to find a nice compactification which contains analogs of the maximal unipotent boundary points studied in Section 6.1. As in that section, it will be convenient to be flexible in our constructions.

In Section 6.3, we will discuss the mirror map, which conjecturally takes a neighborhood of *large radius limit points* of $\mathcal{KM}(V)$ (to be defined in Section 6.2.1) to a neighborhood of a maximally unipotent boundary point of the complex moduli space of V°. This will tie together the similarities that the reader will notice between this section and Section 6.1. The mirror map can be used to identify these neighborhoods with subsets of the SCFT moduli space. It is important to stress at the outset that the mirror map is only locally defined, so at present it is dangerous to try to identify the entire compactifications constructed in this section with a part of the SCFT moduli space. On the other hand, we will see in examples that the compactifications constructed in this section do give a useful geometrical framework for understanding the global structure of the SCFT moduli space, with the caveat that it is to be "corrected" by the mirror map. The problem of giving a precise mathematical description of the SCFT moduli space will be left as one of the exciting mysteries remaining to be explained.

6.2.1. Large Radius Limit Points. Our first step in studying Kähler moduli is to find some nice partial compactifications of the complexified Kähler space $K_{\mathbb{C}}(V)$. For this purpose, fix a maximal dimensional simplicial cone σ whose interior lies in the Kähler cone $K(V)$, and let

$$\mathcal{D}_\sigma = \left(H^2(V, \mathbb{R}) + i\,\mathrm{Int}(\sigma)\right) / \mathrm{im}\, H^2(V, \mathbb{Z}) \subset K_{\mathbb{C}}(V).$$

Note that $\mathcal{D}_\sigma$ is an open subset of $K_{\mathbb{C}}(V)$. Now assume further that σ is generated by a basis $T_1, \ldots, T_r$ of $H^2(V, \mathbb{Z})/\text{torsion}$. Note that the T_i lie in the closure of the Kähler cone. Then $\mathrm{Int}(\sigma) = \{t_1 T_1 + \cdots + t_r T_r : t_1, \ldots, t_r > 0\}$, and the map

$$(6.19) \qquad t_1 T_1 + \cdots + t_r T_r \mapsto (q_1, \ldots, q_r) = (e^{2\pi i t_1}, \ldots, e^{2\pi i t_r})$$

induces a biholomorphism

$$\mathcal{D}_\sigma \simeq (\Delta^*)^r.$$

Using this isomorphism and the inclusion $(\Delta^*)^r \subset \Delta^r$, we get a partial compactification

$$\mathcal{D}_\sigma \subset \mathcal{D}_\sigma^- \simeq \Delta^r.$$

In particular, the origin $0 \in \Delta^r$ becomes a distinguished boundary point denoted $0 \in \mathcal{D}_\sigma^-$.

For $\omega \in H^2(V, \mathbb{R}) + i\,\mathrm{Int}(\sigma)$, having a large imaginary part means that under the map (6.19), the image of ω is close to 0. For this reason, we call $0 \in \mathcal{D}_\sigma^-$ a *large radius limit point*. This is the Kähler analog of a maximally unipotent boundary point. Note that each large radius limit point has canonical coordinates $q_1, \ldots, q_r$ as defined above.

For the nef complete intersections defined in Section 5.5.3, we saw a version of the coordinates $q_1, \ldots, q_r$ in Proposition 5.5.4. There, we let $q_j = e^{t_j}$ and set

$$q^\beta = \prod_j q_j^{\int_\beta T_j} = \exp\left(\int_\beta (t_1 T_1 + \cdots + t_r T_r)\right)$$

for $\beta \in H_2(V, \mathbb{Z})$ (see the discussion following (5.43)). If we replace t_j with $2\pi i t_j$, we get exactly the above variables t_j and q_j, and it follows that aside from an extra factor of $2\pi i$, the formal function $\tilde{I}$ of Proposition 5.5.4 is intrinsically defined on the complexified Kähler space $K_{\mathbb{C}}(V)$. Since $\tilde{I}$ is related to the Picard-Fuchs equations of the mirror V°, we get a hint of the relation between complex and Kähler moduli spaces given by mirror symmetry.

Now that we've defined large radius limit points, we next observe that the different choices of σ give lots of such points. If we denote the corresponding coordinates as $q_1^\sigma, \ldots, q_r^\sigma$ on $\mathcal{D}_\sigma^-$, then we can ask what happens to these coordinates if we change σ. For example, suppose that we have two such cones with $\tau \subset \sigma$. This gives an inclusion $\mathcal{D}_\tau \subset \mathcal{D}_\sigma$, and to relate the corresponding partial compactifications, we express the generators of τ in terms of those of σ:

$$T_i^\tau = \sum_{j=1}^{r} a_{ij} T_j^\sigma, \quad a_{ij} \geq 0.$$

Then the inclusion $\mathcal{D}_\tau \subset \mathcal{D}_\sigma$ extends to a map $\mathcal{D}_\tau^- \to \mathcal{D}_\sigma^-$ given by

$$q_j^\sigma = \prod_{i=1}^{r} (q_i^\tau)^{a_{ij}}.$$

This is reminiscent of (6.4) and (6.6), and the similarity is deeper than one might expect. First, Section 6.1.1 also has cones, namely those generated by the logarithms of the monodromy operators, and (6.4) comes from the inclusion of one cone in another. Furthermore, when we study quantum cohomology in Chapter 8, we will see that the generators T_i of the cone σ considered here are naturally the logarithms of the monodromy of the A-model connection. So the situations are incredibly similar, which is exactly what mirror symmetry predicts.

One potential difference between large radius limit points and maximally unipotent boundary points is that local coordinates are unique for the former but not the latter, because of (6.2). As we will see in Section 6.3.1, this can be corrected by assuming that $g_0, \ldots, g_r$ in (6.1) are integral. This mirrors the above assumption that $T_1, \ldots, T_r$ are integral.

We next discuss Kähler moduli in the special case when $\mathrm{Aut}(V) = \{1\}$ and $K(V)$ is the interior of a rational polyhedral cone in $H^2(V, \mathbb{R})$. Here, the Kähler moduli space is $\mathcal{KM} = K_\mathbb{C}(V)$, and we can construct a toric partial compactification of $\mathcal{KM}$ as follows. The closure $\overline{K(V)}$ is rational polyhedral with respect to the lattice $N = H^2(V, \mathbb{Z})/\text{torsion}$, so that we get the affine toric variety $X_{\overline{K(V)}}$, which we denote X_K for simplicity. Then the isomorphism

$$(6.20) \qquad H^2(X, \mathbb{C})/N \simeq N \otimes \mathbb{C}^* = T_N = \text{the torus of } X_K$$

gives a natural inclusion of open sets

$$(6.21) \qquad \mathcal{KM} = K_\mathbb{C}(V) \subset T_N \subset X_K$$

by the definition of $K_\mathbb{C}(V)$. Using the boundary $B = X_K - T_N$ of the toric variety, we define the partial compactification $\overline{\mathcal{KM}}$ to consist of $\mathcal{KM}$ together with all points of B which are in the closure of the image of $\mathcal{KM}$ in X_K.

The cone $\overline{K(V)}$ determines a point in $0 \in X_K$ which is the fixed point of the torus action and possibly very singular. One can check that $0 \in \overline{\mathcal{KM}}$ since under (6.20), the image of a point with large imaginary part is close to 0. Hence $\overline{\mathcal{KM}}$ is likely to be singular.

We resolve this singularity by standard toric methods. Let Σ_+ be a fan with support $\overline{K(V)}$ such that the toric variety X_{Σ_+} is smooth. Then (6.21) applies with X_{Σ_+} in place of X_K, and we can define $\overline{\mathcal{KM}}$ exactly as before, so that the boundary of $\overline{\mathcal{KM}}$ now consists of points of $X_{\Sigma_+} - T_N$ lying in the closure of the image of $\mathcal{KM}$.

The maximal cones of Σ_+ give distinguished points of X_{Σ_+}. We claim that these are all large radius limit points of $\overline{\mathcal{KM}}$. To see why, note that the maximal cones are all simplicial cones σ with $\mathrm{Int}(\sigma) \subset K(V)$. This is precisely the type of cone used in the definition of large radius limit point. If we regard $\mathcal{KM}$ as a subset of X_{Σ_+}, then applying (6.21) with X_σ in place of X_K shows that

$$\mathcal{D}_\sigma = X_\sigma \cap \mathcal{KM}.$$

Since the affine toric variety X_σ is just $\mathbb{C}^r$, it follows easily that $X_\sigma \cap \overline{\mathcal{KM}}$ is *precisely* the partial compactification $\mathcal{D}_\sigma^-$ discussed earlier.

Hence we get a nice partial compactification with a finite number of large radius limit points. Furthermore, if the imaginary part of $\omega \in H^2(V,\mathbb{R}) + i\,K(V)$ goes to infinity within one of the maximal cones $\sigma \in \Sigma_+$, then the image of ω in $\mathcal{KM}$ approaches the corresponding large radius limit point of $\overline{\mathcal{KM}}$.

This is a very nice picture, but it is still very special. For a general Calabi-Yau manifold, the automorphism group need not be trivial, and the Kähler cone need not be polyhedral. The structure of $K(V)$ can be rather complicated, as we will see in Section 6.2.2 when we study the threefold case in more detail. In general, the best we could hope for is the *Cone Conjecture* [**Morrison3**]. To state this conjecture, let K_+ be the convex hull of $\overline{K(V)} \cap H^2(V,\mathbb{Q})$. Then $K(V) \subset K_+$ by the openness of $K(V)$, so that K_+ consists of $K(V)$ together with the rationally defined part of its boundary.

CONJECTURE 6.2.1. *Let V be a Calabi-Yau manifold of dimension > 2 with Kähler cone $K(V)$. Then there exists a rational polyhedral cone $\Pi \subset K_+$ such that* $\mathrm{Aut}(V) \cdot \Pi = K_+$.

The cone conjecture has been verified in a nontrivial example in [**GM**]. Also, some interesting finiteness statements concerning the action of $\mathrm{Aut}(V)$ on the Kähler cone are corollaries of Conjecture 6.2.1. For example, it implies that $K(V)$ has only finitely many edges modulo $\mathrm{Aut}(V)$. In [**Borcea1**], this was checked for a particular Calabi-Yau threefold V which had a nonpolyhedral Kähler cone and infinite $\mathrm{Aut}(V)$. The cone conjecture also implies that certain classes of Calabi-Yau threefolds admit only finitely many algebraic fiber space structures, and this is known to be true [**Oguiso, OgP**]. Another consequence of the cone conjecture is proved in [**Szendroi**]. A relative version of the cone conjecture is formulated in [**Kawamata2**] and proven there for Calabi-Yau fiber spaces in low dimension.

For us, the key property of the cone conjecture is that we can construct a partial compactification of $\mathcal{KM}$ for *any* Calabi-Yau manifold satisfying the conjecture. Hence we will assume that Conjecture 6.2.1 holds for V. Then Looijenga's work [**Looijenga**] on partial compactifications of quotients of tube domains (such as $H^2(V,\mathbb{R}) + i\,K(V)$) shows that we can find an $\mathrm{Aut}(V)$-invariant fan Σ_+ (possibly infinite) whose support is K_+. We can also assume that $\mathrm{Aut}(V)$ acts freely on Σ_+ and that the maximal cones of Σ_+ are of the form σ as above.

From this data, Looijenga builds a space $\widehat{\mathcal{D}}(\Sigma_+)$ from strata determined by the fan Σ_+. The difficult part of the construction is giving $\widehat{\mathcal{D}}(\Sigma_+)$ the correct topology so that the quotient $\widehat{\mathcal{D}}(\Sigma_+)/\mathrm{Aut}(V)$ is a normal analytic space. This quotient, called a *semi-toric partial compactification*, is the partial compactification $\overline{\mathcal{KM}}$ we want. A description of how this works can be found in [**Morrison3**].

When $\mathrm{Aut}(V)$ acts properly discontinuously and $K(V)$ is a homogeneous self-adjoint cone [**AMRT**], this construction can be done purely within the toric context. For the purposes of exposition, we will work in this special case to see how it relates to what we did earlier. The idea here is that the infinite fan Σ_+ determines a scheme X_{Σ_+} which is locally of finite type, and using (6.21) with X_{Σ_+} in place of X_K, we get an inclusion of open sets

$$K_{\mathbb{C}}(V) \subset X_{\Sigma_+}.$$

Since the fan is invariant under $\mathrm{Aut}(V)$, this inclusion is equivariant with respect to $\mathrm{Aut}(V)$, so that we get an inclusion

$$\mathcal{KM} \subset X_{\Sigma_+}/\mathrm{Aut}(V).$$

A nice picture of what $X_{\Sigma_+}/\mathrm{Aut}(V)$ looks like can be found in [**AMRT**] and is reproduced in [**Morrison3**]. Using the toric boundary of $X_{\Sigma_+}/\mathrm{Aut}(V)$, we get a partial compactification $\overline{\mathcal{KM}}$ exactly as before. Furthermore, since $\mathrm{Aut}(V)$ acts freely on the fan in this case, every maximal cone in the fan gives a large radius limit point with a neighborhood isomorphic to $\mathcal{D}_\sigma^-$. (However, because of the action by $\mathrm{Aut}(V)$, infinitely many different cones of Σ_+ can give the same large radius limit point.) As shown by [**Morrison3**], this picture continues to hold in the more general situation considered by Looijenga.

The upshot of all of this is that we have a nice partial compactification $\overline{\mathcal{KM}}$ containing finitely many large radius limit points. However, there are many such compactifications, because given one, we can always create others by subdividing the fan further. Hence, as with maximally unipotent boundary points, we follow [**Morrison6**] and introduce the following equivalence relation among large radius limit points.

DEFINITION 6.2.2. *Let $\overline{\mathcal{KM}}$ be a smooth normal crossings partial compactification of the Kähler moduli space $\mathcal{KM}$. We say that the large radius limit points $p, q \in \overline{\mathcal{KM}}$ are equivalent if there is a connected set $\Xi \subset \overline{\mathcal{M}} - \mathcal{M}$ containing p and q and a local system $\mathcal{L}$ on the cotangent bundle of $\overline{\mathcal{KM}}$ defined on a neighborhood of Ξ having logarithmic poles on the boundary such that the local system is spanned by the 1-forms $d\log q_k$ for q_k as in (6.19) near each of the points p and q.*

It is easy to see that the multiple large radius points created by further subdividing the fan Σ_+ are equivalent in this sense.

As explained in Section 6.1, mirror symmetry tells us that the Kähler moduli of V should be isomorphic (via the mirror map) to part of the complex moduli of the mirror V° in such a way that each equivalence class of large radius limit points maps to an equivalence class of maximally unipotent boundary points. In fact, the mirror map should preserve the local systems mentioned in the two definitions of equivalence. There is a lot of nonuniqueness in the the construction of $\overline{\mathcal{KM}}$, just as the compactification of the complex moduli of the mirror is not unique. So part of the mirror conjecture is that compatible compactifications can be found.

In practice, one chooses Σ_+ with cones as large as possible, subject to our assumption that X_{Σ_+} is smooth. This makes the equivalence classes smaller, similar to what we did for complex moduli in Section 6.1. However, if we allow for singular compactifications of $\mathcal{KM}$, then there is a canonical smallest one, which is determined by the fan Σ_{SBB} consisting of K_+ and its faces. Since K_+ might have infinitely many faces, toric methods don't apply, but Looijenga's construction still works, and one gets the minimal compactification denoted $\overline{\mathcal{KM}}_{\text{SBB}}$. This construction generalizes the construction of the Satake-Baily-Borel compactification for quotients of bounded symmetric domains, which explains the notation. When $\text{Aut}(V) = 1$ and $\overline{K(V)}$ is rational polyhedral, $\overline{\mathcal{KM}}_{\text{SBB}}$ is the compactification described earlier coming from the affine toric variety X_K of $\overline{K(V)}$.

The existence of a Satake-Baily-Borel style minimal compactification for Kähler moduli implies, via mirror symmetry, that the complex moduli of the mirror should have a similar minimal compactification. Precise conjectures along these lines can be found in [**Morrison3**].

Finally, we turn to a question which may have already occurred to the reader: why do we compactify the complex moduli, but only *partially* compactify the Kähler moduli? The answer is that $\mathcal{KM}$ is actually only a small part of the full Kähler moduli space, and to get the missing pieces, we need to go "beyond the Kähler cone" [**Morrison5**]. The theory of how this works has not been done for general Calabi-Yau manifolds, but extensive studies have been made in two special cases, threefolds and toric hypersurfaces. Hence we now discuss the Kähler moduli of these two very interesting classes of varieties.

6.2.2. Kähler Moduli of Calabi-Yau Threefolds. Let V be a smooth Calabi-Yau threefold. Our goal here is twofold: first, to describe what is known about the structure of the Kähler cone $K(V)$ and its closure $\overline{K(V)}$, and second, to explain how we can enlarge the Kähler moduli space.

In describing the Kähler cone, an important ingredient is the cubic cone $W^* \subset H^2(V, \mathbb{R})$ defined as the set of all classes D with $D^3 = 0$. The basic idea is that away from W^*, $K(V)$ is easy to understand. We will use the work of Wilson [**Wilson2**, **Wilson1**], which relates the structure of $\overline{K(V)}$ to primitive contractions.

DEFINITION 6.2.3. *A* contraction *is a birational morphism* $V \to \overline{V}$ *where* $\overline{V}$ *is a projective normal threefold with* $\rho(\overline{V}) < \rho(V)$ *(where as usual ρ denotes the Picard number). A contraction is called* primitive *if it cannot be factored further into birational morphisms between normal varieties.*

If ϕ is a primitive contraction, then one can prove that up to a multiple, there is a unique numerical class C of 1-cycles on V which is contracted by ϕ. This defines the hyperplane $H = \{D \in H^2(V, \mathbb{R}) : D \cdot C = 0\}$, and it is easy to see that this hyperplane supports a codimension one face of $\overline{K(V)}$. We say that this face *corresponds* to the primitive contraction. These faces determine the structure of $\overline{K(V)}$ away from the cubic cone W^*. More precisely, we have the following result.

PROPOSITION 6.2.4. [**Wilson2**] *Let V be a smooth Calabi-Yau threefold.*

(*i*) *Away from W^*, the cone $\overline{K(V)}$ is locally rational polyhedral.*

(*ii*) *Away from W^*, the codimension one faces of $\overline{K(V)}$ correspond to primitive birational contractions.*

This still allows the Kähler cone to have a fairly complicated structure. For example, it can happen that $\overline{K(V)}$ has infinitely many faces away from W^* which accumulate at W^*. We should also mention that W^* may contain codimension one faces of $\overline{K(V)}$. Such a face is determined by a linear system $|D|$ which maps V to a lower dimensional variety, and the general element D of the linear system lies in the face.

We next describe how to enlarge the Kähler moduli space of V. The idea is to extend the Kähler cone $K(V)$ across certain faces of $\overline{K(V)}$. We will use faces corresponding to primitive contractions, and recall that such contractions ϕ come in three flavors:

- Type I: There is a curve $C \subset V$ so that ϕ is an isomorphism on $V - C$, and ϕ maps C to a point.
- Type II: There is a surface $S \subset V$ so that ϕ is an isomorphism on $V - S$, and ϕ maps S to a point.
- Type III: There is a surface $S \subset V$ so that ϕ is an isomorphism on $V - S$, and ϕ maps S to a curve.

Now suppose that a face of $\overline{K(V)}$ corresponds to a primitive contraction $\phi : V \to \overline{V}$ of Type I. Then we have a (not necessarily irreducible) curve $C \subset V$ which is contracted by ϕ. A basic theorem of threefold geometry states that C can be *flopped* [**Reid2, Kollár**]. This means in particular that there exists a Calabi-Yau threefold V', distinct from V, a curve $C' \subset V'$, and a map $\phi' : V' \to \overline{V}$ which contracts C' and is an isomorphism on $V' - C'$. Since the resulting birational map $V - - \to V'$ is an isomorphism in codimension 1, we have a natural identification $H^2(V) \simeq H^2(V')$. With this identification, $\overline{K(V)}$ and $\overline{K(V')}$ meet along a common face of each, namely the faces supported by the hyperplanes $\{D : D \cdot C = 0\}$ and $\{D' : D' \cdot C' = 0\}$ respectively.

A result of Kawamata [**Kawamata1**] states that the cones $\overline{K(V')}$ for Calabi-Yau threefolds V' (including V itself) which can be obtained from V by a sequence of flops form the chambers of a polyhedral decomposition of the convex hull of their union, which is called the *movable cone* $\mathrm{Mov}(V)$. Since all birational models of V are built from sequences of flops, the group $\mathrm{Bir}(V)$ of birational automorphisms of V acts naturally on $\mathrm{Mov}(V)$. The birational version of the cone conjecture is the *birational cone conjecture* [**Morrison5**], which asserts that there is a rational polyhedral cone Π such that $\mathrm{Bir}(V) \cdot \Pi = \mathrm{Mov}(V)_+$, where $\mathrm{Mov}(V)_+$ is the convex hull of $\mathrm{Mov}(V) \cap H^2(V, \mathbb{Q})$. By a construction similar to what we did above, we can appropriately subdivide $\mathrm{Mov}(V)$ and construct a partial compactification of the Kähler moduli space of *all* of the V' simultaneously. There is somewhat more detail in [**Morrison5**], and we will return to this idea in the context of toric hypersurfaces in Section 6.2.3.

We can enlarge the Kähler moduli space even further. Suppose that ϕ is a Type III contraction containing a ruled surface S over a curve C of genus $g \geq 2$ such that ϕ contracts S to C along the ruling. If we deform the complex structure of V generically, then the generic fiber of $S \to C$ will not remain holomorphic, but $2g - 2$ fibers will deform holomorphically. In fact, the obstruction to deforming a fiber of ϕ to first order is the image of the Kodaira-Spencer class $\rho \in H^1(\Theta_V)$ under a natural map $r : H^1(\Theta_V) \to H^0(K_C)$, and if ρ is generic, then $r(\rho)$ vanishes at $2g - 2$ distinct points. Over these points, the fibers deform to first order [**Wilson2**]. It can also be calculated that there are no higher order obstructions. These $2g - 2$ deformed curves can then be simultaneously flopped. The flopped Calabi-Yau threefold lies in the same complex moduli space, but with a different complex structure. The Kähler structure changes as well. The resulting isomorphism on H^2 can be followed back to the original complex structure, where it becomes the automorphism of $H^2(V)$ given by

$$D \mapsto D + (D \cdot f)\, E, \quad f = \text{class of a fiber}.$$

Since $E \cdot f = -2$, this is a reflection in the hyperplane $\{D : D \cdot f = 0\}$. We then reflect $\mathrm{Mov}(V)$ using the reflections coming from all possible Type III contractions as above. This yields the *reflected movable cone* [**Morrison5**]. If we also enlarge the birational automorphism group to include the above reflections, then we obtain a yet larger Kähler moduli space which uses the reflected movable cone and the enlarged automorphism group just described. Furthermore, a suitable version of the cone conjecture would then allow us to partially compactify this enlarged moduli space.

The Kähler moduli space given by the reflected movable cone is constructed by crossing walls of $\overline{K(V)}$ corresponding to contractions of Types I and III, though we only used Type III contractions where a ruled surface is contracted to a curve of genus $g \geq 2$. At present, there is no proposal for how to enlarge the Kähler moduli space after crossing walls corresponding to Type II and the remaining Type III contractions. However, we will see in Section 6.2.3 how the Kähler moduli space can be enlarged even further when dealing with a Calabi-Yau toric hypersurface.

The final topic we want to discuss is how the Kähler cone behaves when we vary the complex structure of V. In the naive picture drawn in Section 1.1, it looks as if we can vary the complex structure while holding the Kähler structure constant. In the threefold case, this is true generically. More precisely, Wilson [**Wilson2**] has shown that for a generic complex structure on a Calabi-Yau threefold V, the Kähler cone is constant under small deformations of complex structure. This generic rigidity is rather strong. For example, [**Wilson3**] proves that if two Calabi-Yau threefolds V_1 and V_2 are generic in their respective complex moduli spaces and symplectic deformations of each other, then their Kähler cones are the same, and [**Wilson4**] shows that away from the cubic cone W^*, the type (i.e., Type I, II, or III) of a primitive contraction associated to a codimension 1 face of $\overline{K(V_1)}$ is the same as the type of the primitive contraction associated to the corresponding face of $\overline{K(V_2)}$.

However, things are actually more complicated, as hinted in the footnote below the picture in Section 1.1. In particular, there are deformations of V which allow the Kähler cone to vary. Suppose that ϕ is a Type III contraction which maps a ruled surface S to a curve C of genus g. When $g \geq 2$, we were able to reflect the movable cone, as described earlier. But when the base curve has $g = 1$, then the argument given above shows that no fibers deform along with a general deformation of complex structure. This means that for the special complex structure containing E, the cone of effective curves strictly contains the cone of effective curves on the general complex structure, since the class of f is not present for a general complex structure. Dually, the Kähler cone becomes smaller for this special complex structure. It follows that the Kähler moduli space can depend on the complex structure.

6.2.3. Kähler Moduli of Toric Hypersurfaces. Let Δ be a reflexive polytope with Σ a maximal projective subdivision of the normal fan of Δ. If $V \subset X_\Sigma$ is a general Calabi-Yau hypersurface, recall the description of the restriction map $H^2(X_\Sigma) \to H^2(V)$ from Chapter 4. Its kernel is generated by the classes $[D_\rho]$ of divisors associated to edges $\rho \in \Sigma(1)$ spanned by primitive integral vectors which are in the interior of some facet of Δ°. The image of the restriction map is by definition $H^2_{\mathrm{toric}}(V)$. Our goal is to construct a Kähler moduli space for the "toric part" of V. This construction is in a sense "mirror" to the construction of $\overline{\mathcal{M}}_{\text{simp}}$ in Section 6.1.2.

As in Section 6.1.2, we change the problem slightly by excluding integral points of facets of Δ°. We accordingly modify the definition of maximal projective subdivision slightly to avoid these points.

DEFINITION 6.2.5. *Given a reflexive polytope* $\Delta \subset M_{\mathbb{R}}$, *a fan* Σ *in* $N_{\mathbb{R}}$ *is a simplified projective subdivision if it has the following properties:*
- Σ *refines the normal fan of* Δ.
- $\Sigma(1) = (\Delta^\circ \cap N)_0 - \{0\}$.
- X_Σ *is projective and simplicial.*

We can now copy Batyrev's construction from Section 4.1, replacing a maximal projective subdivision Σ by a simplified projective subdivision, which we will again call Σ. Let's show that we get the same class of Calabi-Yau varieties as arise from Batyrev's construction. First, given a maximal projective subdivision Σ, we can remove all of the edges generated by interior points of facets of Δ° to obtain the fan of a blowdown of X_Σ. The resulting fan need not be simplicial, so we may need to further subdivide this fan without adding new edges to get a simplified projective subdivision Σ'. We get a commutative diagram

$$(6.22) \qquad \begin{array}{ccc} X_\Sigma & \overset{\phi}{\dashrightarrow} & X_{\Sigma'} \\ & {\scriptstyle\pi}\searrow \quad \swarrow{\scriptstyle\pi'} & \\ & \mathbb{P}_\Delta & \end{array}$$

By construction, ϕ is an isomorphism except over the points of $\mathbb{P}_\Delta$ associated to the facets of Δ° which contain interior points. Since a generic Calabi-Yau hypersurface $\bar{V} \in |-K_{\mathbb{P}_\Delta}|$ is disjoint from this finite set, we see that its proper transforms under π and π', which are generic Calabi-Yau hypersurfaces in X_Σ and $X_{\Sigma'}$, are isomorphic as claimed. Going the other way is simpler, for any simplified projective subdivision can be refined to a maximal one, and we get isomorphic Calabi-Yau hypersurfaces as before.

Now fix a simplified projective subdivision Σ of the normal fan of Δ and consider $V \subset X_\Sigma$. Because we are excluding interior lattice points in the facets of Δ°, the natural map $H^2(X_\Sigma) \to H^2(V)$ is injective. As already noted, the image is $H^2_{\text{toric}}(V)$, so that

$$H^2(X_\Sigma) \simeq H^2_{\text{toric}}(V).$$

In Section 3.3.3, we saw that the closure of the Kähler cone of X_Σ is $\overline{K(X_\Sigma)} = \text{cpl}(\Sigma)$. Hence we can regard $\text{cpl}(\Sigma)$ as lying in $H^2_{\text{toric}}(V)$. If we define the *toric part of the Kähler cone of* V to be

$$K(V)_{\text{toric}} = K(V) \cap H^2_{\text{toric}}(V),$$

then we have the obvious inclusion $\text{cpl}(\Sigma) \subset \overline{K(V)}_{\text{toric}}$ since ample classes on X_Σ restrict to ample classes on V.

For the moment, suppose that $\text{cpl}(\Sigma) = \overline{K(V)}_{\text{toric}}$ always holds. In this situation, we can easily construct an enlarged toric Kähler moduli space and compactify it using the secondary fan from Section 3.4. If we set $\Xi = (\Delta \cap N)_0 - \{0\}$, then the secondary fan lives naturally in the vector space $A(\Xi) \simeq H^2(X_\Sigma, \mathbb{R}) \simeq H^2_{\text{toric}}(V)$ (this is the notation of Section 3.4). The secondary fan and the lattice given by integer cohomology determine a complete toric variety, which we will temporarily call $\overline{\mathcal{KM}}_{\mathrm{toric}}$. For each simplified projective subdivision Σ, $\mathrm{cpl}(\Sigma)$ is a maximal cone of the secondary fan, and $\mathrm{cpl}(\Sigma) = \overline{K(V)}_{\mathrm{toric}}$ would imply that the corresponding point of $\overline{\mathcal{KM}}_{\mathrm{toric}}$ is a large radius limit point in the sense of Section 6.2.1 (possibly subdividing $\mathrm{cpl}(\Sigma)$ in order to get something smooth).

The problem with this picture is that *equality need not hold* in the inclusion $\mathrm{cpl}(\Sigma) \subset \overline{K(V)}_{\mathrm{toric}}$. The closure $\overline{K(V)}_{\mathrm{toric}}$ of the toric Kähler cone can be strictly larger than $\mathrm{cpl}(\Sigma)$. The reason is that there may exist simplified projective subdivisions $\Sigma' \neq \Sigma$ which give isomorphic Calabi-Yau hypersurfaces. It follows that $\overline{K(V)}_{\mathrm{toric}}$ may contain both $\mathrm{cpl}(\Sigma)$ and $\mathrm{cpl}(\Sigma')$. We will see an example of this later in the section.

In fact, the precise structure of the toric Kähler cone $\overline{K(V)}_{\mathrm{toric}}$ is not known at this time. It is suspected to be rational polyhedral, and this is certainly true in all cases where $\overline{K(V)}_{\mathrm{toric}}$ has been computed. In Conjecture 6.2.8 below, we give a careful description for what we think $\overline{K(V)}_{\mathrm{toric}}$ should look like in general. The essence of the conjecture is that $\overline{K(V)}_{\mathrm{toric}}$ is the union of certain cones $\mathrm{cpl}(\Sigma')$. If correct, this would imply that the space $\mathcal{KM}_{\mathrm{toric}}$ constructed from the secondary fan is a blowup of the "true" compactification of the enlarged toric Kähler moduli space.

Our plan for the rest of the section is, first, to examine the structure of $\overline{K(V)}_{\mathrm{toric}}$, and second, to construct an enlarged Kähler moduli space and then compactify it.

To begin our study of $\overline{K(V)}_{\mathrm{toric}}$, we need to look at how the cones $\mathrm{cpl}(\Sigma)$ fit together as Σ varies over all simplified projective subdivisions. For example, how do we pass from one subdivision to a "nearby" one? This is provided by the language of *linear circuits*, which we now define.

DEFINITION 6.2.6. *Given* $\Xi = (\Delta^\circ \cap N)_0 - \{0\}$, *a linear circuit is a linearly dependent subset* $\Xi' \subset \Xi$ *with the property that any proper subset of* Ξ' *is linearly independent.*

We refer the reader to [**GKZ2**] and the references given there for more details, remarking that the situation in [**GKZ2**] is slightly different since there, affine circuits are studied using affine dependence and triangulations instead of linear dependence and fans respectively.

In what follows, we will use v_i to denote the elements of Ξ. Given a linear circuit $\Xi' \subset \Xi$, we fix a nontrivial linear relation $\sum_i b_i v_i = 0$ among the elements $v_i \in \Xi'$, and decompose $\Xi' = \Xi'_+ \cup \Xi'_-$, where Ξ'_+ denotes the set of elements $v_i \in \Xi'$ which appear with positive coefficient b_i in the linear relation, and Ξ'_- is defined similarly.

For any subset $S \subset \Xi$, we let $\mathcal{C}(S)$ denote the convex cone generated by S. Given a circuit $\Xi' \subset \Xi$ with $|\Xi'_+| \geq 1$, we define $\Sigma_+(\Xi')$ to be the fan whose top dimensional cones are $\mathcal{C}(\Xi' - \{v_i\})$ for all $v_i \in \Xi'_+$. The support of this fan is $\mathcal{C}(\Xi')$. Similarly, we get the fan $\Sigma_-(\Xi')$ provided $|\Xi'_-| \geq 1$. For an example of what this looks like, suppose that $\Xi' = \{v_1, v_2, v_3, v_4\}$, where the linear relation is $v_1 + v_2 = v_3 + v_4$. In this case, $\mathcal{C}(\Xi')$ is a cone with four generators. There are two ways of subdividing this into a pair of simplicial cones: $\Sigma_+(\Xi')$ gives one way, and $\Sigma_-(\Xi')$ the other.

We next describe how a linear circuit interacts with a fan Σ.

DEFINITION 6.2.7. *Let Σ be a projective simplicial fan in $N_{\mathbb{R}}$ with $\Sigma(1) \subset \Xi$, and let $\Xi' \subset \Xi$ be a linear circuit. We say that Σ is supported on Ξ' (or, alternatively, Ξ' is supported by Σ) if the following conditions are satisfied:*

(i) *$\Sigma_+(\Xi')$ is a subfan of Σ.*
(ii) *Let σ be a top dimensional cone of $\Sigma_+(\Xi')$. If there exists a subset $\Pi \subset \Xi'$ such that $\sigma \cup \Pi$ generates a top dimensional cone of Σ, then for all other top dimensional cones $\sigma' \in \Sigma_+(\Xi')$, $\sigma' \cup \Pi$ also generates a cone of Σ.*

Thus, being supported on Ξ' means that the cones of Σ touching $\mathcal{C}(\Xi')$ are determined by $\Sigma_+(\Xi')$ in a particularly strong fashion. An example of a projective subdivision supported on a circuit will be given when we revisit Example 6.1.4.2 later in this section.

Definition 6.2.7 allows us to use linear circuits to modify the fan Σ. More precisely, suppose that Σ is supported on Ξ' and that $|\Xi'_-| \geq 1$. Then we obtain a new simplicial fan $\Sigma' = \mathrm{flip}_{\Xi'}(\Sigma)$ by replacing the simplices of Σ spanned by $\sigma \cup \Pi$, where σ are the cones of $\Sigma_+(\Xi')$ and $\Pi \subset \Xi'$, with the simplices spanned by $\sigma' \cup \Pi$, where σ' are the cones of $\Sigma_-(\Xi')$. For an example of what this looks like, suppose that Σ is supported on the linear circuit $\Xi' = \{v_1, v_2, v_3, v_4\}$ and $v_1 + v_2 = v_3 + v_4$ (this is the example discussed above). In this case, Σ will use one way of dividing $\mathcal{C}(\Xi')$ into two simplicial cones, while the flip Σ' will use the other.

If we start from a simplicial projective fan Σ satisfying $\Sigma(1) \subset \Xi$, any flip Σ' is clearly simplicial and satisfies $\Sigma'(1) \subset \Xi$. Furthermore, using the techniques of [**OdP**], one can show that Σ' is also projective. Thus $\mathrm{cpl}(\Sigma')$ will be a cone in the GKZ decomposition. But when Σ subdivides the normal fan of Δ (as in the definition of simplified projective subdivision), Σ' may fail to do the same. However, if Ξ' is a subset of a proper face of Δ°, then Σ' will also subdivide the normal fan of Δ.

We next consider the geometric significance of the flip $\Sigma' = \mathrm{flip}_{\Xi'}(\Sigma)$. Suppose that Σ is a simplicial projective fan with $\Sigma(1) \subset \Xi$ and that $|\Xi'_+|, |\Xi'_-| \geq 2$. It follows immediately from the definitions that $X_{\Sigma'}$ is obtained from X_Σ by blowing down the subvariety $\cap_{v_i \in \Xi'_-} D_i$, and then blowing up the resulting singularity, replacing it by the subvariety $\cap_{v_i \in \Xi'_+} D_i$ of $X_{\Sigma'}$. We will refer to this birational transformation from X_Σ to $X_{\Sigma'}$ as a *generalized flop*, and will denote it by $\mathrm{flop}_{\Xi'}$. For example, the classic example of a codimension 2 flop can be thought of as the generalized flop associated to the circuit $\{v_1, v_2, v_3, v_4\}$ with $v_1 + v_3 = v_2 + v_4$.

Now let Ξ' be a linear circuit supported on Σ with $|\Xi'_+|, |\Xi'_-| \geq 2$. Also assume that the base locus of $\mathrm{flop}_{\Xi'}$ is disjoint from V, i.e.,

$$\bigcap_{v_i \in \Xi'_-} D_i \cap V = \emptyset$$

for generic Calabi-Yau hypersurfaces $V \subset X_\Sigma$. It follows immediately that $\mathrm{flop}_{\Xi'}$ induces isomorphism $V \simeq V'$ of Calabi-Yau hypersurfaces. Such flops will be called *trivial flops*, and the associated flip $\Sigma' = \mathrm{flip}_{\Xi'}(\Sigma)$ will be called a *trivial flip* of Σ.

When Σ' is a trivial flip of Σ, we get a map

$$H^2(X_{\Sigma'}) \to H^2(V') \simeq H^2(V)$$

which takes Kähler classes to Kähler classes. We therefore conclude that $\mathrm{cpl}(\Sigma') \subset \overline{K(V)}_{\mathrm{toric}}$. This motivates the following definition. Given our Calabi-Yau toric hypersurface $V \subset X_\Sigma$, we consider the cone

$$(6.23) \qquad \widetilde{K(V)}_{\text{toric}} = \bigcup_{\Sigma'} \text{cpl}(\Sigma'),$$

where the union is over all fans Σ' which may be obtained from Σ by a sequence of trivial flips. It follows immediately from the above discussion that

$$\widetilde{K(V)}_{\text{toric}} \subset \overline{K(V)}_{\text{toric}}.$$

CONJECTURE 6.2.8. $\widetilde{K(V)}_{\text{toric}} = \overline{K(V)}_{\text{toric}}$.

To the best of our knowledge, $\widetilde{K(V)}_{\text{toric}}$ coincides with $\overline{K(V)}_{\text{toric}}$ in all known examples. It is also intriguing to note that in these examples, $K(V)_{\text{toric}}$ is always simplicial, even though $K(X_\Sigma)$ need not be. We will illustrate Conjecture 6.2.8 and some ideas which may be relevant to its eventual proof when we revisit Example 6.1.2 below.

For the rest of this section, we will assume that Conjecture 6.2.8 is true.

Let's explore what this says about the toric part of the Kähler moduli. We define the toric part of the (unenlarged) Kähler moduli space to be the quotient

$$(6.24) \qquad \mathcal{KM}_{\text{toric}} = \{\omega \in H^2_{\text{toric}}(V, \mathbb{C}) : \text{Im}(\omega) \in K(V)_{\text{toric}}\}/\text{im}\, H^2_{\text{toric}}(V, \mathbb{Z}).$$

Strictly speaking, we should take the quotient of this by the automorphisms of V. However, $\text{Aut}(V)$ preserves the rational polyhedral cone $\overline{K(V)}_{\text{toric}}$, and each automorphism permutes its minimal generators. Hence $\text{Aut}(V)$ acts via a finite group of automorphisms, so that the true toric Kähler moduli space is a finite quotient of (6.24). This is analogous to what happened in Section 6.1.2, where $\mathcal{M}_{\text{poly}}$ was a finite quotient of $\mathcal{M}_{\text{simp}}$. Just as we found it more convenient to use $\mathcal{M}_{\text{simp}}$, here it is simpler to consider $\mathcal{KM}_{\text{toric}}$ as defined above.

With this definition of the toric Kähler moduli space, we see that the cones $\text{cpl}(\Sigma')$ in (6.23) give a fan subdividing $\overline{K(V)}_{\text{toric}}$. As in Section 6.2.1, the resulting toric variety is a partial compactification of $\mathcal{KM}_{\text{toric}}$, and if we refine further to get a smooth partial compactification, we get lots of large radius limit points. These include the ones coming from $\text{cpl}(\Sigma)$, but there may be more because of the other cones $\text{cpl}(\Sigma')$ which may occur in $\overline{K(V)}_{\text{toric}}$. All of these should correspond to maximally unipotent boundary points on the mirror side.

If we are looking for a *minimal* partial compactification of $\mathcal{KM}_{\text{toric}}$, then it is clear that we should use the affine toric variety given by $\overline{K(V)}_{\text{toric}}$. On the mirror side, we will see later in the chapter that $\overline{\mathcal{M}}_{\text{simp}}$ needs to be blown down a bit in order to get the corresponding minimal compactification of the simplified complex moduli space $\mathcal{M}_{\text{simp}}$, as suggested in Example 6.1.4.2.

We will next discuss some further properties of linear circuits which are useful for doing examples. We first observe that linear circuits give interesting elements of the dual space $A(\Xi)^*$. Since $A(\Xi) = \mathbb{R}^\Xi/M_\mathbb{R}$, we see that $A(\Xi)^*$ is the vector space of linear relations

$$A(\Xi)^* = \{\ell = (\ell_i) \in \mathbb{R}^\Xi : \textstyle\sum_i \ell_i v_i = 0\}.$$

If Ξ' is a linear circuit, we have a relation $\sum_i b_i v_i = 0$, from which we get the decomposition $\Xi' = \Xi'_+ \cup \Xi'_-$. The relation also gives $\ell_{\Xi'} = (b_i) \in A(\Xi)^*$, which is well-defined up to a positive scalar if $\Xi' = \Xi'_+ \cup \Xi'_-$ is fixed. If we think of an element $D \in A(\Xi)$ as a divisor class $D = [\sum_i a_i D_i]$, then

$$\ell_{\Xi'}(D) = \sum_i a_i b_i.$$

We can now characterize $\mathrm{cpl}(\Sigma)$ in terms of the linear circuits supported by Σ. If Σ is supported on Ξ' and $\Sigma' = \mathrm{flip}_{\Xi'}(\Sigma)$, then it is easy to show that $\ell_{\Xi'}(D) \geq 0$ for $D \in \mathrm{cpl}(\Sigma)$, while $\ell_{\Xi'}(D') \leq 0$ for $D' \in \mathrm{cpl}(\Sigma')$. Thus the hyperplane $\ell_{\Xi'} = 0$ separates the interiors of the two cones. Furthermore, one can show that the cone $\mathrm{cpl}(\Sigma)$ can be defined as

$$\mathrm{cpl}(\Sigma) = \{D \in A(\Xi) : \ell_{\Xi'}(D) \geq 0 \text{ for all } \Xi' \text{ supported by } \Sigma\}.$$

In fact, the facets of $\mathrm{cpl}(\Sigma)$ are in one-to-one correspondence with the linear circuits Ξ' such that Σ is supported on Ξ'.

We can also use linear circuits to shrink the GKZ decomposition. This is useful since we're not really interested in the cones $\mathrm{cpl}(\Sigma)$ for which $\Sigma(1)$ is a proper subset of $\Xi = (\Delta^\circ \cap N)_0 - \{0\}$. The idea is as follows. Let Σ be a simplified projective subdivision, and suppose that $\Xi' \subset \Xi$ is any linear circuit with $|\Xi'_-| = 1$, not necessarily supported by Σ. We claim that $\ell_{\Xi'} \geq 0$ on $\mathrm{cpl}(\Sigma)$. To prove this, we relabel Ξ' so that the linear relation is $b_0 v_0 = \sum_{i>0} b_i v_i$, where all of the coefficients are positive. Since $v_0 \in \Xi = \Sigma(1)$, we have $v_0 \in \sigma$ for some maximal cone $\sigma \in \Sigma$. Now take $D = [\sum_i a_i D_i] \in \mathrm{cpl}(\Sigma)$. Since σ is simplicial, we can find $m_\sigma \in M_\mathbb{R}$ such that $\langle m_\sigma, v_i \rangle = -a_i$ for all $v_i \in \sigma$, and note that for *any* $v_i \in \Xi$, we also have $\langle m_\sigma, v_i \rangle \geq -a_i$ since $D \in \mathrm{cpl}(\Sigma)$ (see Section 3.3.3). Then

$$-b_0 a_0 = \langle m_\sigma, b_0 v_0 \rangle = \langle m_\sigma, \sum_i b_i v_i \rangle \geq -\sum_i a_i b_i,$$

and it follows immediately that $\ell_{\Xi'}(D) \geq 0$. To exploit this, recall that the cone

$$A^+(\Xi) = \left\{ [\sum_i a_i D_i] \in A(\Xi) : a_i \geq 0 \text{ for all } i \right\}$$

is the support of the GKZ decomposition. Then we define the *Calabi-Yau cone*[1] to be

$$\mathcal{C}_{\mathrm{CY}} = \{D \in A^+(\Xi) : \ell_{\Xi'}(D) \geq 0 \text{ for all } \Xi' \text{ with } |\Xi'_-| = 1\}.$$

The point is that if Σ is a simplified projective subdivision, then $\mathrm{cpl}(\Sigma) \subset \mathcal{C}_{\mathrm{CY}}$ by the above paragraph. So we need only study the restriction of the GKZ decomposition to the cone $\mathcal{C}_{\mathrm{CY}}$.

We will now illustrate the concepts developed so far in an example we've seen before.

Example 6.1.4.2, revisited. Example 6.1.4 studied the complex moduli of the Calabi-Yau threefold $V \subset X_\Sigma$ coming from the reflexive polytope Δ with vertices $v_1, \dots, v_6$ given by

$$(6.25) \qquad \begin{array}{c} (-1,-2,-3,-7),(1,0,0,0),(0,1,0,0), \\ (0,0,1,0),(0,0,0,1),(0,0,-1,-2) \end{array}$$

We now want to study the the Kähler moduli of the mirror $V^\circ \subset X_{\Sigma^\circ}$. Since we're dealing with V°, we need to switch N and M, Δ and Δ°, etc.

A first observation is that $K(V^\circ) = K(V^\circ)_{\mathrm{toric}}$. This follows from the formulas of Section 4.1, especially (4.4) and (4.7), which imply $h^2(V^\circ) = h^2_{\mathrm{toric}}(V^\circ) = 2$.

---

[1] The relevance of this cone was pointed out to us by B. Sturmfels.

A simplified projective subdivision Σ° of the normal fan of Δ° lives in $M_\mathbb{R}$ and satisfies $\Sigma^\circ(1) = (\Delta \cap M)_0 - \{0\}$. This is precisely the set Ξ of Example 6.1.4.2, and observe that $\Xi = \{v_1, v_2, v_3, v_4, v_5, v_6\}$. Then the cone $\mathrm{cpl}(\Sigma^\circ)$ is part of the GKZ decomposition of Ξ. Earlier, we constructed the secondary fan (6.18) using $\Xi^+ = (\Xi \cup \{0\}) \times \{1\} = (\Delta \cap M)_0 \times \{1\}$. Repeating that construction using Ξ instead gives the GKZ decomposition. Recall that $A(\Xi)^*$ is the vector space of linear relations among the elements of Ξ. In this case, a basis of $A(\Xi)^*$ is given by the rows of the matrix

$$\begin{pmatrix} -2 & -2 & -4 & 1 & 0 & 7 \\ 1 & 1 & 2 & 0 & 1 & -3 \end{pmatrix}$$

Using the dual basis of $A(\Xi)$, the columns of this matrix give vectors $e_{v_1}^*, \dots, e_{v_6}^* \in A(\Xi)$. Then Section 3.4 shows that the GKZ decomposition is given by the following picture:

(6.26)

$$e_{v_5}^*$$
$$e_{v_1}^*, e_{v_2}^*, e_{v_3}^* \quad \sigma_1 \qquad \sigma_2$$
$$\sigma_3 \qquad e_{v_4}^*$$
$$e_{v_6}^*$$

Each edge is labeled with the points $e_{v_i}^*$ which span the edge. There are three maximal cones in the GKZ decomposition, denoted $\sigma_1, \sigma_2, \sigma_3$.

The linear circuits of Ξ are

(6.27)	$\{v_1, v_2, v_3, v_5, v_6\}, \quad \{v_1, v_2, v_3, v_4, v_6\}, \quad \{v_4, v_5, v_6\}, \quad \{v_1, v_2, v_3, v_4, v_5\},$

corresponding to the linear relations

(6.28)
$$v_1 + v_2 + 2v_3 + v_5 - 3v_6 = 0$$
$$-2v_1 - 2v_2 - 4v_3 + v_4 + 7v_6 = 0$$
$$v_4 + v_5 + 2v_6 = 0$$
$$v_1 + v_2 + 2v_3 + 3v_4 + 7v_5 = 0.$$

The linear circuits are most easily found by looking at (6.26). Each circuit is obtained by dividing the plane into two halfplanes along the line determined by one of the edges in the figure. Then Ξ' is the set of vertices not contained in the line, with points of Ξ'_+ on one side of the line and points of Ξ'_- on the opposite side.

It is readily observed that the only linear circuit Ξ' with $|\Xi'_-| = 1$ (using either the relations (6.28) or their negatives) is the first relation with the signs as written. Thus the Calabi-Yau cone $\mathcal{C}_{\mathrm{CY}}$ is the cone inside $A^+(\Xi) = \sigma_1 \cup \sigma_2 \cup \sigma_3$ determined by the single inequality $a_1 + a_2 + 2a_3 + a_5 - 3a_6 \geq 0$, where a general point of $A^+(\Xi)$ is represented by $[\sum_i a_i D_i]$ as usual. This gives $\mathcal{C}_{\mathrm{CY}} = \sigma_1 \cup \sigma_2$.

We will next construct the fans corresponding to σ_1 and σ_2. Let's begin with σ_2. One way to do this would be to take a divisor class in σ_2 and use the convexity conditions for ampleness to determine the fan (this procedure is automated in [**KS**]). However, it is easier to realize that V° comes from the weighted projective space $\mathbb{P}(1, 1, 2, 3, 7)$ (this is how V° was originally constructed). Namely, of the six vertices listed in (6.25), the first five determine a simplex, which gives a fan in $M_{\mathbb{R}}$ by taking cones over faces of the simplex. The associated toric variety is $\mathbb{P}(1,1,2,3,7)$, which is not Fano. However, if we blow-up $\mathbb{P}(1,1,2,3,7)$ by inserting the 1-dimensional cone generated by $(0,0,-1,-2)$, then we get a simplicial fan Σ° whose top dimensional cones are

$$\mathcal{C}(v_2, v_3, v_4, v_5), \; \mathcal{C}(v_1, v_3, v_4, v_5), \; \mathcal{C}(v_1, v_2, v_4, v_5), \; \mathcal{C}(v_1, v_2, v_3, v_4),$$
$$\mathcal{C}(v_2, v_3, v_5, v_6), \; \mathcal{C}(v_1, v_3, v_5, v_6), \; \mathcal{C}(v_1, v_2, v_5, v_6), \; \mathcal{C}(v_1, v_2, v_3, v_6).$$

This fan clearly refines the normal fan of Δ°. One can also check that Σ° is supported on the linear circuits $\Xi'_1 = \{v_1, v_2, v_3, v_5, v_6\}$ and $\Xi'_2 = \{v_1, v_2, v_3, v_4, v_6\}$ (with signs determined by (6.28)). Since the inequalities $\ell_{\Xi'_i}(D) \geq 0$ for $i = 1, 2$ define σ_2, we see that $\mathrm{cpl}(\Sigma^{\circ}) = \sigma_2$.

Furthermore, note that Ξ'_2 has $|(\Xi'_2)_+| = 2$ and $|(\Xi'_2)_-| = 3$, so we get the fan $\Sigma^{\circ\prime} = \mathrm{flip}_{\Xi'_2}(\Sigma^{\circ})$ and the birational map $\mathrm{flop}_{\Xi'_2}$. Using (6.26), one easily checks that $\mathrm{cpl}(\Sigma^{\circ\prime}) = \sigma_1$. Note also that Ξ' is contained in the facet of Δ supported by the hyperplane in $M_{\mathbb{R}}$ defined by the equation

$$\langle m, (-1,-1,-1,1) \rangle = -1$$

corresponding to the vertex $(-1,-1,-1,1) \in \Delta^{\circ}$ (see Example 6.1.4.2). By our earlier remarks, it follows that both Σ° and $\Sigma^{\circ\prime}$ are simplified projective subdivisions.

We will now prove that $\Sigma^{\circ\prime}$ is a trivial flip of Σ°. Since both are simplified projective subdivisions, we are in the situation of (6.22). Furthermore, one can check that $\mathrm{flop}_{\Xi'_2} : X_{\Sigma^{\circ}} - - \to X_{\Sigma^{\circ\prime}}$ is an isomorphism except over the point of $\mathbb{P}_{\Delta^{\circ}}$ corresponding to the vertex $(-1,-1,-1,1) \in \Delta^{\circ}$. Since a generic $\bar{V}^{\circ} \subset \mathbb{P}_{\Delta^{\circ}}$ misses this point, it follows that a generic $V^{\circ} \subset X_{\Sigma^{\circ}}$ misses the base locus of $\mathrm{flop}_{\Xi'_2}$. Hence the general Calabi-Yau hypersurfaces in $X_{\Sigma^{\circ}}$ and $X_{\Sigma^{\circ\prime}}$ are isomorphic and $\Sigma^{\circ\prime}$ is a trivial flip of Σ°.

We have thus shown that $\widetilde{K(V^{\circ})}_{\mathrm{toric}} = \sigma_1 \cup \sigma_2$. The next step is to show that this equals $\overline{K(V^{\circ})}_{\mathrm{toric}}$, i.e., that Conjecture 6.2.8 is true in this case. To begin the proof, let $D \in \overline{K(V^{\circ})}_{\mathrm{toric}}$. First view $V^{\circ} \subset X_{\Sigma^{\circ}}$ and define the curve $C \subset V^{\circ}$ by imposing the equations $x_1 = x_6 = 0$ on V°, using homogeneous coordinates on $X_{\Sigma^{\circ}}$. Then view $V^{\circ} \subset X_{\Sigma^{\circ\prime}}$ and consider the curve $C' \subset V$ defined by $x_1 = x_2 = 0$ using homogeneous coordinates on $X_{\Sigma^{\circ\prime}}$. Since $D \in \overline{K(V^{\circ})}_{\mathrm{toric}}$, we have $C \cdot D \geq 0$ and $C' \cdot D \geq 0$. One can show that these two inequalities are precisely the conditions defining $\sigma_1 \cup \sigma_2$. In fact, the first inequality coincides with $\ell_{\Xi'_1}(D) \geq 0$, where Ξ'_1 is the linear circuit defined above (which is supported on Σ°). The second inequality coincides with $\ell_{\Xi'_3}(D) \geq 0$, where Ξ'_3 is the third circuit in (6.27), with signs as in (6.28). Thus $\overline{K(V^{\circ})}_{\mathrm{toric}} \subset \sigma_1 \cup \sigma_2$, and equality follows.

One approach to proving Conjecture 6.2.8 in general is to produce curves like C, C' above such that the linear functionals given by C, C' coincide with the linear functionals $\ell_{\Xi'}$ defined by appropriate circuits for the various subdivisions used in the definition of $\widetilde{K(V)}_{\mathrm{toric}}$.

Now that we have the Kähler cone $\overline{K(V^{\circ})} = \overline{K(V^{\circ})}_{\mathrm{toric}} = \sigma_1 \cup \sigma_2$, we next discuss the Kähler moduli space of V°. By Section 6.2.1, a partial compactification of $\mathcal{KM}$ is given by the affine toric variety of $\overline{K(V^{\circ})}$, which is easily seen to be smooth. Furthermore, natural coordinates near the large radius limit point are determined by generators of $\overline{K(V^{\circ})}$. The generators D, D' we want are the vectors $D = (-2,1) = e^*_{v_1}$ and $D' = (1,0) = e^*_{v_4}$ in (6.26), and it follows that $D = D_1$ and

$D' = D_4$, where D_i now denotes the toric divisor associated to v_i restricted to V. One can compute the intersection numbers

$$D^3 = 0, \qquad D^2 \cdot D' = 1, \qquad D \cdot D'^2 = 3, \qquad D'^3 = 9,$$

from which we obtain see that the relations $D^3 = 0$, $D'^2 - 3D \cdot D' = 0$ hold in the cohomology of V°. This will be useful below.

The next step is to compare this to the complex moduli of the mirror V. In Example 6.1.4.2, we compactified $\mathcal{M}_{\mathrm{simp}}$ using the secondary fan (6.18). But we now know that we need to combine the cones σ and σ_2 into a single cone, which corresponds to blowing down our previous definition of $\overline{\mathcal{M}}_{\mathrm{simp}}$. When we do this, the Kähler cone $\overline{K(V^\circ)}$ corresponds to a single boundary point of $\overline{\mathcal{M}}_{\mathrm{simp}}$ which can be shown to be maximally unipotent. If fact, if we use the natural coordinates provided by the generators of the cone $\overline{K(V^\circ)}$, then we get two monodromy transformations with associated logarithms N_1, N_2. One can calculate that $N_1^3 = 0$, $N_2^2 - 3N_1 N_2 = 0$.[2] This is enough to prove that the corresponding boundary point of the minimal blowdown is maximally unipotent. Note the similarity with the relations $D^3 = 0$, $D'^2 - 3D \cdot D' = 0$ satisfied by D, D'. This is no accident, for under the mirror map, the logarithms of the monodromy are supposed to match up with cup product with the generators of the corresponding cone in $\overline{K(V^\circ)}$. So we have an explicit example of how mirror symmetry works in this case.

We next turn to the global description of the Kähler moduli space. Our goal is to make this tautologically identical with the global description of the simplified polynomial complex moduli space of the mirror manifold given in Section 6.1.2.

More formally, let $V \subset X_\Sigma$ be a Calabi-Yau toric hypersurface, where Σ is a simplified projective subdivision of the normal fan of the reflexive polytope Δ, and as usual, let $\Xi = (\Delta^\circ \cap N)_0 - \{0\}$. Then we *define* the compactified toric Kähler moduli space $\overline{\mathcal{KM}}_{\mathrm{toric}}$ to be the toric variety associated to the fan in $A(\Xi)$ which is obtained from the secondary fan of $\Xi^+ = (\Delta^\circ \cap N)_0 \times \{1\}$ by replacing collections of cones $\mathrm{cpl}(\Sigma')$ by their unions when they can be related by a sequence of trivial flips (as in Conjecture 6.2.8).

Thus the compactified toric moduli space is essentially the same as the compactified complex moduli space $\mathcal{M}_{\mathrm{simp}}$ of the mirror family after a minimal blowdown. Note that by definition, the compactified Kähler moduli space contains the affine toric variety associated to the Kähler cone. We have also seen in Section 6.2.1 that this toric variety naturally contains the Kähler moduli space. As already mentioned, we are ignoring the role of the automorphism group, which (assuming Conjecture 6.2.8 as always) would give a finite quotient of $\overline{\mathcal{KM}}_{\mathrm{toric}}$.

One nice thing about the definition of $\overline{\mathcal{KM}}_{\mathrm{toric}}$ is the way it effortlessly enlarges the Kähler moduli space and simultaneously compactifies it. For a general Calabi-Yau threefold in Section 6.2.2, we first had to enlarge to the movable cone and then further enlarge to the reflected movable cone. And this only gave a partial compactification. So although it took some effort to describe the toric Kähler moduli space, there is no question that Calabi-Yau toric hypersurfaces are *much* easier to work with than general Calabi-Yau threefolds, not to mention Calabi-Yau varieties of arbitrary dimension.

---

[2]The calculation was done using the methods of [**CdFKM, CFKM**], which are more efficient for this purpose.

The compactified toric Kähler moduli space, together with the moment map discussed in Section 3.3.3, give a mathematical description of the phase structure of the gauged linear sigma model (GLSM) which we outline in Appendix B.5. The GLSM depends on certain parameters, which can be identified with elements of $H^2_{\text{toric}}(X_\Sigma, \mathbb{R})$. This space is naturally identified with the target space of the moment map. As in our discussion of symplectic reduction, there is a symplectic quotient which is obtained as a quotient of a fiber of the moment map over a cohomology class $D \in A^+(\Xi)$. If D is in the cone of $\text{cpl}(\Sigma)$ in the GKZ decomposition of $A^+(\Xi)$, then we proved in Theorem 3.4.2 that the symplectic quotient is just X_Σ. For a general $D \in A(\Xi)$, the symplectic quotient is still defined but need not be a projective toric variety. It follows that the isomorphism classes of the possible symplectic quotients are in one-to-one correspondence with the top dimensional cones in the secondary fan.

The additional ingredient in the GLSM is that a subvariety V of the symplectic quotient is obtained. In this context, the different symplectic quotients corresponding to different components of the secondary fan are called *phases* of the GLSM. If $D \in \text{cpl}(\Sigma)$ for a simplified projective subdivision, then the subvariety V is a Calabi-Yau hypersurface. In general, V need not even be a hypersurface. In the example given in Appendix B.5, there are two phases. The corresponding symplectic quotients are $\mathbb{P}^4$ and $\mathbb{C}^5/\mathbb{Z}_5$, and the associated subvarieties V are the Fermat quintic threefold and the origin. The latter case is an example of a *Landau-Ginzburg orbifold*, also discussed in Appendix B.4.

One consequence of our definition of the toric Kähler moduli space is that if $V \subset X_\Sigma$ and $V^\circ \subset X_{\Sigma^\circ}$ are Batyrev mirrors, then $\overline{\mathcal{KM}}_{\text{toric}}(V)$ and $\overline{\mathcal{M}}_{\text{simp}}(V^\circ)$ are the same space. This may suggest to the reader that the mirror map is the identity, but as we will see in Example 6.2.4.3, the mirror map is not so simple. The actual mirror map will be discussed in detail in Section 6.3.

6.2.4. The Boundary of the Kähler Moduli Space. The discussion in this section is still largely speculative. We will begin with Calabi-Yau threefolds, and then specialize to those which arise as toric hypersurfaces. We take as our starting point the observation that one can often blow down a known Calabi-Yau threefold to a singular variety and then smooth it to get a new Calabi-Yau. More precisely, we have the following definition from [**Morrison9**].

DEFINITION 6.2.9. *Let V be a Calabi-Yau threefold, and let $\phi : V \to \overline{V}$ be a birational contraction. If $\overline{V}$ can be smoothed to a Calabi-Yau threefold $\widetilde{V}$, then the process of going from V to $\widetilde{V}$ is called an* extremal transition.

If ϕ is a primitive birational contraction, then for emphasis, we will sometimes refer to this transition as a *primitive extremal transition*. Recall from Section 6.2.2 that ϕ has Type I, II or III, so that (for example) we can speak of a Type I extremal transition.

In the physics literature, Type I extremal transitions are called *conifold transitions*, since physicists often call a three dimensional node a *conifold*. Conifold transitions are the transitions appearing in what is commonly called "Reid's fantasy". In [**Reid5**], Reid speculates that if the complex moduli spaces of all possible V and $\widetilde{V}$ are glued together along the space of all possible V, then the resulting space may be connected.

Here is a nice example of a conifold transition, taken from the paper [**GMS**].

Example 6.2.4.1. Let $\overline{V} \subset \mathbb{P}^4$ be a generic singular quintic threefold containing the plane in $\mathbb{P}^4$ defined by $x_3 = x_4 = 0$. Such a $\overline{V}$ has equation $x_3 g(x_0, \dots, x_4) + x_4 h(x_0, \dots, x_4) = 0$ for generic quartics g, h. The hypersurface $\overline{V}$ has 16 nodes at the points $x_3 = x_4 = g(x_0, \dots, x_4) = h(x_0, \dots, x_4) = 0$. Consider now the smooth Calabi-Yau complete intersection in $V \subset \mathbb{P}^4 \times \mathbb{P}^1$ defined by

$$y_0 x_4 - y_1 x_3 = 0, \quad y_0 g(x_0, \dots, x_4) + y_1 h(x_0, \dots, x_4) = 0,$$

where y_0, y_1 are homogeneous coordinates on $\mathbb{P}^1$ and $x_0, \dots, x_4$ are again homogeneous coordinates on $\mathbb{P}^4$. It is easy to see that the Kähler cone of V is isomorphic to the Kähler cone of $\mathbb{P}^4 \times \mathbb{P}^1$, i.e., is generated by the pullbacks of the hyperplane classes of $\mathbb{P}^4$ and $\mathbb{P}^1$. The hyperplane class H coming from $\mathbb{P}^4$ spans a face of the Kähler cone and $H^3 = 5 \neq 0$, so the projection map $\mathbb{P}^4 \times \mathbb{P}^1 \to \mathbb{P}^4$ restricts to a primitive contraction $\phi : V \to \overline{V}$. It is of Type I since ϕ contracts the curves lying over the 16 nodes. The singular threefold $\overline{V}$ can be smoothed to a smooth quintic threefold $\widetilde{V} \subset \mathbb{P}^4$, completing the conifold transition.

Our discussion of extremal transitions from V to $\widetilde{V}$ is structured to suggest that it is more natural to simultaneously glue both the complex and Kähler moduli spaces of V and $\widetilde{V}$, rather than just the complex moduli spaces. In other words, we are working with the full SCFT moduli space. As we will see below, identifying the SCFT moduli spaces of V and V° leads to a surprisingly rich structure.

An extremal transition is naturally thought of as a two-step process in the SCFT moduli space: a deformation of Kähler structure with fixed complex structure, followed by a deformation of complex structure with fixed Kähler structure. In more detail, the space of ample classes on $\overline{V}$ pulls back via ϕ to a face F of $K(V)$ by the discussion in Section 6.2.2. We first choose a path in $\overline{\mathcal{KM}}$ approaching a component of the boundary of $\overline{\mathcal{KM}}$ associated to F (keeping the complex structure fixed). Then we glue the product of this boundary component with the complex moduli space of V to the product of the compactified Kähler moduli space for $\widetilde{V}$ with the complex moduli space of $\overline{V}$. Note that the complex moduli space of $\overline{V}$ is a subset of the boundary of the complex moduli space of $\widetilde{V}$. The transition may be completed by smoothing $\overline{V}$, which is done by moving freely in the complex moduli space of $\widetilde{V}$.

It is conjectured [**Morrison9**] that this transition is "mirror" to an extremal transition in the SCFT moduli space of V° performed in the reverse direction. Mirror symmetry conjecturally identifies the path in the SCFT moduli space of V with a path in the SCFT moduli space of V°. If $\widetilde{V}^\circ$ is the mirror of $\widetilde{V}$, then the path describing the transition should have the following description: it starts with a path in complex moduli space of V°, fixing the Kähler structure, and the limiting singular complex structure arises as a primitive contraction of $\widetilde{V}^\circ$, representable as a path in the Kähler moduli space of $\widetilde{V}^\circ$. The transition from V° to $\widetilde{V}^\circ$ is conjectured to be the inverse of an extremal transition.

As noticed in [**BKK, Morrison9**], a nice example of how this works is provided by the Batyrev mirror construction.

Example 6.2.4.2. Suppose that Δ and $\widetilde{\Delta}$ are distinct reflexive polytopes such that $\Delta \subset \widetilde{\Delta}$. For simplicity, we will assume that both have dimension four. We will first use this data to construct an extremal transition from V to $\widetilde{V}$.

Since $\widetilde{\Delta}^\circ \subset \Delta^\circ$, any subdivision $\widetilde{\Sigma}$ of the normal fan of $\widetilde{\Delta}$ can be refined to a subdivision Σ of the normal fan of Δ. This subdivision yields a morphism $\pi : X_\Sigma \to X_{\widetilde{\Sigma}}$. The Calabi-Yau hypersurfaces $V \subset X_\Sigma$ are defined by the equations

$$\sum_{m_i \in \Delta \cap M} \lambda_i t^{m_i} = 0.$$

Similarly, we get Calabi-Yau hypersurfaces $\widetilde{V} \subset X_{\widetilde{\Sigma}}$ defined by equations

$$(6.29) \qquad \sum_{m_i \in \widetilde{\Delta} \cap M} \lambda_i t^{m_i} = 0.$$

Putting $\phi = \pi|_V$, we get a birational contraction $\phi : V \to \overline{V}$, where $\overline{V}$ is defined by (6.29), except that all coefficients λ_i for $m_i \in (\widetilde{\Delta} - \Delta) \cap M$ are zero. The extremal transition is completed by smoothing $\overline{V}$ to $\widetilde{V}$ by allowing arbitrary values of all the λ_i in (6.29). The process of going from V to $\widetilde{V}$ is called a *toric extremal transition*.

Now comes the crucial observation: since $\widetilde{\Delta}^\circ \subset \Delta^\circ$, the identical construction gives a toric extremal transition between the mirrors, now going from $\widetilde{V}^\circ$ and V°. This is exactly as predicted by Morrison's conjecture.

We should also mention that the technique of Example 6.2.4.2 has been applied to the 7555 reflexive polytopes corresponding to hypersurfaces in weighted projective spaces discussed in Section 4.1. It can be shown [**CGGK, ACJM**] that when the toric extremal transitions are used to connect moduli components, the resulting space has all of these 7555 irreducible components lying in the same connected component.

In general, suppose we have an extremal transition from V to $\widetilde{V}$. There is not yet a good mathematical definition of the structure of the SCFT moduli space along the intersection of the two components of the moduli space associated to V and $\widetilde{V}$. The only existing explanation is via physics using type II string theory and massless black holes in the case of Type I extremal transitions [**GMS**]. The type II string theory moduli space is larger than the SCFT moduli space, and these larger moduli spaces get glued together by an extremal transition in a way described by physics as follows. The curves which are to be contracted by $V \to \overline{V}$ correspond to black holes. As we follow the extremal transition to the boundary of the Kähler moduli space of V, the black holes become massless. On the other hand, cohomology classes also correspond to massless particles in type II string theory, and the difference between the Hodge numbers of V and $\widetilde{V}$ are accounted for by these massless black holes. A similar explanation can be given for certain Type III extremal contractions based on the physics explained in [**KMP**]. We will revisit this circle of ideas briefly in Section 12.2.8. (The type II string theory referred to here—with lowercase "t"—has no direct relation to the Types I, II, III of extremal contractions—with uppercase "T".)

We next discuss an example taken from [**CdFKM, KMP**] of an extremal transition which reveals the richly detailed structure of the moduli spaces involved. Surprisingly, it is an example we're already familiar with.

Example 6.2.4.3. We will continue our discussion of the mirror family of anti-canonical hypersurfaces in the toric blowup of $\mathbb{P}(1,1,2,2,2)$ begun in Examples 5.4.2

and 6.1.4.1. We have the reflexive polytope Δ whose polar Δ° satisfies

$$\Xi = (\Delta^\circ \cap N)_0 - \{0\} = \{(-1,-2,-2,-2),(1,0,0,0),(0,1,0,0),$$
$$(0,0,1,0),(0,0,0,1),(0,-1,-1,-1)\}.$$

We label these points $m_1,\ldots,m_6$. The GKZ decomposition from Example 3.7.2 shows that there is a unique simplified projective subdivision Σ with $\Sigma(1) = \Xi$. This gives a Calabi-Yau threefold $V \subset X_\Sigma$. We will construct an extremal transition starting from V, and we will study what this transition looks like on the Kähler moduli space of V.

Using the formulas from Section 4.1, one computes that

$$h^{1,1}(V) = h^{1,1}_{\mathrm{toric}}(V) = 2$$
$$h^{2,1}(V) = 86, \ h^{2,1}_{\mathrm{poly}}(V) = 83,$$

and the Batyrev mirror V° therefore satisfies

$$h^{2,1}(V^\circ) = h^{2,1}_{\mathrm{poly}}(V^\circ) = 2$$
$$h^{1,1}(V^\circ) = 86, \ h^{1,1}_{\mathrm{toric}}(V^\circ) = 83.$$

Let $D_1,\ldots,D_6$ denote the toric divisors associated to $\Sigma(1) = \Xi$, which by $H^2(X_\Sigma) \simeq H^2(V)$ can be thought of as on either X_Σ or V. Then the Kähler cone of V is spanned by the classes of D_1 and D_3. The first six columns of the matrix in Example 6.1.4.1 yield the linear equivalences $D_2 \sim D_1$, $D_4 \sim D_5 \sim D_3$, and $D_6 \sim D_1 - 2D_2$, which are useful in subsequent calculations. In particular, this tells us that $V \in |\sum_i D_i| = |4D_3|$.

We next look for a birational contraction of V. Using the theory developed in Section 6.2.2, we study the faces of the Kähler cone. It will be convenient to use homogeneous coordinates $x_1,\ldots,x_6$ of X_Σ corresponding to $\Sigma(1) = \Xi$. One can check that $|D_1|$ is the pencil defined by sections x_1, x_2, which give a K3 fibration $V \to \mathbb{P}^1$, and $(D_1)^3 = 0$. Hence $|D_1|$ doesn't give a contraction. On the other hand, $|D_3|$ is defined by sections $x_1^2 x_6, x_1 x_2 x_6, x_2^2 x_6, x_3, x_4, x_5$. Since $(D_3)^3 = 8$, we get a primitive contraction $\phi = \phi_{|D_3|}$ to a degree 8 threefold $\overline{V} \subset \mathbb{P}^5$. In fact, $\overline{V}$ is a complete intersection of a quadric and a quartic. To see this, we introduce coordinates $y_0,\ldots,y_5$ on $\mathbb{P}^5$. Using the sections of $|D_3|$ in the order above to define ϕ, we see that $\overline{V}$ lies on the singular rank 3 quadric $y_0 y_2 = y_1^2$, while the section of $|4D_3|$ defining $V \subset X_\Sigma$ is induced by a degree 4 polynomial in the y_i. One can check that ϕ is a primitive contraction of Type III and that it contracts the exceptional surface $x_6 = 0$ to the locus in $\overline{V}$ cut out by $y_0 = y_1 = y_2 = 0$, which is in fact a plane quartic curve. The exceptional divisor $x_6 = 0$ is a ruled surface over this genus 3 curve. The primitive extremal transition is completed by smoothing $\overline{V}$ to a smooth $(2,4)$ complete intersection $\widetilde{V} \subset \mathbb{P}^5$. The relevant Hodge numbers of $\widetilde{V}$ are $h^{1,1}(\widetilde{V}) = 1$ and $h^{2,1}(\widetilde{V}) = 89$.

In terms of Kähler moduli, this transition is described as follows. Since we obtained $\overline{V}$ from a face of the Kähler cone of V, it is clear that $\overline{\mathcal{KM}}(\widetilde{V})$ should be attached to $\overline{\mathcal{KM}}(V)$ along the subset corresponding to this face. However, this isn't quite the full story, since a face of the Kähler cone has *real* codimension one, yet $\overline{\mathcal{KM}}(\widetilde{V}) \subset \overline{\mathcal{KM}}(V)$ has *complex* codimension one. So further work is needed to determine the precise locus where we attach $\overline{\mathcal{KM}}(\widetilde{V})$. We will see below that the mirror tells us what to do.

This is almost a complete picture of the SCFT moduli space. So far, we have looked at the polynomial part of the complex moduli space of V, yet the Hodge numbers of V indicate that there are non-polynomial deformations. The exceptional divisor $x_6 = 0$ will not survive a general deformation of complex structure of V, but since this divisor is a ruled surface over a genus 3 curve, $2g - 2 = 4$ of the fibers will deform with a general deformation of complex structure. On this nearby complex structure, the deformed $|D_3|$ will again induce a primitive contraction, but now it will be of Type I. Furthermore, the 4 isolated curves can be simultaneously flopped, resulting in different complex and Kähler structures. One can show that this flop induces an involution of the SCFT moduli space.

We thus arrive at the following description of the SCFT moduli space of V. For any complex structure in the 86-dimensional complex moduli space, we have a 2-dimensional Kähler moduli space containing a distinguished curve corresponding to the primitive contraction $\phi_{|D_3|}$. Furthermore, in terms of the flop involution of the SCFT moduli space that we constructed above, the 83-dimensional polynomial subspace is fixed by the involution, but it acts nontrivially on the Kähler moduli space by a reflection as in Section 6.2.2. The 89-dimensional complex moduli space of $\widetilde{V}$ contains an 83-dimensional subspace of $(2,4)$ complete intersections whose defining quadric has rank 3 and an 86-dimensional subspace of $(2,4)$ complete intersections whose defining quadric has rank 2. These get identified with the complex moduli space and polynomial moduli space of V, while the Kähler moduli space of $\widetilde{V}$ is identified with a curve in the Kähler moduli space of V (this will be described below).

We next consider the "dual" extremal transition from $\widetilde{V}^\circ$ to V°. We have described V° in Examples 5.4.2 and 6.1.4.1, and $\widetilde{V}^\circ$ can be constructed using the Batyrev-Borisov construction discussed in Section 4.3. When this is done, an extremal transition from $\widetilde{V}^\circ$ to V° can be written down explicitly based on the calculation in [**CdFKM**]. This extremal transition induces a map on moduli $\overline{\mathcal{M}}_{\mathrm{simp}}(\widetilde{V}^\circ) \to \overline{\mathcal{M}}_{\mathrm{simp}}(V^\circ)$. Using the coordinates z_1, z_2 on $\overline{\mathcal{M}}_{\mathrm{simp}}$, one can show [**CdFKM, KMP**] that $\overline{\mathcal{M}}_{\mathrm{simp}}(\widetilde{V}^\circ)$ is attached along the component of the discriminant locus defined by $D_{A \cap \Gamma}(f_\Gamma) = 1 - 4z_2 = 0$ in Example 6.1.4.1. We will give a brief account of this argument in Section 6.3.3. The appearance of the discriminant locus should not be surprising since going from $\widetilde{V}^\circ$ to V° first contracts to a singular variety.

Since the transition from $\widetilde{V}^\circ$ to V° is compatible via mirror symmetry with the transition from V to $\widetilde{V}$, we can finally complete our description of the SCFT moduli space of V. Namely, if we use the mirror map to identify $\overline{\mathcal{M}}_{\mathrm{simp}}(V^\circ)$ with $\overline{\mathcal{KM}}(V)$, then by the previous paragraph, the Kähler moduli space of $\widetilde{V}$ should be identified with the curve in the Kähler moduli space of V defined by the equation $1 - 4z_2 = 0$.

The SCFT moduli space of V° will have the same intricate geometry as that of V. In addition, we can also see the effect of the flop involution as follows. By the proof of Theorem 4.1.5, V° is constructed as a hypersurface in a toric variety X_{Σ° which contains 3 toric divisors D_i, each of which has 2 components D_i', D_i'' after restricting to V°. The divisor classes $D_i' + D_i''$ are part of $H_{\mathrm{toric}}^{1,1}(V^\circ)$, but the individual divisors D_i', D_i'' are not. The involution $D_i' \leftrightarrow D_i''$, $i = 1, 2, 3$ corresponds to the effect of the flop involution in the complex moduli space of V. This involution is trivial on $H_{\mathrm{toric}}^{1,1}(V^\circ)$.

This example has some interesting consequences concerning the mirror map in the toric case. By our secondary fan construction, Batyrev mirrors V and V° give identical moduli spaces $\overline{\mathcal{KM}}_{\mathrm{toric}}(V)$ and $\overline{\mathcal{M}}_{\mathrm{simp}}(V^\circ)$. This might suggest that the mirror map is the identity. However, a closer look at Example 6.2.4.3 reveals that this is wrong. The basic idea is that in order for mirror symmetry to be compatible with extremal transitions, the mirror map must differ from the identity.

Here are the details of how this works. Returning to Example 6.2.4.3, the closure of the Kähler cone of V is $\{t_1 D_3 + t_2 D_1 : t_1, t_2 \geq 0\}$. Note that t_1, t_2 are the coordinates labeled r_1, r_2 in (3.23) in Chapter 3. It follows that the faces of the Kähler cone are defined by $t_j = 0$. Moving to the Kähler moduli space, we now let t_j lie in the upper half plane, and then $q_j = e^{2\pi i t_j}$ is the corresponding local coordinate of $\overline{\mathcal{KM}}(V)$. Hence $\mathrm{Im}(t_j) = 0$ corresponds to $|q_j| = 1$ in the Kähler moduli space of V. Note that this set has real codimension one in $\overline{\mathcal{KM}}(V)$.

On the mirror side, we have the simplified complex moduli of V°, which is given by the same space. The natural coordinates z_1, z_2 described in Example 6.1.4.1 came from the matrix (6.15). The rows of this matrix gave a basis of $A^*(\Xi)$ and the dual basis of $A(\Xi)$, which is where the GKZ decomposition lives. This is precisely the basis D_3, D_1 used above for the Kähler cone of V. Thus the local coordinates z_1, z_2. for $\overline{\mathcal{M}}_{\mathrm{simp}}(V^\circ)$ are the same as the coordinates q_1, q_2 on $\overline{\mathcal{KM}}(V)$ under our identifications of each with a Chow quotient (we will prove this carefully in Section 6.3.3).

To see why the mirror map is *not* $q_i = z_i$, we use the extremal transition from V to $\widetilde{V}$ described in Example 6.2.4.3. As we saw above, the face of the Kähler cone generated by D_3 gives the primitive contraction $\phi = \phi_{|D_3|} : V \to \overline{V}$. In Kähler moduli, this face is defined by $\mathrm{Im}(t_2) = 0$, which gives $|q_2| = 1$. If the mirror map were the identity, then the "dual" extremal transition from $\widetilde{V}^\circ$ to V° would glue in somewhere along the locus $|z_2| = 1$. Yet we saw in Example 6.2.4.3 that this locus is defined by $1 - 4z_2 = 0$, or $z_2 = 1/4$. The discrepancy between these equations shows that the mirror map can't be the identity.

A physicist would explain the discrepancy just noted by saying that the Kähler moduli space "receives instanton corrections". We have seen this phenomenon before, most notably in how instantons (holomorphic maps from a Riemann surface to V) cause the A-model correlation function to differ from the intersection number. We will finally explain the mirror map in Section 6.3.

It is also interesting to note that although $|z_2| = 1$ and $1 - 4z_2$ are not consistent, they at least involve the same variable. This is no accident and is part of a general conjecture of Morrison [**Morrison9**] describing the location of complex and Kähler moduli identifications for extremal transitions.

Sections 12.2.8 and 12.2.10 will discuss further aspects of extremal transitions. We will also see in Section 12.2.9 that extremal transitions can be used to construct mirrors of Calabi-Yau complete intersections in Grassmannians and partial flag manifolds.

6.3. The Mirror Map

In this section, we will describe the *mirror map* between the complex moduli space of a Calabi-Yau manifold V and the Kähler moduli space of its mirror manifold V°. These moduli spaces are presumed to be isomorphic, and the mirror map is an explicit local isomorphism connecting a natural set of coordinates on the Kähler moduli space of V° at a large radius limit point and a natural set of coordinates on the complex moduli space of V near a corresponding maximally unipotent boundary point.

We will start with a general definition of the mirror map, and then apply the definition to the case of toric hypersurfaces. We will conclude with some explicit examples.

6.3.1. Definition of the Mirror Map. The mirror map was written down in [**CdGP**] for the quintic threefold, and other examples with 1-dimensional Kähler moduli were done by essentially the same method [**Font, KT1, KT2, Morrison1**]. The mirror map for higher dimensional moduli was worked out in [**CdFKM, HKTY1, CFKM, HKTY2**], and a physics argument for the mirror map can be found in [**BCOV2**]. In Chapter 8, we will motivate the mirror map mathematically based on quantum cohomology. In Chapter 11, we will give another way to calculate the mirror map, a hint of which is given at the end of Example 6.3.4.1. The purely mathematical definition of the mirror map is due to Morrison, and we will follow the treatment of [**Morrison6**].

To define the mirror map, we begin with a maximally unipotent boundary point in the complex moduli space $\overline{\mathcal{M}}(V)$. Let $g_0,\dots,g_r$ be a basis for $W_2 \subset H^d(V)$ as in Definition 5.2.2 (W_2 is from the monodromy weight filtration and $d = \dim(V)$). Then, as in (6.1), define q_k by

$$(6.30) \qquad \frac{1}{2\pi i}\log q_k = \frac{1}{\langle g_0, \Omega\rangle}\sum_{j=1}^{r}\langle g_j, \Omega\rangle m^{jk},$$

where m^{jk} is the inverse matrix to the matrix m_{ij} from Definition 5.2.2. By Section 6.1.1, $q_i = q_i(p)$ is holomorphic for p in a neighborhood of our maximally unipotent boundary point.

By the discussion following Definition 5.2.2, we can rewrite the formula for q_k as follows. First, note that $y_0 = \langle g_0, \Omega\rangle$ is the unique (up to a constant) solution of the Picard-Fuchs equations which is holomorphic at the maximally unipotent boundary point. Furthermore, if we set

$$y_k = \sum_{j=1}^{r}\langle g_j, \Omega\rangle m^{jk},$$

then by Chapter 5, y_k is a solution of the Picard-Fuchs equations satisfying

$$y_k = y_0\frac{\log z_k}{2\pi i} + \tilde{y}_k,$$

where $\tilde{y}_k$ is holomorphic at the boundary point. Using this notation, (6.30) can be written as

$$\frac{1}{2\pi i}\log q_k = \frac{y_k}{y_0},$$

which implies that q_k is given by

$$(6.31) \qquad q_k = z_k\exp(2\pi i\,\tilde{y}_k/y_0).$$

Note that once y_0 is fixed, y_k is unique up to adding a constant multiple of y_0. Hence q_k is unique up to a constant multiple.

Now let V° be the mirror of V, and assume that the cone conjecture holds for the Kähler cone of V°. Also choose a subdivision Σ_+ of the cone K_+ from Section 6.2.1 which gives a smooth partial compactification $\overline{\mathcal{KM}}(V^\circ)$ (the unenlarged Kähler moduli space.) Then each cone $\sigma \in \Sigma_+$ gives a large radius limit point of $\overline{\mathcal{KM}}(V^\circ)$. By (6.19), the generators T_i of the cone σ determine canonical coordinates q_i, which are unique up to order.

Since $\overline{\mathcal{M}}(V)$ and $\overline{\mathcal{KM}}(V^\circ)$ have the same dimension r (this is part of being mirrors), we can take p in a neighborhood of our maximally unipotent boundary point in $\overline{\mathcal{M}}(V)$ and define

$$(6.32) \qquad M(p) = (q_1(p),\dots,q_r(p)) \in \overline{\mathcal{KM}}(V^\circ),$$

where $q_i(p)$ are the holomorphic functions defined above and we are using the canonical local coordinates of the chosen large radius limit point on the mirror side. It follows that the mirror map M requires *two* sorts of data for its definition:

- The functions $q_1,\dots,q_r$ from (6.30).
- The generators $T_1,\dots,T_r \in H^2(V^\circ)$ of the cone σ which gives the large radius limit point.

The first item is determined purely by the maximally unipotent boundary point of V, while the second is determined by the large radius limit point of V°.

Once the functions $q_1,\dots,q_r$ and cone generators $T_1,\dots,T_r$ are chosen, the map M from (6.30) and (6.32) is defined. For most choices, M will not be the map we want, but—and this gets to the heart of mirror symmetry—if the following conditions are satisfied:

- The classes $g_0,g_1,\dots,g_r$ in W_2 are chosen correctly (or equivalently, the solutions $y_0,y_1,\dots,y_r$ of the Picard-Fuchs equations are chosen correctly), and
- The maximally unipotent boundary and large radius limit points are chosen correctly in compatible compactifications of the moduli spaces,

then the map (6.32) is the mirror map.

To understand how to "choose correctly", we need to explain these two items in more detail. The first goes back to (6.2), where we noted that if we change $g_0,\dots,g_r$ to $g'_j = \sum_{k=0}^r c_{jk}g_k$, then we get new local coordinates $(q'_1,\dots,q'_r)$ with

$$(6.33) \qquad q'_k = e^{2\pi i(c^{k0}/c_{00})}q_k, \quad c^{k0} = \sum_{j=1}^r c_{j0}m'^{jk},$$

where (m'^{jk}) is inverse to the matrix determined by $N_j(g'_k) = m'_{jk}g'_0$.

There are two known methods for dealing with this lack of uniqueness. The first is to use the integral structure [**Morrison6**]. The spaces $W_0 \subset W_2$ in the monodromy weight filtration are defined over $\mathbb{Q}$. If $W_i \cap H^d(V,\mathbb{Z})$ denotes the intersection of W_i with the image of $H^d(V,\mathbb{Z})$ in $H^d(V,\mathbb{Q})$, then we can pick $g_0 \in W_0 \cap H^d(V,\mathbb{Z})$ and $g_1,\dots,g_r \in W_2 \cap H^d(V,\mathbb{Z})$ such that g_0 spans $W_0 \cap H^d(V,\mathbb{Z})$ over $\mathbb{Z}$ and $g_0,g_1,\dots,g_r$ span $W_2 \cap H^d(V,\mathbb{Z})$ over $\mathbb{Z}$. We call $g_0,g_1,\dots,g_r$ a $\mathbb{Z}$-*basis* of $W_0 \subset W_2$.

Given a a $\mathbb{Z}$-basis $g_0,g_1,\dots,g_r$, one easily shows that the numbers m_{jk} defined by $N_j(g_k) = m_{jk}g_0$ are integers. The *integrality conjecture* from Section 5.2.2 asserts that the matrix (m_{jk}) is invertible over $\mathbb{Z}$, or equivalently, that the m^{jk} in (6.30) are integers. This conjecture has been verified for the quintic mirror [**Morrison2**], and other cases of the conjecture are discussed in [**Morrison6**].

We will assume that the integrality conjecture is true. Now suppose we have $\mathbb{Z}$-bases $g_0, g_1, \ldots, g_r$ and $g'_0, g'_1, \ldots, g'_r$ related by $g'_j = \sum_{k=0}^{r} c_{jk} g_k$. It follows that $c_{00} = \pm 1$ and $c_{jk} \in \mathbb{Z}$. Since we are also assuming $m'^{jk} \in \mathbb{Z}$, (6.33) implies $q'_k = q_k$. Hence the functions (6.30) are unique up to order provided we use a $\mathbb{Z}$-basis.

Example 6.3.1.1. Suppose that V is a smooth Calabi-Yau threefold which satisfies the integrality conjecture at a maximally unipotent boundary point, which we will think of as $0 \in \Delta^r$. In the discussion following Proposition 5.6.1, we explained how we can find variables $q_1, \ldots, q_r$ such that the Gauss-Manin connection ∇ is given by

$$
\begin{aligned}
\nabla_{\delta_i} e^0 &= 0 \\
\nabla_{\delta_i} e^k &= \delta_{ik} e^0, \quad 1 \le k \le r \\
\nabla_{\delta_i} e_j &= \sum_{k=1}^{r} Y_{ijk} e^k, \quad 1 \le j \le r \\
\nabla_{\delta_i} e_0 &= e_i,
\end{aligned}
$$
(6.34)

where $\delta_j = 2\pi i q_j\, \partial/\partial q_j$ (see (5.67)). Let us show that $q_1, \ldots, q_r$ determine the mirror map in this case.

In the proof of Proposition 5.6.1, we showed that e_0 is the normalized 3-form Ω and the flat section $e^0 = g_0$ is the integral generator of W_0 such that $\langle g_0, \Omega \rangle = 1$. Furthermore, we also know that $e^1, \ldots, e^r$ are sections of W_2. If we write $q_j = \exp(2\pi i u_j)$, then it follows easily that

$$
g_j = -e^j + u_j e^0, \quad 1 \le j \le r
$$

is a flat section. Since going once counterclockwise around the ith slice of $(\Delta^*)^r$ takes u_j to $u_j + \delta_{ij}$, it follows that $T_i(g_j) = g_j + \delta_{ij} g_0$, and hence $N_i(g_j) = \delta_{ij} g_0$ for $1 \le j \le r$.

Since $e^0, e^1, \ldots, e^r$ are sections of W_2, the same is true for $g_0, g_1, \ldots, g_r$. We also claim that these sections are integral. We already know this for g_0. For the others, note that $\exp(-\sum_i u_i N_i)\, g_j = e^j$ for $1 \le j \le r$. As in the proof of Proposition 5.6.1, the e^j are integral above $0 \in \Delta^r$, which implies that the same is true for the flat section g_j by the discussion in Section 5.1.4. Thus $g_0, g_1, \ldots, g_r$ are a basis of W_2 consisting of flat integral sections such that g_0 spans W_0.

We can now compute the mirror map. Since $N_i(g_j) = \delta_{ij} g_0$, the matrix $(m_{ij}) = (\delta_{ij})$ is the identity, so that the right hand side of (6.30) reduces to $\langle g_j, \Omega \rangle$ (remember that $\langle g_0, \Omega \rangle = 1$). But

$$
\langle g_j, \Omega \rangle = \langle -e^j + u_j e^0, e_0 \rangle = u_j
$$

since $\langle e^j, e_0 \rangle = 0$ for $1 \le j \le r$ by (5.66), and it follows from (6.30) that q_k is indeed the kth coordinate of the mirror map.

In the toric case, there is a second method for picking the correct $g_0, g_1, \ldots, g_r$ in (6.30), which doesn't depend on the integrality conjecture. The basic idea is that toric geometry gives natural choices for the z_i, and by suitably normalizing the constant term of $\tilde{y}_k/y_0$ in (6.31), we get a unique choice for q_k. We will explain how this works in Section 6.3.3.

After computing $q_1, \ldots, q_r$, the other aspect of "choosing correctly" is finding the corresponding large radius limit point of $\overline{\mathcal{KM}}(V^\circ)$. This is potentially non-trivial, for as we saw in Sections 6.1.1 and 6.2.1, the compactifications may be nonunique with possibly many equivalence classes of points to choose from. One approach is to look for suitably minimal compactifications, where one blows down the boundary as much as possible. This may lead to a small number of choices for how to match maximally unipotent boundary points with large radius limit points. But most of the time, the issue is that one needs to have fairly complete knowledge of what the moduli spaces look like. In fact, part of V° being the mirror of V is that *the moduli spaces $\overline{\mathcal{M}}(V)$ and $\overline{\mathcal{KM}}(V^\circ)$ should be understood well enough so that one knows how to match up maximally unipotent boundary points with large radius limit points*. The best example of this is the toric case, where the moduli spaces are essentially the same, so that we know precisely how the boundary points match up. We will explore this in more detail in Section 6.3.3.

Furthermore, once we've picked the correct large radius limit point, we get a cone σ generated by $T_1, \ldots, T_r$. But how are these generators ordered? When we discuss the derivative of the mirror map in Section 6.3.2, we will see that this is essentially determined by the isomorphism $H^{d-1,1}(V) \simeq H^{1,1}(V^\circ)$ from equation (1.5) in Chapter 1.

Once all of the correct choices have been made, the key idea is that the mirror map from $\overline{\mathcal{M}}(V)$ to $\overline{\mathcal{KM}}(V^\circ)$ should be compatible with the A-model and B-model correlation functions. Furthermore, it is conjectured that this compatibility extends to variations of Hodge structure. Over $\overline{\mathcal{M}}(V)$, we have a natural variation of Hodge structure coming from the variation of complex structure on V, and the asymptotic behavior of this variation is what determines whether a boundary point is maximally unipotent. In Chapter 8, we will define the *A-model variation of Hodge structure* on $\overline{\mathcal{KM}}(V^\circ)$, and we will prove that it is maximally unipotent at large radius limit points. Then the mirror conjecture asserts that the mirror map is compatible with these variations of Hodge structure, and this would imply compatibility of the corresponding correlation functions. In Chapter 11, we will see how much of this conjecture can be proved in the toric context.

6.3.2. The Derivative of the Mirror Map. Let V and V° be Calabi-Yau manifolds of dimension $d \geq 3$, and assume that the mirror map has been defined as in Section 6.3.1. Our next task is to relate the following three topics:

- The derivative of the mirror map M.
- The isomorphism $H^{d-1,1}(V) \simeq H^{1,1}(V^\circ)$ from (1.5) in Chapter 1.
- The "correct choice" of $T_1, \ldots, T_r$ once we know $q_1, \ldots, q_r$.

In order to study the mirror map, it is convenient to replace the Kähler moduli space $\mathcal{KM}(V^\circ)$ with the complexified Kähler space

$$K_{\mathbb{C}}(V^\circ) = \big(H^2(V^\circ, \mathbb{R}) + i\, K(V^\circ) \big) / \operatorname{im} H^2(V^\circ, \mathbb{Z}).$$

If we assume the cone conjecture from Section 6.2.1, then this is a discrete cover of $\mathcal{KM}(V^\circ)$. To avoid this, we will simply regard the mirror map as a map

$$M : S \longrightarrow K_{\mathbb{C}}(V^\circ),$$

where S is a neighborhood of a maximally unipotent boundary point of the complex moduli space of V.

A large radius limit point of V° is determined by a cone σ generated by integral classes $T_1, \ldots, T_r \in H^2(V^\circ, \mathbb{C}) = H^{1,1}(V^\circ)$ which lie in the closure of the Kähler cone of V°. This gives coordinates

$$(6.35) \qquad (e^{2\pi i t_1}, \ldots, e^{2\pi i t_r}) \longleftrightarrow \left[\textstyle\sum_j t_j T_j \right] \in K_{\mathbb{C}}(V^\circ).$$

in a neighborhood of the chosen large radius limit point. By (6.32), the mirror map is $M(p) = (q_1(p), \ldots, q_r(p))$ for $p \in S$, and then (6.31) easily implies that M extends to the boundary points and is a local biholomorphism there.

In order to describe the derivative of M, we need to understand the tangent bundles of S and $K_{\mathbb{C}}(V^\circ)$. Let's begin with the latter. Each $T \in H^2(V^\circ, \mathbb{C})$ gives a vector field $\partial/\partial T$ on $H^2(V^\circ, \mathbb{R}) + i\, K(V^\circ)$ which is invariant under translation. Thus we can regard $\partial/\partial T$ as a vector field on $K_{\mathbb{C}}(V^\circ)$. This gives a trivialization

$$T_{K_{\mathbb{C}}(V^\circ)} \simeq H^2(V^\circ, \mathbb{C}) \times K_{\mathbb{C}}(V^\circ).$$

The coordinates (6.35) give vector fields $\partial/\partial t_j$ on $K_{\mathbb{C}}(V^\circ)$ which are naturally identified with $\partial/\partial T_j$. Under the above isomorphism, $\partial/\partial t_j$ is identified with T_j.

In terms of this trivialization, the mirror map $M : S \to K_{\mathbb{C}}(V^\circ)$ induces a map

$$(6.36) \qquad D_M : T_S \to H^2(V^\circ, \mathbb{C}) \times K_{\mathbb{C}}(V^\circ).$$

since M is a local biholomorphism. Since the functions $q_1, \ldots, q_r$ can be used as local coordinates on S, the vector fields $\delta_j = 2\pi i q_j\, \partial/\partial q_j$ on S are a basis of T_S, and then the above formula for M shows that D_M takes δ_j to $\partial/\partial t_j = \partial/\partial T_j$. In terms of (6.36), D_M takes δ_j to T_j.

We next describe the tangent bundle T_S of S. The complex moduli space of V is $\mathcal{M}(V)$, and in a neighborhood S of our maximally unipotent boundary point p_0, we will assume that we have a universal family $\pi : \mathcal{V} \to S$ (strictly speaking, we should go to a finite cover of S). Since V is Calabi-Yau, the Kodaira-Spencer map is an isomorphism [**Bogomolov, Tian, Todorov**]

$$\kappa : T_S \simeq R^1 \pi_* \Theta_{\mathcal{V}/S},$$

where T_S is the tangent bundle of S and $\Theta_{\mathcal{V}/S}$ is the relative tangent bundle.

We also have the Hodge filtration $\mathcal{F}^p$ of $\mathcal{H} = R^d \pi_* \mathbb{C} \otimes_{\mathbb{C}} \mathcal{O}_S$. We are most interested in $\mathcal{F}^d = \pi_* \Omega^d_{\mathcal{V}/S}$ and $\mathcal{F}^{d-1}/\mathcal{F}^d = R^1 \pi_* \Omega^{d-1}_{\mathcal{V}/S}$. Contraction gives a map

$$(6.37) \qquad R^1 \pi_* \Theta_{\mathcal{V}/S} \longrightarrow \underline{\mathrm{Hom}}(\mathcal{F}^d, \mathcal{F}^{d-1}/\mathcal{F}^d),$$

and, as is well known, the composition

$$(6.38) \qquad T_S \longrightarrow \underline{\mathrm{Hom}}(\mathcal{F}^d, \mathcal{F}^{d-1}/\mathcal{F}^d)$$

of κ with (6.37) is the map which sends a vector field X and holomorphic d-form Ω to $[\nabla_X(\Omega)]$, where $[\ldots]$ denotes the class in the quotient $\mathcal{F}^{d-1}/\mathcal{F}^d$.

Since V is Calabi-Yau, we can fix a nonvanishing section Ω of $\mathcal{F}^d$ over S. We normalize it as usual by setting $\widetilde{\Omega} = \Omega/\langle g_0, \Omega \rangle$, where g_0 is from the definition of maximally unipotent monodromy. Then $\widetilde{\Omega}$ induces an isomorphism $\mathcal{F}^d \simeq \mathcal{O}_S$, so that (6.38) gives the isomorphism

$$(6.39) \qquad T_S \simeq R^1 \pi_* \Theta_{\mathcal{V}/S} \simeq R^1 \pi_* \Omega^{d-1}_{\mathcal{V}/S}$$

defined by $X \mapsto [\nabla_X(\widetilde{\Omega})]$. This has the following meaning. Suppose $V = V_p$ for $p \in S$. If we think of X as the infinitesimal deformation taking $V = V_p$ to V_ϵ, then under this deformation, $\widetilde{\Omega}_p$ deforms to $\widetilde{\Omega}_\epsilon = \widetilde{\Omega}_p + \epsilon \nabla_X(\widetilde{\Omega})_p + O(\epsilon^2)$.

We can now bring together the three items discussed earlier. First, if $V = V_p$ for $p \in S$, then combining (6.36) and (6.39) gives an isomorphism

$$(6.40) \qquad H^1(V, \Omega_V^{d-1}) \simeq T_{S,p} \simeq H^2(V^\circ, \mathbb{C}) = H^{1,1}(V^\circ).$$

Thus, the derivative of the mirror map is the d-dimensional version of the isomorphism (1.5) from Chapter 1. Furthermore, we saw above that D_M takes $\delta_j = 2\pi i q_j \partial/\partial q_j$ to T_j. In terms of (6.40), this says that at $V = V_p$,

$$H^1(V, \Omega_V^{d-1}) \simeq H^2(V^\circ, \mathbb{C}) \quad \text{is defined by} \quad 2\pi i q_j \frac{\partial}{\partial q_j} \mapsto T_j.$$

Thus knowing (6.40) is *equivalent* to knowing the "correct choice" for $T_1, \ldots, T_r$!

We conclude our discussion with the following lemma, which will be useful in Chapter 8. The idea is to relate the derivative of the mirror map to the bundle $\mathcal{H} = \mathcal{F}^0 = R^d \pi_* \mathbb{C} \otimes_{\mathbb{C}} \mathcal{O}_S$.

LEMMA 6.3.1. *Let $\widetilde{\Omega}$ be the normalized d-form near a maximally unipotent boundary point p_0. Then:*

(*i*) *In a neighborhood S of p_0, the map*

$$\psi : \mathcal{O}_S \oplus T_S \longrightarrow \mathcal{F}^{d-1} \subset \mathcal{H}$$

defined by $\psi(f, X) = f\widetilde{\Omega} + \nabla_X(\widetilde{\Omega})$ is an isomorphism compatible with (6.39).

(*ii*) *There is a bundle isomorphism*

$$\phi : \mathcal{F}^{d-1} \simeq \left(H^0(V^\circ, \mathbb{C}) \oplus H^2(V^\circ, \mathbb{C}) \right) \times M(S)$$

such that ϕ takes $\widetilde{\Omega}$ to $1 \in H^0(V^\circ, \mathbb{C})$ and is the derivative of the mirror map when restricted to T_S, which is identified with a subset of $\mathcal{F}^{d-1}$ via ψ.

PROOF. The definition of maximally unipotent monodromy implies that $W_0 = W_1$ has dimension 1. Since cup product interacts nicely with the monodromy weight filtration, it follows that Gr_{2d}^W has dimension 1 and $W_{2d-1} = W_{2d-2}$. The arguments used to prove (5.10) then imply that

$$F_{\lim}^d \oplus W_{2d-1} = F_{\lim}^d \oplus W_{2d-2} = H^d(V, \mathbb{C}),$$

where $F_{\lim}^d$ is from the limiting Hodge filtration. We conclude that the Hodge structure on Gr_{2d-2}^W is purely of type $(d-1, d-1)$, and this in turn implies

$$F_{\lim}^{d-1}/F_{\lim}^d \simeq Gr_{2d-2}^W.$$

The definition of maximally unipotent monodromy, together with the duality given by cup product on $Gr_\bullet^W$, imply that if $g \in W_{2d}$ generates Gr_{2d}^W, then the $N_j(g)$ form a basis of Gr_{2d-2}^W. Combining this with (5.3), we see that the lemma is true at p_0 once we go to the canonical extensions. From here, the isomorphism $\mathcal{O}_S \oplus T_S \simeq \mathcal{F}^{d-1}$ follows immediately. The remaining statements of the lemma are now obvious. $\square$

An illustration of this lemma is provided by (6.34), since $e_0 = \widetilde{\Omega}$ and $e_i = \nabla_{\delta_i}(\widetilde{\Omega})$. Thus, if we use the basis 1 of $\mathcal{O}_S$ and δ_i of T_S, then the isomorphism in the first part of Lemma 6.3.1 takes 1 to $\widetilde{\Omega}$ and δ_i to e_i. In particular, (6.34) contains an explicit copy of T_S. Also, the isomorphism in the second part of the lemma takes e_0 to $1 \in H^0(V^\circ, \mathbb{C})$ and e_i to T_i for $1 \leq i \leq r$. In Chapter 8, we will extend ϕ to an isomorphism of $\mathcal{H} = \mathcal{F}^0$ with the trivial bundle given by $\oplus_p H^{2p}(V^\circ, \mathbb{C})$. As we will see, this is part of the Hodge-theoretic formulation of mirror symmetry.

6.3.3. The Mirror Map for Toric Hypersurfaces. For toric hypersurfaces, we can be much more precise about what the mirror map should be, and we can also give an alternate proposal for dealing with the nonuniqueness (6.33).

As usual, we fix a reflexive polytope Δ and let $V \subset X_\Sigma$ be a Calabi-Yau hypersurface corresponding to a simplified projective subdivision Σ of the normal fan of Δ. Also let $V^\circ \subset X_{\Sigma^\circ}$ be a Batyrev mirror of V and set

$$\Xi = (\Delta \cap M)_0 - \{0\} \quad \text{and} \quad \Xi^+ = (\Xi \cup \{0\}) \times \{1\} = (\Delta \cap M)_0 \times \{1\}.$$

Note that $\Sigma^\circ(1) = \Xi$ since Σ° is a simplified projective subdivision of the normal fan of Δ°. Finally, recall that the GKZ decomposition of Ξ and the secondary fan of Ξ^+ live in the vector space $A(\Xi)$.

In order to specify the mirror map, we need more data. The cone $\mathrm{cpl}(\Sigma^\circ)$ is a cone the GKZ decomposition of Ξ. We will subdivide $\mathrm{cpl}(\Sigma^\circ)$ into smooth cones σ. Hence each σ is a simplicial cone whose generators form an integer basis of the lattice.

Our basic claim is that each cone σ gives all the information we need to specify the mirror map. In other words, σ determines corresponding maximally unipotent boundary points and large radius limit points and chooses the constants in (6.33) *without* assuming the integrality conjecture or knowing anything about integer cohomology. Furthermore, by starting with σ, the "correct choice" for $T_1, \dots, T_r$ is made automatically.

Let's begin with the simplified (polynomial) moduli space $\overline{\mathcal{M}}_{\mathrm{simp}}$ constructed in Section 6.1.2 using the secondary fan of Ξ^+ (see (6.12)). Since we want a smooth compactification, we will use a refinement of the the secondary fan which includes the above cones σ refining $\mathrm{cpl}(\Sigma^\circ)$. We denote this compactification by $\overline{\mathcal{M}}_{\mathrm{simp}}$, and assuming Conjecture 6.1.4 (which we do for the rest of the chapter), each σ gives a maximally unipotent boundary point of $\overline{\mathcal{M}}_{\mathrm{simp}}$.

If we fix one of the σ's, we get natural coordinates at the maximally unipotent boundary point as follows. The generators of σ form a basis of the lattice of $A(\Xi)$, denoted N_0 in Section 6.1.2. The dual basis $\ell_1, \dots, \ell_r$ generates the lattice of $A(\Xi)^*$, denoted M_0 in (6.13). This is the lattice of relations among Ξ^+. Thus each basis element can be written $\ell_k = (\ell_{kj})$. Then we saw in (6.14) that

$$(6.41) \qquad z_k = \prod_j \lambda_j^{\ell_{kj}}$$

are the local coordinates at the maximally unipotent boundary point. Recall that the λ_i are coefficients of the Laurent polynomials

$$(6.42) \qquad \sum_{m_i \in (\Delta \cap M)_0} \lambda_i\, t^{m_i} = \lambda_0 + \sum_{m_i \in \Xi} \lambda_i\, t^{m_i},$$

where as usual $m_0 = 0$. This way of writing the local coordinates emphasizes that we are dealing with moduli (actually, simplified moduli, since the by the dominance conjecture, $\overline{\mathcal{M}}_{\mathrm{simp}}$ is a finite cover of the true polynomial moduli space).

The mirror map will be a function of the coordinates $z_1, \dots, z_r$ from (6.41). We will see below that each $\ell_k = (\ell_{kj})$ contributes the sign

$$(6.43) \qquad (-1)^{\ell_{k0}}, \quad k = 1, \dots, r$$

which will play an important role in determining the mirror map.

We now turn to the toric Kähler moduli space $\overline{\mathcal{KM}}_{\mathrm{toric}}(V^\circ)$ discussed in Section 6.2.3. We will assume Conjecture 6.2.8, which implies that up to the action of a finite group, a compactification of the toric Kähler moduli is given by the secondary fan of Ξ^+, provided we combine cones corresponding to trivial flips. In order to have a compactification compatible with the complex moduli of V, we will work with the smooth blow-up $\overline{\mathcal{KM}}_{\text{toric}}(V^\circ)$ given by the refinement of the secondary fan used above. In particular, the cones σ refining $\text{cpl}(\Sigma^\circ)$ give large radius limit points of $\overline{\mathcal{KM}}_{\text{toric}}(V^\circ)$.

For each σ, we get local coordinates $q_1, \dots, q_r$ at the corresponding large radius limit point as follows. First note that we have a natural isomorphism

$$A(\Xi) \simeq H^2_{\text{toric}}(V^\circ, \mathbb{R})$$

by the proof of Theorem 4.1.5. If the generators of σ are $T_1, \dots, T_r$, then the map

$$(\Delta^*)^r \longrightarrow \left(H^2_{\text{toric}}(V^\circ, \mathbb{R}) + i\, K_{\text{toric}}(V^\circ)\right)/\text{im}\, H^2_{\text{toric}}(V^\circ, \mathbb{Z})$$

given by $(e^{2\pi i t_1}, \dots, e^{2\pi i t_r}) \mapsto \left[\sum_{j=1}^r t_j T_j\right]$ extends to an open immersion

$$\Delta^r \subset \overline{\mathcal{KM}}_{\text{toric}}(V^\circ)$$

such that the origin maps to the chosen large radius limit point. Then local coordinates at this point are given by $q_j = e^{2\pi i t_j}$. We will use these coordinates in defining the mirror map.

The compactifications $\overline{\mathcal{M}}_{\text{simp}}(V)$ and $\overline{\mathcal{KM}}_{\text{toric}}(V^\circ)$ are the same, and fixing σ gives compatible maximally unipotent boundary points and large radius limit points. We also have local coordinates $z_1, \dots, z_r$ of the former and $q_1, \dots, q_r$ of the latter. It may not be obvious, but these coordinates are the same. This is because the z_j correspond to a basis of the character group of the torus T_0 contained in the moduli space (this is the notation of Section 6.1.2). Similarly, the q_j correspond to a basis of the 1-parameter subgroups of T_0. Since the two bases involved are dual bases, the z_j and the q_j clearly give the same coordinates. However, in order to keep things straight, we will continue to use z_j for $\overline{\mathcal{M}}_{\text{simp}}(V)$ and q_j for $\overline{\mathcal{KM}}_{\text{toric}}(V^\circ)$.

We can now specify the mirror map. We define $q_k = q_k(z_1, \dots, z_r)$ as in (6.30), and we resolve the lack of uniqueness of q_k as follows. By (6.31), we have $q_k(z_1, \dots, z_r) \sim c\, z_k$ for some constant $c \neq 0$. Then, using (6.33), we can arrange that

$$(6.44) \qquad\qquad q_k(z_1, \dots, z_r) \sim (-1)^{\ell_{k0}}\, z_k,$$

where the sign $(-1)^{\ell_{k0}}$ is as in (6.43). This clearly determines the functions $q_1, \dots, q_r$ uniquely, and defines a unique mirror map (6.32). In terms of the discussion preceding (6.31), this means finding a solution y_k of the Picard-Fuchs equations such that

$$(6.45) \qquad\qquad y_k = y_0\, \frac{\log((-1)^{\ell_{k0}}\, z_k)}{2\pi i} + \tilde{y}_k,$$

where $\tilde{y}_k$ is holomorphic and vanishes at $z_1 = \cdots = z_r = 0$. Then q_k is given by

$$q_k = (-1)^{\ell_{k0}}\, z_k \exp(2\pi i\, \tilde{y}_k/y_0).$$

Although this formula for q_k looks unusual, the sign $(-1)^{\ell_{k0}}$ was present in our very first example of mirror symmetry, as we will now explain.

Example 6.3.3.1. In our discussion of the quintic threefold V in Chapter 2, the mirror map went from the complex moduli of the mirror V° to the Kähler moduli of V. In homogeneous coordinates, V° is defined by

$$\lambda_1\, x_1^5 + \lambda_2\, x_2^5 + \lambda_3\, x_3^5 + \lambda_4\, x_4^5 + \lambda_5\, x_5^5 + \lambda_0\, x_1 x_2 x_3 x_4 x_5 = 0,$$

and the relation between the elements of Ξ^+ is $\ell = (1,1,1,1,1,-5)$. As in (5.36), the moduli coordinate of V° is

$$z = \frac{\lambda_1 \cdots \lambda_5}{\lambda_0^5}.$$

The discussion following (5.36) shows that z is precisely the coordinate called x in Chapter 2. We also have the coordinate q for the Kähler moduli of V.

In Example 6.3.4.1 and its continuation, we will show that the Picard-Fuchs equation of V° has the holomorphic solution

$$y_0(z) = \sum_{n=0}^{\infty} \frac{(5n)!}{(n!)^5}(-z)^n,$$

and there is also the solution

$$y_0 \frac{\log(-z)}{2\pi i} + \frac{5}{2\pi i} \sum_{n=1}^{\infty} \frac{(5n)!}{(n!)^5}\Big[\sum_{j=n+1}^{5n} \frac{1}{j} \Big](-z)^n,$$

which has the required form (6.44). It follows that the toric mirror map for the quintic threefold is given by

$$q = -z \exp\left(\frac{5}{y_0(z)} \sum_{n=1}^{\infty} \frac{(5n)!}{(n!)^5}\Big[\sum_{j=n+1}^{5n} \frac{1}{j} \Big](-z)^n \right).$$

This is the formula given for the mirror map in (2.9). In Chapter 2, we got the minus sign in front of the right hand side by computing the Yukawa coupling of V° and using the fact that the generic V has 2875 lines. Now that we know (6.44), we get the minus sign immediately from $(-1)^{\ell_{k0}} = (-1)^{-5} = -1$.

We can also compare this to the mirror map defined in Section 6.3.1. Since the quintic mirror satisfies the integrality conjecture [**Morrison2**], we can find integral g_0, g_1 with $N(g_1) = g_0$ such that the mirror map of (6.30) is

$$\frac{1}{2\pi i} \log q = \frac{\langle g_1, \Omega \rangle}{\langle g_0, \Omega \rangle}.$$

It is not obvious that this agrees with the above toric formulation. However, as mentioned in Example 5.6.4.1, this follows from the explicit computations given in [**CdGP, Morrison2**]. Furthermore, letting γ_0, γ_1 be Poincaré dual to g_0, g_1, the above formula implies

$$q = \exp\big(2\pi i \textstyle\int_{\gamma_1} \Omega / \int_{\gamma_0} \Omega \big),$$

which is the definition of the mirror map given at the beginning of Section 2.3. Hence we see that for the quintic threefold, all of the various ways of defining the mirror map agree.

One way to understand (6.44) is to change the sign of λ_0: if we replace λ_0 by $-\lambda_0$ in (6.42), then by (6.41), each z_k changes sign by $(-1)^{\ell_{k0}}$. If we had set things up this way, the mirror map would be normalized using $q_k \sim z_k$ instead of (6.44). This method is used in [**BvS**]. On the other hand, [**HKTY1**] uses the same z_k we do, but then introduces new variables $x_k = (-1)^{\ell_{k0}} z_k$ and defines the mirror map in terms of the x_k.

Another way to understand (6.44) is to observe that it is equivalent to specifying the derivative of the mirror map. Using the ideas of Section 6.3.2, the tangent space at the maximally unipotent boundary point can be naturally identified with the space denoted $H^1_{\mathrm{poly}}(V, T_V) \simeq H^{n-2,1}_{\mathrm{poly}}(V)$ in the proof of Theorem 4.1.5, and similarly, the tangent space at the large radius limit point is $H^2_{\mathrm{toric}}(V^\circ)$. In terms of the coordinates z_k and q_k, we have the natural bases $\partial/\partial z_k$ of $H^{n-2,1}_{\mathrm{poly}}(V)$ and $\partial/\partial q_k$ of $H^2_{\mathrm{toric}}(V^\circ)$. In these coordinates, the derivative of the mirror map is the isomorphism

$$H^{n-2,1}_{\mathrm{poly}}(V) \simeq H^2_{\mathrm{toric}}(V^\circ)$$

defined by $\partial/\partial z_k \mapsto \sum_j \frac{\partial q_j}{\partial z_k} \partial/\partial q_j$. At the boundary point $z_1 = \cdots = z_r = 0$, (6.44) tells us that the derivative takes $\partial/\partial z_k$ to $(-1)^{\ell_{k0}} \partial/\partial q_k$. If we untangle the notation, this map takes first order deformations of V determined by monomials to cohomology classes of V° coming from toric divisors, up to the sign $(-1)^{\ell_{k0}}$. Hence this map (or more properly, the version without the signs) is called the *monomial-divisor mirror map* [**AGM1**].

One of the basic conjectures of [**AGM1**, Sect. 4] is that up to an automorphism of order 2, the monomial-divisor mirror map is the derivative of the mirror map. The automorphism of order 2 means that certain signs need to be specified, and this is precisely what we do in (6.44). In our treatment, this conjecture is true by definition, given our toric method for defining the mirror map. However, to really deserve the name "mirror map", the mirror map described here must be compatible with the correlation functions and variations of Hodge structure mentioned at the end of Section 6.3.1. This has yet to be proved in complete generality, so that there is still work to be done.

A final question to ask concerns why (6.44) is the correct way to normalize the mirror map. A justification from physics can be found in [**MP1**]. From a mathematical point of view, the sign $(-1)^{\ell_{k0}}$ is most easily explained using hypergeometric functions, which we will discuss in the next section.

We close by partially computing the mirror map of an example we know well.

Example 6.3.3.2. The mirror V° of an anticanonical hypersurface in the toric blowup of $\mathbb{P}(1,1,2,2,2)$ has been considered several times, most recently in Section 6.2.4. Rather than compute the whole mirror map, we will concentrate on studying the part relevant to the extremal transition studied in Example 6.2.4.3. Recall that there were actually two extremal transitions: the original one, from V to $\widetilde{V}$, and its "dual" transition, from $\widetilde{V^\circ}$ to V°. The original transition involves a face of the Kähler cone of V, while its dual uses part of the discriminant locus of V°. The question is, how do these match up under the mirror map?

Using the local coordinates z_1, z_2 from Example 6.1.4.1 for our maximally unipotent boundary point in $\mathcal{M}_{\mathrm{simp}}$, we will study the mirror map on the locus $z_1 = 0$. The calculation presented here first appeared in [**AGM3**] and has been amplified in [**Morrison9**]. Our goal will be to compute q_2 as a function of z_2, assuming $z_1 = 0$. Since $z_2 = \lambda_1 \lambda_2 / \lambda_6^2$, we get the GKZ operator $\square = \partial_1 \partial_2 - \partial_6^2$, where ∂_i is partial differentiation with respect to λ_i. The calculations leading to (5.39) apply here to give the Picard-Fuchs equation

$$\delta^2 y - 2z_2 \delta(2\delta + 1)y = 0$$

for the periods $y = y(z_2)$, where $\delta = z_2(d/dz_2)$. The general solution is

$$(6.46) \qquad y(z_2) = c_1 + c_2 \log\left(\frac{\sqrt{1 - 4z_2} + 2z_2 - 1}{2z_2}\right).$$

To calculate the mirror map, we need the analytic and logarithmic solutions near $z_2 = 0$. The analytic solutions are just the constants, and we accordingly take $y_0(z_2) = 1$. Since the Taylor expansion of $\sqrt{1 - 4z_2}$ begins $1 - 2z_2 - 2z_2^2 + \cdots$, the function multiplying c_2 in (6.46) has logarithmic growth at $z_2 = 0$. In order that $q_2(z_2) = \exp(2\pi i\, y_1(z_2)/y_0(z_2))$ have the asymptotic behavior predicted by (6.44), we let $c_1 = 1/2$ and $c_2 = 1/(2\pi i)$. This gives

$$y_1(z_2) = \frac{1}{2\pi i} \log\left(\frac{1 - 2z_2 - \sqrt{1 - 4z_2}}{2z_2}\right),$$

and it follows that the mirror map on the locus $z_1 = 0$ is given by

$$(6.47) \qquad q_2(z_2) = \frac{1 - 2z_2 - \sqrt{1 - 4z_2}}{2z_2} = z_2 + 2z_2^2 + 5z_2^3 + \cdots.$$

In $\overline{\mathcal{M}}_{\text{simp}}(V^\circ)$, Example 6.2.4.3 showed that the dual extremal transition occurred along the locus $1 - 4z_2 = 0$, which is part of the discriminant locus. This locus hits $z_1 = 0$ in the point $z_2 = 1/4$. If we substitute this into (6.47), we get $q_2 = 1$. However, $q_2 = \exp(2\pi i\, t_2)$, so that $t_2 \in \mathbb{Z}$ and in particular the imaginary part of t_2 is zero. The point with coordinates q_1, q_2 corresponds to the complexified Kähler class

$$t_1 T_1 + t_2 T_2 = \text{Re}(t_1) T_1 + \text{Re}(t_2) T_2 + i\left(\text{Im}(t_1) T_1 + \text{Im}(t_2) T_2\right) = B + i\,\omega,$$

and $\text{Im}(t_2) = 0$ implies that ω is in the face of the Kähler cone where the original extremal transition occurs. Hence we get exactly the compatibility predicted by mirror symmetry, and we also see why we get $z_2 = 1/4$ rather than $z_2 = 1$, which would be the case if the mirror map were simply $q_2 = z_2$.

In the next section, we will return to this example and compute the full mirror map using hypergeometric functions.

6.3.4. The Mirror Map Via Hypergeometric Functions. In Sections 5.5 and 6.1.3, we used the GKZ hypergeometric system to study Picard-Fuchs equations on the simplified moduli space of a Calabi-Yau toric hypersurface $V \subset X_\Sigma$. Here, we will continue our discussion by using hypergeometric methods to construct the mirror map. We will use the notation of the previous section, and in particular, we will assume that we are at a maximally unipotent boundary point with local coordinates $z_1, \ldots, z_r$ given by (6.41).

The GKZ system on $\overline{\mathcal{M}}_{\text{simp}}$ consists of the differential operators $\square_\ell \lambda_0^{-1}$ discussed in Section 6.1.3, where ℓ is in the lattice M_0 of relations among Ξ^+ (6.13). For $f = \lambda_0 + \sum_{m_i \in \Xi} \lambda_i t^{m_i}$, we also know that the n-form

$$\widetilde{\omega} = \frac{\lambda_0}{f}\frac{dt_1}{t_1} \wedge \cdots \wedge \frac{dt_n}{t_n}$$

is a solution, i.e., $\square_\ell \lambda_0^{-1}\widetilde{\omega} = 0$ for all ℓ.

The form $\widetilde{\omega}$ lives on $X_\Sigma - V$ and its residue on V is the holomorphic $(n-1)$-form Ω. We also have the cycle γ on $T \subset X_\Sigma$ defined by $|t_1| = \cdots = |t_n| = 1$, which gives a cohomology class $g \in H^{n-1}(V)$. It follows that

$$\langle g, \Omega \rangle = \frac{1}{(2\pi i)^n} \int_\gamma \frac{\lambda_0}{f} \frac{dt_1}{t_1} \wedge \cdots \wedge \frac{dt_n}{t_n},$$

where we added the constant $1/(2\pi i)^n$ to make the integral easier to compute. By Chapter 5, we know that this function is a solution of the Picard-Fuchs equations. The above integral is an example of an *Euler integral*, and solutions of the GKZ system of this type are studied in detail in [**GKZ1**].

To see what this solution looks like, we follow [**Batyrev3**] and note that

$$\frac{\lambda_0}{f} = \frac{1}{1 - \sum_{m_i \in \Xi} \lambda_i(-\lambda_0)^{-1} t^{m_i}} = \sum_{k=0}^\infty \left(\sum_{m_i \in \Xi} \lambda_i(-\lambda_0)^{-1} t^{m_i} \right)^k.$$

Suppose that Ξ has s elements. Then substituting this identity into the above integral and integrating term-by-term gives the formula

$$(6.48) \qquad \langle g, \Omega \rangle = \sum_{u_1, \ldots, u_s} \frac{(-u_0)!}{u_1! \cdots u_s!} (-\lambda_0)^{u_0} \lambda_1^{u_1} \cdots \lambda_s^{u_s},$$

where the sum is over all $\ell = (u_1, \ldots, u_s, u_0) \in M_0$ with $u_1, \ldots, u_s \geq 0$. This follows because $\ell \in M_0$ implies $\sum_{i=1}^s u_i m_i = 0$ and $-u_0 = u_1 + \cdots + u_s$. The minus sign in front of λ_0 in (6.48) is significant.

Here is an example of what the series (6.48) looks like in a specific example.

Example 6.3.4.1. For the mirror V° of the quintic threefold, the only relations among Ξ^+ are multiples of $\ell = (1, 1, 1, 1, 1, -5)$. Hence we are summing over $(n, n, n, n, n, -5n)$ for $n \geq 0$, which gives

$$(6.49) \qquad \begin{aligned} \langle g, \Omega \rangle &= \sum_{n=0}^\infty \frac{(-(-5n))!}{n!n!n!n!n!} (-1)^{-5n} \lambda_0^{-5n} \lambda_1^n \lambda_2^n \lambda_3^n \lambda_4^n \lambda_5^n \\ &= \sum_{n=0}^\infty \frac{(5n)!}{(n!)^5} (-z)^n, \end{aligned}$$

where as usual $z = \lambda_1 \lambda_2 \lambda_3 \lambda_4 \lambda_5 / \lambda_0^5$. This is precisely the formula (2.11) for $y_0(z)$ given in Chapter 2.

In general, we can prove that the solution (6.48) is holomorphic at the maximally unipotent boundary point.

PROPOSITION 6.3.2. *If we write the series* (6.48) *in terms of the coordinates* $z_1, \ldots, z_r$ *at the maximally unipotent boundary point, then we get a nonvanishing holomorphic function.*

PROOF. The first step of the proof is to write (6.48) in terms of $z_1, \ldots, z_r$ and show that only nonnegative exponents occur. The z_k come from a basis $\ell_k = (\ell_{kj})$ of the lattice of relations among Ξ^+, so that any ℓ in (6.48) is of the form $\ell = n_1 \ell_1 + \cdots + n_r \ell_r$ for integers $n_1, \ldots, n_r$. It follows that we can write the series as

$$(6.50) \qquad \sum_{n_1, \ldots, n_r} \frac{\left(-\sum_{k=1}^r n_k \ell_{k0} \right)!}{\prod_{j=1}^s \left(\sum_{k=1}^r n_k \ell_{kj} \right)!} ((-1)^{\ell_{10}} z_1)^{n_1} \cdots ((-1)^{\ell_{r0}} z_r)^{n_r},$$

where the sum is over all $n_1, \ldots, n_r$ such that $\sum_{k=1}^r n_k \ell_{kj} \geq 0$ for $j = 1, \ldots, s$. We need to prove that these inequalities imply $n_1, \ldots, n_r \geq 0$, i.e., that every

$\ell = (u_1, \ldots, u_s, u_0) \in M_0$ with $u_1, \ldots, u_s \geq 0$ is a nonnegative integer combination of the $\ell_k = (\ell_{kj})$.

This is actually quite easy to see once we unravel the various interpretations of elements of M_0 and its dual N_0. Over $\mathbb{R}$, we have $N_0 \otimes \mathbb{R} = A(\Xi) = H^2_{\mathrm{toric}}(V^\circ, \mathbb{R})$, and $M_0 \otimes \mathbb{R} = A(\Xi)^*$ is its dual. Each toric divisor D_j on X_{Σ° gives a divisor still called D_j on V, so we get $[D_i] \in H^2_{\mathrm{toric}}(V^\circ, \mathbb{R})$. Since $\Sigma^\circ(1) = \Xi$, $[D_1], \ldots, [D_s]$ give linear functionals on M_0. Furthermore, as a functional on M_0, one can show that $[D_j]$ picks out the jth coordinate of $\ell = (u_1, \ldots, u_s, u_0) \in M_0$. Thus $u_1, \ldots, u_s \geq 0$ simply says that ℓ is in the cone dual to the one generated by the $[D_j]$.

The cone generated by the $[D_j]$ is the effective cone of $A(\Xi)$ and is the support of the GKZ decomposition. Since the cone σ which specifies our maximally unipotent boundary point lies in $\mathrm{cpl}(\Sigma^\circ)$, we see that σ lies in the effective cone, so that its dual contains the dual of the effective cone. Hence ℓ lies in the dual of σ, which is generated by the $\ell_k = (\ell_{kj})$.

We conclude by proving convergence in a neighborhood of the origin. The elementary bound

$$\frac{(u_1 + \cdots + u_s)!}{u_1! \cdots u_s!} \leq s^{u_1 + \cdots + u_s}$$

comes from expanding $(1 + \cdots + 1)^{u_1 + \cdots + u_s}$, and if $u_j = \sum_{k=1}^r n_k \ell_{kj}$ as above, then $\sum_{j=1}^s u_j \leq C \sum_{k=1}^r n_k$ for a suitable constant C independent of the n_k since $n_1 \ldots, n_r \geq 0$. Since the constant term of the series is clearly 1, the proposition follows easily. $\qquad\square$

This proposition shows that the *one* series (6.48) is holomorphic at *every* maximally unipotent boundary point. The proof also suggests an interesting relationship between the GKZ system and the GKZ decomposition. See [**GKZ1**] for further details.

At a maximally unipotent boundary point, the Picard-Fuchs equations have only one holomorphic solution (up to a constant)—this is what the monodromy weight filtration having $\dim(W_0) = 1$ means. Thus Proposition 6.3.2 describes the denominator $\langle g_0, \Omega \rangle$ in our formula (6.30) for the mirror map. For simplicity, we will let y_0 denote the series given in (6.50).

In order to find the kth component q_k of the mirror map, we need to find a solution of the Picard-Fuchs equations satisfying (6.45). However, when dealing with hypergeometric functions, it is convenient to drop the factor $1/(2\pi i)$ in (6.45). Thus we seek a solution y_k such that

$$(6.51) \qquad y_k = y_0 \log \left((-1)^{\ell_{k0}} z_k\right) + \tilde{y}_k, \quad 1 \leq k \leq r,$$

where $\tilde{y}_k$ is holomorphic at the origin with $y_k(0) = 0$. With this normalization, the formula for q_k becomes

$$(6.52) \qquad q_k = (-1)^{\ell_{k0}} z_k \exp(\tilde{y}_k/y_0) \sim (-1)^{\ell_{k0}} z_k, \quad 1 \leq k \leq r$$

since y_0 (resp. y_k) is nonvanishing (resp. vanishing) at the origin. By (6.44), this is precisely what we want.

There are several ways known to find y_k satisfying (6.51). We begin with a variant of the Frobenius method from [**HLY1**] which works nicely for hypergeometric equations. The idea is to replace each n_k in (6.50) with $n_k + \rho_k$, where ρ_k is a real parameter. This means that a factorial such as $\left(\sum_{k=1}^r n_k \ell_{kj}\right)!$ gets replaced with the classical Euler Γ-function

$$\Gamma\left(\sum_{k=1}^{r}(n_k + \rho_k)\ell_{kj} + 1\right).$$

This also affects exponents, so that $((-1)^{\ell_{k0}} z_k)^{n_k}$ becomes $((-1)^{\ell_{k0}} z_k)^{n_k + \rho_k}$. Let the series obtained from (6.50) in this way be denoted $y_0(z; \rho)$, where z stands for $z_1, \ldots, z_r$ and ρ stands for $\rho_1, \ldots, \rho_r$.

If $\partial_{\rho_k} = \partial/\partial \rho_k$ denotes partial differentiation with respect to ρ_k, then the key result is that

$$y_k = \left.(\partial_{\rho_k} y_0(z, \rho))\right|_{\rho=0}, \quad 1 \le k \le r$$

has the desired property (6.51). This follows from the identities

$$\partial_{\rho_k}((-1)^{\ell_{k0}} z_k)^{n_k + \rho_k} = \log((-1)^{\ell_{k0}} z_k) \cdot ((-1)^{\ell_{k0}} z_k)^{n_k + \rho_k}$$

$$\left[\square_\ell, \partial_{\rho_k}\right] = 0.$$

We will omit the details of the proof. Instead, we will give some examples to show how this works in practice.

Example 6.3.4.1, continued. For the quintic mirror example begun earlier, the formula given for $y_0(z)$ in (6.49) means that

$$(6.53) \qquad y_0(z; \rho) = \sum_{n=0}^{\infty} \frac{\Gamma(5n + 5\rho + 1)}{\Gamma(n + \rho + 1)^5} (-z)^{n+\rho}.$$

Then $y_0(z) = y_0(z; 0)$, and the above theory tells us that the logarithmic solution of the GKZ equations is given by

$$y_1(z) = \left.\partial_\rho y_0(z; \rho)\right|_{\rho=0} = y_0(z) \log(-z) + \sum_{n=0}^{\infty} \left.\partial_\rho\left(\frac{\Gamma(5n + 5\rho + 1)}{\Gamma(n + \rho + 1)^5}\right)\right|_{\rho=0} (-z)^n.$$

In order to simplify this, we need to understand the derivative of the Γ-function. Fortunately, we have the following formula for $\Gamma'(s)$ when evaluated at a positive integer:

$$\Gamma'(n + 1) = \Gamma(n + 1)\left(-\gamma + \sum_{j=1}^{n} \frac{1}{j}\right),$$

where γ is Euler's constant. This is proved using induction and the classical fact $\Gamma'(1) = -\gamma$ [**WW**, p. 236]. From here, one easily obtains

$$y_1(z) = y_0(z) \log(-z) + 5 \sum_{n=1}^{\infty} \frac{(5n)!}{(n!)^5}\left[\sum_{j=n+1}^{5n} \frac{1}{j}\right](-z)^n,$$

and then (6.52) gives

$$q = -z \exp\left(\frac{5}{y_0(z)} \sum_{n=1}^{\infty} \frac{(5n)!}{(n!)^5}\left[\sum_{j=n+1}^{5n} \frac{1}{j}\right](-z)^n\right).$$

This is precisely the formula following (2.11) in Chapter 2 for the mirror map of the quintic threefold.

We next switch to the standard Frobenius method, which is based on recurrence relations (and is similar to what is done in [**BvS**]). To simplify notation, let $x = -z$.

Using this variable, the Picard-Fuchs equation (2.8) for the quintic mirror family can be written as

$$\delta^4 y = 5x(5\delta + 1)(5\delta + 2)(5\delta + 3)(5\delta + 4)y,$$

where $\delta = x\,d/dx$ (see (5.37) and the subsequent discussion). Then the standard Frobenius method [**CL**] uses the function

$$\tilde{y}(x;\rho) = \sum_{n=0}^{\infty} a_n(\rho)\, x^{n+\rho}.$$

If L represents the operator given by the Picard-Fuchs equation, then the $a_n(\rho)$ need to be chosen so that $\tilde{y}(x;\rho)$ satisfies the equation

$$(6.54) \qquad\qquad L(\tilde{y}(x;\rho)) = \rho^4\, x^\rho$$

since the indicial equation of the Picard-Fuchs equation is $\rho^4 = 0$ (maximally unipotent monodromy). We get for $n > 0$ the recursion relation

$$(\rho + n)^4 a_n(\rho) = 5(5\rho + 5n - 1)(5\rho + 5n - 2)(5\rho + 5n - 3)(5\rho + 5n - 4)a_{n-1}(\rho).$$

Choosing the solution with $a_0 = 1$, we solve the recursion relation to obtain

$$a_n(\rho) = \frac{\prod_{m=1}^{5n}(5\rho + m)}{\prod_{m=1}^{n}(\rho + m)^5}.$$

This gives

$$(6.55) \qquad\qquad \tilde{y}(x;\rho) = \sum_{n=0}^{\infty} \frac{\prod_{m=1}^{5n}(5\rho + m)}{\prod_{m=1}^{n}(\rho + m)^5}\, x^{n+\rho},$$

and then (6.54) easily implies that the functions

$$(6.56) \qquad\qquad y_i(x) = \frac{1}{i!}\frac{\partial^i \tilde{y}(x;\rho)}{\partial \rho^i}\bigg|_{\rho=0}, \qquad 0 < i < 3$$

satisfy the Picard-Fuchs equation and in fact give a basis of solutions near $x = 0$. Note that we can regard the $y_i(x)$ as the coefficients of the Taylor series of

$$(6.57) \qquad \tilde{y}(x;\rho) = y_0(x) + y_1(x)\,\rho + y_2(x)\,\rho^2 + y_3(x)\,\rho^3 + \cdots .$$

The reader might worry that this differs from our earlier version of the Frobenius method, which used the Γ-function. However, one can easily check that $\tilde{y}(x;\rho)$ is related to the function $y_0(z;\rho)$ defined in (6.53) via

$$y_0(z;\rho) = \tilde{y}(-z;\rho)\,\frac{\Gamma(5\rho + 1)}{\Gamma(\rho + 1)^5}.$$

Classical formulas (or *Mathematica*) give the power series expansion

$$\frac{\Gamma(5\rho + 1)}{\Gamma(\rho + 1)^5} = 1 + \frac{5\pi^2}{3}\rho^2 + \cdots ,$$

and it follows that

$$\frac{\partial \tilde{y}(-z;\rho)}{\partial \rho}\bigg|_{\rho=0} = \frac{\partial y_0(z;\rho)}{\partial \rho}\bigg|_{\rho=0}.$$

Hence both versions of the Frobenius method give the same formula for the logarithmic solution y_1.

The surprise comes when we think about this from a more formal point of view. First replace x with e^{t_1} (so that $\delta = x\,d/dx = d/dt_1$), and then replace ρ by the hyperplane class H in the cohomology ring of $\mathbb{P}^4$. If we also change the index of summation from n to d, then (6.55) becomes the function

$$\tilde{I} = e^{t_1 H} \sum_{d=0}^{\infty} \frac{\prod_{m=1}^{5d}(5H + m)}{\prod_{m=1}^{d}(H + m)^5} e^{dt_1},$$

which was considered in Chapter 5 (see (5.45)). Hence, the strange-seeming cohomology valued function $\tilde{I}$ is completely natural from the point of view of the Frobenius method (we are grateful to the authors of [**LLY**] for explaining this to us). If we expand $\tilde{I}$ in the usual basis of $H^*(\mathbb{P}^4)$, then we get an expression of the form

$$\tilde{I} = y_0(x) + y_1(x)\,H + y_2(x)\,H^2 + y_3(x)\,H^3 + y_4(x)\,H^4,$$

which is just (6.57) with ρ replaced by H (since $H^5 = 0$). It follows that y_0, y_1, y_2, y_3 are the solutions (6.56), while y_4 is not. This justifies the claims made in connection with (5.54) in Section 5.5.3. Furthermore, if we define I_V to be $5H\,\tilde{I}$ as in (5.46), then we see that

$$\begin{aligned}
I_V &= e^{t_1 H} 5H \sum_{d=0}^{\infty} \frac{\prod_{m=1}^{5d}(5H + m)}{\prod_{m=1}^{d}(H + m)^5} e^{dt_1} \\
&= 5H\big(y_0(x) + y_1(x)\,H + y_2(x)\,H^2 + y_3(x)\,H^3\big),
\end{aligned}$$

where y_0, y_1, y_2, y_3 give a basis of solutions of the Picard-Fuchs equation of the quintic mirror. These formulas appeared in Chapter 2 as (2.32) and (2.33). The solutions to the Picard-Fuchs equations were first written down in this form in [**Givental1**]. As indicated at the end of Section 5.5.3, I_V appears in [**LLY**] as $\lim_{\lambda \to 0} HG[\mathcal{I}(\hat{P})](t_1)$, and the above discussion of solving hypergeometric equations makes it clear why the "HG" notation is appropriate. The function $I_V = \lim_{\lambda \to 0} HG[\mathcal{I}(\hat{P})](t_1)$ will play an important role in the treatment of the mirror theorem in Chapter 11.

Example 6.3.4.2. For our final example, we return to the mirror V° of an anticanonical hypersurface in the toric blowup of $\mathbb{P}(1, 1, 2, 2, 2)$. In Example 6.3.3.2, we computed the mirror map assuming $z_1 = 0$. Using hypergeometric functions, we can now determine the map for all z_1 and z_2. We will revisit this mirror map and more in Example 11.2.5.1, using the function $\tilde{I}$ of Section 5.5.3 for this example.

The solution $y_0(z_1, z_2)$ is easy to write down since the generators of the lattice M_0 are the rows $(0, 0, 1, 1, 1, 1, -4)$ and $(1, 1, 0, 0, 0, -2, 0)$ of the matrix (6.15). Then

$$(u_1, u_2, u_3, u_4, u_5, u_6, u_0) = n_1(0, 0, 1, 1, 1, 1, -4) + n_2(1, 1, 0, 0, 0, -2, 0)$$

implies that $u_1 = u_2 = n_1$, $u_3 = u_4 = u_5 = n_2$, $u_6 = n_1 - 2n_2$ and $u_0 = -4n_1$ in (6.48), so that (6.50) gives the series

$$y_0(z_1, z_2) = \sum_{n_1 \geq 2n_2} \frac{(4n_1)!}{(n_1!)^3 (n_2!)^2 (n_1 - 2n_2)!} z_1^{n_1} z_2^{n_2}.$$

The condition $n_1 \geq 2n_2$ shows that $y_0(0, z_2) = 1$, which agrees with what we found in Example 6.3.3.2.

To describe solutions with logarithmic growth, we define $y_0(z_1, z_2; \rho_1, \rho_2)$ as above and compute $y_k(z_1, z_2) = \partial_{\rho_k} y_0(z_1, z_2; \rho_1, \rho_2)\big|_{\rho_1 = \rho_2 = 0}$. This can be done by the methods of the previous example, except that we have to be treat the condition $n_1 \geq 2n_2$ appropriately. The subtle point here is that $y_0(z_1, z_2)$ is really the series

$$y_0(z_1, z_2) = \sum_{n_1, n_2 \geq 0} \frac{\Gamma(4n_1 + 1)}{\Gamma(n_1 + 1)\Gamma(n_2 + 1)^2 \Gamma(n_1 - 2n_2 + 1)} z_1^{n_1} z_2^{n_2}.$$

This agrees with the previous series because the Euler-Γ function has simple poles at $0, -1, -2$, etc. But when we create $y_0(z_1, z_2; \rho_1, \rho_2)$ from this series, it will now have terms for both $n_1 \geq 2n_2$ and $n_1 < 2n_2$, and so will its derivatives with respect to ρ_1 and ρ_2.

Let's illustrate what happens with $y_2 = \partial_{\rho_2} y_0 \big|_{\rho_1 = \rho_2 = 0}$. If we let $\rho = \rho_2$, then it suffices to work with

$$(6.58) \qquad \sum_{n_1, n_2 \geq 0} \frac{\Gamma(4n_1 + 1)}{\Gamma(n_1 + 1)\Gamma(n_2 + \rho + 1)^2 \Gamma(n_1 - 2n_2 - 2\rho + 1)} z_1^{n_1} z_2^{n_2 + \rho}.$$

Differentiating with respect to ρ and setting $\rho = 0$, we get $y_0 \log z_2 + \tilde{y}_2$, where $\tilde{y}_2$ is obtained by differentiating the coefficients of (6.58). This series will have terms of two sorts, those with $n_1 \geq 2n_2$, and those with $n_1 < 2n_2$. The former can be treated by the methods of the previous example, while the latter require some extra work. Here, one uses the identity

$$\Gamma(s)\Gamma(1 - s) = \frac{\pi}{\sin(\pi s)}$$

from [**WW**, p. 239] to obtain

$$\partial_\rho(\Gamma(n_1 - 2n_2 - 2\rho + 1)^{-1}) = 2(-1)^{n_1}(2n_2 - n_1 - 1)!$$

when $n_1 - 2n_2 < 0$. Using this, we get the formula we want:

$$y_2(z_1, z_2) = y_0(z_1, z_2) \log z_2 + 2 \sum_{n_1 < 2n_2} \frac{(4n_1)!(2n_2 - n_1 - 1)!}{(n_1!)^3(n_2!)^2}(-1)^{n_1} z_1^{n_1} z_2^{n_2}$$

$$+ 2 \sum_{n_1 \geq 2n_2} \frac{(4n_1)!}{(n_1!)^3(n_2!)^2(n_1 - 2n_2)!} \left[\sum_{j=1}^{n_1 - 2n_2} \frac{1}{j} - \sum_{j=1}^{n_2} \frac{1}{j} \right] z_1^{n_1} z_2^{n_2}.$$

If we write this as $y_2 = y_0 \log z_2 + \tilde{y}_2$, then the mirror map has $q_2 = z_2 \exp(\tilde{y}_2/y_0)$ by (6.52). This gives an explicit (though very complicated) formula for q_2, and we can work out a similar formula for q_1. Thus we now know the mirror map completely.

As a check of what we did, suppose we set $z_1 = 0$ as in Example 6.3.3.2. Then all of the terms with $n_1 \geq 2n_2$ vanish, and since $y_0(0, z_2) = 1$, we have

$$(6.59) \qquad y_2(0, z_2) = \log z_2 + 2 \sum_{n_2=1}^{\infty} \frac{(2n_2 - 1)!}{(n_2)!^2} z_2^{n_2}.$$

In contrast, Example 6.3.3.2 found that

$$(6.60) \qquad y_2(0, z_2) = \log \left(\frac{1 - 2z_2 - \sqrt{1 - 4z_2}}{2z_2} \right)$$

(this is the formula in the equation preceding (6.47)—the factor $1/(2\pi i)$ disappears because of the way we are normalizing things in this section). These two formulas are equal by (6.46) because the series for $y_2(0, z_2)$ satisfies the Picard-Fuchs equation $\delta^2 y - 2z_2\delta(2\delta + 1)y - 0$ from Example 6.3.3.2. Hence, even though we took a very different approach to finding the mirror map in Example 6.3.3.2, the results we got there agree perfectly with what we found here.

As a final comment, we should mention that the method of computing the mirror map described in the two examples given here has been applied to a large number of toric hypersurfaces [**HKTY1**]. This method (with suitable modifications) has also been applied to many complete intersections [**BvS, HKTY2**].

CHAPTER 7

Gromov-Witten Invariants

As we have seen in Section 1.2, the A-model correlation function on a Calabi-Yau threefold V is expressed in terms of the invariants n_β, which naively are the numbers of rational curves on V which represent the homology class β—see especially (1.7). The goal of this chapter is twofold: to give a rigorous definition of the n_β, and to give a rigorous definition of the related *Gromov-Witten invariants*. In fact, the n_β will be defined in terms of the Gromov-Witten invariants.

We will work in the following general situation. Given a projective algebraic variety X, we fix a homology class $\beta \in H_2(X, \mathbb{Z})$ and cycles $Z_1, \dots, Z_n$ on X. The basic question concerns the structure of the following set of curves:

$$(7.1) \qquad C \subset X \text{ of genus } g, \text{ homology class } \beta, \text{ and } C \cap Z_i \neq \emptyset \text{ for all } i,$$

assuming the Z_i are in general position. We will distinguish between *Gromov-Witten invariants*, which (roughly speaking) are the number of curves (7.1) when there are finitely many, and *Gromov-Witten classes*, which are cohomology classes obtained when the number is infinite.

The Gromov-Witten invariants have their origins in physics, in the *topological sigma model coupled to gravity*. In particular, the genus 0 (sometimes called *tree level*) Gromov-Witten invariants originate from the *topological sigma model*, which is a topological quantum field theory. In particular, Gromov-Witten invariants are by design to be unchanged by deformations of the complex structure of X. A brief description of topological quantum field theories is given in Appendix B.6.

We will see below that (7.1) is a hopelessly naive description of a Gromov-Witten invariant (or class). In order to get a rigorous definition, considerable sophistication is required. There are several ways of doing this, using algebraic geometry or symplectic geometry. The first reasonably general definition was in the context of semi-positive symplectic manifolds: for genus 0 in [**RT1**], with some special cases appearing earlier in [**Ruan2**], and for higher genus in [**RT2**]. Since then there have been several generalizations to general symplectic manifolds, e.g., [**LTi3, Siebert1, Ruan3, FO**]. Algebraically, the invariants were first constructed early on for Grassmannians [**BDW**], then generalized to homogeneous spaces [**LTi1, FP**]. General algebro-geometric definitions are given in [**LTi2**] and [**BF, Behrend**].

Sections 7.1 and 7.2 will describe the main algebraic and symplectic approaches to the definition of Gromov-Witten classes. The multiplicity of definitions may seem a bit daunting, but fortunately they all satisfy the basic axioms of Gromov-Witten classes stated in Section 7.3.1. It is commonly believed that all of these definitions agree with each other in their common domain of validity. In Section 7.3.2, we will see that this has been proved in some cases, though the equivalence is not known in general. So we are in a sense abusing notation by referring to all of these by the same term "Gromov-Witten invariants". Fortunately, as long as what we do involves only the Gromov-Witten axioms, there is no problem, and when we speculate on the consequences of the conjectured equivalence of the various definitions, we will be explicit about which definitions are being discussed.

The chapter will end with some examples of how Gromov-Witten invariants are computed. Then, in Chapter 8, we will use these invariants to construct the *quantum cohomology ring*.

Some excellent introductions to Gromov-Witten theory are available in the literature. We recommend in particular Voisin's book [**Voisin3**], Morrison's Park City notes [**Morrison7**], and Fulton and Pandharipande's Santa Cruz notes [**FP**].

7.1. Definition via Algebraic Geometry

In the algebraic approach to Gromov-Witten theory, Kontsevich [**Kontsevich2**] made the key observation that curves $C \subset X$ in (7.1) should be replaced with n-pointed curves $(C, p_1, \ldots, p_n)$ and holomorphic maps $f : C \to X$. This observation is consistent with the origins of the notion in the nonlinear sigma model (Appendix B.2) and two-dimensional topological quantum field theory coupled to gravity. Then (7.1) tells us to consider maps

$$(7.2) \qquad f : C \to X \text{ such that } f_*[C] = \beta \text{ and } f(p_i) \in Z_i \text{ for } i = 1, \ldots, n.$$

There is an obvious notion of isomorphism of such maps, so that we need to consider their *moduli*. To get a compact moduli space, we will allow certain reducible curves C of genus g. The Deligne-Mumford coarse moduli space $\overline{M}_{g,n}$ of n-pointed stable curves of genus g will play an important role in the theory. A nice introduction to $\overline{M}_{g,n}$ can be found in [**FP**].

The rough idea of a Gromov-Witten class is that the maps in (7.2) give a subset of $\overline{M}_{g,n}$, which in turn gives a cohomology class in $H^*(\overline{M}_{g,n}, \mathbb{Q})$. We use $\mathbb{Q}$ coefficients because $\overline{M}_{g,n}$ exists and is an orbifold of dimension $3g - 3 + n$ whenever $n + 2g \geq 3$. To make this more formal, let $\alpha_i \in H^*(X)$ be the cohomology class dual to the cycle Z_i. Then the *Gromov-Witten class*

$$I_{g,n,\beta}(\alpha_1, \ldots, \alpha_n) \in H^*(\overline{M}_{g,n}, \mathbb{Q})$$

is intuitively supposed to be the cohomology class represented by the set of pointed curves $(C, p_1, \ldots, p_n)$ occurring in (7.2). Thus, Gromov-Witten classes are a system of maps

$$(7.3) \qquad I_{g,n,\beta} : H^*(X, \mathbb{Q})^{\otimes n} \longrightarrow H^*(\overline{M}_{g,n}, \mathbb{Q}).$$

The properties of these maps will be studied in Section 7.3.

Besides cohomology classes, we can also define numerical invariants. This is done as follows: if $I_{g,n,\beta}(\alpha_1, \ldots, \alpha_n)$ is as above, then we get the rational number

$$(7.4) \qquad \langle I_{g,n,\beta} \rangle(\alpha_1, \ldots, \alpha_n) = \int_{\overline{M}_{g,n}} I_{g,n,\beta}(\alpha_1, \ldots, \alpha_n).$$

This is a *Gromov-Witten invariant*. Note that (7.4) vanishes unless the Gromov-Witten class $I_{g,n,\beta}(\alpha_1, \ldots, \alpha_n)$ has a component of top degree in $H^*(\overline{M}_{g,n}, \mathbb{Q})$. Intuitively, $I_{g,n,\beta}(\alpha_1, \ldots, \alpha_n)$ should have top degree when (7.2) consists of finitely many curves, and then the number of such curves is $\langle I_{g,n,\beta} \rangle(\alpha_1, \ldots, \alpha_n)$. However, we will see in Section 7.4 that Gromov-Witten invariants may be fractional or even negative, so that their enumerative significance is not always straightforward.

Everything we've done so far is still extremely naive. To define Gromov-Witten classes, we need to study the moduli of maps (7.2) more carefully. In particular, we need to learn about *stable maps* and *virtual fundamental classes*. We will discuss each of these topics separately.

7.1.1. Stable Maps and Their Moduli.

In looking for an algebraic definition of the Gromov-Witten classes, another key idea from [**Kontsevich2**] was the construction of a compactification of the moduli space of maps (7.2). This is done by means of *stable maps*, which we now define.

DEFINITION 7.1.1. *An n-pointed stable map consists of a connected marked curve $(C, p_1, \ldots, p_n)$ and a morphism $f : C \to X$ satisfying the following properties.*

   (i) The only singularities of C are ordinary double points.

   (ii) $p_1, \ldots, p_n$ are distinct ordered smooth points of C.

   (iii) If C_i is a component of C such that $C_i \simeq \mathbb{P}^1$ and f is constant on C_i, then C_i contains at least 3 special (i.e., nodal or marked) points.

   (iv) If C has arithmetic genus 1 and $n = 0$, then f is not constant.

Given the first two conditions of this definition, the third and fourth are equivalent to the assertion that the data $(f, C, p_1, \ldots, p_n)$ has only finitely many automorphisms. The presence of infinitely many automorphisms is precisely the reason why $\overline{M}_{g,n}$ does not exist if $n + 2g < 3$.

We define families of stable maps as follows.

DEFINITION 7.1.2. *Let X be a projective algebraic variety and let S be a scheme over $\mathbb{C}$. An n-pointed stable map over S is a flat proper morphism $C \to S$ together with n sections $s_1, \ldots, s_n$ and a map $f : C \to X$ such that for each geometric point s of S, the restriction $f_s : C_s \to X$ of f to the geometric fibers of C over s, together with the images of the sections s_i, defines a stable map. Furthermore:*

   (i) We say that $f : C \to X$ has genus g *if for each geometric point s of S, the curve C_s has arithmetic genus g.*

   (ii) Given a homology class $\beta \in H_2(X, \mathbb{Z})$, we say that $f : C \to X$ has class β *if for each geometric point s of S, $(f_s)_*[C_s] = \beta$.*

Then, given $\beta \in H_2(X, \mathbb{Z})$, we can define the contravariant functor

$$\overline{\mathcal{M}}_{g,n}(X, \beta) : (\mathbb{C}\text{-}Schemes) \longrightarrow (Sets)$$

by sending the scheme S to the set of all isomorphism classes of n-pointed stable maps over S of genus g and class β.

In order to solve the moduli problem for stable maps, we need to represent this functor in some sense. The most straightforward approach is to think about the *coarse moduli space* of stable maps $f : C \to X$. The idea of a coarse moduli space is discussed in [**Mumford1**]. Then we have the following basic existence result.

THEOREM 7.1.3. *If X is projective and β is a homology class in $H_2(X, \mathbb{Z})$, then the functor $\overline{\mathcal{M}}_{g,n}(X, \beta)$ has a coarse moduli space $\overline{M}_{g,n}(X, \beta)$ which is a projective scheme over $\mathbb{C}$.*

This theorem was first proved by Alexeev [**Alexeev**]. A fairly explicit construction of $M_{g,n}(X, \beta)$ is given in [**FP**].

A more sophisticated approach to moduli questions involves the use of *algebraic stacks*. For the language of stacks, see [**DM**] or [**BEFFGK**]. For our purposes, it suffices to note that stacks are contravariant functors from the category of schemes to the category of sets which satisfy certain properties, some of which we will make explicit as we go along. Schemes are themselves stacks. To a scheme S, we can associate the representable functor h_S defined by

$$(7.5) \qquad h_S(X) = \mathrm{Hom}(X, S) = \{\text{all morphisms of schemes } f : X \to S\}$$

for any scheme X.

In this language, the basic existence result was stated in [**Kontsevich2**].

THEOREM 7.1.4. *If X is projective and β is a homology class in $H_2(X, \mathbb{Z})$, then the functor $\overline{\mathcal{M}}_{g,n}(X, \beta)$ is an algebraic stack which is proper over $\mathbb{C}$. Furthermore, the stack $\overline{\mathcal{M}}_{0,n}(\mathbb{P}^r, \beta)$ is a smooth stack.*

A proof can be found in [**BM**]. Because of this result, we will use $\overline{\mathcal{M}}_{g,n}(X, \beta)$ to denote the stack determined by the functor. Also, recall that every algebraic stack has an underlying algebraic space. For the stack $\overline{\mathcal{M}}_{g,n}(X, \beta)$, the underlying space is the coarse moduli space $\overline{M}_{g,n}(X, \beta)$. There is a natural morphism of stacks $\overline{\mathcal{M}}_{g,n}(X, \beta) \to \overline{M}_{g,n}(X, \beta)$ which takes an n-pointed stable map over S to its classifying map $S \to \overline{M}_{g,n}(X, \beta)$. Roughly speaking, the existence of this map shows that the stack has more information than the coarse moduli space itself.

Since stacks are a generalization of schemes, we will frequently discuss stacks using the language of schemes. Our desire is to give the reader a geometric feel for $\overline{\mathcal{M}}_{g,n}(X, \beta)$ without having to delve into the technical issues surrounding stacks, at the cost of introducing some imprecision that we trust will not be too distracting. This places some extra demands on the reader who wants to understand our precise meaning in the language of stacks. For instance, we will at times refer to an n-pointed genus g stable map $f : (C, p_1, \ldots, p_n) \to X$ of class β as a *point* of $\overline{\mathcal{M}}_{g,n}(X, \beta)$. This strictly speaking makes no sense, because functors do not have points. Depending on context, we might mean the corresponding point of the algebraic space associated to $\overline{\mathcal{M}}_{g,n}(X, \beta)$, or the corresponding element of $\overline{\mathcal{M}}_{g,n}(X, \beta) (\mathrm{Spec}(\mathbb{C}))$.

Let's say a few words about the special case when $g = 0$ and $X = \mathbb{P}^r$. Here, we can write $\beta = d\ell$, where ℓ is the class of a line. Then $\overline{\mathcal{M}}_{0,n}(\mathbb{P}^r, d\ell)$ is smooth by Theorem 7.1.4, which means that the underlying algebraic space $\overline{M}_{0,n}(\mathbb{P}^r, d\ell)$ is an orbifold. One can show that it is an irreducible, normal projective variety of dimension $rd + r + d + n - 3$. We will frequently write $\overline{M}_{g,n}(\mathbb{P}^r, d)$ in place of $\overline{M}_{g,n}(\mathbb{P}^r, d\ell)$.

7.1.2. From Moduli to Gromov-Witten Classes. Now that we have the moduli space $\overline{M}_{g,n}(X, \beta)$, we get the following natural maps:

$$(7.6) \qquad \begin{aligned} \pi_1 &: \overline{M}_{g,n}(X, \beta) \longrightarrow X^n \\ \pi_2 &: \overline{M}_{g,n}(X, \beta) \longrightarrow \overline{M}_{g,n}. \end{aligned}$$

These maps are easy to define. Given a stable map $f : (C, p_1, \ldots, p_n) \to X$, we get a well-defined n-tuple

$$(f(p_1), \ldots, f(p_n)) \in X^n$$

since $p_1, \ldots, p_n$ is an ordered list of points. This gives π_1 set-theoretically, and it is not hard to check that π_1 is a morphism. As for π_2, first observe that if $f : C \to X$ is a stable map, then C need *not* be a stable curve in the sense of Deligne and

Mumford. However, if $n + 2g \geq 3$, then successively contracting the non-stable components of C gives a stable curve $\widetilde{C}$. Then π_2 is given set-theoretically by sending $f : (C, p_1, \dots, p_n) \to X$ to the isomorphism class of $\widetilde{C}$. It can be seen to be a morphism using the techniques of [**Knudsen**].

We can now explain what the Gromov-Witten classes should look like in the special case when X is smooth and $\overline{M}_{g,n}(X,\beta)$ is an orbifold of dimension

$$(7.7) \qquad (1-g)(\dim X - 3) - \int_\beta \omega_X + n,$$

where ω_X is the canonical class of X. As we will explain in Section 7.1.3, this is the "expected dimension" of the moduli space. Given these hypotheses, the maps (7.6) give natural maps

$$\pi_1^* : H^*(X, \mathbb{Q})^{\otimes n} \longrightarrow H^*(\overline{M}_{g,n}(X, \beta), \mathbb{Q})$$
$$\pi_{2*} : H_*(\overline{M}_{g,n}(X, \beta), \mathbb{Q}) \longrightarrow H_*(\overline{M}_{g,n}, \mathbb{Q}),$$

where we are still assuming $n + 2g \geq 3$. Since all of the spaces involved are orbifolds, Poincaré duality (see Appendix A) and π_{2*} induce the Gysin map

$$\pi_{2!} : H^*(\overline{M}_{g,n}(X, \beta), \mathbb{Q}) \longrightarrow H^{2m+*}(\overline{M}_{g,n}, \mathbb{Q}),$$

where $m = (g-1)\dim X + \int_\beta \omega_X$. Then for $n + 2g \geq 3$ we define the Gromov-Witten class $I_{g,n,\beta}(\alpha_1, \dots, \alpha_n)$ by the formula

$$(7.8) \qquad I_{g,n,\beta}(\alpha_1, \dots, \alpha_n) = \pi_{2!}(\pi_1^*(\alpha_1 \otimes \cdots \otimes \alpha_n)).$$

Note how this gives a precise definition of the cohomology class of the set of curves in (7.2) (remember that α_i is the Poincaré dual of Z_i). One easily checks that $I_{g,n,\beta}(\alpha_1, \dots, \alpha_n)$ is a cohomology class of degree

$$2m + \textstyle\sum_{i=1}^n \deg \alpha_i = 2(g-1)\dim X + 2\int_\beta \omega_X + \sum_{i=1}^n \deg \alpha_i.$$

By (7.4), we also have the Gromov-Witten invariant

$$\langle I_{g,n,\beta} \rangle(\alpha_1, \dots, \alpha_n) = \int_{\overline{M}_{g,n}} I_{g,n,\beta}(\alpha_1, \dots, \alpha_n)$$

whenever $I_{g,n,\beta}(\alpha_1, \dots, \alpha_n)$ has dimension $2(3g - 3 + n)$, the real dimension of $\overline{M}_{g,n}$. Using (7.8), this easily simplifies to

$$(7.9) \qquad \langle I_{g,n,\beta} \rangle(\alpha_1, \dots, \alpha_n) = \int_{\overline{M}_{g,n}(X,\beta)} \pi_1^*(\alpha_1 \otimes \cdots \otimes \alpha_n).$$

Intuitively, this is precisely the number given in (7.2). An important observation is that (7.9) makes sense when $n + 2g < 3$. Thus, although Gromov-Witten classes require $n + 2g \geq 3$, Gromov-Witten invariants are defined for $n, g \geq 0$. In Chapter 11, we will frequently adopt the shorthand

$$\langle \alpha_1, \dots, \alpha_n \rangle_{g,\beta} = \langle I_{g,n,\beta} \rangle(\alpha_1, \dots, \alpha_n)$$

for Gromov-Witten invariants when n is clear from the context.

Everything we've done so far assumes that $\overline{M}_{g,n}(X, \beta)$ is a smooth orbifold of the expected dimension (7.7). A nice example which satisfies these hypotheses is when $g = 0$ and $X = \mathbb{P}^r$. As noted above, $\overline{M}_{0,n}(\mathbb{P}^r, d)$ is an orbifold of dimension $rd + r + d + n - 3$, which is the expected dimension in this case. Full details can be found in [**RM**] or [**FP**].

Unfortunately, definitions (7.8) and (7.9) don't work in general. For example, when $g \geq 1$, and even when $g = 0$ for most manifolds $X \neq \mathbb{P}^r$, the space $\overline{M}_{g,n}(X, \beta)$

may have components whose dimension exceeds the expected dimension. To see how to correct this, we need to highlight the role of the *fundamental class*. In the above situation, we have the fundamental class

$$\xi = [\overline{M}_{g,n}(X,\beta)] \in H_*(\overline{M}_{g,n}(X,\beta),\mathbb{Q}),$$

which under Poincaré duality corresponds to $1 \in H^*(\overline{M}_{g,n}(X,\beta),\mathbb{Q})$. We can rewrite (7.8) in terms of ξ as follows. First note that if $e = \dim \overline{M}_{g,n}(X,\beta)$, then Poincaré duality $PD : H^*(\overline{M}_{g,n}(X,\beta),\mathbb{Q}) \to H_{2e-*}(\overline{M}_{g,n}(X,\beta),\mathbb{Q})$ is cap product with ξ. Since the maps π_1 and π_2 of (7.6) give a map

$$\pi : \overline{M}_{g,n}(X,\beta) \longrightarrow X^n \times \overline{M}_{g,n}$$

such that $\pi_i = p_i \circ \pi$, where p_1 and p_2 are the natural projections, we obtain

$$\pi_{2!}\big(\pi_1^*(\alpha_1 \otimes \cdots \otimes \alpha_n)\big) = \pi_{2!}PD^{-1}\big(\pi_1^*(\alpha_1 \otimes \cdots \otimes \alpha_n) \cap \xi\big)$$
$$= PD^{-1}\pi_{2*}\big(\pi_1^*(\alpha_1 \otimes \cdots \otimes \alpha_n) \cap \xi\big)$$
$$= PD^{-1}p_{2*}\pi_*\big(\pi^*p_1^*(\alpha_1 \otimes \cdots \otimes \alpha_n) \cap \xi\big).$$

where in the last two equalities, PD refers to Poincaré duality for $\overline{M}_{g,n}$. Then the projection formula and (7.8) show that Gromov-Witten classes are given by

$$(7.10) \qquad I_{g,n,\beta}(\alpha_1,\ldots,\alpha_n) = PD^{-1}p_{2*}\big(p_1^*(\alpha_1 \otimes \cdots \otimes \alpha_n) \cap \pi_*(\xi)\big)$$

whenever $n + 2g \geq 3$. Similarly, for $n,g \geq 0$, the formula (7.9) for Gromov-Witten invariant can be written using ξ as

$$\langle I_{g,n,\beta}\rangle(\alpha_1,\ldots,\alpha_n) = \int_\xi \pi_1^*(\alpha_1 \otimes \cdots \otimes \alpha_n).$$

It will be convenient for use in later sections to express the map π_1 from (7.6) in terms of its components. For $1 \leq i \leq n$, we define the *evaluation map*

$$e_i : \overline{M}_{g,n}(X,\beta) \to X$$

by sending a stable map $f : (C,\mathbb{P}_1,\ldots,p_n) \to X$ to $f(p_i)$, or equivalently by composing π_1 with the projection of X^n onto its ith factor. In this notation, the Gromov-Witten invariant is given by

$$(7.11) \qquad \langle I_{g,n,\beta}\rangle(\alpha_1,\ldots,\alpha_n) = \int_\xi e_1^*(\alpha_1) \cup \cdots \cup e_n^*(\alpha_n).$$

The key point is that formulas (7.10) and (7.11) make sense whenever we have a suitable "fundamental class" $\xi \in H_*(\overline{M}_{g,n}(X,\beta),\mathbb{Q})$. A class ξ with the desired properties is usually called the *virtual fundamental class*. However, as we will soon see, the definition is quite subtle.

7.1.3. The Normal Cone. We begin with the *normal cone*, which is a prototypical example of a virtual fundamental class. Suppose that E is a vector bundle of rank r on a smooth variety Y. Given a section $s \in H^0(E)$, let $Z = Z(s) \subset Y$ be the zero scheme of s. As s varies, Z can behave badly, even changing dimension. This is similar to the problem we have with the spaces $\overline{M}_{g,n}(X,\beta)$. However, the procedure in the present case is well known (see [**Fulton1**], or [**Fulton2**] for a more in-depth treatment). If $\mathcal{I}$ is the ideal sheaf of Z in Y, then the *normal cone* to Z in Y is defined by

$$(7.12) \qquad C_Z Y = \underline{\mathrm{Spec}}\Big(\bigoplus_{k=0}^{\infty} \mathcal{I}^k / \mathcal{I}^{k+1} \Big).$$

The normal cone $C_Z Y$ is an affine cone over Z of pure dimension n. The surjection $\mathcal{O}(E^*) \to \mathcal{I}$ given by multiplication by s induces a surjective map

$$\bigoplus_k \mathrm{Sym}^k \left(\mathcal{O}(E^*)/\mathcal{I}\mathcal{O}(E^*) \right) \to \bigoplus_k \mathcal{I}^k / \mathcal{I}^{k+1}.$$

Applying Spec, we get an embedding of $C_Z Y$ in the bundle $E|_Z$ on Z. The normal cone defines a cycle class $[C_Z Y] \in A_n(E|_Z)$. Then the class $s^*[C_Z Y]$ is a class in $A_{n-r}(Z)$. Note that the dimension $n - r$ is what we would get for the dimension of Z if s defines Z as a local complete intersection. Thus, $n - r$ is the "expected dimension" of Z. We also have the following folklore result about $s^*[C_Z Y]$ which can be proved using the results and techniques of [**Fulton2**].

LEMMA 7.1.5. *If $i : Z \hookrightarrow Y$ is the inclusion map, then $i_*(s^*[C_Z Y]) \in A_{n-r}(Y)$ is the Euler class $c_r(E) \cap [Y]$ of E.*

This lemma shows that $s^*[C_Z Y]$ refines the Euler class $c_r(E) \cap [Y]$. The class $s^*[C_Z Y]$ also behaves well as the section s varies. We remark that $s^*[C_Z Y]$ can sometimes be calculated as the Euler class of an associated *excess normal bundle*. We will have occasion later to work in cohomology rather than in the Chow group. In that situation, the Euler class of a rank r bundle E will be taken to be the top Chern class $c_r(E) \in H^{2r}(Y, \mathbb{Z})$. This differs slightly from common usage in topology, where the Euler class is defined as the homology class $c_r(E) \cap [Y]$.

The following example illustrates how normal cones relate to stable maps.

Example 7.1.3.1. Let $V \subset \mathbb{P}^4$ be a smooth quintic threefold. Since a generic V has precisely 2875 lines, any reasonable definition of Gromov-Witten invariant should satisfy $\langle I_{0,0,\ell} \rangle = 2875$ when ℓ is a line on V. (Note that $\langle I_{0,0,\ell} \rangle$ is a function of $n = 0$ arguments, i.e., a number.) We will eventually show that this is the case, but for now, let's see why normal cones are relevant to this example.

First observe that $\overline{M}_{0,0}(\mathbb{P}^4, \ell)$ is the Grassmannian $G(2, 5)$ of lines in $\mathbb{P}^4$. Thus the inclusion $V \subset \mathbb{P}^4$ induces a natural map $\overline{M}_{0,0}(V, \ell) \to \overline{M}_{0,0}(\mathbb{P}^4, \ell) = G(2, 5)$ which sends a stable map to its image line in $\mathbb{P}^4$. Note also that $\dim G(2, 5) = 6$. Example 7.1.5.1 will describe this map as follows. If U is the tautological rank 2 subbundle on the $G(2, 5)$, then the fiber U_ℓ over a line ℓ is the 2-dimensional subspace of $\mathbb{C}^5$ whose projectivization is ℓ. An equation for V gives a section s of the rank 6 bundle $E = \mathrm{Sym}^5 U^*$. Then $\overline{M}_{0,0}(V, \ell)$ is the zero scheme of s.

The above construction produces the class $s^*[C_Z Y] \in A_0(\overline{M}_{0,0}(V, \ell))$, where $C_Z Y$ is the normal cone of $\overline{M}_{0,0}(V, \ell) = Z(s) \subset Y = G(2, 5)$. By Example 7.1.5.1 below, this 0-cycle is the virtual fundamental class of $\overline{M}_{0,0}(V, \ell)$. This is plausible since V has trivial canonical bundle, so that $\overline{M}_{0,0}(V, \ell)$ has expected dimension 0 by (7.7). Hence the virtual fundamental class should be a 0-cycle.

If V is a generic quintic, then one can prove that the zero locus of the section s is a reduced complete intersection, so that $\overline{M}_{0,0}(V, \ell)$ really does have dimension 0. It follows that in the generic case, $\langle I_{0,0,\ell} \rangle$ is precisely the number of lines on V.

In Example 9.1.3.1, we will calculate this number to be 2875, in agreement with Chapter 2.

On the other hand, suppose that $V \subset \mathbb{P}^4$ is the Fermat quintic defined by

$$x_0^5 + x_1^5 + x_2^5 + x_3^5 + x_4^5 = 0.$$

This is a smooth Calabi-Yau threefold with infinitely many lines. To see this, let ζ be any fifth root of unity. Then, for any $i \neq j$, the hyperplane $x_i + \zeta x_j = 0$ intersects V in a cone over a plane Fermat quintic curve, which gives a 1-parameter family of lines lying on V. Furthermore, one can show that the 50 families we get this way contain *all* lines on V and that each line has normal bundle $N = \mathcal{O}_{\mathbb{P}^1}(-3) \oplus \mathcal{O}_{\mathbb{P}^1}(1)$. Hence $\overline{M}_{0,0}(V, \ell)$ is 1-dimensional and has 50 components, each of which is a plane quintic curve. In addition, the normal bundle satisfies $h^0(N) = 2$, which enables one to show that $\overline{M}_{0,0}(V, \ell)$ is nowhere reduced. In fact, each component has multiplicity 2 at its generic point, and multiplicity 5 at each of the 375 special points corresponding to the 375 lines common to more than one cone (for example, the line defined by $x_0 + x_1 = x_2 + x_3 = x_4 = 0$ is of this form). All of this is proved in [**AKa**] by studying deformations.

Although $\overline{M}_{0,0}(V, \ell)$ is complicated, its virtual fundamental class is still the 0-cycle $s^*[C_Z Y] \in A_0(\overline{M}_{0,0}(V, \ell))$ given by the normal cone construction, provided s is induced by the Fermat quintic equation. The structure of the normal cone $C = C_Z Y$ is described in [**ClK**, Example 4.2], where more details are given. Over each of the 50 components in $\overline{M}_{0,0}(V, \ell)$, the normal cone C has a component with fiber dimension 5 and multiplicity 2, and over each of the special points corresponding to the 375 lines which lie on more than one cone, there is a component of C with fiber dimension 6 and multiplicity 5. Thus C has pure dimension 6, as expected.

Turning our attention to the virtual fundamental class $s^*[C]$, one can show that each component of C lying over a component of $\overline{M}_{0,0}(V, \ell)$ contributes 20 to the virtual fundamental class (an Euler class calculation of an excess normal bundle on a plane quintic curve is needed), while each component over the special points contributes 5. By (7.11), the Gromov-Witten invariant $\langle I_{0,0,\ell} \rangle$ is the degree of this 0-cycle, which is $20 \times 50 + 5 \times 375 = 2875$, as expected.

One can check that the calculation just described corresponds precisely to [**AKa**]. In fact, our exposition shows that this correspondence follows from dynamic intersection theory [**Fulton2**, Chapter 11]. The virtual fundamental class in a sense extends the validity of dynamic intersection theory.

7.1.4. The Virtual Fundamental Class. To define the virtual fundamental class of $\overline{M}_{g,n}(X, \beta)$ for arbitrary g, n, X, β, we will need something more general than the normal cone construction. Thus, rather than realizing $\overline{M}_{g,n}(X, \beta)$ as the zero scheme of a section of a vector bundle, we will instead endow $\overline{M}_{g,n}(X, \beta)$ with more structure. In the approach of [**LTi2**], the structure is called a *perfect tangent-obstruction complex*, and in the approach of [**BF**], it is called a *perfect obstruction theory*.

Before we plunge into the details of these definitions, we first indicate why obstructions are relevant to the virtual fundamental class. The basic reason is that obstructions arise naturally when computing the expected dimension (7.7) of the coarse moduli space $\overline{M}_{g,n}(X, \beta)$. The explanation is as follows.

The idea of an "expected dimension" occurs often in algebraic geometry and is perhaps easiest to see in the case of deformations of a compact complex manifold

M. The infinitesimal deformations are given by $H^1(M, \Theta_M)$, and the obstructions lie in $H^2(M, \Theta_M)$. Kuranishi theory [**Kuranishi**] describes the moduli space of M locally as the zero locus of a holomorphic obstruction map

$$(7.13) \qquad U \to H^2(M, \Theta_M),$$

where $U \subset H^1(M, \Theta_M)$ is an open subset. Thus the expected dimension of the moduli space of M is

$$\dim H^1(M, \Theta_M) - \dim H^2(M, \Theta_M).$$

The reasoning is similar in the case of $\overline{M}_{g,n}(X, \beta)$. Let $f : (C, p_1, \dots, p_n) \to X$ be a stable map. The tangent space of the stack $\mathcal{M}_{g,n}(X, \beta)$ at f is the hyperext group

$$\operatorname{Ext}^1_C\big(f^*\Omega^1_X \to \Omega^1_C(\textstyle\sum_{i=1}^n p_i), \mathcal{O}_C\big)$$

while the obstructions live in

$$\operatorname{Ext}^2_C\big(f^*\Omega^1_X \to \Omega^1_C(\textstyle\sum_{i=1}^n p_i), \mathcal{O}_C\big).$$

Here, we are following [**LTi2**]. In [**Kontsevich2**], the tangent space and obstruction space were equivalently written as

$$\mathbb{H}^1\big(C, \Theta'_C \to f^*\Theta_X\big) \quad \text{and} \quad \mathbb{H}^2\big(C, \Theta'_C \to f^*\Theta_X\big),$$

where Θ_X is the tangent sheaf of X and Θ'_C is the sheaf of vector fields on C which vanish at $p_1, \dots, p_n$.

We will see that (7.13) has an analog in this context. Thus, the expected dimension of $\overline{M}_{g,n}(X, \beta)$ is

$$\dim \operatorname{Ext}^1_C\big(f^*\Omega^1_X \to \Omega^1_C(\textstyle\sum_{i=1}^n p_i), \mathcal{O}_C\big) - \dim \operatorname{Ext}^2_C\big(f^*\Omega^1_X \to \Omega^1_C(\textstyle\sum_{i=1}^n p_i), \mathcal{O}_C\big).$$

We look at the long exact sequence for Ext, which begins

$$(7.14) \qquad \begin{aligned} 0 \to \operatorname{Ext}^0_C\big(f^*\Omega^1_X \to \Omega^1_C(\textstyle\sum_{i=1}^n p_i), \mathcal{O}_C\big) &\to \operatorname{Ext}^0_C\big(\Omega^1_C(\textstyle\sum_{i=1}^n p_i), \mathcal{O}_C\big) \to \\ \operatorname{Ext}^0_C\big(f^*\Omega^1_X, \mathcal{O}_C\big) &\to \operatorname{Ext}^1_C\big(f^*\Omega^1_X \to \Omega^1_C(\textstyle\sum_{i=1}^n p_i), \mathcal{O}_C\big) \to \cdots . \end{aligned}$$

Note that $\operatorname{Ext}^0(\Omega^1_C(\textstyle\sum_{i=1}^n p_i), \mathcal{O}_C)$ is the space of infinitesimal automorphisms of $(C, p_1, \dots, p_n)$ (these can exist, since C itself need not be stable). Furthermore, $\operatorname{Ext}^0(f^*\Omega^1_X, \mathcal{O}_C)$ is the space of first order deformations of the map f (with C fixed). The map connecting these terms is seen to be injective, using the stability of f. Thus $\operatorname{Ext}^0_C(f^*\Omega^1_X \to \Omega^1_C(\textstyle\sum_{i=1}^n p_i), \mathcal{O}_C) = 0$. It follows that the expected dimension can be rewritten as

$$\chi(f^*\Theta_X) + \dim \operatorname{Ext}^1\big(\Omega^1_C(\textstyle\sum_{i=1}^n p_i), \mathcal{O}_C\big) - \dim \operatorname{Ext}^0\big(\Omega^1_C(\textstyle\sum_{i=1}^n p_i), \mathcal{O}_C\big),$$

where we have used the isomorphism $\operatorname{Ext}^i_C(f^*\Omega^1_X, \mathcal{O}_C) \simeq H^i(f^*\Theta_X)$. The Hirzebruch-Riemann-Roch theorem gives $\chi(f^*\Theta_X) = -\int_\beta \omega_X + (1 - g)\dim X$. As for the two Ext terms, $\operatorname{Ext}^1(\Omega^1_C(\textstyle\sum_{i=1}^n p_i), \mathcal{O}_C)$ is the space of first order deformations of $(C, p_1, \dots, p_n)$ and $\operatorname{Ext}^0(\Omega^1_C(\textstyle\sum_{i=1}^n p_i), \mathcal{O}_C)$ is the space of infinitesimal automorphisms (recall that $(C, p_1, \dots, p_n)$ need not be stable). Thus the difference is the dimension of the tangent space to $\overline{M}_{g,n}$ (in the sense of orbifolds), which is $3g - 3 + n$. Thus the expected dimension is

$$-\textstyle\int_\beta \omega_X + (1 - g)\dim X + (3g - 3 + n) = (1 - g)(\dim X - 3) - \int_\beta \omega_X + n,$$

in agreement with (7.7).

We can now give a precise definition of virtual fundamental class. As indicated earlier, two ways of doing this are presented in [**LTi2**] and [**BF**]. It is anticipated that these two approaches define the same virtual fundamental class in their common domain of validity. The proof should be straightforward, though it has not appeared in the literature. For this reason, we will sometimes give two proofs of assertions about the virtual fundamental class, one for each approach. Furthermore, when we use a particular approach, we will be explicit about which one we are using. To emphasize the idea of virtual fundamental class, our notation for it will not distinguish between the differing approaches.

We start with the approach of Li and Tian and refer the reader to [**LTi2**] for more details. Let

$$\mathcal{F} : (\mathbb{C}\text{-}Schemes) \longrightarrow (Sets)$$

be a contravariant moduli functor sending a scheme S to the set of all isomorphism classes of geometric objects over S of a certain fixed type. The stack $\overline{\mathcal{M}}_{g,n}(X, \beta)$ is an example of such a moduli functor.

Let S be an affine scheme over $\mathbb{C}$ and let $\mathcal{N}$ be an $\mathcal{O}_S$-module. Let $S_{\mathcal{N}}$ denote the trivial extension of S by $\mathcal{N}$, which means that $S_{\mathcal{N}} = \mathrm{Spec}(\Gamma(\mathcal{O}_S) \oplus \Gamma(\mathcal{N}))$, where $\Gamma(\mathcal{O}_S) \oplus \Gamma(\mathcal{N})$ is the trivial ring extension of $\Gamma(\mathcal{O}_S)$ by $\Gamma(\mathcal{N})$. In particular, we get an infinitesimal extension $S \hookrightarrow S_{\mathcal{N}}$. The *tangent functor $T\mathcal{F}$* of $\mathcal{F}$ is the following collection of functors. To each S as above and element $\alpha \in \mathcal{F}(S)$, we have a contravariant functor

$$T\mathcal{F}(\alpha) : (\mathcal{O}_S\text{-}modules) \longrightarrow (Sets)$$

taking $\mathcal{N}$ to the set of all elements of $\mathcal{F}(S_{\mathcal{N}})$ which restrict to α under the natural "restriction" map $\mathcal{F}(S_{\mathcal{N}}) \to \mathcal{F}(S)$. See [**LTi2**] for a more precise description. Thus the tangent functor encodes the data of all "extensions" of α to all infinitesimal extensions of S. The various functors making up $T\mathcal{F}$ satisfy a natural base change property.

A *tangent-obstruction complex* for $\mathcal{F}$ consists of a complex of functors

$$T^1\mathcal{F} \to T^2\mathcal{F}$$

where the arrow is the zero natural transformation, $T^1\mathcal{F}$ is the tangent functor defined above, and $T^2\mathcal{F}$ gives a reasonable obstruction theory for $\mathcal{F}$. We will not say more about what an obstruction theory is, except to mention that in particular, there is associated to the data $(\alpha, S, \mathcal{N})$ an obstruction class

$$\mathrm{ob}(\alpha, S, \mathcal{N}) \in \Gamma(T^2\mathcal{F}(\alpha) \otimes_{\mathcal{O}_S} \mathcal{N})$$

whose vanishing is necessary and sufficient for the existence of an element $\tilde{\alpha} \in \mathcal{F}(S_{\mathcal{N}})$ extending α. However, an obstruction theory contains more information than this—see [**LTi2**] for more details.

A tangent-obstruction complex is *perfect* if for each (α, S) as above, there is, at least locally on S, a 2-term complex of locally free sheaves of $\mathcal{O}_S$-modules $\mathcal{E}^1 \to \mathcal{E}^2$ such that for any $\mathcal{N}$, $T^i\mathcal{F}(\alpha)(\mathcal{N})$ is the ith sheaf cohomology of the induced complex $\mathcal{E}^{\bullet} \otimes_{\mathcal{O}_S} \mathcal{N}$.

Suppose now that $\mathcal{F}$ is representable by a scheme Z. In the notation of (7.5), this means $\mathcal{F} = h_Z$, where $1_Z \in h_Z(Z) = \mathrm{Hom}(Z, Z)$ corresponds to the universal object $\alpha \in \mathcal{F}(Z)$. For ease of exposition, let us suppose also that there is a perfect tangent-obstruction complex for $\mathcal{F}$ where the $\mathcal{E}^i$ of the definition are all induced from locally free sheaves defined over all of Z (rather than just locally). These sheaves will be denoted by the same symbols $\mathcal{E}^i$. We give a simple example of such a perfect obstruction theory in Example 7.1.4.1 below. For $i = 1, 2$, we let E^i denote the vector bundle with $\mathcal{O}(E^i) = \mathcal{E}^i$. Li and Tian use obstruction theory to construct a (far from unique) *Kuranishi map*

$$F : \widehat{E}^1 \to E^2$$

where $\widehat{E}^1$ is the formal completion of E^1 along its zero section. This Kuranishi map F is a formal relative version of the usual one (7.13). The characteristic property of F is that on each fiber over $z \in Z$, the zero locus of F is formally isomorphic to the completion of Z at z. More precisely, the zero locus of F itself is formally isomorphic to the completion $\widehat{Z}$ of $Z \times Z$ along its diagonal. In particular, the Kuranishi map gives an embedding of $\widehat{Z}$ in $\widehat{E}^1$.

It is now a simple matter to construct a *virtual normal cone*. Consider the normal cone to $\widehat{Z}$ in $\widehat{E}^1$. This is a cone over $\widehat{Z}$. The virtual normal cone $C^{\mathcal{E}^\bullet}$ is the restriction of this cone to $Z \subset \widehat{Z}$. Note that via $\widehat{Z} \subset \widehat{E}^1$, Z is embedded as the zero section of E^1. The virtual normal cone has dimension equal to the rank of E^1.

The Kuranishi map can now be used to give an embedding of $C^{\mathcal{E}^\bullet}$ in E^2. The *virtual fundamental class* is defined to be $s^*[C^{\mathcal{E}^\bullet}]$, where s is the zero section of E^2. It is immediate to see that it is a cycle class on Z of dimension $\mathrm{rk}(E^1) - \mathrm{rk}(E^2)$. This difference can now be identified with the expected dimension of Z (or rather, this is now a definition of the expected dimension). It is obvious from the definition of a perfect tangent-obstruction complex that this difference is completely determined by the tangent-obstruction complex and not by the choice of the $\mathcal{E}^i$. Less obvious, but shown in [**LTi2**], is that the virtual fundamental class itself is independent of the choices of the $\mathcal{E}^i$ and Kuranishi maps.

Example 7.1.4.1. We bring this into down-to-earth terms by returning to the context of a smooth scheme Y, vector bundle E, and section $s \in H^0(E)$. We let $\mathcal{F}$ be the functor represented by $Z = Z(s)$. Said differently,

$$\mathcal{F}(S) = \{\text{all morphisms } f : S \to Y \text{ satisfying } f^*(s) = 0\}.$$

Given an element $f : S \to Y$ of $\mathcal{F}(S)$, we claim that the complex $T^\bullet \mathcal{F}(f)$ defined by the kernel and cokernel of

$$f^* \Theta_Y \longrightarrow f^* \mathcal{O}(E)$$

satisfies the requirements needed for a perfect tangent-obstruction complex. The map in the above complex is the differential of the section $s : Y \to E$ composed with the natural projection $T_E \to E$ of the tangent bundle of E to the normal bundle of the zero section, which is isomorphic E itself. It is clearly induced by a complex defined globally on Z, namely

$$(7.15) \qquad\qquad \Theta_Y|_Z \to \mathcal{O}(E)|_Z,$$

which plays the role of $\mathcal{E}^\bullet$ above. Following the prescription of [**LTi2**], the virtual normal cone $C^{\mathcal{E}^\bullet}$ is seen to coincide with the normal cone $C_Z Y$, with precisely the same embedding into E as was just described. The virtual fundamental class is then just $s^*[C_Z Y]$. This shows that the virtual fundamental class generalizes the normal cone construction given in Section 7.1.3.

So far, we've constructed virtual fundamental classes only in the special case when the functor $\mathcal{F}$ is representable by a scheme and the complex $\mathcal{E}^\bullet$ exists globally.

The paper [**LTi2**] shows that these assumptions can be relaxed. In particular, the construction can be applied to define a virtual fundamental class for certain non-representable functors such as $\overline{\mathcal{M}}_{g,n}(X,\beta)$ which have coarse moduli spaces.

To see how this is done for $\overline{\mathcal{M}}_{g,n}(X,\beta)$, consider a family of n-pointed genus g stable maps $f : \mathcal{C} \to X$ over an affine scheme S (see Definition 7.1.2). The union of the images of the n sections $s_i : S \to \mathcal{C}$ will be denoted as $D \subset \mathcal{C}$. Li and Tian consider the tangent-obstruction complex which associates to f the complex of functors with terms

$$(7.16) \qquad T^i \overline{\mathcal{M}}_{g,n}(X,\beta)(f)(\mathcal{N}) = \mathrm{Ext}^i_{\mathcal{C}/S}\left(f^*\Omega^1_X \to \Omega_{\mathcal{C}/S}(D), \mathcal{O}_{\mathcal{C}} \otimes_{\mathcal{O}_S} \mathcal{N}\right)$$

for $i = 1, 2$. Li and Tian prove that this is a perfect complex. Thus, we can define the *virtual fundamental class of* $\overline{\mathcal{M}}_{g,n}(X,\beta)$ to be the virtual fundamental class of the above perfect tangent-obstruction complex. This class, denoted $[\overline{\mathcal{M}}_{g,n}(X,\beta)]^{\mathrm{virt}}$, has dimension $\mathrm{rk}(E^1) - \mathrm{rk}(E^2)$. Using the display immediately before (7.14), we see that $[\overline{\mathcal{M}}_{g,n}(X,\beta)]^{\mathrm{virt}}$ has the expected dimension (7.7).

We next give a brief treatment of the related construction of the virtual fundamental class due to Behrend and Fantechi [**BF**]. We will not go into depth in describing this natural construction, since it would require a substantial detour into the world of stacks. We continue to refer to [**DM**] or [**BEFFGK**] for the language of stacks, and to [**Vistoli**] for intersection theory on algebraic stacks.

A Deligne-Mumford stack is an algebraic stack with unramified diagonal. The construction of [**BF**] begins with the observation that any Deligne-Mumford stack $\mathcal{M}$ has an *intrinsic normal cone* $\mathcal{C}_{\mathcal{M}}$. This is a stack over $\mathcal{M}$ of pure dimension 0.

We can describe the intrinsic normal cone locally as follows. Since $\mathcal{M}$ has an étale open covering by schemes, we can work over a scheme U. Pick an embedding $U \hookrightarrow W$, where W is smooth. Let $\mathcal{I}$ be the ideal sheaf of U in W, and define the normal cone $C_U W$ by the right hand side of (7.12) as usual. The differentiation map $\mathcal{I} \to \Omega^1_W$ takes f to df and induces a map

$$\bigoplus_k \mathcal{I}^k/\mathcal{I}^{k+1} \to \bigoplus_k \mathrm{Sym}^k\left(\Omega^1_W/\mathcal{I}\Omega^1_W\right).$$

Applying Spec, we get a map

$$T_W|_U \to C_U W$$

since $T_W = \underline{\mathrm{Spec}}(\oplus_{k=0}^{\infty} \mathrm{Sym}^k \Omega^1_W)$. There is a notion of a stack quotient, and the intrinsic normal cone $\mathcal{C}_U$ is defined to be the stack quotient of $C_U W$ by $T_W|_U$. The normal cone $\mathcal{C}_U$ is independent of the choice of W, which implies if we pick an étale covering $\mathcal{M} = \cup_i U_i$ where the U_i are schemes, that these local intrinsic normal cones $\mathcal{C}_{U_i}$ glue together to give $\mathcal{C}_{\mathcal{M}}$.

Replacing the normal cone $C_U W$ in the above construction with the normal sheaf, we also arrive at a construction of the *intrinsic normal sheaf* $\mathcal{N}_{\mathcal{M}}$. It turns out that $\mathcal{C}_{\mathcal{M}}$ embeds in $\mathcal{N}_{\mathcal{M}}$. The embedding is induced locally by the surjection

$$\bigoplus_k \mathrm{Sym}^k(\mathcal{I}/\mathcal{I}^2) \to \bigoplus_k \mathcal{I}^k/\mathcal{I}^{k+1}.$$

Now $\mathcal{M}$ has a cotangent complex $L^{\bullet}_{\mathcal{M}}$ [**Illusie**]. Roughly speaking, $h^0(L^{\bullet}_{\mathcal{M}})$ controls deformations and $h^{-1}(L^{\bullet}_{\mathcal{M}})$ controls obstructions. In [**BF**], an *obstruction theory* for $\mathcal{M}$ roughly speaking denotes a complex of sheaves $\mathcal{E}^{\bullet}$ on $\mathcal{M}$ and a morphism $\mathcal{E}^{\bullet} \to L^{\bullet}_{\mathcal{M}}$ which is an isomorphism on h^0 and a surjection on h^{-1}. To be more precise, we should work in a derived category. Note the similarity with the tangent-obstruction complex of [**LTi2**], when the tangent space $T^1\mathcal{F}$ is fixed but there is some flexibility in the obstructions $T^2\mathcal{F}$.

If $\mathcal{E}^0 \to \mathcal{E}^1$ is a 2-term complex of Abelian sheaves on $\mathcal{M}$, we can form the stack quotient of the action of $\mathcal{E}^0$ on $\mathcal{E}^1$. We have already seen an example of such a quotient in the construction of the intrinsic normal cone above. Now, if $\mathcal{E}^\bullet$ is a complex of arbitrary length, following [**BF**] we can define $h^1/h^0(\mathcal{E}^\bullet)$ to be the stack quotient of the kernel of $\mathcal{E}^1 \to \mathcal{E}^2$ by the cokernel of $\mathcal{E}^{-1} \to \mathcal{E}^0$.

PROPOSITION 7.1.6.

$$h^1/h^0((L_\mathcal{M}^\bullet)^\vee) \simeq \mathcal{N}_\mathcal{M}.$$

PROOF. This follows from [**BF**, pp. 66–69]. The symbol $(L_\mathcal{M}^\bullet)^\vee$ denotes a "dual" in the derived category. $\square$

We can now give the construction of the virtual fundamental class. An obstruction theory $\mathcal{E}^\bullet$ is *perfect* if $h^1/h^0((\mathcal{E}^\bullet)^\vee)$ is smooth over $\mathcal{M}$. It follows from Proposition 7.1.6 and the definition of a perfect obstruction theory that $\mathcal{N}_\mathcal{M}$ embeds in $h^1/h^0((\mathcal{E}^\bullet)^\vee)$, hence $\mathcal{C}_\mathcal{M}$ embeds there as well. Let C be the fiber product of $(E^{-1})^*$ with $\mathcal{C}_\mathcal{M}$ over $h^1/h^0((\mathcal{E}^\bullet)^\vee)$, where $\mathcal{O}(E^{-1}) = \mathcal{E}^{-1}$. This is an ordinary cone contained in the vector bundle $(E^{-1})^*$. The *virtual fundamental class* is then defined to be the intersection C with the zero section of $(E^{-1})^*$.

We now continue our previous example using the approach of [**BF**].

Example 7.1.4.1, revisited. We return to the situation of the zero locus $Z \subset Y$ of a section s of a vector bundle E. Let $\mathcal{E}^\bullet$ be the complex $\mathcal{O}(E^*)|_Z \to (\Omega_Y^1)|_Z$ dual to (7.15), the terms having degrees -1 and 0. Since Y is smooth, we have $L_Y^\bullet \simeq \Omega_Y^1$. There is then a morphism $\mathcal{E}^\bullet \to L_Z^\bullet$ obtained by projecting $\mathcal{E}^\bullet$ onto its degree 0 term, then composing with the functorially defined map $L_Y^\bullet|_Z \to L_Z^\bullet$. This makes $\mathcal{E}^\bullet$ into a perfect obstruction theory. Then $h^1/h^0((\mathcal{E}^\bullet)^\vee)$ becomes the stack quotient of $E|_Z$ by $\Theta_Y|_Z$—compare with (7.15). The virtual fundamental class is then computed as the intersection of a cone C in the bundle $E|_Z$ with the zero section. It follows from the definition of the intrinsic normal cone that $C = C_Z Y$ with the same embedding into $E|_Z$ as given at the beginning of this section. Thus, the virtual fundamental class agrees with $s^*[C_Z Y]$ as claimed.

We now apply this construction to $\overline{M}_{g,n}(X,\beta)$. Following [**Behrend**], we will define a relative version of the virtual fundamental class, which can be simpler to apply. Let $\mathfrak{M}_{g,n}$ be the stack of n-pointed genus g curves which satisfy conditions (i) and (ii) of Definition 7.1.1 (such curves are called *prestable*). Then we have the map of stacks

$$(7.17) \qquad\qquad F : \overline{\mathcal{M}}_{g,n}(X,\beta) \to \mathfrak{M}_{g,n}$$

which forgets the map but doesn't stabilize. A relative version of the above constructions shows that F has an intrinsic normal cone $\mathcal{C}_F$ and sheaf $\mathcal{N}_F$.

Now consider the map of stacks

$$\pi_{n+1} : \overline{\mathcal{M}}_{g,n+1}(X,\beta) \longrightarrow \overline{\mathcal{M}}_{g,n}(X,\beta)$$

which forgets the last point p_{n+1} and contracts any resulting unstable components. As explained in Section 10.1.1 and proved in [**DM**], the universal stable map (in the stack sense) consists of π_{n+1} and the evaluation map

$$e_{n+1} : \overline{\mathcal{M}}_{g,n+1}(X,\beta) \longrightarrow X.$$

The required perfect obstruction theory is $\mathcal{E}^{\bullet} = (R\pi_{n+1*}e_{n+1}^{*}\Theta_X)^{\vee}$. Then $\mathcal{C}_F$ embeds into $h^1/h^0((\mathcal{E}^{\bullet})^{\vee}) = h^1/h^0(R\pi_{n+1*}e_{n+1}^{*}\Theta_X)$. This can again be pulled back to an ordinary cone in an ordinary vector bundle, and intersecting with the zero section gives a *relative virtual fundamental class* $[\overline{\mathcal{M}}_{g,n}(X,\beta)/\mathfrak{M}_{g,n}, R\pi_{n+1*}e_{n+1}^{*}\Theta_X]$ in the bivariant Chow group $A^*(\overline{\mathcal{M}}_{g,n}(X,\beta) \to \mathfrak{M}_{g,n})$. (Bivariant Chow groups are discussed in [**Fulton2**] and [**Vistoli**].) Finally, the *virtual fundamental class of* $\overline{\mathcal{M}}_{g,n}(X,\beta)$ is defined to be

$$[\overline{M}_{g,n}(X,\beta)]^{\text{virt}} = [\mathfrak{M}_{g,n}] \cap [\overline{\mathcal{M}}_{g,n}(X,\beta)/\mathfrak{M}_{g,n}, R\pi_{n+1*}e_{n+1}^{*}\Theta_X]$$

in $A_*(\overline{\mathcal{M}}_{g,n}(X,\beta))$.

Notice that all of our definitions of virtual fundamental classes lie in a Chow group or bivariant Chow group. In the remainder of this chapter, we will frequently identify the virtual fundamental class with its associated homology class, forgetting some information in the process. We do this in order to compare with the symplectic version of the virtual fundamental class, discussed in Section 7.2. We also do this to avoid working with equivariant Chow groups in Chapter 9 and later chapters.

7.1.5. Computing the Virtual Fundamental Class. Rather than give an exhaustive account of how to compute and manipulate virtual fundamental classes, we will instead indicate four situations where they are relatively well understood. We begin with a case where $[\overline{M}_{0,n}(X,\beta)]^{\text{virt}}$ is especially easy to find.

DEFINITION 7.1.7. [**KoM1**] *A smooth projective variety X is said to be* convex *if $H^1(C, f^*T_X) = 0$ for all genus 0 stable maps $f : C \to X$.*

When X is convex, one can prove that $\overline{M}_{0,n}(X,\beta)$ is an orbifold of the expected dimension [**FP**]. In particular, we have the fundamental class $[\overline{M}_{0,n}(X,\beta)] \in H_*(\overline{M}_{0,n}(X,\beta), \mathbb{Q})$. But being convex also implies that the moduli functor of genus 0 stable maps to X is unobstructed. Using either of the approaches described above, we are reduced to a trivial intersection in a trivial bundle, which proves that $[\overline{M}_{0,n}(X,\beta)]^{\text{virt}} = [\overline{M}_{0,n}(X,\beta)]$ in this case. Projective spaces, and more generally homogeneous spaces, are convex.

In Section 7.1.2, we discussed the case when $\overline{M}_{g,n}(X,\beta)$ was an orbifold of the expected dimension. For such an X, we again have $[\overline{M}_{0,n}(X,\beta)]^{\text{virt}} = [\overline{M}_{0,n}(X,\beta)]$. This is because (in the language of [**LTi2**]) the tangent-obstruction complex is induced from the tangent sheaf to the associated moduli stack $\overline{\mathcal{M}}_{g,n}(X,\beta)$ or (in the language of [**BF**]) the perfect obstruction theory is induced from the cotangent sheaf of the stack. The desired equality now follows for reasons similar to the previous paragraph.

We next compute a virtual fundamental class for the quintic threefold.

Example 7.1.5.1. Let $V \subset \mathbb{P}^4$ be a smooth quintic threefold, and let $d > 0$. The goal of this example is to construct the virtual fundamental class $[\overline{M}_{0,0}(V,d\ell)]^{\text{virt}}$. We begin by showing that the natural map $\overline{M}_{0,0}(V,d\ell) \to \overline{M}_{0,0}(\mathbb{P}^4,d)$ is an embedding. To see why this is true, let s be a section of $\mathcal{O}_{\mathbb{P}^4}(5)$ which vanishes on V. Then consider the functor

$$(7.18) \quad F(S) = \{f : C \to \mathbb{P}^4 \text{ is a genus 0 stable map over } S \text{ with } f^*(s) = 0\}/\sim,$$

where $\sim$ denotes isomorphism of stable maps over S. This is the functor associated to $\overline{M}_{0,0}(V,d\ell)$, and it follows that we get an embedding $\overline{M}_{0,0}(V,d\ell) \hookrightarrow \overline{M}_{0,0}(\mathbb{P}^4,d)$.

We can describe this embedding as follows. Let $\mathcal{V}_d$ be the rank $5d + 1$ vector bundle on the stack $\overline{\mathcal{M}}_{0,0}(\mathbb{P}^4, d)$ whose fiber over a stable map $f : C \to \mathbb{P}^4$ is $H^0(C, f^*\mathcal{O}_{\mathbb{P}^4}(5))$. The section s defining V determines a section $\tilde{s}$ of $\mathcal{V}_d$ over $\overline{\mathcal{M}}_{0,0}(\mathbb{P}^4, d)$, and then (7.18) implies that $\overline{\mathcal{M}}_{0,0}(V, d\ell)$ is the zero locus of $\tilde{s}$. We use stacks here because $\mathcal{V}_d$ need *not* be a vector bundle over the orbifold $\overline{M}_{0,0}(\mathbb{P}^4, d)$.

This representation of $\overline{\mathcal{M}}_{0,0}(V, d\ell)$ implies that we are (almost) in the situation of Section 7.1.3. We have a vector bundle $E = \mathcal{V}_d$ over $Y = \overline{\mathcal{M}}_{0,0}(\mathbb{P}^4, d)$ which is smooth (as a stack, though not as a variety), and $Z(\tilde{s}) = \overline{\mathcal{M}}_{0,0}(V, d\ell)$ is the zero locus of the section $\tilde{s}$. Then the normal cone construction gives $\tilde{s}^*[C_Z Y]$. This is a 0-cycle since $\mathcal{V}_d$ has rank $5d + 1$ and $\overline{\mathcal{M}}_{0,0}(\mathbb{P}^4, d)$ has dimension $5d + 1$ by (7.7).

We claim that $\tilde{s}^*[C_Z Y]$ is the virtual fundamental class $[\overline{M}_{0,0}(V, d\ell)]^{\mathrm{virt}}$. To prove this, we will use the approach of Li and Tian. By Example 7.1.4.1, we know that $\tilde{s}^*[C_Z Y]$ is the virtual fundamental class of the perfect tangent-obstruction complex given by the kernel and cokernel of

$$(7.19) \qquad \Theta_{\overline{\mathcal{M}}_{0,0}(\mathbb{P}^4, d)}\big|_{\overline{\mathcal{M}}_{0,0}(V, d\ell)} \to \mathcal{V}_d\big|_{\overline{\mathcal{M}}_{0,0}(V, d\ell)}$$

(see (7.15)). Hence it suffices to prove that the kernel and cokernel of (7.19) coincides with the tangent-obstruction complex defined in (7.16) for $\overline{\mathcal{M}}_{0,0}(V, d\ell)$.

Towards this end, let $f : \mathcal{C} \to V$ be a stable map over a scheme S, with $\pi : \mathcal{C} \to S$ the structure map. Consider the short exact sequence of complexes

$$\begin{array}{ccccccccc}
0 & \to & f^*\mathcal{O}_{\mathbb{P}^4}(-5) & \to & f^*\Omega^1_{\mathbb{P}^4} & \to & f^*\Omega^1_V & \to & 0 \\
& & \downarrow & & \downarrow & & \downarrow & & \\
0 & \to & 0 & \to & \Omega^1_{\mathcal{C}/S} & = & \Omega^1_{\mathcal{C}/S} & \to & 0.
\end{array}$$

We now look at the long exact sequence of relative Ext's.

$$(7.20) \qquad \begin{aligned}
0 &\to \mathrm{Ext}^1_{\mathcal{C}/S}(f^*\Omega^1_V \to \Omega^1_{\mathcal{C}}, \mathcal{O}_{\mathcal{C}}) \to \mathrm{Ext}^1_{\mathcal{C}/S}(f^*\Omega^1_{\mathbb{P}^4} \to \Omega^1_{\mathcal{C}}, \mathcal{O}_{\mathcal{C}}) \to \\
&\quad \mathrm{Ext}^0_{\mathcal{C}/S}(f^*\mathcal{O}_{\mathbb{P}^4}(-5), \mathcal{O}_{\mathcal{C}}) \to \mathrm{Ext}^2_{\mathcal{C}/S}(f^*\Omega^1_V \to \Omega^1_{\mathcal{C}}, \mathcal{O}_{\mathcal{C}}) \to 0,
\end{aligned}$$

where the final $\to 0$ follows from the convexity of $\mathbb{P}^4$.

Let's investigate the nonzero terms of (7.20). The first term is the pullback to S of the tangent space to $\overline{\mathcal{M}}_{0,0}(V, d\ell)$, the second is the pullback to S of the tangent space to $\overline{\mathcal{M}}_{0,0}(\mathbb{P}^4, d)$, the third is the pullback to S of $\mathcal{V}_d$, and the fourth is the obstruction space T^2. It follows from the exactness of (7.20) that we can identify the kernel of (7.19) with $\mathrm{Ext}^1_{\mathcal{C}/S}(f^*\Omega^1_V \to \Omega^1_{\mathcal{C}}, \mathcal{O}_{\mathcal{C}})$ and the cokernel with $\mathrm{Ext}^2_{\mathcal{C}/S}(f^*\Omega^1_V \to \Omega^1_{\mathcal{C}}, \mathcal{O}_{\mathcal{C}})$. This is the tangent-obstruction complex defined in (7.16), and our claim is proved.

The observant reader will notice that we have abused notation slightly in our claim. The bundle $\mathcal{V}_d$ and its Euler class are defined only on the stack $\overline{\mathcal{M}}_{0,0}(V, d)$ and not on the space $\overline{M}_{0,0}(V, d)$. Yet we have defined the virtual fundamental class on the coarse moduli space $\overline{M}_{0,0}(V, d)$. This is because once we have the Euler class on the stack, it can then be pushed forward to the coarse moduli space without regard to the existence of a related bundle on $\overline{M}_{0,0}(V, d)$. We will continue this abuse of notation throughout the book.

The equality $[\overline{M}_{0,0}(V, d\ell)]^{\mathrm{virt}} = \tilde{s}^*[C_Z Y]$ has a very nice consequence. Let $i : \overline{\mathcal{M}}_{0,0}(V, d\ell) \hookrightarrow \overline{\mathcal{M}}_{0,0}(\mathbb{P}^4, d)$ be the embedding described above. Then Lemma 7.1.5 implies that

$$(7.21) \qquad i_*([\overline{M}_{0,0}(V, d\ell)]^{\mathrm{virt}}) = i_*(\tilde{s}^*[C_Z Y]) = c_{5d+1}(\mathcal{V}_d) \cap [\overline{\mathcal{M}}_{0,0}(\mathbb{P}^4, d)].$$

This equation takes place on $\overline{\mathcal{M}}_{0,0}(\mathbb{P}^4, d)$, and pushing forward to $\overline{M}_{0,0}(\mathbb{P}^4, d)$ gives a 0-cycle on $\overline{M}_{0,0}(\mathbb{P}^4, d)$ which represents the Euler class of the bundle $\mathcal{V}_d$. This will lead to a nice formula for the Gromov-Witten invariants studied in Example 7.1.6.1.

In the notation of Example 7.1.3.1, the bundle $\mathrm{Sym}^5 U^*$ on $G(2,5)$ is the bundle $\mathcal{V}_1$ on $\overline{M}_{0,0}(\mathbb{P}^4, 1)$ considered here. Thus the embedding $\overline{M}_{0,0}(V, \ell) \hookrightarrow G(2,5)$ given in that earlier example follows from the above description of $\overline{M}_{0,0}(V, d\ell)$.

The third aspect of computing virtual fundamental classes we wish to discuss concerns "excess dimension". In intersection theory, excess dimension occurs when an intersection has greater dimension than expected. In certain nice situations, it is then possible to compute the intersection class using what's called the excess normal bundle. See [**Fulton2**, Sect. 6.3] for a clear discussion.

The idea of "excess dimension" also applies in our situation. We define the *excess dimension* of $\overline{M}_{g,n}(X, \beta)$ to be the difference between $\dim \overline{M}_{g,n}(X, \beta)$ and the expected dimension (7.7). When the excess dimension is positive, the equality $[\overline{M}_{g,n}(X, \beta)]^{\mathrm{virt}} = [\overline{M}_{g,n}(X, \beta)]$ is clearly impossible, but it is sometimes possible to compute the virtual fundamental class using Euler classes. If $\overline{\mathcal{M}}_{g,n} \to \mathfrak{M}_{g,n}$ is the map of stacks (7.17), then we have the following result.

PROPOSITION 7.1.8. *If the sheaf $R^1\pi_{n+1*}e_{n+1}^*\Theta_X$ which was used to define $[\overline{M}_{g,n}(X, \beta)]^{\mathrm{virt}}$ is locally free of rank e, where e is the excess dimension, then*

$$[\overline{M}_{g,n}(X, \beta)]^{\mathrm{virt}} = c_e(R^1\pi_{n+1*}e_{n+1}^*\Theta_X) \cap [\overline{\mathcal{M}}_{g,n}(X, \beta)/\mathfrak{M}_{g,n}],$$

where $[\overline{\mathcal{M}}_{g,n}(X, \beta)/\mathfrak{M}_{g,n}] \in A^*(\overline{\mathcal{M}}_{g,n}(X, \beta) \to \mathfrak{M}_{g,n})$ is the fundamental class in the sense of [**Fulton2**].

PROOF. See [**Behrend, Getzler1**]. $\qquad\qquad\square$

We will use this proposition in Example 7.4.5.1 when we discuss genus one Gromov-Witten invariants. Note that even when $\overline{M}_{g,n}(X, \beta)$ is smooth, the hypotheses of Proposition 7.1.8 need not hold. The point is that the obstructions for maps in $\overline{M}_{g,n}(X, \beta)$ do not lie in $H^1(C, f^*\Theta_X)$, but rather in its quotient $\mathrm{Ext}^0(f^*\Omega_X^1 \to \Omega_C^1(\sum_i p_i), \mathcal{O}_C)$, which is more subtle to work with. We will see an illustration of this in Section 9.2.3.

Our fourth and final observation concerns the following compatibility between the virtual fundamental classes of $\overline{M}_{g,n}(X, \beta)$ and $\overline{M}_{g,n-1}(X, \beta)$. Consider the commutative diagram

$$
\begin{array}{ccc}
 & & X \\
 & \nearrow^{e_{n+1}} & \uparrow^{e_n} \\
\overline{M}_{g,n+1}(X, \beta) & \xrightarrow{\phi_{n+1}} & \overline{M}_{g,n}(X, \beta) \\
\downarrow^{\pi_{n+1}} & & \downarrow^{\pi_n} \\
\overline{M}_{g,n}(X, \beta) & \xrightarrow{\phi_n} & \overline{M}_{g,n-1}(X, \beta)
\end{array}
$$

where the maps ϕ_i forget the first marked point (and stabilizes), the maps π_i forget the last marked point, and the maps $e_i : \overline{M}_{g,i}(X, \beta) \to X$ are evaluation at the last marked point. By base change, we have

$$R^1\pi_{n+1*}e_{n+1}^*\Theta_X \simeq \phi_n^* R^1\pi_{n*}e_n^*\Theta_X.$$

This implies that

$$(7.22) \qquad [\overline{M}_{g,n}(X,\beta)]^{\mathrm{virt}} = \phi_n^*[\overline{M}_{g,n-1}(X,\beta)]^{\mathrm{virt}}.$$

Said differently, the deformation theory of the nth marked point is unobstructed. See [**Behrend**, Axiom IV]. The compatibility (7.22) will be useful in Chapter 10.

We should point out that the forgetful map $\phi_n : \overline{M}_{g,n}(X,\beta) \to \overline{M}_{g,n-1}(X,\beta)$ doesn't always exist. The problem is that when $\beta = 0$, the process of forgetting a marked point and stabilizing can make the curve vanish. For example, $\phi_3 : \overline{M}_{0,3}(X,0) \to \overline{M}_{0,2}(X,0)$ doesn't exist since $\overline{M}_{0,3}(X,0) = \overline{M}_{0,3} \times X$ and $\overline{M}_{0,2}(X,0) = \emptyset$. Fortunately, one can show without difficulty that ϕ_n exists if either $n + 2g \geq 4$ or $\beta \neq 0$ and $n \geq 1$. The same is true for π_n.

7.1.6. Defining Gromov-Witten Classes and Invariants. Now that we have the virtual fundamental class $\xi = [\overline{M}_{g,n}(X,\beta)]^{\mathrm{virt}}$, we can finally give a rigorous definition of Gromov-Witten classes and invariants. Recall from (7.6) that we have a map $\pi : \overline{M}_{g,n}(X,\beta) \to X^n \times \overline{M}_{g,n}$, and let p_i, $i = 1, 2$, be projection onto the ith factor of $X^n \times \overline{M}_{g,n}$.

DEFINITION 7.1.9. *Let $\beta \in H_2(X,\mathbb{Z})$ be a homology class and $\alpha_1, \dots, \alpha_n \in H^*(X,\mathbb{Q})$ be cohomology classes.*

(i) *If $n + 2g \geq 3$, then the Gromov-Witten class $I_{g,n,\beta}(\alpha_1, \dots, \alpha_n)$ is the cohomology class in $H^*(\overline{M}_{g,n},\mathbb{Q})$ defined by*

$$I_{g,n,\beta}(\alpha_1, \dots, \alpha_n) = PD^{-1}p_{2*}(p_1^*(\alpha_1 \otimes \cdots \otimes \alpha_n) \cap \pi_*(\xi)),$$

where $\xi = [\overline{M}_{g,n}(X,\beta)]^{\mathrm{virt}}$ is the virtual fundamental class of $\overline{M}_{g,n}(X,\beta)$ and PD is Poincaré duality.

(ii) *If $n, g \geq 0$, then the Gromov-Witten invariant $\langle I_{g,n,\beta}\rangle(\alpha_1, \dots, \alpha_n)$ is the rational number defined by*

$$\langle I_{g,n,\beta}\rangle(\alpha_1, \dots, \alpha_n) = \int_\xi e_1^*(\alpha_1) \cup \dots \cup e_n^*(\alpha_n),$$

where the e_i are as in (7.11).

Notice how these definitions are the same as (7.10) and (7.11), except that ξ is now the virtual fundamental class. Also, if $n + 2g \geq 3$, then one can show that

$$\langle I_{g,n,\beta}\rangle(\alpha_1, \dots, \alpha_n) = \int_{\overline{M}_{g,n}} I_{g,n,\beta}(\alpha_1, \dots, \alpha_n).$$

Hence Gromov-Witten invariants are determined by the corresponding Gromov-Witten classes when $n + 2g \geq 3$.

We will study the properties of Gromov-Witten classes and invariants in Section 7.3 and compute some examples in Section 7.4. We should mention that although our treatment of Gromov-Witten classes uses homology and cohomology over $\mathbb{Q}$, many of the papers in the literature use Chow groups instead. A similar definition of a Gromov-Witten invariant is given in [**Behrend**], using the relative version of the virtual fundamental class discussed earlier in Section 7.1.4.

We end this section with a formula from [**Kontsevich2**] for certain Gromov-Witten invariants of the quintic threefold. This result will play an important role in Chapter 11.

Example 7.1.6.1. Let $V \subset \mathbb{P}^4$ be a smooth quintic threefold, and as before let $\ell \in H_2(V, \mathbb{Z})$ be the class of a line. We have seen in Example 7.1.5.1 that the virtual fundamental class $[\overline{M}_{0,0}(V, d\ell)]^{\text{virt}}$ is a 0-cycle. Thus, by Definition 7.1.9, we have

$$\langle I_{0,0,d\ell} \rangle = \int_{\overline{M}_{0,0}(V,d\ell)^{\text{virt}}} 1 = \text{degree of the 0-cycle } [\overline{M}_{0,0}(V, \ell)]^{\text{virt}}.$$

We can compute this degree by working on $\overline{M}_{0,0}(\mathbb{P}^4, d)$ via the embedding $i :$ $\overline{M}_{0,0}(V, d\ell) \hookrightarrow \overline{M}_{0,0}(\mathbb{P}^4, d)$. From (7.21), we know that

$$i_*[\overline{M}_{0,0}(V, d\ell)]^{\text{virt}} = c_{5d+1}(\mathcal{V}_d) \cap [\overline{M}_{0,0}(\mathbb{P}^4, d)],$$

where $\mathcal{V}_d$ is the vector bundle on $\overline{\mathcal{M}}_{0,0}(\mathbb{P}^4, d)$ whose fiber over a genus 0 stable map $f : C \to \mathbb{P}^4$ is $H^0(C, f^*\mathcal{O}_{\mathbb{P}^4}(5))$.

Since $c_{5d+1}(\mathcal{V}_d)$ is a cohomology class of top degree, the degree of the 0-cycle $c_{5d+1}(\mathcal{V}_d) \cap [\overline{M}_{0,0}(\mathbb{P}^4, d)]$ is given by integration over the fundamental class of $\overline{M}_{0,0}(\mathbb{P}^4, d)$. Thus we obtain

$$\langle I_{0,0,d\ell} \rangle = \int_{\overline{M}_{0,0}(\mathbb{P}^4,d)} c_{5d+1}(\mathcal{V}_d).$$

This formula first appeared in [**Kontsevich2**, Sect. 2.2]. We have abused notation slightly, writing $\mathcal{V}_d$ as if it were defined on $\overline{M}_{0,0}(\mathbb{P}^4, d)$, since as already explained, its Euler class is a well-defined 0-cycle on $\overline{M}_{0,0}(\mathbb{P}^4, d)$. In Section 9.2.1, we will show how to compute this integral using equivariant cohomology and localization.

7.2. Definition via Symplectic Geometry

In this section, we give a brief outline of the ideas and techniques used in the symplectic formulations of Gromov-Witten invariants. These definitions are based on the theory of J-holomorphic maps in symplectic geometry and led to the first rigorous treatment of quantum cohomology. Some of the basic ideas appear in [**Witten3**], though the first explicit treatment of genus 0 symplectic Gromov-Witten invariants for semi-positive symplectic manifolds is due to [**Ruan2**] and [**RT1**]. The book [**MS**] can serve as an introduction to this approach. For higher genus invariants in the semi-positive case, definitions were given in [**RT1**] for a fixed complex structure, and then for varying complex structure in [**RT2**]. There are now several definitions of Gromov-Witten invariants for general symplectic manifolds [**LTi3, Siebert1, Ruan3, FO**]. We refer the reader to all of these for the many details that will be omitted from our summary in this section.

7.2.1. Symplectic Manifolds and J-Holomorphic Maps. Let M be a compact symplectic manifold of (real) dimension $2n$. This means that M admits a closed 2-form ω such that ω^n is nondegenerate at every point. Every Kähler manifold has a natural symplectic structure given by its Kähler form, and one of the powerful ideas in symplectic geometry is that any symplectic manifold (M, ω) is *almost Kähler*. In particular, the complex structure and Kähler metric of a Kähler manifold get replaced by an almost complex structure and an associated Riemannian metric which are related to ω.

There are several conditions that can be imposed on an almost complex structure on a symplectic manifold. We formalize two of them in the following definition.

DEFINITION 7.2.1. *Let (M, ω) be a symplectic manifold, and let J be an almost complex structure on M. Then J is tamed by ω if*

$$\omega(X, JX) > 0 \text{ for all } p \in M \text{ and nonzero } X \in T_p M.$$

Furthermore, J is compatible with ω if J is tamed by ω and

$$\omega(JX, JY) = \omega(X, Y) \text{ for all vector fields } X, Y \text{ on } M.$$

Compatibility is equivalent to $g(X, Y) = \omega(JX, Y)$ being a Riemannian metric on M. For either condition, such J's always exist and are all deformations of each other. This implies, for example, that we can define the Chern classes $c_i(TM)$ of the tangent bundle on a symplectic manifold by using *any* ω-tamed J to give the tangent bundle TM a complex structure. If J is tamed by ω, then

$$\mu_J(X, Y) = \frac{1}{2} \left(\omega(X, JY) + \omega(Y, JX) \right).$$

defines a metric, which reduces to g above if J is compatible with ω.

The use of tamed almost complex structures has the advantage of making it easier to achieve "generic" situations. On the other hand, more precise statements can often be made if the stronger condition of compatibility is imposed.

Our main treatment follows the approach initiated in [**Ruan2**]. We will also discuss extensions of this approach [**RT1, RT2**], which develops the theory of Gromov-Witten invariants for symplectic manifolds which are *semi-positive*. This means that for every homology class $\beta \in H_2(M, \mathbb{Z})$ which is represented by a map $f : S^2 \to M$, we never have

$$\int_\beta \omega > 0 \text{ and } 3 - \tfrac{1}{2} \dim_{\mathbb{R}} M \leq \int_\beta c_1(TM) < 0.$$

For example, every Calabi-Yau manifold is semi-positive in its natural symplectic structure, as is $\mathbb{P}^r$. Also note that every smooth projective variety of complex dimension ≤ 3 is semi-positive. We caution the reader that in the literature, semi-positive sometimes has a slightly different meaning.

We next discuss curves lying on a symplectic manifold M. In algebraic geometry, we define curves (and other subvarieties) by equations. This doesn't work in symplectic geometry, so instead we let C be a Riemann surface and consider C^∞ maps

$$f : C \to M$$

which are "holomorphic" in the following sense.

DEFINITION 7.2.2. *Let (M, ω) be symplectic and let J be an almost complex structure tamed by ω. A J-holomorphic map is a C^∞ map $f : C \to M$ from a Riemann surface C to M such that the differential $df : TC \to TM$ is a map of complex vector bundles with respect to the complex structure on C and the almost complex structure on M.*

Note that there is a $\overline{\partial}$-operator which can be applied to maps $f : C \to M$ and is defined by

$$(7.23) \qquad\qquad \overline{\partial}_J = d + J \circ d \circ j_C,$$

where j_C is the (integrable) almost complex structure on C. Note that being J-holomorphic is equivalent to the Cauchy-Riemann equations $\overline{\partial}_J f = 0$ (apply J to the definition of $\overline{\partial}_J$ and use $J^2 = -1$). In particular, if J is integrable, then a J-holomorphic map is the same as a parametrized complex curve in this case.

The basic idea of a symplectic Gromov-Witten invariant is similar to what we did in (7.2): we want to count the number of curves

$$(7.24) \qquad f : C \to M \text{ such that } f_*[C] = \beta \text{ and } f(p_i) \in Z_i \text{ for } i = 1, \dots, n,$$

where as before, $p_1, \dots, p_n$ are fixed points of C, β is a homology class in $H_2(M, \mathbb{Z})$, and $Z_1, \dots, Z_n$ are cycles on M in general position. In the approach of [**Ruan2, RT2**], this is made precise by requiring $f : C \to M$ to be J-holomorphic for a fixed *generic* choice of J (actually, the Cauchy-Riemann equations need to be perturbed). The effect of genericity is to ensure that the relevant moduli spaces have the expected dimension. The resulting invariants will depend only on ω and not on which J we chose. This is consistent with the algebraic case, for we will see in Section 7.3 that the Gromov-Witten invariants from Section 7.1 don't change when we deform the complex structure. From the mirror symmetry point of view, this is reasonable since these invariants appear in A-model correlation functions, which depend only on the Kähler moduli.

One of the nice features of the symplectic approach is that many of the complications of the algebraic case (e.g., moduli spaces being singular or having the wrong dimension) go away when we use a generic almost complex structure. In fact, (7.24) is much closer to the actual definition of Gromov-Witten invariant than it was in the algebraic case. However, in more recent approaches to symplectic Gromov-Witten invariants (for which semi-positivity can be dropped), analogs of the virtual fundamental class are developed. In these situations, it is no longer necessary to achieve the expected dimension.

To describe any of these approaches, we need to understand the relevant moduli spaces. We now turn to this.

7.2.2. Moduli Spaces. We first follow the treatment in [**Ruan2**] (see [**MS**] for an exposition). Here, the idea is to study J-holomorphic maps $f : C \to M$ of class $\beta \in H_2(M, \mathbb{Z})$. We will fix the complex structure on C and the almost complex structure J on M. For technical reasons, we need to assume that f is *simple*, meaning that f doesn't factor as $C \to C' \to M$, where $C \to C'$ has degree > 1. (If f is not simple, we say that it is a *multiple-cover*.) Then we let

$$\mathcal{M}(C, J, \beta) = \{f : C \to M : f \text{ is } J\text{-holomorphic, simple, and } f_*[C] = \beta\}.$$

A first result is that when J is a generic almost complex structure tamed by ω, the set $\mathcal{M}(C, J, \beta)$ has the natural structure of a real manifold of dimension

$$(7.25) \qquad\qquad (1 - g) \dim_{\mathbb{R}} M + 2 \int_\beta c_1(TM).$$

To prove this, one represents $\mathcal{M}(C, J, \beta)$ as the fiber of a map between infinite dimensional manifolds. For generic J, this map has a Fredholm linearization, and then the implicit function theorem implies that the fiber $\mathcal{M}(C, J, \beta)$ is a finite dimensional manifold. The dimension of the fiber is given by the Fredholm index, which by the Hirzebruch-Riemann-Roch theorem gives (7.25). We should mention that similar analyses are used to construct the moduli spaces appearing in Donaldson theory and Seiberg-Witten theory.

The manifold $\mathcal{M}(C, J, \beta)$ also depends nicely on J. More precisely, as we vary J, the different $\mathcal{M}(C, J, \beta)$ are all cobordant provided J is sufficiently generic. However, because $\mathcal{M}(C, J, \beta)$ is usually non-compact, this result is not as strong as one would hope. One complication of this non-compactness will be highlighted in the discussion following Conjecture 7.4.5 in Section 7.4.4.

We next describe an alternate approach to these spaces considered by Ruan and Tian [**RT1**]. Here, one perturbs the notion of J-holomorphic map. Perturbations are needed in part to deal with multiple covers, and as we will note, lead to compactified moduli spaces.

Recall that J-holomorphic means the Cauchy-Riemann equation $\overline{\partial}_J f = 0$. Over the product $C \times M$, let

$$\nu : p_1^* TC \to p_2^* TM$$

be a conjugate linear map, where p_i, $i = 1, 2$ is projection onto the ith factor of $C \times M$. Then consider the *inhomogeneous Cauchy-Riemann equation*

$$(\overline{\partial}_J f)(p) = \nu(p, f(p)), \quad p \in C.$$

We say that $f : C \to M$ is a *perturbed J-holomorphic map* if it satisfies this equation. Now suppose that J and ν are generic and consider the set

$$\mathcal{M}(C, J, \nu, \beta) = \{f : C \to M : \overline{\partial}_J f = \nu, f_*[C] = \beta\}.$$

Notice that we no longer assume that f is simple. This is one of the advantages of working with perturbed J-holomorphic maps. A discussion of how this relates to the homogeneous case (when $\nu = 0$) can be found in Section 2.10 of [**RT1**]. One can show that $\mathcal{M}(C, J, \nu, \beta)$ has a natural manifold structure of dimension (7.25).

So far, the complex structure of the curve C has been fixed. This is the situation in [**RT1**], and it leads to invariants which can be used to define a topological quantum field theory (Appendix B.6) including all genera. If we let the complex structure vary (in physics language, the topological theory is *coupled to gravity*), then we get symplectic analogs of the moduli spaces $\overline{M}_{g,n}(X, \beta)$ considered earlier. Since the spaces $\overline{M}_{g,n}$ are needed, we have to assume that $2g + n \geq 3$. Because of the singularities of $\overline{M}_{g,n}$, however, one has to proceed carefully. In [**RT2**], Ruan and Tian define the space $\mathcal{M}_{g,n}^\mu(J, \nu, \beta)_I$ which, roughly speaking, consists of all perturbed J-holomorphic maps

$$f : (C, p_1, \dots, p_n) \longrightarrow M$$

satisfying $f_*[C] = \beta$. Here, C is smooth of genus g, and the superscript μ signals that we are actually working with a finite cover $\overline{M}_{g,n}^\mu$ of $\overline{M}_{g,n}$. This is typically done by fixing a level structure and is needed in order to get a universal curve. Also, the subscript I denotes that we are restricting to the subset of $\overline{M}_{g,n}^\mu$ consisting of n-pointed curves with trivial automorphism group.

As before, $\mathcal{M}_{g,n}^\mu(J, \nu, \beta)_I$ has a natural manifold structure, and its dimension is given by

$$(7.26) \qquad 2(1 - g)(\tfrac{1}{2} \dim_{\mathbb{R}} M - 3) + 2\int_\beta c_1(TM) + 2n.$$

When M is a projective variety X, we have $\int_\beta c_1(TX) = -\int_\beta \omega_X$, and it follows that this dimension is exactly the expected (real) dimension of $\overline{M}_{g,n}(X, \beta)$ (compare with (7.7)).

We thus have two types of spaces built from perturbed J-holomorphic maps, namely $\mathcal{M}(C, J, \nu, \beta)$ and $\mathcal{M}_{g,n}^\mu(J, \nu, \beta)_I$. Each space has a natural compactification called the *Gromov-Uhlenbeck compactification*. Because (M, ω) is semi-positive, one can prove that these compactifications are obtained by adding finitely many strata of real codimension ≥ 2. We will see below that $\mathcal{M}(C, J, \nu, \beta)$ and $\mathcal{M}_{g,n}^\mu(J, \nu, \beta)_I$ lead to two different (but closely related) notions of Gromov-Witten invariant.

We now describe a third approach to these moduli spaces due to [**LTi3**]. This approach uses almost complex structures J which are compatible with ω. The notion of a *stable C^ℓ map* is introduced. These maps are C^ℓ versions of stable maps $f : C \to M$. The curves C are exactly the same as those in the notion of a stable map, and the marked points are again smooth points. Instead of f being an algebraic morphism, we have the following condition: for each component C_i of C, the composition of f with the normalization map $\widetilde{C_i} \to C_i$ is C^ℓ. Finally, the stability condition is imposed for components C_i such that $f_*[C_i] = 0$, and not just for components on which f is constant. There is a natural notion of equivalence of C^ℓ maps induced by biholomorphisms of C.

There is an also analogous notion of J-holomorphic stable maps in this situation. Here, we require that the maps $\widetilde{C_i} \to C_i \to M$ be J-holomorphic.

If we fix the genus g, the number n of marked points, and the homology class $\beta = f_*[C]$, then the resulting set of equivalence classes of stable C^ℓ maps is denoted by $\overline{\mathcal{F}}^\ell_\beta(M, g, n)$. This space has a natural topology and is the moduli space of interest. Similar spaces are used in the approaches of [**Ruan3, Siebert1**].

7.2.3. Symplectic Gromov-Witten Invariants. The existence of nice compactifications of $\mathcal{M}(C, J, \nu, \beta)$ and $\mathcal{M}^\mu_{g,n}(J, \nu, \beta)_I$ implies that these spaces *almost* have fundamental classes. (The issue is whether the boundary ∂ has a neighborhood U such that ∂ is a deformation retract of U—see Remark 4.3 of [**RT1**].) If we had the correct sort of fundamental class, then we could define Gromov-Witten invariants using formulas similar to (7.10).

Given that we don't have a nice fundamental class, there are two ways we can proceed:

- Use transversality arguments to reduce the definition to counting the number of points in an oriented intersection. This is the approach used in [**RT1, RT2**].
- As with virtual fundamental class discussed in Section 7.1, work in a larger space. In [**Ruan3**], Ruan uses a *virtual neighborhood*, which is obtained by embedding the moduli space into a V-manifold. A related approach can be found in [**Siebert1**], and the spaces $\overline{\mathcal{F}}^\ell_\beta(M, g, n)$ defined above are used in [**LTi3**] in a similar manner.

The second approach is more powerful since it applies to *all* symplectic manifolds, not just semi-positive ones. Nevertheless, our discussion will focus on the first, though we will give a quick mention of further ideas from [**LTi3**] as a representative of the more general techniques.

Let's begin with the invariants coming from $\mathcal{M}(C, J, \nu, \beta)$. Our goal is to define the *mixed Gromov-Witten invariant*

$$(7.27) \qquad \Phi_{g,\beta}(\alpha_1, \dots, \alpha_k \mid \beta_1, \dots, \beta_l),$$

where $\alpha_i, \beta_j \in H^*(M, \mathbb{Z})$ are cohomology classes satisfying

$$(7.28) \qquad \textstyle\sum_{i=0}^k \deg \alpha_i + \sum_{j=0}^l \deg \beta_j = (1 - g) \dim_{\mathbb{R}} M + 2\int_\beta c_1(TM) + 2l.$$

(In [**RT1**], Ruan and Tian use homology classes, so that our α_i and β_j are dual to theirs.) The rough idea is as follows. Fix points $p_1, \dots, p_k$ on C. Then the mixed invariant (7.27) should be the number of maps $f : C \to M$ in $\mathcal{M}(C, J, \nu, \beta)$ such that $f(p_i)$ lies in a cycle dual to α_i and $f(C)$ meets a cycle dual to β_j.

In order to define this precisely, we will represent the cycles dual to α_i and β_j by *pseudo-manifolds* of M. By pseudo-manifold, we mean a continuous map $F : Y \to M$ where Y is stratified so that each stratum has codimension at least two in the next and F is smooth on each stratum. If (F_i, Y_i) and (G_j, Z_j) represent the cycles dual to α_i and β_j, then

$$F = \Pi_{i=1}^{k} F_i \times \Pi_{j=1}^{l} G_j : \Pi_{i=1}^{k} Y_i \times \Pi_{j=1}^{l} Z_j \longrightarrow M^{k+l}$$

is a pseudo-manifold which represents a cycle dual to the cohomology class $\alpha_1 \otimes \cdots \otimes \alpha_k \otimes \beta_1 \otimes \cdots \otimes \beta_l \in H^*(M^{k+l}, \mathbb{Z})$.

On the other hand, we also have the fixed points $p_1, \ldots, p_k \in C$. This gives the *evaluation map*

$$e : \mathcal{M}(C, J, \nu, \beta) \times C^l \longrightarrow M^{k+l}$$

defined by sending $(f : C \to M, q_1, \ldots, q_l) \in \mathcal{M}(C, J, \nu, \beta) \times C^l$ to the point $(f(p_1), \ldots, f(p_k), f(q_1), \ldots, f(q_l)) \in M^{k+l}$.

We thus have two maps to M^{k+l}. For generic J and ν, one can show that the images of these maps are cycles which intersect transversely (in particular, the intersections occur at smooth points of the images). Furthermore, the degree condition (7.28) implies that the images $\mathrm{Im}(F)$ and $\mathrm{Im}(e)$ are cycles of complementary dimension. Hence the intersection consists of finitely many points! Since all of the objects we're dealing with have natural orientations, each $P \in \mathrm{Im}(F) \cap \mathrm{Im}(e)$ has a sign $\epsilon(P) = \pm 1$ determined by the orientations.

DEFINITION 7.2.3. *Let* (M, ω) *be a semi-positive symplectic manifold and* $\beta \in H_2(M, \mathbb{Z})$ *be a homology class. Then, given* $\alpha_i, \beta_j \in H^*(M, \mathbb{Z})$ *satisfying* (7.28), *the* mixed Gromov-Witten invariant *is defined by*

$$\Phi_{g,\beta}(\alpha_1, \ldots, \alpha_k \mid \beta_1, \ldots, \beta_l) = \sum_{P \in \mathrm{Im}(F) \cap \mathrm{Im}(e)} \epsilon(P).$$

Ruan and Tian [**RT1**] show that $\Phi_{g,\beta}(\alpha_1, \ldots, \alpha_k \mid \beta_1, \ldots, \beta_l)$ is independent of the choices of J, ν, the points $p_1, \ldots, p_k \in C$, the complex structure of C, and the pseudo-manifolds representing α_i and β_j. Their paper also includes numerous properties and applications of the mixed invariants.

We can relate this to the algebraic invariants discussed in Section 7.1. Suppose that (M, ω) is a smooth projective variety with its Kähler form. Let

$$\pi : \overline{M}_{g,k+l}(M, \beta) \to \overline{M}_{g,k}$$

be the map which takes a stable map, forgets the map and the last l marked points, and then contracts any resulting unstable components. This is an extension of the map π_2 from (7.6), which is the case $l = 0$. Also let $[pt]$ denote the cohomology class of a point in $\overline{M}_{g,k}$. This represents a fixed curve C with a fixed set of marked points $(p_1, \ldots, p_k)$. Then the invariant $\Phi_{g,\beta}(\alpha_1, \ldots, \alpha_k \mid \beta_1, \ldots, \beta_l)$ is morally the same thing as

$$\int_{[\overline{M}_{g,k+l}(M,\beta)]^{\mathrm{virt}}} \Pi_{i=1}^{k} e_i^*(\alpha_i) \cup \Pi_{j=1}^{l} e_{k+j}^*(\beta_j) \cup \pi^*[pt].$$

However, we are more interested in Gromov-Witten invariants of the type discussed in Section 7.1. For this reason, we turn our attention to the invariants built from $\mathcal{M}''_{g,n}(J, \nu, \beta)_I$. These are the *symplectic Gromov-Witten invariants*

$$(7.29) \qquad\qquad \Psi_{g,n,\beta}(\gamma; \alpha_1, \ldots, \alpha_n),$$

which are defined for cohomology classes $\gamma \in H^*(\overline{M}_{g,n}, \mathbb{Q})$ and $\alpha_i \in H^*(M, \mathbb{Z})$ satisfying

$$\deg \gamma + \sum_{i=0}^{n} \deg \alpha_i = 2(1-g)(\tfrac{1}{2}\dim_{\mathbb{R}} M - 3) + 2\int_\beta c_1(TM) + 2n.$$

The naive idea is that if Z is a cycle in $\overline{M}_{g,n}$ dual to γ and Z_i is a cycle in M dual to α_i, then $\Psi_{g,n,\beta}(\gamma; \alpha_1, \dots, \alpha_n)$ should be the number of perturbed J-holomorphic maps $f : (C, p_1, \dots, p_n) \to M$ such that $(C, p_1, \dots, p_n) \in Z \subset \overline{M}_{g,n}$, $f_*[C] = \beta$, and $f(p_i) \in Z_i$.

The precise definition of (7.29) is similar to what we did in the mixed case, though the details are more complicated because not all classes in $H_*(\overline{M}_{g,n}, \mathbb{Z})$ are represented by pseudo-manifolds, and also because using $\mathcal{M}_{g,n}^\mu(J, \nu, \beta)_I$ means working with a finite cover $\overline{M}_{g,n}^\mu \to \overline{M}_{g,n}$. This allows us to define an invariant $\Psi_{g,n,\beta}^\mu(\gamma; \alpha_1, \dots, \alpha_n)$ which depends on the choice of finite cover. Then, dividing this by the degree of the covering, we get an invariant $\Psi_{g,n,\beta}(\gamma; \alpha_1, \dots, \alpha_n)$ which is independent of μ. Rather than give the details here, we will refer the reader to [**RT2**] for the full definition.

Notice that if we fix $\alpha_1, \dots, \alpha_n$, then knowing $\Psi_{g,n,\beta}(\gamma; \alpha_1, \dots, \alpha_n)$ for all γ determines a unique class $\Psi_{g,n,\beta}(\alpha_1, \dots, \alpha_n) \in H^*(\overline{M}_{g,n}, \mathbb{Q})$ such that

$$\int_{\overline{M}_{g,n}} \gamma \cup \Psi_{g,n,\beta}(\alpha_1, \dots, \alpha_n) = \Psi_{g,n,\beta}(\gamma; \alpha_1, \dots, \alpha_n).$$

We call $\Psi_{g,n,\beta}(\alpha_1, \dots, \alpha_n)$ a *symplectic Gromov-Witten class*. As in the algebraic case, the symplectic classes can be regarded as maps

$$(7.30) \qquad\qquad \Psi_{g,n,\beta} : H^*(M)^{\otimes n} \longrightarrow H^*(\overline{M}_{g,n}, \mathbb{Q}).$$

We will see later that algebraic and symplectic Gromov-Witten classes have similar properties.

There are some nice relations between mixed and symplectic Gromov-Witten invariants. For example, since a mixed invariant fixes the complex structure, it should be no surprise that

$$\Phi_{g,\beta}(\alpha_1, \dots, \alpha_n \,|) = \Psi_{g,n,\beta}([pt]; \alpha_1, \dots, \alpha_n).$$

To state the general relation between these invariants, let $K_{k,l}$ be the cycle in $\overline{M}_{g,k+l}$ obtained by taking the closure of the set

$$\{(C, p_1, \dots, p_{k+l}) \in \overline{M}_{g,k+l} : (C, p_1, \dots, p_k) \text{ is a fixed point in } \overline{M}_{g,k}\}.$$

Then one can show that

$$\Phi_{g,\beta}(\alpha_1, \dots, \alpha_k \,|\, \beta_1, \dots, \beta_l) = \Psi_{g,k+l,\beta}([K_{k,l}]; \alpha_1, \dots, \alpha_k, \beta_1, \dots, \beta_l).$$

This implies that *all* mixed invariants are special cases of the symplectic invariants. A particularly relevant case is when $g = 0$. Since $\overline{M}_{0,3}$ is a single point, it follows that

$$(7.31) \qquad \Phi_{0,\beta}(\alpha_1, \alpha_2, \alpha_3 \,|\, \alpha_4, \dots, \alpha_n) = \Psi_{0,n,\beta}([pt]; \alpha_1, \dots, \alpha_n).$$

A nice explanation of this formula can be found in [**Voisin3**].

We close this section with a quick mention of some of the ideas in the approach of [**LTi3**]. Let $\overline{\mathcal{F}}_\beta^\ell(M, g, n)$ be the moduli space described at the end of Section 7.2.2.

We define a generalized bundle E on $\overline{\mathcal{F}}_\beta^\ell(M, g, n)$ as follows. Given a stable C^ℓ map $f : C \to M$ with C smooth, let $\Lambda_f^{0,1}$ denote the space of continuous sections ν of

$$\mathrm{Hom}(TC, f^*TM)$$

such that $\nu \circ j_C = -J \circ \nu$. These are to be thought of as $f^*(TM)$-valued $(0,1)$ forms on C. Note that the Cauchy-Riemann operator $\overline{\partial}_J$ is in $\Lambda_f^{0,1}$, as is calculated immediately using (7.23), $J^2 = -1$, and $j_C^2 = -1$. The definition of $\Lambda_f^{0,1}$ needs to be slightly modified when C is not smooth (see [**LTi3**] for details) and is compatible with the equivalence which defines $\overline{\mathcal{F}}_\beta^\ell(M, g, n)$. The result is that the spaces $\Lambda_f^{0,1}$ are the fibers of a generalized bundle E on $\overline{\mathcal{F}}_\beta^\ell(M, g, n)$.

By construction, $\overline{\partial}_J$ induces a global section of E. We thus have a section of an infinite dimensional bundle on an infinite dimensional space. The idea is that we would like something that plays the role of the Euler class of a bundle in the finite dimensional case. It is shown in [**LTi3**] that if $\ell \geq 2$, then this section gives E the structure of an *generalized Fredholm orbifold bundle*, a notion introduced in [**LTi3**]. The main result is that to this data is associated an *oriented Euler class*, which is a homology class on the base space of the bundle. This class is supported on the zero locus of a carefully chosen perturbation of the section $\overline{\partial}_J$ of E. In the case of interest, the homology class has dimension given exactly by the expected dimension (7.26). This oriented Euler class plays the role of the virtual fundamental class, and easily leads to a definition of symplectic Gromov-Witten invariants in the familiar manner.

The Gromov-Witten invariants of [**LTi3**] extend the invariants of [**RT2**], which are defined only when M is semi-positive. Note that it is not necessary to assume that $2g + n \geq 3$. There is also a related construction [**Siebert1**]. It is believed that the differing symplectic Gromov-Witten invariants agree with each other, but no proof has been written down as of this writing. In the next section we will discuss to what extent the various symplectic invariants agree with the algebraic Gromov-Witten invariants discussed in Section 7.1.

7.3. Properties of Gromov-Witten Classes

Our next task is to study Gromov-Witten classes and the relations between the various definitions given in the previous two sections. We begin with a description of the properties common to all of the definitions.

7.3.1. Axioms for Gromov-Witten Classes.

In [**KoM1**], Kontsevich and Manin propose a system of axioms for Gromov-Witten classes, and it is now known that both the algebraic classes $I_{g,n,\beta}$ from Section 7.1 and the symplectic classes $\Psi_{g,n,\beta}$ from Section 7.2 satisfy these axioms. In the algebraic case, this is proved in [**BM, LTi2, BF, Behrend**], while in the symplectic case, the proof can be found in [**RT1, RT2, LTi3**]. We will add another axiom to these, the Deformation Axiom, which is proven in the algebraic case in [**LTi2**] and follows from symplectic invariance in the symplectic case.

For simplicity, we will state the axioms for $I_{g,n,\beta}$. Recall from Section 7.1 that if X is a smooth projective variety, then for $g \geq 0$, $n \geq 0$ and $n + 2g \geq 3$, the Gromov-Witten classes are maps

$$I_{g,n,\beta} : H^*(X, \mathbb{Q})^{\otimes n} \longrightarrow H^*(\overline{M}_{g,n}, \mathbb{Q}).$$

Also, for $g, n \geq 0$, the Gromov-Witten invariants are maps

$$\langle I_{g,n,\beta}\rangle : H^*(X, \mathbb{Q})^{\otimes n} \longrightarrow \mathbb{Q}.$$

When $n + 2g \geq 3$, these are related by

$$\langle I_{g,n,\beta}\rangle(\alpha_1, \dots, \alpha_n) = \int_{\overline{M}_{g,n}} I_{g,n,\beta}(\alpha_1, \dots, \alpha_n).$$

In thinking about the axioms given below, the reader should recall the naive interpretation (7.2) of Gromov-Witten classes. Namely, if Z_i is a cycle in X dual to α_i, then $I_{g,n,\beta}(\alpha_1, \dots, \alpha_n)$ should be the cohomology class given by the set of genus g curves $(C, p_1, \dots, p_n) \in \overline{M}_{g,n}$ for which we can find f such that

$$(7.32) \qquad f : (C, p_1, \dots, p_n) \to X \text{ is stable}, f(p_i) \in Z_i, \text{ and } f_*[C] = \beta.$$

With this in mind, we now proceed to the axioms.

Linearity Axiom. The first axiom asserts that $I_{g,n,\beta}$ is linear in each variable. Naively, this is because a sum of cycles is simply their union.

Effectivity Axiom. The next axiom says that for a smooth projective variety X, $I_{g,n,\beta} = 0$ if β is not an effective class. This makes sense because $f_*[C]$ is effective whenever $f : C \to X$ is holomorphic map. In the case of a symplectic manifold (M, ω), this axiom is replaced by the observation that the invariants vanish whenever $\int_\beta \omega < 0$.

Degree Axiom. This axiom asserts that for $\alpha_1, \dots, \alpha_n \in H^*(X, \mathbb{Q})^{\otimes n}$, the cohomology class $I_{g,n,\beta}(\alpha_1, \dots, \alpha_n) \in H^*(\overline{M}_{g,n}, \mathbb{Q})$ has degree

$$2(g - 1) \dim X + 2\int_\beta \omega_X + \sum_{i=1}^n \deg \alpha_i$$

if the α_i are homogeneous classes. When the moduli space $\overline{M}_{g,n}(X, \beta)$ has the expected dimension (7.7) and has a nice fundamental class, this follows from (7.8). In general, the Degree Axiom is a consequence of Definition 7.1.9 because the virtual fundamental class has the expected degree.

The Degree Axiom implies that $I_{g,n,\beta}(\alpha_1, \dots, \alpha_n)$ is a top degree class if and only if

$$(7.33) \qquad \sum_{i=1}^n \deg \alpha_i = 2(1 - g) \dim X - 2\int_\beta \omega_X + 2(3g - 3 + n).$$

It follows that when considering the Gromov-Witten invariant $\langle I_{g,n,\beta}\rangle(\alpha_1, \dots, \alpha_n)$, we can always assume that (7.33) is satisfied (since the invariant is zero otherwise). Using Definition 7.1.9, we see that this is valid for all $n, g \geq 0$.

In the literature, $\deg \alpha_i$ is sometimes replaced by its "algebraic degree" $\frac{1}{2}\deg \alpha_i$, so that if Z is a subvariety of codimension i, then $[Z]$ has degree i. This convention is most useful when dealing with even-dimension cohomology.

Equivariance Axiom. The symmetric group S_n acts naturally on the cohomology groups $H^*(X, \mathbb{Q})^{\otimes n}$ and $H^*(\overline{M}_{g,n}, \mathbb{Q})$. For the latter, the action is given by permuting the points p_i of $(C, p_1, \dots, p_n)$. This axiom asserts that the map

$$I_{g,n,\beta} : H^*(X, \mathbb{Q})^{\otimes n} \to H^*(\overline{M}_{g,n}, \mathbb{Q})$$

is S_n-equivariant. The intuition behind this should be evident from (7.32). For Gromov-Witten invariants, equivariance means that

$$\langle I_{g,n,\beta}\rangle(\alpha_1, \dots, \alpha_i, \alpha_{i+1}, \dots, \alpha_n) =$$
$$(-1)^{\deg \alpha_i \, \deg \alpha_{i+1}} \langle I_{g,n,\beta}\rangle(\alpha_1, \dots, \alpha_{i+1}, \alpha_i, \dots, \alpha_n).$$

Fundamental Class Axiom. If $n + 2g \geq 4$, then we get a natural map $\pi_n :$ $\overline{M}_{g,n} \to \overline{M}_{g,n-1}$ by forgetting the last point. If $[X] \in H^0(X, \mathbb{Q})$ is the fundamental class of X, then this axiom asserts that

$$I_{g,n,\beta}(\alpha_1, \ldots, \alpha_{n-1}, [X]) = \pi_n^* I_{g,n-1,\beta}(\alpha_1, \ldots, \alpha_{n-1}).$$

Using (7.32), this makes sense because $f(p_n) \in X$ puts no condition on p_n. When dealing with a Gromov-Witten invariant, the Fundamental Class Axiom implies

$$(7.34) \qquad \langle I_{g,n,\beta} \rangle (\alpha_1, \ldots, \alpha_{n-1}, [X]) = 0.$$

This follows because the above Gromov-Witten invariant can be nonzero only if $I_{g,n,\beta}(\alpha_1, \ldots, \alpha_{n-1}, [X])$ is a class of top degree. Thus $I_{g,n-1,\beta}(\alpha_1, \ldots, \alpha_{n-1})$ is zero, since $\overline{M}_{g,n-1}$ has smaller dimension. Note that (7.34) holds whenever π_n is defined. The discussion at the end of Section 7.1.5 shows that this is true if either $n + 2g \geq 4$ or $\beta \neq 0$ and $n \geq 1$.

Divisor Axiom. If $n + 2g \geq 4$, then let $\pi_n : \overline{M}_{g,n} \to \overline{M}_{g,n-1}$ be as in the previous axiom. If $\alpha_n \in H^2(X, \mathbb{Q})$, then

$$\pi_{n*} I_{g,n,\beta}(\alpha_1, \ldots, \alpha_{n-1}, \alpha_n) = \left(\int_\beta \alpha_n \right) I_{g,n-1,\beta}(\alpha_1, \ldots, \alpha_{n-1}).$$

To see why this should be true in terms of (7.32), let $f : (C, p_1, \ldots, p_n) \to X$ be stable such that $f_*[C] = \beta$ and $f(p_i) \in Z_i$ for $i = 1, \ldots, n-1$. Then $f(p_n)$ must lie in $f(C) \cap Z_n = \beta \cap Z_n$, which means that there are $\int_\beta \alpha_n$ possible choices for $f(p_n)$. Hence we get the above formula.

For Gromov-Witten invariants, we see that if $\alpha_n \in H^2(X, \mathbb{Q})$ and (7.33) is satisfied, then

$$\langle I_{g,n,\beta} \rangle (\alpha_1, \ldots, \alpha_{n-1}, \alpha_n) = \int_\beta \alpha_n \, \langle I_{g,n-1,\beta} \rangle (\alpha_1, \ldots, \alpha_{n-1}).$$

As with the Fundamental Class Axiom, this holds whenever π_n is defined, so that by Section 7.1.5, this is true if either $n + 2g \geq 4$ or $\beta \neq 0$ and $n \geq 1$.

Point Mapping Axiom. This axiom describes what happens when $\beta = 0$. When $g = 0$, it states that if the α_i are homogeneous cohomology classes, then

$$I_{0,n,0}(\alpha_1, \ldots, \alpha_n) = \begin{cases} \left(\int_X \alpha_1 \cup \cdots \cup \alpha_n \right) [\overline{M}_{g,n}] & \text{if } \sum_{i=1}^n \deg \alpha_i = 2 \dim X \\ 0 & \text{otherwise.} \end{cases}$$

To see why this is reasonable, notice that a map satisfying $f_*[C] = 0$ must be constant. Thus (7.32) reduces to $f(C) \in Z_1 \cap \cdots \cap Z_n$. When $\sum_{i=1}^n \deg \alpha_i = 2 \dim X$, the class has degree 0 by the Degree Axiom. This gives the top formula, and all other classes vanish in this case.

This has a nice consequence for the Gromov-Witten invariants $\langle I_{0,n,0} \rangle$. If we compare the above condition $\sum_{i=1}^n \deg \alpha_i = 2 \dim X$ to (7.33), we see that $n = 3$ is special. More precisely,

$$(7.35) \qquad \langle I_{0,n,0} \rangle (\alpha_1, \ldots, \alpha_n) = \begin{cases} \int_X \alpha_1 \cup \alpha_2 \cup \alpha_3 & \text{if } n = 3 \\ 0 & \text{otherwise.} \end{cases}$$

Unfortunately, this reasoning applies only when $g = 0$. The problem is that $\overline{M}_{g,n}(X, 0) = \overline{M}_{g,n} \times X$ has the expected dimension only when $g = 0$. There are formulas for $I_{g,n,0}$ for all g, but they can be rather complicated. We will see what happens when $g = 1$ in (7.55) below, and the difficulties of the general case are discussed in [**KoM1**, Sect. 2.2.5].

Splitting Axiom. This axiom is a bit complicated to state. First suppose that we have splittings $g = g_1 + g_2$ and $n = n_1 + n_2$ such that $n_i + 2g_i \geq 2$. Notice that if we have stable curves $(C_1, p_1, \ldots, p_{n_1+1})$ and $(C_2, q_1, \ldots, q_{n_2+1})$ of respective genus g_1 and g_2, then the curve C obtained from $C_1 \cup C_2$ by identifying p_{n_1+1} with q_{n_2+1} gives a stable curve $(C, p_1, \ldots, p_{n_1}, q_1, \ldots, q_{n_2})$ of genus g. Thus we get a map

$$(7.36) \qquad \varphi : \overline{M}_{g_1, n_1+1} \times \overline{M}_{g_2, n_2+1} \longrightarrow \overline{M}_{g,n}.$$

Then the splitting axiom asserts that $\varphi^* I_{g,n,\beta}(\alpha_1, \ldots, \alpha_n)$ is given by the formula:

$$\sum_{\beta=\beta_1+\beta_2} \sum_{i,j} g^{ij} \, I_{g_1, n_1+1, \beta_1}(\alpha_1, \ldots, \alpha_{n_1}, T_i) \otimes I_{g_2, n_2+1, \beta_2}(T_j, \alpha_{n_1+1}, \ldots, \alpha_n),$$

where T_i is a homogeneous basis of $H^*(X, \mathbb{Q})$ and (g^{ij}) is the inverse of the matrix (g_{ij}) defined by $g_{ij} = \int_X T_i \cup T_j$. (This means that the cohomology class of the diagonal in $H^*(X \times X, \mathbb{Q})$ is $\sum_{i,j} g^{ij} T_i \otimes T_j$.) Because of the Effectivity Axiom, the above sum is finite.

To see where this formula comes from, first note that the inverse image under φ of the maps (7.32) consists of maps

$$f : (C_1 \cup C_2, p_1, \ldots, p_{n_1}, p, p_{n_1+1}, \ldots, p_n, q) \to X$$

where $f_*[C_1] + f_*[C_2] = \beta$, $f(p_i) \in Z_i$ and $f(p) = f(q)$. The first of these conditions corresponds to the decompositions $\beta_1 + \beta_2 = \beta$ in the above formula, and the last condition means we want $(f(p), f(q))$ to be in the diagonal of $X \times X$. Since the class of the diagonal is $\sum_{i,j} g^{ij} T_i \otimes T_j$, we get the above formula.

Reduction Axiom. Here, we use the fact that gluing together the last two marked points gives a natural map $\psi : \overline{M}_{g-1, n+2} \to \overline{M}_{g,n}$. If Δ_a and g^{ab} are as in the previous axiom, then

$$\psi^* I_{g,n,\beta}(\alpha_1, \ldots, \alpha_n) = \sum_{i,j} g^{ij} \, I_{g-1, n+2, \beta}(\alpha_1, \ldots, \alpha_n, T_i, T_j).$$

This is reasonable because the inverse image under φ of the maps (7.32) consists of maps $f : (C, p_1, \ldots, p_{n+2}) \to X$ of genus $g - 1$ such that $f_*[C] = \beta$, $f(p_i) \in Z_i$ for $i = 1, \ldots, n$ and $f(p_{n+1}) = f(p_{n+2})$. The last condition means that $(f(p_{n+1}), f(p_{n+2}))$ lies in the diagonal of $X \times X$. As above, the class of the diagonal is $\sum_{i,j} g^{ij} T_i \otimes T_j$, and the formula follows.

Deformation Axiom. Let $f : \mathcal{X} \to T$ be a smooth proper map with connected base T, and set $X_t = f^{-1}(t)$ for $t \in T$. Thus, for each $t \in T$ and $\beta_t \in H_2(X_t, \mathbb{Z})$, we get a map

$$I_{g,n,\beta_t}^{X_t} : H^*(X_t, \mathbb{Q})^{\otimes n} \longrightarrow H^*(\overline{M}_{g,n}, \mathbb{Q}).$$

Then, if β_t is a locally constant section of $H_2(X_t, \mathbb{Z})$ and $\alpha_1, \ldots, \alpha_n$ are locally constant sections of $H^*(X_t, \mathbb{Q})$, $I_{g,n,\beta_t}^{X_t}(\alpha_1, \ldots, \alpha_n)$ is constant. The intuition behind this is that Gromov-Witten classes should be invariant under deformation of complex structure. This is clear when we use the symplectic definition of $I_{g,n,\beta}$.

Among these axioms, the Splitting Axiom and the Reduction Axiom are probably the most important. As we will see in Chapter 8, they are closely linked to proving that quantum cohomology is associative. In the papers of Ruan and Tian, these two axioms are known collectively as the *Composition Law*.

Besides these axioms, **[KoM1]** also includes a *Motivic Axiom*, which says that Gromov-Witten classes should be induced by a class of the appropriate degree in the Chow ring of $X^n \times \overline{M}_{g,n}$. This axiom follows since algebraic Gromov-Witten classes are defined using the virtual fundamental class $[\overline{M}_{g,n}(X,\beta)]^{\mathrm{virt}}$.

We should emphasize that the intuitive explanations given for the above axioms are *not* proofs. However, in the symplectic case, some of these "arguments" are fairly close to the actual proofs (with the exception of the Composition Law—see [**RT1, RT2**]). On the other hand, the proofs in the algebraic case [**LTi2, Behrend**] are more sophisticated. As explained in [**BM**], the above axioms reduce to five essential compatibilities between the virtual fundamental classes $[\overline{M}_{g,n}(X,\beta)]^{\mathrm{virt}}$.

7.3.2. Equivalence of the Various Definitions. In Sections 7.1 and 7.2 we saw various definitions of Gromov-Witten classes and invariants. In the algebraic case, we have two definitions since [**LTi2**] and [**BF, Behrend**] give slightly different constructions of the virtual fundamental class. As we noted in Section 7.1.4, these definitions should be equivalent, though the proof has not been written down.

In the symplectic case, the invariants defined in [**LTi3**] extend those of [**RT2**], but we also have the symplectic invariants given in [**Siebert1**]. Again, one expects these definitions to give the same invariants, but this has not yet been proved.

What's really nice is that there has been substantial progress in proving that algebraic and symplectic invariants agree. In particular, the algebraic and symplectic invariants defined in [**LTi2**] and in [**LTi3**] are shown to coincide in [**LTi4**], and the algebraic and symplectic invariants defined in [**Behrend**] and [**Siebert1**] are shown to coincide in [**Siebert2**]. Hence it seems likely that *all* of these definitions agree on their common domain of definition.

There is one situation where this is known to happen. Let $\langle I_{g,n,\beta} \rangle^{\mathrm{LRT}}$ denote the Gromov-Witten invariant defined by either the algebraic construction in [**LTi2**] or the symplectic construction in [**LTi3**] (which generalizes [**RT2**]). Also, let $\langle I_{g,n,\beta} \rangle^{\mathrm{BFS}}$ denote the Gromov-Witten invariant defined by either the algebraic construction in [**Behrend**] (which uses the relative virtual fundamental class in [**BF**]) or the symplectic construction in [**Siebert1**]. Then the following is true.

THEOREM 7.3.1. *Let X be a smooth projective variety acted on transitively by a linear algebraic group. Pick classes $\alpha_1, \ldots, \alpha_n \in H^*(X,\mathbb{Q})$. Then*

$$\langle I_{0,n,\beta} \rangle^{\mathrm{LRT}}(\alpha_1, \ldots, \alpha_n) = \langle I_{0,n,\beta} \rangle^{\mathrm{BFS}}(\alpha_1, \ldots, \alpha_n).$$

In addition, these invariants count the number of pointed maps $f : \mathbb{P}^1 \to X$ satisfying (7.2).

PROOF. We may as well use the respective algebraic definitions of $\langle I_{0,n,\beta} \rangle^{\mathrm{LRT}}$ and $\langle I_{0,n,\beta} \rangle^{\mathrm{BFS}}$ for checking equality. Then Lemma 13 of [**FP**] shows that the counting definition gives the algebraic Gromov-Witten invariant. In particular, the virtual fundamental class coincides with the fundamental class (using *any* algebraic definition of the virtual fundamental class). The group action is used to ensure that we can find cycles Z_i dual to α_i which are in general position.

On the symplectic side, using the transitive group action, Li and Tian show in [**LTi1**] that one gets the same number using the given complex structure J_0 of X and $\nu = 0$. A similar comment applies to the symplectic definition of [**Siebert1**]. $\square$

Theorem 7.3.1 asserts that the invariants are equal, and its proof shows that they really count the number of rational curves of the appropriate type. As we will see in the next section, the enumerative significance of general Gromov-Witten invariants is in general harder to understand.

Finally, we should mention that in addition to the algebraic and symplectic Gromov-Witten invariants discussed so far, there are also related invariants which are of interest. In Chapters 9 and 10, we will encounter two of these, *equivariant Gromov-Witten invariants* and *gravitational correlators*. These play a crucial role in the work of Givental on mirror symmetry, as well as in related ideas and developments.

7.4. Computing Gromov-Witten Invariants, I

Now that we know the definition and properties of Gromov-Witten classes and invariants, it is time to compute some examples. In this section, we will discuss the Gromov-Witten invariants $\langle I_{g,n,\beta}\rangle(\alpha_1,\ldots,\alpha_n)$ for a variety of different spaces X. Most of our examples will have genus $g = 0$ or 1. Given the naive interpretation (7.2) of Gromov-Witten invariants, we will also ask about their enumerative significance. As we will discover, these numbers sometimes have a complicated relation to enumerative geometry.

In Chapter 8, we will see that quantum cohomology is built out of the invariants $\langle I_{0,n,\beta}\rangle$. Chapter 8 will also introduce the *Gromov-Witten potential*, which has some very interesting properties. In particular, it will shed some light on some of the examples presented here. Hence this section is really the first of two sections dealing with the computation and enumerative significance of Gromov-Witten invariants.

7.4.1. Tree-Level Gromov-Witten Classes. When $g = 0$, the classes $I_{0,n,\beta}$ are a *tree-level system of Gromov-Witten classes*. The term "tree-level" refers to the fact that stable curves of genus zero are trees of $\mathbb{P}^1$s (corresponding to tree-like Feynman diagrams in quantum field theory, where the terminology originated). Since we always require that $n + 2g \geq 3$, tree-level classes are defined for $n \geq 3$. The moduli space $\overline{M}_{0,n}$ is an orbifold of dimension $n - 3$, and its cohomology is well-understood. This enables us to say a lot about the $I_{0,n,\beta}$. In particular, we have the following *Reconstruction Theorem* of Kontsevich and Manin.

THEOREM 7.4.1. *Let X be a smooth projective variety with the property that $H^*(X,\mathbb{Q})$ is generated by $H^2(X,\mathbb{Q})$. Also assume that we know the Gromov-Witten invariants $\langle I_{0,3,\beta}\rangle(\alpha_1,\alpha_2,\alpha_3)$ for all $\beta \in H_2(X,\mathbb{Z})$ satisfying $-\int_\beta \omega_X \leq \dim X + 1$ and $\deg\alpha_3 = 2$. Then we can determine all tree-level Gromov-Witten classes $I_{0,n,\beta}(\alpha_1,\ldots,\alpha_n)$ for all $\beta \in H_2(X,\mathbb{Z})$.*

PROOF. We first show that all tree-level classes can be reconstructed from Gromov-Witten invariants. We will use induction on n, and we can assume $n > 3$ since $\overline{M}_{0,3}$ is a point.

To describe the cohomology of $\overline{M}_{0,n}$, notice that the image of the map φ in (7.36) is a divisor in $\overline{M}_{0,n}$. Furthermore, by permuting the points $p_1,\ldots,p_n$, we get other versions of the map φ, which give other divisors in $\overline{M}_{0,n}$. By [**Keel**], the cohomology of $H^*(\overline{M}_{0,n},\mathbb{Q})$ is generated by the classes of these divisors. Using this, [**KoM1**] shows that a cohomology class in $H^*(\overline{M}_{0,n},\mathbb{Q})$ not of top degree is uniquely determined by its intersections with these divisors.

Now let $I_{0,n,\beta}(\alpha_1,\ldots,\alpha_n)$ be one of our classes. If it lies in the top degree, it is $\langle I_{0,n,\beta}\rangle(\alpha_1,\ldots,\alpha_n)$ times the class of a point. If it doesn't, then it is determined by its intersections with the divisors described above. However, when we intersect with the divisor given by the image of φ, we get $\varphi^* I_{0,n,\beta}(\alpha_1,\ldots,\alpha_n)$. By the Splitting Axiom, this is built out of Gromov-Witten classes which have smaller

n. Furthermore, the Equivariance Axiom shows that the same is true for the intersections with the other divisors. It follows that $I_{0,n,\beta}(\alpha_1, \dots, \alpha_n)$ is determined by Gromov-Witten classes with smaller n. By induction, these are determined by Gromov-Witten invariants, and we are done.

The next step is to study the invariants $\langle I_{0,n,\beta}\rangle$. The idea is that certain linear relations among the divisors mentioned above give quadratic relations among the $\langle I_{0,n,\beta}\rangle$ for different n and β. These relations enable one to express any Gromov-Witten invariant in terms of those listed in the statement of the theorem. In particular, the inequality in the statement of the theorem is easy to explain. If the α_i are homogeneous classes, then $\langle I_{0,3,\beta}\rangle(\alpha_1, \alpha_2, \alpha_3) = 0$ unless

$$\sum \deg(\alpha_i) = 2\dim[\overline{M}_{0,n}(X,\beta)]^{\mathrm{virt}} = 2\left(-\int_\beta \omega_X + \dim X\right).$$

Combining this with the obvious inequality $\sum \deg(\alpha_i) \leq 2\dim X + 2$ (still assuming $\deg \alpha_3 = 2$) leads to the desired inequality. The full details of the argument can be found in [**KoM1**]. Below, we indicate how this works when $X = \mathbb{P}^2$. $\square$

7.4.2. Genus Zero Invariants of the Projective Plane. We will compute the tree-level Gromov-Witten invariants for the plane $\mathbb{P}^2$. Here, $\beta = d[\ell]$, where $\ell \subset \mathbb{P}^2$ is a line and $d \geq 0$. To simplify notation, we write $\langle I_{0,n,d}\rangle$ instead of $\langle I_{0,n,\beta}\rangle$. We first compute $\langle I_{0,3,1}\rangle([pt],[pt],[\ell])$. This is easy! Given a map $f : (\mathbb{P}^1, p_1, p_2, p_3) \to \mathbb{P}^2$ such that $f(p_1)$ and $f(p_2)$ are preassigned points and $f(p_3) \in \ell$, we get the following picture:

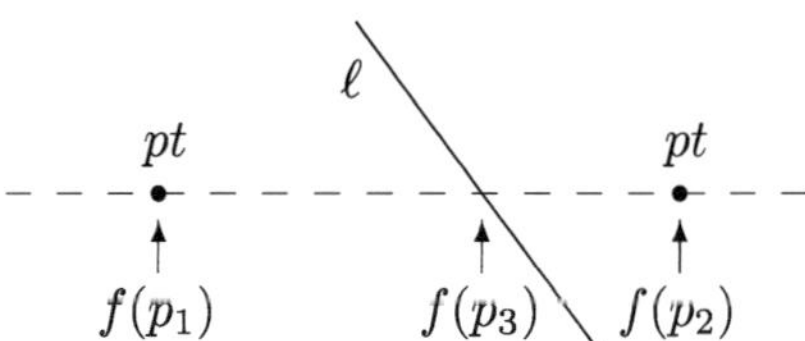

Since the degree is 1, the image of f is the line determined by the two points. Also, since $f(p_1)$ and $f(p_2)$ are fixed and $f(p_3)$ is determined by the intersection $\ell \cap f(\mathbb{P}^1)$, we get a *unique* map. It is then immediate to see that this is the only genus 0 stable map $f : (C, p_1, p_2, p_3) \to \mathbb{P}^2$ with arbitrary source curve C satisfying the required conditions on the p_i. Furthermore, it is straightforward to see that this stable map occurs with multiplicity 1 in the definition (7.9) of the Gromov-Witten invariant. Thus

$$\langle I_{0,3,1}\rangle([pt],[pt],[\ell]) = 1.$$

Furthermore, according to Theorem 7.4.1, this is the *only* invariant we need to compute, since one easily checks that $\langle I_{0,3,1}\rangle([pt],[pt],[\ell])$ is the only Gromov-Witten invariant satisfying the conditions of the theorem.

So what are the other invariants? By the Point Mapping Axiom, we can assume $d \geq 1$. To compute $\langle I_{0,n,d}\rangle(\alpha_1, \dots, \alpha_n)$, we can assume each α_i is $[\mathbb{P}^2]$, $[\ell]$ or $[pt]$. Also, by (7.33), we know that

$$(7.37) \qquad \sum_{i=1}^n \deg \alpha_i = 6d - 2 + 2n.$$

For $n = 0, 1$ this equation has no solutions, and for $n = 2, 3$, one can check that (up to permutation), the only Gromov-Witten invariants are

$$\langle I_{0,2,1}\rangle([pt],[pt]) = \langle I_{0,3,1}\rangle([pt],[pt],[\ell]) = 1.$$

The first equality is by the Divisor Axiom or, more simply, $\langle I_{0,2,1}\rangle([pt],[pt]) = 1$ since there is a unique line through two points in $\mathbb{P}^2$.

If $n \geq 4$, note that $\langle I_{0,n,d}\rangle(\alpha_1,\dots,\alpha_n) = 0$ if some $\alpha_i = [\mathbb{P}^2]$ by the Fundamental Class Axiom. Similarly, if $n \geq 4$ and $\alpha_n = [\ell]$, then the Divisor Axiom shows that

$$\langle I_{0,n,d}\rangle(\alpha_1,\dots,\alpha_{n-1},[\ell]) = d\,\langle I_{0,n-1,d}\rangle(\alpha_1,\dots,\alpha_{n-1}),$$

and we can proceed inductively. The only invariants remaining to compute are the $\langle I_{0,n,d}\rangle([pt],\dots,[pt])$. Since $\deg[pt] = 4$, (7.37) implies that $n = 3d - 1$. Thus we want to compute

$$(7.38) \qquad N_d = \langle I_{0,3d-1,d}\rangle(\underbrace{[pt],\dots,[pt]}_{3d-1 \text{ times}})$$

for $d \geq 1$. Because $\mathbb{P}^2$ has a transitive linear group action, Theorem 7.3.1 shows that N_d gives the number of rational curves of degree d in $\mathbb{P}^2$ passing through $3d-1$ points in general position.

The numbers N_d can alternatively be defined as the degree of the Severi variety of degree d rational plane curves. However, traditional methods in algebraic geometry had previously only yielded the first few numbers. The new feature here is that by interpreting N_d as a Gromov-Witten invariant, one gets the recursion formula

$$(7.39) \qquad N_d = \sum_{\substack{d=d_1+d_2 \\ d_1,d_2>0}} N_{d_1} N_{d_2} \left(d_1^2 d_2^2 \binom{3d-4}{3d_1-2} - d_1^3 d_2 \binom{3d-4}{3d_1-1} \right).$$

Since $N_1 = \langle I_{0,2,1}\rangle([pt],[pt]) = 1$, this relation implies

$$N_2 = 1,\ N_3 = 12,\ N_4 = 620,\ N_5 = 87304,\ \dots$$

This provides an illustration of how stable maps are often easier to work with than embedded curves, even for answering classical questions about embedded curves!

Our proof of (7.39) will use the strategy outlined in the proof of Theorem 7.4.1. We start with the Gromov-Witten class

$$(7.40) \qquad \xi = I_{0,3d,d}(\underbrace{[pt],\dots,[pt]}_{3d-2 \text{ times}},[\ell],[\ell]) \in H^*(\overline{M}_{0,3d},\mathbb{Q})$$

By the Degree Axiom, this is a class of degree $2\dim\overline{M}_{0,3d}-2$, so that its intersection with a divisor is a well-defined rational number.

Fortunately, $\overline{M}_{0,3d}$ has many interesting divisors. We will index the $3d$ marked points on a stable curve in $\overline{M}_{0,3d}$ by the set $\mathcal{I} = \{1,\dots,3d-4,p,q,r,s\}$. In terms of (7.40), this means $\alpha_1 = \cdots = \alpha_{3d-4} = \alpha_p = \alpha_q = [pt]$ and $\alpha_r = \alpha_s = [\ell]$. Now partition $\mathcal{I}$ into disjoint sets A and B. As in the Splitting Axiom, this gives a natural map

$$\varphi_{A,B} : \overline{M}_{0,|A|+1} \times \overline{M}_{0,|B|+1} \longrightarrow \overline{M}_{0,3d}$$

The image of $\varphi_{A,B}$ is a divisor in $\overline{M}_{0,3d}$. Furthermore, we have the fundamental linear equivalence of divisors:

$$(7.41) \qquad \sum_{\substack{r,s\in A\ p,q\in B}} \mathrm{Im}(\varphi_{A,B}) \sim \sum_{\substack{p,r\in A\ q,s\in B}} \mathrm{Im}(\varphi_{A,B}).$$

This follows because of the natural map $\phi : \overline{M}_{0,3d} \to \overline{M}_{0,4}$ which forgets the first $3d - 4$ points and contracts non-stable components. Under this map, the left hand side of (7.41) is the fiber of ϕ over the stable curve:

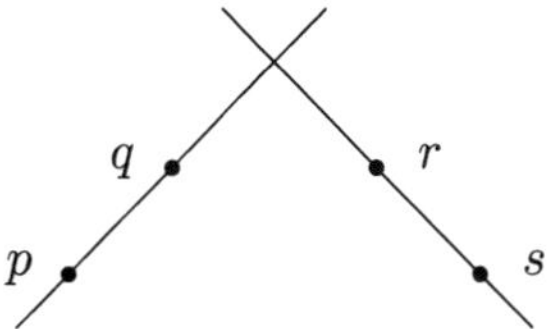

in $\overline{M}_{0,4}$. In a similar way, the right hand side of (7.41) is the fiber of ϕ over the stable curve which has p, r on one component and q, s on the other. The cross-ratio shows that $M_{0,4} \simeq \mathbb{P}^1 - \{0, 1, \infty\}$, and the above stable curve is one of three boundary points of $\overline{M}_{0,4} \simeq \mathbb{P}^1$. Since the boundary points are clearly linearly equivalent, the same is true for the fibers of ϕ over these points, and (7.41) follows.

The next step of the proof is to intersect each side of (7.41) with the Gromov-Witten class ξ defined in (7.40). Let's start with the left hand side of (7.41). The intersection of ξ with $\mathrm{Im}(\varphi_{A,B})$ corresponds to the pullback $\varphi_{A,B}^* \xi$, which we can compute using the Splitting Axiom. In Section 7.3, we stated the Splitting Axiom for one particular partition of the index set $\mathcal{I}$, but by the Equivariance Axiom, it applies to all partitions. Thus, since the class of the diagonal in $\mathbb{P}^2 \times \mathbb{P}^2$ is

$$\Delta = [pt] \otimes [\mathbb{P}^2] + [\ell] \otimes [\ell] + [\mathbb{P}^2] \otimes [pt],$$

the Splitting Axiom and the Equivariance Axiom show that the intersection number $\xi \cdot \mathrm{Im}(\varphi_{A,B})$ is given by the formula

$$\sum_{d=d_1+d_2} \Big(\langle I_{0,|A|+1,d_1} \rangle (\alpha_i, i \in A, [pt]) \cdot \langle I_{0,|B|+1,d_2} \rangle (\alpha_i, i \in B, [\mathbb{P}^2])$$
$$\text{(7.42)} \qquad + \langle I_{0,|A|+1,d_1} \rangle (\alpha_i, i \in A, [\ell]) \cdot \langle I_{0,|B|+1,d_2} \rangle (\alpha_i, i \in B, [\ell])$$
$$+ \langle I_{0,|A|+1,d_1} \rangle (\alpha_i, i \in A, [\mathbb{P}^2]) \cdot \langle I_{0,|B|+1,d_2} \rangle (\alpha_i, i \in B, [pt]) \Big)$$

since $\varphi_{A,B}^* \xi$ is a top degree class.

Although this formula looks complicated, most terms vanish because of the degree condition for Gromov-Witten invariants. For instance, consider the invariant $\langle I_{0,|A|+1,d_1} \rangle (\alpha_i, i \in A, \gamma)$, where γ is one of $[pt], [\ell], [\mathbb{P}^2]$. By our condition on A, we have $\alpha_r = \alpha_s = [\ell]$ and $\alpha_i = [pt]$ otherwise. Hence (7.33) implies that the invariant vanishes unless

$$|A| = 3d_1 + 2 - \tfrac{1}{2} \deg \gamma.$$

Since $|A|$ is fixed, this equation determines d_1 and $\deg \gamma$. Hence at most one term in (7.42) is nonzero.

Now assume that $|A|, |B| \geq 3$. Then the Fundamental Class Axiom implies that $\langle I_{0,|A|+1,d_1} \rangle (\alpha_i, i \in A, [\mathbb{P}^2]) = \langle I_{0,|B|+1,d_2} \rangle (\alpha_i, i \in B, [\mathbb{P}^2]) = 0$. Hence the nonvanishing term must have $\gamma = [\ell]$, which in turn implies $|A| = 3d_1 + 1$ and $|B| = 3d_2 - 1$. Then (7.42) reduces to

$$\langle I_{0,3d_1+2,d_1} \rangle (\underbrace{[pt], \ldots, [pt]}_{3d_1 - 1 \text{ times}}, [\ell], [\ell], [\ell]) \cdot \langle I_{0,3d_2,d_2} \rangle (\underbrace{[pt], \ldots, [pt]}_{3d_2 - 1 \text{ times}}, [\ell]).$$

By the Divisor Axiom and the definition of N_d, this reduces to $d_1^3 d_2 N_{d_1} N_{d_2}$. Thus, we have

$$(7.43) \qquad \xi \cdot \mathrm{Im}(\varphi_{A,B}) = d_1^3 d_2 N_{d_1} N_{d_2} \quad \text{when } |A| = 3d_1 + 1.$$

Note also that $d_1 > 0$, $d_2 > 1$ in this case.

On the other hand, if $|B| = 2$, we leave it to the reader to show that the nonvanishing term of (7.42) is

$$\langle I_{0,3d-1,d-1}\rangle(\underbrace{[pt], \dots, [pt]}_{3d-4 \text{ times}}, [\ell], [\ell], [\ell]) \cdot \langle I_{0,3,1}\rangle([pt], [pt], [\ell]).$$

Using the Divisor Axiom, this reduces to $(d-1)^3 N_{d-1} N_1$, which gives (7.43) with $d_1 = d - 1$ and $d_2 = 1$.

Finally, if $|A| = 2$, the nonvanishing term is

$$\langle I_{0,3,0}\rangle([\ell], [\ell], [\mathbb{P}^2]) \cdot \langle I_{0,3d-1,d}\rangle(\underbrace{[pt], \dots, [pt]}_{3d-2 \text{ times}}, [pt]).$$

The first term in this product is 1 by the Point Mapping Axiom, and the second is N_d. Thus, we get

$$(7.44) \qquad \xi \cdot \mathrm{Im}(\varphi_{A,B}) = N_d \quad \text{when } |A| = 2, \ d_1 = 0, \ d_2 = d.$$

From here, it is easy to see that the intersection of the class ξ with the left hand side of (7.41) is

$$N_d \ + \ \sum_{\substack{d = d_1 + d_2 \\ d_1, d_2 > 0}} N_{d_1} N_{d_2} d_1^3 d_2 \binom{3d-4}{3d_1 - 1}.$$

To see why, consider partitions A, B of the index set $\mathcal{I} = \{1, \dots, 3d - 4, p, q, r, s\}$ with $r, s \in A$ and $p, q \in B$. As we vary over A, B, (7.44) occurs exactly once, when $A = \{r, s\}$, and when $d = d_1 + d_2$ with $d_1, d_2 > 0$, (7.43) occurs exactly $\binom{3d-4}{3d_1 - 1}$ times since $|A| = 3d_1 + 1$ means that besides r and s, A has $3d_1 - 1$ elements chosen from the first $3d - 4$ elements of $\mathcal{I}$.

To complete the proof of (7.39), one shows by a similar argument that the intersection of ξ with the right hand side of (7.41) is

$$\sum_{\substack{d = d_1 + d_2 \\ d_1, d_2 > 0}} N_{d_1} N_{d_2} d_1^2 d_2^2 \binom{3d-4}{3d_1 - 2}.$$

We leave the details to the reader—this is a good exercise in working with the axioms. Given the linear equivalence (7.41), we know that ξ has the same intersection with each side, and the recursion relation (7.39) follows immediately.

In the argument just presented (due to Kontsevich), notice that intersecting (7.41) with ξ gave N_d on one side but not the other. This was crucial for the proof of (7.39), but happens much more generally. In fact, the proof of the Reconstruction Theorem proceeds similarly, by intersecting (7.41) with various Gromov-Witten classes to show how general tree-level invariants can be expressed in terms of the special ones listed in the statement of Theorem 7.4.1.

There are other ways to prove the recursion relation for N_d. For example, in Chapter 8, we will see how the machinery of quantum cohomology automates and simplifies the argument given above. In particular, the original rigorous proof [**RT1**] used the symplectic formulation of quantum cohomology. One can also give a more geometric version of the argument which dispenses with the formality of Gromov-Witten classes but follows the spirit of what we did above (see [**FP**]). Alternatively, one can give a completely self-contained argument using other algebro-geometric methods (see [**Ran1, CH1, CH2, Vakil3**]). Generalizations of the numbers N_d are discussed in [**DI, EK1**].

Finally, we should mention that these methods give nice answers in other varieties with transitive linear group actions. For example, tree-level Gromov-Witten invariants for $\mathbb{P}^3$ and a smooth quadric threefold are computed in [**FP**], and the case of $\mathbb{P}^n$ is discussed in [**Vakil3, GPa**]. Other examples may be found in [**DI**].

7.4.3. Genus Zero Invariants of Calabi-Yau Threefolds. We will discuss tree-level Gromov-Witten invariants for Calabi-Yau threefolds. Here, V will denote a smooth Calabi-Yau threefold. If $\beta \in H_2(V, \mathbb{Z})$ is nonzero, we put $N_\beta = \langle I_{0,0,\beta} \rangle$.

PROPOSITION 7.4.2. *The genus 0 Gromov-Witten invariants of V are completely determined by the N_β and the intersection numbers on V.*

PROOF. We may restrict our attention to those Gromov-Witten invariants $\langle I_{0,n,\beta} \rangle (\alpha_1, \ldots, \alpha_n)$ for which the α_i are homogeneous. By the Degree Axiom, we can assume that $\sum_i \deg \alpha_i = 2n$. Since either α_i is the fundamental class or $\deg \alpha_i \geq 2$, there are two cases to consider.

- One of the α_i is the fundamental class, or
- $\deg \alpha_i = 2$ for all i.

In the first case, the Fundamental Class Axiom, Equivariance Axiom, and Point Mapping Axiom say that the Gromov-Witten invariant is an intersection number or 0. In the second case, the Divisor Axiom gives

$$\langle I_{0,n,\beta} \rangle (\alpha_1, \ldots, \alpha_n) = N_\beta \int_\beta \alpha_1 \cdots \int_\beta \alpha_n$$

if $\beta \neq 0$, while if $\beta = 0$, (7.35) implies that $\langle I_{0,n,0} \rangle = 0$ for $n \neq 3$ and that

$$\langle I_{0,3,0} \rangle (\alpha_1, \alpha_2, \alpha_3) = \int_V \alpha_1 \cup \alpha_2 \cup \alpha_3.$$

This completes the proof of the proposition. $\square$

Example 7.4.3.1. Let $V \subset \mathbb{P}^4$ be a generic quintic threefold and pick a line $\ell \subset V$ (there are 2875 of them). We write $H \in H^2(V, \mathbb{Z})$ for the hyperplane class. Then the only nonzero Gromov-Witten invariants (up to permutation) are

$$\langle I_{0,3,0} \rangle (\alpha_1, \alpha_2, 1) \quad \text{and} \quad \langle I_{0,n,d} \rangle (H, \ldots, H),$$

where for simplicity, we write $\langle I_{0,n,d} \rangle$ instead of $\langle I_{0,n,d[\ell]} \rangle$ in the above. Here, $n, d \geq 0$ are arbitrary.

By the Fundamental Class Axiom, $\langle I_{0,3,0} \rangle (\alpha_1, \alpha_2, 1) = \int_V \alpha_1 \cup \alpha_2$. Putting $N_d = \langle I_{0,0,d} \rangle$, we get $\langle I_{0,n,d} \rangle (H, \ldots, H) = d^n N_d$ by the Divisor Axiom if $d > 0$. Finally, $\langle I_{0,3,0} \rangle (H, H, H) = \int_V H^3 = 5$ by the Point Mapping Axiom.

For the quintic threefold V, the above example reduces finding Gromov-Witten invariants to the problem of determining $N_d - \langle I_{0,0,d} \rangle$, which by Example 7.1.6.1 is given by

$$N_d = \langle I_{0,0,d} \rangle = \int_{\overline{\mathcal{M}}_{0,0}(\mathbb{P}^4, d)} c_{5d+1}(\mathcal{V}_d).$$

The desire, of course, is to compute N_d in terms of rational curves on V. This is still an open problem, due to the Clemens conjecture and the existence of multiple covers. We will discuss each of these and explain their relation to the problem of computing N_d.

As discussed in Chapter 2, one of the striking aspects of mirror symmetry is the predictions it makes concerning rational curves on the quintic threefold V. The difficulty is that we don't know in general if V has finitely many rational curves of a given degree. This leads to the Clemens conjecture, which goes as follows.

CONJECTURE 7.4.3. *Let* $V \subset \mathbb{P}^4$ *be a generic quintic threefold. Then for each degree* $d \geq 1$, *we have:*

(*i*) *There are only finitely many irreducible rational curves* $C \subset V$ *of degree* d.
(*ii*) *These curves, as we vary over* all *degrees, are disjoint from each other.*
(*iii*) *If* $f : \mathbb{P}^1 \to C$ *is the normalization of an irreducible rational curve* $C \subset V$, *then the normal bundle* N_f *is isomorphic to* $\mathcal{O}_{\mathbb{P}^1}(-1) \oplus \mathcal{O}_{\mathbb{P}^1}(-1)$.

Note that it is *not* claimed that the rational curves are smooth. That assertion is false, since 6-nodal rational plane quintic curves $C \subset V$ exist on a generic V [**Vainsencher**]. It is known that there are finitely many rational curves on V of degree less than or equal to 9. Also, as predicted by the conjecture, they are all disjoint and are all are nonsingular with the exception of the plane quintic curves C mentioned above. Furthermore, the smooth ones have normal bundle $\mathcal{O}_{\mathbb{P}^1}(-1) \oplus \mathcal{O}_{\mathbb{P}^1}(-1)$. All of this is proved in [**Katz2, JK1**]. In Section 9.2.3, we will show that the singular quintic curves in V also have the predicted normal bundle, so that Conjecture 7.4.3 is true for $d \leq 9$.

But even if the Clemens conjecture were true for all d, we would still have the problem of *multiple covers*. Suppose that we have an irreducible rational curve $C \subset V$ of degree $\frac{d}{k}$, with normalization map $f : \mathbb{P}^1 \to V$. Composing f with an arbitrary degree k map $g : \mathbb{P}^1 \to \mathbb{P}^1$, we get a stable map $f \circ g : \mathbb{P}^1 \to V$ with $(f \circ g)_*[\mathbb{P}^1] = k\frac{d}{k} = d$. Since g varies in a family, we get a family of stable maps which is positive dimensional for $k > 1$. This shows that $\overline{M}_{0,0}(V, d)$ has positive dimension, even though the expected dimension (7.7) is 0 since V is a Calabi-Yau threefold.

We need to determine the contribution of the above multiple cover family to the virtual fundamental class $[\overline{M}_{0,0}(V, d)]^{\mathrm{virt}}$ and hence to N_d. In Section 7.4.4, we will see that when $C \subset V$ is a smooth rational curve of degree $\frac{d}{k}$ and normal bundle $N \simeq \mathcal{O}_{\mathbb{P}^1}(-1) \oplus \mathcal{O}_{\mathbb{P}^1}(-1)$, the degree k multiple covers contribute k^{-3} to N_d.

For the moment, let's be naive and assume that for all d, V has only finitely many rational curves of degree d, all of which are smooth, disjoint and have normal bundle $\mathcal{O}_{\mathbb{P}^1}(-1) \oplus \mathcal{O}_{\mathbb{P}^1}(-1)$. Then, if n_d is the total number of such curves of degree d, the above discussion suggests that

$$(7.45) \qquad N_d = \sum_{k \mid d} n_{\frac{d}{k}} k^{-3}.$$

since every rational curve of degree $\frac{d}{k}$ contributes k^{-3} to N_d. Using the Divisor Axiom, (7.45) implies

$$(7.46) \qquad \langle I_{0,3,d} \rangle (H, H, H) = d^3 \langle I_{0,0,d} \rangle = d^3 N_d = \sum_{k \mid d} n_{\frac{d}{k}} \left(\frac{d}{k} \right)^3 = \sum_{k \mid d} n_k k^3$$

for $d \geq 1$. This allows us to make contact with the formulas of Chapter 2 as follows. Recall from (2.1) that the A-model correlation function is given by

$$(7.47) \qquad \langle H, H, H \rangle = 5 + \sum_{d=1}^{\infty} n_d d^3 \frac{q^d}{1 - q^d}.$$

Since the right hand side of this equation can be written as

$$5 + \sum_{d=1}^{\infty} \left(\sum_{k \mid d} n_k k^3 \right) q^d,$$

the above formula for $\langle I_{0,3,d} \rangle(H, H, H)$ tells us that

$$(7.48) \qquad \langle H, H, H \rangle = \sum_{d=1}^{\infty} \langle I_{0,3,d} \rangle(H, H, H)\, q^d$$

since $\langle I_{0,3,0} \rangle(H, H, H) = 5$ by Example 7.4.3.1.

This is an important development, for although most of the above equations depend on various assumptions, the formula (7.48) for $\langle H, H, H \rangle$ is rigorously defined. This means that we finally have a firm understanding of the A-model correlation function of the quintic threefold. Furthermore, when Chapter 11 proves the wonderful formulas for $\langle H, H, H \rangle$ given in (2.26), it will use this rigorous version of $\langle H, H, H \rangle$.

We can draw two conclusions from this discussion. The first is that the Gromov-Witten invariants $N_d = \langle I_{0,0,d} \rangle$ are precisely what we need in order to formulate mirror symmetry for the quintic threefold. However, the second conclusion is that we still have a *lot* of work to do, for the problem remains of relating N_d to rational curves on V without all of the assumptions made above. Hence we need to study what the formula (7.45) really means. As we will soon see, the answer is more complicated than one might expect.

7.4.4. Instanton Numbers of Calabi-Yau Threefolds. In order to better understand the relation between N_d and rational curves on the quintic threefold, our strategy will be as follows. We will define *instanton numbers* n_d such that (7.45) is automatically true, and then we will tackle the more subtle problem of relating the n_d to the actual number of degree d rational curves on V.

We will work in the following general situation. Let V be a smooth Calabi-Yau threefold, and let $\beta \in H_2(V, \mathbb{Z})$ be nonzero. Then it is easy to see that there exist unique rational numbers n_β, called *instanton numbers*, such that

$$(7.49) \qquad N_\beta = \langle I_{0,0,\beta} \rangle = \sum_{\beta = k\gamma} n_\gamma k^{-3}.$$

The sum is over all $k > 0$ in $\mathbb{Z}$ and γ in $H_2(V, \mathbb{Z})$ with $\beta = k\gamma$. The existence and uniqueness of the n_β is proved as follows. If β is indivisible in $H_2(V, \mathbb{Z})$, we simply put $n_\beta = \langle I_{0,0,\beta} \rangle$. Otherwise, n_β is inductively defined using (7.49). This proves existence, and uniqueness follows easily.

Given this definition of the instanton number n_β, our first task is to motivate the factor of k^{-3} in (7.49). At a purely formal level, this can be explained because it gives nice formulas. For example, suppose that $\beta \in H_2(V, \mathbb{Z})$ is a nonzero effective class and $\alpha_1, \alpha_2, \alpha_3 \in H^2(X, \mathbb{C})$. Then, as a consequence of (7.49) and the Divisor

Axiom, we have

$$\langle I_{0,3,\beta}\rangle(\alpha_1,\alpha_2,\alpha_3) = N_\beta \int_\beta \alpha_1 \int_\beta \alpha_2 \int_\beta \alpha_3$$

$$(7.50) \qquad = \Big(\sum_{\beta=k\gamma} n_\gamma k^{-3} \Big) \int_\beta \alpha_1 \int_\beta \alpha_2 \int_\beta \alpha_3$$

$$= \sum_{\beta=k\gamma} n_\gamma \int_\gamma \alpha_1 \int_\gamma \alpha_2 \int_\gamma \alpha_3,$$

where the last equality follows because $k^{-1}\int_\beta u = \int_\gamma u$ when $\beta = k\gamma$. Note how this generalizes (7.46).

In Chapter 8, we will formalize the mathematical notion of the A-model correlation functions. For a Calabi-Yau threefold V, the correlation function will be given by

$$\langle \alpha_1, \alpha_2, \alpha_3 \rangle = \sum_\beta \langle I_{0,3,d}\rangle(\alpha_1,\alpha_2,\alpha_3)\, q^d,$$

where $\alpha_i \in H^2(V,\mathbb{C})$ and $q^\beta = e^{2\pi i \int_\beta \omega}$ for a complexified Kähler class ω. Using the Point Mapping Axiom and (7.50), this becomes

$$(7.51) \qquad \langle \alpha_1,\alpha_2,\alpha_3 \rangle = \int_V \alpha_1 \cup \alpha_2 \cup \alpha_3 + \sum_{\beta\neq 0} n_\beta \frac{q^\beta}{1-q^\beta} \int_\beta \alpha_1 \int_\beta \alpha_2 \int_\beta \alpha_3$$

by manipulations similar to the derivation of (7.48). This formula plays an important role in mirror symmetry and has a close relation to quantum cohomology. We will discuss (7.51) in more detail in Section 8.1.2.

There are also deeper reasons for the factor of k^{-3} in (7.49). This was mentioned briefly in Section 7.4.3, and we now give a fuller treatment. Suppose $C \subset V$ is a smooth rational curve with normal bundle $N \simeq \mathcal{O}_{\mathbb{P}^1}(-1) \oplus \mathcal{O}_{\mathbb{P}^1}(-1)$, and set $\gamma = [C] \in H_2(V,\mathbb{Z})$. We want to compute the contribution of degree k multiple covers of C to the Gromov-Witten invariant $N_\beta = \langle I_{0,0,\beta}\rangle$, where $\beta = k\gamma$.

Fix an isomorphism $f : \mathbb{P}^1 \simeq C$. Then a degree k cover $g : \mathbb{P}^1 \to \mathbb{P}^1$ gives a stable map $f \circ g \in \overline{M}_{0,0}(V,\beta)$. Since g naturally lies in $\overline{M}_{0,0}(\mathbb{P}^1,k)$, we get an embedding $\overline{M}_{0,0}(\mathbb{P}^1,k) \to \overline{M}_{0,0}(V,\beta)$. Denote the image of this embedding by $M_{k,C}(V)$. We claim that $M_{k,C}(V)$ is in fact a connected component of $\overline{M}_{0,0}(V,\beta)$.

To see this, consider an arbitrary family $f : \mathcal{C} \to V$ of genus 0 stable maps in the homology class β over a connected parameter space S, such that the stable map over $0 \in S$ is in $M_{k,C}(V)$. We let $\pi : \mathcal{C} \to S$ be the structure morphism. Our assumption on the normal bundle N of C implies the existence of an analytic contraction map $\rho : V \to \overline{V}$ which contracts C to a point p and is an isomorphism off C and p. Consider the composition $\tilde{f} = \rho \circ f$ and note that $\tilde{f}(\pi^{-1}(0))$ is the point p. It follows easily that $\tilde{f}(\pi^{-1}(s))$ is a point for all $s \in S$ by [**CKM**, Lemma 1.5]. Since $f_*[\pi^{-1}(s)] = \beta$, it follows that this point must be p as well. This establishes our claim.

The virtual fundamental class $\overline{M}_{0,0}(V,\beta)^{\mathrm{virt}}$ is 0-dimensional, and by definition, $\langle I_{0,0,\beta}\rangle$ is its degree. The curve C contributes to $\langle I_{0,0,\beta}\rangle$ as follows.

THEOREM 7.4.4. *With the above assumptions on C, let $\beta = k\gamma = k[C]$ and let $M_{k,C}(V)$ be the component of $\overline{M}_{0,0}(V,\beta)$ containing all degree k multiple covers of*

C. Then the contribution to $N_\beta = \langle I_{0,0,\beta} \rangle$ of the portion of the virtual fundamental class $[\overline{M}_{0,0}(V,\beta)]^{\mathrm{virt}}$ supported on $M_{k,C}(V)$ is k^{-3}.

PROOF. Historically, this result was first stated as the assertion that the contribution of degree k multiple covers of C to $\langle I_{0,3,\beta} \rangle (\alpha_1, \alpha_2, \alpha_3)$ is

$$(7.52) \qquad \int_\gamma \alpha_1 \int_\gamma \alpha_2 \int_\gamma \alpha_3$$

when $\alpha_i \in H^2(V, \mathbb{C})$. The equivalence of this assertion and Theorem 7.4.4 follows from the derivation of (7.50). In [**CdGP**], this assertion was implicitly conjectured in the way the three-point function (7.47) was written, and Aspinwall and Morrison sketched a proof in [**AM1**]. Their argument must now be regarded as incomplete, since no direct comparison was made to any of the later definitions of the Gromov-Witten invariant.

We will now sketch a proof due to Voisin [**Voisin2**]. As in Section 7.2, we deform the complex structure of V to a generic almost complex structure J and add a generic inhomogeneous term ν. We consider perturbed J-holomorphic maps $g : \mathbb{P}^1 \to V$ with $g_*[\mathbb{P}^1] = \beta$. In the notation of Section 7.2, these maps form the space $\mathcal{M}(\mathbb{P}^1, J, \nu, \beta)$. If J and ν are sufficiently small and U is a sufficiently small neighborhood of C in V, then the subset $\mathcal{M}^U(\mathbb{P}^1, J, \nu, \beta)$ consisting of maps whose image lies in U is a component (not necessarily connected) of $\mathcal{M}(\mathbb{P}^1, J, \nu, \beta)$. Fixing three points $p_1, p_2, p_3 \in \mathbb{P}^1$, we get the evaluation map

$$\pi_1 : \mathcal{M}^U(\mathbb{P}^1, J, \nu, \beta) \longrightarrow V^3$$

which takes g to $(g(p_1), g(p_2), g(p_3))$. The image of this map is 6-dimensional, has a natural orientation, and can be compactified with a boundary of dimension 4 or less. Hence the image of π_1 gives a well-defined homology class in $H^6(V^3, \mathbb{Z})$. The main result of [**Voisin2**] is that since C is rigidly embedded, this class is precisely the homology class of $C \times C \times C$.

Assuming this result, let $\Gamma_1, \Gamma_2, \Gamma_3$ be Poincaré duals of $\alpha_1, \alpha_2, \alpha_3$ in general position. Since the number of points in which $C \times C \times C$ intersects $\Gamma_1 \times \Gamma_2 \times \Gamma_3$ is given by (7.52) (remember that $[C] = \gamma$), the same is true for $\Gamma_1 \times \Gamma_2 \times \Gamma_3$ intersected with the image of π_1. Then the counting definition of the symplectic Gromov-Witten invariant shows that we get a contribution of (7.52) to $\langle I_{0,3,\beta} \rangle (\alpha_1, \alpha_2, \alpha_3)$.

More recent proofs of Theorem 7.4.4 apply to N_β, showing that degree k covers of C contribute k^{-3}. In Section 9.2.2, we will describe Kontsevich's approach to proving this. We will do the case $d = 2$ in detail but refer to [**Manin2**] for the general case. However, Section 9.2.2 will also present a simplified argument due to Pandharipande. Other proofs of this theorem can be found in [**LLY**] and [**Gathmann2**]. We will discuss the [**LLY**] proof in Section 11.1. $\square$

We next discuss two examples which illustrate the sometimes subtle enumerative meaning of the instanton numbers n_β. Even for the quintic threefold, things are not as simple as one would hope.

Example 7.4.4.1. Let $V \subset \mathbb{P}^4$ be a generic quintic threefold. We will write the instanton numbers n_β as n_d, where $\beta = d[\ell]$ and $\ell \subset V$ is a line. These are the numbers which appear in (7.45) and (7.47) and which are computed by mirror symmetry in the formulas given in Chapter 2.

For $d \leq 9$, the relation between n_d and degree d rational curves is especially nice:

(7.53)   If $d \leq 9$, then n_d = number of rational curves of degree d in V.

This follows from Theorem 7.4.4 since the Clemens conjecture is true for $d \leq 9$.

Let's explain why this is true in more detail. Since Conjecture 7.4.3 holds when $d \leq 9$, the right hand side of (7.53) is finite. Let n'_d denote the number of rational curves of degree d lying in V. As we noted in Section 7.4.3, $d \leq 9$ implies that all rational curves of degree d in V are smooth (except for $d = 5$) with normal bundle $\mathcal{O}_{\mathbb{P}^1}(-1) \oplus \mathcal{O}_{\mathbb{P}^1}(-1)$ [**JK1**].

To prove (7.53), first suppose that $d \neq 5$ and let $f : C' \to V$ be a stable map in $\overline{M}_{0,0}(V, d)$. Since the curves are disjoint as we vary over all degrees ≤ 9, $C = f(C')$ must be a smooth curve of degree $\frac{d}{k}$, where k is the degree of $f : C' \to C$. In the notation of Theorem 7.4.4, f is in the component $M_{k,C}(V)$ of $\overline{M}_{0,0}(V, d)$. This shows that $\overline{M}_{0,0}(V, d)$ is the disjoint union of these components, and then Theorem 7.4.4 implies

$$N_d = \sum_{k \mid d} n'_{\frac{d}{k}} k^{-3}.$$

Given the uniqueness of the n_d, $n'_d = n_d$ follows, and (7.53) is proved for $d \neq 5$.

Next suppose that $d = 5$. By [**Vainsencher**], V contains 17,601,000 6-nodal rational curve of degree 5. Now let $f : C' \to V$ be a stable map in $\overline{M}_{0,0}(V, 5)$. In this case, $f(C')$ can be either a line, a smooth curve of degree 5, or a 6-nodal rational curve of degree 5. This decomposes $\overline{M}_{0,0}(V, 5)$ into three types of corresponding components, and Theorem 7.4.4 applies to the first two types. It remains to consider what happens when $f : C' \to C$, where C is a 6-nodal rational curve in V. One easily sees that f is the normalization map $f : \mathbb{P}^1 \to C$. Furthermore, we will show in Section 9.2.2 that f has normal bundle $N_f = \mathcal{O}_{\mathbb{P}^1}(-1) \oplus \mathcal{O}_{\mathbb{P}^1}(-1)$. This implies that the component of $\overline{M}_{0,0}(V, 5)$ corresponding to C is a smooth point and that its contribution to the virtual fundamental class is 1, consistent with Theorem 7.4.4. Then the argument of the previous paragraph implies $n'_5 = n_5$, completing the proof of (7.53).

Based on (7.53), it is natural to hope that proving the Clemens conjecture for higher d would automatically imply that the corresponding instanton number is the number of rational curves of degree d on the quintic threefold. For a long time, many people believed this, but a recent observation of R. Pandharipande shows that this hope is entirely too naive. In fact, for $d = 10$—the first open case of Conjecture 7.4.3—we have the following conjectural formula for n_{10}:

            If the Clemens conjecture is true for all $d \leq 10$, then
(7.54)
            $n_{10} = 6 \times 17,601,000 +$ number of rational curves of degree 10 in V.

We will prove this carefully in Section 9.2.3. The key point is that double covers of the 17,601,000 6-nodal degree 5 curves in V cause complications. More precisely, we will show in Section 9.2.3 that each such curve C contributes $6\frac{1}{8}$ to N_{10} rather than $\frac{1}{8}$. As we will see, this is because double covers $f : C' \to C$ come in two flavors: those for which f factors through the normalization map, and those for which C' is reducible and f is a local isomorphism in a neighborhood of one of the 6 nodes.

Assuming Conjecture 7.4.3 for $d \leq 10$ and that all C as above contribute $6\frac{1}{8}$ to N_{10}, we can derive (7.54) as follows. First, by definition, we have

$$N_{10} = n_1 10^{-3} + n_2 5^{-3} + n_5 2^{-3} + n_{10}.$$

Then, using $n_d = n'_d$ for $d \leq 9$ and the techniques used to prove (7.53), we obtain

$$N_{10} = n_1 10^{-3} + n_2 5^{-3} + (n_5 - 17,601,600) 2^{-3} + 17,601,000 \cdot 6\tfrac{1}{8} + n'_{10},$$

where n'_{10} is the number of rational curves of degree 10. Then (7.54) follows easily by subtracting these two equations.

Once we prove the Mirror Theorem for the quintic threefold V in Chapter 11, we will be able to compute as many of the instanton numbers n_d as desired. But in light of the previous example, this still leaves us far from knowing the number of rational curves of degree d on V. We first have the nontrivial problem of proving the Clemens conjecture for $d \geq 10$ (see [**JK2**] for some preliminary steps for $10 \leq d \leq 24$), and then we have the difficulty that not all of these curves are smooth. As we saw above, multiple covers of such curves do not always contribute according to Theorem 7.4.4. In fact, it is possible that this problem arises whenever $d \geq 10$ is a multiple of 5. So even though the formulas of Chapter 2 have been completely proved, the quintic threefold still has many unsolved problems to offer.

We next give an example which has even more subtle instanton numbers. Here, we will see that the virtual fundamental class can enter in other ways besides multiple covers.

Example 7.4.4.2. Let $\widehat{V} \subset \mathbb{P}(1,1,1,6,9)$ be the hypersurface of degree 18 defined by

$$x_0^{18} + x_1^{18} + x_2^{18} + x_3^3 + x_4^2 = 0.$$

One can show that $\widehat{V}$ has a unique singular point at $(0,0,0,-1,1)$, and resolving this singularity gives a Calabi-Yau threefold V. The exceptional locus is a divisor $E \simeq \mathbb{P}^2$ lying in V. This is what happens generically, and $\mathbb{P}^2 \subset V$ implies that V has infinitely many rational curves of all degrees.

In order to compute Gromov-Witten invariants, we need to describe some classes on V. Following [**CFKM**], let L and H be the divisor classes on V defined by equations of degrees 1 and 3 respectively. Because of the blow-up, one has $3L + E \in |H|$, and one can show that L and H span $H^2(V, \mathbb{Z})$ and generate the Kähler cone. Furthermore, $\ell = L \cap H$ is a line in $E \simeq \mathbb{P}^2$, and $h = L_1 \cap L_2$ is an elliptic curve, where $L_i \in |L|$ are generic. Then ℓ and h generate $H^4(V, \mathbb{Z}) \simeq H_2(V, \mathbb{Z})$.

In this situation, let $\langle I_{0,3,r,s} \rangle$ denote $\langle I_{0,3,[rh+s\ell]} \rangle$. Then (7.50) becomes

$$\langle I_{0,3,r,s} \rangle (\alpha_1, \alpha_2, \alpha_3) = \sum_{k | \gcd(r,s)} n(\tfrac{r}{k}, \tfrac{s}{k}) \int_{\frac{r}{k}h + \frac{s}{k}l} \alpha_1 \int_{\frac{r}{k}h + \frac{s}{k}l} \alpha_2 \int_{\frac{r}{k}h + \frac{s}{k}l} \alpha_3,$$

where $\alpha_1, \alpha_2, \alpha_3 \in H^2(V, \mathbb{C})$ and $n(r,s)$ is the instanton number for rational curves $C \subset V$ with $C \sim rh + s\ell$.

To see what kinds of numbers we get here, consider

$$\langle I_{0,3,0,2} \rangle (L, L, L) = n(0,1) + n(0,2)2^3.$$

By [**CFKM**], mirror symmetry predicts that $n(0,1) = 3$ and $n(0,2) = -6$, so that $\langle I_{0,3,0,2} \rangle (L, L, L) = -45$. These numbers also follow from Theorem 11.2.16. Hence we have a negative Gromov-Witten invariant.

These numbers can be explained as follows. First, $C \sim \ell$ implies that C is a line in $\mathbb{P}^2$. The moduli space of such C's is the dual $\mathbb{P}^2$. To calculate how this contributes to $n(0,1)$, recall our discussion of the virtual fundamental class from Section 7.1.3. In the situation of Proposition 7.1.8, the construction reduces to finding the degree of the Euler class (or top Chern class) of the obstruction bundle. Furthermore, since we are working with embedded curves, the obstruction space of a line $C \subset \mathbb{P}^2$ is $H^1(N)$, where N is the normal bundle of C. However, $H^1(N)$ is dual to $H^0(N)$ via the pairing

$$H^1(N) \otimes H^0(N) \to H^1(\wedge^2 N) \simeq H^1(T_C^*) = \mathbb{C},$$

so the obstruction bundle is globally $T^*_{\text{moduli}} = T^*_{\mathbb{P}^2}$. It follows that the instanton number is

$$n(0,1) = \int_{\mathbb{P}^2} c_2(T^*_{\mathbb{P}^2}) = (-1)^{\dim(\mathbb{P}^2)}\chi(\mathbb{P}^2) = 3.$$

Similarly, $C \sim 2\ell$ implies that C is a conic in $\mathbb{P}^2$. The moduli space of conics is $\mathbb{P}^5$. Since this is smooth, calculating as above gives

$$n(0,2) = \int_{\mathbb{P}^5} c_5(T^*_{\mathbb{P}^5}) = (-1)^{\dim(\mathbb{P}^5)}\chi(\mathbb{P}^5) = -6.$$

This is the argument given in [**CFKM**], but it is not complete, since the moduli space of stable maps $f : C \to V$ with $f_*[C] = \ell$ is *not* the same as the Hilbert scheme of conics we are working with here.

The computation of $n(0,1)$ can also be done in the symplectic setting—see [**Ruan1**]. The fact that our naive calculation of $n(0,2)$ produces a correct number hints that there may be a generalization of the multiple cover formula for non-isolated curves which allows for a broader geometric description of the n_β in terms of immersed curves rather than in terms of stable maps. Conjecture 7.4.5 below deals with some of the subtleties involved. Further details and calculations for the Calabi-Yau threefold V studied here may be found in [**CFKM**].

We will end our discussion of instanton numbers with a conjectural approach to a symplectic definition of n_β. We fix a symplectic structure on V (e.g., the one defined by a Kähler form). Recall from Section 7.2.2 that $\mathcal{M}(\mathbb{P}^1, J, \beta)$ is the space of simple J-holomorphic maps $f : \mathbb{P}^1 \to V$, where J is a generic almost complex structure tamed by the symplectic form of V. By [**MS**, Chap. 2], each $f \in \mathcal{M}(\mathbb{P}^1, J, \beta)$ is generically immersed in the sense that $\mathbb{P}^1$ has a dense open set U where df is injective and $f^{-1}(f(u)) = \{u\}$ for all $u \in U$. There is a natural action of $\text{PGL}(2, \mathbb{C})$ on $\mathcal{M}(\mathbb{P}^1, J, \beta)$, and we define a *J-holomorphic curve on V of class β* to be an orbit of this action.

We know that $\mathcal{M}(\mathbb{P}^1, J, \beta)$ has real dimension 6 by (7.25). Since this is the dimension of $\text{PGL}(2, \mathbb{C})$ and the action is free, the orbits are the connected components of $\mathcal{M}(\mathbb{P}^1, J, \beta)$. Furthermore, each component $\mathcal{C}$ has a natural orientation, so that we can define the sign $\epsilon(\mathcal{C})$ to be $+1$ or -1 according to whether the map $\phi : \text{PGL}(2, \mathbb{C}) \simeq \mathcal{C}$ is orientation preserving or not, where ϕ is given by the action of $\text{PGL}(2, \mathbb{C})$ on a fixed $f_0 \in \mathcal{C}$. We then get the following conjecture.

CONJECTURE 7.4.5. *Let (V, ω) be a Calabi-Yau threefold with its Kähler form. Fix a generic almost complex structure J tamed by ω. Then:*

(i) *For any homology class $\beta \in H_2(V, \mathbb{Z})$, there are finitely many immersed J-holomorphic curves of genus 0 and class β. These curves are all embedded, and no two intersect (even for different β). Define the integers*

$$n'_\beta = \sum_{C \in \mathcal{M}(\mathbb{P}^1, J, \beta)} \epsilon(C),$$

where the sum is over all connected components C of $\mathcal{M}(\mathbb{P}^1, J, \beta)$.

(ii) *All J-holomorphic maps $C \to V$ of genus 0 are multiple covers of embeddings.*

(iii) *For each β, n_β can be expressed as a sum*

$$n_\beta = n'_\beta + \sum_{\substack{\beta = k\gamma \\ k > 1}} \sum_{C \in \mathcal{M}(\mathbb{P}^1, J, \gamma)} c(k, C),$$

where $c(k, C)$ is an integer determined by the degree k multiple covers of C.

This conjecture says that n_β should be the number of J-holomorphic curves (counted with sign) in the class β, together with certain contributions from multiple covers of degree k when β is divisible by k. In particular, Conjecture 7.4.5 implies that each n_β is an integer. The symplectic definition of $\langle I_{0,0,\beta} \rangle$ shows that n_β is a symplectic invariant, but as we will explain below, it is possible that n'_β is not.

Again, the complication essentially arises from the multiple covers. The spaces $\mathcal{M}(\mathbb{P}^1, J, \beta)$ are not compact as we have remarked in Section 7.2, and multiple covers are needed for the Gromov-Uhlenbeck compactification. As a consequence, it is not even clear that the right hand side of the equality in Conjecture 7.4.5 is a priori a finite sum.

The reader may wonder why we do not conjecture more simply that $n_\beta = n'_\beta$ always. In fact, we know of no examples where this is false. On the other hand, there is an analogous situation in [**Taubes**], which analyzes multiple covers of elliptic curves with trivial normal bundle in symplectic 4-manifolds. In that case, there is a discrete invariant associated to double covers, and the value of this invariant changes the contribution of these covers to the Gromov-Witten invariant. In addition, the space of all J's has certain real codimension 1 loci called *walls* such that when J crosses a wall, the shift in this contribution is an integer. Additional J-holomorphic curves of class 2β appear upon wall-crossing, keeping the Gromov-Witten invariant unchanged. Furthermore, the shift in the contribution of the multiple covers of the double cover is compensated for by the contribution of the multiple covers of the new curves in the class 2β. We are grateful to the authors of [**IP**] for their explanation of both the work of Taubes and of their own paper, which studies the connection between the invariant defined by Taubes and the Gromov-Witten invariant. This discussion led to our formulation of Conjecture 7.4.5.

It is conceivable that a similar situation occurs for rational curves on Calabi-Yau threefolds—there may be walls in the space of all J's, and if the given complex structure of V lies in a wall, then there may two flavors of "generic", depending on which side of the wall a "generic" J is on. As happens in the $g = 1$ situation mentioned in the previous paragraph, crossing a wall could potentially cause some embedded J-holomorphic curves to turn into multiple covers. The hope is to define a discrete invariant $c(k, C)$ for each $k > 1$ and embedded J-holomorphic curve C such that when we cross a wall, the loss in n'_β is compensated for by the gain in $c(k, C)$ in such a way that the sum in Conjecture 7.4.5 remains unchanged.

There is one case where part of Conjecture 7.4.5 can be proved. Suppose that β is *primitive*, meaning that it is not a multiple of another class in $H_2(V,\mathbb{Z})$ by an integer > 1.

PROPOSITION 7.4.6. *When* $\beta \in H_2(V,\mathbb{Z})$ *is nonzero and primitive, then*

$$n_\beta = n'_\beta,$$

where n'_β *is as defined in Conjecture 7.4.5.*

PROOF. This follows from [**Ruan2**]. If we use the Li-Ruan-Tian definition of Gromov-Witten invariant, then $n_\beta = \langle I_{0,0,\beta}\rangle^{\text{LRT}}$ in the notation of Section 7.3.2. By (7.50), the n_β can be recovered from $\langle I_{0,3,\beta}\rangle^{\text{LRT}}$. Since V is semi-positive, we can use the invariants $\Psi_{0,3,\beta}$ in place of the $\langle I_{0,3,\beta}\rangle^{\text{LRT}}$. Let $\alpha_1, \alpha_2, \alpha_3 \in H^2(X,\mathbb{Z})$. In terms of the symplectic Gromov-Witten invariants Ψ and Φ defined in Section 7.2.3, we have

$$\Psi_{0,3,\beta}([pt]; \alpha_1, \alpha_2, \alpha_3) = \Phi_{0,\beta}(a,b,c|)$$

by (7.31).

We can compute $\Phi_{0,\beta}(\alpha_1, \alpha_2, \alpha_3|)$ using the methods of [**Ruan2**] because β primitive implies that all J-holomorphic maps $f : \mathbb{P}^1 \to V$ with $f_*[\mathbb{P}^1] = \beta$ are simple. Thus $\mathcal{M}(\mathbb{P}^1, J, \beta)$ gives *all* J-holomorphic $\mathbb{P}^1$'s in the class β. Then consider the evaluation map

$$\mathcal{M}(\mathbb{P}^1, J, \beta) \times_{\text{PGL}(2,\mathbb{C})} (\mathbb{P}^1)^3 \longrightarrow V^3$$

defined by $(f, p_1, p_2, p_3) \mapsto (f(p_1), f(p_2), f(p_3))$. The closure of the image of this map is a 6-cycle Γ, and by [**MS**, Sect. 7.2], $\Phi_{0,\beta}(\alpha_1, \alpha_2, \alpha_3|)$ is the intersection number $\Gamma \cdot (\Gamma_1 \times \Gamma_2 \times \Gamma_3)$, where as in the proof of Theorem 7.4.4, $\Gamma_1, \Gamma_2, \Gamma_3$ are 4-cycles dual to $\alpha_1, \alpha_2, \alpha_3$.

We know that $\mathcal{M}(\mathbb{P}^1, J, \beta)$ is a union of connected components $\mathcal{C}$. If we fix $f_0 \in \mathcal{C}$, then acting on f_0 by $\text{PGL}(2,\mathbb{C})$ gives an orientation-preserving isomorphism $\mathcal{C} \simeq \epsilon(\mathcal{C})\text{PGL}(2,\mathbb{C})$, where $\epsilon(\mathcal{C})$ determines how to orient $\text{PGL}(2,\mathbb{C})$. Using this, the above map becomes

$$f_0 \times f_0 \times f_0 : \epsilon(\mathcal{C})(\mathbb{P}^1)^3 \simeq \mathcal{C} \times_{\text{PGL}(2,\mathbb{C})} (\mathbb{P}^1)^3 \longrightarrow V^3.$$

Since $f_{0*}[\mathbb{P}^1] = \beta$, it follows that the intersection number of the image of this map with $\Gamma_1 \times \Gamma_2 \times \Gamma_3$ is $\epsilon(\mathcal{C}) \int_\beta \alpha_1 \int_\beta \alpha_2 \int_\beta \alpha_3$. Adding this up over all components gives the required result. $\square$

Similarly, Conjecture 7.4.5 is true for the quintic threefold for all degrees d such that $d \leq 9$, $d \neq 5$. This is because we can already achieve a sufficiently generic situation with J integrable, and then the argument of Example 7.4.4.1 applies to yield $n_d = n'_d$, where the n'_d are defined in Conjecture 7.4.5.

7.4.5. Genus One Invariants. When $g = 1$, we get *elliptic Gromov-Witten invariants* $\langle I_{1,n,\beta}\rangle(\alpha_1, \ldots, \alpha_n)$. These invariants were first discussed in [**BCOV1**] and since then have been studied in various contexts. The examples given below illustrate the often subtle task of determining the enumerative significance of a Gromov-Witten invariant.

Example 7.4.5.1. We begin by studying elliptic Gromov-Witten invariants for the projective plane $\mathbb{P}^2$. If $\ell \subset \mathbb{P}^2$ is a line, we will as usual write $\langle I_{1,n,d}\rangle$ instead of $\langle I_{1,n,d[\ell]}\rangle$.

The first invariant to consider is $\langle I_{1,1,0}\rangle$, which is surprisingly nontrivial. We are dealing with constant maps (the degree is zero), but the Point Mapping Axiom from Section 7.3 doesn't apply. However, as explained in [**KoM1**], there are versions of this axiom for every genus. For $g = 1$ and X arbitrary, one gets

$$(7.55) \qquad \langle I_{1,1,0}\rangle(\alpha) = -\frac{1}{24}\int_X c_{\dim X -1}(T_X)\cup\alpha$$

for $\deg\alpha = 2$. The problem is that the space of n-pointed genus 1 degree 0 maps $f : C \to X$ is $\overline{M}_{1,n}\times X$, which doesn't have the expected dimension. Thus one must compute the virtual fundamental class as in Section 7.1.5. This can be done using Proposition 7.1.8, as will be spelled out in Example 10.1.3.3, and (7.55) follows easily.

Applying (7.55) to $X = \mathbb{P}^2$, we obtain

$$\langle I_{1,1,0}\rangle([\ell]) = -\frac{1}{24}\int_{\mathbb{P}^2} c_1(T_{\mathbb{P}^2})\cup[\ell] = -\frac{1}{8}.$$

So Gromov-Witten invariants can be both fractional *and* negative!

We will consider an interesting related calculation in Example 10.1.3.3 after we generalize Gromov-Witten invariants to *gravitational correlators*.

Fortunately, the other elliptic invariants have a nice enumerative meaning. For $d \geq 1$, define

$$N_d^1 = \langle I_{1,3d,d}\rangle(\underbrace{[pt],\ldots,[pt]}_{3d \text{ times}}).$$

One can prove that N_d^1 is the number of elliptic curves in $\mathbb{P}^2$ of degree d which go through $3d$ generic points. Thus $N_1^1 = N_2^1 = 0$ and $N_3^1 = 1$ since there is a unique (smooth) elliptic curve through 9 generic points in the plane. For $d > 3$, all of the curves counted by N_d^1 are singular. It is easy to show that all of the elliptic Gromov-Witten invariants are determined once we know $\langle I_{1,1,0}\rangle([\ell])$ and N_d^1 for $d \geq 1$.

The main results concerning N_d^1 are various recursion relations. One example can be found in [**Getzler1**]. An elegant recursion was predicted in [**EHX**] using the method of Virasoro constraints which we will briefly discuss in Section 10.1.4. The formula, which was proven in [**Pandharipande2**], says that if N_d is the number of rational curves given by (7.38), then

$$(7.56) \qquad N_d^1 = \frac{1}{12}\binom{d}{3}N_d + \sum_{\substack{d=d_1+d_2\\d_1,d_2>0}} \frac{3d_1^2 d_2 - 2d_1 d_2}{9} N_{d_1} N_{d_2}^1 \binom{3d-1}{3d_1-1}.$$

The proof in [**Pandharipande2**] uses the geometric construction of a relation between codimension 2 strata in $H^4(\overline{M}_{1,4},\mathbb{Q})$ discovered by Getzler, and then applies the WDVV equation to be discussed in Chapter 8. Thus we can compute as many N_d^1 as desired. See [**Ran3**] for an elementary proof.

In the elliptic invariants considered thus far, we have allowed the j-invariant to vary. However, one can also study elliptic curves in $\mathbb{P}^2$ with *fixed* j invariant. More precisely, let $E_{d,j}$ be the number of irreducible elliptic curves of given j-invariant and degree d going through $3d - 1$ generic points in $\mathbb{P}^2$. In [**Ionel, Pandharipande1**], it is shown that

$$
(7.57) \qquad E_{d,j} = \begin{cases} \binom{d-1}{2} N_d & \text{if } j \neq 0, 1728 \\ \frac{1}{3}\binom{d-1}{2} N_d & \text{if } j = 0 \\ \frac{1}{2}\binom{d-1}{2} N_d & \text{if } j = 1728, \end{cases}
$$

where N_d is as above.

Let's try this calculation using Gromov-Witten classes. For simplicity, we will assume that $j \neq 0, 1728$. Since we want to go through $3d - 1$ points in $\mathbb{P}^2$, we begin with the Gromov-Witten class

$$
\xi = I_{1,3d-1,d}(\underbrace{[pt], \dots, [pt]}_{3d-1 \text{ times}}).
$$

By (7.33), this class has degree $6d - 4$, which is two from the top since $\overline{M}_{1,3d-1}$ has dimension $3d - 1$. If we fix j, we get the divisor $Z_j \subset \overline{M}_{1,3d-1}$ consisting of all pointed curves with this j-invariant. Then, naively, we might expect $E_{d,j}$ to be the intersection of ξ and Z_j, i.e.,

$$
(7.58) \qquad E_{d,j} = \xi \cdot Z_j.
$$

We compute $\xi \cdot Z_j$ as follows. We have the natural map $\overline{M}_{1,3d-1} \to \overline{M}_{1,1} \simeq \mathbb{P}^1$ which forgets the last $3d - 2$ points and contracts non-stable components. Then Z_j is simply the fiber over j since $j \neq 0, 1728$. However, the fiber over ∞ consists of all elements of $\overline{M}_{1,3d-1}$ which map to a nodal curve. One easily sees that this fiber is the image of the map $\psi : \overline{M}_{0,3d+1} \to \overline{M}_{1,3d-1}$ which glues together the last two points. It follows that $Z_j \sim \text{Im}(\psi)$, which means that

$$
\xi \cdot Z_j = \xi \cdot \text{Im}(\psi) = \langle I_{1,3d+1,d} \rangle([pt], \dots, [pt], [pt], [\mathbb{P}^2]) +
$$
$$
\langle I_{1,3d+1,d} \rangle([pt], \dots, [pt], [\ell], [\ell]) +
$$
$$
\langle I_{1,3d+1,d} \rangle([pt], \dots, [pt], [\mathbb{P}^2], [pt]),
$$

where the second equality is by the Reduction Axiom (used for the first time). From here, the Fundamental Class and Divisor Axioms easily imply

$$
\xi \cdot Z_j = \xi \cdot \text{Im}(\psi) = d^2 N_d.
$$

Comparing this to (7.57), we see that (7.58) is wrong. The reason is that $E_{d,j}$ counts only irreducible curves in $\mathbb{P}^2$, while $\xi \cdot Z_j$ also includes reducible curves coming from the boundary of $\overline{M}_{1,3d-1}$. In fact, the argument in [**Ionel**] computes $E_{d,j}$ by first doing the above calculation and then (in the symplectic context) carefully taking the boundary into account. We should also mention that in terms of the mixed invariants and symplectic invariants defined in Section 7.2, we have

$$
\xi \cdot Z_j = \Phi_{1,d}([pt] | \underbrace{[pt], \dots, [pt]}_{3d-2 \text{ times}}) = \Psi_{1,3d-1,d}(Z_j; \underbrace{[pt], \dots, [pt]}_{3d-1 \text{ times}})
$$

(see especially the discussion following (7.30)). A discussion of the enumerative significance of these and related elliptic invariants can be found in Remark 2.19 of [**RT2**].

Example 7.4.5.2. In $\mathbb{P}^3$, let ℓ be a line, and then define the numbers

$$N_{a,b} = \langle I_{0,a+b,(a+2b)/4}\rangle(\underbrace{[\ell],\dots,[\ell]}_{a \text{ times}},\underbrace{[pt],\dots,[pt]}_{b \text{ times}})$$

$$N_{a,b}^1 = \langle I_{1,a+b,(a+2b)/4}\rangle(\underbrace{[\ell],\dots,[\ell]}_{a \text{ times}},\underbrace{[pt],\dots,[pt]}_{b \text{ times}}),$$

where we assume that $(a+2b)/4$ is an integer. By Theorem 7.3.1, we know that $N_{a,b}$ gives the number of rational curves in $\mathbb{P}^3$ of degree $d = (a+2b)/4$ passing through a generic lines and b generic points. Recursion relations for the $N_{a,b}$ can be found in [**FP**].

The elliptic invariants $N_{a,b}^1$ can also be computed using recursion relations given in the Appendix to [**Getzler1**]. But, in contrast to the genus zero case, these are not enumerative invariants. For example, one finds that *all* of the $N_{a,b}^1$ computed in the Appendix to [**Getzler1**] are negative, and many are fractional as well. However, if we form the linear combination

$$E_{a,b} = N_{a,b}^1 + \frac{2d-1}{12}N_{a,b},$$

where as above $d = (a+2b)/4$, then [**GeP**] shows that $E_{a,b}$ is the number of elliptic curves in $\mathbb{P}^3$ of degree d which pass through a generic lines and b generic points. Alternate methods for calculating $N_{a,b}^1$ are also discussed in [**Vakil3**, **GPa**] along with a generalization to $\mathbb{P}^n$.

We should also note that it is also possible to study elliptic curves in $\mathbb{P}^3$ with fixed j-invariant. As for $\mathbb{P}^2$, one gets formulas relating the numbers of such curves to the $N_{a,b}$. Precise formulas can be found in [**Ionel**].

Example 7.4.5.3. Finally, we turn to the generic quintic threefold $V \subset \mathbb{P}^4$. Here, we begin with the elliptic analog of (7.51). Let n_d be the instanton number of rational curves on V of degree d, and let e_d be a similar instanton number of "elliptic curves" on V of degree d (the quotation marks will be explained below). We will assume that these numbers are finite and think of the n_d and e_d as counting appropriate curves that are rigidly embedded. Then one has the formula

$$(7.59) \qquad \langle I_{1,1,d}\rangle(H) = \sum_{k|d}\sigma\left(\tfrac{d}{k}\right)ke_k + \tfrac{1}{12}\sum_{k|d}kn_k,$$

where $\sigma\left(\tfrac{d}{k}\right) = \sum_{l|\frac{d}{k}}l$ is the sum of the divisors of $\frac{d}{k}$.

This formula was arrived at through physical reasoning [**BCOV1**], but so far, there is no mathematical proof. However, we can give an intuitive explanation of the formula as follows. We begin with the first sum of (7.59). Let $f : (C,p) \to V$ be a stable curve whose image $E = f(C)$ is an elliptic curve such that $C \to E$ has degree $\frac{d}{k}$ and $E \subset V$ has degree k. Then E is one of our e_k elliptic curves, and since $f(p) \in E \cap H$, there are k choices for $f(p)$. Given $f(p)$, there are $\frac{d}{k}$ choices for $p \in C$, but these are all isomorphic because $C \to E$ is a covering space. It remains to count the number of covering spaces of E of degree $\frac{d}{k}$. This means counting the number of sublattices of $\mathbb{Z}^2$ of index $\frac{d}{k}$, which is well known to equal $\sigma\left(\frac{d}{k}\right)$.

The second sum of (7.59) is more subtle. For instance, consider the case $k = d$. How do the n_d "rational curves" of degree d contribute to $\langle I_{1,1,d}\rangle(H)$? To answer this question, let $C_0 \subset V$ be rational of degree d. As usual, there are d possibilities for $f(p) \in C_0 \cap H$. Given C_0 and $f(p)$, we want to construct genus 1 stable maps $f : (C,p) \to C_0$ which have degree 1. This might seem impossible, except that the definition of stable map allows C to be reducible. This is what has been called a *degenerate instanton* in the physics literature. Here, $C = E \cup C_0$, where E is an elliptic curve joined to C_0 at one point, and $f : C \to C_0$ is the identity on C_0 and constant on E. Once we pick $p \in C$ mapping to $f(p)$, we get a family of stable maps of the sort we're interested in.

When $k|d$ is a proper divisor of d in the second sum of (7.59), things are even more complicated, mainly because of multiple covers. As in the genus zero case, one has rational multiple covers to worry about, and in addition, we have multiple covers arising from the well known degree 2 map from an elliptic curve to $\mathbb{P}^1$. Calculations from physics [**BCOV1**] suggest that the answer is again $\frac{1}{12}$, and this has been proven mathematically when the normal bundle of the underlying rational curve is $\mathcal{O}_{\mathbb{P}^1}(-1) \oplus \mathcal{O}_{\mathbb{P}^1}(-1)$ [**GPa**]. Hence, (7.59) holds in degrees d for which the hypotheses of the Clemens conjecture 7.4.3 holds in all degrees dividing d, together with a similar hypothesis on elliptic curves, assumed disjoint from each other and from the rational curves.

A further complication is that even though e_d is probably finite, it may count more than just smooth elliptic curves on V. This is why we put "elliptic curves" in quotation marks. The best illustration of what's happening is given by e_4, the number of quartic "elliptic curves" on the quintic threefold. In [**BCOV1**], this number is computed to be

$$e_4 = 3721431625.$$

However, by [**ES2**], the number of smooth quartic elliptic curves on a generic quintic threefold is

$$3718024750.$$

This differs from e_4. As explained in [**ES2**], the discrepancy arises because e_4 allows for *singular* curves. In particular, V contains a number of binodal quartic plane curves, and each of these gives rise to a stable map contributing to e_4. These singular curves are counted as follows. If ℓ is one of the 2875 lines on V, then any plane $P \subset \mathbb{P}^4$ containing ℓ intersects V in $\ell \cup C$, where C is a plane quartic. Generically, C is smooth, but it acquires two nodes for a finite number of special planes P. In [**Vainsencher**], it is shown that there are 1185 special planes P. Since

$$3721431625 = 3718024750 + 1185 \times 2875,$$

the numbers are consistent. This gives another example of the care required in understanding the enumerative significance of a Gromov-Witten invariant.

7.4.6. Other Examples. Besides the examples listed so far, there has been a *lot* of other work on Gromov-Witten invariants and enumerative geometry. Rather than continue with an already long list of examples, we will instead indicate some references for further reading.

The enumerative aspect of mirror symmetry goes back to 1991 with the stunning predictions about rational curves on quintic threefolds appearing in [**CdGP**]. This led to a host of papers making enumerative predictions about Calabi-Yau manifolds, including physics papers [**BCOV1, BCOV2, CdFKM, CFKM, HKTY1, HKTY2, BKK, AM2, Font, KT1, KT2**] and mathematics papers [**Katz2, Morrison1, ES2, ES1, LTe, BvS, Voisin1, Meurer, HSS, Givental2, LLY, Givental4, BCFKvS2**]. An nice survey of the enumerative aspects of mirror symmetry can be found in [**Morrison6**].

Although many unsolved problems remain concerning the enumerative geometry of Calabi-Yau manifolds, much of the more recent work has dealt with other types of algebraic varieties. This is partly due to the definition of Gromov-Witten invariants for arbitrary smooth projective varieties, as described in Section 7.1, which led to many of the examples and papers mentioned above. In addition, there have been applications to rational surfaces [**CM, GöP, Vakil1**], Del Pezzo surfaces [**Vakil2**], blow-ups of $\mathbb{P}^n$ [**Gathmann1**] and of convex varieties [**Gathmann2**], complete intersections [**Beauville1**], and nodal genus-2 curves in $\mathbb{P}^2$ [**KQR**]. At the same time, people have been inspired to see what can be done without using Gromov-Witten theory, which resulted in papers by Crauder and Miranda [**CM**], Ran [**Ran1, Ran3**] and Caporaso and Harris [**CH1, CH2, CH3**] (see [**Caporaso**] for a summary). There have also been studies of the Clemens conjecture on quintic threefolds and other related Calabi-Yau threefolds [**JK1, JK2, EJS**]. Taking *all* of these papers into account, we see that there has been a surge of interest in enumerative geometry in recent years. It may have started with mirror symmetry, but it is now taking on a life of its own within algebraic geometry.

This brings us to the end of a long chapter. We have seen that Gromov-Witten invariants provide a nontrivial link to symplectic geometry and play an important role in contemporary algebraic geometry. One of the morals of this chapter is that counting curves is deeper and richer than one might expect. However, there is more to the story, for we have yet to learn about quantum cohomology. As we will see in Chapter 8, this will give further insight into the examples considered here and will complete the circle of ideas needed to understand the mathematical formulation of mirror symmetry.

CHAPTER 8

Quantum Cohomology

In this chapter, we will learn that there are actually two types of quantum cohomology—the so-called *big* and *small* quantum cohomology rings. We will also study the *Gromov-Witten potential*, which may be thought of as the exponential generating function associated to the tree-level Gromov-Witten invariants. The WDVV equation satisfied by the Gromov-Witten potential leads naturally to the associativity of the big quantum cohomology ring and also has some interesting consequences concerning the examples studied in Chapter 7. Another important topic is the Dubrovin formalism, which shows how quantum cohomology gives a variation of Hodge structure on the cohomology ring.

Quantum cohomology is the final ingredient needed in order to understand the mathematical version of mirror symmetry. As we will see in this chapter, it gives a clear statement of the three-point function. Furthermore, it leads to the variation of Hodge structure mentioned in the last paragraph, which is a variation over the Kähler moduli. One of the key ideas of mirror symmetry is that this variation of Hodge structure corresponds, via the mirror map, to the geometric variation of Hodge structure over the complex moduli space of the mirror family. This will be explained in detail when we define mathematical mirror pairs and discuss the Mirror Conjecture at the end of the chapter.

Historically, the idea of quantum cohomology first appeared in the physics literature in the guise of the *chiral ring* of a Calabi-Yau threefold. The chiral ring was understood as a deformation of the ordinary cohomology ring [**LVW**], agreeing with the ordinary cohomology ring at large radius limit points. The quantum product can be understood more generally in the context of topological quantum field theories [**Witten1, Witten2**]. This will be discussed in Appendix B.6. The first mathematical construction of quantum cohomology was for semi-positive symplectic manifolds [**RT1, MS**].

Our approach to quantum cohomology in this chapter will be almost exclusively algebraic. Expository accounts of quantum cohomology can be found in [**Aluffi, FP, MS, Morrison7, Voisin3**].

8.1. Small Quantum Cohomology

The small quantum cohomology ring of a smooth projective variety X is defined using the Gromov-Witten invariants $\langle I_{0,3,\beta} \rangle$ from Chapter 7.

8.1.1. Definition and Properties. Given a basis $T_0 = 1, T_1, \ldots, T_m$ for $H^*(X, \mathbb{Q})$ consisting of homogeneous elements, let $g_{ij} = \int_X T_i \cup T_j$, and then $(g^{ij}) = (g_{ij})^{-1}$ is the inverse matrix. Recall from Chapter 0 that $\omega \in H^2(X, \mathbb{C})$ is a complexified Kähler class if the imaginary part of ω is Kähler.

DEFINITION 8.1.1. *Let ω be a complexified Kähler class on a smooth projective variety X. Then, for $a, b \in H^*(X, \mathbb{C})$, define*

$$a * b = \sum_{i,j} \sum_{\beta \in H_2(X, \mathbb{Z})} \langle I_{0,3,\beta} \rangle (a, b, T_i)\, g^{ij}\, q^{\beta}\, T_j,$$

where $q^{\beta} = e^{2\pi i \int_{\beta} \omega}$. We call $a * b$ the small quantum product *of a and b.*

As written, this definition is admittedly imprecise, as the sum over β could be infinite, so a convergence result is needed. In Section 8.1.3, we will explain several ways to give rigorous meaning to Definition 8.1.1, replacing q^{β} by an appropriate formal expression. This is in fact the common practice. Nevertheless, we prefer to give the above incomplete definition because it makes the small quantum product depend explicitly on Kähler moduli, which is one of the central ideas of quantum cohomology. We ask the reader's indulgence in "suspending disbelief" until we get to Section 8.1.3, at which point the reader will see that all of the claims of this section and the next, suitably interpreted, are perfectly rigorous.

We will use physics notation for "raising indices" and define

$$T^i = \sum_j g^{ij}\, T_j,$$

so that $T^0, \ldots, T^m$ form the dual basis of $T_0, \ldots, T_m$ satisfying $\int_V T^i \cup T_j = \delta_{ij}$. Then we can write the small quantum product more simply as

$$(8.1) \qquad a * b = \sum_{i} \sum_{\beta \in H_2(X, \mathbb{Z})} \langle I_{0,3,\beta} \rangle (a, b, T_i)\, q^{\beta}\, T^i,$$

We will use this notation throughout this chapter and the next.

In Chapter 7, Gromov-Witten invariants were defined for rational cohomology classes. In the above formula for $*$, we extend the definition to complex coefficients by linearity so that $\langle I_{0,3,\beta} \rangle (a, b, T_i)$ makes sense for $a, b \in H^*(X, \mathbb{C})$. Also note that for $\beta \in H_2(X, \mathbb{Z})$ and a, b homogeneous, the degree condition (7.33) implies that the Gromov-Witten invariant $\langle I_{0,3,\beta} \rangle (a, b, T_i)$ in Definition 8.1.1 vanishes unless

$$(8.2) \qquad \deg a + \deg b + \deg T_i = 2 \dim X - 2 \int_{\beta} \omega_X,$$

where ω_X is the canonical class on X. Thus the sum in the above definition is over these β's. One can also check that the binary operation $*$ doesn't depend on which basis T_i we use.

In the case of a Calabi-Yau manifold, we have the following conjecture.

CONJECTURE 8.1.2. *For a Calabi-Yau manifold, the sum in Definition 8.1.1 converges provided the imaginary part of ω is sufficiently large.*

For the time being, we will assume this conjecture. As stated earlier, Section 8.1.3 will discuss how to avoid convergence problems by reinterpreting the q^{β}. Fortunately, there are some projective manifolds where the above sum is known to behave nicely, giving a rigorous meaning to the small quantum product in terms of Kähler moduli.

PROPOSITION 8.1.3. *Let X be a smooth projective variety such that either*

    (i) X is Fano, or

    (ii) X has a transitive action by a semisimple Lie group.

Then the sum in Definition 8.1.1 is finite for any $a, b \in H^2(X, \mathbb{C})$.

PROOF. We need to show that there are only finitely many classes β on X which satisfy (8.2) and are represented by a genus zero stable map. First suppose that X is Fano. Neglecting torsion (which is finite), the β's represented by genus zero stable maps are lattice points in $H_2(X, \mathbb{R})$. Being Fano means that $-\omega_X$ is in the interior of the Kähler cone, so that the linear functional $\beta \mapsto -\int_\beta \omega_X$ is positive on the closure of the cone generated by the effective β's. It follows easily that only finitely many β's can satisfy $-\int_\beta \omega_X < C$ for any constant C. By (8.2), this proves the proposition when X is Fano.

Next suppose that X is a homogeneous space G/P. Then the cohomology of X is known, and the classes β represented by genus zero stable maps are non-negative integer combinations of finitely many classes $\beta_1, \dots, \beta_r$ such that each β_i is the image of a nonconstant map $\mu_i : \mathbb{P}^1 \to X$. By (8.2), it suffices to show that $\int_{\beta_i} \omega_X < 0$, or equivalently, that $\int_{\beta_i} c_1(T_X) > 0$.

The transitive group action implies that T_X is generated by its global sections, which in turn implies that $\mu_i^*(T_X)$ is also generated by global sections. Since $\mu_i^*(T_X)$ is a vector bundle on $\mathbb{P}^1$, $\mu_i^*(T_X) \simeq \oplus_j \mathcal{O}_{\mathbb{P}^1}(a_j)$ where the a_j are all ≥ 0. If we can prove that some $a_j > 0$, then $\int_{\beta_i} c_1(T_X) > 0$ will follow. However, if this weren't true, then $\mu_j^*(T_X)$ would be trivial, and then the natural map $T_{\mathbb{P}^1} \to \mu_i^*(T_X)$ would necessarily be zero since $T_{\mathbb{P}^1} \simeq \mathcal{O}_{\mathbb{P}^1}(2)$. This is impossible since μ_i is nonconstant, and the proposition follows. $\qquad\square$

The second part of the proposition shows that the sum in Definition 8.1.1 is finite for any projective space, Grassmannian, quadric hypersurface, flag manifold, or any product of such varieties.

We next show that $H^*(X, \mathbb{C})$ is a ring under the small quantum product.

THEOREM 8.1.4. *$H^*(X, \mathbb{C})$ is a supercommutative ring with identity under the small quantum product.*

PROOF. By supercommutative, we mean

$$a * b = (-1)^{\deg a \deg b} b * a,$$

similar to cup product. This follows because the Equivariance Axiom from Section 7.3 implies $\langle I_{0,3,\beta} \rangle (a, b, T_i) = (-1)^{\deg a \deg b} \langle I_{0,3,\beta} \rangle (b, a, T_i)$. As for an identity element, let $T_0 = 1 = [X] \in H^0(X, \mathbb{C})$ be the fundamental class of X. This is the identity under cup product, and as we will now show, it is the identity under $*$ as well. We first observe that $\langle I_{0,3,\beta} \rangle (a, T_0, T_i) = 0$ whenever $\beta \neq 0$. This is an easy consequence of the Equivariance and Fundamental Class Axioms. It follows that

$$a * T_0 = \sum_i \langle I_{0,3,0} \rangle (a, T_0, T_i) \, T^i$$
$$= \sum_i \left(\int_X a \cup T_0 \cup T_i \right) T^i$$
$$= a \cup T_0 = a,$$

where the second equality follows from the Point Mapping Axiom and the third from the definition of T^i. Thus T_0 is the identity element for $*$.

The final step is to prove that $*$ is associative. The Splitting Axiom from Section 7.3 will play a key role in the proof. Given homogeneous classes a, b, c in $H^*(X, \mathbb{C})$, we first consider $(a * b) * c$. Expanding via Definition 8.1.1 and using

$T^i = \sum_j g^{ij} T_j$, we get

$$\sum_\beta \sum_{t,u} \left(\sum_{\beta=\beta_1+\beta_2} \sum_{r,s} g^{rs} \langle I_{0,3,\beta_1} \rangle (a,b,T_r) \langle I_{0,3,\beta_2} \rangle (T_s,c,T_t) \right) g^{tu} q^\beta T_u.$$

Now consider the Gromov-Witten class $I_{0,4,\beta}(a,b,c,T_t)$. The Splitting Axiom describes the pullback of this class via the map

$$\varphi : \overline{M}_{0,3} \times \overline{M}_{0,3} \longrightarrow \overline{M}_{0,4} \simeq \mathbb{P}^1$$

defined in (7.36). Note that this pullback vanishes unless

$$I_{0,4,\beta}(a,b,c,T_t) \in H^0(\overline{M}_{0,4}),$$

which henceforth will be assumed. Since φ is the inclusion of a point of $\mathbb{P}^1$, we can regard

$$(8.3) \qquad\qquad \varphi^* I_{0,4,\beta}(a,b,c,T_t)$$

as a number. Then the Splitting Axiom says that (8.3) is *precisely* the quantity in the large parentheses in the above display.

Turning our attention to $a * (b * c)$, we obtain

$$\sum_\beta \sum_{t,u} \left(\sum_{\beta=\beta_1+\beta_2} \sum_{r,s} g^{rs} \langle I_{0,3,\beta_1} \rangle (b,c,T_r) \langle I_{0,3,\beta_2} \rangle (a,T_s,T_t) \right) g^{tu} q^\beta T_u$$

But $\langle I_{0,3,\beta_2} \rangle (a,T_s,T_t) = (-1)^{\deg a \, \deg T_s} \langle I_{0,3,\beta_2} \rangle (T_s,a,T_t)$ by the Equivariance Axiom, and we also have

$$\deg a + \deg T_s + \deg T_t \equiv 0 \bmod 2$$

by (8.2). It follows that

$$\langle I_{0,3,\beta_2} \rangle (a,T_s,T_t) = (-1)^{\deg a(\deg a + \deg T_t)} \langle I_{0,3,\beta_2} \rangle (T_s,a,T_t).$$

Hence the quantity in parentheses in the expression for $a * (b * c)$ is

$$(8.4) \qquad\qquad (-1)^{\deg a(\deg a + \deg T_t)} \, \widetilde{\varphi}^* I_{0,4,\beta}(b,c,a,T_t),$$

where $\widetilde{\varphi}$ is a map similar to φ. Note that $\widetilde{\varphi}$ is the inclusion of another point of $\mathbb{P}^1$, so that it is linearly equivalent to φ.

Comparing the expressions for $(a * b) * c$ and $a * (b * c)$, we see that the theorem will follow immediately once we show that (8.3) and (8.4) are equal. Hence it suffices to show that

$$I_{0,4,\beta}(a,b,c,T_t) = (-1)^{\deg a(\deg a + \deg T_t)} I_{0,4,\beta}(b,c,a,T_t).$$

This is an easy consequence of the Equivariance Axiom and the congruence $\deg a + \deg b + \deg c + \deg T_t \equiv 0 \bmod 2$ obtained from (7.33). $\qquad\square$

Note that the above proof has much in common with the methods used to prove the recursion relation for the Gromov-Witten invariants of $\mathbb{P}^2$ discussed in Section 7.4.2. Here, there were signs to worry about because of the presence of odd-dimensional cohomology classes. An intuitive explanation of associativity can be found in [**Morrison7**].

Besides giving a ring structure, the small quantum product has other properties as well. We first describe the degree of $a * b$. As examples in the next section will reveal, $a * b$ need not be homogeneous, even if a and b are. However, the degrees which appear in $a * b$ aren't entirely arbitrary.

PROPOSITION 8.1.5. *If $a, b \in H^*(X, \mathbb{C})$ are homogeneous, then their small quantum product is a sum $a * b = \sum_l c_l$ of homogeneous classes such that*

$$\deg c_l \equiv \deg a + \deg b \bmod 2.$$

Furthermore, if the class of ω_X in $H^2(X, \mathbb{Z})$ is divisible by an integer r, then

$$\deg c_l \equiv \deg a + \deg b \bmod 2r.$$

PROOF. Since $g^{ij} \neq 0$ only if $\deg T_i + \deg T_j = 2 \dim X$, the degree condition (8.2) implies that in Definition 8.1.1, T_j has a nonzero coefficient only if

$$\deg a + \deg b = \deg T_j - 2 \int_\beta \omega_X.$$

The proposition now follows immediately. $\qquad\square$

We can also describe the interaction of the small quantum product with the intersection pairing on X, which we write as $g(a, b) = \int_X a \cup b$.

PROPOSITION 8.1.6. *For all $a, b, c \in H^*(X, \mathbb{C})$, we have:*

(i) $g(a * b, c) = g(a, b * c) = \sum_{\beta \in H_2(X, \mathbb{Z})} \langle I_{0,3,\beta} \rangle (a, b, c)\, q^\beta.$
(ii) $\int_X a * b = g(a, b).$

PROOF. It is easy to show that $g(T_i * T_j, T_k) = \sum_\beta \langle I_{0,3,\beta} \rangle (T_i, T_j, T_k)\, q^\beta$. Turning to $g(T_i, T_j * T_k)$, supercommutativity of the intersection product implies that

$$g(T_i, T_j * T_k) = (-1)^{\deg T_i (\deg T_j + \deg T_k)} g(T_j * T_k, T_i),$$

and as above we have $g(T_j * T_k, T_i) = \sum_\beta \langle I_{0,3,\beta} \rangle (T_j, T_k, T_i)\, q^\beta$. The first part of the proposition now follows immediately from the Equivariance Axiom. To prove the second part, observe that

$$\int_X a * b = \int_X T_0 \cup (a * b) = g(T_0, a * b) = g(T_0 * a, b) = g(a, b)$$

since T_0 is the identity for cup product and the small quantum product. $\qquad\square$

This proposition implies that $H^*(X, \mathbb{C})$ is a *Frobenius algebra* under $*$ and g (see Definition B.6.2).

Proposition 8.1.6 is also related to a slightly different approach to small quantum cohomology, which begins with the *three-point function* $\langle a, b, c \rangle$ defined by

$$(8.5) \qquad \langle a, b, c \rangle = \sum_{\beta \in H_2(X, \mathbb{Z})} \langle I_{0,3,\beta} \rangle (a, b, c)\, q^\beta.$$

We also call $\langle a, b, c \rangle$ the *A-model correlation function*. It is easy to see that the small quantum product of a and b is the unique element $a * b \in H^*(X, \mathbb{C})$ satisfying

$$g(a * b, c) = \langle a, b, c \rangle \quad \text{for all } c \in H^*(X, \mathbb{C}).$$

In the literature, one often finds quantum cohomology defined this way [**MS, Morrison7, Voisin3**]. Note also that (8.1) and (8.5) imply

$$(8.6) \qquad a * b = \sum_i \langle a, b, T_i \rangle\, T^i$$

which gives another proof that quantum cohomology is determined by the three-point functions. Finally, we should mention that Proposition 8.1.6 implies

$$(8.7) \qquad \langle a, b, c \rangle = \int_X a * b * c,$$

so that the three-point function is expressible completely in terms of the small quantum product.

For a Calabi-Yau manifold, the small quantum product is especially nice.

COROLLARY 8.1.7. *If V is Calabi-Yau, then the small quantum product satisfies $\deg a * b = \deg a + \deg b$. Furthermore, if $a \in H^{p,q}(V)$ and $b \in H^{r,s}(V)$, then $a * b \in H^{p+r,q+s}(V)$.*

PROOF. The first assertion follows immediately from Proposition 8.1.5 since ω_V is trivial. Now suppose that V has dimension d. For the second assertion, it suffices by part (i) of Proposition 8.1.6 to show that

$$\langle I_{0,3,\beta} \rangle(a, b, c) = 0 \quad \text{if} \quad c \notin H^{d-p-r,d-q-s}(V).$$

However, since $\overline{M}_{0,3}$ is a point, Definition 7.1.9 implies that

$$\langle I_{0,3,\beta} \rangle(a, b, c) = \int_{\pi_{1*}(\xi)} a \otimes b \otimes c,$$

where $\pi_1 : \overline{M}_{0,3}(V, \beta) \to V^3$ is the evaluation map and ξ is the virtual fundamental class. Since ξ is algebraic of the expected dimension d by (7.7), one sees that $\pi_{1*}(\xi) \in H_{2d}(V^3)$ is the homology class of an algebraic cycle. The proposition now follows since the above integral is zero unless $a \otimes b \otimes c$ has type (d, d). $\qquad \square$

Corollary 8.1.7 implies that $\oplus_{p=0}^d H^{p,p}(V)$ is a subring of $H^*(V, \mathbb{C})$ endowed with the small quantum product when V is Calabi-Yau. This will be useful when we study the A-variation of Hodge structure later in the chapter.

8.1.2. Examples. We now compute some examples of the small quantum cohomology ring.

Example 8.1.2.1. For $\mathbb{P}^r$, we will use the cohomology basis H^i, $0 \leq i \leq r$, where H is the hyperplane class and the exponent refers to cup product. We claim that

$$(8.8) \qquad H^i * H^j = \begin{cases} H^{i+j} & i+j \leq r \\ q\, H^{i+j-r-1} & i+j \geq r+1, \end{cases}$$

where $q = e^{2\pi i \int_\ell \omega}$ and ℓ is a line in $\mathbb{P}^r$. To prove this, let $\beta = d[\ell]$ for $d \geq 0$. By (8.2), the Gromov-Witten invariant $\langle I_{0,3,d} \rangle(H^i, H^j, H^k)$ (where we write d instead of β) is zero unless

$$(8.9) \qquad i + j + k = r + d(r+1).$$

Since $i, j, k \leq r$, we see that d can only be 0 or 1, and furthermore, $d = 0$ if $i+j \leq r$. From here, it follows easily that $H^i * H^j = H^{i+j}$ when $i + j \leq r$. In particular, when $0 \leq i \leq r$, H^i means the same for both cup product and the small quantum product.

To finish the proof, it suffices to show $H * H^r = q\, H^0$. In this case, we have $d = 1$ and $k = r$ by (8.9). Then one easily obtains

$$H * H^r = \langle I_{0,3,1} \rangle(H, H^r, H^r)\, q\, H^0.$$

As in Section 7.4.2, $\langle I_{0,3,1}\rangle(H, H^r, H^r) = \langle I_{0,3,1}\rangle(H, [pt], [pt]) = 1$ because the two points in $\mathbb{P}^r$ determine a unique line meeting H at a third point. This proves (8.8). Also note that (8.8) gives a nice example of Proposition 8.1.5 since $\omega_{\mathbb{P}^r} = \mathcal{O}_{\mathbb{P}^r}(-r-1)$ is divisible by $r+1$.

Since $H^0 = [\mathbb{P}^r]$ is the identity element and $H^{r+1} = q\, H^0$ for the small quantum product, the small quantum ring of $\mathbb{P}^r$ is isomorphic to

$$\mathbb{C}[H]/\langle H^{r+1} - q\rangle.$$

The usual cohomology ring of $\mathbb{P}^r$, on the other hand, is isomorphic to

$$\mathbb{C}[H]/\langle H^{r+1}\rangle.$$

Since $q \to 0$ as the imaginary part of ω gets large, we see that the small quantum product becomes cup product in the limit.

Example 8.1.2.2. In [**Batyrev2**], a definition is given for the "quantum cohomology ring" of a smooth projective toric variety X. It turns out that this ring agrees with the small quantum cohomology ring if X is Fano, but can be different otherwise. To avoid confusion, Batyrev's ring will be denoted $H_\omega^*(X)$. In [**QR**], it was shown that $H_\omega^*(X)$ agrees with the small quantum cohomology ring for certain projective bundles over projective spaces. In Example 11.2.5.2, we will explain why $H_\omega^*(X)$ agrees with the small quantum cohomology ring in the Fano case (extending the result of [**QR**]) and also give an example in the non-Fano case which shows that $H_\omega^*(X)$ is definitely not the same as the small quantum cohomology ring in general. Example 11.2.5.2 will also show that in any case, $H_\omega^*(X)$ is closely related to the small quantum cohomology ring. We now turn to the definition and properties of $H_\omega^*(X)$. We will see that its properties resemble the properties that we expect from quantum cohomology (as they must, at least in the Fano case).

We put $X - X_\Sigma$, where Σ is the fan in $N_\mathbb{R}$ giving X. We will assume that $v_1, \dots, v_s \in N$ are the primitive integral generators of the 1-dimensional cones in Σ, and for each v_i, we have a torus-invariant divisor $D_i \subset X$.

We begin by recalling the usual description of the cohomology ring of X over $\mathbb{C}$. To v_i and D_i we associate a variable x_i, so that we get the polynomial ring $\mathbb{C}[x_1, \dots, x_s]$. Now consider the following two ideals:

$$P(\Sigma) = \left\langle \sum_{i=1}^s \langle m, v_i\rangle x_i : m \in M \right\rangle$$

$$SR(\Sigma) = \left\langle x_{i_1} \cdots x_{i_k} : \{v_{i_1}, \dots v_{i_k}\} \not\subset \sigma \text{ for all } \sigma \in \Sigma \right\rangle.$$

($SR(\Sigma)$ is the *Stanley-Reisner ideal* of Σ). It is well known [**Fulton3, Oda**] that the map sending $[D_i] \in H^2(X, \mathbb{C})$ to x_i induces an isomorphism

$$(8.10) \qquad H^*(X, \mathbb{C}) \simeq \mathbb{C}[x_1, \dots, x_s]/(P(\Sigma) + SR(\Sigma)).$$

We also need a modification of this due to Batyrev [**Batyrev2**]. Section 3.2.3 defined $\mathcal{P} = \{v_{i_1}, \dots, v_{i_k}\}$ to be a *primitive collection* if $\mathcal{P} \not\subset \sigma$ for all $\sigma \in \Sigma$, but every proper subset of $\mathcal{P}$ does lie in some σ. Then one sees easily that

$$(8.11) \qquad SR(\Sigma) = \left\langle x_{i_1} \cdots x_{i_k} : \{v_{i_1}, \dots v_{i_k}\} \text{ is a primitive collection} \right\rangle.$$

Now consider quantum cohomology. For $\mathbb{P}^r$, we went from $\mathbb{C}[h]/\langle h^{r+1}\rangle$ (cup product) to $\mathbb{C}[h]/\langle h^{r+1} - q\rangle$ (quantum product). In order to generalize this to our toric variety X, we first need some notation. Suppose that $\mathcal{P} = \{v_{i_1}, \dots v_{i_k}\}$ is a primitive collection. Then the sum $v_{\mathcal{P}} = v_{i_1} + \cdots + v_{i_k}$ lies in some cone $\sigma \in \Sigma$, and hence can be written $v_{\mathcal{P}} = c_1 v_{j_1} + \cdots + c_l v_{j_l}$, where $v_{j_1}, \ldots, v_{j_l}$ are the generators of σ and $c_1, \ldots, c_l \geq 0$. This gives the nontrivial relation

$$(8.12) \qquad v_{i_1} + \cdots + v_{i_k} - c_1 v_{j_1} - \cdots - c_l v_{j_l} = 0$$

among $v_1, \ldots, v_s$.

Such relations correspond to homology classes as follows. Since X is smooth, $\mathrm{Pic}(X) = A_{n-1}(X) = H^2(X, \mathbb{Z})$, and it follows that the dual of the exact sequence (3.2) from Chapter 3 is

$$0 \longrightarrow H_2(X, \mathbb{Z}) \longrightarrow \mathbb{Z}^s \longrightarrow N \longrightarrow 0,$$

recalling that $A_{n-1}(X, \mathbb{Z})$ is torsion free if X is smooth. Here, $(\lambda_1, \ldots, \lambda_s) \in \mathbb{Z}^s$ maps to $\sum_{i=1}^s \lambda_i v_i \in N$. It follows that we have a natural isomorphism

$$(8.13) \qquad \{(\lambda_1, \ldots, \lambda_s) \in \mathbb{Z}^s : \sum_{i=1}^s \lambda_i v_i = 0\} \simeq H_2(X, \mathbb{Z}).$$

In particular, the relation (8.12) corresponds to a homology class which we will denote $\beta(\mathcal{P}) \in H_2(X, \mathbb{Z})$ (compare with Lemma 3.3.2). In [**Batyrev1**, Thm. 2.15], Batyrev shows that $\beta(\mathcal{P})$ is an effective 1-cycle.

Now, given a complexified Kähler class $\omega \in H^{1,1}(X) = H^2(X, \mathbb{C})$, define the *quantum Stanley-Reisner ideal*

$$(8.14) \qquad SR_\omega(\Sigma) = \langle x_{i_1} \cdots x_{i_k} - q^{\beta(\mathcal{P})} x_{j_1}^{c_1} \cdots x_{j_l}^{c_l} : \mathcal{P} \text{ is a primitive collection} \rangle$$

where as usual $q^{\beta(\mathcal{P})} = e^{2\pi i \int_{\beta(\mathcal{P})} \omega}$. Then Batyrev defines the ring

$$(8.15) \qquad H_\omega^*(X) = \mathbb{C}[x_1, \ldots, x_s]/(P(\Sigma) + SR_\omega(\Sigma))$$

(see Theorem 5.3 of [**Batyrev2**]). We will call $H_\omega^*(X)$ the *Batyrev quantum ring*.

To see how the ring (8.15) relates to to the usual cohomology ring (8.10), let the imaginary part of ω approach infinity. Then the imaginary part of $\int_{\beta(\mathcal{P})} \omega$ also gets large since $\beta(\mathcal{P})$ is an effective cycle, and it follows that $q^{\beta(\mathcal{P})} \to 0$. Thus the quantum Stanley-Reisner ideal (8.14) turns into the usual Stanley-Reisner ideal (8.11) as the imaginary part of ω goes to infinity. This shows that the usual cohomology ring of X is a limit of $H_\omega^*(X)$.

Let's check how this works for $X = \mathbb{P}^r$. Here, the cone generators are $e_0, \ldots, e_r$, where $e_1, \ldots, e_r$ generate the lattice and $e_0 = -\sum_{i=1}^r e_i$. Thus the ideal $P(\Sigma)$ in the description of $H^*(\mathbb{P}^r, \mathbb{C})$ is

$$P(\Sigma) = \langle x_1 - x_0, \ldots, x_r - x_0 \rangle.$$

Furthermore, note that $\mathcal{P} = \{e_0, \ldots, e_r\}$ is the unique primitive collection. Since $e_0 + \cdots + e_r = 0$, the c_j's in (8.12) are 0. Thus the quantum Stanley-Reisner ideal (8.14) is just

$$SR_\omega(\Sigma) = \langle x_0 \cdots x_r - q^{\beta(\mathcal{P})} \rangle.$$

Putting $P(\Sigma)$ and $SR_\omega(\Sigma)$ together, (8.15) reduces to $\mathbb{C}[x_0]/\langle x_0^{r+1} - q^{\beta(\mathcal{P})} \rangle$. It remains to determine $\beta(\mathcal{P}) \in H_2(\mathbb{P}^r, \mathbb{Z})$. Since $\beta(\mathcal{P})$ is effective and the relation $e_0 + \cdots + e_r = 0$ generates the $\mathbb{Z}$-module of all relations among the e_i, we must have $\beta(\mathcal{P}) = [\ell]$ (see also Lemma 3.3.2 and the sentence following its proof). Hence we recover our earlier description of the small quantum cohomology of $\mathbb{P}^r$.

The ring $H_\omega^*(X)$ has other descriptions as well. If

$$(8.16) \qquad \sum_{i=1}^s a_i v_i = \sum_{i=1}^s b_i v_i, \quad a_i, b_i \geq 0,$$

then $a - b = (a_1 - b_1, \ldots, a_s - b_s)$ gives an element $\beta(a, b) \in H_2(X, \mathbb{Z})$ by (8.13). In [**Batyrev2**], Batyrev shows that the quantum Stanley-Reisner ideal is given by

$$(8.17) \qquad SR_\omega(\Sigma) = \langle x_1^{a_1} \cdots x_s^{a_s} - q^{\beta(a,b)} x_1^{b_1} \cdots x_s^{b_s} : a_i, b_i \text{ as in } (8.16) \rangle.$$

This description of $SR_\omega(\Sigma)$ shows that $H_\omega^*(X)$ depends only on the 1-dimensional cones of Σ and the complexified Kähler class ω. It follows that different fans with the same 1-skeleton have closely related quantum cohomology. For example, let X_1 and X_2 be smooth toric varieties whose fans have the same 1-skeleton, and suppose that their Kähler cones share a common wall in the GKZ decomposition described in Section 3.4. Then X_1 and X_2 are related by a flop and hence might have nonisomorphic (usual) cohomology rings. However, using the Batyrev quantum ring, we can get from one to the other as follows:

$$\begin{array}{ccccccc}
\text{usual} & & \text{quantum} & & \text{quantum} & & \text{usual} \\
H^*(X_1, \mathbb{C}) & \longleftarrow & H_\omega^*(X_1, \mathbb{C}) & \longleftrightarrow & H_\omega^*(X_2, \mathbb{C}) & \longrightarrow & H^*(X_2, \mathbb{C})
\end{array}$$

The arrow on the left tells us that as the imaginary part of ω approaches infinity in the Kähler cone of X_1, the Batyrev quantum ring of X_1 approaches its usual cohomology ring. The arrow on the right is similar, except that now the imaginary part of ω goes to infinity in the Kähler cone of X_2. Finally, the middle arrow is what happens when ω crosses the wall. This is possible because (8.17) allows us to change fans as long as we keep the same 1-dimensional cones.

We can think about the two descriptions of $SR_\omega(\Sigma)$ as follows: (8.17) uses only the 1-skeleton and shows how to extend $H_\omega^*(X)$ over the whole GKZ decomposition, while (8.14) uses the full fan (via the primitive collections) and shows how to take the limit in any particular Kähler cone. See [**Batyrev2**] for further discussion and examples.

Finally, when we let the imaginary part of ω approach infinity, we have taken the limit of the ideal $SR_\omega(\Sigma)$ in a very naive way. The proper way to do this, of course, is via a monomial ordering and a Gröbner basis. In fact, the imaginary part of ω gives a monomial ordering—see the proof of Theorem 5.3 in [**Batyrev2**]. Furthermore, if we partition all possible $\text{Im}(\omega)$'s according to monomial orderings they give, then we get a refinement of the GKZ decomposition called the *Gröbner fan*. This is described in [**Sturmfels1**] and has been used in mirror symmetry to study solutions of hypergeometric equations [**HLY1, HLY2**].

Example 8.1.2.3. Finally, we consider the small quantum cohomology of a smooth Calabi-Yau threefold V. We first show that most of the time, the small quantum product agrees with cup product.

LEMMA 8.1.8. *If V is a smooth Calabi-Yau threefold, then*

$$a * b = a \cup b \quad \text{if either } \deg a \neq 2 \text{ or } \deg b \neq 2.$$

PROOF. First observe that $\langle I_{0,3,\beta} \rangle(\alpha_1, \alpha_2, \alpha_3) = 0$ whenever $\beta \neq 0$ and $\deg \alpha_i \neq 2$ for some i. We proved this for the quintic threefold in Section 7.4.3, and the proof works for any Calabi-Yau threefold. From here, the lemma follows easily. $\square$

It remains to compute the quantum product on $H^2(V, \mathbb{C})$. First suppose that V is a quintic threefold. As usual, we will write $\langle I_{0,3,d} \rangle$ instead of $\langle I_{0,3,\beta} \rangle$. Then the small quantum product is determined uniquely once we know $H * H$, where

$H \in H^2(V, \mathbb{Z})$ is the hyperplane class. By Corollary 8.1.7, $H * H \in H^4(V, \mathbb{C})$, and then Definition 8.1.1 and $\int_V H^3 = 5$ imply that

$$H * H = \frac{1}{5} \sum_{d=0}^{\infty} \langle I_{0,3,d} \rangle (H, H, H) \, q^d \, H^2.$$

We also have the three-point function $\langle H, H, H \rangle$ from (8.5) and (8.7), which here becomes

$$(8.18) \qquad \langle H, H, H \rangle = \int_V H * H * H = \sum_{d=0}^{\infty} \langle I_{0,3,d} \rangle (H, H, H) \, q^d.$$

In Section 7.4.3, we saw that the right hand side of this equation is the A-model correlation function $\langle H, H, H \rangle$ considered in Chapter 2. Hence the three-point function of the quintic threefold is given by the small quantum product. As we saw in (8.6), the converse is also true, for combining the above two formulas gives

$$H * H = \langle H, H, H \rangle \frac{1}{5} H^2.$$

Thus the quantum product is written in terms of the three-point function.

For an arbitrary Calabi-Yau threefold V, the situation is similar if we restrict to classes $a, b, c \in H^2(V, \mathbb{C})$. By (8.5) and (8.7), the three-point function is

$$(8.19) \qquad \begin{aligned} \langle a, b, c \rangle &= \int_V a * b * c = \sum_{\beta} \langle I_{0,3,\beta} \rangle (a, b, c) \, q^\beta \\ &= \langle I_{0,3,0} \rangle (a, b, c) + \sum_{\beta \neq 0} \langle I_{0,0,\beta} \rangle \, q^\beta \int_\beta a \int_\beta b \int_\beta c. \end{aligned}$$

We learned in Section 7.4.4 that there are instanton numbers n_β such that

$$(8.20) \qquad \langle I_{0,0,\beta} \rangle = \sum_{\beta = k\gamma} n_\gamma k^{-3}.$$

which as in (7.51) leads to

$$(8.21) \qquad \langle a, b, c \rangle = \int_V a * b * c = \int_V a \cup b \cup c + \sum_{\beta \neq 0} n_\beta \frac{q^\beta}{1 - q^\beta} \int_\beta a \int_\beta b \int_\beta c.$$

This classic formula for the three-point function of a smooth Calabi-Yau threefold was first given for the quintic [**CdGP**] and appears frequently in the literature [**Witten5, AGM3, Morrison5, Voisin3**]. It makes good sense when the instanton numbers n_β are defined via (8.20), but we know from Section 7.4.4 that n_β need not count the number of rational curves on V with homology class β. (However, Conjecture 7.4.5 asserts that n_β should be closely related to the number of J-holomorphic curves, counted with orientation, in the class β.)

It is very satisfying to see the three-point function of a Calabi-Yau threefold expressed in terms of quantum cohomology. Furthermore, as we saw in Section 8.1.1, the converse is also true: we can construct quantum cohomology in terms of the three-point function.

Example 8.1.2.4. In this example, we will study the effect of a flop on the small quantum cohomology of a Calabi-Yau threefold V. In the terminology of Section 6.2.2, suppose that $\phi : V \to \bar{V}$ is a primitive Type I contraction giving the flop $V \dashrightarrow V'$ such that the proper transform of $C \subset V$ is $C' \subset V'$. We have a natural identification $H^2(V) \simeq H^2(V')$ under which the Kähler cones $K(V)$ and $K(V')$ meet along a common face of each.

In order to compare the quantum cohomologies of V and V', suppose that the reducible curve C consists of rigidly embedded disjoint $\mathbb{P}^1$'s in the homology class β. We will assume further that C is disjoint from all other curves whose classes differ from β.

In this situation, it is known that V and V' have the same Hodge numbers, but may have different intersection forms. Thus the (usual) cohomology rings may be nonisomorphic. However, their small quantum cohomology rings are very closely related, as noted in [**Witten5, AGM3, Morrison5**]. Using (8.21), we write the three-point function of V in the form:

$$\langle a,b,c \rangle = \int_V a \cup b \cup c + n_\beta \frac{q^\beta}{1-q^\beta} \int_\beta a \int_\beta b \int_\beta c + \sum_{\gamma \neq \beta, 0} n_\gamma \frac{q^\gamma}{1-q^\gamma} \int_\gamma a \int_\gamma b \int_\gamma c.$$

In passing to V', the first term on the right may change, as may the second. The third term is unchanged, however. This is because the moduli spaces $\overline{M}_{0,3}(V,\gamma)$ are isomorphic to their counterparts on V', the isomorphisms preserving their virtual fundamental classes. Now, an easy calculation (given in the above references) shows that the sum of the first two terms

$$\int_V a \cup b \cup c + n_\beta \frac{q^\beta}{1-q^\beta} \int_\beta a \int_\beta b \int_\beta c$$

is *unchanged* when we pass from V to V' provided we transform a,b,c via $H^2(V) \simeq H^2(V')$ and β via the Poincaré dual of this map.

From a formal point of view, this says that the three-point functions of V and V' are the same formal power series (under the above identifications), and one easily sees that the same is true for the small quantum cohomology rings (with suitable formal coefficients). However, the small quantum product on V converges (conjecturally) when the imaginary part of ω is sufficiently deep in $K(V)$, and similarly for $K(V')$. Under the map $H^2(V) \simeq H^2(V')$, the Kähler cones are disjoint, so that the quantum cohomology rings V and V' are no longer isomorphic since we are evaluating them at disjoint sets of ω's. What we get instead is an analytic continuation of the three-point function of V to that of V'. The situation is remarkably similar to Example 8.1.2.2, where we studied the effect of a flop on the quantum cohomology of a toric variety. As happened there, quantum cohomology allows us to deform the (usual) cohomology ring of V into the cohomology ring of V'.

All of this concerns a face of the Kähler cone corresponding to a Type I contraction. In [**Wilson4**], the effect of a Type III contraction is studied.

The small quantum product has been an active area of research. In addition to the examples just discussed, one can describe the small quantum cohomology rings of other interesting varieties, including the following:

- **Grassmannians.** The small quantum ring of a Grassmanian was first described in [**Vafa2**], followed by mathematical treatments in [**Witten6, ST**]. Other references are [**MS, FP, Bertram, BCF**]. Topics of interest include quantum versions of the Giambelli and Pieri formulas, relations with the Verlinde algebra, and the Vafa-Intriligator formula.

- **Flag manifolds.** The small quantum ring of a flag variety has a fascinating relation with the Toda lattice. This was first described in [**GK**]. Other references include [**AS, Kim1, Givental3, CiocanF, Kostant**].
- **Fano manifolds.** A lot of work has been done on the small quantum ring of Fano manifolds. These include Grassmannians and flag varieties as well as certain projective hypersurfaces and complete intersections. Papers on this subject include [**ST, Beauville1, TX, CJ**].
- **Rational surfaces.** The small quantum ring of a general rational surface is described in [**CM**]. For a cubic surface, associativity of the quantum product is essentially equivalent to the existence of 27 lines each of which meets 10 others.

8.1.3. Coefficients. In the discussion so far, we have used cohomology with complex coefficients in defining the small quantum product. This is because

$$q^\beta = e^{2\pi i \int_\beta \omega} \in \mathbb{C}$$

when $\omega \in H^2(X, \mathbb{C})$ is a complexified Kähler class. However, the sum used in Definition 8.1.1 is not known to converge. One way to avoid this problem is to regard q^β as a formal variable rather than a number. There are several ways to make this precise.

For a Fano manifold, the sums appearing in Definition 8.1.1 are finite by Proposition 8.1.3. In this case, we can let the coefficients be the semigroup ring

$$R = \mathbb{Q}[q^\beta; \beta \in H_2(X, \mathbb{Z})].$$

This means that each $\beta \in H_2(X, \mathbb{Z})$ gives a formal symbol q^β such that $q^{\beta_1} \cdot q^{\beta_1} = q^{\beta_1 + \beta_2}$, and an element of R is a finite sum of the q^β's with coefficients in $\mathbb{Q}$. An easy example is given by $\mathbb{P}^r$, where we get the ring $R = \mathbb{Q}[q, q^{-1}]$ of Laurent polynomials in q. For any Fano manifold, the small quantum product is defined on $H^*(X, R)$.

If X is an arbitrary projective manifold, this might not work because the sum in Definition 8.1.1 may be infinite. Here, observe that by the Effectivity Axiom, we can assume that β is effective. More precisely, we can assume that β lies in the integral Mori cone $M(X)_\mathbb{Z}$. Then consider

$$\mathcal{R} = \mathbb{Q}[[q^\beta; \beta \in M(X)_\mathbb{Z}]],$$

where the double brackets indicate that we allow infinite sums. The product of two such elements is

$$\sum_{\beta \in M(X)_\mathbb{Z}} a_\beta q^\beta \cdot \sum_{\beta \in M(X)_\mathbb{Z}} b_\beta q^\beta = \sum_{\beta \in M(X)_\mathbb{Z}} \left(\sum_{\beta = \beta_1 + \beta_2} a_{\beta_1} b_{\beta_2} \right) q^\beta.$$

This sum makes sense because for any $\beta \in M(X)_\mathbb{Z}$, the sum in parentheses is finite since the free part of the Mori cone consists of lattice points in a strongly convex cone. Thus $\mathcal{R}$ is a ring, which may be thought of as the formal semigroup ring of the integral Mori cone. An example is given by the quintic threefold, where $\mathcal{R} = \mathbb{Q}[[q]]$ is the ring of formal power series in q. In general, one sees easily that the small quantum product is defined on $H^*(X, \mathcal{R})$.

Finally, when (M, ω) is a symplectic manifold, we can define the small quantum product using Definition 8.1.1, but now it is necessary work with cohomology over the *Novikov ring* $\Lambda(\omega, \mathbb{Q})$. This ring consists of all formal sums

$$\sum_{\beta \in H_2(M,\mathbb{Z})} a_\beta q^\beta, \quad a_\beta \in \mathbb{Q}$$

such that for any $C \in \mathbb{R}$, the set

$$\{\beta : a_\beta \neq 0 \text{ and } \int_\beta \omega < C\}$$

is finite. It is straightforward to show that $\Lambda(\omega, \mathbb{Q})$ is a ring and that the small quantum product is defined on $H^*(X, \Lambda(\omega, \mathbb{Q}))$.

There are other coefficients one can use as well. In the algebraic case, one can replace $\mathcal{R}$ with its ring of invariants under a suitable automorphism group [**Morrison7**], and in the symplectic case, one often restricts the Novikov ring to spherical homology classes [**MS**]. See also [**Getzler1**].

8.2. Big Quantum Cohomology

The quantum cohomology discussed in Section 8.1 doesn't use the full enumerative information provided by the Gromov-Witten invariants $\langle I_{0,n,\beta} \rangle$. This is because the small quantum product is defined in terms of $\langle I_{0,3,\beta} \rangle$, while the more interesting invariants often occur for $n \geq 3$. For example, we saw in Chapter 7 that for $\mathbb{P}^2$, the numbers

$$N_d = \langle I_{0,3d-1,d} \rangle (\underbrace{[pt], \dots, [pt]}_{3d-1 \text{ times}})$$

have some remarkable properties. Yet the small quantum product for $\mathbb{P}^2$ uses only $N_1 = \langle I_{0,3,1} \rangle ([pt], [pt], [\ell]) = 1$. In this section, we will explore a version of quantum cohomology—the so-called *big quantum cohomology*—which takes *all* of the Gromov-Witten invariants into account.

8.2.1. $H^*(X, \mathbb{C})$ **as a Supermanifold.** Before defining the big quantum product, we need to discuss the supermanifold structure of $H^*(X, \mathbb{C})$. Fix a homogeneous basis $T_0 = 1, T_1, \dots, T_m$ of the rational cohomology $H^*(X, \mathbb{Q})$, and for each T_i, introduce a variable t_i of the same degree as T_i. These variables are local coordinates for the supermanifold $H^*(X, \mathbb{C})$. The t_i supercommute, which means that

$$t_i t_j = (-1)^{\deg t_i \deg t_j} t_j t_i,$$

and $t_i^2 = 0$ if t_i has odd degree. Hence the ring $\mathbb{C}[t_0, \dots, t_m]$ is the tensor product of a polynomial ring (the even variables) with an exterior algebra (the odd variables). We define the partial derivative operator $\partial/\partial t_i$ by

$$\frac{\partial}{\partial t_i}(t_i^k \cdot t^\alpha) = k t_i^{k-1} \cdot t^\alpha$$

for any monomial $t^\alpha \in \mathbb{C}[t_0, \dots, t_m]$ not involving t_i. Then we can compute $\partial F/\partial t_i$ for any $F \in \mathbb{C}[t_0, \dots, t_m]$, and it follows that

$$\frac{\partial^2 F}{\partial t_i \partial t_j} = (-1)^{\deg t_i \deg t_j} \frac{\partial^2 F}{\partial t_j \partial t_i}.$$

All of this extends easily to the formal power series ring $\mathbb{C}[[t_0, \dots, t_m]]$.

The reader should consult [**Manin3**] for more about supermanifolds. As discussed in [**Alvarez**, Lect. 2], supercommuting variables arise naturally in quantum physics when studying fermionic systems.

8.2.2. The Gromov-Witten Potential. The key tool used to define the big quantum product is the *Gromov-Witten potential*. As above, we fix a basis $T_0 = 1, T_1, \ldots, T_m$ of the rational cohomology $H^*(X, \mathbb{Q})$. Then put $\gamma = \sum_{i=0}^m t_i T_i$, where t_i is a supercommutative variable with $\deg t_i = \deg T_i$.

DEFINITION 8.2.1. *Let ω be a complexified Kähler class on a smooth projective variety X. Then the* Gromov-Witten potential *is the formal sum*

$$\Phi(\gamma) = \sum_{n=0}^{\infty} \sum_{\beta \in H_2(X, \mathbb{Z})} \frac{1}{n!} \langle I_{0,n,\beta} \rangle (\gamma^n) \, q^\beta,$$

where $q^\beta = e^{2\pi i \int_\beta \omega}$.

In the above sum, when $\beta = 0$ we implicitly have $n \geq 3$ since $\overline{M}_{0,n}(X, 0)$ does not exist if $n \leq 2$. There is some variation in the definition of the Gromov-Witten potential in the literature, as some authors truncate the series by assuming $n \geq 3$ for *all* values of β. This is especially appropriate when we want to relate the Gromov-Witten potential to Gromov-Witten classes.

In order to make sense of this definition, we need to explain what $\langle I_{0,n,\beta} \rangle (\gamma^n)$ means. This is somewhat subtle because of supercommutativity. The idea is to expand

$$\gamma^n = \left(\sum_{i=0}^m t_i T_i \right)^n,$$

regarding both T_i and t_i as supercommuting variables. We will use multinomial notation, where an exponent vector $\alpha = (a_0, \ldots, a_m)$ gives monomials T^α and t^α of total degree $|\alpha| = a_0 + \cdots + a_m$. Also set $\alpha! = a_0! \cdots a_m!$. A useful observation is that each term $t_i T_i$ of γ is even, so that $t_i T_i$ can be commuted as a unit. Then the sign rules of the Gromov-Witten invariants give the formula

$$(8.22) \qquad \frac{1}{n!} \langle I_{0,n,\beta} \rangle (\gamma^n) = \sum_{|\alpha|=n} \epsilon(\alpha) \langle I_{0,n,\beta} \rangle (T^\alpha) \frac{t^\alpha}{\alpha!}.$$

The factor $\epsilon(\alpha) = \pm 1$ is the sign determined by the equation

$$(8.23) \qquad (t_0 T_0)^{a_0} \cdots (t_m T_m)^{a_m} = \epsilon(\alpha) T_0^{a_0} \cdots T_m^{a_m} \, t_0^{a_0} \cdots t_m^{a_m}.$$

Substituting (8.22) into the definition of Φ, we would like to regard Φ as a formal power series in the ring $\mathbb{C}[[t_0, \ldots, t_m]]$ discussed in Section 8.2.1. However, the convergence questions encountered in small quantum cohomology are present in this case as well. To be rigorous, we may need to replace $\mathbb{C}$ with one of the rings R, $\mathcal{R}$ or $\Lambda(\omega, \mathbb{Q})$ from Section 8.1.3. Thus, in the discussion below, we will let $\mathcal{C}$ denote one of the rings

$$\mathbb{C}[[t_0, \ldots, t_m]] \text{ or } R[[t_0, \ldots, t_m]] \text{ or } \mathcal{R}[[t_0, \ldots, t_m]] \text{ or } \Lambda(\omega, \mathbb{Q}))[[t_0, \ldots, t_m]]$$

as appropriate.

Since the Gromov-Witten potential Φ incorporates *all* possible genus zero invariants, we can think of Φ as the generating function of the tree-level Gromov-Witten invariants. Because of the factorials and noncommuting variables, it is more accurate to say that Φ is the *supercommutative exponential generating function* of the $\langle I_{0,n,\beta} \rangle$. From the physics point of view, Φ has a natural interpretation as the genus 0 free energy.

The Gromov-Witten potential is sometimes called the genus 0 Gromov-Witten potential. The *genus g Gromov-Witten potential* can be defined in an analogous fashion using genus g Gromov-Witten invariants.

A simple example of a Gromov-Witten potential is given by an elliptic curve E. Since there are no nonconstant maps $\mathbb{P}^1 \to E$, we only need to consider the Gromov-Witten invariant $\langle I_{0,3,0} \rangle$, which is determined by the Point Mapping Axiom. Let $T_0 = 1 \in H^0(E,\mathbb{C})$, $T_3 = [pt] \in H^2(E,\mathbb{C})$, and pick $T_1, T_2 \in H^1(E,\mathbb{C})$ with $\int_E T_1 \cup T_2 = 1$. Then one computes that

$$(8.24) \qquad \Phi = \tfrac{1}{2} t_0^2 t_3 - t_0 t_1 t_2,$$

The minus sign is due to $\epsilon(\alpha)$ in (8.22). We will discuss some less trivial examples of the Gromov-Witten potential in Section 8.3.

Using the Gromov-Witten potential, we define the big quantum product as follows.

DEFINITION 8.2.2. *Let Φ be the Gromov-Witten potential for a smooth projective variety X. Then define*

$$T_i * T_j = \sum_k \frac{\partial^3 \Phi}{\partial t_i \partial t_j \partial t_k} \, T^k,$$

where g^{kl} is as in Definition 8.1.1 and $T^k = \sum_l g^{kl} T_l$. Extending this linearly gives the big quantum product *on the cohomology $H^*(X,\mathcal{C})$.*

Note that since the Gromov-Witten invariants with $n < 3$ occur as coefficients of terms in Φ which are at most quadratic in the t_i, then their presence or absence in the definition of the Gromov-Witten potential does not affect the definition of the big quantum product.

In the elliptic curve example (8.24), the Gromov-Witten potential is $\Phi = \tfrac{1}{2} t_0^2 t_3 - t_0 t_1 t_2$. Using the cohomology basis T_0, T_1, T_2, T_3 from before, one computes the big quantum product

$$T_1 * T_2 = \frac{\partial^3 \Phi}{\partial t_1 \partial t_2 \partial t_0} \, T_3 = \frac{\partial^3}{\partial t_1 \partial t_2 \partial t_0} \left(\tfrac{1}{2} t_0^2 t_3 - t_0 t_1 t_2 \right) T_3 = T_3$$

since t_1 and t_2 are odd variables. It is easy to see that the big quantum product coincides with ordinary cup product in this case.

The following lemma is useful in working with the third partials of the Gromov-Witten potential.

LEMMA 8.2.3. *For all i,j,k, we have*

$$\frac{\partial^3 \Phi}{\partial t_i \partial t_j \partial t_k} = \sum_{n=0}^{\infty} \sum_{\beta \in H_2(X,\mathbb{Z})} \frac{1}{n!} \langle I_{0,n+3,\beta} \rangle (T_i, T_j, T_k, \gamma^n) \, q^{\beta}.$$

PROOF. Let $\alpha = (a_0, \dots, a_m)$ be an exponent vector with $|\alpha| = n$ such that $\gamma = \alpha - c_i - c_j - c_k$ has nonnegative entries, where c_i, c_j, c_k are the standard basis vectors. If we can prove that

$$\frac{\partial^3}{\partial t_i \partial t_j \partial t_k} \epsilon(\alpha) \langle I_{0,n,\beta} \rangle (T^{\alpha}) \frac{t^{\alpha}}{\alpha!} = \epsilon(\gamma) \langle I_{0,n,\beta} \rangle (T_i, T_j, T_k, T^{\gamma}) \frac{t^{\gamma}}{\gamma!},$$

then the lemma will follow from (8.22). First suppose that t_i, t_j, t_k are odd. We can assume that $a_i = a_j = a_k = 1$, and using the sign rules for $\langle I_{0,n,\beta} \rangle$ and the t_i, we can write

$$\epsilon(\alpha)\langle I_{0,n,\beta}\rangle(T^\alpha)\frac{t^\alpha}{\alpha!} = \epsilon(\gamma + e_i + e_j + e_k)\langle I_{0,n,\beta}\rangle(T_k, T_j, T_i, T^\gamma)\, t_k t_j t_i \frac{t^\gamma}{\gamma!}.$$

Then differentiating three times gives

$$\frac{\partial^3}{\partial t_i \partial t_j \partial t_k}\epsilon(\alpha)\langle I_{0,n,\beta}\rangle(T^\alpha)\frac{t^\alpha}{\alpha!} = \epsilon(\gamma + e_i + e_j + e_k)\langle I_{0,n,\beta}\rangle(T_k, T_j, T_i, T^\gamma)\frac{t^\gamma}{\gamma!}.$$

Hence it suffices to show

$$\epsilon(\gamma + e_i + e_j + e_k)\langle I_{0,n,\beta}\rangle(T_k, T_j, T_i, T^\gamma) = \epsilon(\gamma)\langle I_{0,n,\beta}\rangle(T_i, T_j, T_k, T^\gamma).$$

To prove this, observe that by (8.23), $\epsilon(\gamma) = (-1)^{s(s+1)/2}$, where s is the number of odd variables appearing in t^γ (we can assume the odd variables appear to the first power). Also, (7.33) implies $s + 3 \equiv 0 \bmod 2$ since t_i, t_j, t_k are odd. From here, the desired result follows easily.

The other cases are similar and are left for the reader. $\qquad\square$

8.2.3. The WDVV Equation and Associativity. Our next goal is to show that $H^*(X, \mathcal{C})$ is a ring under the big quantum product. As in Section 8.1, the hard part is associativity. But here, something remarkable happens, for the associative property of big quantum cohomology is equivalent to a certain partial differential equation satisfied by Φ, the so-called WDVV (= Witten-Dijkgraaf-Verlinde-Verlinde) equation [**Witten2, DVV1, DVV2**].

THEOREM 8.2.4. *The Gromov-Witten potential Φ satisfies the equation*

$$\sum_{a,b} \frac{\partial^3 \Phi}{\partial t_i \partial t_j \partial t_a} g^{ab} \frac{\partial^3 \Phi}{\partial t_b \partial t_k \partial t_l} =$$

$$(-1)^{\deg t_i (\deg t_j + \deg t_k)} \sum_{a,b} \frac{\partial^3 \Phi}{\partial t_j \partial t_k \partial t_a} g^{ab} \frac{\partial^3 \Phi}{\partial t_b \partial t_i \partial t_l}$$

for all i, j, k, l.

PROOF. We will only sketch the main ideas since a detailed proof can be found in [**KoM1**]. For simplicity, we will also assume that $H^*(X, \mathbb{C})$ has only even cohomology so that we don't have to worry about signs.

The proof will follow the same strategy used to prove the recursion relation (7.39) for Gromov-Witten invariants of $\mathbb{P}^2$. In place of the class ξ of (7.40), we will use the Gromov-Witten class

$$I_{0,n+4,\beta}(\gamma^n, T_i, T_j, T_k, T_l) \in H^*(\overline{M}_{0,n+4}, \mathcal{C}),$$

where $\gamma = \sum_{i=0}^m t_i T_i$. This class will be evaluated on two linearly equivalent divisors which we now describe. Let the set $\mathcal{I} = \{1, \dots, n, i, j, k, l\}$ index the $n+4$ arguments of $I_{0,n+4,\beta}$ (this is a slight abuse of notation). As in the discussion preceding (7.41), each partition of $\mathcal{I}$ into disjoint sets A and B gives a divisor $\mathrm{im}(\varphi_{A,B}) \subset \overline{M}_{0,n+4}$.

Then define

$$D(i,j|k,l) = \sum_{i,j\in A \; k,l\in B} \text{im}(\varphi_{A,B})$$

$$D(j,k|i,l) = \sum_{j,k\in A \; i,l\in B} \text{im}(\varphi_{A,B})$$

Similar to what we did in (7.41), one can find a linear equivalence

$$(8.25) \qquad\qquad D(i,j|k,l) \sim D(j,k|i,l).$$

Now consider the sum on the left hand side of the WDVV equation. Using $T^k = \sum_l g^{kl} T_l$ and Lemma 8.2.3, this can be written as

$$\sum \frac{1}{n_1! n_2!} \langle I_{0,n_1+3,\beta_1}\rangle (T_i, T_j, T_a, \gamma^{n_1})\, g^{ab}\, \langle I_{0,n_2+3,\beta_2}\rangle (T_b, T_k, T_l, \gamma^{n_2})\, q^{\beta_1+\beta_2},$$

where the sum is over $n_1, n_2, \beta_1, \beta_2, b_2, a, b$. We will break this sum down into smaller sums for which $n_1 + n_2 = n$ and $\beta_1 + \beta_2 = \beta$ are fixed. By commutativity, this smaller sum (corresponding to a particular choice of n and β) can be written as

$$\sum \frac{1}{n_1! n_2!} \langle I_{0,n_1+3,\beta_1}\rangle (\gamma^{n_1}, T_i, T_j, T_a)\, g^{ab}\, \langle I_{0,n_2+3,\beta_2}\rangle (T_b, \gamma^{n_2}, T_k, T_l)\, q^{\beta},$$

where the sum is now over $n_1, n_2, \beta_1, \beta_2, b_2, a, b$ with $n_1 + n_2 = n$ and $\beta_1 + \beta_2 = \beta$. If we expand this sum using (8.22) and apply the Splitting Axiom, one can show that the smaller sum reduces to

$$\int_{D(i,j|k,l)} \frac{1}{n!} I_{0,n+4,\beta}(\gamma^n, T_i, T_j, T_k, T_l)\, q^{\beta},$$

where $D(i,j|k,l)$ is the divisor defined above. Hence the left hand side of the WDVV equation is the sum

$$\sum_{n=0}^{\infty} \sum_{\beta \in H_2(X,\mathbb{Z})} \int_{D(i,j|k,l)} \frac{1}{n!} I_{0,n+4,\beta}(\gamma^n, T_i, T_j, T_k, T_l)\, q^{\beta}.$$

Applying the same process to the right hand side of the WDVV equation, we get the sum

$$\sum_{n=0}^{\infty} \sum_{\beta \in H_2(X,\mathbb{Z})} \int_{D(j,k|i,l)} \frac{1}{n!} I_{0,n+4,\beta}(\gamma^n, T_i, T_j, T_k, T_l)\, q^{\beta}.$$

By the linear equivalence (8.25), the above two sums are equal, and the theorem follows immediately. $\qquad\square$

It is now easy to show that big quantum cohomology is a ring.

THEOREM 8.2.5. *$H^*(X,\mathcal{C})$ is a commutative ring with identity $T_0 = 1$ under the big quantum product.*

PROOF. The big quantum product is supercommutative by the Equivariance Axiom from Section 7.3. We next show that $T_0 = 1$ is the identity not only for

cup product but also for the big quantum product. This is easy, for applying
Lemma 8.2.3 with $T_0 = [X]$ gives

$$(8.26) \quad \frac{\partial^3 \Phi}{\partial t_0 \partial t_j \partial t_k} = \sum_{n=0}^{\infty} \sum_{\beta \in H_2(X,\mathbb{Z})} \frac{1}{n!} \langle I_{0,n+3,\beta} \rangle (T_0, T_j, T_k, \gamma^n) q^{\beta}$$

$$= \langle I_{0,3,0} \rangle (T_0, T_j, T_k) = \int_X T_j \cup T_k = g_{jk},$$

where the second and third equalities are by the Fundamental Class and Point
Mapping Axioms respectively. Since (8.26) implies $T_0 * T_j = T_j$, we see that T_0 is
the identity.

Finally, we need to prove associativity. Using $T^k = \sum_l g^{kl} T_l$ and the definition
of big quantum product, we get

$$(T_i * T_j) * T_k = \sum_{a,b} \sum_{l,s} \frac{\partial^3 \Phi}{\partial t_i \partial t_j \partial t_a} g^{ab} \frac{\partial^3 \Phi}{\partial t_b \partial t_k \partial t_l} g^{ls} T_s.$$

Expanding $(T_j * T_k) * T_i$ in the same way and using the WDVV equation from
Theorem 8.2.4, we see immediately that

$$(T_i * T_j) * T_k = (-1)^{\deg T_i (\deg T_j + \deg T_k)} (T_j * T_k) * T_i,$$

and then associativity follows since $*$ is supercommutative. $\square$

This theorem shows that the WDVV equation implies associativity of the big
quantum product. The converse is also true, so that associativity is equivalent
to the WDVV equation. The is part of the Dubrovin formalism, which will be
discussed in Section 8.4.

We should also mention that big quantum cohomology interacts nicely with the
intersection product $g(a, b) = \int_X a \cup b$. Similar to Proposition 8.1.6, one can show
that the big quantum product satisfies

$$(8.27) \qquad\qquad g(a * b, c) = g(a, b * c)$$

for all $a, b, c \in H^*(X, \mathcal{C})$. It follows that $H^*(X, \mathcal{C})$ is a Frobenius algebra under the
big quantum product, and we also get the formula

$$\int_X T_i * T_j * T_k = \frac{\partial^3 \Phi}{\partial t_i \partial t_j \partial t_k}.$$

There is a Künneth formula for the big quantum product [**KoM2, Kaufmann**].
Also, Section 10.1.1 will discuss an enlargement of the big quantum product called
the *gravitational quantum product*.

8.3. Computing Gromov-Witten Invariants, II

We now revisit some of the examples discussed in Section 7.4 to see how these
examples relate to big quantum cohomology and the Gromov-Witten potential.

8.3.1. Structure of the Gromov-Witten Potential. As we will see below, explicit formulas for the Gromov-Witten potential can be rather complicated.
However, Φ has a structure which makes the formulas easier to understand.

We first show that Φ is homogeneous once we assign appropriate degrees to its variables. Using the notation of the previous section, we have the cohomology basis T_i and the supercommutative variables t_i. By (8.22), a typical term of Φ looks like

$$\langle I_{0,n,\beta}\rangle(T_0^{a_0},\ldots,T_m^{a_m})\frac{t_0^{a_0}}{a_0!}\cdots\frac{t_m^{a_m}}{a_m!}\,q^\beta,$$

where $a_0 + \cdots + a_m = n$. However, we also have the degree condition (7.33), which tells us that

$$a_0 \deg T_0 + \cdots + a_m \deg T_m = 2\dim X - 2\int_\beta \omega_X + 2(3\cdot 0 - 3 + n).$$

Using $a_0 + \cdots + a_m = n$, we can rewrite this in the form

$$a_0(\deg T_0 - 2) + \cdots + a_m(\deg T_m - 2) + 2\int_\beta \omega_X = 2\dim X - 6.$$

Hence the Gromov-Witten potential Φ is homogeneous of degree $2\dim X - 6$ provided we assign the following degrees to $t_0,\ldots,t_m$ and q^β:

(8.28)
$$
\begin{aligned}
t_i &\quad\text{has degree}\quad \deg(T_i) - 2 \\
q^\beta &\quad\text{has degree}\quad 2\int_\beta \omega_X.
\end{aligned}
$$

Thus, for the t_i's, we subtract 2 from their previous degree (so their parity is unchanged). For the q^β's, this requires that they be formal symbols, as in the rings R, $\mathcal{R}$ or $\Lambda(\omega,\mathbb{Q})$ from Section 8.1.3, rather than just numbers.

The homogeneity of the Gromov-Witten potential is very useful in practice. It implies that many terms in the Gromov-Witten potential are zero for degree reasons. Also, when applied to quantum cohomology, homogeneity says that the coefficient of q^β in $T_i * T_j$ has degree $\deg(T_i) + \deg(T_j) + 2\int_\beta \omega_X$, just as we noted for the small quantum product in the proof of Proposition 8.1.5. (However, when interested only in the quantum product, the degree condition is typically easier to use than homogeneity.)

As a simple example, consider the elliptic curve Gromov-Witten potential $\Phi = \frac{1}{2}t_0^2 t_3 - t_0 t_1 t_2$ from (8.24). This is homogeneous of degree -4 provided t_0, t_1, t_2, t_3 have respective degrees $-2, -1, -1, 0$.

A useful property of Φ is that by the Point Mapping Axiom, the terms corresponding to $\beta = 0$ reduce to $\frac{1}{6}\langle I_{0,3,0}\rangle(\gamma^3) = \frac{1}{6}\int_X \gamma^3$. Hence we can decompose the Gromov-Witten potential into "classical" and "quantum" terms (this is the terminology of [**FP**]) as follows:

$$
\begin{aligned}
\Phi &= \Phi_{\text{classical}} + \Phi_{\text{quantum}} \\
&= \frac{1}{6}\int_X \gamma^3 + \sum_n \sum_{\beta\in H_2(X,\mathbb{Z})-\{0\}} \frac{1}{n!}\langle I_{0,n,\beta}\rangle(\gamma^n)\,q^\beta.
\end{aligned}
$$

Note that by the Fundamental Class Axiom, the variable t_0 appears *only* in the classical part $\Phi_{\text{classical}}$ of Φ.

Another nice property of Φ is that by the Divisor Axiom, the variables corresponding to divisors appear in exponential form in Φ_{quantum}. To see why, order the basis elements T_i so that $T_1,\ldots,T_r$ is a basis for $H^2(X,\mathbb{Q})$, and write

$\gamma = t_0 T_0 + \delta + \epsilon$, where $\delta = \sum_{i=1}^{r} t_i T_i$ and $\epsilon = \sum_{i=r+1}^{m} t_i T_i$. Then, using the binomial theorem and the Divisor Axiom, we obtain

$$
\begin{aligned}
\Phi &= \frac{1}{6} \int_X \gamma^3 + \sum_n \sum_{\beta \neq 0} \frac{1}{n!} \langle I_{0,n,\beta} \rangle ((\delta + \epsilon)^n) \, q^\beta \\
&= \frac{1}{6} \int_X \gamma^3 + \sum_{a,b} \sum_{\beta \neq 0} \frac{1}{a!\,b!} \langle I_{0,b,\beta} \rangle (\epsilon^b) \left(\int_\beta \delta \right)^a q^\beta \\
&= \frac{1}{6} \int_X \gamma^3 + \sum_b \sum_{\beta \neq 0} \frac{1}{b!} \langle I_{0,b,\beta} \rangle (\epsilon^b) e^{\int_\beta \delta} \, q^\beta
\end{aligned}
$$
(8.29)

This formula is discussed in [**KoM1**, Prop. 4.4] and has some implications for quantum cohomology (to be discussed in Section 8.5.1).

8.3.2. $\mathbb{P}^2$ Revisited. For $\mathbb{P}^2$, we first compute the Gromov-Witten potential. The cohomology basis is $T_0 = [\mathbb{P}^2] = 1$, $T_1 = [\ell]$ and $T_2 = [pt]$, so that by (8.28), t_0, t_1, t_2 have degrees $-2, 0, 2$ and $q = q^\ell$ has degree -6. As usual, we write Gromov-Witten invariants as $\langle I_{0,n,d} \rangle$. In this notation, $N_d = \langle I_{0,3d-1,d} \rangle (T_2^{3d-1})$ is the number of degree d rational curves passing through $3d - 1$ generic points in $\mathbb{P}^2$.

We know that Φ is homogeneous of degree -2, and the classical part of Φ is $\frac{1}{2}(t_0^2 t_2 + t_0 t_1^2)$. A typical term of the quantum part is

$$
\langle I_{0,a+b,d} \rangle (T_1^a, T_2^b) \frac{t_1^a \, t_2^b}{a!\,b!} \, q^d,
$$

where $2b + (-6)d = -2$ by homogeneity. This gives $b = 3d - 1$, so that the term can be written

$$
\langle I_{0,3d-1,d} \rangle (T_2^{3d-1}) \frac{(dt_1)^a}{a!} \frac{t_2^{3d-1}}{(3d-1)!} \, q^d = N_d \frac{(dt_1)^a}{a!} \frac{t_2^{3d-1}}{(3d-1)!} \, q^d,
$$

where d^a comes from the Divisor Axiom. (This reproduces a calculation done earlier in Section 7.4.2, though here it goes more quickly.) Thus the Gromov-Witten potential of $\mathbb{P}^2$ is

$$
\Phi = \frac{1}{2}(t_0^2 t_2 + t_0 t_1^2) + \sum_{d=1}^{\infty} N_d e^{dt_1} \frac{t_2^{3d-1}}{(3d-1)!} \, q^d.
$$

Now we can have some fun. For indices $(i,j,k,l) = (1,1,2,2)$, the WDVV equation from Theorem 8.2.4 becomes

$$
\Phi_{222} + \Phi_{111}\Phi_{122} = \Phi_{112}^2,
$$
(8.30)

where the subscripts indicate partial derivatives. This is easy to work out since $\Phi_{0jk} = g_{jk}$ by (8.26). To exploit (8.30), note that N_d appears in Φ_{222} in the term

$$
N_d \frac{1}{(3d-4)!} t_2^{3d-4} \, q^d.
$$

This suggests looking at terms containing $t_2^{3d-4} q^d$ in Φ_{112}^2 and $\Phi_{111}\Phi_{122}$. In Φ_{112}, the terms with no t_1 are

$$
N_{d_1} d_1^2 \frac{1}{(3d_1-2)!} t_2^{3d_1-2} q^{d_1},
$$

so that the term containing $t_2^{3d-4}q^d$ in Φ_{112}^2 is

$$\sum_{d_1+d_2=d} N_{d_1} N_{d_2} d_1^2 d_2^2 \frac{1}{(3d_1-2)!} \frac{1}{(3d_2-2)!} t_2^{3d-4} q^d.$$

We leave it to the reader to show similarly that the term containing $t_2^{3d-4}q^d$ in $\Phi_{111}\Phi_{122}$ is

$$\sum_{d_1+d_2=d} N_{d_1} N_{d_2} d_1^3 d_2 \frac{1}{(3d_1-1)!} \frac{1}{(3d_2-3)!} t_2^{3d-4} q^d.$$

From here, the recursion relation

$$N_d = \sum_{d_1+d_2=d} N_{d_1} N_{d_2} \left(d_1^2 d_2^2 \binom{3d-4}{3d_1-2} - d_1^3 d_2 \binom{3d-4}{3d_1-1} \right)$$

follows immediately from (8.30).

It is also easy to determine the big quantum cohomology of $\mathbb{P}^2$. For example, since $\Phi_{011} = 1$, we have

$$T_1 * T_1 = T_2 + \Phi_{111}T_1 + \Phi_{112}T_0.$$

In contrast, the small quantum product for $\mathbb{P}^2$ has $T_1 * T_1 = T_2$, so there is a big difference between the two quantum cohomologies. Note also that associativity of the big quantum product is essentially equivalent to the WDVV equation discussed in Section 8.4. Often, associativity is easier to use in practice. For example, (8.30) follows directly from writing out $(T_1 * T_1) * T_2 = T_1 * (T_1 * T_2)$ and comparing the coefficients of T_0. Hence the recursion relation for N_d is a straightforward consequence of associativity of the big quantum product. As noted in [**FP**], "the quantum formalism has removed any necessity to be clever. One simply writes down the associativity equations, and reads off enumerative information." A systematic approach for doing this is discussed in [**DI**], and applications to $\mathbb{P}^3$ and quadric threefolds can be found in [**FP**]. See also [**Kresch**] for a deeper study of associativity equations and their relation to Kontsevich's First Reconstruction Theorem (Theorem 7.4.1).

For $\mathbb{P}^2$, one can also generalize N_d to the number $N_d(a,b,c)$, which counts the number of rational curves in $\mathbb{P}^2$ of degree d passing through a general points, tangent to b general lines, and tangent to c general lines at specified general points on each line (for $a+b+2c = 3d-1$). These invariants are studied in [**EK1**], which derives recursion relations for the $N_d(a,b,c)$ extending the relations for $N_d(a,b,0)$ given in [**DI**]. Furthermore, one can use the $N_d(a,b,c)$ to define a generalization of the big quantum cohomology of $\mathbb{P}^2$ called the *contact cohomology ring* [**EK2**].

8.3.3. The Quintic Threefold Revisited. We will first study the Gromov-Witten potential of a quintic threefold V. A cohomology basis of $H^*(V,\mathbb{C})$ is given by $T_0 = [V], T_1 = [H], T_2 = [\ell], T_3 = [pt]$, and $\gamma_1, \ldots, \gamma_{204}$, where the γ_i are a basis of $H^3(V,\mathbb{C})$. The corresponding supercommutative variables will be t_0, t_1, t_2, t_3 and $u_1, \ldots, u_{204}$. Also let $q_{ij} = \int_V \gamma_i \cup \gamma_j$. Then the classical part of Φ is

$$\Phi_{\text{classical}} = \frac{1}{2}t_0^2 t_3 + t_0 t_1 t_2 + \frac{5}{6}t_1^3 - \sum_{i<j} q_{ij} t_0 u_i u_j,$$

where the minus sign is due to $\epsilon(\alpha)$ in (8.22).

Turning to the quantum part, we know that Φ and q^β have degree 0 and that t_0 doesn't appear in Φ_{quantum}. Since t_2, t_3 and the u_i have positive degree by (8.28), we see that Φ_{quantum} depends only on t_1. But T_1 is the class of a divisor (the hyperplane section H), so that (8.29) implies

$$(8.31) \qquad \Phi_{\text{quantum}} = \sum_{d \geq 1} \langle I_{0,0,d} \rangle e^{dt_1} q^d,$$

where as usual we write $\langle I_{0,0,d} \rangle$ instead of $\langle I_{0,0,\beta} \rangle$. Thus the Gromov-Witten potential of the quintic threefold is

$$\Phi = \frac{1}{2} t_0^2 t_3 + t_0 t_1 t_2 + \frac{5}{6} t_1^3 - \sum_{i<j} q_{ij} t_0 u_i u_j + \sum_{d \geq 1} \langle I_{0,0,d} \rangle (e^{t_1} q)^d.$$

From here, it is relatively simple to describe the big quantum product of V. It coincides with cup product for all pairs of basis elements except for $T_1 * T_1$, which is given by

$$T_1 * T_1 = \Phi_{111} T_2.$$

Furthermore, one has

$$\int_X T_1 * T_1 * T_1 = \Phi_{111},$$

and, since T_1 is the class of H, Φ_{111} is given by the formula

$$\Phi_{111} = \sum_{d \geq 1} \langle I_{0,0,d} \rangle d^3 (e^{t_1} q)^d = \sum_{d \geq 1} \langle I_{0,3,d} \rangle (H, H, H)(e^{t_1} q)^d.$$

If we replace $e^{t_1} q$ with q in these formulas, we recover *precisely* the small quantum cohomology of the quintic threefold.

More generally, for any smooth Calabi-Yau threefold, we can write the Gromov-Witten potential in an especially nice form. First observe that $\beta \neq 0$ implies

$$(8.32) \qquad \langle I_{0,n,\beta} \rangle(\alpha_1, \dots, \alpha_n) = 0 \quad \text{unless the } \alpha_i \text{ all have degree 2.}$$

Then, if $T_1, \dots, T_r$ is our chosen basis of $H^2(V, \mathbb{C})$, (8.32) implies that Φ_{quantum} is a function of $t_1, \dots, t_r$, and by (8.29), these variables appear in exponential form. Hence Φ_{quantum} can be written

$$\Phi_{\text{quantum}} = \sum_{\beta \neq 0} \langle I_{0,0,\beta} \rangle e^{\int_\beta \delta} q^\beta,$$

where $\delta = \sum_{i=1}^r t_i T_i$. If we drop the factor of q^β and let $\mathbf{q}^\beta = e^{\int_\beta \delta} = \Pi_{i=1}^r e^{t_i \int_\beta T_i}$, then the Gromov-Witten potential is given by the elegant formula

$$(8.33) \qquad \Phi = \Phi_{\text{classical}} + \sum_{\beta \neq 0} \langle I_{0,0,\beta} \rangle \, \mathbf{q}^\beta.$$

Using (8.32) and (8.33), we see that

$$\frac{\partial^3 \Phi}{\partial t_i \partial t_j \partial t_k} = \int_V T_i \cup T_j \cup T_k + \sum_{\beta \neq 0} \langle I_{0,0,\beta} \rangle \int_\beta T_i \int_\beta T_j \int_\beta T_k \, \mathbf{q}^\beta$$

$$(8.34) \qquad\qquad = \langle I_{0,3,0} \rangle(T_i, T_j, T_k) + \sum_{\beta \neq 0} \langle I_{0,3,\beta} \rangle(T_i, T_j, T_k) \, \mathbf{q}^\beta$$

$$= \langle T_i, T_j, T_k \rangle.$$

where the second equality uses the Point Mapping and Divisor Axioms and the third equality uses (8.19). This shows that replacing the Gromov-Witten potential with (8.33) transforms Definition 8.2.2 into the definition of the small quantum product. It follows that there is essentially no difference between the big and small quantum cohomologies for a Calabi-Yau threefold. One can also check that the WDVV equations boil down to the associativity of cup product in this case. This is very different from $\mathbb{P}^2$, where the big quantum product and WDVV equation contain *much* more enumerative information than the small quantum cohomology.

In Section 8.5.1, we will see that for general varieties there is essentially no difference between the big and small quantum products, provided we set the parameters t_i to 0 whenever T_i has degree different from 2.

8.4. Dubrovin Formalism

The algebraic consequence of the Gromov-Witten potential is the big quantum product. In this section, we will explore the main geometric consequence of Φ, which is the *Dubrovin connection*. In Section 8.5, we will use the Dubrovin formalism to construct a naturally occurring variation of Hodge structure on the Kähler moduli of a Calabi-Yau manifold.

Rather than define the Dubrovin connection just for the Gromov-Witten potential Φ, we will follow [**Dubrovin1**] and develop a more general theory which uses an arbitrary potential function F.

As usual, we have the basis $T_0 = 1, \ldots, T_m$ of $H^*(X, \mathbb{C})$ and the associated supercommutative variables $t_0, \ldots, t_m$. We also have the matrix (g_{kl}) of the intersection pairing and its inverse (g^{kl}). Then let $F : H^*(X, \mathbb{C}) \to \mathbb{C}$ be an even formal power series in the t_i. In nice cases, F will converge on an open set of the supermanifold $H^*(X, \mathbb{C})$. Given F, we define the tensor A_{ijk} by

$$(8.35) \qquad A_{ijk} = \frac{\partial^3 F}{\partial t_i \partial t_j \partial t_k}.$$

If we indicate partial derivatives by subscripts, note that $A_{ijk} = F_{kji}$. Following the usual physics convention for "raising indices", we set

$$(8.36) \qquad A_{ij}^k = \sum_l A_{ijl} g^{lk}.$$

We can now define two basic objects: the binary operation

$$T_i * T_j = \sum_k A_{ij}^k T_k$$

and the *Dubrovin connection*

$$\nabla^\lambda_{\frac{\partial}{\partial t_i}} \left(\frac{\partial}{\partial t_j} \right) = \lambda \sum_k A_{ij}^k \frac{\partial}{\partial t_k},$$

where λ is a nonzero complex number.

There is a strong relation between the binary operation $*$ and the connection ∇^λ, where properties of one translate into properties of the other. Here is a brief sketch of three important aspects of this relationship:

- **Torsion and Commutativity.** An easy consequence of (8.35) and (8.36) is the identity

$$A_{ij}^k = (-1)^{\deg t_i \deg t_j} A_{ji}^k.$$

Using this, one can prove without difficulty that the connection ∇^λ has zero torsion (in the supermanifold sense) and that the binary operation $*$ is supercommutative.

- **Curvature and Associativity.** By computing the curvature (in the supermanifold sense) of the connection ∇^λ, one finds that $(\nabla^\lambda)^2$ is a sum of two terms. The first term vanishes since

$$\frac{\partial}{\partial t_l} A_{ij}^k = (-1)^{\deg t_i \deg t_l} \frac{\partial}{\partial t_i} A_{lj}^k$$

by (8.35) and (8.36), and the second term vanishes if and only if

$$\sum_a A_{ij}^a A_{ak}^l = (-1)^{\deg t_i (\deg t_j + \deg t_k)} \sum_a A_{jk}^a A_{ai}^l.$$

We recognize this as the WDVV equation, and then it is easy to show

$$\nabla^\lambda \text{ is a flat connection} \iff F \text{ satisfies the WDVV equation}$$

$$\iff * \text{ is associative.}$$

- **Identity.** We would like T_0 to be the identity for $*$. We have the easy equivalence

$$A_{0ij} = g_{ij} \text{ for all } i,j \iff A_{0i}^j = \delta_{ij} \text{ for all } i,j,$$

and then it is straightforward to prove that

$$T_0 \text{ is the identity for } * \iff A_{0ij} = g_{ij} \text{ for all } i,j$$

$$\iff \nabla^\lambda_{\frac{\partial}{\partial t_0}} \left(\frac{\partial}{\partial t_i} \right) = \lambda \frac{\partial}{\partial t_i} \text{ for all } i.$$

Proofs of these assertions can be found in [**Dubrovin1, KoM1, Manin3**].

From here on, we will always assume that F satisfies the WDVV equation and that $A_{0ij} = g_{ij}$ for all i,j. This means that the Dubrovin connection ∇^λ is flat and torsion free and that $\nabla^\lambda_{\frac{\partial}{\partial t_0}}$ acts as multiplication by λ on $\frac{\partial}{\partial t_i}$. In terms of the binary operation $*$, the consequences for $H^*(X,\mathbb{C})$ are as follows.

PROPOSITION 8.4.1. *If F satisfies the WDVV equation and $A_{0ij} = g_{ij}$ for all i,j, then $H^*(X,\mathbb{C})$ is a Frobenius algebra under $*$ with T_0 as identity.*

PROOF. Our hypothesis implies that $*$ gives a supercommutative algebra with T_0 as identity. So we only need to prove that it is Frobenius. However, another consequence of (8.35) is

$$A_{ijk} = (-1)^{\deg t_i (\deg t_j + \deg t_k)} A_{jki}.$$

We know also that F is an even function in the t_i. These facts imply that

$$g(T_i * T_j, T_k) = g(T_i, T_j * T_k).$$

Thus $H^*(X,\mathbb{C})$ is a Frobenius algebra under $*$. $\qquad\square$

The potential functions Φ and Φ_{small} we studied earlier satisfy the conditions of Proposition 8.4.1. This is why the big and small quantum products make $H^*(X,\mathbb{C})$ into Frobenius algebras.

The construction of the Dubrovin connection defines ∇^λ on the tangent bundle of the supermanifold $M = H^*(X,\mathbb{C})$. If we identify $\frac{\partial}{\partial t_i}$ with T_i, then we get a connection ∇^λ on the trivial bundle $H^*(X, \mathbb{C}) \times M$. One easily checks that this new ∇^λ is given by

$$\nabla^\lambda_{\frac{\partial}{\partial t_i}}(T_j) = \lambda T_i * T_j,$$

so that we can think of the Dubrovin connection as being given by quantum multiplication. This point of view will be useful in the next section.

The Dubrovin formalism has some further aspects, which we now discuss. First note that the theory described so far doesn't use the homogeneity of Φ mentioned in Section 8.3.1. Translated into terms of the Dubrovin connection, homogeneity means that there is an even vector field E, called the *Euler vector field*, which interacts nicely with ∇^λ.

The second thing to observe is that ∇^λ is a one-parameter family of connections ∇^λ. If we regard λ as an affine coordinate for $\mathbb{P}^1$, then [**KoM1, Manin3**] show how ∇^λ extends to a connection $\widehat{\nabla}^\lambda$ on the tangent bundle of $H^*(X, \mathbb{C}) \times \mathbb{P}^1$ with singular points at $\lambda = 0, \infty$. For $X = \mathbb{P}^2$, the Laplace transform of $\widehat{\nabla}^\lambda$ has some surprising connections to the sixth Painlevé equation, as described in [**Manin3**].

The structure provided by the supermanifold $H^*(X, \mathbb{C})$ with the metric g and the even potential function F satisfying the WDVV equation is an example of a *Frobenius manifold* [**Dubrovin1, Dubrovin2**]. This has some interesting consequences. For example, consider the power series expansion of F. Ignoring terms of degree ≤ 2 in the t_i, we can write F in the form

$$F = \sum_{n=3}^{\infty} \sum_{|\alpha|=n} c_\alpha t^\alpha.$$

Then define $\langle I_n \rangle : H^*(X, \mathbb{C})^{\otimes n} \to \mathbb{C}$ by the equation

$$c_\alpha = \frac{\epsilon(\alpha)}{\alpha!} \langle I_n \rangle (T^\alpha), \quad n \geq 3,$$

where we are using the notation of (8.22). This allows us to write F as

$$F = \sum_{n=3}^{\infty} \frac{1}{n!} \langle I_n \rangle (\gamma^n),$$

which should look familiar. In [**Manin3**], Manin proves the remarkable result that for $n \geq 3$, there are maps

$$I_n : H^*(X, \mathbb{C})^{\otimes n} \longrightarrow H^*(\overline{M}_{0,n}, \mathbb{C})$$

which satisfy the Splitting and Equivariance Axioms and determine $\langle I_n \rangle$ in the usual way, i.e.,

$$\langle I_n \rangle (\alpha_1, \dots, \alpha_n) = \int_{\overline{M}_{0,n}} I_n(\alpha_1, \dots, \alpha_n).$$

The I_n, $n \geq 3$, form what Manin calls a *cohomological field theory*, and the $\langle I_n \rangle$ are its *correlation functions*. The conclusion is that any even function F on $H^*(X, \mathbb{C})$ satisfying the WDVV equation leads to a structure surprisingly similar to the Gromov-Witten classes. We should also mention that the Künneth formula for quantum cohomology, which was mentioned in Section 8.2.3, holds more generally for cohomological field theories [**KoM2, Kaufmann**]. One can also define higher genus cohomological field theories [**Manin3, KK**].

Finally, it is important to remember that when discussing the Dubrovin formalism, the function F can be a formal power series. Hence the connections ∇^λ may be formal connections, and for this reason they are sometimes called *formal Dubrovin connections*. In this situation, we must use one of the coefficient rings discussed in Section 8.1.3 in order to define the corresponding ring structure on cohomology.

8.5. The A-Variation of Hodge Structure

In this section, we will show how small quantum cohomology gives a natural variation of Hodge structure on the Kähler moduli space studied in Chapter 6. This will enable us to formulate a precise mathematical definition of mirror pair in Section 8.6.

8.5.1. Big versus Small Quantum Cohomology. In order to study the A-model connection defined in Section 8.5.2, we need to understand the relation between big and small quantum cohomology.

We have our usual setup with the cohomology basis $T_0 = 1, T_1, \ldots, T_m$ of $H^*(X, \mathbb{Q})$, where $T_1, \ldots, T_r$ span $H^2(X, \mathbb{Q})$. The big quantum product, as defined in Definition 8.2.2, is a formal power series in $t_0, t_1, \ldots, t_m$ and $q^\beta = \exp\left(2\pi i \int_\beta \omega\right)$. We begin by observing that the variables $t_1, \ldots, t_r$ appear in exponential form in this series. More precisely, Definition 8.2.2, Lemma 8.2.3, and the manipulations used in (8.29) easily imply the following lemma.

LEMMA 8.5.1. *If we set $\delta = \sum_{i=1}^r t_i T_i$ and $\epsilon = t_0 T_0 + \sum_{i=r+1}^m t_i T_i$, then the big quantum product is given by*

$$T_i * T_j = \sum_k \sum_{n=0}^\infty \sum_\beta \frac{1}{n!} \langle I_{0,n+3,\beta} \rangle (T_i, T_j, T_k, \epsilon^n) \, e^{\int_\beta \delta} q^\beta \, T^k.$$

Using this lemma, there are several ways to compare big and small quantum cohomology. The simplest is to set the variables $t_0, \ldots, t_m = 0$ in the formula for $T_i * T_j$. In the notation of Lemma 8.5.1, we set $\delta = \epsilon = 0$, which gives

$$T_i * T_j \big|_{\delta=\epsilon=0} = \sum_k \sum_\beta \langle I_{0,3,\beta} \rangle (T_i, T_j, T_k) \, q^\beta \, T^k.$$

Since

$$(8.37) \qquad T_i *_{\text{small}} T_j = \sum_k \sum_\beta \langle I_{0,3,\beta} \rangle (T_i, T_j, T_k) \, q^\beta \, T^k,$$

we see that

$$T_i * T_j \big|_{\delta=\epsilon=0} = T_i *_{\text{small}} T_j,$$

so that the small quantum product can be regarded as the restriction of the big quantum product.

However, since we want to use the Dubrovin formalism of Section 8.4, we will take a slightly different approach, where we "restrict to $H^2(X)$". This means the following. The big quantum product is defined (formally) on $H^*(X)$ in terms of $t_0, \ldots, t_m$ and q^β. Hence "restricting to $H^2(X)$" means setting $t_0 = t_{r+1} = \cdots =$

$t_m = 0$ in the formula for $T_i * T_j$. This is equivalent to setting $\epsilon = 0$ in Lemma 8.5.1, which gives

$$(8.38) \qquad T_i * T_j\big|_{\epsilon=0} = \sum_k \sum_\beta \langle I_{0,3,\beta}\rangle(T_i, T_j, T_k)\, e^{\int_\beta \delta} q^\beta\, T^k.$$

We will call this the *restricted big quantum product*. It is similar, though not identical, to the small quantum product (8.37). The similarity is highlighted if we note that

$$e^{\int_\beta \delta} q^\beta = e^{\int_\beta (\delta + 2\pi i\, \omega)}.$$

Thus, the exponential terms in (8.38) and (8.37) both involve the integral over β of an element of $H^2(X)$, in one case $\delta + 2\pi i\,\omega$ and in the other $2\pi i\,\omega$.

Hence we can get these formulas to agree as follows. If we let $\omega = 0$ and set

$$\mathbf{q}^\beta = \exp\left(\int_\beta \delta\right) = \exp\left(\int_\beta \sum_{i=1}^r t_i T_i\right),$$

then the restricted big quantum product becomes

$$(8.39) \qquad T_i * T_j\big|_{\epsilon=0} = \sum_k \sum_\beta \langle I_{0,3,\beta}\rangle(T_i, T_j, T_k)\, \mathbf{q}^\beta\, T^k.$$

This compares very nicely with (8.37). From a formal point of view, we can regard $\mathbf{q}^\beta$ in (8.39) as a formal symbol depending on $\beta, t_1, \ldots, t_r$ in an appropriate coefficient ring, and then we get the same formal series as (8.37) by using a homomorphism of coefficient rings which sends $\mathbf{q}^\beta$ to q^β.

On the other hand, we can also relate (8.39) and (8.37) when regarded as formal functions on $H^2(X)$ in the variables $t_1, \ldots, t_r$. The idea here is that the functions agree provided we set

$$(8.40) \qquad \omega = \frac{1}{2\pi i} \sum_{i=1}^r t_i T_i$$

in (8.37). Hence the convergence assumptions for (8.37) are the same as for (8.39).

With these conventions, the restriction of the big quantum product to $H^2(X)$ is the small quantum product both formally and analytically.

8.5.2. The A-Model Connection. For the rest of the chapter, let V be a Calabi-Yau manifold of dimension $d \geq 3$. Recall from Chapter 1 that the complexified Kähler space of V is the quotient

$$K_{\mathbb{C}}(V) = \{\omega \in H^2(V, \mathbb{C}) : \mathrm{Im}(\omega) \text{ is Kähler}\}/\mathrm{im}\, H^2(V, \mathbb{Z}),$$

where $\mathrm{im}\, H^2(V, \mathbb{Z})$ is the image of the natural map $H^2(V, \mathbb{Z}) \to H^2(V, \mathbb{C})$. We will now explain how the small quantum product $*_{\text{small}}$ gives an interesting connection over $K_{\mathbb{C}}(V)$.

As usual, we use a cohomology basis $T_0, \ldots, T_m$ of $H^*(V, \mathbb{Q})$. Furthermore, the basis elements $T_1, \ldots, T_r$ lying in $H^2(V, \mathbb{Q})$ will be assumed to satisfy:

- $T_1, \ldots, T_r$ are a basis of $\mathrm{im}\, H^2(V, \mathbb{Z})$.
- $T_1, \ldots, T_r$ are in the closure of the Kähler cone.

Since $H^{2,0}(V) = 0$, it is easy to see that such $T_1, \ldots, T_r$ can be found.

Now introduce variables $u_1, \ldots, u_r$ corresponding to $T_1, \ldots, T_r$ and set

$$(8.41) \qquad \omega = \sum_{i=1}^r u_i T_i \in H^2(V, \mathbb{R}) + iK(V).$$

Then, for any $\beta \in H_2(V, \mathbb{Z})$, the function $q^\beta = e^{2\pi i \int_\beta \omega}$ is well-defined on $K_{\mathbb{C}}(V)$ since $T_i \in \operatorname{im} H^2(V, \mathbb{Z})$. In Chapter 6, we used t_i instead of u_i, but the reason for using u_i here will soon become apparent.

Now let $*_{\text{small}}$ be the small quantum product on $H^*(V, \mathbb{C})$. This requires some convergence assumptions, which we will defer until later. Then consider the trivial bundle $\mathcal{H} = H^*(V, \mathbb{C}) \times K_{\mathbb{C}}(V)$ over $K_{\mathbb{C}}(V)$. We define the *A-model connection* ∇ on $\mathcal{H}$ by the rule

$$(8.42) \qquad \nabla_{\frac{\partial}{\partial u_i}}(T_j) = T_i *_{\text{small}} T_j, \quad 1 \le i \le r, \ 0 \le j \le m.$$

Thus ∇ is built from the small quantum product, though it only multiplies by elements of $H^2(V, \mathbb{C})$. Writing out ∇ using the definition of small quantum product gives

$$(8.43) \qquad \nabla_{\frac{\partial}{\partial u_i}}(T_j) = \sum_k \sum_{\beta \in H_2(V,\mathbb{Z})} \langle I_{0,3,\beta}\rangle(T_i, T_j, T_k)\, q^\beta\, T^k.$$

Note that this connection is defined on $K_{\mathbb{C}}(V)$ and that q^β is a nonconstant function on this space. Thus ∇ records the variation of the small quantum product as we vary ω.

Strictly speaking, ∇ is only a formal connection since convergence of the series in (8.43) is still conjectural. For the sake of exposition, however, we will regard ∇ as an honest connection. Section 8.5.3 will make precise convergence assumptions concerning ∇.

The connection ∇ has the following nice property.

PROPOSITION 8.5.2. *The A-model connection ∇ is flat.*

PROOF. This will follow from the flatness of the Dubrovin connection ∇^λ, provided we are careful in how we invoke the theory of Section 8.4.

We will apply the Dubrovin formalism to the Gromov-Witten potential with $q^\beta = 1$. This gives a flat connection ∇^λ on $H^*(V, \mathbb{C})$ defined by

$$(8.44) \qquad \nabla^\lambda_{\frac{\partial}{\partial t_i}}(T_j) = \lambda\, T_i * T_j,$$

where $*$ is the big quantum product and $t_0, \dots, t_m$ are the usual variables for $H^*(V, \mathbb{C})$. When we restrict ∇^λ to $H^2(V, \mathbb{C})$ as in Section 8.5.1, we claim that

$$(8.45) \qquad \nabla = \nabla^\lambda \quad \text{for} \quad \lambda = \frac{1}{2\pi i}.$$

Once we show this, the proposition follows immediately from the flatness of ∇^λ.

To prove (8.45), note that by (8.39), the restriction of the Dubrovin connection can be written

$$\nabla^\lambda_{\frac{\partial}{\partial t_i}}(T_j) = \lambda \sum_k \sum_\beta \langle I_{0,3,\beta}\rangle(T_i, T_j, T_k)\, \mathbf{q}^\beta\, T^k,$$

where we use the variables $t_1, \dots, t_r$ for $H^2(V, \mathbb{C})$. By Section 8.5.1, this becomes

$$(8.46) \qquad \nabla^\lambda_{\frac{\partial}{\partial t_i}}(T_j) = \lambda\, T_i *_{\text{small}} T_j$$

under the substitution $\mathbf{q}^\beta \mapsto q^\beta$, provided q^β is defined as in (8.40) to be

$$\omega = \frac{1}{2\pi i} \sum_{i=1}^r t_i T_i.$$

The A-model connection is defined using variables $u_1, \ldots, u_r$, and comparing the above equation to (8.41), we see that $u_j = (1/2\pi i)\, t_j$. In terms of the coordinates $q_1, \ldots, q_r$ on the complexified Kähler space, this means

$$q_j = \exp(2\pi i\, u_j) = \exp(t_j),$$

and hence

$$(8.47) \qquad \frac{\partial}{\partial t_j} = q_j \frac{\partial}{\partial q_j} \quad \text{and} \quad \frac{\partial}{\partial u_j} = 2\pi i\, q_j \frac{\partial}{\partial q_j} = 2\pi i\, \frac{\partial}{\partial t_j}.$$

From here, (8.45) follows immediately from (8.42) and (8.46), which completes the proof of the proposition. $\qquad\square$

In Chapters 10 and 11, we will find it more convenient to use the variables $t_1, \ldots, t_r$ rather than the $u_1, \ldots, u_r$ used in the definition of the A-model connection. For the rest of this chapter, however, we will continue to use $u_1, \ldots, u_r$ since they relate more directly to the discussion in Chapter 6.

Also, in Chapter 10, we will encounter a twisted form of ∇^λ. This is the *Givental connection* ∇^g, which is defined by

$$\nabla^g_{\frac{\partial}{\partial t_i}}\Big(\sum_{j=0}^{m} a_j T_j \Big) = \hbar \sum_{j=0}^{m} \frac{\partial a_j}{\partial t_i} T_j - \sum_{j=0}^{m} a_j\, T_i *_{\mathrm{small}} T_j,$$

where $\hbar$ is a parameter. For now, we can think of $\hbar$ as a nonzero complex number, but we will see in Section 10.2.3 that $\hbar$ has an intrinsic meaning in $\mathbb{C}^*$-equivariant cohomology. In terms of the Dubrovin connection ∇^λ, one has

$$\nabla^g = \hbar \nabla^\lambda \quad \text{for} \quad \lambda = -\hbar^{-1}.$$

Strictly speaking, ∇^g is not a connection (because of the factor of $\hbar$), but it still makes sense to speak of its flat sections, which are just the flat sections of $\nabla^{-\hbar^{-1}}$.

8.5.3. Asymptotics of the A-Model Connection.

In defining the A-model connection ∇, we used an integral basis $T_1, \ldots, T_r$ lying in the closure of the Kähler cone of V. To study the asymptotics of ∇, let σ be the simplicial cone generated by $T_1, \ldots, T_r$. One easily sees that $\mathrm{Int}(\sigma) \subset K(V)$, so that

$$\mathcal{D}_\sigma = \big(H^2(V, \mathbb{R}) + i\,\mathrm{Int}(\sigma) \big)/\mathrm{im}\, H^2(V, \mathbb{Z}) \subset K_{\mathbb{C}}(V)$$

is an open subset of $K_{\mathbb{C}}(V)$.

As we saw in Section 6.2.1, the map

$$u_1 T_1 + \cdots + u_r T_r \mapsto (q_1, \ldots, q_r) = (e^{2\pi i u_1}, \ldots, e^{2\pi i u_r})$$

induces a biholomorphism $\mathcal{D}_\sigma \simeq (\Delta^*)^r$, and under the inclusion $(\Delta^*)^r \subset \Delta^r$, we can regard $0 \in \Delta^r$ as a large radius limit point in the terminology of Chapter 6. Furthermore, assuming the cone conjecture from Section 6.2.1, this large radius limit point lives naturally on the boundary of a smooth compactification of the Kähler moduli space of V.

The convergence assumption we want to make concerning the small quantum product is that the series in (8.43) converges for all ω whose images under the above map lie in some fixed neighborhood U of $0 \in \Delta^r$. Hence the A-model connection is defined over $U \cap \Delta^r$. However, for simplicity of exposition, we will assume that ∇ is defined over all of Δ^r.

We now apply the methods of Chapter 5 to study the monodromy of the A-model connection. In the partial compactification of D_σ given by Δ^r, the complement $\Delta^r - D_\sigma = \Delta^r - (\Delta^*)^r$ is a divisor with normal crossings $\cup_{j=1}^r D_j$. Let $\mathcal{T}_j$ denote the monodromy about D_j, which is determined by a loop around the origin in the jth factor of $(\Delta^*)^r$. These monodromy transformations are computed as follows.

PROPOSITION 8.5.3. *The A-model connection ∇ of a Calabi-Yau manifold V has regular singular points along $\Delta^r - D_\sigma$. Furthermore, the monodromy transformation $\mathcal{T}_j$ is unipotent and, up to conjugacy, its logarithm $N_j = \log \mathcal{T}_j$ is given by cup product with $-T_j$.*

PROOF. By (8.47), the coordinates $q_j = e^{2\pi i u_j}$ of Δ^r satisfy

$$\frac{\partial}{\partial u_j} = 2\pi i\, q_j \frac{\partial}{\partial q_j},$$

so that (8.43) and $T^l = \sum_s g^{ls} T_s$ imply

$$(8.48) \qquad \nabla_{\frac{\partial}{\partial q_j}}(T_k) = \frac{1}{2\pi i\, q_j} \sum_{l,s} \sum_{\beta \in H_2(V,\mathbb{Z})} \langle I_{0,3,\beta}\rangle(T_j, T_k, T_l)\, g^{ls}\, q^\beta\, T_s.$$

Since β is effective and $T_1, \dots, T_r$ are in the closure of the Kähler cone,

$$q^\beta = q_1^{\int_\beta T_1} \cdots q_r^{\int_\beta T_r}$$

has nonnegative exponents. In fact, we can assume that the exponents are positive whenever $\beta \neq 0$. To see why, suppose that $\int_\beta T_j = 0$ for some j. Then the Divisor Axiom implies

$$\langle I_{0,3,\beta}\rangle(T_j, T_k, T_l) = \left(\int_\beta T_j\right) \langle I_{0,2,\beta}\rangle(T_k, T_l) = 0$$

since $\beta \neq 0$. This allows us to regard the right hand side of (8.48) as a series where q^β has positive exponents in the q_j whenever $\beta \neq 0$ is effective.

By (8.48), the connection matrices of ∇ relative to the basis $T_0, \dots, T_m$ have at worst logarithmic poles along $\Delta^r - D_\sigma$, which by Thm. 4.1 of [**Deligne1**] shows that we have regular singular points.

Now fix j between 1 and r. To compute the monodromy $\mathcal{T}_j$, we work on the "slice" of Δ^r given by $q_i = \text{constant}$ for $i \neq j$. This allows us to regard ∇ as a connection over Δ^* with q_j as coordinate, and the connection matrix of ∇ is given by (8.48) for our fixed j. To compute the residue of this matrix at $q_j = 0$, observe that if $\beta \neq 0$ is effective, then

$$(8.49) \qquad\qquad q^\beta \to 0 \text{ as } q_j \to 0$$

since q^β has positive exponents in the q_j for $\beta \neq 0$.

Then, by (8.49) and (8.48), the residue matrix of $\nabla_{\frac{\partial}{\partial q_j}}$ at $q_j = 0$ is

$$\frac{1}{2\pi i}\left(\sum_l \langle I_{0,3,0}\rangle(T_j, T_k, T_l)\, g^{ls}\right)_{k,s=0,\dots,m} = \frac{1}{2\pi i}\left(\sum_l g^{ls} \int_V T_j \cup T_k \cup T_l\right).$$

We will denote this matrix by $\mathrm{Res}_{q_j=0}(\nabla)$. By the definition of g^{ls}, one sees easily that $\mathrm{Res}_{q_j=0}(\nabla)$ is $1/(2\pi i)$ times the matrix representing the linear map given by cup product with T_j.

To finish the proof, we will use the facts from the second set of bullets following (5.5) in Section 5.1.5. The point is that cup product with T_j is nilpotent, so that

$\mathrm{Res}_{q_j=0}(\nabla)$ is also nilpotent. Hence its only eigenvalue is 0, which implies that the matrix of $\mathcal{T}_j$ is conjugate to $\exp(-2\pi i \mathrm{Res}_{q_j=0}(\nabla))$. This proves that $\mathcal{T}_j$ is unipotent and that its logarithm N_j is conjugate to cup product with $-T_j$. $\square$

Although N_j is conjugate to cup product with $-T_j$, we will see that care is required in interpreting what this means. The situation will become more clear once we understand the canonical extension of $\mathcal{H}$.

Since ∇ has regular singular points, the bundle $\mathcal{H}$ over $(\Delta^*)^r$ has a canonical extension $\bar{\mathcal{H}}$ over Δ^r. The key idea to understanding $\bar{\mathcal{H}}$ is as follows: if s is a multi-valued section of $\mathcal{H}$ which is flat for ∇, then $\tilde{s} = \exp(-\sum_j u_j N_j)\, s$ is a single-valued section of $\mathcal{H}$ which extends naturally to $\bar{\mathcal{H}}$ [**CaK**]. We can thus use sections $\tilde{s}$ to trivialize $\bar{\mathcal{H}}$ over a neighborhood of $0 \in \Delta^r$. (Our signs differ from those in [**CaK**] since we use a counterclockwise monodromy generator.)

One common way to express this is via the connection ∇^c, which is defined by

$$(8.50) \qquad \nabla^c = \nabla + \sum_j N_j\, du_j.$$

An easy computation shows that the ∇^c-flat sections are precisely those of the form $\exp(-\sum_j u_j N_j)\, s$ where s is ∇-flat. Thus ∇^c extends to a connection on $\bar{\mathcal{H}}$, and its flat sections trivialize the canonical extension in a neighborhood of the boundary point $0 \in \Delta^r$.

We can think of our cohomology basis T_k as giving sections of $\mathcal{H}$ over $(\Delta^*)^r$ which trivialize the bundle. Our next task is to relate the T_k to the trivialization given by ∇^c-flat sections.

PROPOSITION 8.5.4. *For each T_k, there is a unique ∇^c-flat section $\tilde{s}_k$ such that $\tilde{s}_k = T_k + $ terms of higher degree and $\tilde{s}_k(0) = T_k$. Furthermore, the matrix of N_j acting on the $\tilde{s}_k$ equals the matrix of cup product with $-T_j$ acting on the T_k.*

PROOF. We first give the proof when $\dim(V) = 3$. For simplicity, we will restrict to even cohomology, so that the cohomology basis can be written as

$$T_0 = 1, T_1, \ldots, T_r, T^1, \ldots, T^r, T^0,$$

where $T^j \in H^4(V, \mathbb{Q})$ are dual to $T_j \in H^2(V, \mathbb{Q})$ for $1 \le j \le r$, and $T^0 \in H^6(V, \mathbb{Q})$ is dual to T_0. These classes are all integral.

We first describe a basis of flat sections of ∇ using the Gromov-Witten potential. We gave a nice formula for Φ at the end of Section 8.3.3, which here can be written

$$(8.51) \qquad \begin{aligned} \Phi &= \frac{(2\pi i)^3}{6} \int_V \left(\sum_{j=1}^r u_j T_j\right)^3 + \sum_{\beta \neq 0} \langle I_{0,0,\beta}\rangle\, q^\beta \\ &= \frac{(2\pi i)^3}{6} \int_V \left(\sum_{j=1}^r u_j T_j\right)^3 + \Phi_{\mathrm{hol}}, \end{aligned}$$

where $q^\beta = q_1^{\int_\beta T_1} \cdots q_r^{\int_\beta T_r}$ and $q_j = \exp(2\pi i u_j)$. As in the proof of Proposition 8.5.3, we can restrict the sum to $\beta \neq 0$ which are effective and satisfy $\int_\beta T_j > 0$ for all j. It follows that Φ_{hol} is holomorphic in the q_j and vanishes at $0 \in \Delta^r$. Then, in terms of Φ, one can easily check that the A-model connection is given by:

$$\nabla_{\frac{\partial}{\partial u_i}}(T^0) = 0$$

$$\nabla_{\frac{\partial}{\partial u_i}}(T^k) = \delta_{ik}T^0, \quad 1 \leq k \leq r$$

$$(8.52) \qquad \nabla_{\frac{\partial}{\partial u_i}}(T_k) = \frac{1}{(2\pi i)^3}\sum_\ell \frac{\partial^3\Phi}{\partial u_i \partial u_k \partial u_\ell}T^\ell, \quad 1 \leq k \leq r$$

$$\nabla_{\frac{\partial}{\partial u_i}}(T_0) = T_i,$$

where we have used (8.47). A straightforward calculation shows that a ∇-flat basis of $\mathcal{H}$ is given by:

$$s^0 = T^0$$

$$s^k = T^k - u_k T^0, \quad 1 \leq k \leq r$$

$$s_k = T_k - \frac{1}{(2\pi i)^3}\sum_\ell \frac{\partial^2\Phi}{\partial u_k \partial u_\ell}T^\ell + \frac{1}{(2\pi i)^3}\frac{\partial\Phi}{\partial u_k}T^0, \quad 1 \leq k \leq r$$

$$(8.53) \qquad s_0 = T_0 - \sum_\ell u_\ell T_\ell - \frac{1}{(2\pi i)^3}\sum_\ell \left(\frac{\partial\Phi}{\partial u_\ell} - \sum_j u_j \frac{\partial^2\Phi}{\partial u_j \partial u_\ell}\right)T^\ell$$

$$+ \frac{1}{(2\pi i)^3}\left(2\Phi - \sum_j u_j \frac{\partial\Phi}{\partial u_j}\right)T^0.$$

These sections are multivalued on $(\Delta^*)^r$ because of the way in which the u_j appear in the "topological" part of Φ, namely $(2\pi i)^3(\int_V (\sum_{j=1}^r u_j T_j)^3)/6$.

The next step is to compute monodromy. The basic observation is that going around the jth slice of $(\Delta^*)^r$ takes u_j to $u_j + 1$. Putting this into the above flat sections enables one to compute the action of $\mathcal{T}_j$ by expanding the partial derivatives of the topological part of Φ in powers of the u_i. Taking the logarithm gives:

$$N_j(s^0) = 0$$

$$N_j(s^k) = -\delta_{jk}s^0, \quad 1 \leq k \leq r$$

$$(8.54) \qquad N_j(s_k) = -\sum_\ell \frac{\partial^3(\Phi/(2\pi i)^3)}{\partial u_j \partial u_k \partial u_\ell}(0)\, s^\ell = -\sum_\ell \int_V T_j \cup T_k \cup T_l \, s^\ell, \ 1 \leq k \leq r$$

$$N_j(s_0) = -s_j.$$

In these equations, it is clear that the coefficients are precisely the entries of the matrix of cup product with $-T_j$ acting on the given cohomology basis.

Once we know how N_j acts on the flat sections of ∇, we can compute what happens when we apply $\exp(-\sum_j u_j N_j)$. Proceeding in this way, we obtain the following ∇^c-flat sections:

$$\tilde{s}^0 = T^0$$

$$\tilde{s}^k = T^k, \quad 1 \leq k \leq r$$

$$(8.55) \qquad \tilde{s}_k = T_k - \frac{1}{(2\pi i)^3}\sum_\ell \frac{\partial^2\Phi_{\text{hol}}}{\partial u_k \partial u_\ell}T^\ell + \frac{1}{(2\pi i)^3}\frac{\partial\Phi_{\text{hol}}}{\partial u_k}T^0, \quad 1 \leq k \leq r$$

$$\tilde{s}_0 = T_0 - \frac{1}{(2\pi i)^3}\sum_\ell \frac{\partial\Phi_{\text{hol}}}{\partial u_\ell}T^\ell + \frac{2}{(2\pi i)^3}\Phi_{\text{hol}}\, T^0$$

These ∇^c-flat sections clearly have the desired form, and they have the value we want at $0 \in \Delta^r$ since Φ_{hol} and its derivatives with respect to $\frac{\partial}{\partial u_j} = 2\pi i q_j \frac{\partial}{\partial q_j}$ vanish at 0. Finally, the operator $\exp(-\sum_j u_j N_j)$ commutes with N_j, so that N_j has the same matrix with respect to (8.53) and (8.55). By (8.54), it follows that N_j acting on (8.55) has the required matrix.

This proves the proposition for the special case of even cohomology on a Calabi-Yau threefold. When V has arbitrary dimension $d \geq 3$, one needs to use the description of the flat sections of ∇ given in Section 10.2.2 in Chapter 10. A complete proof will be given in Corollary 10.2.6. We defer the proof until then. $\square$

This proposition shows that over Δ^r, we can naturally identify the canonical extension $\bar{\mathcal{H}}$ with the trivial bundle $H^*(V, \mathbb{C}) \times \Delta^r$. In particular, the fiber over $0 \in \Delta^r$ is $H^*(V, \mathbb{C})$. Using this, we can now explain the action of the monodromy logarithm N_j. By Proposition 8.5.4, N_j acting on the ∇^c-flat basis $\tilde{s}_k$ is the same as cup product with $-T_j$ acting on the T_k. Since $\tilde{s}_k(0) = T_k$, it follows that *at the fiber over* 0, we have

$$(8.56) \qquad N_j(\alpha) = -T_j \cup \alpha, \quad \alpha \in H^*(V, \mathbb{C}).$$

However, away from $0 \in \Delta^r$, this may no longer hold. For example, when V has dimension 3, the formulas appearing in the proof of Proposition 8.5.4 imply that if we regard T_k, $1 \leq k \leq r$, as a section of $\mathcal{H}$, then

$$N_j(T_k) = -T_j \cup T_k - \frac{1}{(2\pi i)^3} \frac{\partial^2 \Phi_{\mathrm{hol}}}{\partial u_j \partial u_k} T^0, \quad 1 \leq k \leq r.$$

Over most points of Δ^r, this will differ from $-T_j \cup T_k$.

We can also use Proposition 8.5.4 to give $\mathcal{H}$ an integral structure. We begin with the natural integral structure on $H^*(V, \mathbb{C})$, regarded as the fiber over $0 \in \Delta^r$. Then a ∇-flat section s is integral if the value of $\tilde{s} = \exp(-\sum_j u_j N_j)\, s$ at 0 is an integral class in $H^*(V, \mathbb{C})$. The ∇-flat integral sections form a locally constant sheaf $\mathcal{H}_{\mathbb{Z}}$ on $(\Delta^*)^r$, and tensoring with $\mathbb{R}$ gives a locally constant sheaf $\mathcal{H}_{\mathbb{R}}$ of real vector spaces. In particular, there is an action of complex conjugation on $\mathcal{H}$.

An interesting consequence is that if we regard T_k as a section of $\bar{\mathcal{H}}$, then it is neither integral nor real at most fibers. For example, when V has dimension 3, one can check that the ∇-flat sections defined in (8.53) are integral by our choice of the cohomology basis. But these formulas also imply, for example, that

$$T^k = s^k + u_k s^0, \quad 1 \leq k \leq r.$$

Since s^k and s^0 are integral, T^k is usually not even real, much less integral.

Note also that $\mathcal{H}_{\mathbb{Z}}$ is determined by the behavior of $\bar{\mathcal{H}}$ over $0 \in \Delta^r$. Hence it could happen that the integral structure of $\mathcal{H}$ depends on the choice of large radius limit point. We conjecture that this is not the case, i.e., that all such points give the same $\mathcal{H}_{\mathbb{Z}}$. The proof should not be difficult, provided one makes suitable convergence assumptions required for the existence of $\mathcal{H}_{\mathbb{Z}}$.

Since we know the monodromy, we can also determine the monodromy weight filtration of the A-model connection.

PROPOSITION 8.5.5. *For a Calabi-Yau manifold V of dimension $d \geq 3$, the monodromy weight filtration $W_{\bullet}$ of ∇ at $0 \in \Delta^r$ is given by*

$$W_i = \bigoplus_{j \geq 2d - i} H^j(V, \mathbb{C}).$$

PROOF. The filtration $W_\bullet$ is determined by $N = \sum_{i=1}^r a_j N_j$, where $N_j = \log(\mathcal{T}_j)$ and $a_j > 0$. In (8.56), we saw that on the fiber over 0, N_j is cup product with $-T_j$, so that up to a sign, N is cup product with the Kähler class $\omega = \sum_{i=1}^r a_j T_j$ on the fiber over 0. Since

$$\cup \omega^k : H^{d-k}(V, \mathbb{C}) \simeq H^{d+k}(V, \mathbb{C})$$

is an isomorphism by Hard Lefschetz, the procedure described in [**Griffiths2**] for computing the weight filtration easily gives the above description of W_i. In fact, a key property characterizing the monodromy weight filtration is the isomorphism $N^k : W_{d+k}/W_{d+k-1} \simeq W_{d-k}/W_{d-k-1}$, which here is Hard Lefschetz. $\qquad \square$

The weight filtration $W_\bullet$ at 0 can be propagated to a filtration $\mathcal{W}_\bullet$ of the bundle $\bar{\mathcal{H}}$ via ∇^c. When we do this, we get the following result.

COROLLARY 8.5.6. *The monodromy weight filtration* $\mathcal{W}_\bullet$ *of* ∇ *over* Δ^r *is given by the trivial bundle*

$$\mathcal{W}_i = \bigoplus_{j \geq 2d-i} H^j(V, \mathbb{C}) \times \Delta^r.$$

PROOF. At 0, we know that W_i is determined by T_k with $\deg(T_k) \geq 2d - i$. Hence, over Δ^r, it follows that $\mathcal{W}_i$ is determined by $\tilde{s}_k$ with $\deg(T_k) \geq 2d - i$ (this is the notation of Proposition 8.5.4). Since $\tilde{s}_k = T_k +$ terms of higher degree, the corollary now follows easily. $\qquad \square$

Also, using Proposition 8.5.5, we can characterize the monodromy of ∇ as follows.

COROLLARY 8.5.7. *The A-model connection* ∇ *has maximally unipotent monodromy at* $0 \in \Delta^r$.

PROOF. We know that the monodromy is unipotent, so that we need only check conditions (*ii*) and (*iii*) of Definition 5.2.2. Since $H^1(V, \mathbb{C}) = 0$, the previous proposition implies that $\dim W_0 = \dim W_1 = \dim H^{2d}(V, \mathbb{C}) = 1$ and $\dim W_2 = \dim(H^{2d}(V, \mathbb{C}) \oplus H^{2d-2}(V, \mathbb{C})) = 1 + r$. Furthermore, if $g_0 \in H^{2d}(V, \mathbb{C})$ is the Poincaré dual of a point and $g_1, \ldots, g_r \in H^{2d-2}(V, \mathbb{C})$ are dual to $T_1, \ldots, T_r \in H^2(V, \mathbb{C})$ under cup product, then at the fiber over $0 \in \Delta^r$, we have

$$N_i(g_j) = -T_i \cup g_j = -(\textstyle\int_V T_i \cup g_j)g_0 = -\delta_{ij} g_0,$$

where the first equality is by (8.56). The matrix (δ_{ij}) is obviously invertible, so that we have maximally unipotent monodromy. $\qquad \square$

This corollary helps explain why we put so much emphasis on maximally unipotent monodromy in Chapter 5. Under mirror symmetry, the A-model connection should correspond to the Gauss-Manin connection describing the variation of the complex structure of the mirror (see Section 8.6 for a precise statement). Since the A-model connection has nice boundary points with maximally unipotent monodromy, the Gauss-Manin connection of the mirror should have the same property.

The A-model connection also satisfies an analog of the *integrality conjecture* discussed in Section 6.3.1. To see why, note that by choosing $T_1, \ldots, T_r$ to be a basis for $H^2(V, \mathbb{Z})$ in the proof of Corollary 8.5.7, then $g_0, g_1, \ldots, g_r$ is a basis of $H^{2d}(V, \mathbb{Z}) \oplus H^{2d-2}(V, \mathbb{Z})$. Then the integrality conjecture follows immediately from $N_i(g_j) = -\delta_{ij} g_0$. As might be expected, this behavior of the A-model connection is one of the motivations for the integrality conjecture on the mirror side.

Our final comment about the A-model connection concerns the basis $T_0, \ldots, T_m$ of $H^*(V)$. The T_j can be regarded as sections of the canonical extension $\overline{\mathcal{H}}$, but they are not flat for either ∇ or ∇^c, nor are they integral, except at $0 \in \Delta^r$. However, if we look at their behavior on the graded pieces of the weight filtration $\mathcal{W}_\bullet$, then a different picture emerges.

COROLLARY 8.5.8. *If $[T_j]$ denotes the induced section on $Gr_\bullet^{\mathcal{W}}$, then the $[T_j]$ form a basis of flat, integral sections of the canonical extension of $Gr_\bullet^{\mathcal{W}}$.*

PROOF. Corollary 8.5.6 and the definition of ∇ easily imply that $[T_j]$ is flat for the induced connection on $Gr_\bullet^{\mathcal{W}}$. One can also check that the monodromy is trivial on $Gr_\bullet^{\mathcal{W}}$, so that ∇ and ∇^c induce the same connection. This, together with the integrality of T_j above $0 \in \Delta^r$, shows that $[T_j]$ is integral. They obviously form a basis, so that we are done. $\square$

This corollary is interesting for the following reason. In Section 5.6.3, we studied the variation of Hodge structure coming from a family of Calabi-Yau threefolds at a maximally unipotent boundary point. If you look back at the proof of Proposition 5.6.1, you'll see that we began with a basis of $Gr_\bullet^{\mathcal{W}}$ consisting of flat, integral sections and then lifted back to get a basis of $\mathcal{H}$. By Corollary 8.5.8, this is precisely what the T_j do in the A-model case.

The last two corollaries show that the A-model connection is remarkably similar to the Gauss-Manin connection of a family of Calabi-Yau threefolds. This similarity will get deeper once we define the A-model Hodge filtration, which is our next topic of discussion.

8.5.4. The A-Model Hodge Filtration. Our next task is to turn the A-model connection into a variation of Hodge structure. For this, we need something to play the role of the Hodge filtration. As usual, we will assume that V is a Calabi-Yau manifold of dimension $\dim(V) = d \geq 3$. Then set

$$(8.57) \qquad F^p = \bigoplus_{a \leq d-p} H^{a,b}(V).$$

This gives a decreasing filtration $H^*(V, \mathbb{C}) = F^0 \supset F^1 \supset \cdots$. In terms of the bundle $\mathcal{H} = H^*(V, \mathbb{C}) \times K_{\mathbb{C}}(V)$, the subbundles

$$\mathcal{F}^p = F^p \times K_{\mathbb{C}}(V) \subset \mathcal{H}$$

determine a filtration of $\mathcal{F}^0 = \mathcal{H}$.

Our basic claim is that the filtration $\mathcal{F}^\bullet$, together with the A-model connection ∇, form a variation of Hodge structure of weight d, at least near a large radius limit point of the complexified Kähler space $K_{\mathbb{C}}(V)$. To formulate this more precisely, recall that the integral classes $T_1, \ldots, T_r$ generate an open cone $\sigma \subset K(V)$. This gives $(\Delta^*)^r \subset K_{\mathbb{C}}(V)$, and then $0 \in \Delta^r$ is the large radius limit point in question.

In Section 8.5.3, we defined an integral structure $\mathcal{H}_{\mathbb{Z}}$ on $\mathcal{H}$, which in particular gives an action of complex conjugation on $\mathcal{H}$. Then proving that $(\mathcal{H}, \nabla, \mathcal{H}_{\mathbb{Z}}, \mathcal{F}^\bullet)$ is a variation of Hodge structure reduces to showing the following two properties:

- $\nabla(\mathcal{F}^p) \subset \mathcal{F}^{p-1} \otimes \Omega^1_{K_{\mathbb{C}}(V)}$ for every p.
- $\mathcal{F}^p \oplus \overline{\mathcal{F}^{d-p+1}} = \mathcal{H}$ for every p.

The first of these is easy to prove, since the small quantum product is compatible with the Hodge decomposition by Proposition 8.1.7. The argument is as follows. Given $\alpha \in H^{a,b}(V)$ with $a \leq d - p$, we can regard α as a section of $\mathcal{F}^p$. Then

$$\nabla_{\frac{\partial}{\partial u_i}}(\alpha) = T_i *_{\text{small}} \alpha \in H^{a+1,b+1}(V)$$

since $T_i \in H^{1,1}(V)$. It follows that $\nabla_{\frac{\partial}{\partial u_i}}(\alpha)$ is a section of $\mathcal{F}^{p-1}$, which proves Griffiths transversality for ∇.

However, proving the second property, that $\mathcal{F}^p \oplus \overline{\mathcal{F}^{d-p+1}} = \mathcal{H}$, will take more work. For this purpose, we introduce the following pairing: if $\alpha \in H^k(V, \mathbb{C})$ and $\beta \in H^\ell(V, \mathbb{C})$, then

$$(8.58) \qquad S(\alpha, \beta) = \begin{cases} (-1)^{k(k+1)/2} \int_V \alpha \cup \beta & \text{if } k + \ell = 2d \\ 0 & \text{if } k + \ell \neq 2d. \end{cases}$$

If we think of α and β as sections of $\mathcal{H}$, then the above formula for $S(\alpha, \beta)$ still makes sense, so that we can regard S as being defined on $\mathcal{H}$.

The pairing S has the following useful properties.

LEMMA 8.5.9. *Let S be the bilinear form on $\mathcal{H}$ defined above. Then:*

(i) *S has parity $(-1)^d$, i.e., $S(\alpha, \beta) = (-1)^d S(\beta, \alpha)$.*

(ii) *S is flat with respect to ∇, i.e., for any sections α, β of $\mathcal{H}$, we have*

$$\frac{\partial}{\partial u_i} S(\alpha, \beta) = S\big(\nabla_{\frac{\partial}{\partial u_i}}(\alpha), \beta\big) + S\big(\alpha, \nabla_{\frac{\partial}{\partial u_i}}(\beta)\big)$$

for all $1 \leq i \leq r$.

(iii) *The monodromy logarithms N_j are infinitesimal automorphisms of S, i.e., for any $\alpha, \beta \in H^*(V, \mathbb{C})$, we have*

$$S(N_j(\alpha), \beta) + S(\alpha, N_j(\beta)) = 0.$$

(iv) *S takes integer values on $\mathcal{H}_\mathbb{Z}$, i.e, $S(\mathcal{H}_\mathbb{Z}, \mathcal{H}_\mathbb{Z}) \subset \mathbb{Z}$.*

(v) *$S(\mathcal{F}^p, \mathcal{F}^{d-p+1}) = 0$ for all p.*

PROOF. For the first part of the lemma, suppose that $\alpha \in H^k(V, \mathbb{C})$ and $\beta \in H^\ell(V, \mathbb{C})$ with $k + \ell = 2d$. Then $\alpha \cup \beta = (-1)^{k\ell} \beta \cup \alpha$, and the desired equality $S(\alpha, \beta) = (-1)^d S(\beta, \alpha)$ follows easily from the definition of S.

To prove the second part of the lemma, we can assume that $\alpha = T_j$ and $\beta = T_\ell$. Then, using the definition of $\nabla_{\frac{\partial}{\partial u_i}}$ in terms of quantum product with T_i, we are reduced to showing that

$$S(T_i *_{\text{small}} T_j, T_\ell) + S(T_j, T_i *_{\text{small}} T_\ell) = 0.$$

We can assume that $\deg(T_j) = k$ and $\deg(T_\ell) = 2d - k - 2$, since otherwise the above equation is trivially true. Then $\deg(T_i *_{\text{small}} T_j) = k + 2$, so that

$$S(T_i *_{\text{small}} T_j, T_\ell) = (-1)^{(k+2)((k+2)+1)/2} \int_V (T_i *_{\text{small}} T_j) \cup T_\ell$$

$$= -(-1)^{k(k+1)/2} \int_V (T_i *_{\text{small}} T_j) *_{\text{small}} T_\ell,$$

where the second equality follows since $\int_V \alpha *_{\text{small}} \beta = \int_V \alpha \cup \beta$ by Proposition 8.1.6. A similar argument shows that

$$S(T_j, T_i *_{\text{small}} T_\ell) = (-1)^{k(k+1)/2} \int_V T_j *_{\text{small}} (T_i *_{\text{small}} T_\ell).$$

Since T_i has even degree, $T_i *_{\text{small}} T_j = T_j *_{\text{small}} T_i$. Combining this with the associativity of the small quantum product, we get the desired identity.

For the third part of the lemma, note that (ii) implies $S(\mathcal{T}_j(a), \mathcal{T}_j(b)) = S(a, b)$ for any $a, b \in H^*(V, \mathbb{C})$, since $\mathcal{T}_j(a)$ and $\mathcal{T}_j(b)$ are obtained from a and b respectively by parallel translation. Since the logarithm of a unipotent orthogonal transformation is an infinitesimal orthogonal transformation, the desired result follows.

To prove (iv), let s_1, s_2 be integral ∇-flat sections. Then $S(s_1, s_2)$ is constant by part (ii) of the lemma. Setting $\tilde{s}_i = \exp(-\sum_j u_j N_j)\, s_i$, part (iii) implies that $S(\tilde{s}_1, \tilde{s}_2) = S(s_1, s_2)$ is also constant, so that $S(s_1, s_2) = S(\tilde{s}_1(0), \tilde{s}_2(0))$. Then $S(s_1, s_2) \in \mathbb{Z}$ since $\tilde{s}_i(0)$ is integral by the definition of $\mathcal{H}_{\mathbb{Z}}$.

Turning to the final part of the lemma, suppose that $\alpha \in H^{a,b}(V)$, $a \le d - p$, and $\beta \in H^{e,f}(V)$, $e \le p - 1 = d - (d-p+1)$. Then α and β give sections of $\mathcal{F}^p$ and $\mathcal{F}^{d-p+1}$ respectively. Since $\alpha \cup \beta$ has Hodge type $(a+e, b+f)$ with $a + e \le d - 1$, $\alpha \cup \beta$ is not of type (d, d), and then $S(\alpha, \beta) = 0$ follows immediately. $\qquad\square$

One of the key steps in proving that ∇ and $\mathcal{F}^\bullet$ give a variation of Hodge structure is to show that S polarizes a certain mixed Hodge structure. The mixed Hodge structure in question comes from the filtration $F^\bullet$ defined in (8.57) together with the weight filtration from Proposition 8.5.5. One easily checks that

$$(8.59) \qquad F^p \cap \overline{F^{k-p}} \cap W_k = H^{d-p, d-k+p}(V).$$

Since $Gr_k^W = H^{2d-k}(V, \mathbb{C})$, we can regard Gr_k^W as having a pure Hodge structure of weight k such that $(Gr_k^W)^{p, k-p} = H^{d-p, d-k+p}(V)$. It follows that $(W_\bullet, F^\bullet)$ is a mixed Hodge structure.

Recall also that at $0 \in \Delta^r$, the monodromy logarithm N_j is given by cup product with $-T_j$. It follows that $N = \sum_j a_j N_j$, $a_j > 0$, is cup product with $-\omega$, where $\omega = \sum_j a_j T_j$ is a Kähler class. We saw in the proof of Proposition 8.5.5 that $W_\bullet$ is the monodromy weight filtration of N. Then we have the following observation from [**CKS2**].

PROPOSITION 8.5.10. *For any $N = \sum_j a_j N_j$ with $a_j > 0$ for all j, the mixed Hodge structure $(W_\bullet, F^\bullet)$ is polarized by $-N$ and S.*

PROOF. We begin by recalling from [**CaK**] what it means for $(W_\bullet, F^\bullet)$ to be polarized by $-N$ and S. The definition first requires that $W_\bullet$ be the monodromy weight filtration of N (as noted above), that $N(F^p) \subset F^{p-1}$ (obvious from the definitions), and that N is an infinitesimal automorphism of S (this follows from the proof of Lemma 8.5.9). Furthermore, if $k \ge d$, the "primitive part" $P_k \subset Gr_k^W$ is defined by

$$P_k = \ker(L^{k-d+1} : Gr_k^W \to Gr_{2d-k-2}^W).$$

Then the final part of the definition of polarized mixed Hodge structure requires that the pure Hodge structure on P_k be polarized by $S(\cdot, (-N)^{k-d}(\cdot))$ (the minus sign is because our N is the negative of the N used in [**CaK**]). Thus, to prove the proposition, we must show that if $\alpha \in P_k^{p, k-p}$ is nonzero, then

$$(8.60) \qquad S\big(i^{p-(k-p)}\alpha, (-N)^{k-d}(\overline{\alpha})\big) > 0.$$

The key point is that we've set things up so that this becomes the usual polarization property of primitive cohomology. To see how this works, recall that $Gr_k^W = H^{2d-k}(V, \mathbb{C})$ and that $(Gr_k^W)^{p, k-p} = H^{d-p, d-k+p}(V)$. As in the proof of

Proposition 8.5.5, one sees easily that since N is cup product with $-\omega$, the primitive part $P_k \subset Gr_k^W$ is precisely the usual primitive cohomology $H_0^{2d-k}(V, \mathbb{C})$ with respect to ω. Thus, if $\alpha \in P_k^{p,k-p} = H_0^{d-p,d-k+p}(V)$ is nonzero, then the usual Hodge-Riemann bilinear relation (5.1) implies that

$$(-1)^{(2d-k)((2d-k)-1)/2} \int_V \omega^{d-(2d-k)} \wedge i^{(d-p)-(d-k+p)} \alpha \wedge \bar{\alpha} > 0.$$

Using the definition of S, this reduces to (8.60) by easy algebra. □

We can now finally show that we have a variation of Hodge structure.

THEOREM 8.5.11. *Let V be a Calabi-Yau manifold of dimension $d \geq 3$. Then, in a neighborhood of U of $0 \in \Delta^r$, $(\mathcal{H}, \nabla, \mathcal{H}_{\mathbb{Z}}, \mathcal{F}^{\bullet})$ is a variation of Hodge structure over $U \cap (\Delta^*)^r$ of weight d polarized by the bilinear form S.*

PROOF. The philosophy of the proof is similar to how we defined $\mathcal{H}_{\mathbb{Z}}$, where the integral structure over $0 \in \Delta^r$ determined the integral structure of $\mathcal{H}$. In this case, we will use the polarized mixed Hodge structure over $0 \in \Delta^r$ from Proposition 8.5.10 to show that $(\mathcal{H}, \nabla, \mathcal{H}_{\mathbb{Z}}, \mathcal{F}^{\bullet})$ is a variation of Hodge structure near 0.

Our proof will use the powerful machinery of [**CaK, CKS1**], and our notation will follow these papers. We are grateful to the authors of [**CaK**] for explaining the relevance of their work to this situation. Let $\check{D}$ be the flag manifold which classifies filtrations $\widetilde{F}^{\bullet}$ of $H^*(V, \mathbb{C})$ satisfying $\dim(\widetilde{F}^p) = \sum_{a \leq d-p} h^{a,b}(V)$ and $S(\widetilde{F}^p, \widetilde{F}^{d-p+1}) = 0$ for all p. Inside $\check{D}$ is the open set D which classifies those filtrations which induce a polarized Hodge structure of weight d on $H^*(V, \mathbb{C})$.

The filtration $F^{\bullet}$ defined in (8.57) lies in $\check{D}$ since $S(F^p, F^{d-p+1}) = 0$ by Lemma 8.5.9. Then consider the set of filtrations $\exp(-\sum_j u_j N_j) F^{\bullet}$, which is contained in $\check{D}$ since the N_j are infinitesimal automorphisms of S. As defined in [**CaK**], this is a *nilpotent orbit* provided that

- $N_j(F^p) \subset F^{p-1}$.
- $\exp(-\sum_j u_j N_j) F^{\bullet} \in D$ when $\mathrm{Im}(u_j) \gg 0$.

We proved the first bullet in the discussion following the definition (8.57) of the filtration $F^{\bullet}$. For the second bullet, let $N = \sum_j a_j N_j$ with $a_j > 0$. Then N gives the weight filtration $W_{\bullet}$ of Proposition 8.5.5, and $(W_{\bullet}, F^{\bullet})$ is polarized by $-N$ and S by Proposition 8.5.10. It follows from Proposition (4.66) of [**CKS1**] that $\exp(-\sum_j u_j N_j) F^{\bullet}$ is a nilpotent orbit. (We should mention that [**CaK, CKS1**] write the nilpotent orbit as $\exp(\sum_j u_j N_j) F^{\bullet}$ since their N_j is our $-N_j$.)

In more down-to-earth terms, observe that the nilpotent orbit just defined gives a filtration $\widetilde{\mathcal{F}}^{\bullet}$ of $\mathcal{H}$. The key idea, as explained in [**CaK**, p. 74], is as follows. We can regard the filtration $F^{\bullet}$ of $H^*(V, \mathbb{C})$ as lying in the fiber over $0 \in \Delta^r$ of the canonical extension $\bar{\mathcal{H}}$ (this is the notation of Section 8.5.3). Then $\widetilde{\mathcal{F}}^p$ is the subbundle obtained by propagating F^p using ∇^c-flat sections of $\bar{\mathcal{H}}$, where ∇^c is from (8.50). In other words, a ∇^c-flat section $\tilde{s}$ lies in $\widetilde{\mathcal{F}}^p$ if and only if $\tilde{s}(0) \in F^p$.

It follows that close to 0 in $(\Delta^*)^r$, $(\mathcal{H}, \nabla, \mathcal{H}_{\mathbb{Z}}, \widetilde{\mathcal{F}}^{\bullet})$ is a variation of Hodge structure coming from a nilpotent orbit. However, the filtration $\mathcal{F}^{\bullet}$ satisfies Griffiths transversality and agrees with $\widetilde{\mathcal{F}}^{\bullet}$ at $0 \in \Delta^r$. As observed by Deligne [**Deligne2**, Section 2.3], these conditions, together with Theorem (2.8) of [**CaK**], imply that $(\mathcal{H}, \nabla, \mathcal{H}_{\mathbb{Z}}, \mathcal{F}^{\bullet})$ is a variation of Hodge structure in a neighborhood of 0 in $(\Delta^*)^r$. This completes the proof of the theorem. □

In the terminology of [**Morrison7**], we call $(\mathcal{H}, \nabla, \mathcal{H}_\mathbb{Z}, \mathcal{F}^\bullet)$ the *A-variation of Hodge structure*. Given the way we've set things up, it is straightforward to show that the limiting Hodge structure of the A-variation is precisely the mixed Hodge structure $(W_\bullet, F^\bullet)$ considered in Proposition 8.5.10.

For the purposes of mirror symmetry, we need to consider an interesting sub-variation of the A-variation of Hodge structure. Let $\mathcal{H}^{\mathrm{middle}}$ be the bundle coming from $\oplus_{p=0}^d H^{p,p}(V)$, which is the middle part of the Hodge decomposition of $H^*(V, \mathbb{C})$. Note that ∇ preserves $\mathcal{H}^{\mathrm{middle}}$ since the small quantum product is compatible with the Hodge decomposition by Proposition 8.1.7. Then ∇^{middle} will denote the restriction of ∇ to $\mathcal{H}^{\mathrm{middle}}$. Notice also that $\oplus_{p=0}^d H^{p,p}(V)$ has a natural structure over $\mathbb{R}$, which by the procedure described in Section 8.5.3 gives a real structure $\mathcal{H}_\mathbb{R}^{\mathrm{middle}}$. Finally, by abuse of notation, we will let $\mathcal{F}^\bullet$ denote the filtration induced on $\mathcal{H}^{\mathrm{middle}}$ by the A-variation.

With these definitions, it follows easily from Theorem 8.5.11 that we get a real variation of Hodge structure

$$(8.61) \qquad (\mathcal{H}^{\mathrm{middle}}, \nabla^{\mathrm{middle}}, \mathcal{H}_\mathbb{R}^{\mathrm{middle}}, \mathcal{F}^\bullet).$$

This is the *middle A-variation of Hodge structure*, though we will often refer to (8.61) as the *A-variation* for short. All of the above results apply to (8.61). In particular, it follows that (8.61) is polarized by the intersection form

$$(8.62) \qquad S(\alpha, \beta) = (-1)^p \int_V \alpha \cup \beta, \quad \alpha \in H^{p,p}(V), \ \beta \in H^{d-p,d-p}(V).$$

This follows by restricting (8.58) to $\oplus_{p=0}^d H^{p,p}(V)$. We should also mention that $\mathcal{H}^{\mathrm{middle}}$ has a natural structure over $\mathbb{Z}$ when $d = 3$, because $H^{1,1}(V) = H^2(V, \mathbb{C})$ and $H^{2,2}(V) = H^4(V, \mathbb{C})$ in this case. Hence in the threefold case we can replace (8.61) with $(\mathcal{H}^{\mathrm{middle}}, \nabla^{\mathrm{middle}}, \mathcal{H}_\mathbb{Z}^{\mathrm{middle}}, \mathcal{F}^\bullet)$.

We can also describe the behavior of (8.61) at $0 \in \Delta^r$. By Corollary 8.5.7, we have maximally unipotent monodromy, and the limiting mixed Hodge structure over $0 \in \Delta^r$ is $(W_\bullet, F^\bullet)$, where

$$W_k = \bigoplus_{2j \geq 2d-k} H^{j,j}(V), \quad F^p = \bigoplus_{j \leq d-p} H^{j,j}(V).$$

This follows by restriction of the descriptions of $W_\bullet$ and $F^\bullet$ given in Proposition 8.5.5 and (8.57) respectively. Furthermore, these formulas imply $W_{2i} = W_{2i+1}$ for all i and $F^p \oplus W_{2p-2} = \oplus_{j=0}^d H^{j,j}(V)$ for all p. These are precisely the conditions (5.11) for the mixed Hodge structure to be Hodge-Tate. Hence we have proved the following proposition.

PROPOSITION 8.5.12. *The limiting mixed Hodge structure coming from the variation* $(\mathcal{H}^{\mathrm{middle}}, \nabla^{\mathrm{middle}}, \mathcal{H}_\mathbb{R}^{\mathrm{middle}}, \mathcal{F}^\bullet)$ *is Hodge-Tate.*

We conclude this section with a discussion of the three-point function of a Calabi-Yau threefold and how it relates to the A-model connection.

Example 8.5.4.1. Suppose that V is a Calabi-Yau threefold. In Section 8.1.1, we defined the three-point function (or A-model correlation function) $\langle a, b, c \rangle$ for $a, b, c \in H^2(V, \mathbb{C})$ and showed that it can be expressed as

$$\langle a, b, c \rangle = \int_V a *_{\mathrm{small}} b *_{\mathrm{small}} c$$

(see (8.5) and (8.7)). We will discuss two aspects of how $\langle a, b, c \rangle$ relates to the A-model connection ∇^{middle}, which we write as ∇ for short.

We begin by showing how to express $\langle a, b, c \rangle$ in terms of the A-model connection. The key observation is that by definition of ∇, we have

$$\nabla_{\frac{\partial}{\partial u_i}}(T_j *_{\mathrm{small}} T_k) = T_i *_{\mathrm{small}} T_j *_{\mathrm{small}} T_k + \text{terms of lower degree in } H^*(V, \mathbb{C}).$$

Since $1 = T_0$, we easily see that

$$\int_V \nabla_{\frac{\partial}{\partial u_i}} \nabla_{\frac{\partial}{\partial u_j}} \nabla_{\frac{\partial}{\partial u_k}} 1 = \int_V T_i *_{\mathrm{small}} T_j *_{\mathrm{small}} T_k = \langle T_i, T_j, T_k \rangle.$$

This formula can be written in the form

$$(8.63) \qquad \langle T_i, T_j, T_k \rangle = \int_V 1 \cup \nabla_{\frac{\partial}{\partial u_i}} \nabla_{\frac{\partial}{\partial u_j}} \nabla_{\frac{\partial}{\partial u_k}} 1,$$

which is very similar to the formula

$$\int_V \Omega \wedge \nabla_{\theta_1} \nabla_{\theta_2} \nabla_{\theta_3} \Omega$$

for the B-model correlation function or Yukawa coupling defined in Section 5.6. In each case, the correlation function is built in exactly the same way from the connection (A-model or Gauss-Manin), a section of F^3 (1 or Ω), and the intersection pairing. As we will soon see, this similarity is an essential part of mirror symmetry.

The second aspect of the relation between $\langle a, b, c \rangle$ and ∇ is that not only does ∇ determine $\langle a, b, c \rangle$ as just explained, but the converse is also true. This follows from the proof of Proposition 8.5.4. In terms of the Gromov-Witten potential Φ as given in (8.51), we replace (8.34) by

$$(8.64) \qquad \frac{1}{(2\pi i)^3} \frac{\partial^3 \Phi}{\partial u_i \partial u_j \partial u_k} = \langle T_i, T_j, T_k \rangle,$$

and it follows that (8.52) can be written as

$$(8.65) \qquad \begin{aligned} \nabla_{\frac{\partial}{\partial u_i}}(T^0) &= 0 \\ \nabla_{\frac{\partial}{\partial u_i}}(T^k) &= \delta_{ik} T^0, \quad 1 \leq k \leq r \\ \nabla_{\frac{\partial}{\partial u_i}}(T_k) &= \sum_{\ell} \langle T_i, T_k, T_\ell \rangle T^\ell, \quad 1 \leq k \leq r \\ \nabla_{\frac{\partial}{\partial u_i}}(T_0) &= T_i. \end{aligned}$$

Thus ∇ is completely determined by the A-model correlation functions $\langle a, b, c \rangle$.

A special case is where $h^2(V) = 1$. Here, let $H = T_1$ generate $H^2(V, \mathbb{C})$ and $C = T^1 \in H^4(V, \mathbb{C})$ be the dual generator. Also let $u = u_1$. Then, generalizing what we did in Example 8.1.2.3, we have

$$H *_{\mathrm{small}} H = \langle H, H, H \rangle C.$$

Letting $Y(q) = \langle H, H, H \rangle$ denote the three-point function, this becomes

$$(8.66) \qquad H *_{\mathrm{small}} H = Y(q) C.$$

It follows from (8.65) that the matrix of the A-model connection is given by

$$\begin{pmatrix} 0 & 0 & 0 & 0 \\ 1 & 0 & 0 & 0 \\ 0 & Y(q) & 0 & 0 \\ 0 & 0 & 1 & 0 \end{pmatrix}.$$

This matrix (5.62) appeared earlier in the proof of Proposition 5.6.1, which studied Calabi-Yau threefolds with 1-dimensional complex moduli. In particular, the proposition showed that at a maximally unipotent boundary point, the Gauss-Manin connection had the above form, provided that Y was the normalized Yukawa coupling and that we used the mirror map as a local coordinate. As above, this similarity is no accident—it is part of the mirror conjecture.

This example also provides a small preview of Chapter 10. The above matrix implies that $1 = T_0$ satisfies the equation

$$\nabla^2_{\frac{\partial}{\partial u}} \left(\frac{\nabla^2_{\frac{\partial}{\partial u}}}{Y(q)} \right) 1 = 0.$$

In the language of Section 10.3, this is a *quantum differential equation*, and we will see in Example 10.3.2.1 that the above equation implies the relation

$$H *_{\text{small}} H *_{\text{small}} \left(\frac{H *_{\text{small}} H}{Y(q)} \right) = 0$$

in the small quantum cohomology ring of V. This relation is trivial, since the left hand side is an element of $H^8(V)$, which is obviously zero. On the other hand, $Y(q)$ can be extracted from the above quantum differential equation (up to a constant), so that quantum differential equations contain highly nontrivial information about quantum cohomology. Chapter 10 will prove that in general, quantum differential equations similarly give rise to relations in quantum cohomology. We will also give examples to show that many of these relations are nontrivial.

8.6. The Mirror Conjecture

The final task of Chapter 8 is to state a mathematical version of mirror symmetry. For simplicity, we will focus on a local version. Hence the basic idea is that near suitably chosen boundary points of the complex and Kähler moduli spaces, the mirror map discussed in Chapter 6 should turn the Gauss-Manin connection for V into the A-model connection ∇^{middle} for the mirror V°. We begin with a precise statement of what this means in dimension 3.

8.6.1. Mirror Symmetry for Calabi-Yau Threefolds. Assume that V and V° are smooth Calabi-Yau threefolds. As in Section 6.3, we also assume that we have suitably compatible compactifications of the complex moduli space $\mathcal{M}(V)$ of V and the Kähler moduli space $\mathcal{KM}(V^\circ)$ of V°. This means we have:

- A maximally unipotent boundary point p_0 lying on a smooth compactification $\overline{\mathcal{M}}(V)$ of the complex moduli space of V.
- A large radius limit point q_0 lying on a compatible smooth compactification $\overline{\mathcal{KM}}(V^\circ)$ of the Kähler moduli space of the mirror V°.
- The mirror map (6.32), which maps a neighborhood of $p_0 \in \overline{\mathcal{M}}(V)$ to a neighborhood of $q_0 \in \overline{\mathcal{KM}}(V^\circ)$. We will assume that V satisfies the integrality conjecture from Section 5.2.2. By Section 6.3.1, this specifies the mirror map uniquely.

As explained in Section 6.3.1, there are various choices to be made for matching p_0 to q_0 and for specifying the mirror map. We will assume that these choices have been made.

We also have two polarized variations of Hodge structure. On the complex moduli space of V, we have the Gauss-Manin connection and the Hodge filtration on the bundle $\mathcal{H}^V$ given by $H^3(V,\mathbb{C})$. To prevent confusion, the Gauss-Manin connection will be denoted ∇^{GM}. On the mirror side, the Kähler moduli space of V° has the A-model connection ∇^{middle} and the Hodge filtration on the bundle $\mathcal{H}^{V^\circ}$ given by $H^{\mathrm{middle}}(V^\circ) = \oplus_{p=0}^3 H^{p,p}(V^\circ) = \oplus_{p=0}^3 H^{2p}(V^\circ,\mathbb{C})$. Also, $\mathcal{H}^V$ and $\mathcal{H}^{V^\circ}$ have integral structures $\mathcal{H}^V_\mathbb{Z}$ and $\mathcal{H}^{V^\circ}_\mathbb{Z}$ respectively. Finally, we have the following polarizations:

- Since every class $\alpha \in H^3(V,\mathbb{C})$ is primitive, the form

$$Q(\alpha,\beta) = (-1)^{3(3-1)/2} \int_V \alpha \cup \beta = -\int_V \alpha \cup \beta$$

  polarizes $(\mathcal{H}^V, \nabla^{\mathrm{GM}}, \mathcal{H}^V_\mathbb{Z}, \mathcal{F}^\bullet)$ by (5.1).
- In (8.62), we defined the form

$$S(\alpha,\beta) = (-1)^p \int_V \alpha \cup \beta, \quad \alpha \in H^{p,p}(V),\ \beta \in H^{3-p,3-p}(V)$$

  on $\oplus_{p=0}^3 H^{p,p}(V^\circ)$. By Theorem 8.5.11, S polarizes $(\mathcal{H}^{V^\circ}, \nabla^{\mathrm{middle}}, \mathcal{H}^{V^\circ}_\mathbb{Z}, \mathcal{F}^\bullet)$.

In addition to all of this, we also have several distinguished sections:

- For V°, $\mathcal{H}^{V^\circ}$ has the integral generator $[pt]$ of W_0, where $[pt] \in H^{3,3}(V^\circ,\mathbb{Z})$ is the class of a point. There is also the distinguished section $1 = T_0 \in H^0(V,\mathbb{C})$ which generates $\mathcal{F}^3$ by (8.57). Note that $S(1,[pt]) = 1$.
- For V, the bundle $\mathcal{H}^V$ has the section g_0 as in the definition of a maximally unipotent boundary point. Note that g_0 is an integral generator of W_0, and is unique up to ± 1. In addition, there are sections Ω of $\mathcal{F}^3$. To get something canonical, let $y_0 = \langle g_0, \Omega \rangle$. Once g_0 is fixed, then Ω/y_0 is the unique section of $\mathcal{F}^3$ satisfying $Q(\Omega/y_0, g_0) = 1$.

Then we can finally give a precise definition of mirror symmetry for Calabi-Yau threefolds as follows.

DEFINITION 8.6.1. *Given Calabi-Yau threefolds V and V° as above, we say that (V, V°) is a mathematical mirror pair if the mirror map lifts to an isomorphism of the bundles $\mathcal{H}^V$ and $\mathcal{H}^{V^\circ}$ in neighborhoods of $p_0 \in \overline{\mathcal{M}}(V)$ and $q_0 \in \overline{\mathcal{KM}}(V^\circ)$. Furthermore, this isomorphism should preserve the polarized variations of Hodge structure coming from ∇^{GM} and ∇^{middle}, and take the sections Ω/y_0 and g_0 of $\mathcal{H}^V$ to the respective sections 1 and $[pt]$ of $\mathcal{H}^{V^\circ}$.*

Adapting the terminology of [**Morrison7**], a mathematical mirror pair (V, V°) can also be called a *Hodge-theoretic mirror pair*. Definition 8.6.1 is only a local statement, but using the complex and Kähler moduli spaces as discussed in Chapter 6, one can also formulate a more global definition of mathematical mirror pair.

Our next task is to discuss Definition 8.6.1 and give some equivalent formulations. After that, we will discuss what the Mirror Conjecture means.

A first observation is that Definition 8.6.1 assumes that V satisfies the integrality conjecture. As noted at the end of Section 8.5.3, the A-variation of Hodge structure on V° always satisfies the integrality conjecture. Hence our assumption on V is reasonable since the bundle isomorphism of Definition 8.6.1 preserves the integral structure.

A second observation is that there is at most one bundle isomorphism $\mathcal{H}^V \simeq \mathcal{H}^{V^\circ}$ which is compatible with Definition 8.6.1. To see why, note that by [**BG**], $\mathcal{H}^V$ is generated by repeatedly applying $\nabla^{\mathrm{GM}}_{\delta_i}$ to $\widetilde{\Omega} = \Omega/y_0$. On the mirror side, the Lefschetz decomposition shows that $H^{\mathrm{middle}}(V^\circ)$ is generated by $H^{1,1}(V^\circ)$ under cup product, so that the same is true for the small quantum product, at least near q_0. This easily implies that $\mathcal{H}^{V^\circ}$ is generated by repeatedly applying $\nabla^{\mathrm{middle}}_{\delta_i}$ to $T_0 = 1$. The desired uniqueness now follows immediately.

We next discuss Definition 8.6.1 in more down-to-earth terms. This can be done using either correlation functions or potential functions, which will lead to Theorem 8.6.2 and Corollary 8.6.3 below.

We first formulate what a mathematical mirror pair means in terms of correlation functions. On V, we have the B-model correlation functions, also called the normalized Yukawa couplings. Recall from the discussion following Proposition 5.6.1 that there are local coordinates $q_1, \ldots, q_r$ at p_0 such that ∇^{GM} has an especially simple form (5.67), and Example 6.3.1.1 shows that the q_j's determine the mirror map. If $q_j = \exp(2\pi i u_j)$, then according to (5.68), the normalized Yukawa couplings for V are given by

$$
\begin{aligned}
Y_{ijk} &= -\int_V \widetilde{\Omega} \wedge \nabla_{\frac{\partial}{\partial u_i}} \nabla_{\frac{\partial}{\partial u_j}} \nabla_{\frac{\partial}{\partial u_k}} \widetilde{\Omega} \\
&= Q\big(\widetilde{\Omega}, \nabla_{\frac{\partial}{\partial u_i}} \nabla_{\frac{\partial}{\partial u_j}} \nabla_{\frac{\partial}{\partial u_k}} \widetilde{\Omega}\big),
\end{aligned}
$$
(8.67)

where $\widetilde{\Omega} = \Omega/y_0$ is the normalized 3-form, $\nabla = \nabla^{\mathrm{GM}}$ is the Gauss-Manin connection, and the last equality follows from the definition of the polarization Q. These are the B-model correlation functions of V.

On the mirror side, we have $T_1, \ldots, T_r \in H^2(V^\circ, \mathbb{Z})$ such that

$$
(q_1, \ldots, q_r) = (e^{2\pi i u_1}, \ldots, e^{2\pi i u_r}) \mapsto \big[\textstyle\sum_j u_j T_j\big] \in K_{\mathbb{C}}(V^\circ)
$$

gives local coordinates at the large radius limit point q_0. Using these coordinates, (8.7) implies that the A-model correlation functions are given by

$$
\langle T_i, T_j, T_k \rangle = \int_{V^\circ} T_i *_{\mathrm{small}} T_j *_{\mathrm{small}} T_k.
$$
(8.68)

In terms of the A-model connection $\nabla = \nabla^{\mathrm{middle}}$, (8.63) tells us that

$$
\begin{aligned}
\langle T_i, T_j, T_k \rangle &= \int_{V^\circ} 1 \cup \nabla_{\frac{\partial}{\partial u_i}} \nabla_{\frac{\partial}{\partial u_j}} \nabla_{\frac{\partial}{\partial u_k}} 1 \\
&= S\big(1, \nabla_{\frac{\partial}{\partial u_i}} \nabla_{\frac{\partial}{\partial u_j}} \nabla_{\frac{\partial}{\partial u_k}} 1\big),
\end{aligned}
$$
(8.69)

where the last equality uses the definition of the polarization S.

THEOREM 8.6.2. *If V satisfies the integrality conjecture and the mirror map is defined as above, then (V, V°) is a mathematical mirror pair in the sense of Definition 8.6.1 if and only if the B-model correlation functions Y_{ijk} of V and the A-model correlation functions $\langle T_i, T_j, T_k \rangle$ of V° satisfy*

$$
Y_{ijk} = \langle T_i, T_j, T_k \rangle \quad \text{for all } i, j, k.
$$

PROOF. In one direction, the proof is trivial: if we have an isomorphism which takes $\widetilde{\Omega}$ to 1, ∇^{GM} to ∇^{middle}, and Q to S, then the equality of correlation functions follows immediately from (8.67) and (8.69).

Going the other way, assume that the correlation functions match up as desired. We will use the q_j as local coordinates for both the complex moduli of V and the Kähler moduli of V°. The bundle $\mathcal{H}^V$ has basis e_0, $e_j, 1 \leq j \leq r$, $e^j, 1 \leq j \leq r$, e^0 from the discussion following Proposition 5.6.1, and we know that $e_0 = \widetilde{\Omega}$ is our distinguished section. Similarly, $\mathcal{H}^{V^\circ}$ has basis T_0, $T_j, 1 \leq j \leq r$, $T^j, 1 \leq j \leq r$, T^0 from the proof of Proposition 8.5.4, where $T_0 = 1$ is the distinguished section. Then sending $e_i \mapsto T_i$ and $e^i \mapsto T^i$ clearly defines a bundle isomorphism.

To complete the proof, we need to check that the connections are compatible and that the polarizations and integral structures are preserved. Given our assumption that the correlation functions match up, the former is an immediate consequence of (8.52) and (5.67). The polarizations are equally easy to handle, given the definitions of the forms Q and S and the descriptions of e_i, e^i in (5.66) and T_i, T^i in Proposition 8.5.4.

It remains to consider the structure over $\mathbb{Z}$. This is a bit delicate, since the above bases are *not* integral. However, these bases extend to the canonical extension over Δ^r, and over $0 \in \Delta^r$, they are integral. The integral structure on $\mathcal{H}^{V^\circ}$ was defined in Section 8.5.3 by saying that a ∇^{middle}-flat section s is integral if $\exp(-\sum_j u_j N_j)\,s$ is integral above 0. The same is true for $\mathcal{H}^V$, and then it follows easily that the integral structures are preserved by our isomorphism. $\qquad\square$

This theorem is very nice, especially in the way it relates the definition of mathematical mirror pair to the "classical" notion of mirror symmetry introduced in Chapter 1. One drawback of this approach is that when the dimension r of the complex moduli space of V is large, there are potentially many correlation functions to consider, so that the computations become rather complex. One way to simplify things is to use potential functions.

Let's recall how this works. On V, we showed in Lemma 5.6.2 that there is a potential function Φ^{GM}. As explained in Section 5.6.4, this is a function of u_j, where the mirror map is $q_j = e^{2\pi i\, u_j}$. Then, in these variables, Lemma 5.6.2 says that the normalized Yukawa couplings (8.67) are given by

$$(8.70) \qquad Y_{ijk} = \frac{1}{(2\pi i)^3}\frac{\partial^3 \Phi^{\mathrm{GM}}}{\partial u_i \partial u_j \partial u_k} \quad \text{for all } i, j, k.$$

Turning to V°, we have the Gromov-Witten potential Φ^{GW}, which by (8.51) equals

$$\Phi^{\mathrm{GW}} = \frac{(2\pi i)^3}{6}\int_V (\textstyle\sum_{j=1}^r u_j T_j)^3 + \sum_{\beta \neq 0}\langle I_{0,0,\beta}\rangle\, q^\beta,$$

where $q^\beta = q_1^{\int_\beta T_1}\cdots q_r^{\int_\beta T_r}$ and $q_j = \exp(2\pi i\, u_j)$. Similarly, recall that (8.64) says that

$$\langle T_i, T_j, T_k\rangle = \frac{1}{(2\pi i)^3}\frac{\partial^3 \Phi^{\mathrm{GW}}}{\partial u_i \partial u_j \partial u_k} \quad \text{for all } i, j, k.$$

Then Theorem 8.6.2 immediately implies that being a mathematical mirror pair can be formulated in terms of potential functions as follows.

COROLLARY 8.6.3. *If V satisfies the integrality conjecture and the mirror map is defined as above, then (V, V°) is a mathematical mirror pair if and only if*

$$\Phi^{\mathrm{GM}} = \Phi^{\mathrm{GW}}$$

up to quadratic terms in the u_i.

In Section 8.6.2, we will explain how Theorem 8.6.2 and Corollary 8.6.3 apply to the quintic threefold.

In the above discussion of Definition 8.6.1 and its various consequences, the reader should keep in mind the convergence assumptions needed in order for the A-model connection ∇^{middle} to make sense as an actual connection. There are two ways to avoid this difficulty. The first approach is to note that in Theorem 8.6.2, the equation

$$Y_{ijk} = \langle T_i, T_j, T_k \rangle$$

makes perfect sense as a statement about formal power series. So if this equality can be proved, then convergence for $\langle T_i, T_j, T_k \rangle$ follows immediately since Y_{ijk} is known to converge. Then the entire theory of Section 8.5 applies, and everything is fine. Similarly, in the situation of Corollary 8.6.3, it suffices to prove

$$\Phi^{\mathrm{GM}} = \Phi^{\mathrm{GW}}$$

as formal power series, and then everything goes through.

A potentially more interesting approach to the convergence problem would be to create a theory of "formally degenerating variations of Hodge structure", as advocated in [**Morrison7**, Sect. 7.2]. This would require developing a formal version of everything in Section 8.5, and then Definition 8.6.1 would be replaced with a corresponding formal statement.

Now that we have a rigorous definition of mathematical mirror pair, the next step should be to formulate a *Mirror Conjecture*, which would assert every Calabi-Yau threefold V has a mirror V° such that (V, V°) satisfies Definition 8.6.1. Unfortunately, this doesn't work. Rigid Calabi-Yau threefolds V (ones with only trivial deformations, so that $H^{2,1}(V) = 0$) can't have a mirror V° in the sense of Definition 8.6.1, since this would imply $H^{1,1}(V^\circ) \simeq H^{2,1}(V) = 0$, which is impossible. Hence the best we can hope for is a Mirror Conjecture which asserts the existence of a mirror for certain Calabi-Yau threefolds. From this point of view, each of the mirror constructions given in Chapter 4 leads to a Mirror Conjecture. In Section 8.6.4 we will explore in detail what this means for the Batyrev mirror construction.

We close our discussion of mathematical mirror pairs with an interesting question suggested by Corollary 8.6.3. Given any function $\Phi(u_1, \ldots, u_r)$ such that

$$Y_{ijk} = \frac{1}{(2\pi i)^3} \frac{\partial^3 \Phi}{\partial u_i \partial u_j \partial u_k}$$

is holomorphic in $q_j = \exp(2\pi i u_j)$, we can use equations (5.67) to define a connection on a vector bundle, and we also get an obvious Hodge filtration. Using the techniques of Section 8.5.4, one can show that this gives a variation of Hodge structure. We get *lots* of variations this way, because there are lots of possible functions Φ. But which of these variations are geometric, i.e., come from a family of Calabi-Yau threefolds? In other words, what is special about the potential function of a Calabi-Yau variation? Mirror symmetry gives a partial answer, for if Φ comes from a family of Calabi-Yau threefolds, then it should be the Gromov-Witten potential

Φ^{GW} of the mirror. This implies that the potential function Φ has a very particular form. Hence mirror symmetry gives some insight into the general question of which abstract variations of Hodge structure come from geometry.

8.6.2. Mirror Symmetry for the Quintic Threefold. Our next task is to compare the naive version of mirror symmetry for the quintic threefold presented in Chapter 2 with the more sophisticated formulation given in Definition 8.6.1.

In Chapter 2, we stated mirror symmetry for the quintic threefold as

$$(8.71) \qquad 5 + \sum_{d=1}^{\infty} n_d d^3 \frac{q^d}{1 - q^d} = \frac{5}{(1 + 5^5 z) y_0(z)^2} \left(\frac{q}{z} \frac{dz}{dq} \right)^3$$

where the n_d are the instanton numbers of the quintic threefold and z is the moduli coordinate for the quintic mirror (called x in Chapter 2) such that the quintic mirror family has maximally unipotent monodromy at $z = 0$. Furthermore,

$$y_0(z) = \sum_{n=0}^{\infty} \frac{(5n)!}{(n!)^5} (-1)^n z^n$$

$$y_1(z) = y_0(z) \log(-z) + 5 \sum_{n=1}^{\infty} \frac{(5n)!}{(n!)^5} \left[\sum_{j=n+1}^{5n} \frac{1}{j} \right] (-1)^n z^n$$

$$q = \exp(y_1/y_0).$$

Then we have the following nice fact.

THEOREM 8.6.4. *Definition 8.6.1 with $V^\circ =$ the quintic threefold and $V =$ the quintic mirror is equivalent to* (8.71).

PROOF. Given what we now know, this is easy to see. By (8.21), we recognize that the left hand side of (8.71) is the A-model correlation function $\langle H, H, H \rangle$ of the quintic threefold, and in Example 5.6.4.1, we showed that the right hand side is the normalized Yukawa coupling of the quintic mirror. Then (8.71) implies that the series defining the A-model connection converges, and as we've already noted, the quintic mirror family satisfies the integrality conjecture by [**Morrison2**]. Hence the theorem follows from Theorem 8.6.2. $\qquad\square$

In other words, once we prove the formulas in Chapter 2, the full-blown version of mirror symmetry for the quintic threefold presented in Definition 8.6.1 is an immediate consequence. However, the reader should not be deceived by the shortness of the proof just given. A *lot* of work went into what we just said about each side of (8.71). For the A-model, defining the n_d required the definition of Gromov-Witten invariant from Chapter 7, and understanding their enumerative significance was also nontrivial, as explained in Section 7.4.4. Then we had to work out the Hodge theory of the A-model in Section 8.5. On the B-model, we studied the Hodge theory of the mirror family in Chapter 5, and as we saw in Example 5.6.4.1, we needed some extremely explicit monodromy information to prove that the right hand side of (8.71) was the normalized Yukawa coupling.

We next explain how mirror symmetry for the quintic threefold relates to the potential functions, as in Corollary 8.6.3. This will be important, since this is how Chapter 11 will prove mirror symmetry for the quintic threefold.

We begin by describing the A-model and B-model potential functions. These are multivalued functions of q, but rather than writing them in terms of $u_1 = \frac{1}{2\pi i} \log q$, we will instead use $t_1 = \log q$. As mentioned in Section 8.5.1, the variable

$t_1 = 2\pi i\, u_1$ is more natural for the purposes of Chapter 11. For the A-model, this means replacing (8.51) with the Gromov-Witten potential

$$\Phi^{\mathrm{GW}} = \frac{5}{6}t_1^3 + \sum_{d=1}^{\infty}\langle I_{0,0,d}\rangle\, q^d.$$

For the B-model, let Y be the normalized Yukawa coupling given by the right hand side of (8.71). If we think of Y as a function of $t_1 = \log q$, then the potential function is a function $\Phi^{\mathrm{GM}}(t_1)$ whose third derivative is Y. To describe Φ^{GM}, let $z = -x$. If we regard y_0, y_1 as functions of x, then these functions are given by (6.56), and from (6.56) we also get functions y_2, y_3 such that $y_0, \dots, y_3$ give a basis of solutions of the Picard-Fuchs equation (in terms of the variable x) of the quintic mirror. Then the potential function has the following nice formula.

PROPOSITION 8.6.5. *The potential function of the quintic mirror family is*

$$\Phi^{\mathrm{GM}} = \frac{5}{2}\left(\frac{y_1}{y_0}\frac{y_2}{y_0} - \frac{y_3}{y_0}\right)$$

regarded as a function of $t_1 = \log q$ using the mirror map $q = \exp(y_1/y_0)$.

PROOF. We need to show that

$$\frac{d^3}{dt_1^3}\frac{5}{2}\left(\frac{y_1}{y_0}\frac{y_2}{y_0} - \frac{y_3}{y_0}\right) = Y.$$

We did this in Chapter 2, but let's review the argument for completeness. The y_i satisfy the Picard-Fuchs equation derived in Example 5.4.1. Replacing y_i with y_i/y_0 and x with $q = e^{t_1}$, Proposition 5.6.1 shows that the Picard-Fuchs equation becomes

$$\frac{d^2}{dt_1^2}\left(\frac{d^2y/dt_1^2}{Y}\right) = 0.$$

Note also that $t_1 = y_1/y_0$ by the definition of the mirror map. Using this equation and the power series expansions of Y and the y_i, we showed in (2.35) that

$$5\frac{d^2}{dt_1^2}\frac{y_2}{y_0} = Y, \quad 5\frac{d^2}{dt_1^2}\frac{y_3}{y_0} = Yt_1.$$

From here, the argument following (2.40) implies the desired equation. $\square$

Combining this with Corollary 8.6.3, we get the following theorem.

THEOREM 8.6.6. *Definition 8.6.1 with $V^\circ =$ the quintic threefold and $V =$ the quintic mirror is equivalent to the equation*

$$\Phi^{\mathrm{GW}}(\Psi(t_1)) = \frac{5}{2}\left(\frac{y_1}{y_0}\frac{y_2}{y_0} - \frac{y_3}{y_0}\right),$$

where the mirror map is $\Psi(t_1) = y_1/y_0$.

We stated Theorem 8.6.6 in this form since the above equation is the "Mirror Theorem" from [**LLY**] and is how we will prove mirror symmetry for the quintic threefold in Theorem 11.1.1 in Chapter 11.

8.6.3. Higher Dimensions. We next consider the case when V and V° are smooth Calabi-Yau manifolds of dimension $d > 3$. As before, assume that we have corresponding points $p_0 \in \overline{\mathcal{M}}(V)$ and $q_0 \in \overline{\mathcal{KM}}(V^\circ)$. The complication comes when we try to think about the two polarized variations of Hodge structure. We will see that the situation is not quite as nice as the threefold case.

Let's begin with V. Although we have a nice variation of Hodge structure on $H^d(V, \mathbb{C})$, the polarization (5.1) only works on primitive cohomology, which need not be all of $H^d(V, \mathbb{C})$. To remedy this, we fix a Kähler form ω and consider the Lefschetz decomposition. This implies that every $\alpha \in H^d(V, \mathbb{C})$ can be written $\sum_j \omega^j \cup \alpha_{d-2j}$, where $\alpha_{d-2j} \in H_0^{d-2j}(V, \mathbb{C})$ is primitive. Then we have the form

$$Q(\alpha, \beta) = \sum_j (-1)^{d(d-1)/2} (-1)^j \int_V \omega^{2j} \cup \alpha_{d-2j} \cup \beta_{d-2j}.$$

In general, this form is only defined over $\mathbb{R}$ (since ω need not be integral), but it is still a polarization by (5.1). It follows that once we pick a Kähler structure on V, we get a polarized real variation of Hodge structure $(\mathcal{H}^V, \nabla^{\mathrm{GM}}, \mathcal{H}_\mathbb{R}^V, \mathcal{F}^\bullet)$.

On the mirror side, we want to consider the A-variation of Hodge structure on $\oplus_{p=0}^d H^{p,p}(V^\circ)$. Here, while we have the nice polarization form S given by (8.62), the important observation is that in order to define ∇^{middle}, we need to know how $H^{p,p}(V^\circ)$ sits inside $H^{2p}(V, \mathbb{C})$, since quantum cohomology is defined on the latter. This means that we need to fix a complex structure on V°. Furthermore, notice that $\oplus_{p=0}^d H^{p,p}(V^\circ)$ is in general only defined over $\mathbb{R}$. Hence, once we pick a complex structure on V°, we get a polarized real variation of Hodge structure $(\mathcal{H}^{V^\circ}, \nabla^{\mathrm{middle}}, \mathcal{H}_\mathbb{R}^{V^\circ}, \mathcal{F}^\bullet)$.

We also have the distinguished sections $\widetilde{\Omega}$ and g_0 of $\mathcal{H}^V$ and $1 = T_0$ and $[pt]$ of $\mathcal{H}^{V^\circ}$. Then we define higher-dimensional mirror pairs as follows.

DEFINITION 8.6.7. *Given Calabi-Yau manifolds V and V° of dimension d, where $d > 3$, we say that (V, V°) is a* mathematical mirror pair *if we can find a Kähler structure on V and a complex structure on V° such that the mirror map lifts to an isomorphism of the bundles $\mathcal{H}^V$ and $\mathcal{H}^{V^\circ}$ in neighborhoods of $p_0 \in \overline{\mathcal{M}}(V)$ and $q_0 \in \overline{\mathcal{KM}}(V^\circ)$. Furthermore, this isomorphism preserves the polarized real variations of Hodge structure coming from ∇^{GM} and ∇^{middle}, and takes the sections $\widetilde{\Omega}$ and g_0 of $\mathcal{H}^V$ to the respective sections 1 and $[pt]$ of $\mathcal{H}^{V^\circ}$.*

As in the threefold case, a mathematical mirror pair is sometimes called a *Hodge-theoretic mirror pair*. The interesting aspect of this definition is the extra data required—a Kähler structure on V and a complex structure on V°. This data, when added to $\overline{\mathcal{M}}(V)$ and $\overline{\mathcal{KM}}(V^\circ)$, means that we are using the mirror map on the full SCFT moduli space, as pictured in (1.6) in Chapter 1. Notice also that the definition is rather vague about how the extra data is chosen, in contrast to the very specific choice of p_0 and q_0. Hence it is probably best to regard Definition 8.6.7 as only a preliminary version.

Another uncertain aspect of this definition is that the A-variation is always Hodge-Tate at the large radius limit point q_0, while the definition of maximally unipotent monodromy does not imply that the variation of Hodge structure on $H^d(V, \mathbb{C})$ is Hodge-Tate at the corresponding point p_0. Hence Definition 8.6.7 may need strengthening.

In spite of these limitations, Definition 8.6.7 does have a nice relation to d-point functions, which are defined as follows. Suppose that the mirror map is given by $q_1, \dots, q_r$, and as usual, we use the q_j as local coordinates for both complex and Kähler moduli. Also let $q_j = \exp(2\pi i u_j)$, and to simplify notation, let $\delta_j = \partial/\partial u_j$. In this situation, the B-model correlation functions of V are

$$Y_{i_1 \cdots i_d} = (-1)^{d(d-1)/2} \frac{1}{(2\pi i)^d} \int_V \widetilde{\Omega} \wedge \nabla_{\delta_{i_1}} \cdots \nabla_{\delta_{i_d}} \widetilde{\Omega}$$

for $1 \leq i_i, \dots, i_d \leq r$, and similarly the A-model correlation functions of V° are

$$\langle T_{i_1}, \dots, T_{i_d} \rangle = \int_{V^\circ} T_{i_1} *_{\text{small}} \cdots *_{\text{small}} T_{i_d}$$

These correlation functions are related as follows.

PROPOSITION 8.6.8. *If (V, V°) is a mathematical mirror pair in the sense of Definition 8.6.7, then*

$$Y_{i_1 \cdots i_d} = \langle T_{i_1}, \dots, T_{i_d} \rangle \quad \text{for all} \quad 1 \leq i_i, \dots, i_d \leq r.$$

PROOF. As in the proof of Theorem 8.6.2, Definition 8.6.7 implies that

$$(8.72) \qquad Q\big(\widetilde{\Omega}, \nabla_{\delta_{i_1}} \cdots \nabla_{\delta_{i_d}} \widetilde{\Omega}\big) = S\big(1, \nabla_{\delta_{i_1}} \cdots \nabla_{\delta_{i_d}} 1\big),$$

where on the left, ∇ denotes the Gauss-Manin connection, and on the right, it denotes the A-model connection.

To untangle the left hand side, first note that $\widetilde{\Omega}$ is primitive since it is a $(d, 0)$-form, and similarly, the $(0, d)$ component of $\nabla_{\delta_{i_1}} \cdots \nabla_{\delta_{i_d}} \widetilde{\Omega}$ is also primitive. It then follows without difficulty from the definition of Q that the left hand side of (8.72) is the B-model correlation function $Y_{i_1 \cdots i_d}$.

Turning to the right hand side of (8.72), observe that since the T_{i_j} all have degree 2,

$$\nabla_{\delta_{i_1}} \cdots \nabla_{\delta_{i_d}} 1 = T_{i_1} *_{\text{small}} \cdots *_{\text{small}} T_{i_d} + \text{terms of degree} < 2d.$$

Since $1 \in H^0(V^\circ, \mathbb{C})$, the definition of S implies that the right hand side is

$$\int_{V^\circ} 1 \cup \nabla_{\delta_{i_1}} \cdots \nabla_{\delta_{i_d}} 1 = \int_{V^\circ} \nabla_{\delta_{i_1}} \cdots \nabla_{\delta_{i_d}} 1 = \int_{V^\circ} T_{i_1} *_{\text{small}} \cdots *_{\text{small}} T_{i_d},$$

which is precisely the A-model correlation function $\langle T_{i_1}, \dots, T_{i_d} \rangle$. $\qquad\square$

Although we get a nice correspondence of correlation functions when $d > 3$, it is no longer the case that preserving these correlation functions is equivalent to preserving the variations of Hodge structure. This is very different from the threefold case. To see more precisely what this means, let's work out an example.

Example 8.6.3.1. Suppose that V and V° are Calabi-Yau fourfolds. On V°, we are only concerned with classes in $\oplus_{p=0}^4 H^{p,p}(V^\circ)$. Pick a cohomology basis $T_j \in H^{1,1}(V^\circ)$ for $1 \leq i \leq r$ and $T_k \in H^{2,2}(V^\circ)$ for $r + 1 \leq k \leq s$. Also let $T^\ell \in H^{2,2}(V^\circ)$ be the dual basis, so that $\langle T_k, T^\ell \rangle = \delta_{k\ell}$. Then the Degree Axiom, Fundamental Class Axiom and Divisor Axiom show that for $\beta \neq 0$, the only nonzero Gromov-Witten invariants we need to consider are $\langle I_{0,3,\beta} \rangle (T_i, T_j, T_k)$ for $1 \leq i, j \leq r$ and $r + 1 \leq k \leq s$. Define

$$Y_{ijk}^{\text{A}} = \sum_\beta \langle I_{0,3,\beta} \rangle (T_i, T_j, T_k)\, q^\beta T^k,$$

where the superscript "A" stands for "A-model". Then it follows that all of the enumerative information is carried by

$$\nabla_{\delta_i}(T_j) = T_i *_{\text{small}} T_j = \sum_{k=r+1}^{s} Y_{ijk}^{\text{A}} T^k,$$

where ∇ is the A-model connection. In terms of the 4-point function $\langle T_i, T_j, T_a, T_b \rangle$ defined above, note that

$$\langle T_i, T_j, T_a, T_b \rangle = \int_{V^\circ} T_i *_{\text{small}} T_j *_{\text{small}} T_a *_{\text{small}} T_b$$

$$(8.73) \qquad = \int_{V^\circ} (T_i *_{\text{small}} T_j) \cup (T_a *_{\text{small}} T_b)$$

$$= \sum_{k,\ell=r+1}^{s} Y_{ijk}^{\text{A}} Y_{ab\ell}^{\text{A}} \langle T^k, T^\ell \rangle.$$

where the middle equality follows from Proposition 8.1.6.

We now turn to the B-model on the fourfold V. Assume that V satisfies the integrality conjecture at our chosen maximally unipotent boundary point and also that the limiting mixed Hodge structure is Hodge-Tate. Then, as before, we can find coordinates $q_1, \ldots, q_r$ such that if $e_0 = \widetilde{\Omega}$, then $e_i = \nabla_{\delta_i}(e_0)$ for $1 \leq j \leq r$ gives a basis of $\mathcal{W}_6 \cap \mathcal{F}^3$. If we now let e^k be some basis of $\mathcal{W}_4 \cap \mathcal{F}^2$, then we can write

$$\nabla_{\delta_i}(e_j) = \sum_{k=r+1}^{s} Y_{ijk}^{\text{B}} e^k$$

for some functions Y_{ijk}^{B} (where, of course, "B" is for "B-model").

Now suppose that (V, V°) is a mathematical mirror pair. By Section 6.3.2, the derivative of the mirror map is $e_i \mapsto T_i$ for $1 \leq i \leq r$. If we apply Proposition 8.6.8, we get an equality of 4-point functions, so that we can compute the A-model correlation function $\langle T_i, T_j, T_a, T_b \rangle$ in terms of the B-model. But as shown by (8.73), this no longer gives direct knowledge of the Gromov-Witten invariants of V°, which are encoded in the Y_{ijk}^{A}.

As pointed out in [**Morrison7**], what we're really interested in is an equality $Y_{ijk}^{\text{A}} = Y_{ijk}^{\text{B}}$ which would tell us how to find Gromov-Witten invariants using the B-model. But this requires the full isomorphism $\mathcal{H}^V \simeq \mathcal{H}^{V^\circ}$ and in particular uses $e^k \mapsto T^k$ for $r+1 \leq k \leq s$. The latter is equivalent to knowing the isomorphism $H^{2,2}(V) \simeq H^{2,2}(V^\circ)$. Hence we need the 4-dimensional version of the mirror symmetry isomorphisms (1.5) considered in Chapter 1.

However, given a natural choice for an isomorphism $H^{2,2}(V) \simeq H^{2,2}(V^\circ)$, it should be possible to formulate a precise version of Definition 8.6.7 which implies $Y_{ijk}^{\text{B}} = Y_{ijk}^{\text{A}}$. As a consequence, we could predict the Gromov-Witten invariants of V° in terms of the Hodge theory of V. This approach to mirror symmetry in dimension 4, together with some examples, is discussed in [**Morrison7**]. Furthermore, if the isomorphism $H^{2,2}(V) \simeq H^{2,2}(V^\circ)$ preserves the real structure of the limiting mixed Hodge structure, then Definition 8.6.7 is equivalent to the equality $Y_{ijk}^{\text{B}} = Y_{ijk}^{\text{A}}$ for all i, j, k. So there is an analog of Theorem 8.6.2 in the fourfold case, though it is a bit more complicated.

Given the definition of mirror pair for dimension $d > 3$, one can now attempt to formulate a general Mirror Conjecture, as is done in [**Morrison7**]. This will be necessarily vague, however, given the existence of rigid Calabi-Yau manifolds and the lack of a general description of the mirror. Hence we will not state such a general conjecture, though the toric case will be discussed below.

8.6.4. The Toric Mirror Conjecture. As just noted, one of the difficulties in formulating a general Mirror Conjecture is that given V, we do not know how to find V°. In the toric case, the situation is much nicer since we can use the mirror constructions given in Chapter 4. For simplicity, we will restrict to the case of toric hypersurfaces, so that V° will be the Batyrev mirror of V. As we will see, we get conjectures which are more specific in some ways but less specific in others.

We will assume that we are in the situation of Section 4.1. Thus we fix a reflexive polytope Δ of dimension ≥ 3 and let Σ and Σ° be maximal projective subdivisions of the normal fans of Δ and Δ° respectively. Then we get a minimal Calabi-Yau toric hypersurface $V \subset X_\Sigma$ and its Batyrev dual $V^\circ \subset X_{\Sigma^\circ}$.

In order to make some conjectures, Sections 8.6.1 and 8.6.3 suggest that we should begin with the case when V and V° are threefolds. Here, we get the following *Mirror Conjecture for Toric Threefolds*.

CONJECTURE 8.6.9. *If $V \subset X_\Sigma$ and its Batyrev dual $V^\circ \subset X_{\Sigma^\circ}$ are as above and $\dim(V) = \dim(V^\circ) = 3$, then (V, V°) is a mathematical mirror pair in the sense of Definition 8.6.1.*

This is a nice conjecture, but in practice it is not quite what people usually regard as the Toric Mirror Conjecture. This is because the toric case uses different definitions of the moduli spaces and the mirror map.

Let's recall how this works, assuming now that $\dim(V) = d \geq 3$ is arbitrary. In Chapter 6, we constructed the simplified moduli space $\mathcal{M}_{\mathrm{simp}}(V)$ in Section 6.1.2 and the toric Kähler moduli space $\overline{\mathcal{KM}}_{\mathrm{toric}}(V^\circ)$ in Section 6.2.3. To get boundary points, let σ be the cone generated by a basis $T_1, \dots, T_r$ of $H^2_{\mathrm{toric}}(V^\circ, \mathbb{Z})$ lying in the closure of the Kähler cone. We saw in Chapter 6 that σ gives a maximally unipotent boundary point $p_0 \in \mathcal{M}_{\mathrm{simp}}(V)$ (assuming Conjecture 6.1.4) and a large radius limit point $q_0 \in \overline{\mathcal{KM}}_{\mathrm{toric}}(V^\circ)$. The cone σ also determines explicit local coordinates $z_1, \dots, z_r$ and $q_1, \dots, q_r$ about p_0 and q_0 respectively. Further, as explained in Section 6.3.3, we can specify the toric version of the mirror map completely explicitly in a neighborhood of p_0 and q_0.

In order to make some conjectures, we return to the case when V and V° are threefolds. Here, we know that they are smooth by the comments following the proof of Proposition 4.1.3. To further simplify things, we will first assume that

$$(8.74) \qquad H^{1,1}(X_{\Sigma^\circ}) \longrightarrow H^{1,1}(V^\circ)$$

is surjective. In the notation of Section 4.1.3, this means $H^{1,1}_{\mathrm{toric}}(V^\circ) = H^{1,1}(V^\circ)$ and, by the proof of Theorem 4.1.5, is equivalent to $H^{2,1}_{\mathrm{poly}}(V) = H^{2,1}(V)$. It follows that the moduli spaces described in the previous paragraph are the spaces $\overline{\mathcal{KM}}(V^\circ)$ and $\overline{\mathcal{M}}(V)$ considered in Section 8.6.1.

We still have connections ∇^{GM} and ∇^{middle} and the corresponding variations of Hodge structure. Furthermore, as explained in Section 8.6.1, the derivative of the mirror map, the sections $\widetilde{\Omega}$ and 1, and the polarizations give a unique isomorphism $\mathcal{H}^V \simeq \mathcal{H}^{V^\circ}$. But even if this isomorphism preserves the connections ∇^{GM}

and ∇^{middle}, we can't claim that it preserves the integral structures $\mathcal{H}_{\mathbb{Z}}^V$ and $\mathcal{H}_{\mathbb{Z}}^{V^\circ}$. Here is the problem. The mirror map used in Definition 8.6.1 requires the integrality conjecture and hence preserves the integral structure almost by definition. In contrast, the toric definition of the mirror map given in Section 6.3.3 is potentially different and in particular might not preserve the integral structure. Another way to say this is that the monomial-divisor mirror map discussed in Sections 4.1.3 and 6.3.3 is not known to preserve the integral structure. In fact, as far as we know, it might not preserve the real structure.

Conjecturally, of course, the two mirror maps are the same and V satisfies the integrality conjecture. But in order to formulate a conjecture which doesn't depend on such assumptions, we will concentrate on those structures which only involve the complex numbers. Thus, for V, we have the $\mathbb{C}$-*variation of Hodge structure* given by $(\mathcal{H}^V, \nabla^{\text{GM}}, \mathcal{F}^\bullet)$. This is "polarized" by the form Q, which means a strong form of the first Hodge-Riemann bilinear relation, namely $(\mathcal{F}^p)^\perp = \mathcal{F}^{4-p}$ for all p. Similarly, for V°, we have $(\mathcal{H}^{V^\circ}, \nabla^{\text{middle}}, \mathcal{F}^\bullet)$, which as in Section 8.6.1 is polarized by the form S.

Also observe that by Proposition 8.5.5, the monodromy weight filtration on $\mathcal{H}^{V^\circ}$ satisfies $Gr_4^W = H^2(V, \mathbb{C}) = H^{1,1}(V)$ at $0 \in \Delta^r$. Similarly, on $\mathcal{H}^V$, we have $Gr_4^W = H^{2,1}(V)$ at a maximally unipotent boundary point. This is a consequence of the mixed Hodge structure being Hodge-Tate, which as we noted in (5.11), follows from maximally unipotent monodromy.

In this situation, we get the following special case of the *Hodge-Theoretic Toric Mirror Conjecture*.

CONJECTURE 8.6.10. *As above, let V and V° be 3-dimensional Calabi-Yau toric hypersurfaces which are Batyrev mirrors of each other, and assume that (8.74) is onto. Then the mirror map lifts to an isomorphism of the bundles $\mathcal{H}^V$ and $\mathcal{H}^{V^\circ}$ in neighborhoods of $p_0 \in \overline{\mathcal{M}}(V)$ and $q_0 \in \overline{\mathcal{KM}}(V^\circ)$. Furthermore, this isomorphism preserves the polarized $\mathbb{C}$-variations of Hodge structure coming from ∇^{GM} and ∇^{middle}, and takes the sections $\widetilde{\Omega}$ and g_0 of $\mathcal{H}^V$ to the respective sections 1 and $[pt]$ of $\mathcal{H}^{V^\circ}$. Finally, on Gr_4^W at p_0 and q_0, the induced isomorphism is the monomial-divisor mirror map.*

In this case, we can again formulate mirror symmetry in terms of A-model and B-model correlation functions, similar to Theorem 8.6.2. As before, for V° we have the usual the A-model correlation functions of $\langle T_i, T_j, T_k \rangle$. For V, we define the toric B-model correlation functions to be

$$Y_{ijk} = \int_V \widetilde{\Omega} \wedge \nabla_{\delta_i} \nabla_{\delta_j} \nabla_{\delta_k} \widetilde{\Omega},$$

where $\delta_j = 2\pi i q_j \, \partial/\partial q_j$ and $\widetilde{\Omega} = \Omega/y_0$ is the normalized 3-form. This differs from the normalized Yukawa coupling defined in Section 5.6.4 in two ways:

- Here y_0 is any nonvanishing holomorphic solution of the Picard-Fuchs equation for Ω at the maximally unipotent boundary point. This means that we only know y_0 and hence $\widetilde{\Omega}$ up to a constant.
- We are using the toric version of the mirror map, which is potentially different from the mirror map used in defining the normalized Yukawa coupling.

It follows that Y_{ijk} is only defined up to a multiplicative constant, which is why there is no minus sign in the above formula, in contrast to Definition 5.6.3. But, as we pointed out in Section 5.6.4, our computational techniques only determine Y_{ijk}

up to a constant, unless we make an extremely detailed study of the monodromy. Fortunately, there is still a potential function in this case (the proof of Lemma 5.6.2 still applies), so that there is in essence only one constant to worry about.

From here, it is easy to formulate a version of Theorem 8.6.2, which asserts that Conjecture 8.6.10 is equivalent to

$$(8.75) \qquad Y_{ijk} = \langle T_i, T_j, T_k \rangle \quad \text{for all } i, j, k,$$

where we are allowed to multiply the Y_{ijk} by a constant c (independent of i, j, k) in order to achieve equality. We will omit the details of the straightforward proof,

Here is an example of what Conjecture 8.6.10 looks like when the moduli spaces in question have dimension two.

Example 8.6.4.1. An example considered many times is the toric resolution V of a degree 8 hypersurface in $\mathbb{P}(1,1,2,2,2)$ and its Batyrev mirror V°. We know that V has two-dimensional toric Kähler moduli (which in this case coincides with the number of Kähler moduli), so that (8.74) is surjective. We begin by writing down one of the A-model correlation functions for V.

In Example 6.2.4.3, we saw that a basis of the Kähler cone of V is generated by classes denoted D_3 and D_1. From [**CdFKM**], these classes can be described as follows: $D_3 = H$ is the proper transform of an ample class on $\mathbb{P}(1,1,2,2,2)$ defined by an equation of degree 2, and $D_1 = L$ is the proper transform of a divisor defined by an equation of degree 1. If $E \subset V$ is the exceptional locus of the map $V \to \bar{V}$, then $H \sim 2L + E$. As we noted in Example 6.2.4.3, $H^3 = 8$ and $L^3 = 0$.

The cone σ generated by H, L (which here is the whole Kähler cone) determines a large radius limit point for V and a maximally unipotent boundary point for V°. Local coordinates for the Kähler moduli of V are $q_j = \exp(2\pi i \, u_j)$, where $\sigma = \{u_1 H + u_2 L : u_1, u_2 \geq 0\}$. If we let h, ℓ be the dual basis of $H_2(V, \mathbb{Z})$, then $\beta \in H_2(V, \mathbb{Z})$ can be written $\beta = ah + b\ell$. Since $H^3 = 8$ and $\int_\beta H = a$, it follows that (8.21) gives the equation

$$(8.76) \qquad \langle H, H, H \rangle = 8 + \sum_{(a,b) \neq (0,0)} n(a,b) \, \frac{a^3 \, q_1^a q_2^b}{1 - q_1^a q_2^b},$$

where $n(a,b)$ is the instanton number $n_{ah+b\ell}$ defined in Section 7.4.4. Naively, $n(a,b)$ is the number of rational curves on V in the homology class $ah + b\ell$, but as we saw in Example 7.4.4.1, the relation between instanton numbers and rational curves is more subtle than one first suspects.

Let's use Conjecture 8.6.10 to obtain a conjectural formula for $\langle H, H, H \rangle$. This will give predictions for all of the instanton numbers $n(a,b)$ simultaneously. First observe that the way we've set up the notation, we are using the Toric Mirror Conjecture with V and V° interchanged. We hope this will not cause too much confusion. In the discussion below, we will use ∇ for the Gauss-Manin connection of V°. Then, according to (8.75), the Toric Mirror Conjecture predicts that

$$\langle H, H, H \rangle = Y_{111} = \int_{V^\circ} \widetilde{\Omega} \wedge \nabla_{\delta_1} \nabla_{\delta_1} \nabla_{\delta_1} \widetilde{\Omega},$$

where $\delta_1 = 2\pi i q_1 \frac{\partial}{\partial q_1}$.

To compute the integral on the right hand side, we begin with z_1, z_2, which are the moduli coordinates of V°. Using these variables, we computed the Yukawa couplings

$$K^{ij} = \int_{V^\circ} \Omega \wedge \nabla^i_{z_1 \frac{\partial}{\partial z_1}} \nabla^j_{z_2 \frac{\partial}{\partial z_2}} \Omega, \quad i + j = 3$$

in Example 5.6.2.1. We need to normalize Ω by dividing by y_0 and then switch to the mirror coordinates q_1, q_2. Since

$$\delta_1 = 2\pi i q_1 \frac{\partial}{\partial q_1} = \frac{q_1}{z_1}\frac{\partial z_1}{\partial q_1} z_1 \frac{\partial}{\partial z_1} + \frac{q_1}{z_2}\frac{\partial z_2}{\partial q_1} z_2 \frac{\partial}{\partial z_2},$$

inserting the above equations into the formula for $\langle H, H, H \rangle$ yields

$$\langle H, H, H \rangle = \frac{(2\pi i)^3}{y_0^2}\left[\left(\frac{q_1}{z_1}\frac{\partial z_1}{\partial q_1}\right)^3 K^{30} + 3\left(\frac{q_1}{z_1}\frac{\partial z_1}{\partial q_1}\right)^2\left(\frac{q_1}{z_2}\frac{\partial z_2}{\partial q_1}\right)K^{21}\right.$$

$$\left. + 3\left(\frac{q_1}{z_1}\frac{\partial z_1}{\partial q_1}\right)\left(\frac{q_1}{z_2}\frac{\partial z_2}{\partial q_1}\right)^2 K^{12} + \left(\frac{q_1}{z_2}\frac{\partial z_2}{\partial q_1}\right)^3 K^{03}\right].$$

In Example 5.6.2.1, we computed the K^{ij} explicitly up to a single constant $c \neq 0$, and in Example 6.3.4.2, we computed y_0 and q_2 explicitly as functions of z_1, z_2. The same can be done for q_1, so that the above formula allows us to write $\langle H, H, H \rangle$ in terms of q_1, q_2.

Let's first determine the constant c. We know that $q_j = z_j + \cdots$ by the toric normalization of the mirror map (6.44), and by Example 5.6.2.1,

$$K^{30} = \frac{c}{(1 - 256z_1)^2 - 512^2 z_1^2 z_2},$$

so that $K^{30} = c + \cdots$. Since $y_0 = 1 + \cdots$, the above formula for $\langle H, H, H \rangle$ implies

$$\langle H, H, H \rangle = (2\pi i)^3 c + \cdots.$$

Since $\langle H, H, H \rangle = 8 + \cdots$, we must have

$$c = \frac{8}{(2\pi i)^3},$$

which is similar to what happened in Chapter 2. Using this value for c, we then get the expansion

$$\langle H, H, H \rangle = 8 + 640\frac{q_1}{1 - q_1} + 10032\frac{2^3 q_1^2}{1 - q_1^2} + 640\frac{q_1 q_2}{1 - q_1 q_2} +$$

$$288384\frac{3^3 q_1^3}{1 - q_1^3} + 72224\frac{2^3 q_1^2 q_2}{1 - q_1^2 q_2} + 10979984\frac{4^3 q_1^4}{1 - q_1^4} + \cdots.$$

As many terms as desired can be computed, depending on one's patience and available computing power. Comparing this to (8.76), we get the instanton numbers $n(a, b)$. One can show that $n(a, b) = n(a, a - b)$ when $a \geq 1$ (e.g., $n(1, 1) = n(1, 0) = 640$) and $n(a, b) = 0$ when $b > a$ except for $n(0, 1) = 4$. Hence the series for $\langle H, H, H \rangle$ determines $n(a, b)$ for all $(a, b) \neq (0, 1)$. Further computations with this and other examples of Calabi-Yau toric hypersurfaces can be found in **[CdFKM, HKTY1]**.

In Chapter 11, we will show in Example 11.2.5.1 that the above formulas follow from Theorem 11.2.16. Also, using the detailed monodromy calculations in **[CdFKM]**, one can actually prove that the integrality conjecture is satisfied in this case and that Y_{111} is the normalized Yukawa coupling in the strict sense of Section 5.6.4 once we set $c = 8/(2\pi i)^3$.

We next turn to the case where V and its Batyrev mirror V° are still threefolds, but we no longer assume that the map (8.74) is surjective. In this case, we have to deal with the simplified moduli space $\overline{\mathcal{M}}_{\mathrm{simp}}(V)$ from Section 6.1.2 and the toric Kähler moduli space $\overline{\mathcal{KM}}_{\mathrm{toric}}(V^\circ)$ from Section 6.2.3.

One way to formulate a conjecture would be as follows. We have the bundles $\mathcal{H}^V$ and $\mathcal{H}^{V^\circ}$ on the full moduli spaces, and by restriction, we get bundles

$$\mathcal{H}^V_{\mathrm{simp}} = \text{restriction of } \mathcal{H}^V \text{ to } \overline{\mathcal{M}}_{\mathrm{simp}}(V)$$
$$\mathcal{H}^{V^\circ}_{\mathrm{toric}} = \text{restriction of } \mathcal{H}^{V^\circ} \text{ to } \overline{\mathcal{KM}}_{\mathrm{toric}}(V^\circ).$$

The restricted bundles inherit natural polarized variations of Hodge structure. Then the *Hodge-Theoretic Toric Mirror Conjecture* would assert that there is a bundle isomorphism $\mathcal{H}^V_{\mathrm{simp}} \simeq \mathcal{H}^{V^\circ}_{\mathrm{toric}}$ which preserves the polarized $\mathbb{C}$-variations of Hodge structure (as described in Conjecture 8.6.10). However, we do not state this formally as a conjecture, because a bundle isomorphism $\mathcal{H}^V_{\mathrm{simp}} \simeq \mathcal{H}^{V^\circ}_{\mathrm{toric}}$ requires that we know an isomorphism $H^{2,1}(V) \simeq H^{1,1}(V^\circ)$, yet in the toric case, we only have the isomorphism

$$H^{2,1}_{\mathrm{poly}}(V) \simeq H^{1,1}_{\mathrm{toric}}(V^\circ)$$

coming from the monomial-divisor mirror map discussed in Sections 4.1.3 and 6.3.3. Recall that $H^{2,1}_{\mathrm{poly}}(V)$ is the subspace of $H^{2,1}(V) \simeq H^1(V, \Theta_V)$ corresponding to deformations obtained by varying the defining equation of $V \subset X_\Sigma$. So the problem is that we don't yet understand how to define $\mathcal{H}^V_{\mathrm{simp}} \simeq \mathcal{H}^{V^\circ}_{\mathrm{toric}}$.

However, we can formulate a fairly precise version of the Toric Mirror Conjecture provided we replace $\mathcal{H}^V_{\mathrm{simp}}$ and $\mathcal{H}^{V^\circ}_{\mathrm{toric}}$ with some slightly smaller bundles. To see what this looks like, we begin with V°. Let $i : V^\circ \hookrightarrow X_{\Sigma^\circ}$ be the inclusion map and define

$$H^*_{\mathrm{toric}}(V^\circ) = \mathrm{im}\big(i^* : H^*(X_{\Sigma^\circ}) \to H^*(V^\circ)\big).$$

Note that $H^*_{\mathrm{toric}}(V^\circ)$ consists of cohomology classes of type (p,p) since $H^{p,q}(X_{\Sigma^\circ}) = 0$ for $p \neq q$. It follows that restricting to $H^*_{\mathrm{toric}}(V^\circ)$ gives the subbundle

$$H^*_{\mathrm{toric}}(V^\circ) \times \overline{\mathcal{KM}}_{\mathrm{toric}}(V^\circ) \subset H^{\mathrm{middle}}(V^\circ) \times \overline{\mathcal{KM}}_{\mathrm{toric}}(V^\circ) = \mathcal{H}^{V^\circ}_{\mathrm{toric}}.$$

We will denote this subbundle by $\mathcal{TH}^{V^\circ}_{\mathrm{toric}}$, where the $\mathcal{T}$ stands for the toric subring we are using.

When X_{Σ° is convex (e.g., $X_{\Sigma^\circ} = \mathbb{P}^4$) and V is ample, the argument of Proposition 4 in [**Pandharipande3**] shows that $H^*_{\mathrm{toric}}(V^\circ)$ is a subring of $H^*(V^\circ)$ with respect to the small quantum product. We don't know if this is true in general, so we will further assume

$$(8.77) \qquad H^*_{\mathrm{toric}}(V^\circ) \text{ is a subring of } H^*(V^\circ) \text{ under } *_{\mathrm{small}}.$$

Under this assumption, it follows that the A-model connection ∇^{middle} induces a connection on $\mathcal{TH}^{V^\circ}_{\mathrm{toric}}$. We claim that we in fact get a polarized variation of Hodge structure. To prove this, we will assume that X_{Σ° is smooth. Then we have the following lemma.

LEMMA 8.6.11. *If V° is a Calabi-Yau toric hypersurface in the smooth toric variety X_{Σ°, then the restriction of cup product is nondegenerate on $H^*_{\mathrm{toric}}(V^\circ)$.*

PROOF. To prove the lemma, it suffices to show that

$$\int_{V^\circ} i^*(\alpha) \cup i^*(\beta) = 0 \text{ for all } \beta \in H^*(X_{\Sigma^\circ}) \Longrightarrow i^*(\alpha) = 0.$$

However, by the projection formula, we know that

$$\int_{V^\circ} i^*(\alpha) \cup i^*(\beta) = \int_{X_{\Sigma^\circ}} \alpha \cup \beta \cup [V],$$

and [**HLY1**, (3.36)] implies that for $\alpha \in H^*(X_{\Sigma^\circ})$, we have

$$\alpha \cup [V] = 0 \text{ in } H^*(X_{\Sigma^\circ}) \iff i^*(\alpha) = 0 \text{ in } H^*_{\mathrm{toric}}(V^\circ).$$

These two facts easily imply the required nondegeneracy. $\square$

Assuming (8.77), we can repeat the construction of Section 8.5 with $\mathcal{H}^{V^\circ}_{\mathrm{toric}}$ replaced by $T\mathcal{H}^{V^\circ}_{\mathrm{toric}}$. The polarization comes from the lemma just proved, and as explained in Section 8.5.3, we get an integral structure on $T\mathcal{H}^{V^\circ}_{\mathrm{toric}}$ since $H^*_{\mathrm{toric}}(V^\circ)$ is naturally defined over $\mathbb{Z}$.

On the mirror side, we need to find a corresponding variation of Hodge structure. Here, we go back to the proof of Theorem 4.1.5, where we had the exact sequence

$$\longrightarrow \oplus_i H^1(V \cap D_i, \mathbb{C}) \longrightarrow H^3(V, \mathbb{C}) \longrightarrow Gr_3^W H^3(Z_f, \mathbb{C}) \longrightarrow 0.$$

In this sequence, the D_i are the toric divisors of X_{Σ° and $Z_f = T \cap V$ is the affine hypersurface in the torus $T \subset X_{\Sigma^\circ}$. In the course of the proof, we eventually proved

$$h^{2,1}(Gr_3^W H^3(Z_f, \mathbb{C})) = h^{2,1}_{\mathrm{poly}}(V).$$

This suggests that there is a natural isomorphism

(8.78) $$H^{2,1}_{\mathrm{poly}}(V) \simeq H^{2,1}(Gr_3^W H^3(Z_f, \mathbb{C}))$$

induced by $H^{2,1}_{\mathrm{poly}}(V) \subset H^{2,1}(V) \to H^{2,1}(Gr_3^W H^3(Z_f, \mathbb{C}))$, where the last map comes from the above exact sequence. Since a proof of this has not yet been written down, we will simply assume that (8.78) is an isomorphism.

Given this assumption, the natural way to create a variation of Hodge structure using $H^{2,1}_{\mathrm{poly}}(V)$ is using the natural polarized variation of Hodge structure on $Gr_3^W H^3(Z_f, \mathbb{C})$ over the moduli space $\overline{\mathcal{M}}_{\mathrm{simp}}(V)$. We will denote the resulting bundle by $T\mathcal{H}^V_{\mathrm{simp}}$, where the T now stands for the torus we are intersecting with.

We can now state the *Hodge-Theoretic Toric Mirror Conjecture* for threefolds.

CONJECTURE 8.6.12. *As above, let V and V° be 3-dimensional Calabi-Yau toric hypersurfaces which are Batyrev mirrors of each other, and assume (8.77) and (8.78) and that X_{Σ° is smooth. Then the mirror map lifts to an isomorphism of the bundles $T\mathcal{H}^V_{\mathrm{simp}}$ and $T\mathcal{H}^{V^\circ}_{\mathrm{toric}}$ in neighborhoods of $p_0 \in \overline{\mathcal{M}}(V)_{\mathrm{simp}}$ and $q_0 \in \overline{\mathcal{KM}}(V^\circ)_{\mathrm{toric}}$. Furthermore, this isomorphism preserves the polarized $\mathbb{C}$-variations of Hodge structure coming from ∇^{GM} and ∇^{middle}, and takes the sections $\widetilde{\Omega}$ and g_0 of $\mathcal{H}^V$ to the respective sections 1 and $[\mathrm{pt}]$ of $\mathcal{H}^{V^\circ}$. Finally, on Gr_4^W at p_0 and q_0, the induced isomorphism is the monomial-divisor mirror map.*

Given the number of assumptions made in the statement of this conjecture, we should regard it as "work-in-progress". But we are hopeful that Conjecture 8.6.12 is reasonably close to what is true. One encouraging fact is that when (8.74) is surjective, this conjecture reduces to Conjecture 8.6.10.

Finally, we turn to the case when V and V° have dimension > 3. Here, we run into some immediate problems when trying to formulate a Toric Mirror Conjecture. The main problem is that a minimal Calabi-Yau V may be singular when $\dim(V) > 3$. Yet we have defined Gromov-Witten invariants only for smooth projective varieties. Presumably this is only a technical difficulty—eventually, one hopes that Gromov-Witten theory should be available for a wider class of varieties. Furthermore, if the bundle isomorphism $\mathcal{H}^V_{\mathrm{simp}} \simeq \mathcal{H}^{V^\circ}_{\mathrm{toric}}$ preserves the Hodge filtrations, then we would have

$$(8.79) \qquad h^{p,q}(V) = h^{q,q}(V^\circ), \quad p + q = \dim(V).$$

We proved this for $q = 1$ in Chapter 4, but for other q, the best result known is the formula

$$h^{p,q}_{\mathrm{st}}(V) = h^{q,q}_{\mathrm{st}}(V^\circ), \quad p + q = \dim(V)$$

of (4.16) relating the string theoretic Hodge numbers of V and V° [**BD**]. When V and V° are smooth, this reduces to (8.79), but when V is singular, $h^{p,q}_{\mathrm{st}}(V)$ may differ from $h^{p,q}(V)$.

Of course, we can avoid this difficulty by assuming that V and V° are smooth, though this would limit the scope of the conjecture. But then we would still need to know how to define $\mathcal{H}^V_{\mathrm{simp}} \simeq \mathcal{H}^{V^\circ}_{\mathrm{toric}}$. This requires more than just the equality of Hodge numbers (8.79)—we would need actual isomorphisms $H^{p,q}(V) \simeq H^{q,q}(V^\circ)$ for $p + q = \dim(V)$. This is the same problem we encountered in Section 8.6.3.

For these reasons, we will not state a Hodge-Theoretic Toric Mirror Conjecture in the general case. It is possible that some version of Conjecture 8.6.12 is true in higher dimensions, though more work is needed in order to state a precise conjecture. According to the *Hodge-Theoretic Mirror Symmetry Conjecture* stated in [**Morrison7**, Lect. 8.2], the ultimate version of the conjecture should involve certain sub-variations of Hodge structure such as $\mathcal{TH}^{V^\circ}_{\mathrm{toric}}$. However, finding a precise version of this conjecture is still an open problem in algebraic geometry.

8.6.5. Conclusion. In this section, we have given a rigorous definition of mathematical mirror pair (Definitions 8.6.1 and 8.6.7) and have considered various versions of the Hodge-Theoretic Toric Mirror Conjecture (Conjectures 8.6.10 and 8.6.12). These show that we have a good idea of what mirror symmetry means for threefolds, though more work needs to be done on Conjecture 8.6.12. But once we get into higher dimensions, there are large chunks of the conjectures which are not specified. We are a long way from a definitive Hodge-theoretic version of the Mirror Conjecture, even in the toric case.

However, several "Mirror Theorems" have appeared recently in the literature. These include a complete proof of mirror symmetry for the quintic threefold, as well as results for Calabi-Yau toric hypersurfaces and more generally Calabi-Yau toric complete intersections. In the latter case, one uses the Batyrev-Borisov mirror construction from Section 4.3.[1] Also, Givental states "Mirror Theorems" which apply to certain toric complete intersections which aren't Calabi-Yau, and the "Mirror Principle" of [**LLY**] applies to many situations beyond the case of a Calabi-Yau threefold and its mirror.

---

[1]Discussions and examples of Calabi-Yau complete intersections in toric varieties can be found in [**BvS, LTe, HKTY2, Givental2, Givental4**] and in the references listed in Section 7.4.6.

In Chapter 11, we will see that although these "Mirror Theorems" make no direct reference to the Hodge-theoretic versions of mirror symmetry discussed here, they still share the basic idea of transforming the hard problem of computing Gromov-Witten invariants into something simpler. In the case of dimension > 3, it is not clear how these results relate to Hodge theory. Thus, just as we don't yet have the definitive "Hodge-Theoretic Mirror Conjecture" in dimension > 3, neither do we have the definitive "Mirror Theorem" in this case. The hope is that algebraic geometers, as they continue to explore the mathematics of mirror symmetry, will prove more general "Mirror Theorems" and formulate more precise "Mirror Conjectures" which will eventually converge to a broad understanding of the phenomenon of mirror symmetry.

This chapter has brought us to a clearer conception of what the mathematical version of mirror symmetry should look like. It is disappointing not to have a precise statement of the "Mirror Conjecture" when $\dim(V) > 3$, but this shouldn't deter us from examining the "Mirror Theorems" which have been proved so far. With this goal in mind, we will next develop some of the needed machinery in Chapters 9 and 10, and then study some "Mirror Theorems" in Chapter 11.

CHAPTER 9

Localization

In this chapter, we define equivariant cohomology and explain the powerful method of localization. The ideas and techniques we learn here will not only complete some of the work begun in Chapter 7, but also provide some of the tools needed to prove the Mirror Theorem in Chapter 11.

We begin in Section 9.1 with the definition and basic properties of equivariant cohomology. We will also discuss the localization theorem [**AB**] and its various corollaries. The basic idea is that when a smooth variety has a group action, cohomology classes on the variety "localize" to classes on the fixed point locus, which allow the calculations to proceed readily. We will see that the equivariant Euler class of the normal bundle of the fixed point set plays an important role. As applications, we will describe localization for $\mathbb{P}^r$ and then use localization to determine the number of lines on a generic quintic threefold. In general, localization was first introduced in enumerative geometry using torus actions on Hilbert schemes in the paper [**ES2**].

The main work of the chapter begins with Section 9.2, which applies localization to the moduli space $\overline{M}_{0,n}(\mathbb{P}^r, d)$. The projective space $\mathbb{P}^r$ admits a natural action of the algebraic torus $T = (\mathbb{C}^*)^{r+1}$, which induces an action of T on $\overline{M}_{0,n}(\mathbb{P}^r, d)$. Following [**Kontsevich2**], we will learn how to describe the components of the fixed point set using certain labeled graphs Γ and how to compute the equivariant normal bundle of the corresponding components of the fixed point set. Then, in Sections 9.2.2 and 9.2.3, we will continue the study of Gromov-Witten invariants of a Calabi-Yau threefold begun in Section 7.4.4. We first compute the contribution of degree d multiple covers of a rigidly embedded smooth rational curve in the threefold, and then we specialize to the quintic threefold and study the more subtle question of double covers of a nodal rational curve of degree 5. This will give a rigorous proof of the claims made about the instanton number n_{10} in Chapter 7.

The final section of the chapter gives a very quick introduction to equivariant Gromov-Witten invariants and localization of the virtual fundamental class.

9.1. The Localization Theorem

We begin by defining *equivariant cohomology*. We give here a topological construction, and refer to [**EG**] for an algebraic construction. Let G be a compact connected Lie group, classified by the principal G-bundle $EG \to BG$ with EG, whose total space EG is contractible. This G-bundle is uniquely determined up to homotopy equivalence. The example we will use is the algebraic torus $G = (\mathbb{C}^*)^n$, in which case $BG = (\mathbb{CP}^\infty)^n$, and $EG = \pi_1^* S \otimes \cdots \otimes \pi_n^* S$, where $\pi_i : BG \to \mathbb{CP}^\infty$ is the ith projection and S is the tautological bundle on $\mathbb{CP}^\infty$ whose sheaf of sections is $\mathcal{O}_{\mathbb{CP}^\infty}(-1)$.

9.1.1. Equivariant Cohomology. If now X is a topological space with a G-action, put $X_G = X \times_G EG$, which is itself a bundle over BG with fiber X. For future use, we let $i_X : X \hookrightarrow X_G$ denote the inclusion of a fiber.

DEFINITION 9.1.1. *The* equivariant cohomology *of X is defined to be*

$$H_G^*(X) = H^*(X_G),$$

where $H^*(X_G)$ is the ordinary cohomology of X_G.

Equivariant cohomology enjoys many of the usual properties of ordinary cohomology, such as existence of flat equivariant pullbacks and proper equivariant pushforwards. These operations satisfy the usual properties of the corresponding operations in ordinary cohomology [**AB**]. Note that $H_G^*(\text{point}) = H^*(BG)$. By pullback via $X \to$ point, we see that in general $H_G^*(X)$ is an $H^*(BG)$ module. Thus $H^*(BG)$ may be regarded as the coefficient ring for equivariant cohomology. Note also that $i_X : X \to X_G$ induces a "forgetful" map $i_X^* : H_G^*(X) \to H^*(X)$.

Our primary interest is when $G = (\mathbb{C}^*)^n$. In this case, let $M(G)$ be the character group of the torus G. For each $\rho \in M(G)$, we get a 1-dimensional vector space $\mathbb{C}_\rho$ with a G-action given by ρ. If $L_\rho = (\mathbb{C}_\rho)_G$ is the corresponding line bundle over BG, then the assignment $\rho \mapsto -c_1(L_\rho)$ defines an isomorphism $\psi : M(G) \simeq H^2(BG)$, which in turn induces a ring isomorphism $\mathrm{Sym}(M(G)) \simeq H^*(BG)$. We call $\psi(\rho)$ the *weight* of ρ.

In particular, if ρ_i is the character of $G = (\mathbb{C}^*)^n$ defined by $\rho_i(t_1, \ldots, t_n) = t_i$, then we let λ_i denote the weight of ρ_i. Thus we get an isomorphism

$$(9.1) \qquad H_G^*(\text{point}) = H^*(BG) \simeq \mathbb{C}[\lambda_1, \ldots, \lambda_n].$$

We denote the line bundle L_{ρ_i} by $\mathcal{O}(-\lambda_i)$, so that $\lambda_i = c_1(\mathcal{O}(\lambda_i))$. Note that under the identification (9.1), the map i_{point}^* can be thought of as a "nonequivariant limit" which maps all λ_i to 0. More generally, for any X, the map i_X^* can be thought of as a nonequivariant limit, which in particular maps all λ_i to 0.

In general, for a G-space X, an *equivariant vector bundle* is a vector bundle E over X such that the action of G on X lifts to an action of E which is linear on fibers. In this situation, E_G is a vector bundle over X_G, and the *equivariant Chern classes* $c_k^G(E) \in H_G^*(X)$ are defined to be the ordinary Chern classes $c_k(E_G)$. If E has rank r, then the top Chern class $c_r^G(E)$ is called the *equivariant Euler class* of E and is denoted

$$\mathrm{Euler}_T(E) \in H_G^*(X).$$

In Chapter 7, the Euler class was a homology class (the Poincaré dual of the top Chern class), but from here on, we will find it more convenient to regard the Euler class (ordinary or equivariant) as a cohomology class.

Example 9.1.1.1. The diagonal action of $G = (\mathbb{C}^*)^n$ on $\mathbb{C}^n$ gives an equivariant vector bundle E over $X = $ point such that $E_G \simeq \mathcal{O}(\lambda_1) \oplus \cdots \oplus \mathcal{O}(\lambda_1)$. Thus $\lambda_1, \ldots, \lambda_n$ are the *weights* of this representation. Since $\lambda_i = c_1(\mathcal{O}(\lambda_i))$, it follows that

$$(9.2) \qquad c_k^G(E) = \sigma_k(\lambda_1, \ldots, \lambda_n) \in \mathbb{C}[\lambda_1, \ldots, \lambda_n],$$

where σ_k is the kth elementary symmetric function.

9.1.2. The Theorem of Atiyah and Bott. The next notion that is needed is the localization theorem. Suppose that we have an action of a torus $T = (\mathbb{C}^*)^n$ on a smooth manifold X. By [**Iversen**], the fixed point locus X^T is a union of smooth connected components Z_j. Let $i_j : Z_j \hookrightarrow X$ be the inclusion, and let N_j denote the normal bundle of Z_j in X. Since N_j is an equivariant vector bundle, it has an equivariant Euler class

$$\mathrm{Euler}_T(N_j) \in H_T^*(Z_j).$$

The equivariant inclusion $i_j : Z_j \to X$ induces $i_j^* : H_T^*(X) \to H_T^*(Z_j)$. In addition, since Z_j is a submanifold of X, we also have a Gysin map

$$i_{j!} : H_T^*(Z_j) \longrightarrow H_T^*(X).$$

This is because the induced map $(Z_j)_T \to X_T$ has finite approximations by embeddings of submanifolds. The Gysin map has the property that for any $\alpha \in H_T^*(Z_j)$,

$$(9.3) \qquad\qquad i_j^* \circ i_{j!}(\alpha) = \alpha \cup \mathrm{Euler}_T(N_j).$$

Proofs of these assertion can be found in [**Audin**, App. A to Chap. 6]. Finally, note that $H_T^*(Z_j)$ is a module over $H^*(BT) = \mathbb{C}[\lambda_1, \dots, \lambda_n]$. If

$$\mathcal{R}_T \simeq \mathbb{C}(\lambda_1, \dots, \lambda_n)$$

is the field of fractions of $H^*(BT) \simeq \mathbb{C}[\lambda_1, \dots, \lambda_n]$, then we get the *localization* $H_T^*(Z_j) \otimes \mathcal{R}_T$. An important observation [**AB**] is that $\mathrm{Euler}_T(N_j)$ is an invertible element in $H_T^*(Z_j) \otimes \mathcal{R}_T$.

With this set-up, we have the following *Localization Theorem* proved in [**AB**].

PROPOSITION 9.1.2. *There is an isomorphism*

$$H_T^*(X) \otimes \mathcal{R}_T \xrightarrow{\sim} \bigoplus_j H_T^*(Z_j) \otimes \mathcal{R}_T$$

induced by the map $\alpha \mapsto \big(i_j^*(\alpha)/\mathrm{Euler}_T(N_j)\big)_j$. *Furthermore, the inverse is induced by* $(\alpha_j)_j \mapsto \sum_j i_{j!}(\alpha_j)$. *In particular, for any* $\alpha \in H_T^*(X) \otimes \mathcal{R}_T$, *we have*

$$\alpha = \sum_j i_{j!}\Big(\frac{i_j^*(\alpha)}{\mathrm{Euler}_T(N_j)}\Big).$$

Let's see what the localization theorem says about $\mathbb{P}^r$.

Example 9.1.2.1. Consider the action of $T = (\mathbb{C}^*)^{r+1}$ on $X = \mathbb{P}^r$ given by

$$(t_0, \dots, t_r) \cdot (x_0, \dots, x_r) = (t_0^{-1} x_0, \dots, t_r^{-1} x_r).$$

The inverses have been chosen so that $(t_0, \dots, t_r)$ acts on the homogeneous form $x_j \in H^0(\mathcal{O}_{\mathbb{P}^r}(1))$ as multiplication by t_j.

We begin by defining the equivariant hyperplane class. First note that $\mathbb{P}_T^r$ is the projectivization of the vector bundle $\mathcal{O}(-\lambda_0) \oplus \cdots \oplus \mathcal{O}(-\lambda_r)$ over BT. Thus, $\mathbb{P}_T^r = \mathbb{P}(E_T^*)$, where E^* is the dual of the bundle E defined in Example 9.1.1.1 and $\mathbb{P}$ denotes projectivization. This gives the tautological line bundle $\mathcal{O}_{\mathbb{P}_T^r}(1)$, and we let

$$p - c_1(\mathcal{O}_{\mathbb{P}_T^r}(1)) \in H_T^*(\mathbb{P}^r).$$

One can check that p is the equivariant Chern class $c_1^T(\mathcal{O}_{\mathbb{P}^r}(1))$. For this reason, we refer to p as the *equivariant hyperplane class.*

We next compute the equivariant cohomology of $\mathbb{P}^r$. The standard formula for the cohomology of a projective bundle (e.g., [**GH**, p. 606]) implies

$$(-p)^r - c_1(E_T^*)(-p)^{r-1} + c_2(E_T^*)(-p)^{r-2} + \cdots + c_r(E_T^*) = 0.$$

Note that the bundle denoted T in [**GH**] is $\mathcal{O}_{\mathbb{P}_T^r}(-1)$ in our notation, so that $c_1(T) = -p$. Since $c_k(E_T^*) = (-1)^k c_k^T(E) = (-1)^k \sigma_k(\lambda_0, \ldots, \lambda_r)$ by (9.2), it follows that

$$(9.4) \qquad \prod_{j=0}^{r} (p - \lambda_j) = 0$$

in $H_T^*(\mathbb{P}^r)$. We conclude that the equivariant cohomology of $\mathbb{P}^r$ is

$$H_T^*(\mathbb{P}^r) = \mathbb{C}[p, \lambda_0, \ldots, \lambda_r] \Big/ \prod_{j=0}^{r} (p - \lambda_j).$$

Observe that by formally putting $\lambda_i = 0$, we recover the usual description of the ordinary cohomology of $\mathbb{P}^r$, replacing p by the ordinary hyperplane class $H = i_{\mathbb{P}^r}^* p$.

To apply the localization theorem, we first need to understand the fixed points of the T-action. We have $r+1$ fixed points q_j, ordered as usual, so that the jth coordinate of q_j is nonzero, all other coordinates being 0. Note that $(q_j)_T = \mathbb{P}(\mathcal{O}(-\lambda_j))$, so that $\mathcal{O}_{\mathbb{P}_T^r}(-1)$ restricts to $\mathcal{O}(-\lambda_j)$ on $(q_j)_T$. Accordingly, the tautological bundle $\mathcal{O}_{\mathbb{P}_T^r}(1)$ restricts to $\mathcal{O}(\lambda_j)$. This implies that

$$(9.5) \qquad i_j^*(p) = \lambda_j.$$

Said differently, the T-action on $\mathbb{C} q_j$ is has weight $-\lambda_j$ because $(t_0, \ldots, t_r) \cdot q_j = t_j^{-1} q_j$ by definition. More generally, any element of $H_T^*(\mathbb{P}^r)$ can be written as a polynomial $F(p)$ in p with coefficients in $\mathbb{C}[\lambda_0, \ldots, \lambda_r]$, and (9.5) implies

$$(9.6) \qquad i_j^*(F(p)) = F(\lambda_j).$$

Since q_j is a point, the normal bundle N_j is given by the tangent space

$$N_j = T_{q_j} \mathbb{P}^r = \operatorname{Hom}(\mathbb{C} q_j, \mathbb{C}^{r+1}/\mathbb{C} q_j).$$

Hence the representation of T on N_j has weights $\lambda_j - \lambda_k$ for $k \neq j$, so that

$$(9.7) \qquad \operatorname{Euler}_T(N_j) = \prod_{k \neq j} (\lambda_j - \lambda_k).$$

One now computes that the isomorphism of Proposition 9.1.2 is given by

$$(9.8) \qquad p \mapsto \left(\frac{\lambda_j}{\prod_{i \neq j}(\lambda_j - \lambda_i)} \right)_{j=0}^{r}.$$

To describe the inverse of this map, let

$$(9.9) \qquad \phi_j = \prod_{k \neq j} (p - \lambda_k).$$

Then (9.6) implies

$$(9.10) \qquad i_k^*(\phi_j) = \begin{cases} \operatorname{Euler}_T(N_j) & \text{if } k = j \\ 0 & \text{otherwise.} \end{cases}$$

Hence the final assertion of Proposition 9.1.2 tells us that

$$\phi_j = \sum_{k=0}^{r} i_{k!}\left(\frac{i_k^*(\phi_j)}{\prod_{\ell\neq k}(\lambda_k - \lambda_\ell)}\right) = i_{j!}\left(\frac{i_j^*(\phi_j)}{\prod_{\ell\neq j}(\lambda_j - \lambda_\ell)}\right) = i_{j!}(1).$$

Using $H_T^*(q_j) = \mathbb{C}[\lambda_0,\ldots,\lambda_r]$ and the projection formula $i_{j!}(i_j^*(\alpha)\cup 1) = \alpha\cup i_{j!}(1)$, it follows easily that the inverse of (9.8) is given by

$$(\alpha_j)_{j=0}^{r} \mapsto \sum_{j=0}^{r} \alpha_j \cup \phi_j = \sum_{j=0}^{r} \alpha_j\,\phi_j.$$

Thus $\phi_0,\ldots,\phi_r$ form a basis of $H_T^*(\mathbb{P}^r) \otimes \mathcal{R}_T$ as a vector space over $\mathcal{R}_T$.

Finally, take an arbitrary element of $H_T^*(\mathbb{P}^r) \otimes \mathcal{R}_T$ and write it as above as a polynomial $F(p)$ in p with coefficients in $\mathcal{R}_T = \mathbb{C}(\lambda_0,\ldots,\lambda_r)$. Then, using (9.6) again, the final assertion of the localization theorem gives the identity

$$F(p) = \sum_{j=0}^{r} F(\lambda_j)\frac{\prod_{k\neq j}(p - \lambda_k)}{\prod_{k\neq j}(\lambda_j - \lambda_k)}.$$

This shows that Proposition 9.1.2 can be regarded as a far-reaching generalization of the Lagrange interpolation formula.

From Proposition 9.1.2, we obtain an integration formula by pushing forward to a point. For any variety X with a T-action, the trivial map $X \to$ point induces an equivariant projection map $\pi_X : X_T \to BT$. The pushforward map $\pi_{X!}$ is given by integration along the fiber and in this context is called the *equivariant integral*. Accordingly, we will often write $\pi_{X!}$ as

$$\int_{X_T} : H_T^*(X) \longrightarrow H^*(BT).$$

Using the equivariant integral, we get the following corollary of Proposition 9.1.2.

COROLLARY 9.1.3. *For any $\alpha \in H_T^*(X) \otimes \mathcal{R}_T$, we have*

$$\int_{X_T} \alpha = \sum_{j} \int_{(Z_j)_T} \left(\frac{i_j^*(\alpha)}{\mathrm{Euler}_T(N_j)}\right).$$

Proposition 9.1.2 and Corollary 9.1.3 gives rise to the notion of *localization*. By this, we mean the procedure of equivariantly restricting a class to each fixed point component and dividing by the equivariant normal bundle of that component. This reduces the calculation of an equivariant integral of a class to the sum over all fixed point components of the equivariant integrals of the restrictions of the class.

Since we eventually want to apply this theory to the orbifold $\overline{M}_{0,n}(\mathbb{P}^r, d)$, we will actually need a slight variant of Corollary 9.1.3 for smooth stacks. As mentioned in Chapter 7, the reader should consult [**Vistoli**] for intersection theory on algebraic stacks and [**BEFFGK**] for an introduction to stacks.

Hence, suppose that an orbifold X is the underlying variety of a smooth stack. This means that X admits local charts U/H, where as in Appendix A, U is smooth and H is a small subgroup of $\mathrm{GL}(n,\mathbb{C})$ acting on U. When we have a T-action on the corresponding smooth stack, we can locally (on T and X) realize this as an action on U. In this way, we can still make sense of fixed point loci Z_j, inclusions $i_j : Z_j \hookrightarrow X$, and equivariant normal bundles by locally working in U. Then we get the following stack version of Corollary 9.1.3.

CoROLLARY 9.1.4. *Let X be an orbifold which is the variety underlying a smooth stack with a T-action. If $\alpha \in H_T^*(X) \otimes \mathcal{R}_T$, then*

$$\int_{X_T} \alpha = \sum_j \int_{(Z_i)_T} \left(\frac{i_j^*(\alpha)}{a_i \mathrm{Euler}_T(N_i)} \right),$$

where a_i is the order of the group H occurring in a local chart at the generic point of Z_i.

9.1.3. Polynomials in Chern Classes. For many applications, we can formulate Corollary 9.1.3 in more down-to-earth terms. Suppose that we have a vector bundle E on a smooth variety X with a T-action, and we want to compute $\int_X P(c_k(E))$, where P is a polynomial in the Chern classes of E. We can put this into an equivariant context by replacing $c_k(E)$ by the equivariant Chern class $c_k^T(E) \in H_G^*(X)$ defined in Section 9.1.1 and pulling back via the inclusion i_{point}. To see that this gives $\int_X P(c_k(E))$, consider the commutative diagram

(9.11)
$$\begin{array}{ccc} X & \longrightarrow & \mathrm{point} \\ i_X \downarrow & & \downarrow i_{\mathrm{point}} \\ X_T & \longrightarrow & BT \end{array}$$

This implies $i_{\mathrm{point}}^* \circ \int_{X_T} = \int_X \circ\, i_X^*$. Hence

$$i_{\mathrm{point}}^* \int_{X_T} P(c_k^T(E)) = \int_X i_X^* P(c_k^T(E)) = \int_X P(c_k(E)),$$

as claimed above. The key point is that $\int_{X_T} P(c_k^T(E))$ is an equivariant integral which can be computed by localization. More precisely, by combining the above equation with Corollary 9.1.3, we will get an equation which computes $\int_X P(c_k(E))$ in terms of the behavior of E on the fixed point set of the T-action on X.

To see what this says more explicitly, let's see what happens when the fixed point set X^T is finite. For each point $Z_j \in X^T$, the restriction $E|_{Z_j}$ decomposes into characters of T, say $\chi_j^1, \dots, \chi_j^s$. Each χ_j^k is a linear combination of the basic characters $\rho_1 \dots, \rho_n$ already defined. Here we are using additive notation for characters, i.e., $(\rho_1 + \rho_2)(t) = \rho_1(t)\rho_2(t)$. Let us write $\chi_j^k = \ell_j^k(\rho_1, \dots, \rho_n)$ for this linear combination, which we express symbolically as $\chi_j^k = \ell_j^k(\rho)$.

Since ρ_i has weight λ_i, it follows that the weight of $\chi_j^k = \ell_j^k(\rho)$ is $\ell_j^k(\lambda)$, where $\ell_j^k(\lambda)$ denotes the linear combination of $\lambda_1, \dots, \lambda_n$ obtained by substituting λ_i for ρ_i in $\ell_j^k(\rho)$. Hence the action of T on $E|_{Z_j}$ has weights $\ell_j^k(\lambda)$, which implies

$$i_j^*(c_k^T(E)) = c_k^T(E|_{Z_j}) = \sigma_k(\ell_j^1(\lambda), \dots, \ell_j^s(\lambda)).$$

Furthermore, since Z_j is a point, the normal bundle N_j is the tangent space $T_{Z_j} X$. If the characters of the T-action on $T_{Z_j} X$ are $t_j^1(\rho), \dots, t_j^d(\rho)$, where $d = \dim X$, then

(9.12)
$$\mathrm{Euler}_T(N_j) = \prod_k t_k^j(\lambda).$$

We can now apply i_{point}^* to the right hand side of the formula of Corollary 9.1.3. Keeping the notation as above, we arrive at the following statement.

PROPOSITION 9.1.5. *Suppose that T acts on X with isolated fixed points Z_j and that E is an equivariant vector bundle on X. If $P(c_k(E))$ is a polynomial in the Chern classes of E, then*

$$\int_X P(c_k(E)) = \sum_j \frac{P(\sigma_k(\ell_j^1(\lambda), \dots, \ell_j^s(\lambda)))}{\prod_k t_k^j(\lambda)},$$

where σ_k is the kth elementary symmetric function.

This is the formulation used for enumerative applications in [**ES2**]. If we repeat the argument in the case of a general Z_j, we are led to the formulation appearing in [**Kontsevich2**].

We can also generalize Proposition 9.1.5 to smooth stacks. This is done by replacing Corollary 9.1.3 by Corollary 9.1.4 in the above argument. The result is that if we describe the stack by an orbifold, then we must divide each term in the sum given in Proposition 9.1.5 by the order of the automorphism group of the object associated with the corresponding Z_j. We will illustrate this in Section 9.2.2 below.

Example 9.1.3.1. We calculate the number of lines on a general quintic threefold $V \subset \mathbb{P}^4$ using Proposition 9.1.5. In Example 7.1.5.1, we showed how to realize the scheme of lines on V as the zero locus of a section of the vector bundle $\mathrm{Sym}^5 U^*$ on the Grassmannian $G(2,5)$ of lines in $\mathbb{P}^4$. Here, U is the tautological rank 2 subbundle on the Grassmannian, so that the fiber U_ℓ over a line ℓ is the 2-dimensional subspace of $\mathbb{C}^5$ whose projectivization is ℓ. Then an equation for $V \subset \mathbb{P}^4$ induces a section s of the rank 6 bundle $\mathrm{Sym}^5 U^*$, and Example 7.1.6.1 shows that the number of lines is the degree of the Euler class $c_6(\mathrm{Sym}^5 U^*)$. This is the Gromov-Witten invariant

$$\langle I_{0,0,1} \rangle = \int_{G(2,5)} c_6(\mathrm{Sym}^5 U^*).$$

We will use Proposition 9.1.5 with $E = \mathrm{Sym}^5 U^*$ and $P(c_k(E))$ equal to the Euler class of $\mathrm{Sym}^5 U^*$, using the natural action of $(\mathbb{C}^*)^5$ on $\mathbb{C}^5$ given in coordinates by

$$(\lambda_1, \dots, \lambda_5) \cdot (x_1, \dots, x_5) = (\lambda_1^{-1} x_1, \dots, \lambda_5^{-1} x_5).$$

This induces an action of $T = (\mathbb{C}^*)^5$ on $G(2,5)$ with 10 isolated fixed points L_I corresponding to the 10 coordinate lines in $\mathbb{P}^4$. Each L_I is indexed by $I \subset \{1, \dots, 5\}$ with $|I| = 2$, so that L_I is defined by the equations $x_j = 0$ for all $j \notin I$. The normal bundle of L_I is just the tangent space to $G(2,5)$ at L_I, whose equivariant Euler class we need. As in (9.12), we need to decompose the T-action on the tangent space into characters, replace each ρ_i by λ_i, and multiply together.

The tangent bundle of the Grassmannian is $\mathrm{Hom}(U, \mathbb{C}^5/U)$. The key observation is that at $L_I \in G(2,5)$, the restriction $U|_{L_I}$ has characters $-\rho_i$ for $i \in I$. It follows that the characters of the T-action on the tangent space at L_I are

$$\{\rho_i - \rho_j \mid i \in I, \ j \notin I\}.$$

Hence the tangent space of $G(2,5)$ at L_I has equivariant Euler class

$$\prod_{i \in I, \ j \notin I} (\lambda_i - \lambda_j).$$

The other ingredient we need is the equivariant Euler class of $i_I^*(\mathrm{Sym}^5 U^*)$, where $i_I : \{L_I\} \hookrightarrow G(2,5)$ is the inclusion. This is computed using the characters of $U^*|_{L_I}$, which as above are ρ_i for $i \in I$. Hence the equivariant Euler class of $i_I^*(\mathrm{Sym}^5 U^*)$ is

$$\prod_{a=0}^{5}(a\lambda_{i_1} + (5-a)\lambda_{i_2}),$$

where we have written $I = \{i_1, i_2\}$.

Proposition 9.1.5 now implies

$$\int_{G(2,5)} c_6(\mathrm{Sym}^5 U^*) = \sum_{|I|=2} \frac{\prod_{a=0}^{5}(a\lambda_{i_1} + (5-a)\lambda_{i_2})}{\prod_{i\in I}\prod_{j\notin I}(\lambda_i - \lambda_j)},$$

where as above $I = \{i_1, i_2\}$. All of the λ_i cancel as they must, and the answer is computed to be 2875, the number of lines on a generic quintic threefold.

The calculation in the example just completed involves the simplification of a sum of 10 rational expressions in the λ_i, each of which has both numerator and denominator equal to a product of 6 linear binomials in the λ_i. While this may be tedious to carry out by hand, it is very fast by computer. This method applies to give rapid calculations in the Hilbert scheme of lines, conics, and twisted cubics in projective spaces, as these are all smooth spaces with well understood torus actions. For example, the calculation of twisted cubics on complete intersection Calabi-Yau threefolds is given in [**ES2**]. There are two ingredients needed for these calculations. The first ingredient is to find a smooth moduli space for the projective curves in question together with a T-action whose fixed point locus and weights are computable. The second ingredient is to represent the number in question as the degree of a polynomial in the Chern classes of an equivariant vector bundle whose weights at the fixed points are computable.

9.2. Localization in $\overline{M}_{0,n}(\mathbb{P}^r, d)$

To find the number of degree d rational curves on the quintic threefold, it was Kontsevich's insight [**Kontsevich2**] that the method of Section 9.1.3 could be applied by using the moduli space of stable maps described in Section 7.1.1. The space $\overline{M}_{0,n}(\mathbb{P}^r, d)$ is not smooth, but $\overline{\mathcal{M}}_{0,n}(\mathbb{P}^r, d)$ is a smooth stack, so that Corollary 9.1.4 applies.

9.2.1. Kontsevich's Approach. The natural action of $T = (\mathbb{C}^*)^{r+1}$ on $\mathbb{P}^r$ induces a T-action on $\overline{M}_{0,n}(\mathbb{P}^r, d)$. The fixed point loci and equivariant normal bundles of the fixed point loci were worked out in [**Kontsevich2**] (see [**GPa**] for the case $g > 0$, including the localization formula for virtual normal bundles). We quickly sketch the main ideas and results.

A T-fixed point of $\overline{M}_{0,n}(\mathbb{P}^r, d)$ consists of a stable map $(f, C, p_1, \ldots, p_n)$ where each component C_i of C is either mapped by f to a T-fixed point of $\mathbb{P}^r$ or else multiply covers a coordinate line. In addition, each marked point p_i, each node of C, and each ramification point of f is mapped to a T-fixed point of $\mathbb{P}^r$. This implies that if C_i is a degree d_i cover of a coordinate line via f, then homogeneous coordinates on C_i and the coordinate line can be chosen so that the cover is given by $(x_0, x_1) \mapsto (x_0^{d_i}, x_1^{d_i})$.

From this, it follows immediately that the fixed point components Z_j of the T-action on $\overline{M}_{0,n}(\mathbb{P}^r, d)$ can be described by combinatorial data. Let $q_0, \ldots, q_r$ denote the usual T-fixed points of $\mathbb{P}^r$. To a stable map $(f, C, p_1, \ldots, p_n)$ we associate a tree Γ whose vertices v are in one-to-one correspondence with the connected components C_v of $f^{-1}(\{q_0, \ldots, q_r\})$. Thus each C_v is either a point of C or a connected union of irreducible components of C. The edges e of Γ correspond to those irreducible components C_e of C which are mapped by f onto some coordinate line $\ell_e \subset \mathbb{P}^r$.

The tree Γ also has various labels. We associate to each vertex v the number i_v defined by $f(C_v) = q_{i_v}$ and the set S_v consisting of those i for which the marked point p_i is in C_v. In addition, we associate to each edge e the degree d_e of the map $f|_{C_e} : C_e \to \ell_e$. Then the connected components of $\overline{M}_{0,n}(\mathbb{P}^r, d)^T$ are in one-to-one correspondence with connected trees Γ with labels i_v, S_v, and d_e satisfying the following conditions:

- If an edge e contains vertices v and v', then $i_v \neq i_{v'}$ and ℓ_e is the coordinate line joining q_{i_v} and $q_{i_{v'}}$.
- $\sum_e d_e = d$.
- $\{1, \ldots, n\} = \coprod_v S_v$.

We will refer to the data (Γ, i_v, S_v, d_e) as the *graph* of the stable map, and we usually will abbreviate all this data by the symbol Γ. Recall that the *valence* of the vertex v, denoted $\mathrm{val}(v)$, is the number of edges connected to v.

The stable maps with fixed graph Γ naturally define a substack

$$\overline{\mathcal{M}}_\Gamma \subset \overline{\mathcal{M}}_{0,n}(\mathbb{P}^r, d).$$

To study $\overline{\mathcal{M}}_\Gamma$, suppose we have $(f, C, p_1, \ldots, p_n) \in \overline{\mathcal{M}}_\Gamma$. For each vertex v such that C_v is a curve, note that C_v has $n(v) = |S_v| + \mathrm{val}(v)$ special points—there are $|S_v|$ marked points lying on C_v and $\mathrm{val}(v)$ nodes where C_v meets an irreducible component C_e for an edge e containing v. In fact, $(f|_{C_v}, C_v)$ plus the $n(v)$ points just described form a stable curve, so that we get an element of $\overline{M}_{0,n(v)}$.

A more careful description of this is as follows. Given a collection $\{C_v\} \in \prod_{\dim C_v = 1} \overline{M}_{0,n(v)}$ of $n(v)$-pointed genus 0 stable curves, we construct a curve C by linking C_v to $C_{v'}$ via a new curve $C_e \simeq \mathbb{P}^1$ at $0, \infty \in \mathbb{P}^1$ whenever there is an edge e containing v and v' (this can be done systematically). Then define $f : C \to \mathbb{P}^r$ by contracting each C_v to q_{i_v} and letting $f|_{C_e}$ be $(x_0, x_1) \mapsto (x_0^{d_e}, x_1^{d_e})$. This gives a morphism

$$\psi_\Gamma : \prod_{\dim C_v = 1} \overline{M}_{0,n(v)} \longrightarrow \overline{\mathcal{M}}_\Gamma.$$

Note that $n(v)$ can be defined for all vertices v and that $\dim C_v = 1 \Leftrightarrow C_v$ contains a component of C contracted by $f \Leftrightarrow n(v) \geq 3$.

We define $\overline{M}_\Gamma$ to be the above product (and we let $\overline{M}_\Gamma$ be a point if there are no contracted components). The map $\psi_\Gamma : \overline{M}_\Gamma \to \overline{\mathcal{M}}_\Gamma$ described above is not an isomorphism, but rather a finite morphism. In terms of stacks, there is a finite group of automorphisms A_Γ acting on $\overline{M}_\Gamma$ such that the quotient (in the stack sense) is $\overline{\mathcal{M}}_\Gamma$. The group A_Γ fits into an exact sequence

$$(9.13) \qquad 0 \longrightarrow \prod_e \mathbb{Z}/d_e\mathbb{Z} \longrightarrow A_\Gamma \longrightarrow \mathrm{Aut}(\Gamma) \longrightarrow 0,$$

where $\mathrm{Aut}(\Gamma)$ is the group of automorphisms of Γ which preserve the labels. We can explain this sequence as follows. The subgroup $\prod_e \mathbb{Z}/d_e\mathbb{Z}$ arises from the covering transformations of the degree d_e covers. The quotient map arises as follows. Each $g \in A_\Gamma$ takes a collection $\{C_v\} \in \overline{M}_\Gamma$ of $n(v)$-pointed genus 0 stable curves to a different collection $\{C'_v\}$ such that $f = \psi_\Gamma(\{C_v\})$ and $f' = \psi_\Gamma(\{C'_v\})$ are isomorphic stable maps. This isomorphism identifies the connected components of $f^{-1}(q_i)$ and $(f')^{-1}(q_i)$. In particular, each C_v is identified with some C'_w such that $i_v = i_w$. It is easy to see that the permutation of the set of vertices $v \mapsto w$ is independent of the choice of the C_v and induces an automorphism $\alpha(g)$ of Γ which preserves labels. The map $g \mapsto \alpha(g)$ defines the map $A_\Gamma \to \mathrm{Aut}(\Gamma)$ in (9.13), which is then easily seen to be exact.

As a consequence, when Corollary 9.1.4 is applied to $\overline{M}_{0,n}(\mathbb{P}^r, d)$, the number a_Γ appearing in the denominator of the term corresponding to the fixed component $\overline{\mathcal{M}}_\Gamma$ is just the order of A_Γ.

Example 9.2.1.1. Let us compute the fixed point components of the standard T-action on $\overline{M}_{0,0}(\mathbb{P}^1, 2)$. The possible graphs Γ are:

$$
\overset{\textstyle 1 \qquad 1}{\underset{\textstyle 0 \quad\ 1 \quad\ 0}{\bullet\!-\!\bullet\!-\!\bullet}}
\qquad\qquad
\overset{\textstyle 1 \qquad 1}{\underset{\textstyle 1 \quad\ 0 \quad\ 1}{\bullet\!-\!\bullet\!-\!\bullet}}
\qquad\qquad
\overset{\textstyle 2}{\underset{\textstyle 0 \quad\ 1}{\bullet\!-\!\bullet}}
$$

In these graphs, we have placed d_e above the edge e and i_v below the vertex v. The S_v are empty in this case since there are no marked points.

This gives three components of the fixed point set of T acting on $\overline{M}_{0,0}(\mathbb{P}^1, 2)$. The corresponding T-fixed stable maps are as follows in the three respective cases:

(*i*) The curve C has two components, $C = C_1 \cup C_2$. The map f restricts on each component to an isomorphism $f|_{C_i} : C_i \overset{\sim}{\to} \mathbb{P}^1$, and $f(C_1 \cap C_2) = q_1$.

(*ii*) The curve C has two components, $C = C_1 \cup C_2$. The map f restricts on each component to an isomorphism $f|_{C_i} : C_i \overset{\sim}{\to} \mathbb{P}^1$, and $f(C_1 \cap C_2) = q_0$.

(*iii*) The curve C is irreducible, and coordinates (z_0, z_1) on $C \simeq \mathbb{P}^1$ can be chosen so that $f(z_0, z_1) = (z_0^2, z_1^2)$.

In each case, we have $A_\Gamma \simeq \mathbb{Z}_2$, hence $a_\Gamma = 2$. In cases (*i*) and (*ii*), the nontrivial automorphism arises from the obvious automorphism of Γ, while in case (*iii*), the nontrivial automorphism arises from the double cover.

Example 9.2.1.2. We compute the fixed point components of the standard T-action on $\overline{M}_{0,1}(\mathbb{P}^2, 2)$. The T-action on $\mathbb{P}^2$ has three fixed points

$$q_0 = (1, 0, 0), \quad q_1 = (0, 1, 0), \quad \text{and} \quad q_2 = (0, 0, 1).$$

Accordingly, the possible graphs Γ are:

$$
\overset{\textstyle \{1\}\ 1 \qquad 1}{\underset{\textstyle i \quad\ j \quad\ k}{\bullet\!-\!\bullet\!-\!\bullet}}
\qquad\qquad
\overset{\textstyle 1\ \{1\}\ 1}{\underset{\textstyle i \quad\ j \quad\ k}{\bullet\!-\!\bullet\!-\!\bullet}}
\qquad\qquad
\overset{\textstyle \{1\}\ 2}{\underset{\textstyle i \quad\ j}{\bullet\!-\!\bullet}}
$$

$$
i \neq j,\ j \neq k \qquad\qquad i \neq j,\ j \neq k \qquad\qquad i \neq j
$$

where we have added the label S_v above the vertex v whenever S_v is nonempty.

Note that the second graph corresponds to a stable map with source curve $C = C_1 \cup C_2 \cup C_3$ with C_2 intersecting each of C_1 and C_3 in a point and no other intersections. The curve C_1 is mapped isomorphically by f onto the line joining q_i and q_j, the curve C_2 is mapped to the point q_j, and the curve C_3 is mapped isomorphically by f onto the line joining q_j and q_k. The marked point p_1 is a smooth point of C_2. Note that f is stable since C_2 contains three special points, namely p_1, $C_1 \cap C_2$, and $C_2 \cap C_3$. The existence of a contracted component can be recognized immediately from the middle vertex v of the graph, since $n(v) = 1 + 2 = 3$.

To carry out calculations in $\overline{M}_{0,n}(\mathbb{P}^r, d)$, we use Corollary 9.1.4. Hence we need to describe the normal bundles of the components of the fixed point set. In this context, we denote the normal bundle of $\overline{\mathcal{M}}_\Gamma$ by N_Γ. References for the calculation of its equivariant Euler class include [**Kontsevich2, Givental4, GPa**] (the reference [**GPa**] includes the generalization to $\overline{M}_{g,n}(\mathbb{P}^r, d)$, which requires a localization formula for the equivariant virtual fundamental class).

Recall from Section 7.1.4 that the tangent space to the stack $\overline{\mathcal{M}}_{0,n}(\mathbb{P}^r, d)$ at the stable map $(f, C, p_1, \ldots, p_n)$ is the hyperext group

$$\mathrm{Ext}^1_C\big(f^*\Omega^1_{\mathbb{P}^r} \to \Omega^1_C(D), \mathcal{O}_C\big),$$

where $D = \sum_i p_i$. To understand this group, we use the long exact sequence for Ext, which gives

$$
\begin{aligned}
0 &\longrightarrow \mathrm{Hom}(\Omega^1_C(D), \mathcal{O}_C) \longrightarrow H^0(C, f^*T_{\mathbb{P}^r}) \longrightarrow \\
& \mathrm{Ext}^1(f^*\Omega^1_{\mathbb{P}^r} \to \Omega^1_C(D), \mathcal{O}_C) \longrightarrow \mathrm{Ext}^1(\Omega^1_C(D), \mathcal{O}_C) \longrightarrow 0,
\end{aligned}
$$
(9.14)

as well as $\mathrm{Ext}^2(f^*\Omega^1_{\mathbb{P}^r} \to \Omega^1_C(D), \mathcal{O}_C) = 0$, since $H^1(C, f^*T_{\mathbb{P}^r}) = 0$ by the convexity of $\mathbb{P}^r$. For more general g and X, the exact sequence (9.14) must be replaced by a similar exact sequence which calculates the tangent-obstruction complex, as explained in Section 7.1.4

At a T-fixed point of $\overline{\mathcal{M}}_\Gamma \subset \overline{M}_{0,n}(\mathbb{P}^r, d)$, we can find the weights of the T-action on the tangent space using (9.14). Then we can determine the weights of the T-action on the normal bundle N_Γ by looking at the T-action on the terms in (9.14) and restricting to the "moving part". Then multiplying these weights together gives the equivariant Euler class of N_Γ in the usual way.

Before we can give the formula for $\mathrm{Euler}_T(N_\Gamma)$, we first need some notation. Following [**Kontsevich2**], we define a *flag* F to be a pair (v, e) such that e is an edge containing the vertex v. We put $i(F) = v$, and we also let $j(F)$ denote the vertex of e different from e. Given a flag $F = (v, e)$, we put

$$\omega_F = \frac{\lambda_{i(F)} - \lambda_{j(F)}}{d_e}.$$

This is the element of $H^2(BT)$ corresponding to the weight of the T-action on the tangent space of the component C_e of C associated to the edge e at the point p_F lying over i_v. We also let e_F be the first Chern class of the bundle on $\overline{\mathcal{M}}_\Gamma$ whose fiber is the cotangent space to the component associated to v at p_F.

If v is a vertex with $\mathrm{val}(v) = 1$, then the unique flag containing v will be denoted by $F(v)$. Similarly, if $\mathrm{val}(v) = 2$, then the two flags containing v will be denoted by $F_1(v), F_2(v)$.

THEOREM 9.2.1. *The equivariant Euler class of the normal bundle N_Γ is a product of contributions from the flags, vertices, and edges. More precisely,*

$$\mathrm{Euler}_T(N_\Gamma) = e^F_\Gamma \, e^v_\Gamma \, e^e_\Gamma,$$

where e_Γ^{F}, e_Γ^{v} and e_Γ^{e} are defined by the formulas

$$e_\Gamma^{\mathrm{F}} = \prod_{n(i(F)) \geq 3} (\omega_F - e_F) \Big/ \prod_{j \neq i(F)} (\lambda_{i(F)} - \lambda_j)$$

$$e_\Gamma^{\mathrm{v}} = \prod_v \prod_{j \neq i_v} (\lambda_{i_v} - \lambda_j) \prod_{\substack{\mathrm{val}(v)=2 \\ S_v = \emptyset}} (\omega_{F_1(v)} + \omega_{F_2(v)}) \Big/ \prod_{\substack{\mathrm{val}(v)=1 \\ S_v = \emptyset}} \omega_{F(v)}$$

$$e_\Gamma^{\mathrm{e}} = \prod_e \frac{(-1)^{d_e}(d_e!)^2 (\lambda_i - \lambda_j)^{2d_e}}{d_e^{2d_e}} \prod_{\substack{a+b=d_e \\ k \neq i,j}} \left(\frac{a\lambda_i + b\lambda_j}{d_e} - \lambda_k \right).$$

PROOF. Complete proofs can be found in [**Kontsevich2**] and [**GPa**]. We will instead sketch a small part of the proof which involves Lemma 9.2.2 below. This lemma will be useful later.

We begin with the observations that $\mathrm{Hom}(\Omega_C^1(D), \mathcal{O}_C)$ is the space of infinitesimal automorphisms of $(C, p_1, \dots, p_n)$ and $H^0(C, f^*T_{\mathbb{P}^r})$ is the space of first order deformations of the map f. Hence we see that

$$H^0(C, f^*T_{\mathbb{P}^r})/\mathrm{Hom}(\Omega_C^1(D), \mathcal{O}_C).$$

is the subspace $S_{\Gamma,f}$ of the tangent space to $\overline{M}_\Gamma$ corresponding to deformations of the stable map f for which the moduli of the points $p_1, \dots, p_n$ are fixed. Combining this with (9.14) gives the exact sequence

$$0 \longrightarrow S_{\Gamma,f} \longrightarrow T_f \overline{\mathcal{M}}_{0,n}(\mathbb{P}^r, d) \xrightarrow{\kappa} \mathrm{Ext}^1(\Omega_C^1(D), \mathcal{O}_C) \longrightarrow 0.$$

However, we also have the map

$$\mu : \mathrm{Ext}^1(\Omega_C^1(D), \mathcal{O}_C) \longrightarrow H^0(C, \underline{\mathrm{Ext}}^1(\Omega_C^1(D), \mathcal{O}_C))$$

coming from the local-global spectral sequence for Ext. Since C is a curve, the spectral sequence degenerates at E_2, which shows that μ is surjective. But the sheaf $\underline{\mathrm{Ext}}^1(\Omega_C^1(D), \mathcal{O}_C)$ is a torsion sheaf supported at the nodes of C, since $\Omega_C^1(D)$ is locally free at the smooth points of C. It follows immediately that

$$\underline{\mathrm{Ext}}^1(\Omega_C^1(D), \mathcal{O}_C) \simeq \underline{\mathrm{Ext}}^1(\Omega_C^1, \mathcal{O}_C),$$

so that we can regard μ as a surjective map

$$\mu : \mathrm{Ext}^1(\Omega_C^1(D), \mathcal{O}_C) \longrightarrow H^0(C, \underline{\mathrm{Ext}}^1(\Omega_C^1, \mathcal{O}_C)).$$

By [**DM**], $H^0(C, \underline{\mathrm{Ext}}^1(\Omega_C, \mathcal{O}_C))$ is the space of first order smoothings of C, forgetting both the map f and the marked points p_i. The argument given in [**DM**] is still valid in this context, even though the curve C need not be stable.

This discussion shows that $\kappa^{-1}(\mu^{-1}(\{0\})) \subset T_f \overline{\mathcal{M}}_{0,n}(\mathbb{P}^r, d)$ is the space of first order deformations of f which preserve the nodes. Since the first order deformations of f coming from $T_f \overline{M}_\Gamma$ obviously preserve the nodes, we see that $T_f \overline{M}_\Gamma \subset \kappa^{-1}(\mu^{-1}(\{0\}))$. Thus $\mu \circ \kappa$ induces a surjection

$$N_{\Gamma,f} \longrightarrow H^0(C, \underline{\mathrm{Ext}}^1(\Omega_C, \mathcal{O}_C)).$$

This map is compatible with the natural action of T on $H^0(C, \underline{\mathrm{Ext}}^1(\Omega_C, \mathcal{O}_C))$.

We now state an important lemma from [**Kontsevich2**] which computes this space explicitly. Let q_i be a node of C, and let C_1^i, C_2^i denote the two components of C containing the point q_i.

LEMMA 9.2.2. *There is a natural isomorphism*

$$H^0(C, \underline{\mathrm{Ext}}^1(\Omega_C, \mathcal{O}_C)) \simeq \bigoplus_i \left(T_{q_i} C_1^i \otimes T_{q_i} C_2^i\right)$$

respecting the T-actions.

PROOF. Since the calculation is purely local, we let C be a local (analytic) curve defined in a local surface S by the equation $xy = 0$, where (x, y) are local coordinates on S. The components of C will be denoted by C_1 and C_2, and let $q = (0,0)$. We compute the $\underline{\mathrm{Ext}}^1$ using the standard exact sequence

$$(9.15) \qquad 0 \to I_C/I_C^2 \to \Omega_S^1|_C \to \Omega_C^1 \to 0,$$

where the first map sends $[f]$ to $df|_C$ and the second map is pullback to C. The long exact sequence for $\underline{\mathrm{Ext}}^\bullet(\cdot, \mathcal{O}_C)$ applied to (9.15) gives

$$(9.16) \qquad \underline{\mathrm{Ext}}^1(\Omega_C^1, \mathcal{O}_C) \simeq \mathrm{coker}\left(T_S|_C \xrightarrow{h} (I_C/I_C^2)^*\right),$$

where $(I_C/I_C^2)^*$ denotes $\underline{\mathrm{Hom}}(I_C/I_C^2, \mathcal{O}_C)$. One computes that $h(\partial/\partial x)$ is the map taking the generator xy of I_C/I_C^2 to y, while $h(\partial/\partial y)$ takes xy to x.

Intrinsically, this leads to the isomorphism

$$(9.17) \qquad T_p C_1 \otimes T_p C_2 \simeq \underline{\mathrm{Ext}}^1(\Omega_C^1, \mathcal{O}_C)$$

defined by

$$(9.18) \qquad v_1 \otimes v_2 \mapsto [f \mapsto v_1 v_2(f)|_C],$$

using the identification (9.16). To make sense of this, we interpret v_1 and v_2 as the restrictions to C_1 and C_2 of vector fields on S, with each v_i tangent to C_i. We restrict the resulting function $v_1 v_2 f$ to C. The assignment $f \mapsto v_1 v_2(f)|_C$ is an element of $(I_C/I_C^2)^*$ for given vector fields v_1, v_2. Its value modulo the image of h is easily seen to depend only on $v_1 \otimes v_2$ at the point p. In terms of the local coordinates x, y, (9.18) sends the generator $(\partial/\partial x) \otimes (\partial/\partial y)$ to the class of the element of $(I_C/I_C^2)^*$ taking xy to 1. This calculation shows that (9.18) indeed defines an isomorphism (9.17).

The naturality of (9.17) shows that the isomorphism is compatible with the natural T-actions on each side. $\qquad \square$

In terms of proving Theorem 9.2.1, Lemma 9.2.2 is used in the formula for $\mathrm{Euler}_T(N_\Gamma)$ in two places:

- The terms $\omega_F - e_F$ occurring in e_Γ^{F} arise from the nodes of C where a contracted component meets a component mapped to $\mathbb{P}^r$ with positive degree.
- The terms $\omega_{F_1(v)} + \omega_{F_2(v)}$ occurring in e_Γ^{v} arise from the nodes where two components meet, each of which is mapped to $\mathbb{P}^r$ with positive degree.

To complete the proof, one needs to understand the other terms in e_Γ^{F}, e_Γ^{v} and e_Γ^{e}. For example, one can show the following:

- The denominator of e_Γ^{v} arises from the weights of the T-action on the infinitesimal automorphisms in $\mathrm{Hom}(\Omega_C^1, \mathcal{O}_C)$.
- The denominator of e_Γ^{F} (resp. the first factor of e_Γ^{v}) arises from a contribution of the tangent space to $\mathbb{P}^r$ at $q_{i(F)}$ (resp. $q_{i(v)}$).

These follow without difficulty from (9.14). However, we've said nothing about the terms appearing in e_Γ^{e}, so that a lot of work remains to prove the theorem. As already mentioned, full details can be found in [**Kontsevich2, GPa**]. $\qquad \square$

In the formula for $\mathrm{Euler}_T(N_\Gamma)$ given in Theorem 9.2.1, it should be clear that there are many cancelations between the three factors e_Γ^{F}, e_Γ^{v} and e_Γ^{e}. Therefore this formula is not necessarily the most efficient way to calculate in specific examples. The advantage of Theorem 9.2.1 is that it gives an easily automated algorithm for calculating $\mathrm{Euler}_T(N_\Gamma)$. We will work out a nontrivial example of these formulas (and verify their correctness) in Section 9.2.2 below.

We should mention that the identification

$$H^0(C, \underline{\mathrm{Ext}}^1(\Omega_C^1, \mathcal{O}_C)) \simeq \bigoplus_i (T_{q_i} C_1^i \otimes T_{q_i} C_2^i)$$

given in Lemma 9.2.2 appeared implicitly in [**Mumford2, HM**]. We will use Lemma 9.2.2 in Section 9.2.2 below. This lemma also plays a central role in the proofs of the Mirror Theorem described in Chapter 11.

The T-representation associated to the right hand side of Lemma 9.2.2 arises naturally in computations on more general spaces $\overline{M}_{0,n}(X, \beta)$, even though the lemma does not apply as stated. This follows from the interplay between the tangent space to $\overline{\mathcal{M}}_{0,n}(X, \beta)$ and the virtual fundamental class arising from the long exact sequence for $\mathrm{Ext}^\bullet(f^*\Omega_{\mathbb{P}^r}^1 \to \Omega_C^1(D), \mathcal{O}_C)$.

Example 9.2.1.3. One of the first applications of the theory developed here was the calculation of the Gromov-Witten invariant $N_4 = \langle I_{0,0,4} \rangle$ of the quintic threefold [**Kontsevich2**]. Recall from Example 7.1.6.1 that

$$N_4 = \int_{\overline{\mathcal{M}}_{0,0}(\mathbb{P}^4, 4)} \mathrm{Euler}(\mathcal{V}_4),$$

where $\mathcal{V}_4$ is the vector bundle over $\overline{\mathcal{M}}_{0,0}(\mathbb{P}^4, 4)$ whose fiber at $(f, C) \in \overline{\mathcal{M}}_{0,0}(\mathbb{P}^4, 4)$ is $H^0(C, f^*\mathcal{O}_{\mathbb{P}^4}(5))$.

For the action of $T = (\mathbb{C}^*)^5$ on $\mathbb{P}^4$, one can write down all possible graphs $\overline{\mathcal{M}}_\Gamma$ and compute the corresponding equivariant Euler classes $\mathrm{Euler}_T(N_\Gamma)$. One can also compute the weights of the restriction of $\mathrm{Euler}(\mathcal{V}_4)$ by explicitly representing cohomology classes as Čech cocycles. We will explain how this can be done in Section 9.2.2. When we then apply Proposition 9.1.5, the resulting computation is large, but as noted in [**Kontsevich2**], "during 5 minutes on Sun" yields the answer

$$N_4 = \frac{15517926796875}{64}.$$

In terms of instanton numbers, one has $N_4 = n_1 \, 4^{-3} + n_2 \, 2^{-3} + n_4$, and using the known values $n_1 = 2875$ and $n_2 = 609250$, Kontsevich then obtained

$$n_4 = 242467530000.$$

This was the first rigorous computation of n_4, confirming the prediction [**CdGP**] of mirror symmetry. As we showed in Example 7.4.4.1, this means that a generic quintic threefold contains exactly 242467530000 rational quartic curves.

9.2.2. Multiple Covers and Gromov-Witten Invariants. Suppose that V is a Calabi-Yau threefold and that $C \subset V$ is a smooth infinitesimally rigid rational curve. Now that we know about localization, we can describe Kontsevich's approach to the calculation of the contribution of degree d covers of C to the Gromov-Witten invariant $\langle I_{0,0,d[C]} \rangle$. According to Theorem 7.4.4, the answer is $1/d^3$.

We fix an isomorphism $\mathbb{P}^1 \simeq C$, and our rigidity assumption implies that the normal bundle of $C \subset V$ is $N \simeq \mathcal{O}_{\mathbb{P}^1}(-1) \oplus \mathcal{O}_{\mathbb{P}^1}(-1)$. The virtual fundamental class $[\overline{M}_{0,0}(V, d[C])]^{\mathrm{virt}}$ is a 0-cycle, and by definition, $\langle I_{0,0,d[C]} \rangle$ is its degree. We want to compute the contribution to $\langle I_{0,0,d[C]} \rangle$ of the portion of the virtual fundamental class supported on the variety $M_{d,C}(V)$ of degree d multiple covers of C. In the discussion preceding Theorem 7.4.4, we proved that $M_{d,C}(V)$ is a connected component of $\overline{M}_{0,0}(V, d[C])$.

The isomorphism $\mathbb{P}^1 \simeq C$ induces an isomorphism $M_{d,C}(V) \simeq \overline{M}_{0,0}(\mathbb{P}^1, d)$, and the latter is the space that we will use for our calculations. It is easy to see that $\dim \overline{M}_{0,0}(\mathbb{P}^1, d) = 2d - 2$. The first step in the proof is to represent the restriction of $[\overline{M}_{0,0}(V, d[C])]^{\mathrm{virt}}$ to $\overline{M}_{0,0}(\mathbb{P}^1, d)$ as the Euler class of a suitable bundle.

This is done as follows. Consider a stable map $f \in \overline{M}_{0,0}(\mathbb{P}^1, d)$. Using the exact sequence defining the normal bundle of C in V

$$0 \to f^* T_{\mathbb{P}^1} \to f^* T_V \to f^*(\mathcal{O}_{\mathbb{P}^1}(-1) \oplus \mathcal{O}_{\mathbb{P}^1}(-1)) \to 0,$$

we note that the hypotheses of Proposition 7.1.8 are satisfied, since

$$h^1(C, f^* T_V) = h^1(C, f^*(\mathcal{O}_{\mathbb{P}^1}(-1) \oplus \mathcal{O}_{\mathbb{P}^1}(-1))) = 2d - 2,$$

where the last equality comes from Riemann-Roch. Then the proposition implies that the restriction of the virtual fundamental class is just the Euler class of the rank $2d - 2$ obstruction bundle $R^1 \pi_{1*} e_1^*(\mathcal{O}_{\mathbb{P}^1}(-1) \oplus \mathcal{O}_{\mathbb{P}^1}(-1))$, where the universal stable map is given by the family $\pi_1 : \overline{M}_{0,1}(\mathbb{P}^1, d) \to \overline{M}_{0,0}(\mathbb{P}^1, d)$ of curves and the universal map is $e_1 : \overline{M}_{0,1}(\mathbb{P}^1, d) \to \mathbb{P}^1$.

It follows that Theorem 7.4.4 from Chapter 7 can be reformulated as follows.

THEOREM 9.2.3. *Let $C \subset V$ be a rigidly embedded smooth rational curve in a Calabi-Yau threefold. Then the contribution of degree d multiple covers of C to the Gromov-Witten invariant $\langle I_{0,0,d[C]} \rangle$ is*

$$\int_{\overline{\mathcal{M}}_{0,0}(\mathbb{P}^1, d)} \mathrm{Euler}\left(R^1 \pi_{1*} e_1^*(\mathcal{O}_{\mathbb{P}^1}(-1) \oplus \mathcal{O}_{\mathbb{P}^1}(-1))\right) = \frac{1}{d^3}.$$

PROOF. The idea of the proof is to compute the integral using localization of stacks (Corollary 9.1.4) for the usual action of $T = (\mathbb{C}^*)^2$ on $\mathbb{P}^1$.

First consider the special case $d = 2$. In Example 9.2.1.1, we identified the three fixed point components of the T-action on $\overline{\mathcal{M}}_{0,0}(\mathbb{P}^1, 2)$. Label the graphs depicted in Example 9.2.1.1 consecutively as Γ_1, Γ_2, Γ_3. For each graph Γ_i, we will compute the weights of the T-actions on the normal bundle N_{Γ_i} and on the obstruction bundle $R^1 \pi_{1*} e_1^*(\mathcal{O}_{\mathbb{P}^1}(-1) \oplus \mathcal{O}_{\mathbb{P}^1}(-1))$ restricted to Z_{Γ_i}.

We begin with Γ_1, which is the labeled graph

$$\overset{\textstyle 1 \qquad\quad 1}{\underset{\textstyle 0 \qquad 1 \qquad 0}{\bullet\!\!-\!\!-\!\!-\!\!\bullet\!\!-\!\!-\!\!-\!\!\bullet}}$$

The corresponding fixed point component Z_{Γ_1} is a single point, corresponding to a stable map $f_1 : C_1 \cup C_2 \to \mathbb{P}^1$ such that $f_1|_{C_i}$ is an isomorphism and $f_1(p) = q_1$, where $\{p\} = C_1 \cap C_2$.

We first compute the equivariant Euler class. There are 4 flags, 3 vertices, and 2 edges. Using the formulas given in Theorem 9.2.1, we obtain

$$e_{\Gamma_1}^{\Gamma} = 1/\left((\lambda_0 - \lambda_1)^2(\lambda_1 - \lambda_0)^2\right)$$
$$e_{\Gamma_1}^{v} = \left((\lambda_0 - \lambda_1)^2(\lambda_1 - \lambda_0)\right)(2\lambda_1 - 2\lambda_0)/(\lambda_0 - \lambda_1)^2$$
$$e_{\Gamma_1}^{e} = (\lambda_0 - \lambda_1)^4,$$

and then

$$(9.19) \qquad\qquad \text{Euler}_T(N_{\Gamma_1}) = 2(\lambda_1 - \lambda_0)^2$$

follows immediately. To give more insight into where these formulas come from, we will verify (9.19) by calculating the tangent weights in a more elementary way.

To describe the 2-dimensional tangent space to $\overline{\mathcal{M}}_{0,0}(\mathbb{P}^1, 2)$ at f_1, consider the substack $\overline{\mathcal{M}}(\mathbb{P}^1, 2)_1 \subset \overline{\mathcal{M}}_{0,0}(\mathbb{P}^1, 2)$ of 0-pointed stable maps $f : C_1 \cup C_2 \to \mathbb{P}^1$ such that $f|_{C_i}$ has degree 1. Note that $f_1 \in \overline{\mathcal{M}}(\mathbb{P}^1, 2)_1$. There is a morphism $\overline{\mathcal{M}}(\mathbb{P}^1, 2)_1 \to \mathbb{P}^1$ taking f to $f(C_1 \cap C_2) \in \mathbb{P}^1$. This morphism has degree 2 (as a map of stacks) due to the automorphism switching C_1 and C_2. The fixed point $Z_{\Gamma_1} = \{f_1\}$ is in this manner mapped to $q_1 \in \mathbb{P}^1$. The automorphism corresponds to an automorphism of the graph Γ_1 and will contribute an extra factor of 2 in the denominator when we apply Proposition 9.1.5.

The tangent space to $\overline{\mathcal{M}}(\mathbb{P}^1, 2)_1$ at f_1 gives a tangent direction in $T_{f_1}\overline{\mathcal{M}}_{0,0}(\mathbb{P}^1, 2)$ which simply moves the point q_1. Hence the weight of the T-action on this direction is the same as the tangent weight to $\mathbb{P}^1$ at q_1. This weight is easily computed to be $\lambda_1 - \lambda_0$ by our usual methods.

To find the remaining weight on $T_{f_1}\overline{\mathcal{M}}_{0,0}(\mathbb{P}^1, 2)$, note that by a suitable diagram chase involving the maps used in the proof of Theorem 9.2.1, the quotient

$$T_{f_1}\overline{\mathcal{M}}_{0,0}(\mathbb{P}^1, 2) / T_{f_1}\overline{\mathcal{M}}(\mathbb{P}^1, 2)_1$$

is naturally isomorphic to

$$H^0(C_1 \cup C_2, \underline{\text{Ext}}^1(\Omega_{C_1 \cup C_2}, \mathcal{O}_{C_1 \cup C_2})).$$

Then Lemma 9.2.2 implies that the quotient is isomorphic to

$$T_p C_1 \otimes T_p C_2.$$

All of these isomorphisms preserve the natural T-action. But each $T_p C_i$ is isomorphic to $T_{q_1}\mathbb{P}^1$ as a T-representation, so that the remaining weight is $2(\lambda_1 - \lambda_0)$. Hence the weights of $T_f\overline{\mathcal{M}}_{0,0}(\mathbb{P}^1, 2)$ are $\lambda_1 - \lambda_0$ and $2(\lambda_1 - \lambda_0)$, which implies that the equivariant Euler class is $2(\lambda_1 - \lambda_0)^2$, in agreement with (9.19).

We now turn to the obstruction bundle $R^1\pi_{1*}e_1^*(\mathcal{O}_{\mathbb{P}^1}(-1) \oplus \mathcal{O}_{\mathbb{P}^1}(-1))$. To calculate its T-weights, we need only find the weights of $R^1\pi_{1*}e_1^*(\mathcal{O}_{\mathbb{P}^1}(-1))$ and duplicate the answer. The restriction of $f_1^*\mathcal{O}_{\mathbb{P}^1}(-1)$ to C_i is just $\mathcal{O}_{C_i}(-1)$ for $i = 1, 2$, and then the normalization map induces the short exact sequence

$$0 \longrightarrow f_1^*\mathcal{O}_{\mathbb{P}^1}(-1) \longrightarrow \mathcal{O}_{C_1}(-1) \oplus \mathcal{O}_{C_2}(-1) \longrightarrow f_1^*\mathcal{O}_{q_1}(-1) \longrightarrow 0.$$

Taking cohomology gives

$$H^1(C_1 \cup C_2, f_1^*\mathcal{O}_{\mathbb{P}^1}(-1)) = H^0(\mathbb{P}^1, \mathcal{O}_{q_1}(-1)),$$

which is 1-dimensional. But the action of T on $\mathcal{O}_{\mathbb{P}^1}(-1)$ at q_1 is $-\rho_1$, so that the weight is $-\lambda_1$. Hence the weights of the obstruction bundle at f_1 are $-\lambda_1$, $-\lambda_1$.

The calculations at the fixed point Z_{Γ_2} associated to Γ_2 are entirely analogous. We just take the results of the calculation for Γ_1 and switch λ_0, λ_1, which gives

$$\begin{aligned}
\text{Tangent weights} : &\quad \lambda_0 - \lambda_1, \; 2(\lambda_0 - \lambda_1) \\
\text{Obstruction weights} : &\quad -\lambda_0, \; -\lambda_0.
\end{aligned}$$

Finally, we turn our attention to Γ_3. There are 2 flags, 2 vertices, and 1 edge. By Theorem 9.2.1, we get

$$e_{\Gamma_3}^{\mathrm{F}} = 1/\left((\lambda_0 - \lambda_1)(\lambda_1 - \lambda_0)\right)$$

$$e_{\Gamma_3}^{\mathrm{v}} = (\lambda_1 - \lambda_0)(\lambda_0 - \lambda_1)\Big/\left(\left(\frac{\lambda_0 - \lambda_1}{2}\right)\left(\frac{\lambda_1 - \lambda_0}{2}\right)\right)$$

$$e_{\Gamma_3}^{\mathrm{e}} = \frac{2^2(\lambda_0 - \lambda_1)^4}{2^4},$$

so that $\mathrm{Euler}_T(N_{\Gamma_3}) = (\lambda_0 - \lambda_1)(\lambda_1 - \lambda_0)$. One can also obtain this result by explicit calculations as we did for Γ_1.

We also need the weights of the obstruction bundle at the fixed point Γ_3. Here, the fixed point is represented by the degree 2 map $f_3 : \mathbb{P}^1 \to \mathbb{P}^1$ defined by $f_3(z_0, z_1) = (z_0^2, z_1^2)$. Following the same format we did for Γ_1, it suffices to compute the weight of the T action on $H^1(\mathbb{P}^1, f_3^* \mathcal{O}_{\mathbb{P}^1}(-1))$ and duplicate the answer.

Consider the nonzero cohomology class of $H^1(\mathbb{P}^1, f_3^* \mathcal{O}_{\mathbb{P}^1}(-1))$ represented by the Čech cocycle

$$\frac{1}{z_0 z_1} \in Z^0(U_0 \cap U_1, f_3^* \mathcal{O}_{\mathbb{P}^1}(-1)),$$

where $U_i = \{z \in \mathbb{P}^1 : z_i \neq 0\}$. The variable $x_i \in H^0(\mathbb{P}^1, \mathcal{O}_{\mathbb{P}^1}(1))$ has weight λ_i, so that z_i has weight $\frac{1}{2}\lambda_i$ since $x_i = z_i^2$ via f_3. It follows that $1/(z_0 z_1)$ has weight $-(\lambda_0 + \lambda_1)/2$. Hence the obstruction weights are $-(\lambda_0 + \lambda_1)/2$, $-(\lambda_0 + \lambda_1)/2$.

We now have of the ingredients needed to apply Proposition 9.1.5. Thus the integral

$$\int_{\overline{\mathcal{M}}_{0,0}(\mathbb{P}^1,2)} \mathrm{Euler}\left(R^1 \pi_{1*} e_1^*(\mathcal{O}_{\mathbb{P}^1}(-1) \oplus \mathcal{O}_{\mathbb{P}^1}(-1))\right)$$

is equal to the sum

$$\frac{(\lambda_1)(\lambda_1)}{2(\lambda_1 - \lambda_0)(2(\lambda_1 - \lambda_0))} + \frac{(\lambda_0)(\lambda_0)}{2(\lambda_0 - \lambda_1)(2(\lambda_0 - \lambda_1))}$$

$$+ \frac{(-(\lambda_0 + \lambda_1)/2)(-(\lambda_0 + \lambda_1)/2)}{2(\lambda_1 - \lambda_0)(\lambda_0 - \lambda_1)} = \frac{1}{8}.$$

Note that each term has an extra factor of 2 in the denominator, due to the order 2 automorphisms at each of the fixed points, as noted in Example 9.2.1.1. This completes the proof of Theorem 9.2.3 when $d = 2$.

For general d, [**Manin2**] gives a proof which combines the techniques discussed here with clever combinatorial summing techniques. Instead, we will prove the theorem for arbitrary d using a simplification due to R. Pandharipande. The idea is to change the torus action on $\mathcal{O}_{\mathbb{P}^1}(-1) \oplus \mathcal{O}_{\mathbb{P}^1}(-1)$ to simplify the calculation. We first restrict the torus to the subtorus $T' = \mathbb{C}^* \times \{1\} \subset T = (\mathbb{C}^*)^2$. We let $\hbar$ denote the restriction of λ_0 to T'. We will use the action of T' on $\overline{\mathcal{M}}_{0,0}(\mathbb{P}^1, d)$ obtained by restricting the action of T, but will use a different action of T' on $\mathcal{O}_{\mathbb{P}^1}(-1) \oplus \mathcal{O}_{\mathbb{P}^1}(-1)$.

We equip $\mathcal{O}_{\mathbb{P}^1}(-1) \oplus \mathcal{O}_{\mathbb{P}^1}(-1)$ with a T'-action as follows. We put

$$t \cdot (l(x_0, x_1), m(x_0, x_1)) = (l(tx_0, x_1), m(x_0, t^{-1}x_1)).$$

Here, l and m are homogeneous of degree -1 in (x_0, x_1). Note that this action is compatible with the action of T' on $\mathbb{P}^1$.

We will now calculate the integral of $\mathrm{Euler}(R^1\pi_{1*}e_1^*(\mathcal{O}_{\mathbb{P}^1}(-1)\oplus\mathcal{O}_{\mathbb{P}^1}(-1)))$ on $\overline{\mathcal{M}}_{0,0}(\mathbb{P}^1,d)$ using the above T'-action and Corollary 9.1.4. It is easy to see that the fixed point loci of the T'-action on $\overline{\mathcal{M}}_{0,0}(\mathbb{P}^1,d)$ coincide with the fixed point loci of the T-action. Let Γ be a graph corresponding to a fixed point component $\overline{\mathcal{M}}_\Gamma$ of $\overline{\mathcal{M}}_{0,0}(\mathbb{P}^1,d)$. We claim that if Γ has more than one edge, then its contribution to the localization formula vanishes.

This is straightforward to see. Let $f : C \to \mathbb{P}^1$ be a typical stable map in the component $\overline{\mathcal{M}}_\Gamma$ corresponding to Γ, and decompose C into components C_i. Let $\{r_i\}$ be the set of nodes of C. If there is more than one edge in Γ, then there exists at least one node. Then we have the normalization exact sequence

$$0 \longrightarrow f^*(\mathcal{O}_{\mathbb{P}^1}(-1)\oplus\mathcal{O}_{\mathbb{P}^1}(-1)) \longrightarrow \bigoplus_i (f|_{C_i})^*(\mathcal{O}_{\mathbb{P}^1}(-1)\oplus\mathcal{O}_{\mathbb{P}^1}(-1))$$
$$\longrightarrow \bigoplus_i f^*(\mathcal{O}_{\mathbb{P}^1}(-1)\oplus\mathcal{O}_{\mathbb{P}^1}(-1))_{r_i} \longrightarrow 0.$$

From the above exact sequence, we see that $H^1(C, f^*(\mathcal{O}_{\mathbb{P}^1}(-1)\oplus\mathcal{O}_{\mathbb{P}^1}(-1)))$ contains the subspace $\oplus_i H^0(C, f^*(\mathcal{O}_{\mathbb{P}^1}(-1)\oplus\mathcal{O}_{\mathbb{P}^1}(-1))_{r_i})$. Since $f \in \overline{\mathcal{M}}_\Gamma$, we know that $f(r_i)$ is either q_0 or q_1. If $f(r_i) = q_1$, then $1/x_1$ is a local basis for $\mathcal{O}_{\mathbb{P}^1}(-1)$ at q_1. Thus T' acts trivially on the first $f^*\mathcal{O}_{\mathbb{P}^1}(-1)_{r_i}$ factor. Similarly, if $f(r_i) = q_0$, $1/x_0$ is a basis for $\mathcal{O}_{\mathbb{P}^1}(-1)$ at q_0, hence T' acts trivially on the second $f^*\mathcal{O}_{\mathbb{P}^1}(-1)_{r_i}$ factor. In either case, we see that the existence of a node implies that one of the T'-weights of $\oplus_i H^0(C, f^*(\mathcal{O}_{\mathbb{P}^1}(-1)\oplus_{\mathbb{P}^1}\mathcal{O}(-1))_{r_i})$ is zero. This weight contributes multiplicatively to the restriction of the equivariant Euler class of the obstruction bundle. It follows that the fixed point component $\overline{\mathcal{M}}_\Gamma$ contributes 0, which proves our claim.

Therefore we are reduced to a single graph Γ with one edge corresponding to a degree d map, namely

$$\overset{d}{\underset{\underset{0\qquad 1}{}}{\bullet\!\!-\!\!-\!\!-\!\!\bullet}}$$

For this graph, $\overline{\mathcal{M}}_\Gamma$ is a single point with automorphism group $A_\Gamma = \mathbb{Z}/d\mathbb{Z}$, and as usual, the corresponding map $f : \mathbb{P}^1 \to \mathbb{P}^1$ is defined by $f(z_0, z_1) = (z_0^d, z_1^d)$.

We will compute the weights of the restriction of the obstruction bundle using the open cover U_0, U_1 from the argument for $d = 2$ given earlier. For general d, one can show that a basis for $H^1(\mathbb{P}^1, f^*(\mathcal{O}_{\mathbb{P}^1}(-1)\oplus\mathcal{O}_{\mathbb{P}^1}(-1)))$ is given by the Čech cocycles

$$(9.20)\qquad\qquad \left(\frac{1}{z_0^i z_1^{d-i}}, 0\right), \quad \left(0, \frac{1}{z_0^i z_1^{d-i}}\right), \quad 1 \le i \le d-1.$$

We know that $x_0, x_i \in H^0(\mathbb{P}^1, \mathcal{O}_{\mathbb{P}^1}(1))$ have weights $\hbar,\ 0$. Since $x_i = z_i^d$ via f, it follows that z_0, z_1 have weights $\hbar/d,\ 0$. Hence the cocycles in (9.20) have T'-weights $-i\hbar/d$ and $(d-i)\hbar/d$ for $1 \le i \le d-1$, and thus the equivariant Euler class of the obstruction bundle contributes $(-1)^{d-1}((d-1)!)^2\hbar^{2d-2}/d^{2d-2}$ in the localization theorem.

We also need to compute the equivariant Euler class of the normal bundle N_Γ. Using Theorem 9.2.1 with $\lambda_0 = \hbar$ and $\lambda_1 = 0$, we obtain

$$e_\Gamma^{\mathrm{F}} = 1/((\hbar - 0)(0 - \hbar)) = -\hbar^{-2}$$

$$e_\Gamma^{\mathrm{v}} = (\hbar - 0)(0 - \hbar) \Big/ \left(\left(\frac{\hbar - 0}{d}\right)\left(\frac{0 - \hbar}{d}\right)\right) = d^2$$

$$e_\Gamma^{\mathrm{e}} = \frac{(-1)^d (d!)^2 (\hbar - 0)^{2d}}{d^{2d}} = \frac{(-1)^d (d!)^2 \hbar^{2d}}{d^{2d}},$$

from which we conclude that

$$\mathrm{Euler}_{T'}(N_\Gamma) = (-1)^{d-1}(d!)^2 \hbar^{2d-2}/d^{2d-2}.$$

Finally, since we are dealing with stacks, we also have the factor a_Γ arising from automorphisms of $\overline{\mathcal{M}}_\Gamma$. The exact sequence (9.13) shows that $a_\Gamma = d$ in this case. This gives an extra factor of d in the denominator. Putting everything together, Corollary 9.1.5 implies that the integral in question is equal to

$$\frac{(-1)^{d-1}\left((d-1)!\right)^2 \hbar^{2d-2}/d^{2d-2}}{d \cdot (-1)^{d-1}(d!)^2 \hbar^{2d-2}/d^{2d-2}} = \frac{1}{d^3},$$

as claimed. This completes the proof of the theorem. $\qquad\square$

An entirely different proof of Theorem 9.2.3 is given in [**LLY**, Corollary 3.5]. This proof uses ideas similar to those used in the proof of the Mirror Theorem given in Section 11.1. We will explain this proof in Example 11.1.7.1.

9.2.3. Double Covers of 6-Nodal Quintics and n_{10}. When we studied rational curves on a quintic threefold $V \subset \mathbb{P}^4$ in Section 7.4.4, we showed that for $d \leq 9$, the instanton number n_d is the number of rational curves of degree d on V. A crucial part of the argument is Theorem 9.2.3, which tells us how multiple covers of rigidly embedded smooth rational curves contribute to $N_d = \langle I_{0,0,d} \rangle$. However, as pointed out to us by R. Pandharipande, Theorem 9.2.3 does *not* apply to double covers of nodal rational curves of degree 5 lying in V. This is similar to pathologies about multiple covers of nodal curves noticed in [**Katz3, BL**]. Instead of contributing $2^{-3} = \frac{1}{8}$ as predicted by Theorem 9.2.3, we will prove below that each such curve contributes $6\frac{1}{8}$ to $N_{10} = \langle I_{0,0,10} \rangle$.

Let's first explain why V contains nodal rational curves of degree 5. If $P \subset \mathbb{P}^4$ is a plane, the curve $C = V \cap P$ is a plane quintic curve of arithmetic genus 6. We want to find rational curves among these C. The curve C would be rational if it had 6 nodes. Note that C has a node at a point p if and only if the plane P is simply tangent to V at p. It is one condition on P to be tangent to V. Hence we expect 6 conditions on P to be tangent to V at 6 points. Since the curves C are parametrized by the Grassmannian $G(3,5)$ of projective planes $P \subset \mathbb{P}^4$, and $G(3,5)$ has dimension 6, we expect a generic quintic threefold V to contain a finite nonzero number of 6-nodal rational plane quintic curves C. This naive dimension count can in fact be made rigorous, and the number of such curves (including multiplicity) is known to be $17,601,000$ [**Vainsencher**]. Below, we will show that the multiplicities are generically all 1, so that a generic quintic threefold has precisely $17,601,000$ 6-nodal plane quintic curves.

We now let C_n be one of these nodal rational plane quintics on a generic quintic threefold V. By a *degree d multiple cover of C_n*, we mean a genus 0 stable map $f : C \to C_n$ with $f_*[C] = d[C_n] = 5d$. We can also think of a multiple cover as a map $f : C \to V$ factoring through C_n. We want to find the contribution of the multiple covers to the Gromov-Witten invariant N_{5d}. A naive interpretation of the multiple cover formula from Theorem 9.2.3 would suggest a contribution of $1/d^3$. We show that this is already false for $d = 2$. There are three steps involved:

- First, we will compute the normal bundle of the normalization of C_n.
- Next, we will show that the double covers of C_n are a union of connected components of $\overline{M}_{0,0}(V, 10)$.
- Finally, we will prove that the restriction of the virtual fundamental class $[\overline{M}_{0,0}(V, 10)]^{\mathrm{virt}}$ to these components has total degree $6\frac{1}{8}$.

We begin with the first of these steps, which will tell us that C_n is rigidly embedded in V. Note that we used this fact earlier in Section 7.4.4.

LEMMA 9.2.4. *If V is a generic quintic threefold and $g : \mathbb{P}^1 \to C_n$ is the normalization of $C_n \subset V$, then g has normal bundle $N_g = \mathcal{O}_{\mathbb{P}^1}(-1) \oplus \mathcal{O}_{\mathbb{P}^1}(-1)$ in V.*

PROOF. Since N_g is a rank 2 bundle on $\mathbb{P}^1$, we have $N_g = \mathcal{O}_{\mathbb{P}^1}(a) \oplus \mathcal{O}_{\mathbb{P}^1}(b)$ for some a, b, and the adjunction formula implies $a + b = -2$ since V is Calabi-Yau. Hence $a = b = -1$ if and only if $h^0(\mathbb{P}^1, N_g) = 0$.

We will adapt the argument of [**Katz2**, p. 152–153] to this situation. Let $\mathbb{P} = \mathbb{P}(H^0(\mathbb{P}^4, \mathcal{O}_{\mathbb{P}^4}(5)))$ be the projective space of quintics in $\mathbb{P}^4$, and let M be the moduli space of genus 0 irreducible rational plane quintic curves in $\mathbb{P}^4$. Note that M is fibered over $G(3, 5)$ with fibers isomorphic to open subsets of $\overline{M}_{0,0}(\mathbb{P}^2, 5)$, so that M is irreducible of dimension 20. Let

$$I = \{(D, V) \in M \times \mathbb{P} : D \subset V\}$$

be the incidence correspondence. By the Riemann-Roch theorem, $h^0(D, \mathcal{O}_D(5)) = 20$ for $D \in M$. Since each curve $D \in M$ is a complete intersection in $\mathbb{P}^4$, it is projectively normal. We can then compute that the fibers of the projection $\pi_1 : I \to M$ are projective spaces of dimension 105. Hence I is irreducible of dimension 125, which is also the dimension of $\mathbb{P}$.

Now let

$$I_1 = \{(D, V) \in I : h^0(\mathbb{P}^1, N_g) \geq 1\},$$

where $g : \mathbb{P}^1 \to D \subset V$ is the normalization map. By the above description of N_g, we see that $(D, V) \in I - I_1$ if and only if $N_g = \mathcal{O}_{\mathbb{P}^1}(-1) \oplus \mathcal{O}_{\mathbb{P}^1}(-1)$. We will show in a moment that $I - I_1$ is nonempty. Assuming this, then by the irreducibility of I, we have that $\dim I_1 \leq 124$. Hence $\pi_2(I_1)$ must be a proper subvariety of $\mathbb{P}$, and for any $V \in \mathbb{P} - \pi_2(I_1)$ we conclude that the assertion of the lemma holds for all plane rational quintics on V.

So we are reduced to writing down a single example, which we now do. Let (z_0, z_1) be homogeneous coordinates on $\mathbb{P}^1$, and consider the curve D which is the image of the map

$$(9.21) \qquad \begin{aligned} &(z_0, z_1) \mapsto \\ &(z_0^5 + z_1^5, z_0 z_1(z_0^3 + z_0^2 z_1 + z_0 z_1^2), z_0 z_1(z_0^3 + 3z_0^2 z_1 + 2z_0 z_1^2 + z_1^3), 0, 0). \end{aligned}$$

The curve D lies in the quintic threefold $F = 0$, where

$$(9.22) \qquad \begin{aligned} F(x_0, \dots, x_4) = &\, x_0^3 x_1^2 + 3x_0^2 x_1^3 + 20x_0 x_1^4 + 61x_1^5 - x_0^3 x_1 x_2 + x_0^2 x_1^2 x_2 \\ &- 24x_0 x_1^3 x_2 - 25x_1^4 x_2 - 3x_0^2 x_1 x_2^2 + 2x_0 x_1^2 x_2^2 - 40x_1^3 x_2^2 \\ &+ x_0^2 x_2^3 + 4x_0 x_1 x_2^3 + 35x_1^2 x_2^3 - x_0 x_2^4 - 10x_1 x_2^4 + x_2^5 \end{aligned}$$

and $x_0, \ldots, x_4$ are coordinates on $\mathbb{P}^4$. A general quintic containing D has equation of the form

$$(9.23) \qquad F + x_3 G + x_4 H = 0,$$

where G, H are each homogeneous of degree 4.

From the normal bundle exact sequence

$$(9.24) \qquad 0 \longrightarrow T_{\mathbb{P}^1} \longrightarrow g^* T_V \longrightarrow N_g \to 0,$$

we see that $h^0(\mathbb{P}^1, N_g) = 0$ if and only if $h^0(\mathbb{P}^1, g^* T_V) = 3$. By Riemann-Roch, this happens if and only if $H^1(\mathbb{P}^1, g^* T_V) = 0$. We will compute $H^1(\mathbb{P}^1, g^* T_V)$ using a computational device from [**Katz1**], which we generalize slightly. Suppose that $f : C \to \mathbb{P}^4$ with C a rational curve, possibly reducible. Put $\mathcal{L} = f^* \mathcal{O}_{\mathbb{P}^4}(1)$. If $\widetilde{F} = 0$ is an equation for V, let $\widetilde{F}_j$ denote the pullback via f of the partial derivative of $\widetilde{F}$ with respect to x_j. Note that $\widetilde{F}_j \in H^0(\mathbb{P}^1, 4\mathcal{L})$. Now consider the exact sequences

$$0 \longrightarrow \mathcal{O}_{\mathbb{P}^4} \longrightarrow \mathcal{O}(1)^5 \longrightarrow T_{\mathbb{P}^4} \to 0$$

and

$$0 \longrightarrow T_V \longrightarrow T_{\mathbb{P}^4}|_V \longrightarrow \mathcal{O}_V(5) \longrightarrow 0.$$

We can pull back these sequences via f, and combine them to get a map

$$\Phi_V : H^0(\mathbb{P}^1, \mathcal{L})^5 \longrightarrow H^0(\mathbb{P}^1, 5\mathcal{L})$$

given by

$$(s_0, \ldots, s_4) \mapsto \sum_{j=0}^{4} s_j \widetilde{F}_j.$$

An easy diagram chase shows that $H^1(C, q^* T_V) \simeq \mathrm{coker}\ \Phi_V$. Here, we have used $h^1(C, \mathcal{O}_C) = h^1(C, g^* T_{\mathbb{P}^4}) = 0$, where the latter equality follows from the convexity of $\mathbb{P}^4$.

The normal bundle can then be computed directly using linear algebra (and a computer). For example, letting $G = x_0^4$ and $H = x_1^4$, the matrix of Φ_V is an integer valued 30×26 matrix which we compute has rank 26. The resulting quintic $F + x_3 G + x_4 H = 0$ is singular, but the generic quintic of the form (9.23) is smooth, so by semicontinuity of the rank, a smooth V can be found as well. $\qquad \square$

This lemma implies that when V is generic, each 6-nodal rational plane quintic curve occurs with multiplicity 1, so that V has exactly $17,601,000$ such curves.

We next turn to the second step of our study of double covers of C_n, where we show that such covers give rise to 7 connected components of $\overline{M}_{0,0}(V, 10)$. Let $M_{2,C_n}(V) \subset \overline{\mathcal{M}}_{0,0}(V, 10)$ be the subspace classifying all stable 0-pointed genus 0 double covers $f : C \to C_n$. Then we have the following result.

PROPOSITION 9.2.5. $M_{2,C_n}(V)$ *consists of 7 connected components of the moduli space* $\overline{M}_{0,0}(V, 10)$. *Furthermore:*

(i) *One component is isomorphic to* $\overline{M}_{0,0}(\mathbb{P}^1, 2)$ *and consists of double covers which factor through the normalization.*

(ii) *The other 6 components are smooth points of* $\overline{M}_{0,0}(V, 10)$ *and correspond to double covers* $f : C_1 \cup C_2 \to C_n$ *such that* $f|_{C_i}$ *is the normalization map and f is a local isomorphism above one of the nodes of* C_n.

PROOF. We start with double covers which factor through the normalization map $g : \mathbb{P}^1 \to C_n$. Let $\mathcal{M}_1$ be the stack of stable double covers $f : C \to C_n$ which factor through g, and consider the associated moduli space $M_1 \subset \overline{M}_{0,0}(V, 10)$. It is immediate to see that these are the maps $f \in \overline{M}_{0,0}(V, 10)$ which factor as $f = g \circ h$ for some $h : C \to \mathbb{P}^1$ in $\overline{M}_{0,0}(\mathbb{P}^1, 2)$. We claim that M_1 is a component of $\overline{M}_{0,0}(V, 10)$. To see this, we analyze $\overline{M}_{0,0}(V, 10)$ infinitesimally at $f \in M_1$. Since $\overline{M}_{0,0}(\mathbb{P}^1, 2)$ is smooth (in the orbifold sense) and complete of dimension 2, then to show that M_1 is a component, it suffices to show that the tangent space to $\overline{M}_{0,0}(V, 10)$ at such f also has dimension 2. As usual, this tangent space is $\mathrm{Ext}^1(f^*\Omega_V^1 \to \Omega_C^1, \mathcal{O}_C)$.

First suppose that $(f, C) \in M_1$ with C irreducible. Then $C \simeq \mathbb{P}^1$, and the long exact sequence for $\mathrm{Ext}^\bullet(f^*\Omega_V^1 \to \Omega_{\mathbb{P}^1}^1, \mathcal{O}_{\mathbb{P}^1})$ simplifies to

$$0 \to H^0(\mathbb{P}^1, T_{\mathbb{P}^1}) \longrightarrow H^0(f^*T_V) \longrightarrow \mathrm{Ext}^1(f^*\Omega_V^1 \longrightarrow \Omega_{\mathbb{P}^1}^1, \mathcal{O}_{\mathbb{P}^1}) \longrightarrow 0$$

and the isomorphism $\mathrm{Ext}^2(f^*\Omega_V^1 \to \Omega_{\mathbb{P}^1}^1, \mathcal{O}_{\mathbb{P}^1}) \simeq H^1(f^*T_V)$. Using Lemma 9.2.4 and (9.24), we see that $g^*T_V \simeq \mathcal{O}_{\mathbb{P}^1}(2) \oplus \mathcal{O}_{\mathbb{P}^1}(-1) \oplus \mathcal{O}_{\mathbb{P}^1}(-1)$, which implies that

$$f^*T_V = \mathcal{O}_{\mathbb{P}^1}(4) \oplus \mathcal{O}_{\mathbb{P}^1}(-2) \oplus \mathcal{O}_{\mathbb{P}^1}(-2).$$

This immediately implies $\mathrm{Ext}^1(f^*\Omega_V^1 \to \Omega_{\mathbb{P}^1}^1, \mathcal{O}_{\mathbb{P}^1})$ has dimension 2.

It is also possible to have double covers $(f, C) \in M_1$ where C has two components, each mapping onto C_n with degree 1. For these curves, we can again compute that $\mathrm{Ext}^1(f^*\Omega_V^1 \to \Omega_C^1, \mathcal{O}_C)$ has dimension 2. We omit the details. This completes the proof of part (i) of the proposition.

Now suppose that $f : C \to C_n$ is a degree 2 cover which does not factor through the normalization map g. Write $C = \cup_i C_i$, and let d_i be the degree of f on each C_i, so that $\sum_i d_i = 2$. Note that there can be at most 2 components mapping to C_n with positive degree. If a tree of components C_i were contracted by f, (i.e., if $d_i = 0$ for all curves C_i in the tree), then it could meet at most 2 other components, contradicting stability. Hence C can have only one or two components. If it had just one component, then C would be normal, and f would factor through the normalization g. So $C = C_1 \cup C_2$ is a union of two smooth rational components and $d_1 = d_2 = 1$. It follows that $f|_{C_i}$ is identified with the normalization map g, and we only have to describe how to glue these two maps together. If f maps the node of C to a smooth point of C_n, it is immediate to see that f factors through g. Similarly, it is immediate to see that if f maps the node of C to a node of C_n, then f factors through g if and only if the branches of C near its node are mapped to the same branch of C_n at the image node. Thus, when f doesn't factor through g, it must map to both branches of C_n at the image node. In other words, f is a local isomorphism above the node.

We conclude that there are precisely 6 double covers of C_n which don't factor through the normalization, one for each node of C_n. Let $p_1, \ldots p_6$ denote the nodes, and let $f_i : C = C_1 \cup C_2 \to C_n \subset V$ be the double cover which restricts to the normalization map on each C_i, takes the node of C to p_i, and takes C_1 and C_2 to opposite branches of C_n at p_i.

The above argument shows that $M_{2,C_n}(V)$ consists of the 6 maps $f_1, \ldots, f_6$ together with the component $\overline{M}_{0,0}(\mathbb{P}^1, 2)$ of double covers which factor through the normalization. Hence, to complete the proof of the proposition, we need only show that the f_i are smooth points of $\overline{M}_{0,0}(V, 10)$, or equivalently, that the tangent space $\mathrm{Ext}^1(f_i^*\Omega_V^1 \to \Omega_C^1, \mathcal{O}_C)$ has dimension 0 at these points. By the usual exact sequence, it then suffices to show that the map

$$(9.25) \qquad \mathrm{Ext}^1(\Omega^1_C, \mathcal{O}_C) \to H^1(C, f_i^* T_V)$$

is surjective. Let K be the kernel of $f_1^* \Omega^1_V \to \Omega^1_C$. Using local coordinates, it is easy to see that K is locally free of rank 2. Then surjectivity of (9.25) is equivalent to $H^1(C, K^*) = 0$. We now turn to this computation.

We will prove that $H^1(C, K^*) = 0$ by an indirect argument. Consider the incidence correspondence

$$(9.26) \qquad \mathcal{I} = \{(p, D, V) \in \mathbb{P}^4 \times M \times \mathbb{P} : p \text{ is a node of } D \subset V\}.$$

where as in the proof of Lemma 9.2.4, M is the moduli space of genus 0 irreducible rational plane quintic curves in $\mathbb{P}^4$.

Observe that the set of irreducible rational curves with a node at a fixed point is irreducible. This follows because the curves with a node at, say, $(1,0,0,0,0)$ can be parametrized by the map

$$(z_0, z_1) \mapsto (\alpha_0(z_0, z_1), z_0 z_1 \alpha_1(z_0, z_1), z_0 z_1 \alpha_2(z_0, z_1), z_0 z_1 \alpha_3(z_0, z_1), z_0 z_1 \alpha_4(z_0, z_1))$$

for some homogeneous forms α_i with α_0 of degree 5 and α_i of degree 3 for $1 \le i \le 4$. In the above map, we have arbitrarily let $(1,0)$ and $(0,1)$ be the two points mapping to the node. Note that (9.21) is a special case of this construction. The set of all α_i form an irreducible variety, and the desired irreducibility follows.

Using this, we can prove that the incidence correspondence $\mathcal{I}$ defined in (9.26) is irreducible. This follows from the previous paragraph by projecting onto $\mathbb{P}^4 \times M$ and then onto $\mathbb{P}^4$.

From here, the argument used in proving Lemma 9.2.4 reduces us to finding one example of a $(p, D, V) \in \mathcal{I}$ such that the $H^1(C, K^*) = 0$. We take D to be as in (9.22), letting $p = (1,0,0,0,0)$. This node is the image of the two points $(1,0)$ and $(0,1)$ under the parametrization (9.21). We can use *Macaulay* [**BS**] to explicitly construct the curve C as a line-pair in $\mathbb{P}^2$, construct the map f as the restriction of an explicit map $\mathbb{P}^2 \to \mathbb{P}^4$, then chase through the constructions to build the sheaf K^*. We then calculate that $H^1(C, K^*) = 0$, as desired. Note that *Macaulay* computes in nonzero characteristic, but we can conclude the vanishing generically anyway by upper-semicontinuity of cohomology. $\square$

Finally, we compute the degree of the restriction of the virtual fundamental class $[\overline{M}_{0,0}(V, 10)]^{\mathrm{virt}}$ to the components described in Proposition 9.2.5.

THEOREM 9.2.6. *If C_n is a nodal rational plane quintic curve in the quintic threefold $V \subset \mathbb{P}^4$, then the contribution of double covers of C_n to the Gromov-Witten invariant $N_{10} = \langle I_{0,0,10} \rangle$ is $6\frac{1}{8}$.*

PROOF. We begin with the component $\overline{M}_{0,0}(\mathbb{P}^1, 2)$ given by double covers which factor through the normalization. Here, one simply follows the proof of Theorem 9.2.3. The tangent and obstruction weights are unchanged from those computed earlier, so that this component contributes $1/8$ to N_{10}.

Turning to the 6 components represented by the maps $f_1, \ldots, f_6$, the key observation is that these are smooth points of the expected dimension. It follows by the discussion following Definition 7.1.7 that $[\overline{M}_{0,0}(V, 10)]^{\mathrm{virt}}$ restricts to the usual fundamental class of a point, which is just the point with coefficient 1. Hence each of the 6 components contributes 1 to N_{10}, and the theorem is proved. $\square$

9.3. Equivariant Gromov-Witten Invariants

We now give a brief discussion of equivariant Gromov-Witten invariants, beginning with the case when $g = 0$ and X is convex for simplicity.

9.3.1. Genus Zero Invariants For X Convex. Suppose X is smooth and convex. As in Section 7.1.5, this implies $[\overline{M}_{0,n}(X,\beta)]^{\text{virt}} = [\overline{M}_{0,n}(X,\beta)]$. We can then define genus 0 equivariant Gromov-Witten invariants as follows.

Suppose that X is equipped with an action by an algebraic group G. Then $h \in G$ acts naturally on $\overline{M}_{0,n}(X,\beta)$ by the formula $h \cdot f = \tau_h \circ f$, where $\tau_h : X \to X$ is the automorphism of X associated to h. In Section 9.2, we analyzed this action in detail when $X = \mathbb{P}^r$ and $G = (\mathbb{C}^*)^{r+1}$.

Since $[\overline{M}_{0,n}(X,\beta)]^{\text{virt}} = [\overline{M}_{0,n}(X,\beta)]$, the definition of Gromov-Witten invariant given in Section 7.1.6 simplifies to

$$\langle I_{0,n,\beta}\rangle(\alpha_1,\dots,\alpha_n) = \int_{\overline{M}_{0,n}(X,\beta)} e_1^*(\alpha_1) \cup \cdots \cup e_n^*(\alpha_n),$$

where $\alpha_i \in H^*(X)$ and

$$e_i : \overline{M}_{0,n}(X,\beta) \longrightarrow X$$

is the evaluation map which sends a stable map $f : (C, p_1, \dots, p_n) \to X$ to $f(p_i)$ (see Section 7.1.2).

To make an equivariant version of this, first observe that e_i is equivariant with respect to the actions of G on $\overline{M}_{0,n}(X,\beta)$ and X, so that we have an equivariant pullback map

$$e_i^* : H_G^*(X) \longrightarrow H_G^*(\overline{M}_{0,n}(X,\beta)).$$

Furthermore, since X is convex, we know that $\overline{M}_{0,n}(X,\beta)$ is an orbifold, which means that we have an equivariant integral

$$\int_{\overline{M}_{0,n}(X,\beta)_G} : H_G^*(\overline{M}_{0,n}(X,\beta)) \longrightarrow H^*(BG).$$

We now have all of the ingredients needed for the definition.

DEFINITION 9.3.1. *Suppose that X is smooth and convex, and let $\alpha_1, \dots, \alpha_n \in H_G^*(X)$. Then the expression*

$$\langle I_{0,n,\beta}\rangle_G(\alpha_1,\dots,\alpha_n) = \int_{\overline{M}_{0,n}(X,\beta)_G} e_1^*(\alpha_1) \cup \cdots \cup e_n^*(\alpha_n) \in H^*(BG)$$

is a genus 0 equivariant Gromov-Witten invariant.

We can relate this definition to ordinary Gromov-Witten invariants as follows. From Section 9.1.1, we have the "nonequivariant limit" map $i_{\text{point}}^* : H^*(BG) \to H^*(\text{point}) = \mathbb{C}$, and we also have $i_X^* : H_G^*(X) \to H^*(X)$. Then the identity

$$i_{\text{point}}^*\langle I_{0,n,\beta}\rangle_G(\alpha_1,\dots,\alpha_n) = \langle I_{0,n,\beta}\rangle(i_X^*\alpha_1,\dots,i_X^*\alpha_n)$$

is proved using the commutative diagram (9.11) from Section 9.1.3. Thus equivariant Gromov-Witten invariants are refinements of ordinary Gromov-Witten invariants which carry more information coming from the group action.

Next suppose that E is an equivariant vector bundle over the smooth convex variety X. If $Y \subset X$ is a smooth subvariety defined by the vanishing of a generic section of E, then it is possible to define equivariant versions of certain Gromov-Witten invariants of Y, even though Y itself has no group action. Rather than discuss the general case, we will comment on one example of particular interest to us.

Example 9.3.1.1. Suppose that V is a quintic threefold. From Example 7.1.6.1, we have the Gromov-Witten invariant

$$N_d = \langle I_{0,0,d} \rangle = \int_{\overline{\mathcal{M}}_{0,0}(\mathbb{P}^4,d)} \mathrm{Euler}(\mathcal{V}_d)$$

where $\mathcal{V}_d$ is the vector bundle over $\overline{\mathcal{M}}_{0,0}(\mathbb{P}^4, d)$ whose fiber at $(f, C) \in \overline{\mathcal{M}}_{0,0}(\mathbb{P}^4, d)$ is $H^0(C, f^*\mathcal{O}_{\mathbb{P}^4}(5))$. Since both $\overline{\mathcal{M}}_{0,0}(\mathbb{P}^4, d)$ and $\mathcal{V}_d$ have natural actions by $T = (\mathbb{C}^*)^5$, we see that the equivariant version of this invariant is given by

$$\langle I_{0,0,d} \rangle_T = \int_{\overline{\mathcal{M}}_{0,0}(\mathbb{P}^4,d)_T} \mathrm{Euler}_T(\mathcal{V}_d) \in H^*(BT).$$

One way to view the computation of N_4 described in Example 9.2.1.3 is that we applied Corollary 9.1.4 to compute the equivariant Gromov-Witten invariant $\langle I_{0,0,4} \rangle_T$ and then took the nonequivariant limit to get the desired answer N_4.

We will not make explicit use of equivariant Gromov-Witten invariants in what follows, though they appear implicitly in some of the formulas in Chapter 11. However, Chapter 11 will definitely use the strategy described above of working with certain equivariant classes and then obtaining interesting results in the nonequivariant limit.

9.3.2. Localization of Virtual Fundamental Classes. To define equivariant Gromov-Witten invariants for an arbitrary smooth projective variety X and any genus $g \geq 0$, we need an equivariant version of the virtual fundamental class. In the case when X has a $\mathbb{C}^*$-action, this has been worked out in [**GPa**], provided that one uses the equivariant Chow group $A_*^{\mathbb{C}^*}(X)$ of X.

There are two key points to the theory. The first, as one might expect, is to define an equivariant virtual fundamental class $[\overline{M}_{g,n}(X, \beta)]_{\mathbb{C}^*}^{\mathrm{virt}} \in A_*^{\mathbb{C}^*}(\overline{M}_{g,n}(X, \beta))$. From here, one gets equivariant Gromov-Witten classes without difficulty. But there is also a second key ingredient, which is a localization formula. This is no surprise, since we've already seen how localization plays an important role in calculating contributions to Gromov-Witten invariants.

The basic idea is that the construction of [**BF, Behrend**] associates an equivariant virtual fundamental class

$$[Y]_{\mathbb{C}^*}^{\mathrm{virt}} \in A_*^{\mathbb{C}^*}(Y)$$

to any algebraic scheme Y with a $\mathbb{C}^*$-action and $\mathbb{C}^*$-equivariant perfect obstruction theory. When we apply this to $Y = \overline{M}_{g,n}(X, \beta)$ with the $\mathbb{C}^*$-action inherited from the action on X, the perfect obstruction theory used to define the ordinary virtual fundamental class inherits a natural $\mathbb{C}^*$-action, which gives $[\overline{M}_{g,n}(X, \beta)]_{\mathbb{C}^*}^{\mathrm{virt}} \in A_*^{\mathbb{C}^*}(\overline{M}_{g,n}(X, \beta))$.

Also, as explained in [**GPa**], the equivariant virtual fundamental class localizes nicely. In the above situation of a $\mathbb{C}^*$ action on Y with $\mathbb{C}^*$ equivariant perfect obstruction theory, the components of the fixed point locus Z_j inherit a $\mathbb{C}^*$-fixed perfect obstruction theory, leading to an equivariant virtual fundamental class in

$A_*^{\mathbb{C}^*}(Z_i)$. The *virtual normal bundle* N_j^{vir} to Z_j is obtained from the "moving part" of the virtual tangent space determined by the obstruction theory. Then the localization formula [**GPa**] for the equivariant virtual fundamental class is

$$[X]_{\mathbb{C}^*}^{\mathrm{virt}} = \sum_j i_{j!}\left(\frac{[Z_j]_{\mathbb{C}^*}^{\mathrm{virt}}}{\mathrm{Euler}_{\mathbb{C}^*}(N_j^{\mathrm{virt}})}\right),$$

where as usual, $\mathrm{Euler}_{\mathbb{C}^*}(N_j^{\mathrm{virt}})$ is the equivariant Euler class of N_j^{virt}.

It is straightforward to generalize this to $G = (\mathbb{C}^*)^n$. This localization plays a crucial role in Theorem 11.2.16. The localization methods described in [**GPa**] also have some nice applications to Gromov-Witten invariants, including the multiple cover contributions to genus 1 Gromov-Witten invariants mentioned in Example 7.4.5.3.

CHAPTER 10

Quantum Differential Equations

This chapter will introduce an extension of Gromov-Witten invariants called *gravitational correlators* and use them to define *quantum differential equations*. In Section 10.1, we define the gravitational correlators and study their properties. Section 10.2 then uses gravitational correlators to explicitly describe the flat sections of the cohomology bundle with respect to the A-model connection constructed in Chapter 8. These flat sections also allow us in Section 10.3 to define a formal $H^*(X)$-valued function J on $H^*(X)$. This function may be viewed as a generating function built up from some of the more interesting gravitational correlators. In particular, the A-model correlation functions of a Calabi-Yau threefold can be extracted completely from J. Quantum differential operators are then defined to be differential operators annihilating J. We will prove that these operators give rise to relations in quantum cohomology, which explains the intrinsic interest of J. This function also plays a central role in Givental's approach to the Mirror Theorem, which will be discussed in Section 11.2.

10.1. Gravitational Correlators

In this section, we discuss the *gravitational correlators* and their properties. These are invariants of smooth algebraic varieties and include among them the Gromov-Witten invariants studied in Chapter 7. We define the gravitational correlators in Section 10.1.1 and study their properties in Section 10.1.2. The section concludes with a series of examples.

10.1.1. Definition of the Correlators. Before we can define the gravitational correlators, we need to recall some facts about stable maps. Given a smooth algebraic variety X and $\beta \in H_2(X, Z)$, we have the moduli stack $\overline{\mathcal{M}}_{g,n}(X, \beta)$ from Section 7.1.1, with associated coarse moduli space $\overline{M}_{g,n}(X, \beta)$. There are also evaluation maps

$$e_i : \overline{\mathcal{M}}_{g,n}(X, \beta) \longrightarrow X, \quad i = 1, \ldots, n$$

given by $e_i(f, p_1, \ldots, p_n) = f(p_i)$, as well as the map

$$\pi_{n+1} : \overline{\mathcal{M}}_{g,n+1}(X, \beta) \longrightarrow \overline{\mathcal{M}}_{g,n}(X, \beta)$$

defined by forgetting the last marked point and contracting unstable components.

Since $\overline{M}_{g,n}(X, \beta)$ is not a fine moduli space for $\overline{\mathcal{M}}_{g,n}(X, \beta)$, it is not the parameter space for a universal curve. However, we claim that π_{n+1} is a stack-theoretic version of the universal curve, with $e_{n+1} : \overline{\mathcal{M}}_{g,n+1}(X, \beta) \to X$ as the universal stable map. To see why this is true, we will use the same symbols e_{n+1} and π_{n+1} for the corresponding maps formed from the coarse moduli spaces $\overline{M}_{g,n+1}(X, \beta)$

and $\overline{M}_{g,n}(X,\beta)$. We observe that for any $(f,C,p_1,\ldots,p_n) \in \overline{M}_{g,n}(X,\beta)$, there is a natural isomorphism

$$(10.1) \qquad C/\mathrm{Aut}(f) \simeq \pi_{n+1}^{-1}(f),$$

where $\mathrm{Aut}(f)$ denotes the finite group of automorphisms of the stable map f. This is defined by sending $p \in C$ to $(f : C \to X, p_1,\ldots,p_n,p)$ if p is smooth and distinct from $p_1,\ldots,p_n$. However, if $p = p_i$ for some i, then p maps to the stable map $(f : C \cup \mathbb{P}^1 \to X, p_1,\ldots,p_i',\ldots,p_n,p_{n+1}')$, where $\mathbb{P}^1$ is attached to C at p_i and p_i',p_{n+1}' are distinct points of $\mathbb{P}^1$ different from the attaching point. Finally, we leave the case when p is a node to the reader. Since we can identify f with the restriction of e_{n+1} to $\pi_{n+1}^{-1}(f)$, our claim is proved. In the corresponding stack-theoretic statement, the stack takes the automorphisms of f into account, so that stack-theoretically, π_{n+1} is the universal curve.

The description of the fibers of π_{n+1} given in (10.1) shows that π_{n+1} has tautological sections $s_i : \overline{M}_{g,n}(X,\beta) \to \overline{M}_{g,n+1}(X,\beta)$ defined by

$$s_i(f,C,p_1,\ldots,p_n) = (f : C \cup \mathbb{P}^1 \to X, p_1,\ldots,p_i',\ldots,p_n,p_{n+1}').$$

In terms of (10.1), this is the map which sends $(f,C,p_1,\ldots,p_n)$ to $p_i \in C$.

Now let $\mathcal{L}_i$ denote the "cotangent line at the ith marked point", i.e., the line bundle on $\overline{M}_{g,n}(X,\beta)$ whose fiber over the stable map $(f : C \to X, p_1,\ldots,p_n)$ is the cotangent space $T_{p_i}^*C$. To make this rigorous, let ω_{n+1} be the relative dualizing sheaf of π_{n+1}. Then $\mathcal{L}_i$ is defined to be the sheaf $s_i^*\omega_{n+1}$. Note that $\mathcal{L}_i$ is a line bundle on the stack $\overline{\mathcal{M}}_{g,n}(X,\beta)$, though it need not be locally free on $\overline{M}_{g,n}(X,\beta)$.

Besides the the line bundles $\mathcal{L}_i$ and the evaluation maps e_i, we will also need the virtual fundamental class $[\overline{M}_{g,n}(X,\beta)]^{\mathrm{virt}}$. By Section 7.1.4, this homology class has degree $2(1-g)(d-3) - 2\int_\beta \omega_X + 2n$, where d is the complex dimension of X. We can now make a definition.

DEFINITION 10.1.1. *Given classes $\gamma_1,\ldots\gamma_n \in H^*(X)$ and nonnegative integers d_i for each $i = 1,\ldots,n$, the invariant*

$$\langle \tau_{d_1}\gamma_1,\ldots,\tau_{d_n}\gamma_n \rangle_{g,\beta} = \int_{[\overline{M}_{g,n}(X,\beta)]^{\mathrm{virt}}} \prod_{i=1}^{n} \left(c_1(\mathcal{L}_i)^{d_i} \cup e_i^*(\gamma_i) \right)$$

is a gravitational correlator.

Note that gravitational correlators are only defined when $\overline{M}_{g,n}(X,\beta)$ is defined. This puts restrictions on the triples (g,n,β).

The integral used in Definition 10.1.1 treats $\mathcal{L}_i$ as if it were a line bundle on the moduli space $\overline{M}_{g,n}(X,\beta)$. Strictly speaking, this integral should be evaluated on the stack $\overline{\mathcal{M}}_{g,n}(X,\beta)$. But if we think of the integrand as being pushed forward to $\overline{M}_{g,n}(X,\beta)$, then the integral makes perfect sense when evaluated on $\overline{M}_{g,n}(X,\beta)$. Here and in the remainder of the book, we will adopt this convention, which is similar in spirit to the convention used for the bundles $\mathcal{V}_d$ and their Euler classes in Example 7.1.5.1.

Our terminology for gravitational correlators is motivated by similar terminology used in physics. As outlined in Appendix B.2, the nonlinear sigma model associates (a BRST cohomology class of) an operator $\mathcal{O}_\gamma$ to each cohomology class γ. The operators $\mathcal{O}_\gamma$ yield other physical operators $\mathcal{O}_{i,\gamma}$ for each integer $i \geq 0$ called *gravitational descendants*. The relationship between the gravitational correlators and the gravitational descendants is given by the equation

$$(10.2) \qquad \langle \tau_{d_1}\gamma_1, \ldots, \tau_{d_n}\gamma_n \rangle_{g,\beta} = \frac{\langle \mathcal{O}_{d_1,\gamma_1}, \ldots, \mathcal{O}_{d_n,\gamma_n} \rangle_{g,\beta}}{\prod_{j=1}^{n} d_j!}.$$

The left hand side of (10.2) is the mathematical gravitational correlator from Definition 10.1.1. The expression $\langle \mathcal{O}_{d_1,\gamma_1}, \ldots, \mathcal{O}_{d_n,\gamma_n} \rangle_{g,\beta}$ on the right hand side of (10.2) refers to the part of the physical correlation function $\langle \mathcal{O}_{d_1,\gamma_1}, \ldots, \mathcal{O}_{d_n,\gamma_n} \rangle$ which is computed using maps $f : \Sigma \to X$ with Σ of genus g and $f_*[X] = \beta$ and the gravitational descendant fields. The idea of a physical correlation function is sketched in Appendix B.1. The factors in the denominator of the right hand side of (10.2) are merely a matter of convention.

An important observation is that the Gromov-Witten invariants are precisely those gravitational correlators with all $d_i = 0$. If we abbreviate $\tau_0\gamma$ as γ, then this means that in the notation of Chapter 7,

$$\langle \gamma_1, \ldots, \gamma_n \rangle_{g,\beta} = \langle I_{g,n,\beta} \rangle (\gamma_1, \ldots, \gamma_n).$$

In this chapter, we will use the simpler notation $\langle \gamma_1, \ldots, \gamma_n \rangle_{g,\beta}$ for Gromov-Witten invariants. Another abbreviation we will use frequently is to write $\tau_d 1$ as τ_d, where $1 \in H^0(X)$ is the fundamental class of X.

There are also useful combinations of the gravitational correlators, yielding formal functions. We first recall our notation from Chapter 8. We have a basis $T_0 = 1, T_1, \ldots, T_m$ for $H^*(X, \mathbb{Q})$. We also put $\gamma = \sum_i t_i T_i$ as in the definition of the Gromov-Witten potential given in Section 8.2.2. If ω is a complexified Kähler class on X, then we define the *genus g couplings*

$$(10.3) \qquad \langle\langle \tau_{d_1}\gamma_1, \ldots, \tau_{d_n}\gamma_n \rangle\rangle_g = \sum_{k=0}^{\infty} \sum_{\beta} \frac{1}{k!} \langle \tau_{d_1}\gamma_1, \ldots, \tau_{d_n}\gamma_n, \underbrace{\gamma, \ldots, \gamma}_{k \text{ times}} \rangle_{g,\beta} q^\beta,$$

where as usual $q^\beta = e^{2\pi i \int_\beta \omega}$. We sometimes abbreviate the gravitational correlator inside the sum on the right hand side as

$$\langle \tau_{d_1}\gamma_1, \ldots, \tau_{d_n}\gamma_n, (\gamma)^k \rangle_{g,\beta}$$

when there is no risk of confusion.

Similar to what we did in Sections 8.1.3 and 8.2.2, we can interpret the expression (10.3) as having coefficients in $\mathcal{C}$, where $\mathcal{C}$ is one of the rings

$$\mathbb{C}[[t_0, \ldots, t_m]] \text{ or } R[[t_0, \ldots, t_m]] \text{ or } \mathcal{R}[[t_0, \ldots, t_m]] \text{ or } \Lambda(\omega, \mathbb{Q}))[[t_0, \ldots, t_m]]$$

as appropriate. Thus the gravitational correlators and genus g couplings are well-defined formal objects.

It is illuminating to rephrase this using a *genus g gravitational Gromov-Witten potential*. We introduce supercommuting variables t_d^j for $d \geq 0$ and $0 \leq j \leq m$ with $\deg t_d^j = \deg T_j$ such that t_0^j is our usual variable t_j. We then modify the definition of the Gromov-Witten potential Φ (Definition 8.2.1) to obtain the genus g gravitational Gromov-Witten potential Φ_g^{grav}. We do this by replacing the expression $\gamma = \sum_i t_i T_i$ by the formal expression

$$\gamma = \sum_{d=0}^{\infty} \sum_{j=0}^{m} t_d^j \tau_d T_j,$$

then using genus g maps in place of genus 0 maps and expanding as in Section 8.2.2. More precisely, we have the following definition.

DEFINITION 10.1.2. *Let ω be a complexified Kähler class on a smooth projective variety X. Then the* genus g gravitational Gromov-Witten potential *is the formal sum*

$$\Phi_g^{\mathrm{grav}}(\gamma) = \sum_{n=0}^{\infty} \sum_{\beta \in H_2(X,\mathbb{Z})} \frac{1}{n!} \langle \gamma^n \rangle_{g,\beta} \, q^\beta,$$

where $q^\beta = e^{2\pi i \int_\beta \omega}$.

In the above sum, we only sum over pairs (n,β) for which $\overline{M}_{g,n}(X,\beta)$ exists. Note that

$$(10.4) \qquad \langle\langle \tau_{d_1} T_{i_1}, \dots, \tau_{d_n} T_{i_n} \rangle\rangle_g = \frac{\partial \Phi_g^{\mathrm{grav}}}{\partial t_{d_1}^{i_1} \cdots \partial t_{d_n}^{i_n}} \Big|_{t_d^j = 0 \text{ for all } d > 0}$$

where we use the sign convention for partial derivatives with respect to odd variables from Section 8.2.1. In the special case where $g = 0$ and all of the d_i are 0, then we can write

$$\langle\langle T_{j_1}, \dots, T_{j_n} \rangle\rangle_0 = \frac{\partial^n \Phi}{\partial t_{j_1} \cdots \partial t_{j_n}},$$

where Φ is the ordinary Gromov-Witten potential.

It follows immediately that the big quantum product (Definition 8.2.2) can be rewritten in the form

$$(10.5) \qquad T_i * T_j = \sum_k \langle\langle T_i, T_j, T_k \rangle\rangle_0 \, T^k,$$

where we have used g^{kl} to raise indices, putting $T^k = \sum_l g^{kl} T_l$. As explained in Chapter 8, this comes from the Gromov-Witten potential Φ via the Dubrovin formalism. If we replace Φ by Φ_0^{grav}, we get a new associative product on $H^*(X)$ defined by

$$T_i *_g T_j = \sum_k \frac{\partial^3 \Phi_0^{\mathrm{grav}}}{\partial t_i \partial t_j \partial t_k} T^k.$$

This is the *gravitational quantum product*. The coefficients in the gravitational quantum product can be viewed as belonging to one of the rings

$$\mathbb{C}[[t_d^j]] \quad \text{or} \quad R[[t_d^j]] \quad \text{or} \quad \mathcal{R}[[t_d^j]] \quad \text{or} \quad \Lambda(\omega,\mathbb{Q}))[[t_d^j]]$$

as appropriate.

10.1.2. Properties of the Correlators. The gravitational correlators just defined have properties similar to the properties of Gromov-Witten invariants described in Section 7.3. We only mention the ones we will use. These follow from properties of the virtual fundamental class [**BM, Behrend**].

Degree Axiom. This axiom states that for homogeneous classes γ_i, the gravitational correlator

$$\langle \tau_{d_1} \gamma_1, \dots, \tau_{d_n} \gamma_n \rangle_{g,\beta}$$

can be nonzero only if

$$\sum_{i=1}^n (2d_i + \deg \gamma_i) = 2(1-g) \dim X - 2\int_\beta \omega_X + 2(3g - 3 + n).$$

This follows easily from Definition 10.1.1 and the formula for the dimension of the virtual fundamental class.

Fundamental Class Axiom. As in Chapter 7, $\pi_n : \overline{M}_{g,n}(X,\beta) \to \overline{M}_{g,n-1}(X,\beta)$ is the map forgetting the last point and contracting unstable components. This map is defined if either $n + 2g \geq 4$ or $\beta \neq 0$ and $n \geq 1$. Let $\mathcal{L}'_i$ denote the line bundle on $\overline{M}_{g,n-1}(X,\beta)$ whose fiber is the cotangent space to the universal curve at the ith marked point. Then we have the identity

$$(10.6) \qquad c_1(\mathcal{L}_i) = \pi_n^* c_1(\mathcal{L}'_i) + \widetilde{D}_{(i,n|1,\ldots,i-1,i+1,\ldots n-1)},$$

where the term $\widetilde{D}_{(i,n|1,\ldots,i-1,i+1,\ldots n-1)}$ in (10.6) denotes the closure of the class of stable maps whose source has components C_1, C_2, with points $\{i,n\}$ on C_1, and points $\{1,\ldots,i-1,i+1,\ldots n-1\}$ on C_2, with f constant on C_1. Note also that the virtual fundamental class of $\overline{M}_{g,n}(X,\beta)$ is the pullback of the virtual fundamental class of $\overline{M}_{g,n-1}(X,\beta)$, as we observed in (7.22). We combine this fact together with (10.6) and the calculation in [**Witten3**] to arrive at

$$\langle \tau_{d_1}\gamma_1, \ldots, \tau_{d_{n-1}}\gamma_{n-1}, 1 \rangle_{g,\beta} =$$

$$\sum_{i=1}^{n-1} \langle \tau_{d_1}\gamma_1, \ldots, \tau_{d_{i-1}}\gamma_{i-1}, \tau_{d_i-1}\gamma_i, \tau_{d_{i+1}}\gamma_{i+1}, \ldots, \tau_{d_{n-1}}\gamma_{n-1} \rangle_{g,\beta},$$

where any term involving τ_{-1} is taken to be zero.

Divisor Axiom. Next, let D be a divisor. Then (10.6) and compatibility of the virtual fundamental classes leads to

$$\langle \tau_{d_1}\gamma_1, \ldots, \tau_{d_{n-1}}\gamma_{n-1}, D \rangle_{g,\beta} = \left(\textstyle\int_\beta D\right)\langle \tau_{d_1}\gamma_1, \ldots, \tau_{d_{n-1}}\gamma_{n-1} \rangle_{g,\beta}$$

$$+ \sum_{j=1}^{n-1} \langle \tau_{d_1}\gamma_1, \ldots, \tau_{d_{j-1}}\gamma_{j-1}, \tau_{d_j-1}(D \cup \gamma_j), \tau_{d_{j+1}}\gamma_{j+1}, \ldots, \tau_{d_{n-1}}\gamma_{n-1} \rangle_{g,\beta}.$$

As with the Fundamental Class Axiom, this axiom is valid if either $n + 2g \geq 4$ or $\beta \neq 0$ and $n \geq 1$.

Splitting Axiom. To make this axiom easier to state, we shift viewpoint a bit and extend the definition of Gromov-Witten classes to cover gravitational terms as well. Using the same notation as Definition 7.1.9, if $2g + n \geq 3$ and $\gamma_1, \ldots, \gamma_n \in H^*(X,\mathbb{Q})$, we call the element of $H^*(\overline{M}_{g,n},\mathbb{Q})$ given by

$$I_{g,n,\beta}(\tau_{d_1}\gamma_1, \ldots, \tau_{d_n}\gamma_n) =$$

$$(10.7) \qquad PD^{-1}p_{2*}\left(\prod_{i=1}^{n} c_1(\mathcal{L}_i)^{d_i} \cup p_1^*(\gamma_1 \otimes \cdots \otimes \gamma_n) \cap \pi_*(\xi) \right)$$

a *gravitational class*, where ξ is the virtual fundamental class and

$$PD : H^*(\overline{M}_{g,n},\mathbb{Q}) \simeq H_{6g-6+2n-*}(\overline{M}_{g,n},\mathbb{Q})$$

is the Poincaré duality isomorphism. Here, we are using our convention that we evaluate (10.7) by understanding the expression in parentheses as a class on $\mathcal{M}_{g,n}(X,\beta)$ which is then pushed forward to $\overline{M}_{g,n}(X,\beta)$. With this notation, note that

$$(10.8) \qquad \langle \tau_{d_1}\gamma_1, \ldots, \tau_{d_n}\gamma_n \rangle_{g,\beta} = \int_{\overline{M}_{g,n}} I_{g,n,\beta}(\tau_{d_1}\gamma_1, \ldots, \tau_{d_n}\gamma_n).$$

As in Section 7.3, we consider maps

$$\varphi : \overline{M}_{g_1,n_1+1}(X,\beta) \times \overline{M}_{g_2,n_2+1}(X,\beta) \longrightarrow \overline{M}_{g,n}(X,\beta)$$

obtained by gluing together the sources of two stable maps to form a new stable map. Then the Splitting Axiom is the assertion that

$$\varphi^* I_{g,n,\beta}(\tau_{d_1}\gamma_1,\ldots,\tau_{d_n}\gamma_n) = \sum_{i,\beta=\beta_1+\beta_2} I_{g_1,n_1+1,\beta_1}(\tau_1\gamma_1,\ldots,\tau_{d_{n_1}}\gamma_{n_1},T_i) \otimes$$
$$I_{g_2,n_2+1,\beta_2}(T^i,\tau_{d_{n_1+1}}\gamma_{n_1+1},\ldots,\tau_n\gamma_n).$$

Dilaton Axiom. Our final axiom, which has no analog in Gromov-Witten theory, is the identity

$$\langle \tau_1,\tau_{d_1}\gamma_1,\ldots,\tau_{d_n}\gamma_n\rangle_{g,\beta} = (2g-2+n)\langle\tau_{d_1}\gamma_1,\ldots,\tau_{d_n}\gamma_n\rangle_{g,\beta}.$$

To interpret this, we identify the universal curve $C_{g,n}(X,\beta) \to \overline{M}_{g,n}(X,\beta)$ with the map $\pi_{n+1} : \overline{M}_{g,n+1}(X,\beta) \to \overline{M}_{g,n}(X,\beta)$ used earlier in the Fundamental Class Axiom. With this identification, we see that the sheaf $\mathcal{L}_1$ restricts on the fiber of π_{n+1} over a stable map $(f,C,p_1,\ldots,p_n)$ to the sheaf of dualizing differentials on C which vanish at the p_i. The Dilaton Axiom follows from the compatibility of the virtual fundamental classes with respect to π_{n+1} and the fact that sheaf $\mathcal{L}_1$ has degree $2g-2+n$ on each fiber.

Although gravitational correlators seem to be more complicated than Gromov-Witten invariants, in the genus 0 case it was observed by [**Dubrovin2**] that genus 0 gravitational correlators can be calculated in terms of the Gromov-Witten invariants $\langle\gamma_1,\ldots,\gamma_n\rangle_{0,\beta}$. Dubrovin's proof uses the Divisor Axiom and Lemma 10.2.2 below. We will instead describe briefly a different method due to R. Pandharipande. The idea is to add 3 auxiliary marked points p_{n+1},p_{n+2},p_{n+3}, pull back the integrand of the definition of gravitational correlator to $\overline{M}_{0,n+3}(X,\beta)$, and then cup it with the class

$$\phi = e_{n+1}^*(H) \cup e_{n+2}^*(H),$$

where H is an ample class on X. Forgetting the first $n-1$ marked points and the map, we get a map $s : \overline{M}_{0,n+3}(X,\beta) \to \overline{M}_{0,4}$. We then cup ϕ with both sides of the pullback via s of the linear equivalence $D(n,n+1|n+2,n+3) \sim D(n,n+3|n+1,n+2)$ described in (8.25). Using the Divisor, Fundamental Class, and Splitting Axioms, one obtains a recursive scheme for calculating the genus 0 gravitational correlators in terms of genus 0 Gromov-Witten invariants. An example of this method is sketched below in Example 10.1.3.1. Another approach to the reduction of gravitational correlators is given in [**KoM3**].

10.1.3. Examples of the Correlators. We now give a few examples to illustrate the calculation of gravitational correlators. For convenience, when $X = \mathbb{P}^r$ and $\beta = d[\ell]$, where $\ell \subset \mathbb{P}^r$ is a line, we will sometimes write $\overline{M}_{g,n}(\mathbb{P}^r,d)$ in place of $\overline{M}_{g,n}(\mathbb{P}^r,\beta)$, and $\langle\tau_1\gamma_1,\ldots,\tau_n\gamma_n\rangle_{g,d}$ in place of $\langle\tau_1\gamma_1,\ldots,\tau_n\gamma_n\rangle_{g,\beta}$.

Example 10.1.3.1. We will compute some gravitational correlators for $X = \mathbb{P}^1$. As indicated in Section 7.1.2, $\overline{M}_{0,n}(\mathbb{P}^1,d)$ has the expected dimension $2d+n-2$, so it is not necessary to insert a virtual fundamental class into the computations.

We start by computing degree 1 invariants. Note that $\overline{M}_{0,1}(\mathbb{P}^1,1) \simeq \mathbb{P}^1$, where the isomorphism takes $f : (C,p) \to \mathbb{P}^1$ to $f(p)$. Under this isomorphism, $\mathcal{L}_1$ is identified with the cotangent bundle of $\mathbb{P}^1$, yielding the one-point gravitational correlator $\langle \tau_1 \rangle_{0,1} = -2$. Next, let H be the hyperplane class of $\mathbb{P}^1$ (the class of a point). Then we have the Gromov-Witten invariant

$$\langle H \rangle_{0,1} = \langle H, H, H \rangle_{0,1} = \langle I_{0,3,1} \rangle (H, H, H) = \langle I_{0,3,1} \rangle (pt, pt, pt) = 1,$$

where the first equality is by the Divisor Axiom and the last is obvious since there is a unique automorphism of $\mathbb{P}^1$ sending $0, 1, \infty$ to any three distinct points p_1, p_2, p_3.

We can now compute the gravitational correlator $\langle H, \tau_1 \rangle_{0,1}$ in two ways. First, the Dilaton Axiom gives

$$\langle H, \tau_1 \rangle_{0,1} = -\langle H \rangle_{0,1} = -1.$$

Another way is to use the Divisor Axiom to obtain

$$\langle H, \tau_1 \rangle_{0,1} = (1) \cdot \langle \tau_1 \rangle_{0,1} + \langle H \rangle_{0,1} = -2 + 1 = -1.$$

We next compute the invariants $\langle \tau_{2d}, 1 \rangle_{0,d}$ and $\langle \tau_{2d-1} H, 1 \rangle_{0,d}$ for any d using the reduction process mentioned at the end of Section 10.1.2. To determine $\langle \tau_{2d-1} H, 1 \rangle_{0,d}$, the first step is to note that the Fundamental Class Axiom implies

$$\langle \tau_1 H, 1 \rangle_{0,1} = \langle H \rangle_{0,1} = 1.$$

Now, for general d, take 3 extra marked points, and choose H as the ample class. We consider the class

$$c_1(\mathcal{L}_1)^{2d-1} \cup e_1^*(H) \cup e_3^*(H) \cup e_4^*(H),$$

which we cup with both sides of the pullback of the linear equivalence $D(2,3|4,5) \sim D(2,5|3,4)$ via the map $s : \overline{M}_{0,5}(\mathbb{P}^1, d) \to \overline{M}_{0,4}$. We then integrate over $\overline{M}_{0,5}(\mathbb{P}^1, d)$ using the Splitting Axiom. Many terms are seen to be zero, and the nonzero terms give

$$(10.9) \qquad \langle \tau_{2d-1} H, 1, H, H \rangle_{0,d} \langle 1, H, 1 \rangle_{0,0} = \langle \tau_{2d-1} H, 1, 1, 1 \rangle_{0,d-1} \langle H, H, H \rangle_{0,1}.$$

Note that $\langle 1, H, 1 \rangle_{0,0} = 1$ by the Point Mapping Axiom of Gromov-Witten theory, and $\langle H, H, H \rangle_{0,1} = 1$ as noted above. Then the Divisor and Fundamental Class Axioms reduce (10.9) to

$$d^2 \langle \tau_{2d-1} H, 1 \rangle_{0,d} = \langle \tau_{2d-3} H, 1 \rangle_{0,d-1}.$$

This recursion, together with our calculation of $\langle \tau_1 H, 1 \rangle_{0,1}$, gives

$$(10.10) \qquad\qquad\qquad \langle \tau_{2d-1} H, 1 \rangle_{0,d} = \frac{1}{(d!)^2}.$$

Turning our attention to $\langle \tau_{2d}, 1 \rangle_{0,d}$, the Fundamental Class Axiom implies

$$\langle \tau_2, 1 \rangle_{0,1} = \langle \tau_1 \rangle_{0,1} = -2.$$

Then, for general d, we consider the class $c_1(\mathcal{L}_1)^{2d} \cup e_3^*(H) \cup e_4^*(H)$, cup with the same linear equivalence as before, then integrate and use the Splitting Axiom. We arrive at

$$\langle \tau_{2d}, 1, H, H \rangle_{0,d} \langle 1, H, 1 \rangle_{0,0} = \langle \tau_{2d}, 1, 1, 1 \rangle_{0,d-1} \langle H, H, H \rangle_{0,1},$$

from which we obtain

$$d^2 \langle \tau_{2d}, 1 \rangle_{0,d} + 2d \langle \tau_{2d-1} H, 1 \rangle_{0,d} = \langle \tau_{2d-2}, 1 \rangle_{0,d-1}$$

by the Divisor and Fundamental Class Axioms. This recursion, together with our formula for $\langle \tau_2, 1 \rangle_{0,1}$ and (10.10) can be solved to give

$$(10.11) \qquad \langle \tau_{2d}, 1 \rangle_{0,d} = \frac{-2}{(d!)^2} \left(1 + \frac{1}{2} + \cdots + \frac{1}{d} \right).$$

This calculation of (10.10) and (10.11) is due to Pandharipande. These formulas are implicit in [**Givental2**], as we will see in Example 11.2.1.1.

Example 10.1.3.2. Let V be a Calabi-Yau threefold. This implies that the virtual fundamental class $[\overline{M}_{0,n}(V, \beta)]^{\mathrm{virt}}$ has dimension n. We will compute all of the gravitational correlators of the form $\langle \tau_d \gamma, 1 \rangle_{0,\beta}$. The Degree Axiom shows that $\langle \tau_d \gamma, 1 \rangle_{0,\beta}$ is nonzero only for $2d + \deg \gamma = 4$. Hence we need only consider $\langle \tau_2, 1 \rangle_{0,\beta}$, $\langle \tau_1 D, 1 \rangle_{0,\beta}$ for a divisor D, and $\langle C, 1 \rangle_{0,\beta} = 0$ for a curve C. We may also assume $\beta \neq 0$.

Recall from Section 7.4.4 that we have Gromov-Witten invariants $\langle I_{0,0,\beta} \rangle = \sum_{k | \beta} n_{\frac{\beta}{k}} k^{-3}$, where the n_β are the instanton numbers defined in Section 7.4.4. As in Chapter 7, we put $N_\beta = \langle I_{0,0,\beta} \rangle$.

Now we can compute the desired gravitational correlators. We begin with $\langle \tau_2, 1 \rangle_{0,\beta}$. Here, the Fundamental Class Axiom and the Dilaton Axiom imply

$$\langle \tau_2, 1 \rangle_{0,\beta} = \langle \tau_1 \rangle_{0,\beta}$$
$$= -2N_\beta,$$

and similar arguments show that if D is a divisor, then

$$\langle \tau_1 D, 1 \rangle_{0,\beta} = \langle D \rangle_{0,\beta}$$
$$= \left(\textstyle\int_\beta D \right) N_\beta.$$

Finally, if C is the class of a curve, the Fundamental Class Axiom tells us that $\langle C, 1 \rangle_{0,\beta} = 0$. We will use these computations in Section 10.3.2.

Here is an example where the virtual fundamental class plays a crucial role.

Example 10.1.3.3. Let X be arbitrary of dimension d. We will evaluate $\langle \tau_1 \rangle_{1,0}$. A genus 1 stable map $f : (C, p) \to X$ with $f_*[C] = 0$ is a constant map. It follows that (C, p) is a stable 1-pointed curve, and then we easily get an isomorphism

$$\overline{M}_{1,1}(X, 0) \simeq X \times \overline{M}_{1,1}$$

which identifies $f : (C, p) \to X$ with $(f(C), (C, p))$. The expected dimension is 1, while the actual dimension of $\overline{M}_{1,1}(X, 0)$ is $d + 1$. Thus the excess dimension is d. Since $f^* \Theta_X \simeq \mathcal{O}_C^d$, we have $h^1(C, f^* \Theta_X) = d$. This shows that the hypotheses of Proposition 7.1.8 are satisfied. In the notation of that proposition, we need to compute the dth Chern class of $R^1 \pi_{2*} e_2^* \Theta_X$. Using the above isomorphism, one easily gets

$$R^1 \pi_{2*} e_2^* \Theta_X \simeq p_2^* R^1 \pi_{2*} \mathcal{O}_{\overline{M}_{1,2}} \otimes p_1^* \Theta_X,$$

where p_i are the two projections of $X \times \overline{M}_{1,1}$ and the π_2 on the right hand side is the usual map $\overline{M}_{1,2} \to \overline{M}_{1,1}$ which forgets the second point. For simplicity, let $\mathcal{L} = R^1 \pi_{2*} \mathcal{O}_{\overline{M}_{1,2}}$, which is a line bundle on $\overline{M}_{1,1}$. Then Proposition 7.1.8

and standard facts about Chern classes imply that the virtual fundamental class $\xi = [\overline{M}_{1,0}(X,0)]^{\mathrm{virt}}$ is given by

$$\xi = c_d(p_2^*\mathcal{L} \otimes p_1^*\Theta_X) \cap [\overline{M}_{1,0}(X,0)]$$

(10.12)

$$= \left(p_1^*c_d(X) + p_2^*c_1(\mathcal{L}) \cup p_1^*c_{d-1}(X)\right) \cap [\overline{M}_{1,0}(X,0)],$$

where the $c_k(X) = c_k(\Theta_X)$ are the Chern classes of the cotangent bundle.

We can also describe the line bundle $\mathcal{L}_1$ in this situation. If $\widetilde{\mathcal{L}}_1$ denotes the corresponding bundle on $\overline{M}_{1,1}$, then $\mathcal{L}_1 = p_2^*(\widetilde{\mathcal{L}}_1)$ since no curves get contracted in the map $\overline{M}_{1,1}(X,0) \to \overline{M}_{1,1}$. It is also known that $\int_{\overline{M}_{1,1}} c_1(\widetilde{\mathcal{L}}_1) = \frac{1}{24}$ [**Mumford2**]. We can now compute

$$\langle \tau_1 \rangle_{1,0} = \int_{[\overline{M}_{1,1}(X,0)]^{\mathrm{virt}}} c_1(\mathcal{L}_1)$$

$$= \int_{X \times \overline{M}_{1,1}} p_2^*c_1(\widetilde{\mathcal{L}}_1) \cup \left(p_1^*c_d(X) + p_2^*c_1(\mathcal{L}) \cup p_1^*c_{d-1}(X)\right)$$

$$= \int_{\overline{M}_{1,1}} c_1(\widetilde{\mathcal{L}}_1) \int_X c_d(X)$$

$$= \frac{1}{24} \int_X c_d(X),$$

where the second equality uses (10.12) and the third follows since $c_1(\widetilde{\mathcal{L}}_1) \cup c_1(\mathcal{L}) = 0$ for dimension reasons.

In the literature, (10.12) is usually stated differently. By Serre duality, $\mathcal{L} = R^1\pi_{2*}\mathcal{O}_{\overline{M}_{1,2}}$ is dual to $\pi_{2*}\omega_2$, the push-forward of the relative dualizing sheaf of $\pi_2 : \overline{M}_{1,2} \to \overline{M}_{1,1}$. It is customary to denote this sheaf by $\mathcal{H}$, and then, if we regard the virtual fundamental class as a cohomology class, (10.12) becomes

$$[\overline{M}_{1,0}(X,0)]^{\mathrm{virt}} = p_1^*c_d(X) - p_2^*c_1(\mathcal{H}) \cup p_1^*c_{d-1}(X).$$

This formula was explained in [**BCOV1**] and reinterpreted in the context of stable maps in [**Katz3**]. See also [**Getzler1**].

We can now recompute $\langle \tau_1 \rangle_{1,0}$ with a pleasantly surprising result. First note that since $\mathcal{L}_1 = p_2^*(\widetilde{\mathcal{L}}_1)$, the gravitational class $I_{1,1,0}(\tau_1)$ defined in (10.7) becomes

$$I_{1,1,0}(\tau_1) = PD^{-1}p_{2*}\left((c_1(\mathcal{L}_1) \cup p_1^*(1)) \cap \xi\right)$$

$$= c_1(\widetilde{\mathcal{L}}_1) \cup PD^{-1}p_{2*}\left(p_1^*(1) \cap \xi\right)$$

$$= c_1(\widetilde{\mathcal{L}}_1) \cup I_{1,1,0}(1),$$

where the second equality follows from the projection formula and $I_{1,1,0}(1)$ is a Gromov-Witten class in $H^0(\overline{M}_{1,1})$. Then (10.8) implies that

$$\langle \tau_1 \rangle_{1,0} = \int_{\overline{M}_{1,1}} c_1(\widetilde{\mathcal{L}}_1) \cup I_{1,1,0}(1) = \deg(I_{1,1,0}(1)) \int_{\overline{M}_{1,1}} c_1(\widetilde{\mathcal{L}}_1) = \frac{1}{24}\deg(I_{1,1,0}(1)),$$

where the $\deg(I_{1,1,0}(1))$ is defined using the isomorphism $\deg : H^0(\overline{M}_{1,1}) \simeq \mathbb{C}$ which takes 1 to 1.

To compute $\deg(I_{1,1,0}(1))$, let $\psi : \overline{M}_{0,3} \to \overline{M}_{1,1}$ be the map which glues together the first two marked points to produce a nodal genus 1 curve. Since $\overline{M}_{0,3}$ is a single point, we get

$$\deg(I_{1,1,0}(1)) = \int_{\overline{M}_{0,3}} \psi^*I_{1,1,0}(1).$$

Combining the above two equations and using the Reduction and Point Mapping Axioms from Chapter 7, we obtain

$$\langle \tau_1 \rangle_{1,0} = \frac{1}{24} \int_{\overline{M}_{0,3}} \psi^* I_{1,1,0}(1)$$

$$= \frac{1}{24} \int_{\overline{M}_{0,3}} g^{ij} I_{0,3}(1, T_i, T_j)$$

$$= \frac{1}{24} \sum g^{ij} g_{ij}.$$

Since $g^{ij} = -g^{ji}$ if T_i and T_j are both odd and $g^{ij} = g^{ji}$ if at least one is even, this leads to $\langle \tau_1 \rangle_{1,0} = \frac{1}{24} \sum_i (-1)^i b_i(X)$, where $b_i(X)$ is the ith Betti number of X. Comparing our two computations of $\langle \tau_1 \rangle_{1,0}$, we arrive at the familiar formula

$$\int_X c_d(X) = \sum_{i=0}^{d} (-1)^i b_i(X).$$

10.1.4. The Virasoro Conjecture. We close this section with a brief description of an intriguing conjecture concerning gravitational correlators. Define the *gravitational potential* to be the infinite sum

$$\Phi_{\text{grav}} = \sum_{g=0}^{\infty} \kappa^{2g-2} \Phi^g_{\text{grav}},$$

where κ is a parameter. The *partition function* is then defined as

$$Z = e^{\Phi_{\text{grav}}}.$$

The *Virasoro conjecture* asserts that there is a Virasoro algebra (B.27) of formal differential operators in the t_d^j which annihilate Z [**EHX**].

More precisely, the Virasoro algebra contains explicit differential operators L_n for $n \geq -1$ in the variables t_d^j which satisfy $[L_n, L_m] = (n - m)L_{n+m}$. Then the equation $L_{-1}Z = 0$ follows from the Fundamental Class Axiom [**Witten3**], and one can also derive the equation $L_0 Z = 0$ from a combination of the Degree, Divisor, and Dilaton Axioms [**Hori**]. It is conjectured that $L_n Z = 0$ for all $n \geq -1$.

It is sometimes possible to extend this algebra of differential operators to a central extension of the full Virasoro algebra L_n, $n \in \mathbb{Z}$. See Appendix B.3 for the description of this central extension, especially (B.27). The central charge of this representation is the Euler characteristic $\chi(X)$ of X. It is *not* conjectured that $L_n Z = 0$ for $n < -1$. In fact, by combining the commutation relation $[L_n, L_{-n}] = 2nL_0 + (n^3 - n)\chi(X)/12$ together with $Z \neq 0$, one can show that $L_n Z = L_{-n} Z = 0$ is not possible for *any* $n > 1$.

In the case where X is a point, this conjecture becomes the Witten conjecture [**Witten3**], which has been proven by Kontsevich [**Kontsevich1**]. One of the key ideas in Kontsevich's proof is that the partition function can be represented as a *matrix integral*, an integral over a space of Hermitian matrices. This leads to an intrinsic action of the Virasoro algebra. In physics, there is reason to believe this more general partition functions can be similarly described as a *matrix model*, which would lead to a natural action of a Virasoro algebra.

This method has been applied to predict rational and elliptic Gromov-Witten invariants for projective spaces. For $\mathbb{P}^2$, the method yields (7.39) for rational invariants and (7.56) for elliptic invariants.

The Virasoro algebra proposed by [**EHX**] can work only if every class in $H^*(X)$ is of Hodge type (p,p) (see [**Borisov2**]). A modified pair of Virasoro algebras was proposed in the general case by S. Katz, and it was shown in [**EJX**] that this leads to correct predictions in examples. In formulating this conjecture, the Degree Axiom is replaced by a similar axiom which uses Hodge types.

It was also shown in [**EX**] that the Virasoro conjecture, together with an additional hypothesis, leads to relations among genus g gravitational descendants, called *topological recursion relations* (TRRs), since they reduce gravitational descendants involving τ_k to descendants involving τ_{k-1}, leading to a recursion. The TRRs described in [**EX**] are sufficient to determine the $g = 2$ Gromov-Witten invariants for $\mathbb{P}^2$. It was pointed out to us by E. Getzler that the methods of [**Getzler1**, **Getzler2**] can be applied to give the same TRR, thereby proving the correctness of the TRR derived in [**EX**]. We will give a TRR for $g = 0$ in Lemma 10.2.2 below.

To study this conjecture, [**LiuT**] writes $L_n Z$ as $(\sum_{g \geq 0} \kappa^{2g-2}\Psi_{g,n})Z$, where the $\Psi_{g,n}$ are formal expressions in the t_d^j and q^β (but not κ). One can show that $\Psi_{g,n}$ depends only on gravitational correlators of genus $\leq g$. The main result of [**LiuT**] is that $\Psi_{0,n} = 0$ for all $n \geq -1$, provided that $H^*(X)$ only has (p,p) Hodge types.

10.2. The Givental Connection

In Section 8.4, we showed that any potential function on the supermanifold $H^*(X)$ leads naturally to a family of formal flat connections ∇^λ on the tangent bundle of $H^*(X,\mathbb{C})$. When the potential function is the Gromov-Witten potential, we get some especially interesting connections which are closely related to the A-model connection studied in Section 8.5.

We will actually study a twisted version of the Dubrovin connection called the *Givental connection* ∇^g. In Section 10.2.1, we will describe how a basis for the ∇^g-flat sections can be written down in terms of the gravitational correlators. We will discuss how this relates to the A-model connection of Chapter 8, and in Section 10.2.3 will also explain how the parameter $\hbar$ in the Givental connection relates to equivariant cohomology.

The flat sections constructed in this section will be used in Section 10.3 to define Givental's function J_X, which in Section 10.3 will lead to quantum differential equations and relations in quantum cohomology.

10.2.1. Flat Sections. Given a smooth projective variety X, we let Φ denote the Gromov-Witten potential and $*$ the big quantum product. But rather than use the Dubrovin connection of Φ defined in Section 8.4, we will instead follow [**Givental2**] and modify this connection slightly. We identify $\partial/\partial t_i$ with T_i as in Section 8.4, and think of the connection as being defined on a trivial cohomology bundle over $H^*(X,\mathbb{C})$. We then define the *Givental connection* ∇^g by

$$(10.13) \qquad \nabla^g_{\frac{\partial}{\partial t_i}}\left(\sum_j a_j T_j\right) = \hbar \sum_j \frac{\partial a_j}{\partial t_i}T_j - \sum_j a_j\, T_j * T_i,$$

where $\hbar$ is a parameter. In Section 10.2.3 we will identify $\hbar$ with a generator of $H^2(B\mathbb{C}^*)$. This will be useful in Section 11.1.

Even though ∇ is not a connection (because of $\hbar$), we still use standard terminology for connections and in particular the notion of a flat section. As noted in Section 8.5.2, the Givental connection relates to the Dubrovin connection ∇^λ via the formula $\nabla^g = \hbar \nabla^{-\hbar^{-1}}$. Hence the flat sections of (10.13) are the same as those of the Dubrovin connection ∇^λ with parameter $\lambda = -\hbar^{-1}$.

For each index $i = 0, \ldots, m$, we define the formal section

$$(10.14) \qquad s_a = T_a + \sum_{n=0}^{\infty} \hbar^{-(n+1)} \sum_j \langle\langle \tau_n T_a, T_j \rangle\rangle_0 \, T^j.$$

Notice that $\langle\langle \tau_n T_a, T_j \rangle\rangle_0$ is a formal power series in $t_0, \ldots, t_m$. If we use the formal symbol "c" to denote $c_1(\mathcal{L}_1)$, then we can write (10.14) formally as

$$(10.15) \qquad s_a = T_a + \sum_j \left\langle\left\langle \frac{T_a}{\hbar - c}, T_j \right\rangle\right\rangle_0 T^j,$$

since a formal expansion of the denominator in powers of $c/\hbar$ leads to the expression in (10.14) after replacing a power c^n by τ_n.

The sections s_a behave very nicely in this situation.

PROPOSITION 10.2.1. *The sections s_a form a basis for the ∇^g-flat sections. By (10.13), this is equivalent to the equation*

$$\hbar \frac{\partial s_a}{\partial t_i} = T_i * s_a, \qquad a, i = 0, \ldots m,$$

where $*$ denotes the big quantum product.

Before giving the proof, we need the following result from [**Witten3**]. In the language of Section 10.1.4, this lemma is a topological recursion relation.

LEMMA 10.2.2. *For $d_1, d_2, d_3 \geq 0$ and $0 \leq j_1, j_2, j_3 \leq m$, we have*

$$\langle\langle \tau_{d_1+1} T_{j_1}, \tau_{d_2} T_{j_2}, \tau_{d_3} T_{j_3} \rangle\rangle_0 = \sum_a \langle\langle \tau_{d_1} T_{j_1}, T_a \rangle\rangle_0 \, \langle\langle T^a, \tau_{d_2} T_{j_2}, \tau_{d_3} T_{j_3} \rangle\rangle_0,$$

where as usual, T^a denotes the dual basis of T_a with respect to cup product.

PROOF. Consider for $n \geq 3$ the map from $\overline{M}_{0,n}(X, \beta)$ to a point, which we write as $\pi : \overline{M}_{0,n}(X, \beta) \to \overline{M}_{0,3}$ to take advantage of the functorial nature of the spaces of stable maps. The map π can be thought of as the map taking $(f, C, p_1, \ldots, p_n)$ to the curve obtained from C by deleting $p_4, \ldots, p_n$ and contracting all unstable components. Let $\mathcal{L}_1'$ be the (trivial) bundle on $\overline{M}_{0,3}$ whose definition is analogous to that of the bundle $\mathcal{L}_1$. We compute that

$$(10.16) \qquad \pi^* \mathcal{L}_1' = \mathcal{L}_1 - \sum_{K \cup L = \{4, \ldots, n\}} D_{(\{1\} \cup K | \{2,3\} \cup L)},$$

where $D_{(I|J)}$ is the closure of the set of stable maps f whose source contains the points of I in one component on which f is constant, and the points of J in another component as in the picture

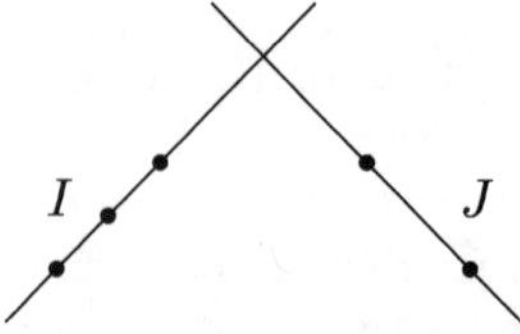

We now multiply (10.16) by $\prod_a c_1(\mathcal{L}_a)^{d_a} \cup e_a^*(T_{j_a})$, integrate against the virtual fundamental class, and use the Splitting Axiom. This gives

$$\langle \tau_{d_1+1} T_{j_1}, \tau_{d_2} T_{j_2}, \tau_{d_3} T_{j_3}, \tau_{d_4} T_{j_4} \ldots, \tau_{d_n} T_{j_n} \rangle_{0,\beta} =$$

$$(10.17) \qquad \sum_{\substack{K \cup L = \{4,\ldots,n\} \\ \beta = \beta_1 + \beta_2}} \sum_a \pm \langle \tau_{d_1} T_{j_1}, \tau_{d_{k_1}} T_{j_{k_1}}, \ldots, T_a \rangle_{0,\beta_1} \times$$

$$\langle T^a, \tau_{d_2} T_{j_2}, \tau_{d_3} T_{j_3}, \tau_{d_{l_1}} T_{j_{l_1}}, \ldots \rangle_{0,\beta_2},$$

where $K = \{k_1, \ldots\}$ and $L = \{l_1, \ldots\}$. The sign in (10.17) arises from the possibly different ordering of the odd cohomology classes between the left and right hand sides of the equation.

Upon expanding using (10.3), it can now readily be seen that (10.17) is a coefficient in the desired equality

$$\langle\langle \tau_{d_1+1} T_{j_1}, \tau_{d_2} T_{j_2}, \tau_{d_3} T_{j_3} \rangle\rangle_0 = \sum_a \langle\langle \tau_{d_1} T_{j_1}, T_a \rangle\rangle_0 \langle\langle T^a, \tau_{d_2} T_{j_2}, \tau_{d_3} T_{j_3} \rangle\rangle_0.$$

Note that the reordering of odd cohomology classes leading to the sign in (10.17) is precisely the same reordering as is done in comparing the variables on the right hand side of the statement of the lemma with the left hand side. $\qquad\square$

We can now prove Proposition 10.2.1, which asserts that the s_a are flat sections.

PROOF OF PROPOSITION 10.2.1. We begin with the observation

$$\frac{\partial}{\partial t_i} \langle\langle \tau_{d_1} \gamma_1, \ldots, \tau_{d_n} \gamma_n \rangle\rangle_0 = \langle\langle T_i, \tau_{d_1} \gamma_1, \ldots, \tau_{d_n} \gamma_n \rangle\rangle_0,$$

which follows from (10.3) by an argument similar to what we did in the proof of Lemma 8.2.3.

Using this, we compute from (10.14) that

$$\hbar \frac{\partial s_a}{\partial t_i} = \sum_{n=0}^{\infty} \sum_j \hbar^{-n} \langle\langle T_i, \tau_n T_a, T_j \rangle\rangle_0 \, T^j,$$

while from (10.14) and (10.5) we have

$$T_i * s_a = \sum_j \langle\langle T_i, T_a, T_j \rangle\rangle_0 \, T^j \; +$$

$$(10.18) \qquad \sum_{n=0}^{\infty} \sum_{j,c} \hbar^{-(n+1)} \langle\langle \tau_n T_a, T_j \rangle\rangle_0 \langle\langle T_i, T^j, T_c \rangle\rangle_0 \, T^c.$$

Lemma 10.2.2 implies that the second term on the right hand side of (10.18) is

$$\sum_{n=0}^{\infty} \sum_c \hbar^{-(n+1)} \langle\langle T_i, \tau_{n+1} T_a, T_c \rangle\rangle_0 \, T^c.$$

After substituting this back into (10.18), we obtain the desired result. $\qquad\square$

One consequence of Proposition 10.2.1 is that

$$\hbar \frac{\partial s_a}{\partial t_0} = s_a$$

since T_0 is the identity for the big quantum product. It follows that s_a depends on t_0 via a multiplicative factor of $e^{t_0/\hbar}$. Because of this, we will often ignore the dependence on t_0 since we can easily restore it by inserting a factor of $e^{t_0/\hbar}$.

In the sequel, we will sometimes restrict the base manifold of the bundle and connection ∇^g. In particular, we will restrict to the subspace

$$M = H^0(X, \mathbb{C}) \oplus H^2(X, \mathbb{C}) \subset H^*(X, \mathbb{C}).$$

in Section 10.3. If $T_1, \ldots, T_r$ form a basis of $H^2(X, \mathbb{Q})$, then the variables for M are $t_0, t_1, \ldots, t_r$. To see what ∇^g looks like in this case, first note that by Section 8.5.1, the restriction of the big quantum product to $H^2(X, \mathbb{C})$ is precisely $*_{\text{small}}$, provided we let $q^\beta = 1$ in (10.3) and use the conventions explained in the discussion surrounding (8.39). Then, since $T_0 = 1$ is the identity for both big and small quantum cohomology, it follows that the restriction of ∇^g to M is given by

$$(10.19) \qquad \nabla^g_{\frac{\partial}{\partial t_i}} \Big(\sum_{j=0}^m a_j T_j \Big) = \hbar \sum_{j=0}^m \frac{\partial a_j}{\partial t_i} T_j - \sum_{j=0}^m a_j T_i *_{\text{small}} T_j$$

for $i = 0, 1, \ldots, r$. In particular, the sections s_a, when restricted to M, satisfy

$$\hbar \frac{\partial s_a}{\partial t_i} = T_i *_{\text{small}} s_a, \quad a = 0, \ldots, m, \ i = 0, \ldots, r.$$

Once restricted to M, the formula for s_a simplifies nicely.

PROPOSITION 10.2.3. *Let $\delta = \sum_{i=1}^r t_i T_i$. Then the restriction of s_a to M can be written as*

$$s_a = e^{t_0/\hbar} \Big(e^{\delta/\hbar} \cup T_a + \sum_{\beta \neq 0} \sum_{j=0}^m e^{\int_\beta \delta} \Big\langle \frac{e^{\delta/\hbar} \cup T_a}{\hbar - c}, T_j \Big\rangle_{0,\beta} T^j \Big).$$

PROOF. First, we noted above that t_0 appears via the multiplicative factor $e^{t_0/\hbar}$. Hence it suffices to prove the proposition when $t_0 = 0$. Using $\delta = \sum_{i=1}^r t_i T_i$, the gravitational coupling (10.3) becomes

$$\langle\langle \tau_n T_a, T_j \rangle\rangle_0 = \sum_{k=0}^\infty \sum_\beta \frac{1}{k!} \langle \tau_n T_a, T_j, (\delta)^k \rangle_{0,\beta},$$

where we set $q^\beta = 1$ as explained above. Note that we are using the notation introduced in the discussion following (10.3).

Since δ is a divisor, we can simplify this expression using the Divisor Axiom. When $\beta \neq 0$, an easy induction on k shows that we have the identity

$$(10.20) \qquad \langle \tau_n T_a, T_j, (\delta)^k \rangle_{0,\beta} = \sum_{\mu+\nu=k} \frac{k!}{\mu! \nu!} \Big(\int_\beta \delta \Big)^\mu \langle \tau_{n-\nu}(T_a \cup \delta^\nu), T_j \rangle_{0,\beta},$$

where we set $\tau_{n-\nu}(T_a \cup (\gamma)^{k-\nu}) = 0$ if $n - \nu < 0$. We need a different identity for $\beta = 0$, since $\overline{M}_{0,2}(X, 0)$ does not exist. Here, we use the Divisor Axiom to obtain

$$(10.21) \qquad \langle \tau_n T_a, T_j, (\delta)^k \rangle_{0,0} = \langle \tau_{n-k+1} \big(T_a \cup \delta^{k-1} \big), T_j, \delta \rangle_{0,0}.$$

This clearly vanishes if $k > n + 1$, and if $k = n + 1$, it reduces to the integral $\int_X T_a \cup T_j \cup \delta^k$ by the Point Mapping Axiom for Gromov-Witten invariants. Finally, if $k < n+1$, we have the identification $\overline{M}_{0,3}(X, 0) \simeq X$ which sends (f, C, p_1, p_2, p_3) to $f(C) \in X$. Also, $\overline{M}_{0,4}(X, 0) \to \overline{M}_{0,3}(X, 0)$ is the projection $X \times \overline{M}_{0,4} \to X$, which makes it easy to show that $\mathcal{L}_1$ is trivial on $\overline{M}_{0,3}(X,0)$ and hence has zero Chern class. Thus (10.21) vanishes when $k < n+1$, and we conclude that

$$(10.22) \qquad \langle \tau_n T_a, T_j, (\delta)^k \rangle_{0,0} = \left(\int_X T_a \cup T_j \cup \delta^k \right) \delta_{k,n+1}.$$

Using (10.20) and (10.22), our formula for $\langle\langle \tau_n T_a, T_j \rangle\rangle_0$ simplifies to

$$\frac{1}{(n+1)!} \int_X T_a \cup T_j \cup \delta^{n+1} + \sum_{\beta \neq 0} \sum_{k=0}^{\infty} \sum_{\mu+\nu=k} \frac{1}{\mu!\nu!} \left(\int_\beta \delta \right)^\mu \langle \tau_{n-\nu}(T_a \cup \delta^\nu), T_j \rangle_{0,\beta}$$

$$= \frac{1}{(n+1)!} \int_X T_a \cup T_j \cup \delta^{n+1} + \sum_{\beta \neq 0} \sum_{\nu=1}^{\infty} e^{\int_\beta \delta} \frac{1}{\nu!} \langle \tau_{n-\nu}(T_a \cup \delta^\nu), T_j \rangle_{0,\beta}.$$

If we insert this expression into the definition of s_a, we obtain

$$(10.23) \qquad \begin{aligned} s_a = T_a + \sum_j \Big(\sum_{n=0}^{\infty} h^{-(n+1)} \frac{1}{(n+1)!} \int_X T_a \cup T_j \cup \delta^{n+1} \Big) T^j + \\ \sum_{\beta \neq 0} \sum_j e^{\int_\beta \delta} \Big(\sum_{n \geq \nu} h^{-(n+1)} \frac{1}{\nu!} \langle \tau_{n-\nu}(T_a \cup \delta^\nu), T_j \rangle_{0,\beta} \Big) T^j. \end{aligned}$$

This looks complicated, but using the identity

$$(10.24) \qquad \sum_j \left(\int_X \phi \cup T_j \right) T^j = \phi \quad \text{for all } \phi \in H^*(X, \mathbb{C}),$$

the large expression in parentheses in the first line of (10.23) simplifies to $T_a \cup e^{\delta/h}$. As for the second line, the large expression in parentheses can be simplified by setting $k = n - \nu$. This implies $h^{-(n+1)} = h^{-(k+1)}h^{-\nu}$, and then the expression becomes

$$\sum_{k=0}^{\infty} h^{-(k+1)} \langle \tau_k(T_a \cup e^{\delta/h}), T_j \rangle_{0,\beta}.$$

In the symbolic notation introduced in (10.15), this equals

$$\left\langle \frac{T_a \cup e^{\delta/h}}{h - c}, T_j \right\rangle_{0,\beta},$$

and from here, the proposition follows immediately. $\qquad \square$

10.2.2. Flat Sections of the A-Model Connection. Suppose that V is a Calabi-Yau manifold of dimension d. In Section 8.5.2, we defined the A-model connection by the equation

$$\nabla_{\frac{\partial}{\partial u_j}}(T_k) = T_j *_{\text{small}} T_k, \quad 1 \leq j \leq r,\ 0 \leq k \leq m,$$

where $u_j = \frac{1}{2\pi i} t_j$. If we restrict the Givental connection ∇^g to $H^2(V, \mathbb{C})$, (10.19) implies that $-2\pi i \nabla$ is precisely ∇^g for $h = -2\pi i$. For the remainder of Section 10.2.2, we fix this value of h.

It follows that the sections s_a defined in Section 10.2.1 restrict to flat sections of the A-model connection ∇. We will use these sections using the following general notation: given a cohomology class $T \in H^*(V, \mathbb{C})$, let

$$s(T) = e^{\delta/h} \cup T + \sum_{\beta \neq 0} \sum_{j=0}^{m} e^{\int_\beta \delta} \left\langle \frac{e^{\delta/h} \cup T}{h - c}, T_j \right\rangle_{0,\beta} T^j.$$

By Proposition 10.2.3, we see that $s(T_a)$ is the section s_a, and it follows easily that $s(T)$ is a flat formal section of ∇ for any $T \in H^*(V, \mathbb{C})$. Following the practice explained in Section 8.5.1, we will let $q^\beta = e^{\int_\beta \delta}$, so that $s(T)$ can be written

$$(10.25) \qquad s(T) = e^{\delta/\hbar} \cup T + \sum_{\beta \neq 0} \sum_{j=0}^{m} q^\beta \left\langle \frac{e^{\delta/\hbar} \cup T}{\hbar - c}, T_j \right\rangle_{0,\beta} T^j.$$

Using this formula for $s(T)$, we can easily determine the monodromy of ∇. We will assume that we are in the situation of Section 8.5.3, where $T_1, \ldots, T_r$ are integral classes lying in the closure of the Kähler cone of V. Furthermore, we have variables $q_j = e^{t_j} = e^{2\pi i u_j}$ such that ∇ is a connection in a neighborhood of the origin in $(\Delta^*)^r$. We are making the same convergence assumptions as in Section 8.5.3.

Now let $\mathcal{T}_j$ be the monodromy transformation given by going around the jth factor in $(\Delta^*)^r$.

THEOREM 10.2.4. *The monodromy $\mathcal{T}_j$ and its logarithm N_j act on the flat sections of the A-model connection as follows:*

$$\mathcal{T}_j(s(T)) = S\big(e^{-T_j} \cup T\big)$$
$$N_j(s(T)) = -S(T_j \cup T).$$

PROOF. Since $q_j = e^{t_j}$, $\mathcal{T}_j$ takes t_j to $t_j + 2\pi i$. We saw above that ∇ is a multiple of the Givental connection for $\hbar = -2\pi i$, so that the action of $\mathcal{T}_j$ can be written $t_j \mapsto t_j - \hbar$.

We need to see what effect this has on (10.25). Since $\delta = \sum_{i=1}^{r} t_i T_i$, $\mathcal{T}_j$ takes δ to $\delta + 2\pi i\, T_j = \delta - \hbar T_j$. This has no effect on $q^\beta = e^{\int_\beta \delta}$ since T_j and β are integral. However, δ also appears in $e^{\delta/\hbar}$, and here the effect of $\mathcal{T}_j$ is given by

$$e^{\delta/\hbar} \longmapsto e^{(\delta - \hbar T_j)/\hbar} = e^{\delta/\hbar - T_j} = e^{\delta/\hbar} \cup e^{-T_j}.$$

This last equality uses the fact that T_j commutes with δ since T_j has even degree. Inserting this into (10.25), we immediately get the desired formula for $\mathcal{T}_j(s(T))$, and then the formula for $N_j(s(T))$ follows by taking logarithms. $\qquad\square$

Theorem 10.2.4 shows that the monodromy logarithm N_j is essentially given by cup product with $-T_j$. More precisely, $N_j(s(T_a)) = -s(T_j \cup T_a)$ implies that the matrix of cup product with $-T_j$ relative to the cohomology basis $\{T_a\}$ is exactly the matrix of N_j relative to the basis of flat sections $\{s(T_a)\}$. This is a fact we used frequently in Chapter 8.

We can also use Theorem 10.2.4 to study the canonical extension of ∇. Recall from Section 5.1.4 that since $s(T)$ is a flat section of ∇, the section

$$\tilde{s}(T) = \exp\left(-\tfrac{1}{2\pi i}\textstyle\sum_j \log(q_j) N_j\right) s(T)$$

extends naturally to the canonical extension. Then Theorem 10.2.4 gives the following nice formula for $\tilde{s}(T)$.

COROLLARY 10.2.5. *Given $T \in H^*(V, \mathbb{C})$, the section $\tilde{s}(T)$ is given by*

$$\tilde{s}(T) = T + \sum_{\beta \neq 0} \sum_{j=0}^{m} q^\beta \left\langle \frac{T}{\hbar - c}, T_j \right\rangle_{0,\beta} T^j.$$

PROOF. This is easy. First observe that $\log(q_j) = t_j$ and $\hbar = -2\pi i$ imply that

$$\tilde{s}(T) = \exp\left(\hbar^{-1}\sum_j t_j N_j\right) s(T)$$
$$= s\left(\exp\left(\hbar^{-1}\sum_j t_j(-T_j)\right) \cup T\right)$$
$$= s\left(e^{-\delta/\hbar} \cup T\right),$$

where the second equality follows from $N_j(s(T)) = s((-T_j) \cup T)$. Recall also that $\delta = \sum_j t_j T_j$. However, by (10.25), we have

$$s\left(e^{-\delta/\hbar} \cup T\right) = e^{\delta/\hbar} \cup \left(e^{-\delta/\hbar} \cup T\right) + \sum_{\beta \neq 0}\sum_{j=0}^{m} q^\beta \left\langle \frac{e^{\delta/\hbar} \cup (e^{-\delta/\hbar} \cup T)}{\hbar - c}, T_j \right\rangle_{0,\beta} T^j$$

$$= T + \sum_{\beta \neq 0}\sum_{j=0}^{m} q^\beta \left\langle \frac{T}{\hbar - c}, T_j \right\rangle_{0,\beta} T^j.$$

Putting these together, the corollary follows immediately. $\qquad\square$

Notice how the formula for $\tilde{s}(T)$ involves only $q^\beta = \Pi_i q_i^{\int_\beta T_i}$. As usual, we can assume that β is effective, and then $\int_\beta T_i \geq 0$ since T_i is in the closure of the Kähler cone. It follows that $\tilde{s}(T)$ extends holomorphically to 0.

Using the above corollary, we can finally prove Proposition 8.5.4 from Chapter 8. We restate the proposition as follows.

COROLLARY 10.2.6. *For each homogeneous $T \in H^*(V, \mathbb{C})$, the section $\tilde{s}(T)$ satisfies $\tilde{s}(T) = T +$ terms of higher degree and $\tilde{s}(T)(0) = T$. Furthermore, the matrix of N_j acting on the $\tilde{s}(T_a)$ equals the matrix of cup product with $-T_j$ acting on the T_a.*

PROOF. The final statement of the corollary follows immediately from the comments made after the proof of Theorem 10.2.4. For the other assertions, first note that unwinding the symbolic notation used in Corollary 10.2.5 gives the formula

$$\tilde{s}(T) = T + \sum_{\beta \neq 0}\sum_{j=0}^{m}\sum_{n=0}^{\infty} q^\beta \hbar^{-(n+1)} \langle \tau_n T, T_j \rangle_{0,\beta} T^j.$$

Since V is Calabi-Yau, the Degree Axiom implies that all terms are 0 except for those j's which satisfy

$$2n + \deg T + \deg T_j = 2d - 2, \quad d = \dim(V).$$

Since $\deg T^j = 2d - \deg T_j$, we have $\deg T^j = \deg T + 2n + 2$. Hence, in the above formula for $\tilde{s}(T)$, any nonzero term in the summation must have degree $> \deg T$.

Finally, we need to show that the value of $\tilde{s}(T)$ is T when $q_i = 0$ for all i. We noted earlier that in the above formula for $\tilde{s}(T)$, we can restrict to those $\beta \neq 0$ such that the exponents of

$$q^\beta = \Pi_i q_i^{\int_\beta T_i}$$

are all nonnegative. Hence it suffices to show that for each such β, at least one exponent is positive. But this is obvious, for otherwise we would have $\int_\beta T_i = 0$ for all i. This is impossible since $\beta \neq 0$ and the T_i are a basis of $H^2(V, \mathbb{C})$. $\qquad\square$

When we first introduced the symbolic expression $\frac{T}{\hbar-c}$ in (10.15), it might have seemed rather artificial. But given how it simplified the proofs of Theorem 10.2.4 and Corollary 10.2.5, this symbolic notation has more than proved its worth. We will encounter this notation often in Section 10.3.

In the case of a Calabi-Yau threefold, we studied the flat sections of the A-model connection in Chapter 8. Let's compare those formulas with the ones derived here.

Example 10.2.2.1. Let V be a Calabi-Yau threefold. In Section 8.5.3, we constructed the flat sections of ∇ using the Gromov-Witten potential

$$\Phi = \frac{(2\pi i)^3}{6} \int_V (\textstyle\sum_{j=1}^r u_j T_j)^3 + \sum_{\beta \neq 0} N_\beta \, q^\beta$$

$$= \frac{(2\pi i)^3}{6} \int_V (\textstyle\sum_{j=1}^r u_j T_j)^3 + \Phi_{\mathrm{hol}}$$

from (8.51) (as usual, $N_\beta = \langle I_{0,0,\beta} \rangle$). Here, we are restricting to even cohomology.

In proof of Proposition 8.5.4, we gave formulas for the flat sections in (8.53), and in (8.55), we applied $\exp(-\sum_j u_j N_j) = \exp(-\frac{1}{2\pi i} \sum_j \log(q_j) N_j)$ to these flat sections. In particular, we obtained the formula

$$(10.26) \qquad \tilde{s}_0 = T_0 - \frac{1}{(2\pi i)^3} \sum_j \frac{\partial \Phi_{\mathrm{hol}}}{\partial u_j} T^j + \frac{2}{(2\pi i)^3} \Phi_{\mathrm{hol}} \, T^0$$

from (8.55). In this equation, $T_0 = 1$ and T_j, $1 \le j \le r$, is a basis of $H^2(V, \mathbb{C})$, and then $T^j \in H^4(V, \mathbb{C})$, $1 \le j \le r$, and $T^0 \in H^6(V, \mathbb{C})$ are dual classes.

In the language of Section 8.5.3, $\tilde{s}_0$ is a ∇^c-flat section, and we showed that its value at 0 is T_0. But Corollary 10.2.5 asserts that the section $\tilde{s}(T_0)$ is also ∇^c-flat, and Corollary 10.2.6 shows that its value at 0 is also T_0. It follows that $\tilde{s}_0 = \tilde{s}(T_0)$.

To see what this implies, let's write out $\tilde{s}(T_0)$ using the above cohomology basis. Since $T_0 = 1$, the formula given in Corollary 10.2.5 becomes

$$\tilde{s}(T_0) = T_0 + \sum_{\beta \neq 0} q^\beta \left(\left\langle \frac{1}{\hbar - c}, T^0 \right\rangle_{0,\beta} T_0 + \sum_j \left\langle \frac{1}{\hbar - c}, T^j \right\rangle_{0,\beta} T_j + \right.$$

$$(10.27) \qquad\qquad \left. \sum_j \left\langle \frac{1}{\hbar - c}, T_j \right\rangle_{0,\beta} T^j + \left\langle \frac{1}{\hbar - c}, T_0 \right\rangle_{0,\beta} T^0 \right).$$

Let's see what happens when we compare coefficients of (10.26) and (10.27). For the coefficient of T^0, first observe that

$$\left\langle \frac{1}{\hbar - c}, T_0 \right\rangle_{0,\beta} = \sum_{n=0}^{\infty} \hbar^{-(n+1)} \langle \tau_n, 1 \rangle_{0,\beta} = \hbar^{-3} \langle \tau_2, 1 \rangle_{0,\beta}$$

since all of the other terms vanish by the Degree Axiom. Then, comparing the coefficients of T^0 and using the above formula for Φ, we obtain the identity

$$2 \frac{1}{(2\pi i)^3} N_\beta = \hbar^{-3} \langle \tau_2, 1 \rangle_{0,\beta}.$$

Since $\hbar = -2\pi i$, this implies $\langle \tau_2, 1 \rangle_{0,\beta} = -2N_\beta$, exactly as in Example 10.1.3.2.

Turning to the other terms in the summation in (10.27), one sees easily that the coefficients of T_0 and T_j vanish by the Degree and Fundamental Class Axioms. Finally, the equality of the coefficients of T^j in (10.26) and (10.27) is equivalent to $\langle \tau_1, T_j \rangle_{0,\beta} = -(\int_\beta T_j) N_\beta$, which follows from the Divisor and Dilaton Axioms.

10.2.3. The Parameter $\hbar$. We close this section with a brief explanation of how the parameter $\hbar$ arises naturally from equivariant cohomology. Our discussion is taken from [**Givental2**].

We consider $\mathbb{P}^1$ with the $\mathbb{C}^*$-action $t \cdot (x_0, x_1) = (t^{-1}x_0, x_1)$, and we also let $\hbar \in H^2(B\mathbb{C}^*)$ be the corresponding generator, so that $H^*(B\mathbb{C}^*) = \mathbb{C}[\hbar]$. We can compare this with the $(\mathbb{C}^*)^2$-action given in Example 9.1.2.1 by the substitution $\lambda_0 \mapsto \hbar$ and $\lambda_1 \mapsto 0$. It follows that $H^*_{\mathbb{C}^*}(\mathbb{P}^1) \simeq \mathbb{C}[p, \hbar]/\langle p(p - \hbar)\rangle$.

Let $\mathcal{R}_{\mathbb{C}^*} = \mathbb{C}(\hbar)$ be the field of fractions of $H^*(B\mathbb{C}^*)$ (this is the notation of Section 9.1.2). Then $\{p/\hbar, (\hbar - p)/\hbar\}$ is a basis for $H^*_{\mathbb{C}^*}(\mathbb{P}^1) \otimes_{H^*(B\mathbb{C}^*)} \mathcal{R}_{\mathbb{C}^*}$ as a vector space over $\mathcal{R}_{\mathbb{C}^*}$. This basis is especially nice relative to the localization theorem from Section 9.1.2. (Note that $\mathcal{R}_{\mathbb{C}^*}$ can be replaced with the smaller ring $\mathbb{C}[\hbar, \hbar^{-1}]$.)

In Chapter 11, we will use the following important idea. Given a stable map $f : (\mathbb{P}^1, p_1, \ldots, p_n) \to X$, we set $\beta = f_*[\mathbb{P}^1] \in H_2(X, \mathbb{Z})$. Then the graph of f gives a natural element of $\overline{M}_{0,n}(X \times \mathbb{P}^1, (\beta, 1))$. One advantage this space has over $\overline{M}_{0,n}(X, \beta)$ is that $\mathbb{C}^*$ acts on $X \times \mathbb{P}^1$ by using the trivial action on X and the above action on $\mathbb{P}^1$.

Because of the $\mathbb{C}^*$-action, we can use the equivariant Gromov-Witten invariants defined in Section 9.3. To simplify the exposition, we will assume that X is smooth and convex. Then $X \times \mathbb{P}^1$ is also convex, so that Definition 9.3.1 applies. Then, given $\gamma \in H^*_{\mathbb{C}^*}(X \times \mathbb{P}^1)$, we get the equivariant invariant

$$\langle I_{0,n,(\beta,1)}\rangle_{\mathbb{C}^*}(\gamma, \ldots, \gamma) = \int_{\overline{M}_{0,n}(X \times \mathbb{P}^1, (\beta,1))_{\mathbb{C}^*}} e_1^*(\gamma) \cup \cdots \cup e_n^*(\gamma) \in H^*(B\mathbb{C}^*).$$

We will find it convenient to regard this as an element of the field of fractions $\mathcal{R}_{\mathbb{C}^*} = \mathbb{C}(\hbar)$. Using these invariants, we define the equivariant potential function

$$\mathcal{F}^1(\gamma) = \sum_{n=0}^{\infty} \sum_{\beta \in H_2(X)} \frac{1}{n!} \langle I_{0,n,(\beta,1)}\rangle_{\mathbb{C}^*}(\gamma, \ldots, \gamma)\, q^\beta.$$

To make this into a formal power series, we introduce some variables. Note that

$$H^*_{\mathbb{C}^*}(X \times \mathbb{P}^1) \otimes_{H^*(B\mathbb{C}^*)} \mathcal{R}_{\mathbb{C}^*} = H^*(X) \otimes_{\mathbb{C}} H^*_{\mathbb{C}^*}(\mathbb{P}^1) \otimes_{H^*(B\mathbb{C}^*)} \mathcal{R}_{\mathbb{C}^*}.$$

It follows that given a cohomology basis T_i of $H^*(X)$, $\{T_j \otimes p/\hbar,\ T_i \otimes (\hbar - p)/\hbar\}$ is a basis of the localized cohomology over $\mathcal{R}_{\mathbb{C}^*}$. Then put

$$\gamma = \sum_j t_j T_j \otimes p/\hbar + \sum_i \tau_i T_i \otimes (\hbar - p)/\hbar$$

into the formula for $\mathcal{F}^1$. We can think of t_j and τ_i as taking values in $\mathcal{R}_{\mathbb{C}^*}$.

Once we express $\mathcal{F}^1$ in terms of the variables t_j, τ_i, some nice things happen. First, consider the second derivative

$$\Phi_{ij} = \frac{\partial^2 \mathcal{F}^1}{\partial \tau_i \partial t_j}.$$

We will think of $\Phi = (\Phi_{ij})$ as a matrix-valued formal function. Then Givental shows that the transpose of Φ gives flat sections of ∇^g.

PROPOSITION 10.2.7. *If Φ^* denotes the transpose of Φ, then for all j,*

$$\hbar \frac{\partial}{\partial t_j} \Phi^* = s_j * \Phi^*,$$

where s_j is the flat section defined in (10.15).

PROOF. The proof is similar to the proof of Proposition 10.2.1, although it is somewhat simpler because we are dealing with equivariant Gromov-Witten invariants rather than gravitational correlators. Details can be found in [**Givental2**]. □

Another nice aspect of Φ_{ij} is that it can be naturally interpreted in terms of gravitational correlators. To see this, let

$$\psi_{ij} = \left\langle\!\left\langle \frac{T_j}{\hbar - c}, T_i \right\rangle\!\right\rangle_0.$$

By (10.15), this is the coefficient of T^i in the expansion of s_j in the dual basis, where as usual $\{T^i\}$ is the dual basis to $\{T_i\}$. We can regard ψ_{ij} as a function of the t_i and $\hbar$, i.e.,

$$\psi_{ij} = \psi_{ij}(t_0, \dots, t_m, \hbar).$$

Then this relates to the above function Φ_{ij} as follows.

LEMMA 10.2.8.

$$\Phi_{ij} = -\sum_{k,\ell} \hbar^{-2} \psi_{ik}(t_0, \dots, t_m, \hbar)\, g^{kl}\, \psi_{j\ell}(\tau_0, \dots, \tau_m, \hbar).$$

PROOF. This lemma is proved in [**Givental2**], so we will only give a brief sketch of the proof. The point is that we compute the left hand side using localization, just as in Chapter 9. The $\mathbb{C}^*$-action on $X \times \mathbb{P}^1$ induces an action on the space of stable maps. One sees that the fixed point locus consists of stable maps $f : C \to X \times \mathbb{P}^1$ with $C = C' \cup C_0 \cup C_\infty$ which satisfy $f(C_0) \subset X \times \{0\}$, $f(C_\infty) \subset X \times \{\infty\}$, and $f(C') \subset \{x\} \times \mathbb{P}^1$ for some point $x \in X$. The components C_0 and C_∞ could be empty. The component C' meets C_0 and C_∞ at one point each, and the $\hbar - c$ factor in the definition of ψ_{ij} arises from the application of Lemma 9.2.2 to these two nodes. □

Finally, we should mention that these ideas can be used to give a different proof of Proposition 10.2.1 which deduces the result from Lemma 10.2.8 and Proposition 10.2.7, again using localization.

10.3. Relations in Quantum Cohomology

In this section, we use the flat sections s_a from (10.14) to study quantum cohomology. We will regard ∇^g as a formal connection on the trivial bundle $M \times H^*(X, \mathbb{C})$ over the base $M = H^0(X, \mathbb{C}) \oplus H^2(X, \mathbb{C})$. Thus the variables are $t_0, t_1, \dots, t_r$, and ∇^g is given by the small quantum product as in (10.19).

10.3.1. Givental's Function J. Following Givental, we define the function

$$(10.28) \qquad\qquad J = \sum_j \langle s_j, 1 \rangle\, T^j,$$

where $\langle \alpha, \beta \rangle = \int_X \alpha \cup \beta$ denotes the usual intersection pairing on cohomology. If we use equation (10.14), the definition of J can also be written as

$$(10.29) \qquad J = J(t_0, \dots, t_m, \hbar^{-1}) = 1 + \sum_{n=0}^{\infty} \sum_{a=0}^{m} \hbar^{-(n+1)} \langle\!\langle \tau_n T_a, 1 \rangle\!\rangle_0\, T^a.$$

We showed in Section 8.5.1 that the variables $t_1, \dots, t_r$ appear in big quantum cohomology as exponentials $q_j = e^{t_j}$. Then, when we regard the small quantum product as the restriction of the big quantum product, the series giving $*_{\text{small}}$ is a formal power series in the q_j. For simplicity, we will write $*_{\text{small}}$ as $*$ in what follows.

Let $P(\hbar\partial/\partial t, e^t, \hbar)$ be a formal power series in the quantities

$$\hbar\partial/\partial t_0, \dots, \hbar\partial/\partial t_r, e^{t_0}, \dots, e^{t_r}, \hbar.$$

For definiteness, we can and will assume that the exponential terms are all to the left of the derivatives, although this is not needed in Theorem 10.3.1 below.

If $H^*(X, \mathcal{C})$ is cohomology with coefficients in $\mathcal{C} = \mathbb{C}[[q_1, \dots, q_r]]$, then we denote by $P(T, q, 0)$ the formal power series in $H^*(X, \mathcal{C})$ obtained from $P(\hbar\partial/\partial t, e^t, \hbar)$ by the substitutions

$$\hbar\partial/\partial t_j \longmapsto T_j, \quad e^{t_j} \longmapsto q_j, \quad \hbar \longmapsto 0$$

and replacing composition of differential operators with quantum product.

The key result is the following.

THEOREM 10.3.1. *Suppose $P(\hbar\partial/\partial t, e^t, \hbar)J = 0$, where P is a formal power series with notation as above. Then the relation $P(T, q, 0) = 0$ holds in small quantum cohomology.*

PROOF. The following proof was suggested by B. Kim. Introduce the *dual Givental connection* $\widetilde{\nabla}^g$, which is defined by

$$\widetilde{\nabla}^g_{\frac{\partial}{\partial t_i}}\left(\sum_j a_j T_j\right) = \hbar \sum_j \frac{\partial a_j}{\partial t_i} T_j + \sum_j a_j\, T_i * T_j.$$

A quick calculation verifies the identity

$$(10.30) \qquad \hbar\frac{\partial}{\partial t_i}\langle G, H\rangle = \langle \nabla^g_{\frac{\partial}{\partial t_i}} G, H\rangle + \langle G, \widetilde{\nabla}^g_{\frac{\partial}{\partial t_i}} H\rangle$$

for $H^*(X, \mathcal{C})$-valued functions of $t_0, \dots, t_m$. One also calculates that

$$(10.31) \qquad \widetilde{\nabla}^g_{\partial/\partial t_{i_1}} \cdots \widetilde{\nabla}^g_{\partial/\partial t_{i_k}} 1 = T_{i_1} * \cdots * T_{i_k} + \hbar\, A(T, q, \hbar)$$

for some formal power series A in the T_i, q_i and $\hbar$.

The expression PJ is defined to be $\sum_j P\langle s_j, 1\rangle T^j$. Since $T_0, \dots, T_m$ are linearly independent, $PJ = 0$ implies that $P(\hbar\partial/\partial t, e^t, \hbar)\langle s_j, 1\rangle = 0$ for all j. Proposition 10.2.1 also implies that $\nabla^g_{\partial/\partial t_i} s_j = 0$ for all j. Then, by repeated application of (10.30) and (10.31), one can show that

$$0 = P\langle s_j, 1\rangle = \langle s_j, P(T, q, 0)\rangle + \hbar\langle s_j, B(T, q, \hbar)\rangle, \quad j = 0, \dots, m$$

for some formal power series $B(T, q, \hbar)$. This is an identity in $\hbar$, which allows us to set $\hbar = 0$ and obtain

$$\langle s_j, P(T, q, 0)\rangle = 0, \quad j = 0, \dots, m.$$

Since the s_j form a basis, we conclude that the relation $P(T, q, 0) = 0$ holds using the small quantum product on $H^*(X, \mathcal{C})$. $\square$

This theorem motivates the following definition.

DEFINITION 10.3.2. *A differential operator $P(\hbar\partial/\partial t, e^t, \hbar)$ satisfying the hypothesis of Theorem 10.3.1 is called a* quantum differential operator. *The equation $P(\hbar\partial/\partial t, e^t, \hbar)Y = 0$ (which is satisfied by $Y = J$) is called a* quantum differential equation.

Another way to think about Theorem 10.3.1 is as follows. If we let $P_{\widetilde{\nabla}^g}$ denote the expression obtained from $P(\hbar\partial/\partial t, e^t, \hbar)$ by the substitution

$$\hbar\frac{\partial}{\partial t_i} \longmapsto \widetilde{\nabla}^g_{\frac{\partial}{\partial t_i}},$$

then (10.30) implies that

$$(10.32) \qquad\qquad P\langle s_j, 1\rangle = \langle s_j, P_{\widetilde{\nabla}^g}1\rangle$$

since s_j is a flat section for ∇^g. Thus, for the Givental function J, we have the equivalences

$$(10.33) \qquad PJ = 0 \iff P\langle s_j, 1\rangle = 0 \text{ for all } j \iff P_{\widetilde{\nabla}^g}1 = 0.$$

The first equivalence is from the proof of Theorem 10.3.1, and the second follows from (10.32) since the s_j form a basis. This explains the role of "1" in the formula (10.28) for J.

Note that if $P(T, q) = 0$ is a relation in small quantum cohomology, then $P(\hbar\partial/\partial t, e^t)$ need not be a quantum differential operator (see Example 10.3.2.1). But if the relation $P(T, q) = 0$ is of at most second order, then it is easy to see that $P(\hbar\partial/\partial t, e^t)$ is a quantum differential operator, provided the exponential terms are placed to the left of the derivatives. Also, Example 10.3.1.2 below will show that in some special cases, quantum differential equations can be derived directly from relations in quantum cohomology. Finally, we should also mention that in Chapter 11, we will use Theorem 10.3.1 to calculate the quantum cohomology of certain toric varieties in Example 11.2.5.2.

Our next task is to simplify the formula for J so that we can compute some examples of quantum differential operators. The following lemma records two formulas for J which will be used frequently.

LEMMA 10.3.3. *The Givental J-function of X is given by the following two formulas:*

$$J = e^{(t_0+\delta)/\hbar}\left(1 + \sum_{\beta\neq 0}\sum_{a=0}^{m} q^\beta\left\langle\frac{T_a}{\hbar-c}, 1\right\rangle_{0,\beta}T^a\right)$$

$$= e^{(t_0+\delta)/\hbar}\left(1 + \sum_{\beta\neq 0} q^\beta PD^{-1}e_{1*}\left(\frac{1}{\hbar-c}\cap[\overline{M}_{0,2}(X,\beta)]^{\mathrm{virt}}\right)\right),$$

where $\delta = \sum_{i=1}^r t_i T_i$ and $q^\beta = e^{\int_\beta \delta}$. In the second equality, PD is Poincaré duality and $e_1 : \overline{M}_{0,2}(X,\beta) \to X$ is evaluation at the first marked point.

PROOF. Recall that J is built from the flat sections s_a, and since we are restricting to $M = H^0(X,\mathbb{C}) \oplus H^2(X,\mathbb{C})$, Proposition 10.2.3 implies that

$$s_a = e^{t_0/\hbar}\left(e^{\delta/\hbar}\cup T_a + \sum_{\beta\neq 0}\sum_{j=0}^{m} q^\beta\left\langle\frac{e^{\delta/\hbar}\cup T_a}{\hbar-c}, T_j\right\rangle_{0,\beta}T^j\right).$$

Hence the Givental function J is given by

$$J = \sum_{a=0}^{m} \langle s_a, 1 \rangle \, T^a$$

$$(10.34) \qquad = e^{t_0/\hbar} \Big(\sum_{a=0}^{m} \langle e^{\delta/\hbar} \cup T_a, 1 \rangle \, T^a +$$

$$\sum_{a=0}^{m} \sum_{\beta \neq 0} \sum_{j=0}^{m} q^\beta \Big\langle \frac{e^{\delta/\hbar} \cup T_a}{\hbar - c}, T_j \Big\rangle_{0,\beta} \langle T^j, 1 \rangle \, T^a \Big).$$

The summation in the second line of (10.34) is easy to simplify:

$$\sum_{a=0}^{m} \langle e^{\delta/\hbar} \cup T_a, 1 \rangle \, T^a = \sum_{a=0}^{m} \langle e^{\delta/\hbar}, T_a \rangle \, T^a = e^{\delta/\hbar},$$

where the last equality is by (10.24). The third line of (10.34) is easier, for one easily sees that

$$\langle T^j, 1 \rangle = \begin{cases} 1 & \text{if } T_j = T_0 = 1 \\ 0 & \text{otherwise.} \end{cases}$$

Hence only the terms with $j = 0$ contribute to the third line of (10.34). It follows that

$$(10.35) \qquad J = e^{t_0/\hbar} \Big(e^{\delta/\hbar} + \sum_{\beta \neq 0} \sum_{a=0}^{m} q^\beta \Big\langle \frac{e^{\delta/\hbar} \cup T_a}{\hbar - c}, 1 \Big\rangle_{0,\beta} T^a \Big).$$

We can simplify this further by observing that although the cohomology basis $\{T_a\}$ and its dual basis $\{T^a\}$ appear in (10.35), J is independent of which basis we use. So we can replace $\{T_a\}$ with any other basis. Furthermore, we could even use a nonhomogeneous basis, and since we're using formal coefficients, we can use the basis given by $\{e^{-\delta/\hbar} \cup T_a\}$. Since this has dual basis $\{e^{\delta/\hbar} \cup T^a\}$, we see that J can be written as

$$J = e^{t_0/\hbar} \Big(e^{\delta/\hbar} + \sum_{\beta \neq 0} \sum_{a=0}^{m} q^\beta \Big\langle \frac{e^{\delta/\hbar} \cup (e^{-\delta/\hbar} \cup T_a)}{\hbar - c}, 1 \Big\rangle_{0,\beta} e^{\delta/\hbar} \cup T^a \Big),$$

which simplifies to the first of our desired formulas for J.

To prove the second formula, we observe that

$$(10.36) \qquad \Big\langle \frac{T_a}{\hbar - c}, 1 \Big\rangle_{0,\beta} = \int_{[\overline{M}_{0,2}(X,\beta)]^{\mathrm{virt}}} \frac{e_1^*(T_a)}{\hbar - c}.$$

By the projection formula for e_1, this is the same as

$$\int_X T_a \cup PD^{-1} \Big(e_{1*} \Big(\frac{1}{\hbar - c} \cap [\overline{M}_{0,2}(X,\beta)]^{\mathrm{virt}} \Big) \Big).$$

If we apply (10.24) with $\phi = PD^{-1}(e_{1*}(1/(\hbar - c)) \cap [\overline{M}_{0,2}(X,\beta)]^{\mathrm{virt}})$ and use (10.36), we obtain

$$\sum_{a=0}^{m} \Big\langle \frac{T_a}{\hbar - c}, 1 \Big\rangle_{0,\beta} T^a = PD^{-1} \Big(e_{1*} \Big(\frac{1}{\hbar - c} \cap [\overline{M}_{0,2}(X,\beta)]^{\mathrm{virt}} \Big) \Big).$$

This shows that the second formula for J in the statement of the lemma is an immediate consequence of the first. $\qquad \square$

An important consequence of Lemma 10.3.3 and the vanishing of $\langle T_a, 1 \rangle_{0,\beta}$ (by the Fundamental Class Axiom) is that

$$(10.37) \qquad J = e^{(t_0+\delta)/\hbar}\left(1 + o(\hbar^{-1})\right),$$

as noted in [**Givental4**]. This fact is crucial to the statement of Theorem 11.2.2 in Section 11.2.

We now give two examples which give formulas for J and illustrate how quantum differential equations give relations in quantum cohomology. We will give some further examples in Section 10.3.2.

Example 10.3.1.1. Let's compute J for $X = \mathbb{P}^1$. If H is the hyperplane section (which is a point), then $\delta = t_1 H$ and $\int_\beta \delta = dt_1$ for $\beta = d \in H^2(\mathbb{P}^1, \mathbb{Z}) = \mathbb{Z}$. Then Lemma 10.3.3 implies

$$J = e^{(t_0+t_1 H)/\hbar}\left(1 + \sum_{d=1}^{\infty} q^d \left(\left\langle \frac{H}{\hbar - c}, 1 \right\rangle_{0,d} 1 + \left\langle \frac{1}{\hbar - c}, 1 \right\rangle_{0,d} H\right)\right).$$

Using (10.10) and (10.11), this becomes

$$J = e^{(t_0+t_1 H)/\hbar}\left(1 + \sum_{d=1}^{\infty} q^d \left(\hbar^{-2d}\frac{1}{(d!)^2}1 + \hbar^{-(2d+1)}\left[\frac{-2}{(d!)^2}\left(1 + \frac{1}{2} + \cdots + \frac{1}{d}\right)\right]H\right)\right)$$

$$= e^{(t_0+t_1 H)/\hbar}\left(1 + \sum_{d=1}^{\infty} \left(\frac{q}{\hbar^2}\right)^d \frac{1}{(d!)^2}\left(1 - 2\left(\frac{H}{\hbar}\right)\left(1 + \frac{1}{2} + \cdots + \frac{1}{d}\right)\right)\right)$$

It is straightforward to verify that J is annihilated by $(\hbar d/dt_1)^2 - e^{t_1}$, so that $(\hbar d/dt_1)^2 - e^{t_1}$ is a quantum differential operator. Then Theorem 10.3.1 yields the small quantum cohomology relation $H^2 = q$ after the substitutions $\hbar d/dt_1 \mapsto H$ and $e^{t_1} \mapsto q$, in agreement with Example 8.1.2.1.

We also get a very interesting formula for J as follows. Since $H^2 = 0$, the above formula for J can be written as

$$J = e^{(t_0+t_1 H)/\hbar} \sum_{d=0}^{\infty} q^d \frac{1}{\left(d!\hbar^{d-1}(\sum_{j=1}^{d} 1/j)H + d!\hbar^d\right)^2},$$

and then, using $H^2 = 0$ again, we obtain

$$J = e^{(t_0+t_1 H)/\hbar} \sum_{d=0}^{\infty} q^d \frac{1}{\left((H + \hbar)(H + 2\hbar)\cdots(H + d\hbar)\right)^2}.$$

We will see in Section 11.2.1 that this is a special case of Givental's version of the Mirror Theorem.

Example 10.3.1.2. Consider n-dimensional projective space $X = \mathbb{P}^n$ and let H be the hyperplane class of $\mathbb{P}^n$. We take our cohomology basis to be $1, H, \ldots, H^n$, with variables t_0, t_1 corresponding to cohomology classes $1, H$. If we let $q = e^{t_1}$, then the small quantum product satisfies $H^{n+1} = e^{t_1}$ by Example 8.1.2.1.

Let's show that this relation leads to a quantum differential equation. The absence of any t_0, t_1 dependence in the quantum powers H^i for $i \leq n$ implies that the analysis used to obtain (10.31) leads in this case to the more precise identity

$$\left(\widetilde{\nabla}_{\frac{d}{dt_1}}\right)^{n+1} 1 = H^{n+1}$$

and since $H^{n+1} = e^{t_1}$ in small quantum cohomology, this becomes

$$\left(\tilde{\nabla}_{\frac{d}{dt_1}}\right)^{n+1} 1 = e^{t_1}.$$

By (10.33), it follows that

$$(10.38) \qquad \left(\hbar \frac{d}{dt_1}\right)^{n+1} - e^{t_1}$$

is a quantum differential operator.

One can also go the other way. Proposition 11.2.1 asserts that for $\mathbb{P}^n$, Givental's function J is given by the formula

$$(10.39) \qquad J = e^{(t_0+t_1 H)/\hbar} \sum_{d=0}^{\infty} e^{dt_1} \frac{1}{\big((H+\hbar)(H+2\hbar)\cdots(H+d\hbar)\big)^{n+1}}.$$

For now, we note two things about this formula. First, for $n = 1$, it agrees with the computation in Example 10.3.1.1 (once we set $q = e^{t_1}$) and second, for general n, one can show that the right hand side of (10.39) is annihilated by the differential operator (10.38). Hence, once we prove (10.39) in Chapter 11, it will follow that (10.38) is a quantum differential operator, and then Theorem 10.3.1 will imply the relation $H^{n+1} = e^{t_1} = q$ in the small quantum cohomology ring of $\mathbb{P}^n$.

10.3.2. Calabi-Yau Threefolds. For the rest of the section, we will assume that V is a Calabi-Yau threefold. We begin by explaining what the Givental J-function looks like in terms of the Gromov-Witten invariants

$$N_\beta = \langle I_{0,0,\beta} \rangle$$

discussed in Section 7.4.4. As in Example 10.2.2.1, we restrict to even cohomology, and we order the cohomology basis of $H^{\text{even}}(V)$ so that $T_0 = 1, T_1, \ldots, T_r$ generate $H^2(V)$, and $T^j \in H^4(V, \mathbb{C})$, $1 \le j \le r$, and $T^0 \in H^6(V, \mathbb{C})$ are dual classes. Also let $\delta = \sum_{i=1}^{r} t_i T_i$.

We now give a formula for J. Here, Lemma 10.3.3 gives the formula

$$J = e^{(t_0+\delta)/\hbar} \left(1 + \sum_{\beta \ne 0} q^\beta \Big(\hbar^{-2} \sum_{a=1}^{r} \langle \tau_1 T_a, 1 \rangle_{0,\beta} T^a + \hbar^{-3} \langle \tau_2, 1 \rangle_{0,\beta} T^0 \Big)\right).$$

This follows easily using the Degree and Fundamental Class Axioms. In Example 10.1.3.2, we showed that

$$\langle \tau_1 T_a, 1 \rangle_{0,\beta} = \big(\textstyle\int_\beta T_a\big) N_\beta$$

$$\langle \tau_2, 1 \rangle_{0,\beta} = -2N_\beta,$$

so that the above formula simplifies to

$$(10.40) \qquad J = e^{(t_0+\delta)/\hbar} \left(1 + \sum_{\beta \ne 0} q^\beta \Big(\hbar^{-2} N_\beta \big(\textstyle\sum_{a=1}^{r} (\int_\beta T_a) T^a\big) - 2\hbar^{-3} N_\beta T^0\Big)\right).$$

However, (10.24) easily implies that the Poincaré dual of $\beta \in H_2(V, \mathbb{C})$ is the class $\sum_{a=1}^{r} \big(\int_\beta T_a\big) T^a \in H^4(V, \mathbb{C})$. If we abuse notation slightly and let β denote this class in $H^4(V, \mathbb{C})$, then we can rewrite our formula for the Givental function of a Calabi-Yau threefold as

$$(10.41) \qquad J = e^{(t_0+\delta)/\hbar} \left(1 + \hbar^{-2} \sum_{\beta \ne 0} N_\beta q^\beta \beta - 2\hbar^{-3} \sum_{\beta \ne 0} N_\beta q^\beta\, pt\right).$$

If an independent method can be given for computing J, then by comparing co-efficients with the above formula, the N_β can be computed. Givental asserts such a method for toric complete intersections in [**Givental4**]. We will illustrate this method for the quintic threefold in Example 11.2.1.3.

We next explain how the Givental J function relates to the Gromov-Witten potential and the small quantum product. The result is as follows.

PROPOSITION 10.3.4. *If Φ is the Gromov-Witten potential of a Calabi-Yau threefold V, then the Givental J-function of V is given by*

$$J = e^{t_0/\hbar}\left(1 + \hbar^{-1}\sum_{a=1}^{r} t_a T_a + \hbar^{-2}\sum_{a=1}^{r}\frac{\partial\Phi}{\partial t_a} T^a + \hbar^{-3}\left(\sum_{a=1}^{r} t_a\frac{\partial\Phi}{\partial t_a} - 2\Phi\right)T^0\right).$$

Furthermore, for any $1 \leq i, j \leq r$, we have

$$\frac{\partial^2 J}{\partial t_i \partial t_j} = \hbar^{-2} e^{(t_0+\delta)/\hbar}\, T_i * T_j.$$

PROOF. Recall from (8.33) that the Gromov-Witten potential is given by

$$\Phi = \frac{1}{6}\int_V \delta^3 + \sum_{\beta\neq 0} N_\beta\, q^\beta = \frac{1}{6}\int_V \delta^3 + \Phi_{\text{hol}},$$

where $\delta = \sum_{a=1}^{r} t_a T_a$. Using this and (10.40), one sees easily that J becomes

$$J = e^{(t_0+\delta)/\hbar}\left(1 + \hbar^{-2}\sum_{a=1}^{r}\frac{\partial\Phi_{\text{hol}}}{\partial t_a} T^a - 2\hbar^{-3}\Phi_{\text{hol}}\, T^0\right),$$

and multiplying this out gives

$$J = e^{t_0/\hbar}\left(1 + \hbar^{-1}\delta + \hbar^{-2}\frac{1}{2}\delta^2 + \hbar^{-3}\frac{1}{6}\delta^3 + \right.$$

$$\left. \hbar^{-2}\sum_{a=1}^{r}\frac{\partial\Phi_{\text{hol}}}{\partial t_a} T^a + \hbar^{-3}\delta\sum_{a=1}^{r}\frac{\partial\Phi_{\text{hol}}}{\partial t_a} T^a - 2\hbar^{-3}\Phi_{\text{hol}}\, T^0\right).$$

From here, we leave it to the reader to show that the desired formula for J follows using the following identities:

$$\delta\sum_{a=1}^{r}\frac{\partial\Phi_{\text{hol}}}{\partial t_a} T^a = \left(\sum_{a=1}^{r} t_a\frac{\partial\Phi_{\text{hol}}}{\partial t_a}\right)T^0$$

(10.42)
$$\frac{1}{6}\delta^3 = \frac{1}{6}\left(\int_V \delta^3\right)T^0 = \frac{1}{3}\sum_{a=1}^{r} t_a\frac{\partial}{\partial t_a}\frac{1}{6}\left(\int_V \delta^3\right)T^0$$

$$\frac{1}{2}\delta^2 = \frac{1}{2}\sum_{a=1}^{r}\left(\int_V \delta^2 \cup T_a\right)T^a = \sum_{a=1}^{r}\frac{\partial}{\partial t_a}\frac{1}{6}\left(\int_V \delta^3\right)T^a.$$

Note that the second line uses the Euler formula and the third uses (10.24).

To prove the formula for $\partial^2 J/\partial t_i \partial t_j$, we differentiate the formula for J given in the statement of the proposition with respect to t_i and t_j. This easily gives

$$\frac{\partial^2 J}{\partial t_i \partial t_j} = e^{t_0/\hbar}\left(\hbar^{-2}\sum_{a=1}^{r}\frac{\partial^3\Phi}{\partial t_i \partial t_j \partial t_a}T^a + \hbar^{-3}\left(\sum_{a=1}^{r} t_a\frac{\partial^3\Phi}{\partial t_i \partial t_j \partial t_a}\right)T^0\right).$$

Using an identity similar to the first line of (10.42), we obtain

$$\frac{\partial^2 J}{\partial t_i \partial t_j} = \hbar^{-2} e^{(t_0+\delta)/\hbar} \sum_a \frac{\partial^3 \Phi}{\partial t_i \partial t_j \partial t_a} T^a$$
$$= \hbar^{-2} e^{(t_0+\delta)/\hbar} T_i * T_j.$$

The second equality follows the discussion following the proof of Proposition 10.2.1, which explained why the small quantum product is the restriction of the big quantum product to $M = H^0(V) \oplus H^2(V)$. $\square$

As an application of this proposition, let's compute an interesting quantum differential operator.

Example 10.3.2.1. Suppose that the Calabi-Yau threefold V satisfies $h^2(V) = 1$. Here, let H denote the generator of $H^2(V)$ and let $C \in H^4(V)$ be the dual generator. The point class will be denoted by $pt \in H^6(V, \mathbb{Z})$, so we have $H \cup C = pt$. As usual, the variables t_0, t_1 correspond to $1, H$, and we let $q = e^{t_1}$.

In Section 8.5.4, we studied the three-point function $Y(q) = \langle H, H, H \rangle$ and noted the relation

$$H * H = Y(q) \, C$$

in (8.66). Using Proposition 10.3.4, we see that

$$\left(\hbar \frac{d}{dt_1} \right)^2 J = e^{(t_0+t_1 H)/\hbar} H * H = e^{(t_0+t_1 H)/\hbar} Y(q) \, C.$$

This easily implies that

$$\left(\hbar \frac{d}{dt_1} \right)^2 \left(\frac{(\hbar \frac{d}{dt_1})^2 J}{Y(q)} \right) = \left(\hbar \frac{d}{dt_1} \right)^2 e^{(t_0+t_1 H)/\hbar} C = e^{(t_0+t_1 H)/\hbar} H \cup H \cup C = 0$$

since $H^8(V) = 0$, so that

$$(10.43) \qquad \left(\hbar \frac{d}{dt_1} \right)^2 \left(\frac{(\hbar \frac{d}{dt_1})^2}{Y(q)} \right)$$

is a quantum differential operator. By Theorem 10.3.1, this gives the relation

$$H * H * (H * H/Y) = 0$$

in the small quantum cohomology ring of V.

This above relation is rather trivial since $H^8(V) = 0$. But the differential equation (10.43) is still very interesting, mainly because of its striking similarity to the Picard-Fuchs equation (5.63) derived in Section 5.6 for the Yukawa coupling of a Calabi-Yau threefold. This is no coincidence and is a key part of mirror symmetry. We will pursue the relation between Picard-Fuchs equations and quantum differential equations in Theorem 10.3.5 below.

We can also use this example to illustrate the comment made in Section 10.3.1 that the converse of Theorem 10.3.1 may fail, i.e., not all relations in quantum cohomology give rise to quantum differential equations. For the Calabi-Yau threefold V considered here, the relation $H * H * H * H = 0$ holds in quantum cohomology, yet $(\hbar d/dt_1)^4 J \neq 0$. This is because a simple calculation using (10.30) gives the equation

$$\left(\hbar\frac{d}{dt_1}\right)^4 \langle s_j, 1\rangle = \hbar^2 \frac{d^2 Y}{dt_1^2}\langle s_j, C\rangle + 2\hbar\frac{dY}{dt_1}\langle s_j, pt\rangle.$$

Since C and pt are independent cohomology classes and Y is nonconstant, it follows easily that $(\hbar d/dt_1)^4\langle s_j, 1\rangle$ can't vanish for all j.

We conclude this section by explaining how Theorem 10.3.1 relates to the A-model connection ∇ discussed in Example 10.2.2.1. Here, recall that the natural variables are u_j, which are related to t_j by $t_j = 2\pi i u_j$. Now consider an operator of the form

$$P_\nabla = P(\nabla_{\frac{\partial}{\partial u_1}}, \ldots, \nabla_{\frac{\partial}{\partial u_r}}, q_1, \ldots, q_r),$$

where $q_j = e^{2\pi i u_j}$. We will assume that P_∇ is a polynomial in the connection terms with coefficients which are formal power series in the q_j. We will also assume that the q_j are all to the left of the connection terms.

The operator P_∇ can be applied to any section of the bundle $\mathcal{H}$ on which the connection ∇ is defined. In particular, we have the section given by $1 = T_0$, and we say that P_∇ is a *Picard-Fuchs operator* if

$$P_\nabla 1 = 0.$$

This terminology is inspired by mirror symmetry, for in the Mirror Conjectures discussed in Section 8.6, the section 1 corresponds to the normalized 3-form $\widetilde{\Omega}$ on the mirror, and the usual Picard-Fuchs operators are those which annihilate $\widetilde{\Omega}$.

Given an operator P_∇, write it as

(10.44) $$P_\nabla = \sum_\alpha A_\alpha(q)\nabla^\alpha,$$

where $\alpha = (a_1, \ldots, a_r)$ is a multi-index and

$$\nabla^\alpha = \nabla_{\frac{\partial}{\partial u_1}}^{a_1} \cdots \nabla_{\frac{\partial}{\partial u_r}}^{a_r}.$$

Setting $|\alpha| = \sum_i a_i$ as usual, we let $m = \max_\alpha\{|\alpha| : A_\alpha(q) \neq 0\}$ be the order of P_∇. Then define the $\hbar$-*homogenization* of P_∇ to be the differential operator

(10.45) $$P(\hbar\partial/\partial t, e^t, \hbar) = \sum_\alpha \left(\frac{\hbar}{2\pi i}\right)^{m-|\alpha|} A_\alpha(e^t)\left(\hbar\frac{\partial}{\partial t}\right)^\alpha,$$

where $(\hbar\partial/\partial t)^\alpha$ has the obvious meaning. (We will see below why the $2\pi i$ is necessary.) Also set

$$P_m(T, q) = \sum_{|\alpha|=m} A_\alpha(q)\,T^\alpha,$$

where T^α is formed using $T_1, \ldots, T_r$ under the small quantum product in $H^*(V)$. We can think of $P_m(T, q)$ as a quantum version of the principal part (or characteristic form) of the operator P_∇.

We can relate these objects as follows.

THEOREM 10.3.5. *Let V be a Calabi-Yau threefold and let P_∇ be an operator of order m as in* (10.44). *If $P(\hbar\partial/\partial t, e^t, \hbar)$ is defined by* (10.45), *then:*

(*i*) *P_∇ is a Picard-Fuchs operator if and only if $P(\hbar\partial/\partial t, e^t, \hbar)$ is a quantum differential operator. In other words,*

$$P_\nabla 1 = 0 \iff P(\hbar\partial/\partial t, e^t, \hbar)\, J = 0.$$

(*ii*) *If P_∇ is Picard-Fuchs, then $P_m(T, q) = 0$ in small quantum cohomology.*

PROOF. We begin by noting a special property of quantum differential equations in this case. Let J be as in (10.41), where $\hbar$ is regarded as a variable, and let J_0 denote the formula obtained from J by replacing $\hbar$ by a nonzero complex number $\hbar_0$. Then we claim that

$$(10.46) \qquad P(\hbar\partial/\partial t, e^t, \hbar)\, J = 0 \iff P(\hbar_0\partial/\partial t, e^t, \hbar_0)\, J_0 = 0.$$

To see this, first note since $P(\hbar\partial/\partial t, e^t, \hbar)$ is homogeneous in $\hbar$, $P(\hbar\partial/\partial t, e^t, \hbar) = \hbar^m \widetilde{P}$, where $\widetilde{P}$ is an operator not involving $\hbar$. Also note that $\hbar$ appears homogeneously in the coefficients of J. To see what this means, we multiply out (10.40) to obtain

$$J = 1 + \hbar^{-1} \sum_{j=1}^{r} A_j T_j + \hbar^{-2} \sum_{j=1}^{r} B_j T^j + \hbar^{-3} C\, T^0,$$

where A_j, B_j, C are functions of $q_1, \ldots, q_r$. Using $P(\hbar\partial/\partial t, e^t, \hbar) = \hbar^m \widetilde{P}$, it follows easily that $P(\hbar\partial/\partial t, e^t, \hbar)\, J = 0$ is equivalent to

$$\widetilde{P}\, 1 = \widetilde{P}\, A_j = \widetilde{P}\, B_j = \widetilde{P}\, C = 0 \quad \text{for all } j.$$

Since $\hbar_0 \neq 0$, this is also equivalent to $P(\hbar_0\partial/\partial t, e^t, \hbar_0)\, J_0 = 0$, and (10.46) follows.

We can now prove the theorem. By (10.46), we can specialize $\hbar$ to any nonzero constant. Hence, for the rest of the proof, we can assume $\hbar = 2\pi i$ without any loss of generality. We will soon see why this is the correct choice.

By (10.33), we know that

$$(10.47) \qquad P(\hbar\partial/\partial t, e^t, \hbar)\, J = 0 \iff P_{\widetilde{\nabla}^g} 1 = 0.$$

where $\widetilde{\nabla}^g$ is the dual Givental connection and $P_{\widetilde{\nabla}^g}$ is defined by the substitution

$$\hbar\frac{\partial}{\partial t_j} \longmapsto \widetilde{\nabla}^g_{\frac{\partial}{\partial t_j}}.$$

Now comes the key observation: since $\hbar = 2\pi i$, the dual Givental connection $\widetilde{\nabla}^g$ is precisely $2\pi i \nabla$. It follows that

$$\widetilde{\nabla}^g_{\frac{\partial}{\partial t_j}} = 2\pi i \nabla_{\frac{\partial}{\partial t_j}} = \nabla_{\frac{\partial}{\partial u_j}},$$

where the second equality uses $t_j = 2\pi i\, u_j$. Since $\hbar = 2\pi i$, comparing (10.44) and (10.45) shows that $P_{\widetilde{\nabla}^g}$ is precisely the operator P_∇, and then (10.47) implies

$$P(\hbar\partial/\partial t, e^t, \hbar)\, J = 0 \iff P_\nabla 1 = 0.$$

This proves the first part of the theorem, and the second part now follows immediately from Theorem 10.3.1. $\qquad\square$

This theorem shows how the A-model connection of a Calabi-Yau threefold is deeply connected to small quantum cohomology. But an even deeper connection is to the B-model of the mirror, as predicted by mirror symmetry. In the next chapter, we will prove some special cases of the Mirror Theorem, which will finally link together all of the amazing mathematics we've been studying.

CHAPTER 11

The Mirror Theorem

The goal of this chapter is to prove some versions of the the Mirror Theorem, with special emphasis on the quintic threefold. We also touch on some interesting topics associated with the techniques in the proof. As noted in Chapters 1 and 2, the proofs involve several new concepts not discussed previously.

We learned in Chapter 8 that there are several ways to formulate a "Mirror Theorem". The versions in Section 8.6 were stated in terms of variations of Hodge structure. Here, we will take a different point of view and describe two closely related approaches to the Mirror Theorem, both of which are based on equivariant intersection theory in the space of stable maps. The historically first approach, due to Givental [**Givental2**], features the gravitational correlators defined in Chapter 10. These correlators lead to the Givental J-function defined in Section 10.3.1, and we will see that J plays an important role in Givental's version of the Mirror Theorem. Givental's methods extend to toric complete intersections [**Givental4**], homogeneous spaces [**Kim2**], and in part to higher genus [**Givental5**]. The approach of Lian, Liu, and Yau [**LLY**], on the other hand, emphasizes the interplay between the linear and nonlinear sigma models. Mathematically, the nonlinear sigma model is understood in terms of intersection theory on moduli spaces of stable maps, while the linear sigma model is understood in terms of intersection theory on the projective spaces N_d of tuples of homogeneous forms of degree d on $\mathbb{P}^1$. The main results are expressed in terms of the equivariant intersection theory of N_d. The method also applies to certain non-compact varieties. We will explain the [**LLY**] approach first, as fewer technical details are needed in the development.

In Section 11.1, we outline the proof of the Mirror Theorem for the quintic threefold given in [**LLY**], stated here as Theorem 11.1.1. By Section 8.6.2, this implies the Hodge-theoretic version of mirror symmetry from Chapter 8. In the approach of [**LLY**], equivariant versions of Gromov-Witten invariants are encoded in a sequence of equivariant cohomology classes $\hat{Q}_d$ on the projective spaces N_d. These are shown to be related to equivariant cohomology classes $\hat{P}_d$ which are associated to hypergeometric functions which give periods on the mirror family. The proof relies on a direct comparison of $\hat{P}_d$ and $\hat{Q}_d$. This is implemented using the concepts of linked Euler data (Definition 11.1.3), itself based on the gluing lemma (Lemma 11.1.2) for multiplicative cohomology classes, and mirror transformations (Definition 11.1.10). We will also explain how this method applies to the multiple cover calculation made in Section 9.2.2.

In Section 11.2, we turn to Givental's approach to the Mirror Theorem. Here, we will focus on the case of a complete intersection $X \subset \mathbb{P}^n$, which we represent as the zero locus of a section of $V = \bigoplus_{i=1}^{\ell} \mathcal{O}_{\mathbb{P}^n}(a_i)$, $a_i > 0$. In this situation, we define two formal cohomology-valued functions J_V and I_V. The function J_V is closely related to the Givental J-function of the complete intersection, while I_V is the function studied in Section 5.5.3. In Section 11.2.1, we state Theorem 11.2.2, which shows how to compute $J_\mathcal{V}$ in terms of $I_\mathcal{V}$ when $\sum_{i=1}^{\ell} a_i \leq n+1$. For $\mathcal{V} = \mathcal{O}_{\mathbb{P}^4}(5)$, this Mirror Theorem implies mirror symmetry for the quintic threefold. More generally, in Section 11.2.2, we will consider a complete intersection $X \subset Y$ defined by the vanishing of a section of a vector bundle $\mathcal{V}$ on Y which is a direct sum of line bundles. Here, we will give the general definition of $J_\mathcal{V}$ and discuss the *Quantum Hyperplane Section Principle*, which conjecturally computes $J_\mathcal{V}$. We will then prove Theorem 11.2.2, first for $\mathcal{V} = 0$ in Section 11.2.3, and then for general $\mathcal{V}$ in Section 11.2.4. The proof uses localization in equivariant cohomology, applied to equivariant versions of $J_\mathcal{V}$ and $I_\mathcal{V}$. When $\mathcal{V} = 0$, the localization formula (Corollary 9.1.4) from Chapter 9 leads to a recursion relation for the localizations of $J_\mathcal{V}$ that can be explicitly summed to yield $J_\mathcal{V}$. For general $\mathcal{V}$, we don't quite get a recursion, so that more work is needed. The chapter will end with a discussion of Givental's version of the Mirror Theorem for toric complete intersections.

11.1. The Mirror Theorem for the Quintic Threefold

In this section, we sketch the proof of the Mirror Theorem for the quintic threefold, following [**LLY**].

11.1.1. Statement of the Theorem. Let $V \subset \mathbb{P}^4$ be a generic quintic threefold. We begin by recalling the Gromov-Witten invariant $\langle I_{0,0,d} \rangle$. Fix a positive integer d and consider the maps

$$\overline{M}_{0,1}(\mathbb{P}^4, d) \xrightarrow{e_1} \mathbb{P}^4$$
$$\downarrow \pi_1$$
$$\overline{M}_{0,0}(\mathbb{P}^4, d)$$

where π_1 forgets the marked point and e_1 is the usual evaluation map. Let $\mathcal{V} = \mathcal{O}_{\mathbb{P}^4}(5)$, and put

$$\mathcal{V}_d = \pi_{1*}e_1^*(\mathcal{V}),$$

a vector bundle on $\overline{\mathcal{M}}_{0,0}(\mathbb{P}^4, d)$. One easily checks that this agrees with the bundle defined in Example 7.1.5.1. It follows that the above Gromov-Witten invariant is given by

$$\langle I_{0,0,d} \rangle = \int_{\overline{\mathcal{M}}_{0,0}(\mathbb{P}^4,d)} c_{5d+1}(\mathcal{V}_d).$$

In Chapter 7, this was denoted N_d, but here we will instead use K_d (as in [**LLY**]) since N_d will have a different meaning below. Also, $c_{5d+1}(\mathcal{V}_d)$ is the Euler class of $\mathcal{V}_d$, which here will be denoted $\mathrm{Euler}(\mathcal{V}_d)$. In this notation, the above formula is written

$$K_d = \int_{\overline{M}_{0,0}(\mathbb{P}^4,d)} \mathrm{Euler}(\mathcal{V}_d).$$

Notice that we wrote this as an integral over the moduli space $\overline{M}_{0,0}(\mathbb{P}^4, d)$ and not the stack $\overline{\mathcal{M}}_{0,0}(\mathbb{P}^4, d)$. This is because although $\mathcal{V}_d$ is only vector bundle over the stack, its Euler class $\mathrm{Euler}(\mathcal{V}_d)$ is a well-defined 0-cycle on $\overline{M}_{0,0}(\mathbb{P}^4, d)$, so that the above integral makes perfect sense. In what follows, we will often abuse notation by speaking of $\mathcal{V}_d$ as if it were a vector bundle on $\overline{M}_{0,0}(\mathbb{P}^4, d)$.

We also know from Chapter 7 that K_d can be written

$$K_d = \sum_{k|d} n_{\frac{d}{k}} k^{-3},$$

where the n_d are the instanton numbers discussed in Section 7.4.4. Furthermore, by Section 8.3.3, the Gromov-Witten potential of the quintic threefold is

$$\Phi(T) = \frac{5}{6}T^3 + \sum_{d>0} K_d\, e^{dT},$$

so that Φ''' is the A-model correlation function. The goal of Section 11.1 is to show that Φ is determined by the Hodge theory of the quintic mirror family as specified by mirror symmetry, thereby proving the formulas for n_d given in Chapter 2.

In order to state the result proved in [**LLY**], consider the formal $H^*(\mathbb{P}^4,\mathbb{C})$-valued expression

$$e^{-Ht/\hbar} \sum_{d\geq 0} \frac{\prod_{m=0}^{5d}(5H - m\hbar)}{\prod_{m=1}^{d}(H - m\hbar)^5} e^{dt},$$

where H is the hyperplane class. If we expand this in terms of powers of H, we obtain

$$5H\left(y_0 - y_1\frac{H}{\hbar} + y_2\frac{H^2}{\hbar^2} - y_3\frac{H^3}{\hbar^3}\right).$$

for certain functions $y_i = y_i(t)$. We saw in Chapter 6 that the y_i are a basis of solutions of the hypergeometric differential equation

$$\left[\left(\frac{d}{dt}\right)^4 - 5e^t\left(5\frac{d}{dt}+1\right)\left(5\frac{d}{dt}+2\right)\left(5\frac{d}{dt}+3\right)\left(5\frac{d}{dt}+4\right)\right]y = 0$$

for the periods of the quintic mirror, as can readily be checked by the Frobenius method. This was explained in the continuation of Example 6.3.4.1. Note that the y_i coincide with the coefficients of the function I from Section 5.5.3 associated to the set $\mathcal{A} = \Delta \cap M$, where Δ is the convex hull of the set (4.13) of primitive integral generators of the standard fan for $\mathbb{P}^4$. By Proposition 5.5.4, we know that the y_i satisfy the GKZ system associated to $\mathcal{A}$.

We put $\Psi(t) = \dfrac{y_1}{y_0}$. We have already noted in Section 2.6.2 that the mirror map is defined by $q = \exp(\Psi(t))$. Then the *Mirror Theorem* of [**LLY**] goes as follows.

THEOREM 11.1.1. *If Φ is the Gromov-Witten potential of the quintic threefold, then*

$$\Phi(\Psi(t)) = \frac{5}{2}\left(\frac{y_1}{y_0}\frac{y_2}{y_0} - \frac{y_3}{y_0}\right).$$

This formulation of the mirror conjecture for the quintic was first stated in [**Kontsevich3**]. In Example 11.2.1.3, we will show that Theorem 11.1.1 follows from the results of [**Givental2**], though the first fully detailed proof appeared in [**LLY**]. We explained in Section 2.6.3 how this version of the mirror theorem leads to "classical" mirror symmetry (2.26). Also recall from Proposition 8.6.5 that the right hand side of the equation in the above theorem is the "potential function" for the variation of Hodge structure of the quintic mirror. As noted in Section 8.6.2, Theorem 11.1.1 implies that the quintic mirror and quintic threefold form a mathematical mirror pair in the sense of Defintion 8.6.1.

The remainder of this section will be devoted to proving Theorem 11.1.1, and we will sketch a second proof of this theorem in Section 11.2.

11.1.2. Linear Models of Stable Moduli. We next introduce a "linearized" version of the space of stable maps which will play a key role in what follows. In physics, this is analogous to going from a nonlinear sigma model to a linear sigma model. We will develop the theory for hypersurfaces of degree ℓ in $\mathbb{P}^n$ and later specialize to the quintic threefold in $\mathbb{P}^4$.

We begin by compactifying the space of degree d maps $\mathbb{P}^1 \to \mathbb{P}^n$. The usual moduli space $\overline{M}_{0,0}(\mathbb{P}^n, d)$ doesn't do this, because a generic element of $\overline{M}_{0,0}(\mathbb{P}^n, d)$ only determines such a map only up to reparametrization. Hence we will instead use the space

$$M_d = \overline{M}_{0,0}(\mathbb{P}^1 \times \mathbb{P}^n, (1, d)).$$

Here, the generic element is a map $\mathbb{P}^1 \to \mathbb{P}^1 \times \mathbb{P}^n$ whose image has degree $(1, d)$. The key observation is that reparametrizing the source doesn't change the image, which is the graph of a degree d map $\mathbb{P}^1 \to \mathbb{P}^n$. Hence M_d compactifies the space of such maps, as claimed.

For our second compactification, note that a degree d map $\mathbb{P}^1 \to \mathbb{P}^n$ can be represented by an $(n+1)$-tuple $(\alpha_0, ..., \alpha_n)$ of degree d homogeneous polynomials on $\mathbb{P}^1$. This leads to the projective space

$$N_d = \mathbb{P}(H^0(\mathbb{P}^1, \mathcal{O}_{\mathbb{P}^1}(d))^{n+1}),$$

where $\mathbb{P}$ denotes the projective space of 1-dimensional subspaces. The role of N_d in the linear sigma model is explained in Appendix B.5 in the case $n = 4$.

Both M_d and N_d admit natural torus actions. In what follows, we will make heavy use of Sections 9.1 and 9.2, so that the reader may wish to review these sections before proceeding. Proposition 9.1.2 (and corollaries), Example 9.1.2 and Section 9.2.1 are especially important.

Now let G be the torus $G = \mathbb{C}^* \times T$, where $T = (\mathbb{C}^*)^{n+1}$. We first describe the G-action on $\mathbb{P}^1 \times \mathbb{P}^n$ we will use. Let $t \in \mathbb{C}^*$ and $(t_0, ..., t_n) \in T = (\mathbb{C}^*)^{n+1}$. Then

$$(t, t_0, \dots, t_n) \cdot \big((w_0, w_1), (x_0, \dots, x_n)\big) = \big((t^{-1}w_0, w_1), (t_0^{-1}x_0, \dots, t_n^{-1}x_n)\big).$$

As in Example 9.1.2.1, the inverses have been chosen so the action on homogeneous forms induced by pullback via the G-action will have positive exponents.

We next describe the G-action $M_d = \overline{M}_{0,0}(\mathbb{P}^1 \times \mathbb{P}^n, (1, d))$. Here, we use the induced action coming from the G-action on $\mathbb{P}^1 \times \mathbb{P}^n$. This action is closely related to the action of $(\mathbb{C}^*)^{r+1}$ on $\overline{M}_{0,n}(\mathbb{P}^r, d)$ studied in Section 9.2. Recall that we gave a careful description of the components of the fixed point locus. Here, we are interested in the components of the fixed point locus of the action of G on M_d, and we get an analogous description, with the following modifications:

- Contracted components of a G-fixed stable map must map to a G-fixed point of $\mathbb{P}^1 \times \mathbb{P}^n$.
- Noncontracted components must cover a G-fixed curve in $\mathbb{P}^1 \times \mathbb{P}^n$. These curves are of the form $\mathbb{P}^1 \times \{q_i\}$ and $\{p\} \times \ell$, where $q_i \in \mathbb{P}^n$ is T-fixed, p is 0 or ∞, and $\ell \subseteq \mathbb{P}^n$ is a coordinate line.
- If $\sum'$ denotes the sum over components covering $\mathbb{P}^1 \times \{q_i\}$ and $\sum''$ denotes the sum over components covering $\{p\} \times \ell$, then $\sum' d_i = 1$ and $\sum'' d_i = d$, where d_i are the degrees of the the coverings.

These components of the fixed point locus of G are analogous to those occurring in the proof of Lemma 10.2.8.

Finally, we describe how G acts on N_d. Here, the action is given by

$$(t, t_0, \dots, t_n) \cdot (\alpha_0(w_0, w_1), \dots, \alpha_n(w_0, w_1)) =$$
$$(t_0^{-1}\alpha_0(t^{-1}w_0, w_1), \dots, t_n^{-1}\alpha_n(t^{-1}w_0, w_1)).$$

This gives a linearized action of G on the projective space N_d. The G-fixed points of N_d are the points

$$p_{i,r} = (0, \dots, 0, w_0^r w_1^{d-r}, 0, \dots, 0),$$

where the nonzero monomial occurs in the ith location. The inclusion of $p_{i,r}$ in N_d will be denoted by $i_{p_{i,r}}$.

Let $\lambda_0, \dots, \lambda_n$ generate $H^*(BT)$ as in Section 9.1.1, and similarly introduce a generator $\hbar$ for $H^*(B\mathbb{C}^*)$. This generator is called α in [**LLY**], but here we adopt the notation of [**Givental2**] to keep notation consistent throughout the chapter (although the sign of $\hbar$ is different in [**Givental2**], corresponding to different conventions regarding the $\mathbb{C}^*$-action).

We can describe the equivariant cohomology $H_G^*(N_d)$ in terms of λ_i and $\hbar$. The answer is similar to the one given in Example 9.1.2.1. By regarding $p_{i,r} = (0, \dots, 0, w_0^r w_1^{d-r}, 0, \dots, 0)$ as a vector in $H^0(\mathbb{P}^1, \mathcal{O}_{\mathbb{P}^1}(d))^{n+1}$, we diagonalize the action of G on $H^0(\mathbb{P}^1, \mathcal{O}_{\mathbb{P}^1}(d))^{n+1}$. Given how G acts on N_d, the weight of G acting on $p_{i,r}$ is clearly $-(\lambda_i + r\hbar)$. Hence we get the same formulas as in Example 9.1.2.1, provided the weights $\{\lambda_i\}$ are replaced by $\{\lambda_i + r\hbar\}$. In particular, $(N_d)_G$ is the projectivization of the bundle $\oplus_{i,r}\mathcal{O}(-\lambda_i - r\hbar)$ over BG, and if κ is the equivariant hyperplane class of N_d, then in place of (9.4), we have the relation

$$(11.1) \qquad \prod_{i,r}(\kappa - \lambda_i - r\hbar) = 0$$

in $H_G^*(N_d)$. Furthermore, as a special instance of (9.9), we put

$$(11.2) \qquad \phi_{p_{i,r}} = \prod_{(j,s)\neq(i,r)} (\kappa - (\lambda_j + s\hbar)),$$

since the $\lambda_j + s\hbar$ are the weights of the linearized G-action on N_d. Recall that the $\phi_{p_{i,r}}$ are a basis of $H_G^*(N_d)$ as a module over $H^*(BG)$.

We note that any class $Q \in H_G^*(N_d)$ can be represented by a polynomial in $\kappa, \hbar$, and λ_i. Our discussion together with (9.5) shows that the class $i_{p_{i,r}}^*(Q)$ is the element of $H^*(BG) = \mathbb{C}[\hbar, \lambda_i]$ obtained by substituting $\lambda_i + r\hbar$ for κ. For this reason, it is reasonable to also denote $i_{p_{i,r}}^*(Q)$ by $Q(\lambda_i + r\hbar)$. Note that this is well-defined, by virtue of the relation (11.1). Similar notation will also be used when Q is an element of an appropriate localization (in the sense of commutative algebra) of the cohomology ring.

As a simple example of how the localization theorem applies to this situation, let's show that

$$(11.3) \qquad Q(\lambda_i + r\hbar) = i_{p_{i,r}}^*(Q) = \int_{(N_d)_G} \phi_{p_{i,r}} \cup Q,$$

where $\int_{(N_d)_G}$ is the equivariant integral defined immediately before Corollary 9.1.3. To see this, first note that if $N_{i,r}$ is the normal bundle of $p_{i,r}$ and $\text{Euler}_G(N_{i,r})$ is its equivariant Euler class, then (9.7) implies

$$(11.4) \qquad \operatorname{Euler}_G(N_{i,r}) = \prod_{(j,s) \neq (i,r)} (\lambda_i + r\hbar - (\lambda_j + s\hbar)).$$

Furthermore, adapting (9.10) to N_d, we easily see that

$$(11.5) \qquad i^*_{p_{j,s}}(\phi_{p_{i,r}}) = \begin{cases} \operatorname{Euler}_G(N_{i,r}) & \text{if } (j,s) = (i,r) \\ 0 & \text{otherwise.} \end{cases}$$

We can now apply Corollary 9.1.3 to conclude that

$$\int_{(N_d)_G} \phi_{p_{i,r}} \cup Q = \frac{i^*_{p_{i,r}}(\phi_{p_{i,r}} \cup Q)}{\operatorname{Euler}_G(N_{i,r})} = i^*_{p_{i,r}}(Q).$$

We should point out that the localization theorems from Chapter 9 use localizations (in the sense of commutative algebra) of equivariant cohomology. As in Section 9.1.2, let $\mathcal{R}_T$ be the field of fractions of $H^*(BT) \simeq \mathbb{C}[\lambda_i]$. Then the tensor product $H^*_G(N_d) \otimes_{H^*(BT)} \mathcal{R}_T$ will be denoted $\mathcal{R}_T H^*_G(N_d)$. This is the localization of $H^*_G(N_d)$ where nonzero polynomials in the λ_i are inverted. Similarly, if $\mathcal{R}_G$ is the field of fractions of $H(BG) \simeq \mathbb{C}[\lambda_i, \hbar]$, then we get the localization $\mathcal{R}_G H^*_G(N_d)$ where nonzero polynomials in the λ_i and $\hbar$ are inverted.

We now assert that there is a natural G-equivariant morphism

$$\varphi : M_d \longrightarrow N_d$$

described as follows. Consider a map $f : \mathbb{P}^1 \to \mathbb{P}^1 \times \mathbb{P}^n$ of bidegree $(1, d)$. This map represents an element of M_d. Reparametrizing the source $\mathbb{P}^1$ as needed, we can represent $f \in M_d$ by a map of the form

$$(11.6) \qquad (w_0, w_1) \mapsto ((w_1, w_0), (\alpha_0(w_0, w_1), ..., \alpha_n(w_0, w_1)),$$

where the α_i are homogeneous forms of degree d in the coordinates (w_0, w_1) of $\mathbb{P}^1$. With this representation of f, we define $\varphi(f) = (\alpha_0, ..., \alpha_n)$. We summarize this by saying that specifying a parametrized degree d map is equivalent to specifying an $(n+1)$-tuple of homogeneous degree d forms on $\mathbb{P}^1$ without a common factor.

The peculiar-looking switch of (w_0, w_1) to (w_1, w_0) arises because φ is required to be $\mathbb{C}^*$-equivariant. For $t \in \mathbb{C}^*$, $t \cdot f$ is given by

$$(w_0, w_1) \mapsto ((t^{-1}w_1, w_0), (\alpha_0, \dots, \alpha_n)) = ((w_1, tw_0), (\alpha_0, \dots, \alpha_n)).$$

To calculate $\varphi(t \cdot f)$, we must put $t \cdot f$ in the standard form (11.6). We do this by replacing w_0 by $t^{-1}w_0$. The result is

$$\varphi(t \cdot f) = (\alpha_0(t^{-1}w_0, w_1), ..., \alpha_n(t^{-1}w_0, w_1)),$$

which equals $t \cdot \varphi(f)$ as claimed.

For an arbitrary stable map f, the description of $\varphi(f) = (\alpha_0, \dots, \alpha_n)$ comes from studying contracted components. Let $p_1 : \mathbb{P}^1 \times \mathbb{P}^n \to \mathbb{P}^1$ and $p_2 : \mathbb{P}^1 \times \mathbb{P}^n \to \mathbb{P}^n$ be the two projection maps, and suppose that $p_1 \circ f$ contracts a component to a point $p \in \mathbb{P}^1$, while $p_2 \circ f$ has degree d' on this component. In this case, one can show that the α_i have a common factor $L(w_0, w_1)^{d'}$, where L is a linear function vanishing at p. We will use this fact in the proof of Theorem 11.1.4 below. For a more explicit formula for φ, see [**Givental2**] or [**LLY**]. These references also prove that φ is a morphism on M_d (the proof given in [**LLY**] is due to J. Li). In [**Givental2**], what is actually constructed is a map $L_d \to N_d$, where L_d is closely related to M_d. We will define L_d during the proof of Lemma 11.2.12.

11.1.3. Euler Data. Now fix an integer $\ell \geq 1$ and set $\mathcal{V} = \mathcal{O}_{\mathbb{P}^n}(\ell)$. As in Section 11.1.1, we have a diagram

$$
\begin{array}{ccc}
\overline{M}_{0,1}(\mathbb{P}^n, d) & \xrightarrow{\;e_1\;} & \mathbb{P}^n \\
\big\downarrow{\scriptstyle \pi_1} & & \\
\overline{M}_{0,0}(\mathbb{P}^n, d) & &
\end{array}
$$

(11.7)

and we define

$$
\mathcal{V}_d = \pi_{1*} e_1^*(\mathcal{V}).
$$

This is a bundle on the stack $\overline{\mathcal{M}}_{0,0}(\mathbb{P}^n, d)$, though as in Section 11.1.1 we will abuse notation and regard $\mathcal{V}_d$ as a bundle on $\overline{M}_{0,0}(\mathbb{P}^n, d)$.

We now define some important equivariant cohomology classes. Consider the natural projection $\pi : M_d \to \overline{M}_{0,0}(\mathbb{P}^n, d)$ and define

$$
\chi_d^{\mathcal{V}} = \mathrm{Euler}_G(\pi^*(\mathcal{V}_d)) \in H_G^*(M_d).
$$

Using the above map $\varphi : M_d \to N_d$, we define

(11.8)
$$
\hat{Q}_d = \varphi_!(\chi_d^{\mathcal{V}}) \in H_G^*(N_d),
$$

the equivariant pushforward of $\chi_d^{\mathcal{V}}$. We can regard $\hat{Q}_d$ as the "linearization" of $\chi_d^{\mathcal{V}}$. We next define

(11.9)
$$
\hat{P}_d = \prod_{m=0}^{\ell d} (\ell \kappa - m \hbar) \in H_G^*(N_d).
$$

The definitions of $\hat{P}_d$ and $\hat{Q}_d$ are for the bundle $\mathcal{V} = \mathcal{O}_{\mathbb{P}^n}(\ell)$. In the special case when $\mathcal{V} = \mathcal{O}_{\mathbb{P}^4}(5)$, the equivariant cohomology classes $\hat{P}_d$ and $\hat{Q}_d$ will play a central role in the proof of Theorem 11.1.1.

The paper [**LLY**] defines $\hat{P}_d$ and $\hat{Q}_d$ for a much wider class of bundles $\mathcal{V}$ than considered here. If we think of $\mathcal{O}_{\mathbb{P}^n}(\ell)$ as associated to a hypersurface, then the types of bundles considered in [**LLY**] include not only those associated to complete intersections in $\mathbb{P}^n$ but also *concavex* bundles. We will define concavex bundles in the remarks at the end of Section 11.1.5 below.

The way we've set things up, the bundle $\mathcal{V} = \mathcal{O}_{\mathbb{P}^n}(\ell)$ gives rise to the infinite sequence of bundles $\{\mathcal{V}_d\}_{d=1}^{\infty}$. We also define $\mathcal{V}_d$ for $d = 0$ by letting $\mathcal{V}_0$ denote the bundle $\mathcal{V}$ on $\mathbb{P}^n$. It is important to note that although these bundles live on different spaces, there are still nontrivial relations among them. To see how this works, pick $0 < r < d$ and define $\overline{M}(r, d - r)$ by the pullback diagram

$$
\begin{array}{ccc}
\overline{M}(r, d-r) & \xrightarrow{\;p_1\;} & \overline{M}_{0,1}(\mathbb{P}^n, r) \\
{\scriptstyle p_2}\big\downarrow & & \big\downarrow{\scriptstyle e_1} \\
\overline{M}_{0,1}(\mathbb{P}^n, d-r) & \xrightarrow{\;e_1\;} & \mathbb{P}^n
\end{array}
$$

(11.10)

where e_1 is the usual evaluation map and the p_i are projections. We can think of $\overline{M}(r, d-r)$ as consisting of pairs of 1-pointed stable curves $((f_1, C_1, p_1), (f_2, C_2, p_2))$ such that $f_1(p_1) = f_2(p_2)$. Then we have a diagram

$$
\begin{array}{ccc}
\overline{M}(r, d-r) & \xrightarrow{\;\psi\;} & \overline{M}_{0,0}(\mathbb{P}^n, d) \\
{\scriptstyle e_1}\big\downarrow & & \\
\mathbb{P}^n & &
\end{array}
$$

(11.11)

where e_1 is the evaluation map (well-defined by the definition of $\overline{M}(r, d - r)$). Also, the horizontal map takes $((f_1, C_1, p_1), (f_2, C_2, p_2))$ to the 0-pointed stable map (f, C), where C is the curve obtained from $C_1 \cup C_2$ by identifying p_1 and p_2, and $f|_{C_i} = f_i$ (again well-defined by how we defined $\overline{M}(r, d - r)$).

When $r = 0$ or $r = d$, we must modify the definition of $\overline{M}(r, d - r)$. Since $\overline{M}_{0,1}(\mathbb{P}^n, 0)$ is empty, we replace $e_1 : \overline{M}_{0,1}(\mathbb{P}^n, 0) \to \mathbb{P}^n$ by $1_{\mathbb{P}^n} : \mathbb{P}^n \to \mathbb{P}^n$ in (11.10). For $r = 0$, this implies $\overline{M}(0, d) = \overline{M}_{0,1}(\mathbb{P}^n, d)$, where in (11.10), p_1 is identified with the evaluation map e_1 and p_2 is the identity. Also, in (11.11), note that ψ is identified with the map $\pi_1 : \overline{M}_{0,1}(\mathbb{P}^n, d) \to \overline{M}_{0,0}(\mathbb{P}^n, d)$ forgetting the marked point. The case when $r = d$ is similar.

Given this setup, we get the following *gluing lemma*.

LEMMA 11.1.2. *If b is any multiplicative equivariant characteristic class for vector bundles (such as the equivariant Euler class), then on $\overline{M}(r, d - r)$ we have*

$$e_1^* b(\mathcal{V}) \psi^* b(\mathcal{V}_d) = p_1^* \pi_1^* b(\mathcal{V}_r) p_2^* \pi_1^* b(\mathcal{V}_{d-r}).$$

(To make sense of this when $r = 0$ or $r = d$, we must interpret $\pi_1^* b(\mathcal{V}_0)$ as $b(\mathcal{V}_0) = b(\mathcal{V})$ whenever it occurs.)

PROOF. The lemma is trivial when $r = 0, d$. Now suppose $0 < r < d$ and let $((f_1, C_1, p_1), (f_2, C_2, p_2))$ map to (f, C) as above. Then we have the exact sequence

$$0 \longrightarrow f^* \mathcal{V} \longrightarrow f_1^* \mathcal{V} \oplus f_2^* \mathcal{V} \longrightarrow \mathcal{V}|_p \longrightarrow 0,$$

where $p = f_1(p_1) = f_2(p_2)$. This in turn gives the exact sequence

$$0 \longrightarrow H^0(C, f^* \mathcal{V}) \longrightarrow H^0(C_1, f_1^* \mathcal{V}) \oplus H^0(C_2, f_2^* \mathcal{V}) \longrightarrow \mathcal{V}|_p \longrightarrow 0.$$

The lemma now follows without difficulty. $\qquad\qquad\square$

The gluing lemma gives an identity which relates the equivariant Euler classes $\mathrm{Euler}_G(\mathcal{V}_d) \in H_G^*(\overline{M}_{0,0}(\mathbb{P}^n, d))$. Since $\hat{Q}_d \in H_G^*(N_d)$ is defined using $\mathrm{Euler}_G(\mathcal{V}_d)$, it is reasonable to ask if these classes satisfy a similar identity. But classes in $H_G^*(N_d)$ are easy to study. The fixed point set is simple—it consists of the isolated points $p_{i,r}$—so that by the localization theorem, the class $\hat{Q}_d$ is uniquely determined by the $i_{p_{i,r}}^*(\hat{Q}_d)$. This suggests that we seek an identity for the $i_{p_{i,r}}^*(\hat{Q}_d)$ which is similar to Lemma 11.1.2.

We will make this precise in Definition 11.1.3 below. But before we can state the definition, we need the map $\overline{} : N_d \to N_d$ defined by

$$(\alpha_0(w_0, w_1), \dots, \alpha_n(w_0, w_1)) \mapsto (\alpha_0(w_1, w_0), \dots, \alpha_n(w_1, w_0)).$$

This map is clearly T-equivariant, but it is not $\mathbb{C}^*$-equivariant. In [**LLY**], it is shown that $\overline{} : H_G^*(N_d) \to H_G^*(N_d)$ satisfies

$$\overline{\kappa} = \kappa - d\hbar$$

(11.12)
$$\overline{\hbar} = -\hbar$$

$$\overline{\lambda_i} = \lambda_i.$$

We also recall from Example 9.1.2.1 that the fixed points of the action of G (or T) on $\mathbb{P}^n$ are denoted q_i.

DEFINITION 11.1.3. *Let $\Omega \in \mathcal{R}_T H_T^*(\mathbb{P}^n)$ be an invertible class. A sequence $Q = \{Q_d\}$ with $Q_d \in \mathcal{R}_T H_G^*(N_d)$, $d > 0$ is called an Ω-Euler data if for all $r = 0, ..., d$ and $i = 0, ..., n$, we have*

$$i_{q_i}^*(\Omega)i_{p_{i,r}}^*(Q_d) = \overline{i_{p_{i,0}}^*(Q_r)}\, i_{p_{i,0}}^*(Q_{d-r}),$$

where $Q_0 = \Omega$. The set consisting of all Ω-Euler data is denoted A^Ω.

We noted earlier that $i_{p_{i,r}}^*(Q_d) = Q_d(\lambda_i + r\hbar)$, and we know from (9.6) that $i_{q_i}^*(\Omega) = \Omega(\lambda_i)$. Hence the above equation can be written

$$\Omega(\lambda_i)Q_d(\lambda_i + r\hbar) = \overline{Q_r(\lambda_i)}Q_{d-r}(\lambda_i).$$

Note that $\overline{\Omega} = \Omega$ since $\Omega \in \mathcal{R}_T H_T^*(\mathbb{P}^n)$. So this equation is trivial when $r = 0$.

Below, we will see that the sequences $\hat{Q} = \{\hat{Q}_d\}$ and $\hat{P} = \{\hat{P}_d\}$ defined in (11.8) and (11.9) are Euler data for $\Omega = \mathrm{Euler}_T(\mathcal{V}) = \mathrm{Euler}_T(\mathcal{O}_{\mathbb{P}^n}(\ell)) = \ell p$, where p is the equivariant hyperplane class. The name "Euler data" was chosen in part because $\hat{Q}$ is defined using the equivariant Euler class.

We first give the easy proof that $\hat{P}$ is an ℓp-Euler data.

Example 11.1.3.1. We first note that $\Omega = \ell p$ is invertible in $\mathcal{R}_T H_G^*(\mathbb{P}^n)$. This can be seen by localization (Proposition 9.1.2), since the restrictions $i_{q_i}^*(\Omega) = \Omega(\lambda_i) = \ell\lambda_i$ are all invertible. Also recall from (11.9) that

$$\hat{P}_d = \prod_{m=0}^{\ell d} (\ell\kappa - m\hbar).$$

Then $\hat{P} = \{\hat{P}_d\}$ being an ℓp-Euler data is equivalent to the identity

$$\ell\lambda_i \prod_{m=0}^{\ell d} (\ell(\lambda_i + r\hbar) - m\hbar) = \prod_{m=0}^{\ell r} (\ell\lambda_i + m\hbar) \prod_{m=0}^{\ell(d-r)} (\ell\lambda_i - m\hbar),$$

which is immediately verified.

It will take more work to show that $\hat{Q}$ is an ℓp-Euler data.

THEOREM 11.1.4. *The sequence $\hat{Q} = \{\hat{Q}_d\} = \{\varphi_! \chi_d^\mathcal{V}\}$ is an ℓp-Euler data.*

PROOF. We need to show that

$$(11.13) \qquad \ell\lambda_i\, \hat{Q}_d(\lambda_i + r\hbar) = \overline{\hat{Q}_r(\lambda_i)}\, \hat{Q}_{d-r}(\lambda_i) \quad \text{for all } r, i.$$

We begin by computing $i_{p_{i,r}}^*(\hat{Q}_d)$, which is given by

$$(11.14) \qquad \hat{Q}_d(\lambda_i + r\hbar) = \int_{(N_d)_G} \phi_{p_{i,r}}\hat{Q}_d = \int_{(M_d)_G} \varphi^*(\phi_{p_{i,r}})\chi_d^\mathcal{V}.$$

The first equality is from (11.3), while the second equality follows from the definition of $\hat{Q}_d$ and the projection formula. Our strategy will be to compute the integral on the right hand side of (11.14) using the localization formula from Corollary 9.1.3.

We first identify which G-fixed components in M_d contribute to this integral. Let $i_\Gamma : \overline{M}_\Gamma \subset M_d$ denote the component corresponding to the graph Γ. Then, when we apply Corollary 9.1.3 to this integral, $\overline{M}_\Gamma$ will contribute a term which has

$$i_\Gamma^*(\varphi^*(\phi_{p_{i,r}})) = (\varphi|_{\overline{M}_\Gamma})^*\big(i_{\varphi(\overline{M}_\Gamma)}^*(\phi_{p_{i,r}})\big)$$

as a factor. Since $\varphi(\overline{M}_\Gamma)$ must be one of the fixed points $p_{j,s}$, the term corresponding to $\overline{M}_\Gamma$ vanishes if $(j,s) \neq (i,r)$ by the definition of $\phi_{p_{i,r}}$ from (11.2).

It remains to determine the components of the G-fixed point set which map to $p_{i,r}$ under φ. For this purpose, we will use a variant of the map ψ from (11.11). Given a point of $\overline{M}(r, d - r)$ represented by $f_j : (C_j, p_j) \to \mathbb{P}^n$ with $f_1(p_1) = f_2(p_2) = q \in \mathbb{P}^n$, define $C = C_0 \cup C_1 \cup C_2$, with $C_0 = \mathbb{P}^1$, and $0 \in \mathbb{P}^1$ is glued to $p_1 \in C_1$ while $\infty \in \mathbb{P}^1$ is glued to $p_2 \in C_2$. Then define $f : C \to \mathbb{P}^1 \times \mathbb{P}^n$ by

$$f|_{C_0}(z) = (z, q)$$
$$f|_{C_1}(z) = (0, f_1(z))$$
$$f|_{C_2}(z) = (\infty, f_2(z)).$$

Then $f \in M_d$, so that this procedure defines a map

$$(11.15) \qquad \bar{\psi} : \overline{M}(r, d - r) \longrightarrow M_d.$$

Let $\{F_k\}$ be the set of T-fixed components of $\overline{M}_{0,1}(\mathbb{P}^n, k)$ whose marked point is mapped to q_i. Then $F_r \times F_{d-r} \subset \overline{M}(r, d - r)$, and we will identify this set with its image under $\bar{\psi}$. Thus we will write $F_r \times F_{d-r} \subset M_d$.

We claim that the $F_r \times F_{d-r}$, as we vary over all possible F_r and F_{d-r}, are precisely the components of the G-fixed point set which map to $p_{i,r}$ under φ. To prove this, take $f \in F_r \times F_{d-r}$. By definition, $p_1 \circ f$ contracts C_0, $p_2 \circ f$ has degree r on C_1, and $p_2 \circ f$ has degree $d - r$ on C_2. Then the description of $\varphi(f)$ given in Section 11.1.2 shows that $\varphi(f) = p_{i,r}$. Using the more detailed description of φ given in [**LLY**], one can show further that $F_r \times F_{d-r}$ is a component of the G-fixed point set and that all components in $\varphi^{-1}(p_{i,r})$ arise this way.

It follows that when we compute the integral on the right in (11.14) using localization, we need only use components of the form $F_r \times F_{d-r}$. We will first treat the case where $r \neq 0, d$. The following notation will be useful. Let $\mathcal{L}_{k,1}$ denote the line bundle on $\overline{M}_{0,1}(\mathbb{P}^n, k)$ whose fiber at $f : (C, p) \to \mathbb{P}^n$ is the cotangent space $T_p^*(C)$, as defined at the beginning of Section 10.1.1. Also, for any k and component F_k, we let $N(F_k)$ denote the normal bundle of F_k in $\overline{M}_{0,1}(\mathbb{P}^n, k)$. Finally, we let $\pi_1 : \overline{M}_{0,1}(\mathbb{P}^n, k) \to \overline{M}_{0,0}(\mathbb{P}^n, k)$ be the map in (11.7) which was used in the definition of the bundle $\mathcal{V}_k$. Then, applying Corollary 9.1.4 to the integral on the right hand side of (11.14), we claim that

$$\hat{Q}_d(\lambda_i + r\hbar) = \int_{(M_d)_G} \varphi^*(\phi_{p_{ir}}) \chi_d^\vee$$

$$= -\hbar^{-2}(\ell\lambda_i)^{-1} i_{p_{i,r}}^*(\phi_{i,r}) \prod_{j \neq i}(\lambda_i - \lambda_j) \times$$

$$(11.16) \qquad \sum_{F_r} \int_{(F_r)_G} \frac{\pi_1^* \mathrm{Euler}_G(\mathcal{V}_r)}{a_r \mathrm{Euler}_G(N(F_r))(\hbar - c_1^G(\mathcal{L}_{r,1}))} \times$$

$$\sum_{F_{d-r}} \int_{(F_{d-r})_G} \frac{\pi_1^* \mathrm{Euler}_G(\mathcal{V}_{d-r})}{a_{d-r} \mathrm{Euler}_G(N(F_{d-r}))(-\hbar - c_1^G(\mathcal{L}_{d-r,1}))},$$

where a_k is the order of the finite group of automorphisms associated to F_k. (In [**LLY**], the a_k are included in the definition of integration over orbifolds and hence don't appear explicitly in their formulas.) We now outline this calculation, omitting many of the details.

The automorphisms fixing a generic stable map $f \in F_r \times F_{d-r}$ can only arise from automorphisms of $f|_{C_1} \in F_r$ and $f|_{C_2} \in F_{d-r}$, so that the group of automorphisms needed in applying the localization formula has order $a_r a_{d-r}$. This explains the factors of a_r and a_{d-r} in the denominators in (11.16).

We next need the equivariant Euler class of the normal bundle $N(F_r \times F_{d-r})$ of $F_r \times F_{d-r}$ in M_d. This computation is the most technical part of the argument, since there are several contributions to $\mathrm{Euler}_G(N(F_r \times F_{d-r}))$. We will be sketchy at times and refer the reader to [**LLY**] for a more complete proof. To study this normal bundle, let $M^i(k) \subset \overline{M}_{0,1}(\mathbb{P}^n, k)$ be the set of stable maps which take the marked point to $q_i \in \mathbb{P}^n$. Note that the normal bundle of this embedding is the pullback of $T_{q_i}\mathbb{P}^n$ via e_1. Then consider the following diagram:

$$F_r \times F_{d-r} \;\longrightarrow\; M^i(r) \times M^i(d-r) \;\longrightarrow\; \overline{M}_{0,1}(\mathbb{P}^n, r) \times \overline{M}_{0,1}(\mathbb{P}^n, d-r)$$
$$\downarrow$$
$$\overline{M}(r, d-r)$$
$$\downarrow \bar{\psi}$$
$$M_d$$

where $\bar{\psi}$ is from (11.15). The definition of $\overline{M}(r, d-r)$ given in (11.10) shows that $e_1^{-1}(q_i) = M^i(r) \times M^i(d-r) \subset \overline{M}(r, d-r)$, where e_1 is the evaluation map from (11.11). Hence the normal bundle of this embedding is also the pullback of $T_{q_i}\mathbb{P}^n$.

If we then compute the normal bundle of $F_r \times F_{d-r} \subset M_d$, the above diagram implies that in the equivariant K-group of $F_r \times F_{d-r}$, we have

$$N(F_r \times F_{d-r}) = N(F_r) + N(F_{d-r}) - 2T_{q_i}\mathbb{P}^n + T_{q_i}\mathbb{P}^n + N(\bar{\psi}).$$

In this equation, we have omitted various pullbacks for clarity. When we take equivariant Euler classes, we obtain (again omitting the pullbacks)

$$\mathrm{Euler}_G(N(F_r \times F_{d-r})) = \frac{\mathrm{Euler}_G(N(F_r))\mathrm{Euler}_G(N(F_{d-r}))}{\prod_{j \neq i}(\lambda_i - \lambda_j)}\mathrm{Euler}_G(N(\bar{\psi}))$$

since $\mathrm{Euler}_G(T_{q_i}\mathbb{P}^n) = \prod_{j \neq i}(\lambda_i - \lambda_j)$ by (9.7). This appears in the denominator in the localization theorem, so that we can explain three more factors in (11.16).

It remains to consider the normal bundle of the map $\bar{\psi}$. First, we consider node deformations, which we studied in Lemma 9.2.2. For $p_1 = C_0 \cap C_1$, the lemma tells us to look at $T_0 C_0 \otimes T_{p_1} C_1$. Since $C_0 \simeq \mathbb{P}^1$ equivariantly and $\mathcal{L}_{r,1}$ is defined using the cotangent bundle of C_1, the weights are $\hbar$ and $-c_1^G(\mathcal{L}_{r,1})$. This explains the factor of $\hbar - c_1^G(\mathcal{L}_{r,1})$ in the denominator in (11.16), and deforming $p_2 = C_0 \cap C_2$ similarly leads to the factor of $-\hbar - c_1^G(\mathcal{L}_{d-r,1})$ in the other denominator.[1] Finally, we study the deformations of $f|_{C_0}$. We know $C_0 = \mathbb{P}^1$, and the definitions of $\overline{M}(r, d-r)$ and $\bar{\psi}$ show that such f's take 0 and ∞ to $(0, q)$ and (∞, q) respectively, for some $q \in \mathbb{P}^n$. We can deform this map by allowing the first coordinates of $(0, q)$ and (∞, q) to vary. We identify these normal deformations to the infinitesimal automorphisms of $\mathbb{P}^1$ modulo those preserving 0 and ∞. Using standard coordinates (z_0, z_1) on $\mathbb{P}^1$, this is given by the vector fields $z_0 \partial/\partial z_1$ and $z_1 \partial/\partial z_0$, which gives a factor of $-\hbar^2$. Since this appears in the denominator, we get the factor of $-\hbar^{-2}$ in (11.16).

The final ingredient in using the localization formula is to identify the restriction of $\varphi^*(\phi_{i,r})\chi_d^{\mathcal{V}} = \varphi^*(\phi_{i,r})\pi^*(\mathrm{Euler}_G(\mathcal{V}_d))$ to $F_r \times F_{d-r}$. Since $\varphi(F_r \times F_{d-r}) = p_{i,r}$, the restriction of $\psi^*(\psi_{i,r})$ is $i_{p_{i,r}}^*(\psi_{i,r})$, which is one of the factors in (11.16). As for

---

[1]This is related to Lemma 10.2.8 and explains the $\hbar - c$ in formulas such as (10.15).

the restriction of $\pi^*(\mathrm{Euler}_G(\mathcal{V}_d))$, the gluing lemma (Lemma 11.1.2) implies that on $F_r \times F_{d-r}$, we get the identity

$$\mathrm{Euler}_G(\mathcal{V})\mathrm{Euler}_G(\mathcal{V}_d) = \mathrm{Euler}_G(\mathcal{V}_r)\mathrm{Euler}_G(\mathcal{V}_{d-r}),$$

where we have omitted the pullbacks for convenience. Since $\mathrm{Euler}_G(\mathcal{V}) = \ell p$, we see that the restriction of $\chi_d^{\mathcal{V}}$ to $F_r \times F_{d-r}$ is

$$(\ell p)^{-1}\pi_1^*\mathrm{Euler}_G(\mathcal{V}_r)\,\pi_1^*\mathrm{Euler}_G(\mathcal{V}_{d-r}).$$

These factors appear in the numerator in the localization formula and complete our explanation of (11.16).

Performing a similar analysis for $r = 0$ leads to the formula

$$(11.17) \quad \hat{Q}_d(\lambda_i) = -\hbar^{-1}i_{p_{i,0}}^*(\phi_{p_{i,0}}) \sum_{F_d} \int_{(F_d)_G} \frac{\pi_1^*\mathrm{Euler}_G(\mathcal{V}_d)}{a_d\mathrm{Euler}_G(N(F_d))(-\hbar - c_1^G(\mathcal{L}_{d,1}))}.$$

We can now prove the theorem. If we consider the above formula with d replaced by r and apply (11.12), we obtain

$$\overline{\hat{Q}_r(\lambda_i)} = \hbar^{-1}\overline{i_{p_{i,0}}^*(\phi_{p_{i,0}})} \sum_{F_r} \int_{(F_r)_G} \frac{\pi_1^*\mathrm{Euler}_G(\mathcal{V}_r)}{a_r\mathrm{Euler}_G(N(F_r))(\hbar - c_1^G(\mathcal{L}_{r,1}))},$$

and furthermore, (11.17) with d replaced with $d - r$ gives

$$\hat{Q}_{d-r}(\lambda_i) = -\hbar^{-1}i_{p_{i,0}}^*(\phi_{p_{i,0}}) \sum_{F_{d-r}} \int_{(F_{d-r})_G} \frac{\pi_1^*\mathrm{Euler}_G(\mathcal{V}_{d-r})}{a_{d-r}\mathrm{Euler}_G(N(F_{d-r}))(-\hbar - c_1^G(\mathcal{L}_{d-r,1}))}.$$

If we multiply these two equations and compare the result with ℓp multiplied by (11.16), we see that our desired equation (11.13) reduces to the identity

$$i_{p_{i,r}}^*(\phi_{i,r}) \prod_{j \neq i}(\lambda_i - \lambda_j) = \overline{i_{p_{i,0}}^*(\phi_{p_{i,0}})}\, i_{p_{i,0}}^*(\phi_{p_{i,0}}).$$

However, recall from (11.5) that $i_{p_{i,r}}^*(\phi_{p_{i,r}})$ is just $\mathrm{Euler}_G(N_{i,r})$, which was computed in (11.4). Hence the above identity reduces to

$$\prod_{(j,s)\neq(i,r)} (\lambda_i - \lambda_j + (r-s)\hbar) \prod_{j\neq i}(\lambda_i - \lambda_j) =$$

$$\prod_{(j,s)\neq(i,0)} (\lambda_i - \lambda_j - s\hbar) \prod_{(j,s)\neq(i,0)} (\lambda_i - \lambda_j + s\hbar),$$

where on the left hand side, j ranges from 0 to d, while on the right hand side, j ranges from 0 to $d - r$ in the first product and 0 to r in the second. This identity is easily verified, which completes the proof when $r \neq 0, d$.

Since the Euler identity is trivial when $r = 0$ (see the discussion following Definition 11.1.3), we only have to consider $r = d$. Here, one derives a formula for $\hat{Q}_d(\lambda_i + d\hbar)$ similar to the formula for $\hat{Q}_d(\lambda_i)$ given in (11.17), and comparing these formulas easily gives the desired identity. Details can be found in [**LLY**]. $\qquad\square$

11.1.4. Linked Euler Data. We now turn to an important concept. Let $\mathcal{S}$ be the set of sequences $P = \{P_d\}_{d=1}^{\infty}$ with $P_d \in \mathcal{R}_G H_G^*(N_d)$ for all d, and note that $i_{p_{i,0}}^*$ maps $\mathcal{R}_G H_G^*(N_d)$ to $\mathcal{R}_G$.

DEFINITION 11.1.5. *Two sequences $P, Q \in \mathcal{S}$ are linked if $i_{p_{i,0}}^*(P_d - Q_d) \in \mathcal{R}_G$ vanishes at $\hbar = (\lambda_i - \lambda_j)/d$ for all $i \neq j$, $d > 0$.*

To make this definition precise, we need to explain what $\hbar = (\lambda_i - \lambda_j)/d$ means. Define the subgroup $G' \subset G$ by

$$G' = \{(t, t_0, \ldots, t_n) : t^d = t_i/t_j\}.$$

The inclusion $G' \hookrightarrow G$ induces a map $BG' \to BG$, which in turn induces a pullback on cohomology $\mathrm{res}_{ij} : H^*(BG) \to H^*(BG')$. It is easy to see from our discussion in Section 9.1.1 that res_{ij} is induced by restriction of characters. This implies $\mathrm{res}_{ij}(\hbar) = \mathrm{res}_{ij}((\lambda_i - \lambda_j)/d)$, which we sometimes abbreviate more simply as $\hbar = (\lambda_i - \lambda_j)/d$, understanding that we are interpreting these symbols as living in G'-equivariant cohomology. Similarly, if a space X has a G-action, we get a restriction map $\mathrm{res}_{ij} : H_G^*(X) \to H_{G'}^*(X)$. Then we get the composition

$$(11.18) \qquad H_G^*(N_d) \xrightarrow{i_{p_{i,0}}^*} H_G^*(p_{i,0}) \xrightarrow{\mathrm{res}_{ij}} H_{G'}^*(p_{i,0}),$$

and the condition of Definition 11.1.5 means that after tensoring with $\mathcal{R}_G$, we have $\mathrm{res}_{ij} \circ i_{p_{i,0}}^*(P_d - Q_d) = 0$ for all $i \neq j$, d. For later purposes, we note that the above composition equals

$$(11.19) \qquad H_G^*(N_d) \xrightarrow{\mathrm{res}_{ij}} H_{G'}^*(N_d) \xrightarrow{i_{p_{i,0}}^*} H_{G'}^*(p_{i,0}),$$

so that being linked also means that $i_{p_{i,0}}^* \circ \mathrm{res}_{ij}(P_d - Q_d) = 0$ for all $i \neq j$, d after tensoring with $\mathcal{R}_G$.

The subgroup G' is of interest because it fixes the space of multiple covers of the line $\overline{q_i q_j} \subset \mathbb{P}^n$. Consider the line $L_{i,j} \subset N_d$ given parametrically by

$$L_{i,j} = \{(0, \ldots, 0, a\, w_0^d, 0, \ldots, 0, b\, w_1^d, 0, \ldots, 0)\},$$

where $a\, w_0^d$ is in the jth position and $b\, w_1^d$ is in the ith position. Note that $p_{i,0} \in L_{i,j}$, corresponding to $a = 0$. A general point of $L_{i,j}$ is the image under φ of a degree d multiple cover of the line $\overline{q_i q_j} \subset \mathbb{P}^n$. The subgroup G' fixes $L_{i,j}$ pointwise, which implies $H_{G'}^*(L_{i,j}) = H^*(L_{i,j}) \otimes H^*(BG')$. It follows that if $u \in L_{i,j}$ is any point, the restriction map

$$H_{G'}^*(L_{i,j}) \longrightarrow H_{G'}^*(u)$$

is independent of u. Thus the maps in the diagram

$$H_G^*(N_d) \xrightarrow{\mathrm{res}_{ij}} H_{G'}^*(N_d) \longrightarrow H_{G'}^*(L_{i,j})$$
$$\searrow^{i_u^*} \qquad \downarrow$$
$$H_{G'}^*(u)$$

are also independent of u. Since (11.19) is the special case when $u = p_{i,0}$, we see that in the "restriction" map $i_{p_{i,0}}^* \circ \mathrm{res}_{ij}$ in the discussion following (11.19), we can replace $p_{i,0}$ by any point $u \in L_{i,j}$. Hence P and Q are linked if and only if they agree on multiple covers of coordinate lines in $\mathbb{P}^n$.

We now state a key result.

THEOREM 11.1.6. *At $\hbar = (\lambda_i - \lambda_j)/d$, $i \neq j$, we have*

$$i_{p_{i,0}}^*(\hat{Q}_d) = \prod_{m=0}^{\ell d} \left(\ell\lambda_i - m\frac{(\lambda_i - \lambda_j)}{d} \right).$$

In particular, the Euler data $\hat{P}$ and $\hat{Q}$ are linked.

PROOF. We first recall that

$$\hat{P}_d = \prod_{m=0}^{\ell d} (\ell\kappa - m\hbar).$$

By (11.18), we can think of $\hbar = (\lambda_i - \lambda_j)/d$ in terms of the composition $\mathrm{res}_{ij} \circ i^*_{p_{i,0}}$. Then $\mathrm{res}_{ij} \circ i^*_{p_{i,0}}(\hbar) = (\lambda_i - \lambda_j)/d$, and Section 11.1.2 implies $i^*_{p_{i,0}}(\kappa) = \lambda_i$. Hence

$$\mathrm{res}_{ij} \circ i^*_{p_{i,0}}(\hat{P}_d) = \prod_{m=0}^{\ell d} \left(\ell\lambda_i - m\frac{(\lambda_i - \lambda_j)}{d}\right),$$

and the final part of the theorem will follow once we compute $\mathrm{res}_{ij} \circ i^*_{p_{i,0}}(\hat{Q}_d)$.

We proved above that the restriction of $\hat{Q}_d$ can also be represented as the composition $i^*_u \circ \mathrm{res}_{ij}(\hat{Q}_d)$ for any $u \in L_{i,j}$. In particular, we can choose

$$u = (0, \ldots, 0, w_0^d, 0, \ldots, 0, w_1^d, 0, \ldots, 0) \in L_{i,j} \subset N_d,$$

where w_0^d and w_1^d are in the jth and i^{th} positions respectively.

The strategy of the proof is that since $\hat{Q}_d = \varphi_!(\chi_d^\vee)$, we can do the calculation on M_d. For u as above, consider the map $f : \mathbb{P}^1 \to \mathbb{P}^1 \times \mathbb{P}^n$ be defined by

$$f(w_0, w_1) = ((w_1, w_0), (0, \ldots, 0, w_0^d, 0, \ldots, 0, w_1^d, 0, \ldots, 0)).$$

This is an element of M_d, and we clearly have $\varphi(f) = u$. The key fact is that the map $\varphi : M_d \to N_d$ is a local isomorphism in a neighborhood of these points. Now consider the commutative diagram:

$$
\begin{array}{ccccc}
H_G^*(M_d) & \xrightarrow{\mathrm{res}_{ij}} & H_{G'}^*(M_d) & \xrightarrow{i_f^*} & H_{G'}^*(f) \\
\varphi_! \downarrow & & \varphi_! \downarrow & & \varphi_! \downarrow \\
H_G^*(N_d) & \xrightarrow{\mathrm{res}_{ij}} & H_{G'}^*(N_d) & \xrightarrow{i_u^*} & H_{G'}^*(u)
\end{array}
$$

Since the vertical map on the right is an isomorphism, it follows that it suffices to compute $i_f^* \circ \mathrm{res}_{ij}(\chi_d^\vee)$.

Recall that $\chi_d^\vee$ is the equivariant Euler class of $\pi^* V_d$, so that over f, we need to compute the weights of the G'-action on $H^0(\mathbb{P}^1, f^* \mathcal{O}_{\mathbb{P}^n}(\ell))$. A basis for these sections is given by $w_0^m w_1^{\ell d - m}$ for $0 \le m \le \ell d$. Since f has degree d, we see that w_0 has weight λ_j/d while w_1 has weight λ_i/d. This is similar to what happened in Chapter 9 in the discussion following (9.20). Thus the section $w_0^m w_1^{\ell d - m}$ has weight

$$m\left(\frac{\lambda_j}{d}\right) + (\ell d - m)\frac{\lambda_i}{d} = \ell\lambda_i - m\left(\frac{\lambda_i - \lambda_j}{d}\right).$$

Multiplying these weights together gives the desired formula for $i^*_{p_{i,0}}(\hat{Q}_d)$ at $\hbar = (\lambda_i - \lambda_j)/d$. $\square$

We also need a uniqueness result.

THEOREM 11.1.7. *Suppose that P, Q are linked Ω-Euler data with the property that $\deg_\hbar i^*_{p_{i,0}}(P_d - Q_d) \le (n+1)d - 2$ for all $0 \le i \le n$ and $d \ge 1$. Then $P = Q$.*

PROOF. We will give only a brief sketch of the proof. A complete argument requires the Reciprocity Lemma [**LLY**, Lemma 2.4], which studies the consequences of being Euler data. The proof begins by noting that $\{\kappa^s\} = \{1, \kappa, \kappa^2, \ldots\}$ generates

$H_G^*(N_d)$ as an $H^*(BG)$ module. Since the intersection pairing is nondegenerate, it suffices to show that

$$(11.20) \qquad \int_{(N_d)_G} \kappa^s(P_d - Q_d) = 0 \quad \text{for all } s \geq 0.$$

Let L_s denote the integral on the left hand side of the equation. The definition of Euler data implies that P_d and Q_d depend polynomially on $\hbar$, which implies that L_s is a polynomial in $\hbar$. This will be important below.

The proof is by induction on d, so that we assume $P_j = Q_j$ for $0 \leq j \leq d-1$ (recall that $P_0 = Q_0 = \Omega$). If we compute L_s by localization (Corollary 9.1.3), we get a sum over the fixed points $p_{i,r}$. Since P is an Euler data, we can express $i^*_{p_{i,r}}(P_d)$ in terms of $P_1, \ldots, P_{d-1}$ when $0 < r < d$, and the same is true for Q. By our induction hypothesis, it follows that only the terms with $r = 0, d$ can contribute to the integral of (11.20). Using [**LLY**, Lemma 2.4], one can show that the localization formula for L_s reduces to

$$(11.21) \qquad L_s = \sum_{i=0}^{n} \left(\frac{\lambda_i^s A_i(\hbar)}{\hbar^d} + \frac{(\lambda_i + d\hbar)^s A_i(-\hbar)}{(-\hbar)^d} \right),$$

where

$$(11.22) \qquad A_i(\hbar) = \frac{(-1)^d}{d! \prod_{j \neq i}(\lambda_i - \lambda_j)} \frac{i^*_{p_{i,0}}(P_d - Q_d)}{\prod_{j \neq i} \prod_{s=1}^{d}(\lambda_i - (\lambda_j + s\hbar))}.$$

The denominator is explained in part by the formula

$$\text{Euler}_G(N_{p_{i,0}}) = i^*_{p_{i,0}}(\phi_{p_{i,0}}) = \prod_{(j,s) \neq (i,0)} (\lambda_i - (\lambda_j + s\hbar)).$$

The formulas (11.21) and (11.22) have $\hbar$ in both numerator and denominator. But P, Q are linked, so that $i^*_{p_{i,0}}(P_d - Q_d)$ vanishes at $\hbar = (\lambda_i - \lambda_j)/d$ for all $j \neq i$. The factor $\lambda_i - (\lambda_j + d\hbar)$ in the denominator also vanishes at $\hbar = (\lambda_i - \lambda_j)/d$, hence cancels with a factor in $i^*_{p_{i,0}}(P_d - Q_d)$. Furthermore, using [**LLY**, Lemma 2.4] and the induction hypothesis, one can show that the other factors containing $\hbar$ in the denominator also cancel in (11.22). This shows that A_i is a polynomial in $\hbar$. Then (11.22) and the degree bound given in the statement of the theorem imply that

$$\deg_\hbar A_i \leq (n+1)d - 2 - nd = d - 2.$$

Since L_s is a polynomial in $\hbar$ for all s, (11.21) implies that $\hbar^d$ must divide

$$(11.23) \qquad \sum_{i=0}^{n} \left(\lambda_i^s A_i(\hbar) + (-1)^d(\lambda_i + d\hbar)^s A_i(-\hbar) \right).$$

We can now show $A_i = 0$ for all i. Otherwise, let m be the minimum degree in $\hbar$ of the nonzero terms of the A_i, and write

$$A_i = a_i \hbar^m + b_i \hbar^{m+1} + \cdots, \qquad a_i \neq 0 \text{ for some } i.$$

The degree bound on A_i implies $m + 1 < d$, so that the coefficients of $\hbar^m$, $\hbar^{m+1}$ in (11.23) must vanish. If $d + m$ is even, the vanishing of the coefficient of $\hbar^m$ implies

$$\sum_{i=0}^{n} 2\lambda_i^s a_i = 0,$$

from which we conclude $a_i = 0$ for all i, a contradiction. When $d + m$ is odd, one uses the coefficient of $\hbar^{m+1}$ in a similar way. $\qquad\square$

11.1.5. Hypergeometric Functions and Mirror Transformations.

We begin with some formal functions which take values in equivariant cohomology. Let S_0 denote the set of sequences $B = \{B_d\}_{d=1}^{\infty}$ such that $B_d \in \mathcal{R}_G H_G^*(\mathbb{P}^n)$ for all d.

We now associate a formal function to any $B \in S_0$. This is a generalization of hypergeometric functions.

DEFINITION 11.1.8. *For $B = \{B_d\} \in S_0$ and $\Omega \in \mathcal{R}_T H_G^*(\mathbb{P}^n)$ invertible, we define*

$$HG[B](t) = e^{-pt/\hbar}\left(\Omega + \sum_{d=1}^{\infty} \frac{B_d e^{dt}}{\prod_{k=0}^{n} \prod_{m=1}^{d}(p - \lambda_k - m\hbar)} \right),$$

where $p \in H_G^*(\mathbb{P}^n)$ is the equivariant hyperplane class of $\mathbb{P}^n$.

If we take the nonequivariant limit $\lambda_i \to 0$, we get the formal expression

$$\lim_{\lambda_i \to 0} HG[B](t) = e^{-Ht/\hbar}\left(\Omega + \sum_{d=1}^{\infty} \frac{\bar{B}_d e^{dt}}{\prod_{k=0}^{n} \prod_{m=1}^{d}(H - m\hbar)} \right),$$

where $H \in H^2(\mathbb{P}^n)$ is the usual hyperplane class and $\bar{B}_d = \lim_{\lambda_i \to 0} B_d$. This is a formal function in $H_{\mathbb{C}^*}^*(\mathbb{P}^n)$ tensored with $\mathbb{C}(\hbar)$. Since $\mathbb{C}^*$ acts trivially on $\mathbb{P}^n$, we get a formal function in $H^*(\mathbb{P}^n) \otimes \mathbb{C}(\hbar)$, where $\hbar$ is now just a parameter. We've seen formal functions of this sort in earlier chapters—Givental's J-function for $X = \mathbb{P}^n$ from Chapter 10 is an obvious example. Furthermore, we will see below in Section 11.1.7 that the formal functions $HG[B](t)$ play a key role in the proof of the Mirror Theorem for the quintic threefold.

Our next task is to relate the $HG[B](t)$ to the Euler data we've been studying. Given $\Omega \in \mathcal{R}_T H_G(\mathbb{P}^n)$ invertible as in Definition 11.1.8, we let $\mathcal{A}^{\Omega}$ denote the set of all Ω-Euler data. We also have $\mathcal{A}^{\Omega} \subset S$, where S is the set of sequences $Q = \{Q_d\}_{d=1}^{\infty}$ such that $Q_d \in \mathcal{R}_G H_G^*(N_d)$ for all $d \geq 1$.

It's not obvious that S has anything to do with S_0—for the latter, the entire sequence lies in $\mathcal{R}_G H_G^*(\mathbb{P}^n)$, while for the former, each term lies in the equivariant cohomology of a different space. Fortunately, there is a natural way to get from S to S_0. We have for each d the G-equivariant map defined by

(11.24) $\qquad I_d : \mathbb{P}^n \to N_d, \qquad (a_0, \dots, a_n) \mapsto (a_0 w_1^d, \dots, a_n w_1^d).$

Notice that I_d takes the fixed point $q_i \in \mathbb{P}^n$ to the fixed point $p_{i,0} \in N_d$. Then we define the map $\mathcal{I} : S \to S_0$ by $\mathcal{I}(P)_d = I_d^*(P_d)$.

Using $\mathcal{I}$, any Ω-Euler data Q gives $\mathcal{I}(Q) \in S_0$, which in turn determines the formal hypergeometric function $HG[\mathcal{I}(Q)](t)$. Let's work this out in a familiar example.

Example 11.1.5.1. For $\Omega = \ell p$, we have the ℓp Euler data $\hat{P}$ defined in (11.9). To compute $\mathcal{I}(\hat{P})$, we first note that $I_d^*(\kappa) = p$, where $\kappa \in H_G^*(N_d)$ and $p \in H_G^*(\mathbb{P}^n)$ are the equivariant hyperplane classes. This follows immediately since $I_d : \mathbb{P}^n \to N_d$ is an equivariant embedding of projective spaces.

Then, using the definition (11.9) of $\hat{P}$, we obtain

$$I_d^*(\hat{P}_d) = I_d^*\left(\prod_{m=0}^{\ell d} (\ell\kappa - m\hbar) \right) = \prod_{m=0}^{\ell d} (\ell p - m\hbar).$$

From this, we get the formal function

$$(11.25) \qquad HG[\mathcal{I}(\hat{P})](t) = e^{-pt/\hbar} \sum_{d=0}^{\infty} \frac{\prod_{m=0}^{\ell d}(\ell p - m\hbar)}{\prod_{m=1}^{d}\prod_{k=0}^{n}(p - \lambda_k - m\hbar)} e^{dt},$$

which coincides (up to the sign of $\hbar$) with the function $I_{\mathcal{V}}$ which was considered previously in [**Givental2**] to describe the quantum cohomology of degree ℓ hypersurfaces in $\mathbb{P}^n$. In Section 11.1.7, we will see that when $\mathcal{V} = \mathcal{O}_{\mathbb{P}^4}(5)$, the nonequivariant limit $\lim_{\lambda_i \to 0} HG[\mathcal{I}(\hat{P})](t)$ gives the solutions y_0, y_1, y_2, y_3 of the Picard-Fuchs equation of the quintic mirror which we studied in Section 6.3.4.

Besides $\hat{P}$, the other important ℓp-Euler data is $\hat{Q}$ from Theorem 11.1.4. Here, we get the formal function

$$HG[\mathcal{I}(\hat{Q})](t) = e^{-pt/\hbar}\left(\Omega + \sum_{d=1}^{\infty} \frac{I_d^*(\hat{Q}_d)e^{dt}}{\prod_{k=0}^{n}\prod_{m=1}^{d}(p - \lambda_k - m\hbar)} \right).$$

We can compute $I_d^*(\hat{Q}_d)$ as follows.

LEMMA 11.1.9. *Let* $\phi^d = \prod_{k=0}^{n}\prod_{m=1}^{d}(p - \lambda_k - m\hbar) \in H_G^*(\mathbb{P}^n)$ *denote the denominator of the coefficient of* e^{dt} *in* $HG[\mathcal{I}(\hat{Q})](t)$. *Then:*

(i) *If* π_1 *and* e_1 *are as in* (11.7) *and* $\mathcal{L}_{d,1}$ *is as in the proof of Theorem 11.1.4, then*

$$I_d^*(\hat{Q}_d) = \phi^d \, e_{1!}\left(\frac{\pi_1^*\mathrm{Euler}_G(\mathcal{V}_d)}{\hbar(\hbar + c_1^G(\mathcal{L}_{d,1}))} \right).$$

(ii) $\deg_\hbar I_d^*(\hat{Q}_d) \leq (n+1)d - 2.$

PROOF. We will prove part (i) by showing that for each $i = 0, \ldots, n$, both sides have equal localizations at q_i. For the left hand side, we have $I_d(q_i) = p_{i,0}$, so that

$$i_{q_i}^* I_d^*(\hat{Q}_d) = i_{p_{i,0}}^*(\hat{Q}_d) = \hat{Q}_d(\lambda_i).$$

If we combine this with (11.17), we obtain

$$i_{q_i}^* I_d^*(\hat{Q}_d) = i_{p_{i,0}}^*(\phi_{p_{i,0}}) \sum_{F_d} \int_{(F_d)_G} \frac{\pi_1^*\mathrm{Euler}_G(\mathcal{V}_d)}{a_d\mathrm{Euler}_G(N(F_d))(\hbar(\hbar + c_1^G(\mathcal{L}_{d,1})))}.$$

However, if we let $\phi_{q_i} = \prod_{j\neq i}(p - \lambda_j)$ as in (9.9), then (11.4) and the definition of ϕ^d easily imply

$$i_{p_{i,0}}^*(\phi_{p_{i,0}}) = i_{q_i}^*(\phi_{q_i})\, i_{q_i}^*(\phi^d),$$

so that the above formula becomes

$$(11.26) \quad i_{q_i}^* I_d^*(\hat{Q}_d) = i_{q_i}^*(\phi_{q_i})\, i_{q_i}^*(\phi^d) \sum_{F_d} \int_{(F_d)_G} \frac{\pi_1^*\mathrm{Euler}_G(\mathcal{V}_d)}{a_d\mathrm{Euler}_G(N(F_d))(\hbar(\hbar + c_1^G(\mathcal{L}_{d,1})))}.$$

Turning our attention to the right hand side, we let

$$\alpha^d = \frac{\pi_1^*\mathrm{Euler}_G(\mathcal{V}_d)}{\hbar(\hbar + c_1^G(\mathcal{L}_{d,1}))},$$

so that we need to compute the localization of $\phi^d e_{1!}(\alpha_d)$ at q_i. We do this by adapting (11.14) to $\mathbb{P}^n$, which gives

$$i_{q_i}^*(\phi^d e_{1!}(\alpha_d)) = \int_{(\mathbb{P}^n)_G} \phi_{q_i} \phi^d\, e_{1!}(\alpha_d) = \int_{(\overline{M}_{0,1}(\mathbb{P}^n,d))_G} e_1^*(\phi_{q_i} \phi^d)\, \alpha_d.$$

However, if we compute the integral on the right using localization, the factor of ϕ_{q_i} implies that we need only consider components of the fixed point set which map to q_i under e_1. These are precisely the components $\{F_d\}$ used in the proof of Theorem 11.1.4, and then the localization formula, applied to this integral, leads immediately to the right hand side of (11.26). This proves part (i).

For part (ii) of the lemma, we know that $\hat{Q}_d$ and hence $I_d^*(\hat{Q}_d)$ are polynomials in $\hbar$, since $\hat{Q}$ is an Euler data. However, the definition of ϕ^d shows that $\deg_\hbar \phi^d = (n+1)d$. Then the desired degree bound on $I_d^*(\hat{Q}_d)$ follows from part (i). $\qquad\square$

Notice that the bound for $\deg_\hbar I_d^*(\hat{Q}_d)$ proved in the second part of the lemma is precisely the degree bound needed for the uniqueness theorem (Theorem 11.1.7).

Our eventual goal is to relate $HG[\mathcal{I}(\hat{P})](t)$ and $HG[\mathcal{I}(\hat{Q})](t)$ in the case of the quintic threefold. But neither these functions nor the Euler data $\hat{P}$ and $\hat{Q}$ are equal. So we need a method for transforming from one Euler data to another. This leads to the following definition.

DEFINITION 11.1.10. *An invertible map* $\mu : \mathcal{A}^\Omega \to \mathcal{A}^\Omega$ *is called a* mirror transformation *if* $\mu(P)$ *is linked to* P *for all* $P \in \mathcal{A}^\Omega$.

As we will soon see, the mirror map discussed in Section 11.1.1 can be described as a mirror transformation in the sense of this definition.

Our goal here is to construct a special class of mirror transformations μ with the property that we can easily relate $HG[\mathcal{I}(P)](t)$ and $HG[\mathcal{I}(\mu(P))](t)$. We begin by considering some elementary transformations of the hypergeometric functions $HG[B](t)$.

LEMMA 11.1.11. *Given any* $B \in \mathcal{S}_0$, *and any* $f, g \in e^t \mathcal{R}_T[[e^t]]$, *then there exists a unique* $\widetilde{B} \in \mathcal{S}_0$ *such that*

$$e^{f/\hbar} HG[B](t + g) = HG[\widetilde{B}](t).$$

PROOF. This is a straightforward formal power series calculation [**LLY**]. $\qquad\square$

Using this lemma, we can now describe the class of mirror transformations we are interested in.

PROPOSITION 11.1.12. *Given* $f, g \in e^t \mathcal{R}_T[[e^t]]$, *there is a mirror transformation* $\mu : \mathcal{A}^\Omega \to \mathcal{A}^\Omega$ *with the property that*

$$e^{f/\hbar} HG[\mathcal{I}(P)](t + g) = HG[\mathcal{I}(\mu(P))](t).$$

for all Euler data $P \in \mathcal{A}^\Omega$.

PROOF. Given f, g, the map $B \mapsto \widetilde{B}$ from Lemma 11.1.11 defines a transformation $\mu_0 : \mathcal{S}_0 \to \mathcal{S}_0$. In this notation, the previous lemma implies that

$$e^{f/\hbar} HG[B](t + g) = HG[\mu_0(B)](t)$$

for any $B \in \mathcal{S}_0$. To prove the proposition, it suffices to show that μ_0 lifts to a mirror transformation μ such that $\mathcal{I}(\mu(P)) = \mu_0(\mathcal{I}(P))$.

Recall that $\mathcal{I} : \mathcal{S} \to \mathcal{S}_0$ is defined by $\mathcal{I}(P)_d = I_d^*(P_d)$, for I_d as in (11.24). To define μ, we need to construct a map going the other way, which we do as follows. The isomorphism given by Proposition 9.1.2 for N_d can be written in the form

$$\mathcal{R}_G H_G^*(N_d) \overset{\oplus i_{p_{i,r}}^*}{\longrightarrow} (\mathcal{R}_G)^{(N_d)^G},$$

where $(N_d)^G$ as usual denotes the G-fixed point set. This implies that in order to specify an element $\omega \in \mathcal{R}_G H_G^*(N_d)$, we merely need to specify its restrictions $i_{p_{i,r}}^*(\omega)$ for all i, r. Accordingly, we define $\mathcal{L} : \mathcal{S}_0 \to \mathcal{S}$ uniquely by the conditions

$$(11.27) \qquad i_{p_{i,r}}^*(\mathcal{L}(B)_d) = (i_{q_i}^*(\Omega))^{-1} \overline{i_{q_i}^*(B_r)}\, i_{q_i}^*(B_{d-r}), \quad 0 \le i \le n,\; 0 \le r \le d$$

for any $B \in \mathcal{S}_0$. In these equations, $B_0 = \Omega$.

The map $\mathcal{L}$ has some nice properties. First, we have $\mathcal{I} \circ \mathcal{L} = \mathrm{id}_{\mathcal{S}_0}$. To see this, note that $I_d(q_i) = p_{i,0}$ and (11.27) for $r = 0$ imply

$$i_{q_i}^*(\mathcal{I}(\mathcal{L}(B))_d) = i_{q_i}^*(I_d^*(\mathcal{L}(B)_d))$$
$$= i_{p_{i,0}}^*(\mathcal{L}(B)_d) = (i_{q_i}^*(\Omega))^{-1} \overline{i_{q_i}^*(B_0)}\, i_{q_i}^*(B_d) = i_{q_i}^*(B_d),$$

where the last equality follows since $B_0 = \Omega \in \mathcal{R}_T H_T^*(\mathbb{P}^n)$. By localization in $\mathcal{R}_T H_G^*(\mathbb{P}^n)$, it follows that $\mathcal{I}(\mathcal{L}(B)) = B$. A second nice property of $\mathcal{L}$ is that for any $B \in \mathcal{S}_0$, $\mathcal{L}(B)$ satisfies the Euler identity in Definition 11.1.3, so if $\mathcal{L}(B)_d \in \mathcal{R}_T H_G^*(N_d)$ for all d, then $\mathcal{L}(B)$ is an Euler data. Finally, it is also immediate from the definition of $\mathcal{L}$ that if $Q \in \mathcal{A}^\Omega \subset \mathcal{S}$, then $Q = \mathcal{L} \circ \mathcal{I}(Q)$.

Now that we have $\mathcal{L} : \mathcal{S}_0 \to \mathcal{S}$ and $\mathcal{I} : \mathcal{S} \to \mathcal{S}_0$, the map $\mu_0 : \mathcal{S}_0 \to \mathcal{S}_0$ extends to $\mu = \mathcal{L} \circ \mu_0 \circ \mathcal{I} : \mathcal{S} \to \mathcal{S}$. In the language of [**LLY**], μ is the *Lagrange lift* of μ_0. When μ is defined this way, note that $\mathcal{I} \circ \mu = \mu_0 \circ \mathcal{I}$ follows immediately from $\mathcal{I} \circ \mathcal{L} = \mathrm{id}_{\mathcal{S}_0}$. Hence it remains to show that μ is a mirror transformation. Given $P \in \mathcal{A}^\Omega$, the condition that P and $\mu(P)$ are linked (i.e., that $i_{p_{i,0}}^*(P_d - \mu(P)_d)$ vanishes at $\hbar = (\lambda_i - \lambda_j)/d$) follows immediately from the explicit power series computation used to establish Lemma 11.1.11. Finally, we need to prove that $\mu(P)$ is an Ω-Euler data. The previous paragraph shows that $\mu(P)$ automatically satisfies the Eulerity condition, so that the final step is to show $\mu(P)_d \in \mathcal{R}_T H_G^*(N_d)$. Here, we refer the reader to [**LLY**, Lemma 2.15] for the proof. $\square$

We now have almost everything we need to prove the Mirror Theorem. But before beginning the proof, we should explain how Sections 11.1.3, 11.1.4 and 11.1.5 relate to [**LLY**]. The bundle $\mathcal{O}_{\mathbb{P}^n}(\ell)$ we've been using is a special case of a *convex* bundle, which is T-equivariant bundle $\mathcal{V}$ such that $\mathrm{Euler}_T(\mathcal{V})$ is invertible in $\mathcal{R}_T H_T^*(\mathbb{P}^n)$ and $H^1(C, f^*\mathcal{V}) = 0$ for every 0-pointed genus 0 stable map $f : C \to \mathbb{P}^n$. Other examples of convex bundles are $\mathcal{V} = \oplus_i \mathcal{O}_{\mathbb{P}^n}(\ell_i)$, $\ell_i > 0$. Similarly, [**LLY**] defines a bundle $\mathcal{V}$ to be *concave* if the above conditions hold, except that we now require $H^0(C, f^*\mathcal{V}) = 0$ for every 0-pointed genus 0 stable map f. Finally, a direct sum of a convex bundle and a concave bundle is called *concavex*.

Given a concavex bundle $\mathcal{V}$ on $\mathbb{P}^n$, we define bundles $\mathcal{V}_d$ on $\overline{M}_{0,0}(\mathbb{P}^n, d)$ as follows. When $\mathcal{V}$ is convex, we use the formula $\mathcal{V}_d = \pi_{1*}e_1^*(\mathcal{V})$, just as in the discussion following (11.7). For a concave bundle, however, we set $\mathcal{V}_d = R^1\pi_{1*}e_1^*(\mathcal{V})$. Finally, if $\mathcal{V}$ is concavex, $\mathcal{V}_d$ is the direct sum of these constructions applied to the convex and concave parts of $\mathcal{V}$.

Finally, suppose that $\mathcal{V}$ is a concavex bundle of the form $\oplus_i \mathcal{O}_{\mathbb{P}^n}(\ell_i)$, $\ell_i \neq 0$. Using the $\mathcal{V}_d$, we define $\hat{Q}_d$ by (11.8), just as before. Furthermore, we can also generalize (11.9) and define

$$(11.28) \qquad \hat{P}_d = \prod_{\ell_i > 0} \prod_{m=0}^{\ell_i d} \left(\ell_i \kappa - m\hbar \right) \times \prod_{\ell_i < 0} \prod_{m=1}^{-\ell_i d - 1} \left(\ell_i \kappa + m\hbar \right).$$

The important point is that the results we proved for $\hat{Q}$ and $\hat{P}$ when $\mathcal{V} = \mathcal{O}_{\mathbb{P}^n}(\ell)$ remain valid when $\mathcal{V} = \oplus_i \mathcal{O}_{\mathbb{P}^n}(\ell_i)$—see [**LLY**] for the details.

11.1.6. Critical Bundles. Before we can prove the Mirror Theorem for the quintic threefold $V \subset \mathbb{P}^4$, we need to explain how the Gromov-Witten invariants $K_d = \langle I_{0,0,d} \rangle$ of V relate to what we've been doing so far. In Theorem 11.1.13, we will show that the K_d appear naturally in the hypergeometric function $HG[\mathcal{I}(\hat{Q})](t)$ for the bundle $\mathcal{V} = \mathcal{O}_{\mathbb{P}^4}(5)$. We will also discuss a subclass of concavex bundles (the *critical bundles*) which satisfy a modified version of this theorem.

Recall from Section 11.1.1 that the Gromov-Witten invariants K_d of V are given by

$$K_d = \int_{\overline{M}_{0,0}(\mathbb{P}^4, d)} \mathrm{Euler}(\mathcal{V}_d).$$

In terms of the K_d, the Gromov-Witten potential Φ of the quintic threefold is written

$$\Phi = \frac{5t^3}{6} + \sum_{d=1}^{\infty} K_d \, e^{dt}.$$

Then we have the following result.

THEOREM 11.1.13. *Let $\mathcal{V} = \mathcal{O}_{\mathbb{P}^4}(5)$ and define $\hat{Q}_d$ as in (11.8). Then, in the nonequivariant limit $\lambda_i \to 0$, we have*

$$\lim_{\lambda_i \to 0} HG[\mathcal{I}(\hat{Q})](t) = 5H \left(1 - t\frac{H}{\hbar} + \frac{\Phi'}{5}\frac{H^2}{\hbar^2} - \frac{t\Phi' - 2\Phi}{5}\frac{H^3}{\hbar^3} \right),$$

where H is the hyperplane class in $H^*(\mathbb{P}^4)$.

PROOF. Since $\hat{Q}$ is an Euler data for $\Omega = 5p$, the nonequivariant limit of Definition 11.1.8 gives

$$\lim_{\lambda_i \to 0} HG[\mathcal{I}(\hat{Q})](t) = e^{-Ht/\hbar} \left(5H + \sum_{d=1}^{\infty} \frac{\lim_{\lambda_i \to 0} I_d^*(\hat{Q}_d)}{\prod_{m=1}^{d}(H - m\hbar)^5} e^{dt} \right).$$

Now take the coefficient of e^{dt} in this expression and write it in terms of the basis $1, H, \ldots, H^4$ of $H^*(\mathbb{P}^4)$:

$$e^{-Ht/\hbar} \frac{\lim_{\lambda_i \to 0} I_d^*(\hat{Q}_d)}{\prod_{m=1}^{d}(H - m\hbar)^5} = A_0 + A_1 H + A_2 H^2 + A_3 H^3 + A_4 H^4.$$

We can determine the A_i by multiplying this equation by H^k and integrating over $\mathbb{P}^4$. This gives

$$A_{4-k} = \int_{\mathbb{P}^4} H^k e^{-Ht/\hbar} \frac{\lim_{\lambda_i \to 0} I_d^*(\hat{Q}_d)}{\prod_{m=1}^{d}(H - m\hbar)^5}.$$

We will leave it to the reader to check that the theorem is now a consequence of the following formulas:

$$(11.29) \qquad \int_{\mathbb{P}^4} H^k e^{-Ht/\hbar} \frac{\lim_{\lambda_i \to 0} I_d^*(\hat{Q}_d)}{\prod_{m=1}^d (H - m\hbar)^5} = \begin{cases} \hbar^{-3}(2 - dt)K_d & k = 0 \\ \hbar^{-2} dK_d & k = 1 \\ 0 & k > 1. \end{cases}$$

The case $k = 0$ of (11.29) appeared explicitly in [**LLY**], and the cases $k \geq 1$ are implicit in [**LLY**].

To prove (11.29), we use the formula for $I_d^*(\hat{Q}_d)$ given in Lemma 11.1.9. Taking the nonequivariant limit of this as $\lambda_i \to 0$, we obtain

$$\lim_{\lambda_i \to 0} I_d^*(\hat{Q}_d) = \prod_{m=1}^d (H - m\hbar)^5 e_{1!}\left(\frac{\pi_1^* \mathrm{Euler}(\mathcal{V}_d)}{\hbar(\hbar + c_1(\mathcal{L}_{d,1}))} \right).$$

If we insert this into the integral on the left hand side of (11.29), we see that the integral simplifies to

$$
\begin{aligned}
& \int_{\mathbb{P}^4} H^k e^{-Ht/\hbar} e_{1!}\left(\frac{\pi_1^* \mathrm{Euler}(\mathcal{V}_d)}{\hbar(\hbar + c_1(\mathcal{L}_{d,1}))} \right) \\
(11.30) \qquad = & \int_{\overline{M}_{0,1}(\mathbb{P}^4, d)} e_1^* H^k e^{-e_1^* Ht/\hbar} \frac{\pi_1^* \mathrm{Euler}(\mathcal{V}_d)}{\hbar(\hbar + c_1(\mathcal{L}_{d,1}))} \\
= & \int_{\overline{M}_{0,0}(\mathbb{P}^4, d)} \mathrm{Euler}(\mathcal{V}_d)\, \pi_{1!}\left(e_1^* H^k \frac{e^{-e_1^* Ht/\hbar}}{\hbar(\hbar + c_1(\mathcal{L}_{d,1}))} \right).
\end{aligned}
$$

Since $\mathrm{Euler}(\mathcal{V}_d)$ has top degree, this integral is just K_d times the degree 0 component of the second factor in the integrand. Since π_1 has relative dimension 1, this is seen to be the degree of

$$\begin{cases} \pi_{1!}(\hbar^{-3}(-e_1^* Ht - c_1(\mathcal{L}_{d,1}))) & k = 0 \\ \pi_{1!}(\hbar^{-2} e_1^* H) & k = 1 \\ 0 & k > 1, \end{cases}$$

which simplifies to

$$\begin{cases} \hbar^{-3}(-dt + 2) & k = 0 \\ \hbar^{-2} d & k = 1 \\ 0 & k > 1 \end{cases}$$

by integration over a fiber of π_1. This establishes (11.29), and theorem is proved. $\qquad \square$

This theorem is actually valid in a more general context. An examination of the proof shows that a key step is the argument following (11.30), where we use the fact that $\mathrm{Euler}(\mathcal{V}_d)$ has top degree on $\overline{M}_{0,0}(\mathbb{P}^4, d)$. This leads to the following definition.

DEFINITION 11.1.14. *A concavex bundle $\mathcal{V}$ on $\mathbb{P}^n$ is called critical if the induced bundle $\mathcal{V}_d$ on $\overline{M}_{0,0}(\mathbb{P}^n, d)$ has rank equal to $(n+1)d + n - 3 = \dim \overline{M}_{0,0}(\mathbb{P}^n, d)$ for all $d > 0$.*

When $\mathcal{V}$ is critical, $\mathrm{Euler}(\mathcal{V}_d)$ is a top degree class on $\overline{M}_{0,0}(\mathbb{P}^n, d)$, so that we can define

$$K_d = \int_{\overline{M}_{0,0}(\mathbb{P}^n,d)} \mathrm{Euler}(\mathcal{V}_d).$$

generalizing what we did for the quintic threefold. In this situation, Lemma 11.1.9 still applies since $\mathcal{V}$ is concavex, and the argument of Theorem 11.1.13 also applies since $\mathcal{V}$ is critical. Hence we have proved that

$$(11.31) \qquad \int_{\mathbb{P}^n} H^k e^{-Ht/\hbar} \frac{\lim_{\lambda_i \to 0} I_d^*(\hat{Q}_d)}{\prod_{m=1}^d (H - m\hbar)^{n+1}} = \begin{cases} \hbar^{-3}(2 - dt)K_d & k = 0 \\ \hbar^{-2} d\, K_d & k = 1 \\ 0 & k > 1. \end{cases}$$

See[**LLY**] for the details. We will see below that this has interesting applications.

In Section 11.1.3, we considered bundles of the form $\mathcal{V} = \mathcal{O}_{\mathbb{P}^n}(\ell)$. The reader can easily check that $\mathcal{V} = \mathcal{O}_{\mathbb{P}^4}(5)$ is the only critical bundle of this type. However, when $\mathcal{V} = \oplus_i \mathcal{O}_{\mathbb{P}^n}(\ell_i)$, there are other critical bundles of interest. Here is a particularly relevant example.

Example 11.1.6.1. Consider the concave bundle on $\mathbb{P}^1$ given by

$$\mathcal{V} = \mathcal{O}_{\mathbb{P}^1}(-1) \oplus \mathcal{O}_{\mathbb{P}^1}(-1).$$

Here, $\mathcal{V}_d$ is defined by

$$\mathcal{V}_d = R^1 \pi_{1*} e_1^* \big(\mathcal{O}_{\mathbb{P}^1}(-1) \oplus \mathcal{O}_{\mathbb{P}^1}(-1) \big),$$

and for $f \in \overline{M}_{0,0}(\mathbb{P}^1, d)$, one easily checks that

$$\mathrm{rank}\, \mathcal{V}_d = \dim H^1(\mathbb{P}^1, f^*\mathcal{O}_{\mathbb{P}^1}(-1) \oplus f^*\mathcal{O}_{\mathbb{P}^1}(-1)) = 2d - 2 = \dim \overline{M}_{0,0}(\mathbb{P}^1, d).$$

It follows that $\mathcal{V}$ is critical. Then (11.31) shows that the nonequivariant limit of $HG[\mathcal{I}(\hat{Q})](t)$ involves

$$K_d = \int_{\overline{M}_{0,0}(\mathbb{P}^1,d)} \mathrm{Euler}\big(R^1\pi_{1*}e_1^*(\mathcal{O}_{\mathbb{P}^1}(-1) \oplus \mathcal{O}_{\mathbb{P}^1}(-1))\big).$$

This is *exactly* the integral computed in Theorem 9.2.3, which computed the multiple cover contributions of rigidly embedded smooth rational curves in Calabi-Yau threefolds. We will soon apply the methods of [**LLY**] to give a second proof of this theorem.

We should also mention that [**LLY**] contains a complete list of all critical bundles on $\mathbb{P}^n$ which are direct sums of line bundles.

11.1.7. Proof of Theorem 11.1.1. It is now time to prove the Mirror Theorem for the quintic threefold. For the rest of the section, we will assume that $\mathcal{V} = \mathcal{O}_{\mathbb{P}^4}(5)$, and we will use $\hat{P}$ and $\hat{Q}$ as defined in Section 11.1.3. The key step in the proof will be to construct two mirror transformations μ and ν such that $\mu(\hat{P}) = \nu(\hat{Q})$. This will imply a relation between the hypergeometric functions $HG[\mathcal{I}(\hat{P})](t)$ and $HG[\mathcal{I}(\hat{Q})](t)$, which in the nonequivariant limit $\lambda_i \to 0$ will give the desired result.

In order to construct the desired mirror transformations, we first need to look at $HG[\mathcal{I}(\hat{P})](t)$ more closely. If we take the nonequivariant limit $\lambda_i \to 0$ of (11.25), we get

$$
\lim_{\lambda_i \to 0} HG[\mathcal{I}(\hat{P})](t) = e^{-Ht/\hbar} \sum_{d=0}^{\infty} \frac{\prod_{m=0}^{5d}(5H - m\hbar)}{\prod_{m=1}^{d}(H - m\hbar)^5} e^{dt}
$$

(11.32)

$$
= 5H\left(y_0 - y_1\frac{H}{\hbar} + y_2\frac{H^2}{\hbar^2} - y_3\frac{H^3}{\hbar^3}\right).
$$

These are the functions y_0, y_1, y_2, y_3 discussed in Section 11.1.1. They appear in the mirror map $\Psi(t) = y_1/y_0$ and also in the statement of Theorem 11.1.1.

We now construct mirror transformations μ and ν by using Proposition 11.1.12 twice. For μ, we let

$$
(11.33)\quad f_1 = -(\log y_0)\hbar + \frac{h_2}{y_0}\sum_{k=0}^{4}\lambda_k \in e^t\mathcal{R}_T[[e^t]], \quad h_2 = \sum_{d=1}^{\infty}\frac{(5d)!}{(d!)^5}\left(\sum_{m=1}^{d}\frac{1}{m}\right)e^{dt},
$$

and $g_1 = 0$, and then Proposition 11.1.12 gives a mirror transformation μ such that

$$
(11.34)\qquad e^{f_1/\hbar}HG[\mathcal{I}(\hat{P})](t) = HG[\mathcal{I}(\mu(\hat{P})](t).
$$

For our second mirror transformation, let $f_2 = 0$ and $g_2 = \Psi(t) - t \in e^t\mathcal{R}[[e^t]]$, where $\Psi(t) = y_1/y_0$ is the mirror map. Applying the proposition a second time, we obtain a mirror transformation ν such that

$$
(11.35)\qquad HG[\mathcal{I}(\nu(\hat{Q}))](t) = HG[\mathcal{I}(\hat{Q})](\Psi(t)).
$$

We can relate the mirror transforms $\mu(\hat{P})$ and $\nu(\hat{Q})$ as follows.

THEOREM 11.1.15. $\mu(\hat{P}) = \nu(\hat{Q})$.

PROOF. We will follow [**LLY**]. We know that $\hat{P}, \hat{Q}$ are linked Euler data by Theorem 11.1.6. Furthermore, since μ and ν are mirror transformations, the pairs $\hat{P}, \mu(\hat{P})$ and $\hat{Q}, \nu(\hat{Q})$ are also linked. By transitivity, we conclude that $\mu(\hat{P}), \nu(\hat{Q})$ are linked Euler data. Thus, the theorem will follow from Theorem 11.1.7 once we prove the degree bound

$$
\deg_\hbar i_{p_{i,0}}^*(\mu(P)_d - \nu(Q)_d) \le 5d - 2.
$$

Since $i_{p_{i,0}} = I_d \circ i_{q_i}$, it suffices to show that

$$
\frac{I_d^*(\mu(P)_d - \nu(Q)_d)}{\prod_{k=0}^{4}\prod_{m=1}^{d}(p - \lambda_k - m\hbar)} \equiv 0 \bmod \hbar^{-2}.
$$

However, by Definition 11.1.8, this is equivalent to

$$
HG[\mathcal{I}(\nu(\hat{Q}))](t) \equiv HG[\mathcal{I}(\mu(\hat{P}))](t) \bmod \hbar^{-2},
$$

and then, using (11.34) and (11.35), we are reduced to proving

$$
(11.36)\qquad e^{f_1/\hbar}HG[\mathcal{I}(\hat{P})](t) \equiv HG[\mathcal{I}(\hat{Q})](\Psi(t)) \bmod \hbar^{-2}.
$$

To prove this, we first expand $HG[\mathcal{I}(\hat{P})](t)$ in powers of $\hbar^{-1}$ to obtain

$$
HG[\mathcal{I}(\hat{P})](t) = 5p\left(y_0 - h^{-1}\left(py_1 + h_2\sum_{k=0}^{4}\lambda_k\right) + O(\hbar^{-2})\right).
$$

Then (11.33) and some algebra show that

$$
e^{f_1/\hbar}HG[\mathcal{I}(\hat{P})](t) = 5p - \hbar^{-1}5p^2\frac{y_1}{y_0} + O(\hbar^{-2}).
$$

Turning our attention to the right hand side of (11.36), we first observe that $\deg_\hbar I_d^*(\hat{Q}_d) \leq 5d - 2$ by part (ii) of Lemma 11.1.9. Then Definition 11.1.8 implies

$$HG[\mathcal{I}(\hat{Q})](t) = e^{-pt/\hbar}\big(5p + O(\hbar^{-2})\big)$$
$$= 5p - \hbar^{-1}5p^2 t + O(\hbar^{-2}).$$

Since $\Psi = y_1/y_0$, we see that

$$HG[\mathcal{I}(\hat{Q})](\Psi(t)) = 5p - \hbar^{-1}5p^2\frac{y_1}{y_0} + O(\hbar^{-2}).$$

From here, the desired equation (11.36) follows immediately by comparing the above expressions for $e^{f_1/\hbar}HG[\mathcal{I}(\hat{P})](t)$ and $HG[\mathcal{I}(\hat{Q})](\Psi(t))$. The completes the proof of the theorem. $\qquad\square$

Note that the expansions of $HG[\mathcal{I}(\hat{P})](t)$ and $HG[\mathcal{I}(\hat{Q})](t)$ given in the above proof explain how the mirror transformations μ and ν were chosen: the basic idea was to make sure that the congruence (11.36) was satisfied.

COROLLARY 11.1.16. *In the nonequivariant limit $\lambda_i \to 0$, we have*

$$\lim_{\lambda_i \to 0} HG[\mathcal{I}(\hat{Q})](\Psi(t)) = \frac{1}{y_0} \lim_{\lambda_i \to 0} HG[\mathcal{I}(\hat{P})](t).$$

PROOF. If we combine Theorem 11.1.15 with (11.34) and (11.35) and take the nonequivariant limit, we obtain

$$\lim_{\lambda_i \to 0} HG[\mathcal{I}(\hat{Q})](\Psi(t)) = \lim_{\lambda_i \to 0} e^{f_1/\hbar}HG[\mathcal{I}(\hat{P})](t).$$

However, we know from (11.33) that $f_1 = -(\log y_0)\hbar + h_2 \sum_{k=0}^{4} \lambda_k$. This implies $\lim_{\lambda_i \to 0} e^{f_1/\hbar} = 1/y_0$, and the corollary follows. $\qquad\square$

It is now easy to prove Theorem 11.1.1. By Theorem 11.1.13, we have

$$\lim_{\lambda_i \to 0} HG[\mathcal{I}(\hat{Q})](\Psi(t)) =$$
$$5H\left(1 - \Psi(t)\frac{H}{\hbar} + \frac{\Phi'(\Psi(t))}{5}\frac{H^2}{\hbar^2} - \frac{\Psi(t)\Phi'(\Psi(t)) - 2\Phi(\Psi(t))}{5}\frac{H^3}{\hbar^3}\right),$$

and by (11.32), we also have

$$\lim_{\lambda_i \to 0} HG[\mathcal{I}(\hat{P})](t) = 5H\left(y_0 - y_1\frac{H}{\hbar} + y_2\frac{H^2}{\hbar^2} - y_3\frac{H^3}{\hbar^3}\right).$$

Combining these equations with Corollary 11.1.16 gives

$$5H\left(1 - \Psi(t)\frac{H}{\hbar} + \frac{\Phi'(\Psi(t))}{5}\frac{H^2}{\hbar^2} - \frac{\Psi(t)\Phi'(\Psi(t)) - 2\Phi(\Psi(t))}{5}\frac{H^3}{\hbar^3}\right)$$
$$= 5H\left(1 - \frac{y_1}{y_0}\frac{H}{\hbar} + \frac{y_2}{y_0}\frac{H^2}{\hbar^2} - \frac{y_3}{y_0}\frac{H^3}{\hbar^3}\right).$$

Equating coefficients, an easy calculation yields

$$\Phi(\Psi(t)) = \frac{5}{2}\left(\frac{y_1}{y_0}\frac{y_2}{y_0} - \frac{y_3}{y_0}\right).$$

This completes the proof of the Mirror Theorem for the quintic threefold.

The proof just given is really nice, but it gets even better when we realize that the work done in earlier in the section gives other significant results without much more effort. Several examples of this sort appear in [**LLY**], but we will give only one which is relevant to our purposes.

Example 11.1.7.1. As in Example 11.1.6.1, consider $\mathcal{V} = \mathcal{O}_{\mathbb{P}^1}(-1) \oplus \mathcal{O}_{\mathbb{P}^1}(-1)$. By (11.31), we know that

$$(11.37) \qquad \int_{\mathbb{P}^1} e^{-Ht/\hbar} \frac{\lim_{\lambda_i \to 0} I_d^*(\hat{Q}_d)}{\prod_{m=1}^d (H - m\hbar)^2} = \hbar^{-3}(2 - dt)K_d,$$

where

$$K_d = \int_{\overline{M}_{0,0}(\mathbb{P}^1, d)} \mathrm{Euler}\big(R^1 \pi_{1*} e_1^*(\mathcal{O}_{\mathbb{P}^1}(-1) \oplus \mathcal{O}_{\mathbb{P}^1}(-1))\big).$$

We can compute K_d as follows. We know that $\hat{P}$ and $\hat{Q}$ are linked Euler data, and Lemma 11.1.9 implies $\deg_\hbar I_d^*(\hat{Q}_d) \leq 2d - 2$. However, in this case (11.28) shows that

$$\hat{P}_d = \prod_{m=1}^{d-1} (-\kappa + m\hbar)^2 = \prod_{m=1}^{d-1} (\kappa - m\hbar)^2.$$

Thus $I_d^*(\hat{P}_d) = \Pi_{m=1}^{d-1}(p - m\hbar)^2$, which also has degree $\leq 2d - 2$. It follows that the degree condition of Theorem 11.1.7 is satisfied in this case, so that $\hat{P} = \hat{Q}$. Thus

$$\begin{aligned}
\int_{\mathbb{P}^1} e^{-Ht/\hbar} \frac{\lim_{\lambda_i \to 0} I_d^*(\hat{Q}_d)}{\prod_{m=1}^d (H - m\hbar)^2} &= \int_{\mathbb{P}^1} e^{-Ht/\hbar} \frac{\lim_{\lambda_i \to 0} I_d^*(\hat{P}_d)}{\prod_{m=1}^d (H - m\hbar)^2} \\
&= \int_{\mathbb{P}^1} e^{-Ht/\hbar} \frac{\prod_{m=1}^{d-1}(H - m\hbar)^2}{\prod_{m=1}^d (H - m\hbar)^2} \\
&= \int_{\mathbb{P}^1} e^{-Ht/\hbar} \frac{1}{(H - d\hbar)^2} \\
&= \hbar^{-3}(2 - dt)\, d^{-3},
\end{aligned}$$

where the last equality uses $H^2 = 0$ in $H^*(\mathbb{P}^1)$. Comparing this to (11.37) implies $K_d = d^{-3}$, which completes our second proof of Theorem 9.2.3.

This example shows that we need to broaden our idea of what a "Mirror Theorem" is, for the same ideas and techniques which lead to mirror symmetry for the quintic threefold simultaneously lead to other interesting results. The paper [**LLY**] represents an important step towards understanding the larger context of mirror symmetry.

We will see in the next section that compared to [**LLY**], Givental's approach to the Mirror Theorem has a parallel but different structure. He considers two cohomology-valued formal functions $I_\mathcal{V}$ (11.38) and $J_\mathcal{V}$ (11.52), then changes variables so that a uniqueness theorem applies. See Theorem 11.2.2 and [**Givental2**]. His method will be illustrated for the quintic threefold in Example 11.2.1.3, and further comments comparing the two approaches will be made near the end of Section 11.2.4.

11.2. Givental's Approach

In this section, we explain Givental's formulation of the Mirror Theorem for complete intersections in projective spaces (Theorem 11.2.2). The technique involves the construction of formal cohomology-valued functions $I_{\mathcal{V}}$ and $J_{\mathcal{V}}$, where $J_{\mathcal{V}}$ is a variant of the function J discussed in Section 10.3. The Mirror Theorem is then the assertion that $I_{\mathcal{V}} = J_{\mathcal{V}}$ after a change of variables.

The approach taken in [**Givental2**] is explained in more detail in [**BDPP, Pandharipande3**], which can also serve as useful references for this section. The case of the quintic is discussed in [**Givental6**]. We will sketch a slightly different proof of Givental's theorem for nef complete intersections in $\mathbb{P}^n$, combining ideas appearing in [**Givental2, Givental4, Kim2, Pandharipande3**], to which we refer the reader. Some of the needed ideas introduced in [**Givental2**] have already been explained in Chapters 9 and 10. We will also discuss toric complete intersections and give some illustrative examples.

11.2.1. The Mirror Theorem for Nef Complete Intersections in $\mathbb{P}^n$.

We consider a smooth complete intersection $X \subset \mathbb{P}^n$ of ℓ hypersurfaces of degrees $a_1, \dots, a_\ell$. We put $\mathcal{V} = \oplus_{i=1}^{\ell} \mathcal{O}_{\mathbb{P}^n}(a_i)$, so that X is defined by the vanishing of a global section of $\mathcal{V}$. We will also assume that the anticanonical class of X is nef, which in this context is equivalent to $\sum_{i=1}^{\ell} a_i \leq n + 1$.

Our first goal is to give a careful statement of Theorem 11.2.2, which relates the two cohomology-valued formal functions $I_{\mathcal{V}}$ and $J_{\mathcal{V}}$ defined below. Of these two functions, $I_{\mathcal{V}}$ is the more elementary. Following [**Givental2, Kim2**], we define

$$I_{\mathcal{V}} = I_{\mathcal{V}}(t_0, t_1, \hbar^{-1})$$

$$(11.38) \qquad = e^{(t_0 + t_1 H)/\hbar} \, \mathrm{Euler}(\mathcal{V}) \sum_{d=0}^{\infty} e^{dt_1} \frac{\prod_{i=1}^{\ell} \prod_{m=1}^{a_i d} (a_i H + m\hbar)}{\prod_{m=1}^{d} (H + m\hbar)^{n+1}},$$

where $H \in H^2(\mathbb{P}^n)$ is the hyperplane class. The notation $I_{\mathcal{V}}(t_0, t_1, \hbar^{-1})$ indicates that we will sometimes think of $I_{\mathcal{V}}$ in terms of its expansion as a formal series in powers of $\hbar^{-1}$. If $X = \mathbb{P}^n$, so that $\mathcal{V}$ is trivial, then $I_{\mathcal{V}}$ will be denoted as $I_{\mathbb{P}^n}$.

The formal function $J_{\mathcal{V}}$, on the other hand, is more complicated to define. In Chapter 10, we studied the Givental function $J = J_X$. If $\dim X \geq 3$, then $H^2(X)$ is generated by $i^* H$, and then Lemma 10.3.3 gives two formulas for J_X:

$$
J_X = e^{(t_0 + t_1 i^* H)/\hbar} \left(1 + \sum_{d=1}^{\infty} \sum_{a=0}^{m} e^{dt_1} \left\langle \frac{T_a}{\hbar - c}, 1 \right\rangle_{0,d} T^a \right)
$$

$$
(11.39)
$$

$$
= e^{(t_0 + t_1 i^* H)/\hbar} \left(1 + \sum_{d=1}^{\infty} e^{dt_1} PD^{-1} e_{1*} \left(\frac{1}{\hbar - c} \cap [\overline{M}_{0,2}(X, d)]^{\mathrm{virt}} \right) \right).
$$

Here, T_a is a cohomology basis of $H^*(X)$, T^a is the dual basis, PD is Poincaré duality, $e_1 : \overline{M}_{0,2}(X, d) \to X$ is the evaluation map, and

$$
(11.40) \qquad \left\langle \frac{T_a}{\hbar - c}, 1 \right\rangle_{0,d} = \sum_{k=0}^{\infty} \hbar^{-(k+1)} \langle \tau_k T_a, 1 \rangle_{0,d}
$$

is the symbolic notation for gravitational correlators used in Chapter 10. Previously, we wrote the Givental function simply as J, but since we need to distinguish between X and $\mathbb{P}^n$, we will now write it as J_X. The function J_X plays a central role in the theory of quantum differential equations (see Theorem 10.3.1).

Notice that we can't directly compare $I_\mathcal{V}$ and J_X since one takes values in $H^*(\mathbb{P}^n)$ while the other take values in $H^*(X)$. For this reason, we will introduce a variant of $J_\mathcal{V}$ which takes values in $H^*(\mathbb{P}^n)$. Then Givental's theorem will explain how $I_\mathcal{V}$ and $J_\mathcal{V}$ are related when X is nef.

Our strategy for explaining the theorem will be as follows. We will begin with the case $\mathcal{V} = 0$, for here, Givental's theorem is the simple assertion that $I_{\mathbb{P}^n} = J_{\mathbb{P}^n}$ (see Proposition 11.2.1 below). Then we will define $J_\mathcal{V}$ for general bundles on $\mathbb{P}^n$ and state Givental's Mirror Theorem (Theorem 11.2.2). Proofs will be deferred until Sections 11.2.3 and 11.2.4. We will also explain how $J_\mathcal{V}$ relates to J_X and show that Theorem 11.2.2 implies mirror symmetry for the quintic threefold.

At this point, we can make contact with the approach of [**LLY**]. Givental's function $I_\mathcal{V}$ reduces for $t_0 = 0$ to the nonequivariant limit $\lim_{\lambda_i \to 0} HG[\mathcal{I}(\hat{P})](t_1)$ in the notation of [**LLY**], up to a change of sign in $\hbar$. In comparing calculations in this section with corresponding calculations in [**LLY**], we will no longer mention this difference in sign explicitly. Furthermore, in the general case, the content of Theorem 11.2.2 is that $I_\mathcal{V}$ and $J_\mathcal{V}$ coincide after a change of variables. In the approach of [**LLY**], the corresponding statement is the equality of $HG[\mathcal{I}(\hat{P})](t_1)$ and $HG[\mathcal{I}(\hat{Q})](t_1)$ after the same change of variables.

We are now ready to state Givental's version of the Mirror Theorem in the special case of projective space itself, $X = \mathbb{P}^n$. We define $I_{\mathbb{P}^n}$ by (11.38) with $\ell = 0$, and define $J_{\mathbb{P}^n}$ as in (11.39).

PROPOSITION 11.2.1. *For projective space $\mathbb{P}^n$, we have $I_{\mathbb{P}^n} = J_{\mathbb{P}^n}$.*

We will prove this in Section 11.2.3 below. However, in Chapter 10 we verified Proposition 11.2.1 when $n = 1$, as we now recall.

Example 11.2.1.1. When $X = \mathbb{P}^1$, the first line of (11.39) simplifies to

$$J_{\mathbb{P}^1} = e^{(t_0+t_1 H)/\hbar}\left(1 + \sum_{d=1}^{\infty}\left(\frac{e^{dt_1}}{\hbar^{2d}}\langle\tau_{2d-1}H, 1\rangle_{0,d}1 + \frac{e^{dt_1}}{\hbar^{2d+1}}\langle\tau_{2d}, 1\rangle_{0,d}H\right)\right)$$

by the Divisor Axiom from Section 10.1.2. Also, (11.38) for $n = 1$ and $\ell = 0$ is

$$I_{\mathbb{P}^1} = e^{(t_0+t_1 H)/\hbar}\sum_{d=0}^{\infty}e^{dt_1}\frac{1}{\left((H+\hbar)(H+2\hbar)\cdots(H+d\hbar)\right)^2}.$$

We computed $\langle\tau_{2d-1}H, 1\rangle_{0,d}$ and $\langle\tau_{2d}, 1\rangle_{0,d}$ in Example 10.1.3.1, and we used these formulas in Example 10.3.1.1 to show that $I_{\mathbb{P}^1} = J_{\mathbb{P}^1}$.

From the point of view of Proposition 11.2.1, this process can be reversed. Once we know $I_{\mathbb{P}^1} = J_{\mathbb{P}^1}$, expanding $I_{\mathbb{P}^1}$ in powers of H (note $H^2 = 0$ in $H^*(\mathbb{P}^1)$) easily gives formulas for $\langle\tau_{2d-1}H, 1\rangle_{0,d}$ and $\langle\tau_{2d}, 1\rangle_{0,d}$. In this way, we can regard Proposition 11.2.1 as computing the gravitational correlators of $\mathbb{P}^1$.

Our next task is to define $J_\mathcal{V}$ when $\mathcal{V} = \oplus_{i=1}^{\ell}\mathcal{O}_{\mathbb{P}^n}(a_i)$ as above. We first need some notation. Given any $d > 0$, the space of sections $H^0(C, f^*(\mathcal{V}))$ forms a vector bundle $\mathcal{V}_{d,k}$ over $\overline{\mathcal{M}}_{0,k}(\mathbb{P}^n, d)$ as $f : C \to \mathbb{P}^n$ varies in $\overline{\mathcal{M}}_{0,k}(\mathbb{P}^n, d)$. More precisely, let $\pi_{k+1} : \overline{\mathcal{M}}_{0,k+1}(\mathbb{P}^n, d) \to \overline{\mathcal{M}}_{0,k}(\mathbb{P}^n, d)$ be the map forgetting the $(k+1)^{\text{st}}$ marked point and let $e_{k+1} : \overline{\mathcal{M}}_{0,k+1}(\mathbb{P}^n, d) \to \mathbb{P}^n$ be evaluation at the $(k+1)^{\text{st}}$ point. Then

$$\mathcal{V}_{d,k} = \pi_{k+1*}e_{k+1}^*(\mathcal{V}).$$

In particular, $\mathcal{V}_{d,0} = \pi_{1*}e_1^*(\mathcal{V})$ is precisely the bundle denoted $\mathcal{V}_d$ earlier in the chapter. One can also show that $\mathcal{V}_{d,k} = \pi_k^*(\mathcal{V}_{d,k-1})$ when $k > 0$.

Now fix i between 1 and k. For each $f : C \to \mathbb{P}^n$ in $\overline{\mathcal{M}}_{0,k}(\mathbb{P}^n, d)$, we can consider those sections of $H^0(C, f^*(\mathcal{V}))$ which vanish at the ith marked point. These form the fibers of a bundle $\mathcal{V}'_{d,k,i}$ on $\overline{\mathcal{M}}_{0,k}(\mathbb{P}^n, d)$ fitting into the exact sequence

$$(11.41) \qquad 0 \longrightarrow \mathcal{V}'_{d,k,i} \longrightarrow \mathcal{V}_{d,k} \longrightarrow e_i^*(\mathcal{V}) \longrightarrow 0.$$

This implies that

$$(11.42) \qquad \mathrm{Euler}(\mathcal{V}_{d,k}) = \mathrm{Euler}(\mathcal{V}'_{d,k,i}) \cup e_i^*(\mathrm{Euler}(\mathcal{V})).$$

In particular, we get the bundle $\mathcal{V}'_{d,2,1}$ on $\overline{\mathcal{M}}_{0,2}(\mathbb{P}^n, d)$, and using this bundle, we define $J_{\mathcal{V}}$ by the formula

$$
\begin{aligned}
(11.43) \qquad J_{\mathcal{V}} = {}& e^{(t_0 + t_1 H)/\hbar}\,\mathrm{Euler}(\mathcal{V}) \times \\
& \left(1 + \sum_{d=1}^{\infty}\sum_{b=0}^{\infty} \frac{e^{dt_1}}{\hbar^{b+1}} e_{1!}\big(c_1(\mathcal{L}_1)^b \cup \mathrm{Euler}(\mathcal{V}'_{d,2,1})\big)\right).
\end{aligned}
$$

In terms of the symbolic notation used above in (11.39), we can write this as

$$(11.44) \qquad J_{\mathcal{V}} = e^{(t_0 + t_1 H)/\hbar}\,\mathrm{Euler}(\mathcal{V})\left(1 + \sum_{d=1}^{\infty} e^{dt_1} e_{1!}\left(\frac{\mathrm{Euler}(\mathcal{V}'_{d,2,1})}{\hbar - c}\right)\right).$$

We note two important properties of $J_{\mathcal{V}}$. First, when $\mathcal{V} = 0$, the corresponding complete intersection is $X = \mathbb{P}^n$. Since $\mathbb{P}^n$ is convex, the virtual fundamental class of $\overline{\mathcal{M}}_{0,2}(\mathbb{P}^n, d)$ coincides with the usual fundamental class, so that the second formula of (11.39) simplifies to

$$J_{\mathbb{P}^n} = e^{(t_0 + t_1 H)/\hbar}\left(1 + \sum_{d=1}^{\infty} e^{dt_1} e_{1!}\left(\frac{1}{\hbar - c}\right)\right).$$

Since the Euler class of the zero bundle is 1, this agrees with (11.44) when $\mathcal{V} = 0$. Thus $J_{\mathcal{V}} = J_{\mathbb{P}^n}$ when $\mathcal{V} = 0$.

For the second property, consider the large expression in parenthesis in (11.43) and observe that the coefficient of $\hbar^{-1}$ contains a factor of

$$\mathrm{Euler}(\mathcal{V}) \cup e_{1!}\big(\mathrm{Euler}(\mathcal{V}'_{d,2,1})\big) = e_{1!}\big(\mathrm{Euler}(\mathcal{V}_{d,2})\big) = 0.$$

We have used (11.42) in the first equality. The second equality follows since e_1 factors through the map $\pi_2 : \overline{\mathcal{M}}_{0,2}(\mathbb{P}^n, d) \to \overline{\mathcal{M}}_{0,1}(\mathbb{P}^n, d)$ forgetting the second marked point: we saw above that $\mathcal{V}_{d,2} = \pi_2^*(\mathcal{V}_{d,1})$, so that

$$\pi_{2!}\big(\mathrm{Euler}(\mathcal{V}_{d,2})\big) = \pi_{2!}\pi_2^*\big(\mathrm{Euler}(\mathcal{V}_{d,1})\big) = 0$$

since the fibers of π_2 have positive dimension. This implies that

$$(11.45) \qquad J_{\mathcal{V}} = e^{(t_0 + t_1 H)/\hbar}\,\mathrm{Euler}(\mathcal{V})\big(1 + o(\hbar^{-1})\big),$$

which is similar to the property (10.37) of the Givental J-function.

We can now state Givental's Mirror Theorem for complete intersections in $\mathbb{P}^n$.

THEOREM 11.2.2. *Let $X \subset \mathbb{P}^n$ be a complete intersection of ℓ hypersurfaces of degree a_i with $\sum_{i=1}^{\ell} a_i \leq n + 1$. If $\mathcal{V} = \oplus_{i=1}^{\ell} \mathcal{O}(a_i)$, then the formal cohomology-valued functions $I_{\mathcal{V}}$ and $J_{\mathcal{V}}$ coincide after a triangular weighted homogeneous change of variables:*

$$t_0 \mapsto t_0 + f(e^{t_1})\hbar + h(e^{t_1}), \quad t_1 \mapsto t_1 + g(e^{t_1}),$$

where f, g, h are weighted homogeneous power series such that $\deg f = \deg g = 0$, $\deg h = 1$, and $\deg e^{t_1} = n + 1 - \sum_i a_i$.

As we observed earlier, the condition $\sum_{i=1}^{\ell} a_i \leq n+1$ says that the anticanonical class of X is nef. In particular, the largest possible value of $\sum_{i=1}^{\ell} a_i$ is $n + 1$, which occurs when X is an elliptic curve, a K3 surface, or a Calabi-Yau manifold of dimension ≥ 3. Notice also that this theorem includes Proposition 11.2.1 as a special case (when $\mathcal{V} = 0$). We will sketch a proof of Theorem 11.2.2 in Section 11.2.4.

Another useful comment is that the power series f, g, h in Theorem 11.2.2 are uniquely determined from the coefficients of $I_{\mathcal{V}}$. To see this, apply the change of variables $t_0 \mapsto t_0 + f(e^{t_1})\hbar + h(e^{t_1})$ and $t_1 \mapsto t_1 + g(e^{t_1})$ to $J_{\mathcal{V}}$. Ignoring the factor of $e^{t_0/\hbar}$ in $I_{\mathcal{V}}$ and $J_{\mathcal{V}}$ after making this substitution in $J_{\mathcal{V}}$, (11.45) gives

$$J_{\mathcal{V}} = e^f \operatorname{Euler}(\mathcal{V})\big(1 + h\,\hbar^{-1}1 + (t_1 + g)\hbar^{-1}H + o(\hbar^{-1})\big).$$

Equating this with $I_{\mathcal{V}}$, we see that

$$
\begin{aligned}
e^f &\longleftrightarrow \text{ coefficient of } 1 \text{ in } \tilde{I}_{\mathcal{V}} \\
(11.46) \qquad e^f h &\longleftrightarrow \text{ coefficient of } \hbar^{-1}1 \text{ in } \tilde{I}_{\mathcal{V}} \\
e^f(t_1 + g) &\longleftrightarrow \text{ coefficient of } \hbar^{-1}H \text{ in } \tilde{I}_{\mathcal{V}},
\end{aligned}
$$

where $\tilde{I}_{\mathcal{V}}$ is (11.38) without the factor of $\operatorname{Euler}(\mathcal{V})$ (hence $I_{\mathcal{V}} = \operatorname{Euler}(\mathcal{V}) \cup \tilde{I}_{\mathcal{V}}$). It follows that $I_{\mathcal{V}}$ determines the coordinate change used in the Mirror Theorem, as claimed. In particular, if

$$I_{\mathcal{V}} = e^{(t_0 + t_1 H)/\hbar} \operatorname{Euler}(\mathcal{V})\big(1 + o(\hbar^{-1})\big)$$

holds already with the existing variables t_0, t_1, then we must have $f = g = h = 0$ and $I_{\mathcal{V}} = J_{\mathcal{V}}$. Here is an example where this occurs.

Example 11.2.1.2. Let $X \subset \mathbb{P}^n$ be a hypersurface of degree $\ell < n$. The assumption on ℓ easily leads to $I_{\mathcal{V}} = e^{(t_0 + t_1 H)/\hbar} \operatorname{Euler}(\mathcal{V})\big(1 + o(\hbar^{-1})\big)$, so that by the above discussion, we have $f = g = h = 0$ in Theorem 11.2.2. Thus

$$J_{\mathcal{V}} = I_{\mathcal{V}} = \ell H e^{(t_0 + t_1 H)/\hbar}\left(1 + \sum_{d=1}^{\infty} e^{dt_1} \frac{(\ell H + \hbar)(\ell H + 2\hbar)\cdots(\ell H + d\ell\hbar)}{\big((H + \hbar)(H + 2\hbar)\cdots(H + d\hbar)\big)^{n+1}}\right).$$

We will soon see that this has some interesting consequences for the small quantum cohomology ring of X.

Before we can fully understand Theorem 11.2.2, we need to explain the relation between J_X and $J_{\mathcal{V}}$. For this, we will use the Gysin map $i_! : H^*(X) \to H^{*+2\ell}(\mathbb{P}^n)$ induced by the inclusion $i : X \hookrightarrow \mathbb{P}^n$. Then we have the following result.

PROPOSITION 11.2.3. *Let $i : X \hookrightarrow \mathbb{P}^n$ be the complete intersection determined by $\mathcal{V} = \oplus_{i=1}^{\ell} \mathcal{O}_{\mathbb{P}^n}(a_i)$, and assume that $\dim X \geq 3$. Then $J_{\mathcal{V}} = i_!(J_X)$.*

PROOF. Our assumption $\dim X \geq 3$ implies that $H^2(\mathbb{P}^n) \simeq H^2(X)$. Thus both $J_{\mathcal{V}}$ and J_X are formal functions of t_0 and t_1. Note also that $i : X \hookrightarrow \mathbb{P}^n$ induces an inclusion $j : \overline{\mathcal{M}}_{0,2}(X, d) \hookrightarrow \overline{\mathcal{M}}_{0,2}(\mathbb{P}^n, d)$ such that the diagram

$$
\begin{array}{ccc}
\overline{\mathcal{M}}_{0,2}(X, d) & \xrightarrow{\;j\;} & \overline{\mathcal{M}}_{0,2}(\mathbb{P}^n, d) \\
e_1 \downarrow & & \downarrow e_1 \\
X & \xrightarrow{\;i\;} & \mathbb{P}^n
\end{array}
$$

commutes, where e_1 is the evaluation map.

The key fact here is that we can compute the virtual fundamental class of $\overline{\mathcal{M}}_{0,2}(X,d)$. By definition, X is the zero locus of a section s of $\mathcal{V}$. But s induces a section $\tilde{s}$ of the bundle $\mathcal{V}_{d,k}$ defined above. As in Example 7.1.5.1, it follows that the zero section of $\tilde{s}$ (in the stack sense) can be naturally identified with $\overline{\mathcal{M}}_{0,2}(X,d)$. This allows us to construct $[\overline{M}_{0,2}(X,d)]^{\mathrm{virt}}$ using the normal cone construction for the vector bundle $\mathcal{V}_{d,2}$. Generalizing (7.21), we then obtain

$$(11.47) \qquad j_*\big([\overline{M}_{0,2}(X,d)]^{\mathrm{virt}}\big) = \mathrm{Euler}(\mathcal{V}_{d,2}) \cap [\overline{M}_{0,2}(\mathbb{P}^n,d)].$$

Thus the virtual fundamental class of $\overline{\mathcal{M}}_{0,2}(X,d)$ refines the Euler class of $\mathcal{V}_{d,2}$.

We can now prove the proposition. We begin with the second formula for J_X given in (11.39). Using the projection formula and the above commutative diagram, one easily sees that

$$i_!(J_X) = e^{(t_0+t_1 H)/\hbar}\left(1 + \sum_{d=1}^{\infty} e^{dt_1} e_{1!} PD^{-1} j_*\left(\frac{1}{\hbar - c} \cap [\overline{M}_{0,2}(X,d)]^{\mathrm{virt}}\right)\right).$$

In the symbolic notation $1/(\hbar - c)$, the c refers to the Chern class $c_1(\mathcal{L}_1)$ of the bundle $\mathcal{L}_1$ on $\overline{\mathcal{M}}_{0,2}(X,d)$. This is the pullback via j of the bundle on $\overline{\mathcal{M}}_{0,2}(\mathbb{P}^n,d)$ also denoted $\mathcal{L}_1$. It follows by the projection formula for j that

$$j_*\left(\frac{1}{\hbar - c} \cap [\overline{M}_{0,2}(X,d)]^{\mathrm{virt}}\right) = \frac{1}{\hbar - c} \cap j_*\big([\overline{M}_{0,2}(X,d)]^{\mathrm{virt}}\big)$$

$$= \frac{1}{\hbar - c} \cap \big(\mathrm{Euler}(\mathcal{V}_{d,2}) \cap [\overline{M}_{0,2}(\mathbb{P}^n,d)]\big)$$

$$= \frac{\mathrm{Euler}(\mathcal{V}_{d,2})}{\hbar - c} \cap [\overline{M}_{0,2}(\mathbb{P}^n,d)],$$

so that our formula for $i_!(J_X)$ simplifies to

$$i_!(J_X) = e^{(t_0+t_1 H)/\hbar}\left(1 + \sum_{d=1}^{\infty} e^{dt_1} e_{1!}\left(\frac{\mathrm{Euler}(\mathcal{V}_{d,2})}{\hbar - c}\right)\right).$$

However, (11.42) implies

$$\mathrm{Euler}(\mathcal{V}_{d,2}) = \mathrm{Euler}(\mathcal{V}'_{d,2,1}) \cup e_1^*\big(\mathrm{Euler}(\mathcal{V})\big),$$

If we insert this into the formula just derived for $i_!(J_X)$ and use the projection formula for e_1, we obtain

$$i_!(J_X) = e^{(t_0+t_1 H)/\hbar}\,\mathrm{Euler}(\mathcal{V})\left(1 + \sum_{d=1}^{\infty} e^{dt_1} e_{1!}\left(\frac{\mathrm{Euler}(\mathcal{V}'_{d,2,1})}{\hbar - c}\right)\right),$$

which is precisely the definition of $J_{\mathcal{V}}$. This proves the proposition. $\square$

Now that we know $i_!(J_X) = J_{\mathcal{V}}$, we next study whether or not any information gets lost in going from J_X to $J_{\mathcal{V}}$. We will use the following two observations:

- Since X is a complete intersection of ample hypersurfaces, the formula $i_!(i^*(\alpha)) = \alpha \cup \mathrm{Euler}(\mathcal{V})$ shows that $i_!$ is injective on the the image of i^*. Note also that $i^*(H^*(\mathbb{P}^n)) \subset H^{\mathrm{even}}(X)$.
- The Degree Axiom from Section 10.1.2 shows that J_X involves only the even cohomology of X.

When $\dim X \geq 3$ is odd, $H^{\mathrm{even}}(\mathbb{P}^n) \to H^{\mathrm{even}}(X)$ is surjective by the Lefschetz theorem. The first bullet shows that $i_!$ is injective on $H^{\mathrm{even}}(X)$, and combining this with the second bullet, we see that J_X can be recovered from $J_\mathcal{V} = i_!(J_X)$.

On the other hand, when $\dim X \geq 3$ is even, we will write $\dim X = 2m$. Here, J_X still takes values in $H^{\mathrm{even}}(X)$, but $i^* : H^{\mathrm{even}}(\mathbb{P}^n) \to H^{\mathrm{even}}(X)$ need not be surjective. The problem is that $H^{2m}(X)$ has the Lefschetz decomposition

$$H^{2m}(X) = \mathbb{C} \cdot i^*(H^m) \oplus H_0^{2m}(X),$$

where $H_0^{2m}(X)$ is the primitive cohomology. The image of i^* completely misses $H_0^{2m}(X)$, yet J_X will usually have a nonzero component in this space. There are various notations for the image of $i^* : H^*(\mathbb{P}^n) \to H^*(X)$. In the terminology of Section 8.6.4, we call the image $H^*_{\mathrm{toric}}(X)$ since $\mathbb{P}^n$ is a toric variety. In [**Givental3**], the image is denoted $H^*(\mathcal{V})$, which is notation we will use here. Using the above decomposition of $H^{2m}(X)$, we get

$$H^{\mathrm{even}}(X) = H^*(\mathcal{V}) \oplus H_0^{2m}(X)$$

since X is a complete intersection. This induces a decomposition

$$(11.48) \qquad\qquad J_X = J_X^\mathcal{V} + J_X^0,$$

where $J_X^\mathcal{V}$ takes values in $H^*(\mathcal{V})$ and J_X^0 is the "primitive part" of J_X. To relate this to $i_!(J_X)$, recall that $H_0^{2m}(X)$ is the kernel of $i_! : H^{2m}(X) \to H^{2m+2\ell}(\mathbb{P}^n)$. Hence, when we apply $i_!$ to (11.48), we get $J_\mathcal{V} = i_!(J_X^\mathcal{V})$. We conclude that for an even-dimensional complete intersection, $J_\mathcal{V}$ reflects that portion of J_X which comes from $\mathbb{P}^n$.

We can unify these observations by noting that the decomposition $J_X = J_X^\mathcal{V} + J_X^0$ from (11.48) makes sense for any dimension (thus $J_X^0 = 0$ when $\dim X$ is odd). Then the above discussion shows that for all X, the expression $J_X^\mathcal{V}$ can be recovered from $J_\mathcal{V} = i_!(J_X^\mathcal{V})$. Hence the Mirror Theorem tells us how to compute some of the genus 0 gravitational correlators of X associated to cohomology classes coming from the ambient space.

One can also state a version of the Mirror Theorem which takes place entirely within $H^*(\mathcal{V})$. If we write $I_\mathcal{V} = \mathrm{Euler}(\mathcal{V})\, \tilde{I}_\mathcal{V}$, then $i^*(\tilde{I}_\mathcal{V})$ and $J_X^\mathcal{V}$ are formal functions which take values in $H^*(\mathcal{V})$. Also note that $i_!$ is injective on $H^*(\mathcal{V})$ and takes $i^*(\tilde{I}_\mathcal{V})$, $J_X^\mathcal{V}$ to $I_\mathcal{V}$, $J_\mathcal{V}$ respectively. It follows that Theorem 11.2.2 is equivalent to the assertion that $i^*(\tilde{I}_\mathcal{V})$ and $J_X^\mathcal{V}$ become equal after a coordinate change of the appropriate type. This is how Givental states the Mirror Theorem in [**Givental3**], although he allows the ambient space to be a toric variety. We will discuss this more general case in Section 11.2.5. We should also mention that our approach, which uses $H^*(\mathbb{P}^n)$ rather than $H^*(\mathcal{V})$, is based on [**Kim2**].

For an example of what the Mirror Theorem looks like in practice, let's consider the quintic threefold.

Example 11.2.1.3. The quintic threefold $V \subset \mathbb{P}^4$ corresponds to $\mathcal{V} = \mathcal{O}_{\mathbb{P}^4}(5)$, so that $\mathrm{Euler}(\mathcal{V}) = 5H$. Here, H is the hyperplane class in $H^*(\mathbb{P}^4)$. Then

$$I_\mathcal{V} = e^{(t_0 + t_1 H)/\hbar}\, 5H \sum_{d=0}^{\infty} e^{dt_1} \frac{(5H + \hbar) \cdots (5H + 5d\hbar)}{((H + \hbar) \cdots (H + d\hbar))^5}.$$

If we compare this with (11.32), we see the same formula, except for the sign of $\hbar$ and a factor of $e^{t_0/\hbar}$. It follows that

$$I_{\mathcal{V}} = e^{t_0/\hbar}\, 5H\left(y_0 + y_1\frac{H}{\hbar} + y_2\frac{H^2}{\hbar^2} + y_3\frac{H^3}{\hbar^3}\right).$$

where y_0, y_1, y_2, y_3 are the basis of solutions of the Picard-Fuchs equation of the quintic mirror discussed in Section 11.1.1.

The Mirror Theorem tells us that $I_{\mathcal{V}}$ equals $J_{\mathcal{V}}$ after a suitable change of variables, and by (11.46), we can read off f, g, h from $I_{\mathcal{V}}$. Hence the above formula for $I_{\mathcal{V}}$ implies

$$e^f = y_0 \Longrightarrow f = \log(y_0)$$
$$e^f h = 0 \Longrightarrow h = 0$$
$$e^f(t_1 + g) = y_1 \Longrightarrow g = \frac{y_1}{y_0} - t_1,$$

which gives $t_0 \mapsto t_0 + f\hbar = t_0 + \log(y_0)\hbar$ and $t_1 \mapsto t_1 + g = y_1/y_0$. Note that the latter is precisely the mirror map used in Section 11.1.1. Now write $J_{\mathcal{V}} = J_{\mathcal{V}}(t_0, t_1)$ and $I_{\mathcal{V}} = I_{\mathcal{V}}(t_0, t_1)$. Then Theorem 11.2.2 implies

$$J_{\mathcal{V}}(t_0 + \log(y_0)\hbar, y_1/y_0) = I_{\mathcal{V}}(t_0, t_1).$$

If we let $s = y_1/y_0$ and note that $e^{(t_0 + \log(y_0)\hbar)/\hbar} = y_0 e^{t_0/\hbar}$, then the above equation simplifies to

$$J_{\mathcal{V}}(t_0, s) = \frac{1}{y_0} I_{\mathcal{V}}(t_0, t_1).$$

Combining this with the above formula for $I_{\mathcal{V}}$, we obtain

$$(11.49) \qquad J_{\mathcal{V}}(t_0, s) = e^{t_0/\hbar} 5H\left(1 + s\frac{H}{\hbar} + \frac{y_2}{y_0}\frac{H^2}{\hbar^2} + \frac{y_3}{y_0}\frac{H^3}{\hbar^3}\right).$$

However, we have some nice formulas for J_V from Section 10.3.2. We will use the cohomology basis of $H^*(V)$ given by $T_0 = 1$, $T_1 = i^*H$, $T^1 = i^*H^2/5$ and $T^0 = i^*H^3/5$. Note that T_0 and T^0 are dual, and similarly for T_1 and T^1. If $\Phi(s)$ is the Gromov-Witten potential of the quintic threefold, then Proposition 10.3.4 implies that

$$J_V(t_0, s) = e^{t_0/\hbar}\left(1 + \hbar^{-1}s\,T_1 + \hbar^{-2}\Phi'(s)\,T^1 + \hbar^{-3}\big(s\Phi'(s) - 2\Phi(s)\big)T^0\right).$$

If we apply $i_!$ and use Proposition 11.2.3, this becomes

$$(11.50) \qquad J_{\mathcal{V}}(t_0, s) = e^{t_0/\hbar}\, 5H\left(1 + s\frac{H}{\hbar} + \frac{\Phi'(s)}{5}\frac{H^2}{\hbar^2} + \frac{s\Phi'(s) - 2\Phi(s)}{5}\frac{H^3}{\hbar^3}\right).$$

Now let $s = \Psi(t_1)$. Then equating (11.50) with the formula for $J_{\mathcal{V}}(t_0, s)$ given in (11.49), we easily get

$$\Phi(\Psi(t_1)) = \frac{5}{2}\left(\frac{y_1}{y_0}\frac{y_2}{y_0} - \frac{y_3}{y_0}\right).$$

This is precisely the Mirror Theorem for the quintic threefold which we discussed in Section 11.1.7. As we noted in Theorem 8.6.4, this implies the Hodge-theoretic version of mirror symmetry stated in Section 8.6.

It remains to complete the discussion begun in Chapter 2. Recall that Section 2.6.2 gave the above formulas for $I_{\mathcal{V}}$ and $J_{\mathcal{V}}$ and discussed how Theorem 11.2.2

relates to mirror symmetry. Looking back at this section, one of the key formulas was (2.34), which stated that when $t_0 = 0$ and $\hbar = 1$, we have

$$(I_V/y_0)'' = (5 + 2875q + \cdots)(H^3 + sH^4),$$

where $'$ denotes d/ds. We then showed that the series in parentheses, namely $5 + 2875q + \cdots$, was the normalized Yukawa coupling Y of the quintic mirror. Thus

$$(11.51) \qquad (I_V/y_0)'' = Y(H^3 + sH^4).$$

Let's use this to show how Givental's Mirror Theorem gives the "classical" formulation of mirror symmetry for the quintic threefold. As we saw above, Theorem 11.2.2 implies $I_V/y_0(t_0, t_1) = J_V(t_0, s)$. It follows that

$$(I_V/y_0)'' = J_V'',$$

where $' = d/ds$. But Proposition 10.3.4 implies

$$J_V'' = \hbar^{-2} e^{(t_0 + sH)/\hbar}\, H * H,$$

where $*$ is the small quantum product in $H^*(V)$. Strictly speaking, we should write $i^*(H)$ instead of H, but this makes the notation too cluttered. Now recall from (8.66) that $H * H = \langle H, H, H \rangle\, C$, where $C \in H^4(V)$ is the class of a line in V. Thus

$$J_V'' = \hbar^{-2} e^{(t_0 + sH)/\hbar} \langle H, H, H \rangle\, C.$$

Now let $t_0 = 0$ and $\hbar = 1$. Since $i_!(C) = H^3 \in H^6(\mathbb{P}^4)$, it follows that

$$J_V'' = i_!(J_V)'' = e^{sH} \langle H, H, H \rangle\, H^3 = \langle H, H, H \rangle\, (H^3 + sH^4)$$

in $H^*(\mathbb{P}^4)$. If we compare this to (11.51), we immediately see that $Y = \langle H, H, H \rangle$. Hence we recover the version of mirror symmetry discussed in Chapter 2.

Also note that mirror symmetry for the quintic threefold can be regarded as one way of computing the instanton numbers n_d of V. A different way of extracting the n_d from the Mirror Theorem is explained in [**Givental2**].

We conclude our discussion of the Mirror Theorem by explaining another way to relate J_V and J_X. The idea is to define a modified "quantum product" on $H^*(\mathbb{P}^n)$ [**Givental2, Kim2**] which can be used to construct J_V in the same way that J arises from the ordinary small quantum product, (10.15), and (10.28). This product is carefully written down and analyzed in [**Pandharipande3**], where the product is denoted by $*_X$. For us, the most interesting property of $*_X$ is the following.

PROPOSITION 11.2.4. *Suppose that $X \subset \mathbb{P}^n$ is a complete intersection. Then the map $i^* : H^*(\mathbb{P}^n) \otimes \mathbb{C}[[q]] \to H^*(X) \otimes \mathbb{C}[[q]]$ is a ring homomorphism, where $H^*(X)$ has the small quantum product $*$ and $H^*(\mathbb{P}^n)$ has the modified quantum product $*_X$.*

PROOF. This is proved in [**Pandharipande3**, Prop. 4], where it is attributed to T. Graber. $\square$

Furthermore, the function J_V yields relations in the ring $(H^*(\mathbb{P}^n) \otimes \mathbb{C}[[q]], *_X)$ just as in Section 10.3.1, where we saw that J_X yields relations in the small quantum cohomology ring of X.

PROPOSITION 11.2.5. *Suppose $P(\hbar\partial/\partial t_1, e^{t_1}, \hbar)J_V = 0$, where P is a polynomial in $\hbar\partial/\partial t_1$ whose coefficients are power series in $e^{t_1}, \hbar$. Then the relation $P(H, q, 0) = 0$ holds in $H^*(\mathbb{P}^n) \otimes \mathbb{C}[[q]]$ with the product $*_X$.*

PROOF. This is proven in [**Pandharipande3**, Lemma 3]. $\square$

Combining Propositions 11.2.4 and 11.2.5, we get the following useful corollary.

COROLLARY 11.2.6. *Suppose* $P(\hbar\partial/\partial t_1, e^{t_1}, \hbar)J_{\mathcal{V}} = 0$, *where* P *is a formal power series as above. Then the relation* $P(i^*(H), q, 0) = 0$ *holds in the small quantum cohomology ring of* X.

PROOF. From Proposition 11.2.5, we first conclude that $P(H, q, 0) = 0$ in $H^*(\mathbb{P}^n)$ with the $*_X$ product. By Proposition 11.2.4, we get $i^*(P(H, q, 0)) = P(i^*(H), q, 0)$, the latter expression being evaluated in the small quantum cohomology ring of X. This expression is therefore zero as well. $\square$

Here is an example of how we can combine Theorem 11.2.2 and Corollary 11.2.6.

Example 11.2.1.2, continued. Let $X \subset \mathbb{P}^n$ be a hypersurface of degree $\ell < n$. In the first part of this example, we explained how the Mirror Theorem for $\mathcal{V} = \mathcal{O}_{\mathbb{P}^n}(\ell)$ gives the equality

$$J_{\mathcal{V}} = I_{\mathcal{V}} = \ell H e^{(t_0 + t_1 H)/\hbar}\left(1 + \sum_{d>0}^{\infty} e^{dt_1} \frac{(\ell H + \hbar)(\ell H + 2\hbar)\cdots(\ell H + d\ell\hbar)}{\big((H + \hbar)(H + 2\hbar)\cdots(H + d\hbar)\big)^{n+1}}\right).$$

An easy calculation shows that $I_{\mathcal{V}}$ is annihilated by the operator

$$\left(\hbar\frac{d}{dt_1}\right)^n - e^{t_1}\ell\left(\ell\hbar\frac{d}{dt_1} + \hbar\right)\cdots\left(\ell\hbar\frac{d}{dt_1} + (\ell - 1)\hbar\right).$$

Since $J_{\mathcal{V}} = I_{\mathcal{V}}$, this operator also annihilates $J_{\mathcal{V}}$. Then Corollary 11.2.6 implies

$$H^n - q\ell\,(\ell H)^{\ell-1} = 0$$

in the small quantum cohomology ring of the hypersurface X. This agrees with the relation found in [**Beauville1**].

11.2.2. The Quantum Hyperplane Section Principle. So far, we've discussed the Mirror Theorem for a complete intersection $X \subset \mathbb{P}^n$. We now want to put this in a broader context via the *Quantum Hyperplane Section Principle* for a complete intersection. This principle is suggested in [**Givental3**] and formalized in [**Kim2**], where the quantum hyperplane section principle is explained for complete intersections in a generalized flag variety. This principle is similar in spirit to though broader in scope than the quantum restriction formula for restricting linear sigma model correlation functions to anticanonical hypersurfaces [**MP1**].

We will formulate the principle in a general setting where it has not yet been proven. Let Y be a smooth algebraic variety, and let $X \subset Y$ be a complete intersection, the zero locus of a vector bundle $\mathcal{V} = \oplus_{i=1}^{\ell}\mathcal{L}_i$, where the $\mathcal{L}_i$ are convex line bundles on Y. Here, *convex* has the same meaning it did in Section 11.1.5. Thus $H^1(C, f^*(\mathcal{L}_i)) = 0$ for all 0-pointed genus 0 stable maps $f : C \to Y$.

Given $\beta \in H_2(Y, \mathbb{Z})$, we now generalize some of the constructions of Section 11.2.1. In particular, we let $\mathcal{V}_{\beta,k}$ be the vector bundle over $\overline{\mathcal{M}}_{0,k}(Y, \beta)$ whose fiber over f is the space of sections of $H^0(C, f^*(\mathcal{V}))$. Letting $\pi_k : \overline{\mathcal{M}}_{0,k}(Y, \beta) \to \overline{\mathcal{M}}_{0,k-1}(Y, \beta)$ be the map forgetting the kth marked point, then $\pi_k^*\mathcal{V}_{\beta,k-1} = \mathcal{V}_{\beta,k}$. Also, for each i between 1 and k, the sections of $H^0(C, f^*(\mathcal{V}))$ which vanish at the ith marked point form the fibers of a bundle $\mathcal{V}'_{\beta,k,i}$ on $\overline{\mathcal{M}}_{0,n}(Y, \beta)$. This bundle fits into the exact sequence

$$0 \longrightarrow \mathcal{V}'_{\beta,k,i} \longrightarrow \mathcal{V}_{\beta,k} \longrightarrow e_i^*\mathcal{V} \longrightarrow 0.$$

For simplicity, we will assume that Y is convex. We now define the formal function $J_{\mathcal{V}}$ in $H^*(Y)$ as follows:

$$(11.52) \qquad J_{\mathcal{V}} = e^{(t_0 + \delta)/\hbar} \mathrm{Euler}(\mathcal{V}) \left(1 + \sum_{\beta \neq 0} q^\beta \, e_{1!} \left(\frac{\mathrm{Euler}(\mathcal{V}'_{\beta,2,1})}{\hbar - c} \right) \right),$$

where e_1 as usual is the evaluation map and $1/(\hbar - c)$ is the symbolic notation used in (11.44). We are using the convention $\delta = \sum_{i=1}^r t_i T_i$, where the T_i range over a basis for $H^2(Y)$, and we set $q^\beta = e^{\int_\beta \delta}$. When Y is the convex variety $\mathbb{P}^n$, notice how this reduces to (11.44). Also, for Y general (but still convex) and $\mathcal{V} = 0$, one can show that $J_{\mathcal{V}} = J_Y$ without difficulty.

We put $\mathcal{L}_j(\beta) = \int_\beta c_1(\mathcal{L}_j)$, and define a second cohomology valued formal function

$$\overline{I}_{\mathcal{V}} = e^{(t_0 + \delta)/\hbar} \, \mathrm{Euler}(\mathcal{V}) \times$$

$$\left(1 + \sum_{0 \neq \beta \in M(Y)} q^\beta \prod_{j=1}^\ell \frac{\prod_{m=-\infty}^{\mathcal{L}_j(\beta)} (c_1(\mathcal{L}_j) + m\hbar)}{\prod_{m=-\infty}^{0} (c_1(\mathcal{L}_j) + m\hbar)} \, e_{1!} \left(\frac{1}{\hbar - c} \right) \right).$$

As in Section 5.5.3, we note that all but finitely many terms cancel in the product.

We now state the quantum hyperplane section principle. We will assume that the Kähler cone of T is simplicial, generated by classes T_i. We introduce corresponding variables t_i and $q_i = e^{t_i}$ as usual. We define the degree of the q_i via

$$c_1(X) - c_1(\mathcal{V}) = \sum_{i=1}^r (\deg q_i) \, T_i.$$

Note that if $Y = \mathbb{P}^n$ and $\mathcal{V} = \oplus_{i=1}^\ell \mathcal{O}(a_i)$, then $\deg q = n + 1 - \sum_{i=1}^\ell a_i$.

CONJECTURE 11.2.7. *As above, let $X \subset Y$ be a complete intersection corresponding to $\mathcal{V}$. Assume that Y is convex and that $\deg q_i \geq 0$ for all i. Then the formal vector-valued functions $\overline{I}_{\mathcal{V}}$ and $J_{\mathcal{V}}$ coincide after a unique triangular weighted homogeneous change of variables:*

$$t_0 \mapsto t_0 + f_0(q)\hbar + h(q), \qquad t_i \mapsto t_i + f_i(q)$$

where $h, f_0, \ldots, f_k$ are weighted homogeneous power series and $\deg f_0 = \deg f_i = 0$, $\deg h = 1$.

Let's first see what this conjecture says when $Y = \mathbb{P}^n$. We begin with Proposition 11.2.2, which asserts $I_{\mathbb{P}^n} = J_{\mathbb{P}^n}$. By (11.38),

$$(11.53) \qquad I_{\mathbb{P}^n} = e^{(t_0 + t_1 H)/\hbar} \sum_{d=0}^\infty q^d \frac{1}{\prod_{m=1}^d (H + m\hbar)^{n+1}},$$

where $q = e^{t_1}$. In the remarks following (11.44), we noted that $J_{\mathbb{P}^n} = J_{\mathcal{V}}$ for $\mathcal{V} = 0$. Then (11.44) implies

$$(11.54) \qquad J_{\mathbb{P}^n} = e^{(t_0 + t_1 H)/\hbar} \left(1 + \sum_{d=1}^\infty e^{dt_1} e_{1!} \left(\frac{1}{\hbar - c} \right) \right).$$

This follows since the Euler class of the trivial bundle is 1. Comparing these two expressions, we conclude that

$$(11.55) \qquad e_{1!} \left(\frac{1}{\hbar - c} \right) = \frac{1}{\prod_{m=1}^d (H + m\hbar)^{n+1}}.$$

Now substitute this into the above formula for $\overline{I}_V$ and notice that $L_j(\beta)$ is just $a_j d$ in this case. It now follows easily that $\overline{I}_V$ coincides with I_V as defined in (11.38).

This shows that for $\mathbb{P}^n$, the Quantum Hyperplane Section Principle stated in Conjecture 11.2.7 is equivalent to the Mirror Theorem from Theorem 11.2.2, once we prove Proposition 11.2.1. This is typical of how Conjecture 11.2.7 works: combining (11.55) with the Quantum Hyperplane Section Principle tells us what the Mirror Conjecture should be for complete intersections in $\mathbb{P}^n$. This same reasoning applies more generally, so that once we know the right hand side of (11.55) for Y convex, we immediately know that the corresponding Mirror Theorem should be for complete intersections in Y.

As evidence for Conjecture 11.2.7, we note from the above discussion that proving Theorem 11.2.2 will verify the conjecture for $Y = \mathbb{P}^n$. A proof is given for complete intersections in generalized flag manifolds in [**Kim2**], and an application to the virtual numbers of rational curves in Calabi-Yau complete intersection threefolds in Grassmannians and complete flag manifolds is given in [**BCFKvS1**]. In general, the Quantum Hyperplane Section Principle appears to give sensible results for a variety of ambient spaces. In addition to complete flag manifolds studied in [**Kim1, BCFKvS1**], the Quantum Hyperplane Section Principle produces verifiable results for certain varieties of nets of quadrics (see [**Tjøtta**]).

11.2.3. Proof of Proposition 11.2.1. We now sketch a proof of Proposition 11.2.1, which asserts that

$$I_{\mathbb{P}^n} = J_{\mathbb{P}^n},$$

where $I_{\mathbb{P}^n}$ and $J_{\mathbb{P}^n}$ are given in (11.53) and (11.54). Our treatment follows the conventions of [**Kim2**], which was inspired by [**Givental4**]. The key step in the proof will be the recursion given in Lemma 11.2.9. In Section 11.2.4, we will prove the Mirror Theorem (Theorem 11.2.2) by combining the methods used here with some important new ideas.

Our strategy will be to create equivariant versions of $I_{\mathbb{P}^n}$ and $J_{\mathbb{P}^n}$ and prove they are equal by the localization techniques similar to what we did in Chapter 9 and Section 11.1. We will use the standard action of $T = (\mathbb{C}^*)^{n+1}$ on $\mathbb{P}^n$, as described in Example 9.1.2.1. This induces an action on $\overline{M}_{0,2}(\mathbb{P}^n, d)$ such that the evaluation map $e_1 : \overline{M}_{0,2}(\mathbb{P}^n, d) \to \mathbb{P}^n$ is equivariant.

We begin by defining an equivariant version of the form (11.54) of $J_{\mathbb{P}^n}$:

$$(11.56) \qquad J_T = e^{(t_0 + t_1 p)/\hbar}\left(1 + \sum_{d=1}^{\infty} q^d e_{1!}\left(\frac{1}{\hbar - c_T}\right)\right),$$

where c_T is a symbol for the equivariant Chern class $c_1^T(\mathcal{L}_1)$. We also set $q = e^{t_1}$, and as usual, p is the equivariant hyperplane class of $\mathbb{P}^n$. The expression J_T takes values in $H_T^*(\mathbb{P}^n)$, and we can recover $J_{\mathbb{P}^n}$ by taking the nonequivariant limit $\lambda_i \to 0$, $p \to H$, $c_T \to c_1(\mathcal{L}_1)$, or more precisely, by $J_{\mathbb{P}^n} = i_{\mathbb{P}^n}^*(J_T)$, where $i_{\mathbb{P}^n}$ is as defined in Section 9.1.1.

It is convenient to drop the exponential factor in (11.56) and put

$$S = 1 + \sum_{d=1}^{\infty} q^d e_{1!}\left(\frac{1}{\hbar - c_T}\right),$$

so that $J_T = e^{(t_0 + t_1 p)/\hbar} S$.

We next decompose S into pieces which are easier to study. We let $Z_i = \int_{(\mathbb{P}^n)_T} S \cup \phi_i$, where $\phi_i = \prod_{k \neq i}(p - \lambda_k)$ is the basis of $H_T^*(\mathbb{P}^n) \otimes_{H^*(BT)} \mathcal{R}_T$ from Example 9.1.2.1. The important fact is that the Z_i are sufficient to determine S, and hence J_T and ultimately $J_{\mathbb{P}^n}$, since we can recover S via $S = \sum_{i=0}^n Z_i \cup \phi^i$, where ϕ^i is the basis dual to ϕ_i.

The projection formula implies

$$\int_{(\mathbb{P}^n)_T} e_{1!}\left(\frac{1}{\hbar - c_T}\right) \cup \phi_i = \int_{\overline{M}_{0,2}(\mathbb{P}^n, d)_T} \frac{e_1^*(\phi_i)}{\hbar - c_T},$$

so that we can write Z_i in the form

$$(11.57) \qquad Z_i = 1 + \sum_{d=1}^{\infty} q^d \int_{\overline{M}_{0,2}(\mathbb{P}^n, d)_T} \frac{e_1^*(\phi_i)}{\hbar - c_T}.$$

We now note a simplification. Since $\dim \overline{M}_{0,2}(\mathbb{P}^n, d) = (n+1)d + n - 1$, any cohomology class on $\overline{M}_{0,2}(\mathbb{P}^n, d)$ of degree less than $(n+1)d + n - 1$ has vanishing integral. Note in addition that ϕ_i has cohomological degree n. When we expand $1/(\hbar - c_T)$ as a geometric series in powers of $c_T/\hbar$ and substitute into (11.57), we conclude that the terms involving c_T^k can only contribute to the expansion of (11.57) for $k \geq (n+1)d - 1$. Thus, we may harmlessly truncate the geometric series. For reasons that will become clearer soon, we truncate somewhat less than we are able to, beginning our expansion with c_T^d (which we can do, since $d \leq (n+1)d - 1$). Using the identity

$$(11.58) \qquad \sum_{k=d}^{\infty} \frac{c_T^k}{\hbar^{k+1}} = \left(\frac{c_T}{\hbar}\right)^d \frac{1}{\hbar - c_T},$$

we obtain in place of (11.57) the alternative form

$$(11.59) \qquad Z_i = Z_i(q, \hbar) = 1 + \sum_{d=1}^{\infty} \left(\frac{q}{\hbar}\right)^d \int_{\overline{M}_{0,2}(\mathbb{P}^n, d)_T} \frac{e_1^*(\phi_i)\, c_T^d}{\hbar - c_T}.$$

We will evaluate the integrals appearing in Z_i using localization on $\overline{M}_{0,2}(\mathbb{P}^n, d)$ in the form of Corollary 9.1.4. Each integral

$$\int_{\overline{M}_{0,2}(\mathbb{P}^n, d)_T} \frac{e_1^*(\phi_i)\, c_T^d}{\hbar - c_T}$$

will be a sum over the components of the fixed point set, which by Section 9.2.1 are described by certain labeled graphs Γ.

Let Γ be a graph corresponding to a fixed point component $\overline{M}_\Gamma$ of $\overline{M}_{0,2}(\mathbb{P}^n, d)$, and as in Section 9.2.1, we have the inclusion map $i_\Gamma : \overline{M}_\Gamma \to \overline{M}_{0,2}(\mathbb{P}^n, d)$. Our first task is to identify those components which contribute to the above integral under localization. Recall that the fixed points of T acting on $\mathbb{P}^n$ are q_j for $0 \leq j \leq n$. Then observe that $i_\Gamma^*(e_1^*(\phi_i))$ vanishes unless $e_1(\overline{M}_\Gamma) = q_i$. This follows because if $e_1(\overline{M}_\Gamma) = q_j$ for $j \neq i$, then

$$i_\Gamma^*(e_1^*(\phi_i)) = i_{q_j}^*(\phi_i) = 0,$$

where the last equality is by (9.10). This implies that when we localize the above integral, we can restrict the sum to those Γ for which $e_1(\overline{M}_\Gamma) = q_i$. (Note the similarity to what we did at the beginning of the proof of Theorem 11.1.4.)

Since the equivariant hyperplane class p restricts to λ_i over q_i, (9.10) implies that $i_\Gamma^*(e_1^*\phi_i) = i_{q_i}^*(\phi_i) = \prod_{k\neq i}(\lambda_i - \lambda_k)$. We also use the symbol c_Γ to denote the restriction of the symbol c_T to Z_Γ. More precisely, we define

$$\frac{1}{\hbar - c_\Gamma} = \sum_{k=0}^{\infty} \hbar^{-(k+1)} c_1^T (i_\Gamma^* \mathcal{L}_1)^k.$$

Finally, we let a_Γ be the order of the automorphism group of a typical stable map with graph Γ. Then Corollary 9.1.4 gives

$$(11.60) \qquad Z_i(q, \hbar) = 1 + \sum_{d=1}^{\infty} \left(\frac{q}{\hbar}\right)^d \sum_\Gamma^{(i)} \int_{(\overline{M}_\Gamma)_T} \frac{c_\Gamma^d \prod_{k\neq i}(\lambda_i - \lambda_k)}{a_\Gamma(\hbar - c_\Gamma)\mathrm{Euler}_T(N_\Gamma)},$$

where $\sum_\Gamma^{(i)}$ means that we sum only over those Γ satisfying $e_1(\overline{M}_\Gamma) = q_i$.

We can further reduce the number of components $\overline{M}_\Gamma$ to consider as follows. If $(f, C, p_1, p_2) \in \overline{M}_\Gamma$, then $e_1(\overline{M}_\Gamma) = q_i$ implies that $e_1(f) = f(p_1) = q_i$. We will let $C' \subset C$ be the component of C containing p_1. We then distinguish the following 2 cases:

Type A: The component $C' \subset C$ containing p_1 is mapped by f onto a curve.

Type B: The component $C' \subset C$ containing p_1 is mapped by f onto a point.

We claim that Type B graphs Γ do not contribute to Z_i, since $c_\Gamma^d = 0$ when Γ is of Type B.

To prove that $c_\Gamma^d = 0$ for Type B graphs, we need to recall the description of $\overline{M}_\Gamma$ given in Section 9.2.1. The graph Γ has vertices which correspond to connected components of $f^{-1}(\{q_0, \dots, q_n\})$. Since Γ has Type B, we know that C' lies in some connected component C_v of $f^{-1}(q_i)$, where v is the corresponding vertex of Γ. As in Section 9.2.1, the number $n(v)$ is defined to be the number of marked points in C_v together with the number of nodes in C_v which connect C_v to a component C_e of C on which f has degree $d_e \geq 1$. Since we are working on $\overline{M}_{0,2}(\mathbb{P}^n, d)$, there are at most 2 marked points. Furthermore, there at most d nodes of the above sort since $\sum_e d_e = d$ by the definition of Γ. It follows that $n(v) \leq d + 2$.

The number $n(v)$ is important because, as explained in Section 9.2.1, $\overline{M}_{0,n(v)}$ is a factor in the product which gives $\overline{M}_\Gamma$. Our strategy will be to study the restriction of c_Γ^d to $\overline{M}_{0,n(v)}$. Since $n(v) \leq d+2$, it follows that $\dim \overline{M}_{0,n(v)} = n(v) - 3 \leq d - 1$. Furthermore, c_Γ in this case is induced from the equivariant Chern class $c_1^T(\mathcal{L}_1)$ of the corresponding bundle $\mathcal{L}_1$ on $\overline{M}_{0,n(v)}$. However, T acts trivially on $\overline{M}_{0,n(v)}$, so that $H_T^*(\overline{M}_{0,n(v)}) \simeq H^*(\overline{M}_{0,n(v)}) \otimes H^*(BT)$. In addition, T acts trivially on $\mathcal{L}_1$, which implies $c_1^T(\mathcal{L}_1) = c_1(\mathcal{L}_1) \otimes 1$ under the above isomorphism. Since $c_1(\mathcal{L}_1)$ has degree 2 and $\dim \overline{M}_{0,n(v)} \leq d - 1$, we have $c_1(\mathcal{L}_1)^d = 0$. This proves that $c_\Gamma^d = 0$, as claimed.

Since graphs of Type B don't contribute to the integral, we let $\sum_\Gamma^A$ denote the sum over Type A graphs. We continue to assume that $e_1(\overline{M}_\Gamma) = q_i$ even though our notation suppresses the index i. Then (11.60) becomes

$$(11.61) \qquad Z_i(q, \hbar) = 1 + \sum_{d=1}^{\infty} \sum_\Gamma^A \left(\frac{q}{\hbar}\right)^d \int_{(\overline{M}_\Gamma)_T} \frac{c_\Gamma^d \prod_{k\neq i}(\lambda_i - \lambda_k)}{a_\Gamma(\hbar - c_\Gamma)\mathrm{Euler}_T(N_\Gamma)}.$$

Using this, we can prove the following lemma.

LEMMA 11.2.8 (Regularity Lemma). *The expression Z_i is an element of the ring $\mathbb{Q}(\lambda_i, \hbar)[[q]]$. The coefficient of each q^d is a rational function of λ_i and $\hbar$ which is regular at each $\hbar = (\lambda_i - \lambda_j)/k$, for all $j \neq i$ and $k \geq 1$.*

PROOF. We need to examine the coefficient of q^d in (11.61). For this purpose, suppose that Γ has type A. The factor of $\mathrm{Euler}_T(N_\Gamma)$ in the denominator is in $H_T^*(\overline{M}_\Gamma)$ and hence is rational in the λ_i. So it remains to analyze c_Γ.

Recall that $c_\Gamma = c_1^T(i_\Gamma^*(\mathcal{L}_1))$. In Section 10.1.1, we defined $\mathcal{L}_1$ to be the line bundle on $\overline{M}_{0,2}(\mathbb{P}^n, d)$ whose fiber at (f, C, p_1, p_2) is the cotangent space $T_{p_1}^* C$. In order to compute $c_1^T(i_\Gamma^*(\mathcal{L}_1))$, we need to compute the weights of T acting on $T_{p_1}^* C$ when $(f, C, p_1, p_2) \in M_\Gamma$. As above, let $C' \subset C$ be the irreducible component containing p_1. Since Γ has Type A, we know that C' is mapped to the line $\overline{q_i q_j}$ with degree d'. At $q_i \in \overline{q_i q_j} \simeq \mathbb{P}^1$, the cotangent space $T_{q_i}^* \mathbb{P}^1$ has weight $\lambda_i - \lambda_j$ (this follows by the techniques used in Example 9.1.2.1), and since $f|_{C'}$ has degree d', the T-action on $T_{p_1}^* C = T_{p_1}^* C'$ has weight $(\lambda_j - \lambda_i)/d'$. It follows that $c_\Gamma = (\lambda_j - \lambda_i)/d'$, which tells us that

$$\int_{(\overline{M}_\Gamma)_T} \frac{c_\Gamma^d \prod_{k \neq i}(\lambda_i - \lambda_k)}{a_\Gamma(\hbar - c_\Gamma)\mathrm{Euler}_T(N_\Gamma)} = \int_{(\overline{M}_\Gamma)_T} \frac{((\lambda_j - \lambda_i)/d')^d \prod_{k \neq i}(\lambda_i - \lambda_k)}{a_\Gamma(\hbar - (\lambda_j - \lambda_i)/d')\mathrm{Euler}_T(N_\Gamma)}$$

in this case. From here, the desired result is now easily verified. $\square$

The lemma just proved shows that we can substitute $\hbar = (\lambda_j - \lambda_i)/d$ in Z_j and obtain a well-defined expression. This is needed for the recursion lemma, which goes as follows.

LEMMA 11.2.9 (Recursion Lemma). *For integers $i \neq j$ with $0 < i, j < n$, put*

$$C_{i,j}(d) = \frac{1}{d! \prod_{(k,m) \neq (j,d), k \neq i} (\lambda_i - \lambda_k + m(\lambda_j - \lambda_i)/d)},$$

where m ranges from 1 to d in the product in the denominator. Then

$$Z_i(q, \hbar) = 1 + \sum_{j \neq i} \sum_{d=1}^{\infty} \left(\frac{q}{\hbar}\right)^d \frac{C_{i,j}(d)}{\lambda_i - \lambda_j + d\hbar} Z_j\left(\frac{q}{\hbar} \frac{\lambda_j - \lambda_i}{d}, \frac{\lambda_j - \lambda_i}{d}\right).$$

PROOF. This recursion will be derived from (11.61). As noted in the discussion following (11.57), we could have truncated the geometric series for $1/(\hbar - c_\Gamma)$ to begin with the term $c_\Gamma^{(n+1)d-1}/\hbar^{(n+1)d}$ in (11.58). Had we done so, we would have been led to the recursion appearing in [**Givental2**]. The choice we made here leads to the recursion relation appearing in [**Kim2**] and avoids the change of variables for q used in [**Givental2**].

The graphs Γ occurring in (11.61) are all of Type A. For $(f, C, p_1, p_2) \in \overline{M}_\Gamma$, we know that $f(p_1) = q_i$, and if $C' \subset C$ is the component containing p_1, then there is $j \neq i$ such that $f|_{C'} : C' \to \overline{q_i q_j}$ is a cover of degree d'. In this situation, the definition of Γ implies that $C = C' \cup C''$, with C'' corresponding to a graph Γ'' and $p_2 \in C''$. If we let $p = C' \cap C''$, then $q_j = f(p)$. Furthermore, if we regard p as the first marked point on C'', then $(f|_{C''}, C'', p, p_2)$ is a 2-pointed stable map with graph Γ'' in $\overline{M}_{0,2}(\mathbb{P}^n, d'')$, where $d = d' + d''$. For these graphs, we claim that

$$a_\Gamma = d' a_\Gamma''$$

$$c_\Gamma = \frac{\lambda_j - \lambda_i}{d'}$$

$$\text{(11.62)} \qquad \text{Euler}_T(N_\Gamma) = \text{Euler}_T(N_{\Gamma''})\left(\frac{\lambda_j - \lambda_i}{d'} - c_{\Gamma''}\right) \times$$

$$\left(\prod_{m,k}^{(i)} m\left(\frac{\lambda_j - \lambda_i}{d'}\right) + \lambda_i - \lambda_k\right),$$

where $\prod_{m,k}^{(i)}$ is a product over $0 \le m \le d' - 1$ and $0 \le k \le n$ with the restriction $(m, k) \ne (0, i)$.

The formula for a_Γ arises because we have a cyclic group of order d' of automorphisms of $f|_{C'}$, as in (9.13) and the subsequent discussion. We proved the formula for c_Γ in the proof of Lemma 11.2.8. Finally, the calculation of $\text{Euler}_T(N_\Gamma)$ proceeds by filtering the equivariant normal bundle N_Γ. The factor of $\text{Euler}_T(N_{\Gamma''})$ arises from deformations of $f|_{C''}$. The factor

$$\frac{\lambda_j - \lambda_i}{d'} - c_{\Gamma''}$$

arises from the application of Lemma 9.2.2 to the node p, similar to what we did in the proof of Theorem 11.1.4. The last factor comes from deformations of $f|_{C'}$ which fix the point p. The deformations are given by the vector fields $v_{m,k} = z_i^{d'-m} z_j^m \partial/\partial x_k$ with $0 \le m < d'$. Here, $x_0, \ldots, x_n$ are coordinates on $\mathbb{P}^n$ and z_i, z_j are coordinates on C' such that $f|_{C'}$ is given by $x_i = z_i^{d'}$, $x_j = z_j^{d'}$. The weight of $v_{m,k}$ is clearly

$$(d' - m)\frac{\lambda_i}{d'} + m\frac{\lambda_j}{d'} - \lambda_k = m\left(\frac{\lambda_j - \lambda_i}{d'}\right) + \lambda_i - \lambda_k.$$

Note that the vector fields $v_{d',k}$ do not fix p and hence do not contribute. Also, the vector field $v_{0,i}$ is (up to a scalar) the image of the vector field $z_i \partial/\partial z_i$ of C' fixing p_1 and p. This comes from an automorphism of the stable map, so that we must also exclude $v_{0,i}$. Then multiplying the remaining weights gives the final factor in the formula for $\text{Euler}_T(N_\Gamma)$.

We now substitute (11.62) into (11.61) and calculate that Z_i is given by

$$\text{(11.63)} \qquad Z_i = 1 + \sum_{j \ne i} \sum_{d'=1}^{\infty} \frac{(q/\hbar)^{d'}\left((\lambda_j - \lambda_i)/d'\right)^{d'} \prod_{k \ne i}(\lambda_i - \lambda_k)}{(\prod_{k \ne j}(\lambda_j - \lambda_k))(\hbar - (\lambda_j - \lambda_i)/d')(d') \prod_{m,k}^{(i)}(m(\lambda_j - \lambda_i)/d' + \lambda_i - \lambda_k)}$$

$$\times \left(1 + \sum_{d''=1}^{\infty} \sum_{\Gamma''}{}^{(j)}\left(\frac{q}{\hbar}\right)^{d''} \int_{(\overline{M}_{\Gamma''})_T} \frac{((\lambda_j - \lambda_i)/d')^{d''} \prod_{k \ne j}(\lambda_j - \lambda_k)}{\text{Euler}_T(N_{\Gamma''})a_{\Gamma''}((\lambda_j - \lambda_i)/d' - c_{\Gamma''})}\right).$$

In obtaining (11.63), we have inserted cancelling factors of $\prod_{k \ne j}(\lambda_j - \lambda_k)$ and also used $d = d' + d''$.

We can now prove the desired formula for Z_i by noting the following, starting with (11.63). The last line of (11.63) is just

$$Z_j\big(q(\lambda_j - \lambda_i)/(d'\hbar), (\lambda_j - \lambda_i)/d'\big),$$

as can be seen from (11.60). As for the first line, note that the large fraction contains

$$\left(\frac{q}{\hbar}\right)^{d'} \frac{1}{\lambda_i - \lambda_j + d'\hbar}$$

as a factor. Thus the lemma reduces to proving that the remaining factor is $C_{i,j}(d')$. In other words, we need to show that

$$(11.64) \qquad C_{i,j}(d') = \frac{((\lambda_j - \lambda_i)/d')^{d'} \prod_{k \neq i}(\lambda_i - \lambda_k)}{\left(\prod_{k \neq j}(\lambda_j - \lambda_k)\right) \prod_{m,k}^{(i)} (m(\lambda_j - \lambda_i)/d' + \lambda_i - \lambda_k)}.$$

To prove this, put $A(m,k) = m(\lambda_j - \lambda_i)/d' + \lambda_i - \lambda_k$, and note the following three identities involving $A(m,k)$:

$$(d')!\left(\frac{\lambda_j - \lambda_i}{d'}\right)^{d'} = \prod_{m=1}^{d'} A(m,i)$$
$$\lambda_i - \lambda_k = A(0,k)$$
$$\lambda_j - \lambda_k = A(d',k).$$

This enables one to express both sides of (11.64) in terms of the $A(m,k)$, together with a factor of $(d')!$ in the denominator. After a little cancelling of some of the $A(m,k)$'s on the right hand side, we achieve the desired result. $\square$

At a first glance, it may not be obvious that the identity proved in Lemma 11.2.9 is a recursion. This becomes clearer when we write Z_i in the form

$$Z_i(q,\hbar) = \sum_{d=0}^{\infty} Z_{i,d}(\hbar)\, q^d,$$

where $Z_{i,0}(\hbar) = 1$. Then for $d \geq 1$, Lemma 11.2.9 implies that

$$Z_{i,d}(\hbar) = \sum_{\substack{d=d'+d'' \\ d' \geq 1}} \frac{C_{i,j}(d')}{\lambda_i - \lambda_j + d'\hbar} \left(\frac{\lambda_j - \lambda_i}{d'\hbar}\right)^{d''} Z_{j,d''}\left(\frac{\lambda_j - \lambda_i}{d'}\right)$$

Since $d' \geq 1$, it follows that $Z_{i,d}$ is determined by the $Z_{j,d''}$ for $d'' < d$. Then, since Lemma 11.2.9 also implies $Z_{i,0} = 1$, we see that Z_i is uniquely characterized by the identity in the lemma.

Now that we have the recursion lemma, we can finally prove Proposition 11.2.1.

PROOF OF PROPOSITION 11.2.1. The plan of the proof is to create an equivariant version I_T of $I_{\mathbb{P}^n}$ and show that it leads to the above recursion. Recall from (11.53) that $I_{\mathbb{P}^n}$ is given by

$$I_{\mathbb{P}^n} = e^{(t_0 + t_1 H)/\hbar} \sum_d \frac{q^d}{((H + \hbar)\cdots(H + d\hbar))^{n+1}}.$$

Consider the equivariant expression

$$I_T = e^{(t_0 + t_1 p)/\hbar} \sum_{d=0}^{\infty} q^d \frac{1}{\prod_{j=0}^{n} \prod_{m=1}^{d}(p - \lambda_j + m\hbar)},$$

whose nonequivariant limit is $I_{\mathbb{P}^n}$. Our goal is to prove $I_T = J_T$. We begin by decomposing I_T the same way we did J_T. Hence we drop the exponential factor and let

$$S' = \sum_{d=0}^{\infty} q^d \frac{1}{\prod_{j=0}^{n} \prod_{m=1}^{d} (p - \lambda_j + m\hbar)},$$

and then we set $Z_i' = \int_{(\mathbb{P}^n)_T} S' \cup \phi_i$. We compute Z_i' as follows. Adapting what we did in (11.3) to $\mathbb{P}^n$, we see that $Z_i' = i_{q_i}^*(S') = S'(\lambda_i)$. This implies

$$Z_i' = 1 + \sum_{d=1}^{\infty} q^d \frac{1}{\prod_{j=0}^{n} \prod_{m=1}^{d} (\lambda_i - \lambda_j + m\hbar)}.$$

It is clear from this equation that Z_i' lies in $\mathbb{Q}(\lambda, \hbar)[[q]]$ and satisfies the regularity of Lemma 11.2.8. Furthermore, a direct calculation shows that Z_i' satisfies the recursion of Lemma 11.2.9. By the uniqueness noted in the discussion following the proof of the lemma, we get $Z_i = Z_i'$ for all i.

To finish the proof, we recall from the discussion following (11.56) that

$$J_T = e^{(t_0 + t_1 p)/\hbar} \sum_{i=0}^{n} Z_i \cup \phi^i.$$

Since similar reasoning implies

$$I_T = e^{(t_0 + t_1 p)/\hbar} \sum_{i=0}^{n} Z_i' \cup \phi^i,$$

we conclude that $I_T = J_T$. Then the desired result $I_{\mathbb{P}^n} = J_{\mathbb{P}^n}$ is obtained by taking the nonequivariant limit $\lambda_i \to 0$, $p \to H$. $\qquad\square$

11.2.4. Proof of Theorem 11.2.2. In order to prove the Mirror Theorem stated in Section 11.2.1, we will generalize the steps used in the proof of Proposition 11.2.1. However, the recursion we get in this case (Lemma 11.2.11) is significantly weaker, so that we will also need Lemmas 11.2.12, 11.2.14, and 11.2.15 below. Our treatment is based on [**Givental2, Pandharipande3**], though we use some conventions from [**Kim2, Givental4**]. Implicit in our discussion is a proof of Conjecture 11.2.7 for complete intersections in projective space.

Just as in the proof of Proposition 11.2.1, we begin by giving an equivariant version of $J_{\mathcal{V}}$. From the definition of $J_{\mathcal{V}}$ given in (11.44), we define

$$(11.65) \qquad J_T = e^{(t_0 + pH)/\hbar} \, \mathrm{Euler}_T(\mathcal{V}) \left(1 + \sum_{d=1}^{\infty} q^d e_{1!}\left(\frac{\mathrm{Euler}_T(\mathcal{V}_{d,2,1}')}{(\hbar - c_T)} \right) \right).$$

Since the Euler class of the trivial bundle is 1, this reduces to (11.56) when $\mathcal{V} = 0$. Note that T acts naturally on $\mathcal{V} = \oplus_{i=1}^{\ell} \mathcal{O}_{\mathbb{P}^n}(a_i)$, so that $\mathrm{Euler}_T(\mathcal{V}_{d,2,1}')$ makes sense.

As in Section 11.2.3, we drop the exponential factor and put

$$(11.66) \qquad S_{\mathcal{V}} = S_{\mathcal{V}}(q, \hbar) = \mathrm{Euler}_T(\mathcal{V}) \left(1 + \sum_{d=1}^{\infty} q^d e_{1!}\left(\frac{\mathrm{Euler}_T(\mathcal{V}_{d,2,1}')}{(\hbar - c_T)} \right) \right).$$

We also put

$$\tilde{S}_{\mathcal{V}} = 1 + \sum_{d>0} q^d e_{1!}\left(\frac{\mathrm{Euler}_T(\mathcal{V}_{d,2,1}')}{(\hbar - c_T)} \right),$$

so that $S_\mathcal{V} = \mathrm{Euler}_T(\mathcal{V})\tilde{S}_\mathcal{V}$. As we have seen in Example 11.1.3.1, p is invertible after tensoring with the field of rational functions in the λ_i, hence the same is true of $\mathrm{Euler}_T(\mathcal{V}) = (\prod_{i=1}^\ell a_i)p^\ell$. In particular, $\tilde{S}_\mathcal{V}$ is uniquely determined by $S_\mathcal{V}$, and can be expressed as

$$\tilde{S}_\mathcal{V} = (\mathrm{Euler}_T(\mathcal{V}))^{-1}S_\mathcal{V},$$

where we again allow rational functions in the λ_i to make sense of this. In the rest of this chapter, we will frequently adopt similar notation, using a tilde over a symbol to denote division by $\mathrm{Euler}_T(\mathcal{V})$.

Finally, we put

$$Z_{i,\mathcal{V}} = \int_{(\mathbb{P}^n)_T} \tilde{S}_\mathcal{V} \cup \phi_i = i_{q_i}^*(\tilde{S}_\mathcal{V}).$$

We want to study whether $Z_{i,\mathcal{V}}$ has the same behavior as the function Z_i studied in Section 11.2.3. The first step is easy, for it is straightforward to show that (11.57) generalizes to

$$(11.67) \qquad Z_{i,\mathcal{V}} = 1 + \sum_{d=1}^\infty q^d \int_{\overline{M}_{0,2}(\mathbb{P}^n,d)_T} \frac{\mathrm{Euler}_T(\mathcal{V}'_{d,2,1})\, e_1^*(\phi_i)}{\hbar - c_T}.$$

However, the truncation we did in (11.58) and (11.59) no longer works in general because of the presence of $\mathrm{Euler}_T(\mathcal{V}'_{d,2,1})$. Hence, instead of using localization to go from (11.59) to (11.60), we instead apply localization to the integrals appearing in (11.67). This easily gives

$$(11.68) \qquad Z_{i,\mathcal{V}} = 1 + \sum_{d=1}^\infty q^d \sum_\Gamma^{(i)} \int_{(\overline{M}_\Gamma)_T} \frac{i_\Gamma^*(\mathrm{Euler}_T(\mathcal{V}'_{d,2,1}))\prod_{k\neq i}(\lambda_i - \lambda_k)}{a_\Gamma(\hbar - c_\Gamma)\mathrm{Euler}_T(N_\Gamma)},$$

where $\sum_\Gamma^{(i)}$ has the same meaning it did in (11.60).

Before continuing the proof, we will pause to make an interesting observation about truncation. The key point is that the Mirror Theorem requires a change of variables because we can't truncate in general. For example, truncation does work if $\mathcal{V} = \oplus_{i=1}^\ell \mathcal{O}(a_i)$ satisfies $\sum_{i=1}^\ell a_i < n$. The straightforward verification uses the fact that $\mathcal{V}'_{d,2,1}$ has rank $d\sum_{i=1}^\ell a_i < dn$. Thus $d + \mathrm{rank}(\mathcal{V}'_{d,2,1}) < d + nd$, so $d + \mathrm{rank}(\mathcal{V}'_{d,2,1}) \leq (n+1)d - 1$, as required by an argument analogous to the one preceding (11.58). In this case, the proof below simplifies considerably, becoming much more like what we did in Section 11.2.3. The result is that when $\sum_{i=1}^\ell a_i < n$, the Mirror Theorem asserts that $I_\mathcal{V} = J_\mathcal{V}$. In other words, Theorem 11.2.2 holds with $f_0 = f_1 = h = 0$ in this situation. Note that we saw a special case of this in Example 11.2.1.2. In general, when $\sum_{i=1}^\ell a_i \leq n+1$, we will see how the lack of truncation complicates the recursion and forces us to change variables.

Returning to the proof of the Mirror Theorem, we note the following result of [**Pandharipande3**], which generalizes Lemma 11.2.8.

LEMMA 11.2.10 (Regularity Lemma). *The expression $Z_{i,\mathcal{V}}$ is an element of the ring $\mathbb{Q}(\lambda_i,\hbar)[[q]]$. The coefficient of each q^d is a rational function of λ_i and $\hbar$ which is regular at each $\hbar = (\lambda_i - \lambda_j)/k$, for all $j \neq i$ and $k \geq 1$.*

PROOF. For a graph Γ appearing in $\sum_\Gamma^{(i)}$, we need to study the integral

$$(11.69) \qquad \int_{(\overline{M}_\Gamma)_T} \frac{i_\Gamma^*(\mathrm{Euler}_T(\mathcal{V}'_{d,2,1}))\prod_{k\neq i}(\lambda_i-\lambda_k)}{a_\Gamma(\hbar-c_\Gamma)\mathrm{Euler}_T(N_\Gamma)}.$$

In Section 11.2.3, we divided the graphs Γ into Type A and Type B graphs. For graphs of Type A, the proof of Lemma 11.2.8 shows that $c_\Gamma = (\lambda_j - \lambda_i)/d'$ for some $d' \geq 1$ and $j \neq i$. Hence for these components, the above integral becomes

$$\int_{(\overline{M}_\Gamma)_T} \frac{i_\Gamma^*(\mathrm{Euler}_T(\mathcal{V}'_{d,2,1}))\prod_{k\neq i}(\lambda_i-\lambda_k)}{a_\Gamma(\hbar-(\lambda_j-\lambda_i)/d')\mathrm{Euler}_T(N_\Gamma)}.$$

As in Lemma 11.2.8, this expression is clearly rational in $\lambda_i, \hbar$ and is regular at $\hbar = (\lambda_i - \lambda_j)/k$ for $k \geq 1$.

Now suppose that Γ is of Type B. In this case, we showed in the discussion preceding Lemma 11.2.8 that $c_\Gamma^d = 0$. Thus, when we expand the factor of $\hbar - c_\Gamma$ in the denominator of (11.69) into a geometric series, we can ignore all terms which involve $c_\Gamma^k/\hbar^{k+1}$ for $k \geq d$. Thus, we only need consider those terms with $k < d$. Each such term is a rational function of the λ_i times a positive power of $\hbar^{-1}$ and hence is clearly a rational function of $\hbar$ and the λ_i satisfying the claimed regularity. $\qquad\square$

As in Section 11.2.3, this lemma shows that we can substitute $\hbar = (\lambda_i - \lambda_j)/d$ in $Z_{j,\mathcal{V}}$ and obtain a well-defined expression.

LEMMA 11.2.11 (Recursion Lemma). *Let* $\mathcal{V} = \oplus_{i=1}^\ell \mathcal{O}_{\mathbb{P}^n}(a_i)$, *and for integers* $i \neq j$ *with* $0 \leq i,j \leq n$, *put*

$$C_{i,j}^{\mathcal{V}}(d) = \frac{\prod_{k=1}^\ell \prod_{m=1}^{a_k d}(a_k\lambda_i + m(\lambda_j-\lambda_i)/d)}{d!\prod_{(k,m)\neq(j,d),k\neq i}(\lambda_i-\lambda_k+m(\lambda_j-\lambda_i)/d)},$$

where m ranges from 1 to d in the product in the denominator. Then the difference

$$R_i = Z_{i,\mathcal{V}}(q,\hbar) - \sum_{j\neq i}\sum_{d=1}^\infty \left(\frac{q}{\hbar}\right)^d \frac{C_{i,j}^{\mathcal{V}}(d)}{\lambda_i-\lambda_j+d\hbar} Z_{j,\mathcal{V}}\left(\frac{q}{\hbar}\frac{\lambda_j-\lambda_i}{d},\frac{\lambda_j-\lambda_i}{d}\right)$$

is in $\mathcal{R}_T[\hbar^{-1}][[q]]$.

PROOF. The identity

$$\frac{1}{\hbar-c_\Gamma} = \left(\frac{c_\Gamma}{\hbar}\right)^d\frac{1}{\hbar-c_\Gamma} + \sum_{k=0}^{d-1}c_\Gamma^k\hbar^{-(k+1)}$$

shows that we can rewrite (11.68) as

$$Z_{i,\mathcal{V}} = 1 + \sum_{d=1}^\infty\sum_\Gamma{}^{(i)}\left(\frac{q}{\hbar}\right)^d\int_{(\overline{M}_\Gamma)_T}\frac{c_\Gamma^d i_\Gamma^*(\mathrm{Euler}_T(\mathcal{V}'_{d,2,1}))\prod_{k\neq i}(\lambda_i-\lambda_k)}{a_\Gamma(\hbar-c_\Gamma)\mathrm{Euler}_T(N_\Gamma)}$$

$$+ \text{ terms which are polynomial in } \hbar^{-1}.$$

Furthermore, we noted in the proof of Lemma 11.2.10 that $c_\Gamma^d = 0$ for Type B graphs. Hence the above formula becomes

$$Z_{i,\mathcal{V}} = 1 + \sum_{d=1}^\infty\sum_\Gamma{}^A\left(\frac{q}{\hbar}\right)^d\int_{(\overline{M}_\Gamma)_T}\frac{c_\Gamma^d i_\Gamma^*(\mathrm{Euler}_T(\mathcal{V}'_{d,2,1}))\prod_{k\neq i}(\lambda_i-\lambda_k)}{a_\Gamma(\hbar-c_\Gamma)\mathrm{Euler}_T(N_\Gamma)}$$

$$+ \text{ terms which are polynomial in } \hbar^{-1},$$

where as in (11.61), $\sum_{\Gamma}^{A}$ means that we sum only over graphs Γ of Type A with $e_1(\overline{M}_\Gamma) = q_i$. Working modulo polynomials in $\hbar^{-1}$, the above equation is the same as (11.61), except for the factor $i_\Gamma^*(\mathrm{Euler}_T(\mathcal{V}'_{d,2,1}))$ in the numerator. Hence, if we follow the argument of Lemma 11.2.9, we are led to an expression analogous to (11.63), with some extra factors in the numerator. These factors can be identified with the numerator of $C_{i,j}^{\mathcal{V}}(d')$ by computing the weights of the T-action on the bundle $\mathcal{V}'_{d,2,1}$ (we omit the details of this argument). It follows that R_i is zero modulo polynomials in $\hbar^{-1}$, and the lemma is proved. $\qquad\square$

If one compares this lemma to the $\mathcal{V} = 0$ version given in Lemma 11.2.9, we see that $R_i = 1$ in that case. So for $\mathcal{V}$ trivial, we get a genuine recursion, where the coefficient of q^d in Z_i is determined by the coefficients of $q^{d'}$ in the Z_j for $d' < d$. But when $\mathcal{V}$ is arbitrary, we only know that R_i is a polynomial in $\hbar^{-1}$. Hence we don't get a recursion in the same way we did when $\mathcal{V} = 0$. Thus, to complete the proof of Theorem 11.2.2, something new is needed. The basic idea is that once we supplement the regularity and recursion of Lemmas 11.2.10 and 11.2.11 with the "double construction" given in Lemma 11.2.12 below, we will have enough properties to force uniqueness. This will finally be achieved in Lemma 11.2.14.

LEMMA 11.2.12 (Double Construction Lemma). *Let z be a variable and write $\mathcal{V} = \oplus_{j=1}^{\ell} \mathcal{O}_{\mathbb{P}^n}(a_j)$. Then the expression*

$$W(z, \hbar) = \sum_{i=0}^{n} \frac{\lambda_i \sum_j a_j}{\prod_{j \neq i}(\lambda_i - \lambda_j)} e^{\lambda_i z} Z_{i,\mathcal{V}}(qe^{z\hbar}, \hbar) Z_{i,\mathcal{V}}(q, -\hbar)$$

is in $H^*(BT)[\hbar][[q, z]]$.

PROOF. Recall the map $\varphi : M_d \to N_d$ and the bundle $\mathcal{V}_d$ discussed in Section 11.1.2, where $M_d = \overline{M}_{0,0}(\mathbb{P}^1 \times \mathbb{P}^n, (1, d))$. Here, we will use a variant of this construction. We will consider $\overline{M}_{0,2}(\mathbb{P}^1 \times \mathbb{P}^n, (1, d))$, which has the evaluation maps $e_1, e_2 : \overline{M}_{0,2}(\mathbb{P}^1 \times \mathbb{P}^n, (1, d)) \to \mathbb{P}^1 \times \mathbb{P}^n$. Following [**Givental2**], we define

$$L_d = e_1^{-1}(\{0\} \times \mathbb{P}^n) \cap e_2^{-1}(\{\infty\} \times \mathbb{P}^n).$$

Let i_{L_d} be the inclusion of L_d in $\overline{M}_{0,2}(\mathbb{P}^1 \times \mathbb{P}^n, (1, d))$. There is a map $\mu : L_d \to N_d$ defined by $\mu = \varphi \circ \pi_1 \circ \pi_2 \circ i_{L_d}$, where π_k is the map induced by forgetting the kth marked point, for $k = 1, 2$. This map is discussed in [**Givental2**], where N_d is denoted by L'_d.

Also recall the map $\pi : M_d \to \overline{M}_{0,0}(\mathbb{P}^n, d)$ induced by the natural projection $\mathbb{P}^1 \times \mathbb{P}^n \to \mathbb{P}^n$. We get from this a map $\widetilde{\pi} : L_d \to \overline{M}_{0,0}(\mathbb{P}^n, d)$ defined by $\widetilde{\pi} = \pi \circ \pi_1 \circ \pi_2 \circ i_{L_d}$. Then define $\widetilde{\chi}_d^{\mathcal{V}} = \mathrm{Euler}_G(\widetilde{\pi}^* \mathcal{V}_d)$. We assert the identity

$$(11.70) \qquad \sum_{d=0}^{\infty} q^d \int_{(L_d)_G} e^{\mu^*(\kappa)z} \, \widetilde{\chi}_d^{\mathcal{V}} = \sum_{i=0}^{n} \frac{\lambda_i \sum_j a_j}{\prod_{j \neq i}(\lambda_i - \lambda_j)} e^{\lambda_i z} Z_{i,\mathcal{V}}(qe^{z\hbar}, \hbar) Z_{i,\mathcal{V}}(q, -\hbar).$$

Once we prove (11.70), the desired conclusion follows, since $\mu^*(\kappa)$ and $\mathrm{Euler}_G(\mathcal{V}_d)$ are in $H_G^*(M_d)$ and hence have polynomial dependence on $\hbar$ and the λ_i.

We will prove (11.70) using the localization formula of Corollary 9.1.4. Since the argument is very similar to the proof of Theorem 11.1.4, we will only sketch the computation, omitting most of the details. A complete proof of (11.70) in the case when $\sum_j a_j = n + 1$ can be found in [**Pandharipande3**, Lemma 7].

We begin by examining the coefficient

$$(11.71) \qquad \int_{(L_d)_G} e^{\mu^*(\kappa)z}\, \widetilde{\chi}_d^{\mathcal{V}}$$

of q^d in the left hand side of (11.70). The first step is to identify the fixed point components of L_d. These are very similar to the fixed point components which arose in the proof of Theorem 11.1.4.

First, fix an i with $0 \le i \le n$ and let q_i be the corresponding T-fixed point of $\mathbb{P}^n$. For each r with $0 < r \le d$, let F'_r be one of the components of $\overline{M}_{0,2}(\mathbb{P}^n, r)^T$ consisting of stable maps for which the first marked point is mapped to the fixed point $q_i \in \mathbb{P}^n$. If in addition $r < d$, choose a component F'_{d-r} of $\overline{M}_{0,2}(\mathbb{P}^n, d-r)^T$ consisting of stable maps for which the first marked point is mapped to $q_i \in \mathbb{P}^n$. We now describe a component of $(L_d)^G$ which is isomorphic to $F'_r \times F'_{d-r}$. We will abuse notation by denoting this component by $F'_r \times F'_{d-r}$. A typical map $f \in F'_r \times F'_{d-r}$ has source curve $C = C_0 \cup C_1 \cup C_2$ with $C_0 = \mathbb{P}^1$ with $f : C \to \mathbb{P}^1 \times \mathbb{P}^n$ given by

$$f|_{C_0}(z) = (z, q_i)$$
$$f|_{C_1}(z) = (0, f_1(z))$$
$$f|_{C_2}(z) = (\infty, f_2(z)).$$

The curve C_1 contains two marked points (p'_1, p_1) and $f_1 : (C_1, p'_1, p_1) \to \mathbb{P}^n$ is an element of F'_r. In addition, C_2 contains two marked points (p'_2, p_2) and the map $f_2 : (C_2, p'_2, p_2) \to \mathbb{P}^n$ is in F'_{d-r}. Finally, we attach $0 \in C_0$ to $p'_1 \in C_1$ and we attach $\infty \in C_0$ to $p'_2 \in C_2$. Then the map $f : (C, p_1, p_2) \to \mathbb{P}^1 \times \mathbb{P}^n$ is an element of L_d, and the set of all such maps is the component $F'_r \times F'_{d-r}$ that we set out to describe. We leave the simpler description of the components corresponding to $r = 0$ or $r = d$ to the reader.

We now describe the contribution of this component to the integral (11.71) by localization. We begin with the restrictions of the terms in the integrand to $F'_r \times F'_{d-r}$. Note that $\mu(F'_r \times F'_{d-r})$ is the G-fixed point $p_{i,r}$ of N_d, as follows from the discussion in Theorem 11.1.4. We thus see that $\mu^*(\kappa)$ restricts to $\lambda_i + r\hbar$ on $F'_r \times F'_{d-r}$ by the discussion in Section 11.1.2.

We next compute the restriction of $\widetilde{\chi}_d^{\mathcal{V}}$ to $F'_r \times F'_{d-r}$. Letting ρ_1 and ρ_2 be the projections onto the first and second factors of $F'_r \times F'_{d-r}$ respectively, we have the exact sequence

$$0 \to \rho_1^*(\mathcal{V}'_{r,2,1}|_{F'_r}) \oplus \rho_2^*(\mathcal{V}'_{d-r,2,1}|_{F'_{d-r}}) \to (\widetilde{\pi}^*\mathcal{V}_d)|_{F'_r \times F'_{d-r}} \to \mathcal{V}_{q_i} \otimes_{\mathbb{C}} \mathcal{O}_{F'_r \times F'_{d-r}} \to 0.$$

To see where this exact sequence comes from, we consider its fibers. The fiber over $f \in F'_r \times F'_{d-r}$ of the map $(\widetilde{\pi}^*\mathcal{V}_d)|_{F'_r \times F'_{d-r}} \to \mathcal{V}_{q_i} \otimes_{\mathbb{C}} \mathcal{O}_{F'_r \times F'_{d-r}}$ is the restriction map

$$H^0(C, f^*\mathcal{V}) \longrightarrow H^0(C_0, f|_{C_0}^*\mathcal{V}) = \mathcal{V}_{q_i}.$$

The kernel of this map consists of pairs of sections of $f_1^*\mathcal{V}$ and $f_2^*\mathcal{V}$ which vanish at their respective marked points p'_1 and p'_2. These are the fibers of $\rho_1^*(\mathcal{V}'_{r,2,1}|_{F'_r})$ and $\rho_2^*(\mathcal{V}'_{d-r,2,1}|_{F'_{d-r}})$ respectively. This exact sequence shows that $\widetilde{\chi}_d^{\mathcal{V}}$ restricts to

$$\lambda_i(\textstyle\sum_j a_j)\,\mathrm{Euler}_G(\mathcal{V}'_{r,2,1}|_{F'_r})\,\mathrm{Euler}_G(\mathcal{V}'_{d-r,2,1}|_{F'_{d-r}}),$$

since $\mathcal{V}_{q_i} = (\oplus_j \mathcal{O}_{\mathbb{P}^n}(a_j))_{q_i}$ has weight $\lambda_i \sum_j a_j$.

Finally, we compute the equivariant Euler class of the normal bundle. The result is

$$\mathrm{Euler}_G(N(F'_r \times F'_{d-r})) = \prod_{j \neq i}(\lambda_i - \lambda_j) \times$$

$$\frac{(\hbar - c_1^T(\mathcal{L}_{r,1}))\mathrm{Euler}_G(N(F'_r))}{\prod_{j \neq i}(\lambda_i - \lambda_j)} \frac{(-\hbar - c_1^T(\mathcal{L}_{d-r,1}))\mathrm{Euler}_G(N(F'_{d-r}))}{\prod_{j \neq i}(\lambda_i - \lambda_j)}.$$

(The reasoning for not doing the obvious cancelation will soon become clear.) The calculation is quite similar to the calculation of $\mathrm{Euler}_G(N(F_r \times F_{d-r}))$ in the proof of Theorem 11.1.4. The only difference is that $\mathrm{Euler}_G(N(F_r \times F_{d-r}))$ had a factor of $-\hbar^2$ due to deformations of automorphisms of $\mathbb{P}^1$ which deform 0 and ∞. This factor is not present in $\mathrm{Euler}_G(N(F'_r \times F'_{d-r}))$, since the composition with the projection to the $\mathbb{P}^1$ factor of $\mathbb{P}^1 \times \mathbb{P}^n$ of all maps in L_d always fixes 0 and ∞ by definition of L_d.

Putting all of this together, we see that the contribution of $F'_r \times F'_{d-r}$ to the integral (11.71) is

$$\frac{\lambda_i \sum_j a_j}{\prod_{j \neq i}(\lambda_i - \lambda_j)} e^{(\lambda_i + r\hbar)z} \times$$

$$\int_{(F'_r)_G} \frac{i^*_{F'_r}(\mathrm{Euler}_G(\mathcal{V}'_{r,2,1})) \prod_{j \neq i}(\lambda_i - \lambda_j)}{a_r(\hbar - c_{F'_r})\mathrm{Euler}_G(N(F'_r))} \times$$

$$\int_{(F'_{d-r})_G} \frac{i^*_{F'_{d-r}}(\mathrm{Euler}_G(\mathcal{V}'_{d-r,2,1})) \prod_{j \neq i}(\lambda_i - \lambda_j)}{a_{d-r}(-\hbar - c_{F'_{d-r}})\mathrm{Euler}_G(N(F'_{d-r}))}.$$

In the above, a_r and a_{d-r} are the orders of the relevant finite automorphism groups, and $c_{F'_r}$ and $c_{F'_{d-r}}$ are the restrictions of $c_1^T(\mathcal{L}_{r,1})$ and $c_1^T(\mathcal{L}_{d-r,1})$ to F'_r and F'_{d-r} respectively.

By Corollary 9.1.4, it follows that the integral (11.71) is obtained by summing the above contribution over all F'_r and F'_{d-r} and then summing from $i = 0$ to n. Then multiplying by q^d and summing over all $d \geq 0$ gives the left hand side of (11.70).

In order to relate this to the right hand side of (11.70), recall the formula for $Z_{i,\mathcal{V}}$ given in (11.68):

$$Z_{i,\mathcal{V}}(q,\hbar) = 1 + \sum_{d=1}^{\infty} q^d \sum_{\Gamma}^{(i)} \int_{(\overline{M}_\Gamma)_T} \frac{i^*_\Gamma(\mathrm{Euler}_T(\mathcal{V}'_{d,2,1})) \prod_{j \neq i}(\lambda_i - \lambda_j)}{a_\Gamma(\hbar - c_\Gamma)\mathrm{Euler}_T(N_\Gamma)}.$$

As explained in (11.60), $\sum_\Gamma^{(i)}$ is the sum over all components $\overline{M}_\Gamma$ in the fixed point locus of $\overline{M}_{0,2}(\mathbb{P}^n, d)$ which map to q_i under e_1. If we switch from d to r, we can then write the above formula as

$$Z_{i,\mathcal{V}}(q,\hbar) = 1 + \sum_{r=1}^{\infty} q^r \sum_{F'_r} \int_{(F'_r)_T} \frac{i^*_{F'_r}(\mathrm{Euler}_T(\mathcal{V}'_{r,2,1})) \prod_{j \neq i}(\lambda_i - \lambda_j)}{a_r(\hbar - c_{F'_r})\mathrm{Euler}_T(N(F'_r))}.$$

Furthermore, we can replace T with $G = \mathbb{C}^* \times T$ since $\mathbb{C}^*$ acts trivially on $\mathbb{P}^n$. If we now expand the right hand side of (11.70) using this formula for $Z_{i,\mathcal{V}}$, one easily sees that the result is precisely the expression for the left hand side of (11.70) described in the previous paragraph. Hence we have proved (11.70), and the lemma follows as explained earlier. $\square$

The formula for $Z_{i,\nu}$ just given shows that $Z_{i,\nu}$ can naturally be regarded as a formal power series in $\hbar^{-1}$. Hence it is surprising that the expression $W(z,\hbar)$ defined in Lemma 11.2.12 is a polynomial in $\hbar$. As we will see, this will play an important role in the proof of the uniqueness lemma (Lemma 11.2.14).

We next unify the properties of the $Z_{i,\nu}$ as expressed in the previous three lemmas. Hence we will consider collections $\{Z_i\}_{i=0}^n$ such that $Z_i \in \mathbb{Q}(\lambda_j)[[\hbar^{-1}, q]]$ for all i. Then, following [**Pandharipande3**], we make the following definition.

DEFINITION 11.2.13. *We say that a collection $\{Z_i\}_{i=0}^n$ is of class $\mathcal{P}$ if*

(i) $Z_i(0, \hbar) = 1$ *for all i.*
(ii) Z_i *satisfies the regularity condition satisfied by $Z_{i,\nu}$ in Lemma 11.2.10.*
(iii) Z_i *satisfies the recursion condition satisfied by $Z_{i,\nu}$ in Lemma 11.2.11.*
(iv) *If we define $W(z,\hbar)$ as in Lemma 11.2.12 using Z_i in place of $Z_{i,\nu}$, then $W(z,\hbar)$ lies in $H^*(BT)[\hbar][[q,z]]$.*

Note that $\{Z_{i,\nu}\}_{i=0}^n$ is of class $\mathcal{P}$ by the three lemmas mentioned in Definition 11.2.13. We can now state the crucial uniqueness lemma.

LEMMA 11.2.14 (Uniqueness Lemma). *Suppose that $\{Z_i\}_{i=0}^n$ and $\{Z_i'\}_{i=0}^n$ are of class $\mathcal{P}$. If for all $0 \le i \le n$, we have $Z_i \equiv Z_i' \bmod \hbar^{-2}$ when these expressions are expanded formally in powers of $\hbar^{-1}$, then $Z_i = Z_i'$ for all i.*

PROOF. We will follow the proof of [**Givental4, Kim2**]. Put $Z_i = \sum_d Z_{i,d} q^d$, and write Z_i' similarly. We know that $Z_{i,0} = Z_{i,0}' = 1$ since $\{Z_i\}$ and $\{Z_i'\}$ are of class $\mathcal{P}$. Hence we can inductively suppose for some $d_0 > 0$ that $Z_{i,d} = Z_{i,d}'$ for $0 \le d < d_0$ and $0 \le i \le n$. Being of class $\mathcal{P}$ implies that the recursion of Lemma 11.2.11 applies to both $\{Z_i\}$ and $\{Z_i'\}$. Combining this with our inductive hypotheses, we see that for all i, $Z_{i,d_0}' - Z_{i,d_0}$ can be expanded as a polynomial in $\hbar^{-1}$. Put

$$D_i(\hbar) = Z_{i,d_0}' - Z_{i,d_0} = \hbar^{-2r_i}\left(A_i \hbar^{-1} + B_i + O(\hbar)\right)$$

where $A_i, B_i \in \mathbb{C}(\lambda_j)$ and $r_i \ge 0$. Since $Z_i \equiv Z_i' \bmod \hbar^{-2}$, $Z_{i,d_0}' = Z_{i,d_0}$ will follow once we show that r_i can be taken to be 0 for all i.

Hence we need to prove that $r_i \ge 1$ implies $A_i = B_i = 0$. For this purpose, let W and W' be the expressions built from $\{Z_i\}$ and $\{Z_i'\}$ respectively using the formula of Lemma 11.2.12. Then our inductive assumption easily implies that the coefficient of q^{d_0} in $W' - W$ is given by

$$\sum_{i=0}^n \frac{\lambda_i \sum_j a_j}{\prod_{j \ne i}(\lambda_i - \lambda_j)}\, e^{\lambda_i z}\left(e^{d_0 \hbar z} D_i(\hbar) + D_i(-\hbar)\right).$$

If we substitute the above formula for D_i into this expression and simplify, we see that the coefficient of q^{d_0} in $W' - W$ has the form

$$\sum_{i=0}^n \frac{\lambda_i \sum_j a_j}{\prod_{j \ne i}(\lambda_i - \lambda_j)}\, e^{\lambda_i z}\hbar^{-2r_i}\left(A_i d_0 z + 2B_i + O(\hbar)\right).$$

Since $\{Z_i\}$ and $\{Z_i'\}$ are of class $\mathcal{P}$, we know that W and W' are polynomials in $\hbar$. Thus $\hbar^{-2r_i}(A_i d_0 z + 2B_i)$ is a polynomial in $\hbar$, and it follow $r_i \ge 1$ implies $A_i = B_i = 0$. This completes the proof of the lemma. $\qquad\square$

Before proving the Mirror Theorem, we need two more ingredients. First, we need to assign weights to the variables which occur. We assign the weight 1 to each of the variables $\lambda_i, p, \hbar$, and weight $n + 1 - \sum a_i$ to q. Below, when we use the term "homogeneous", it will be with respect to these weights. Then we have the following lemma which shows that being of class $\mathcal{P}$ is unaffected by certain coordinate changes.

LEMMA 11.2.15. *Consider formal power series $f(q) \in q\mathbb{Q}[[q]]$, and $g(q), h(q) \in qH^*(BT)[[q]]$, each homogeneous with degrees $\deg f = \deg g = 0, \deg h = 1$. Then, if $\{Z_i\}_{i=0}^n$ is of class $\mathcal{P}$, so is $\{Z_i''\}_{i=0}^n$, where*

$$Z_i''(q, \hbar) = e^{(f(q)+h(q)/\hbar+g(q)\lambda_i/\hbar)} Z_i(qe^{g(q)}, \hbar).$$

PROOF. For the case when $\sum_j a_j = n + 1$, a careful proof can be found in [**Pandharipande3**, Lemma 10]. That argument easily generalizes to $\sum_j a_j \leq n+1$, so that we will omit the details. $\square$

At last we sketch a proof of Givental's Mirror Theorem for nef complete intersections in $\mathbb{P}^n$.

PROOF OF THEOREM 11.2.2. We need to show that $I_\mathcal{V}$ and $J_\mathcal{V}$ agree after an appropriate homogeneous change of variables. As in the case when $\mathcal{V} = 0$, our strategy will be to create equivariant versions of $I_\mathcal{V}$ and $J_\mathcal{V}$ and then prove an equivariant version of the theorem.

The equivariant version of $J_\mathcal{V}$ is J_T from (11.65). Dropping the exponential from J_T gave $S_\mathcal{V}$ in (11.66), and then dropping $\mathrm{Euler}_T(\mathcal{V})$ gave $\tilde{S}_\mathcal{V}$. Finally, we obtained $Z_{i,\mathcal{V}}$ by the formula

$$Z_{i,\mathcal{V}} = \int_{(\mathbb{P}^n)_T} \tilde{S}_\mathcal{V} \cup \phi_i = i_{q_i}^*(\tilde{S}_\mathcal{V}).$$

We know that J_T is uniquely determined by the $Z_{i,\mathcal{V}}$, and $\{Z_{i,\mathcal{V}}\}_{i=0}^n$ is of class $\mathcal{P}$ by Lemmas 11.2.10, 11.2.11 and 11.2.12. Note also that with the above weights, $\tilde{S}_\mathcal{V}$ is homogeneous of weight 0. This can be seen from (11.66) and three easy facts: the bundle $\mathcal{V}'_{d,2,1}$ has rank $\sum_{j=1}^{\ell} a_j d$, the fibers of $e_1 : \overline{M}_{0,2}(\mathbb{P}^n, d) \to \mathbb{P}^n$ have dimension $(n + 1)d - 1$, and a typical term in the expansion of $1/(\hbar - c_T)$ is $c_T^k/\hbar^{k+1}$. We leave the details to the reader.

Turning our attention to the equivariant version of $I_\mathcal{V}$, we define

$$(11.72) \qquad \tilde{S}'_\mathcal{V}(q, \hbar) = 1 + \sum_{d=1}^{\infty} q^d \frac{\prod_{j=1}^{\ell} \prod_{m=1}^{a_j d}(a_j p + m\hbar)}{\prod_{j=0}^{n} \prod_{m=1}^{d}(p - \lambda_j + m\hbar)},$$

and then let

$$I_T = e^{(t_0+t_1 p)/\hbar} \, \mathrm{Euler}_T(\mathcal{V}) \, \tilde{S}'_\mathcal{V}(q, \hbar).$$

The nonequivariant limit of I_T as $\lambda_i \to 0$ is clearly $I_\mathcal{V}$. Also observe that $\tilde{S}'_\mathcal{V}$ is homogeneous of weight 0. Finally, we define

$$Z'_{i,\mathcal{V}} = \int_{(\mathbb{P}^n)_T} \tilde{S}'_\mathcal{V} \cup \phi_i = i_{q_i}^*(\tilde{S}'_\mathcal{V}).$$

One can verify that $\{Z'_{i,\mathcal{V}}\}_{i=0}^n$ is of class $\mathcal{P}$. We omit the details of the straightforward but somewhat lengthy verification.

In order to compare I_T and J_T, first note that

$$J_T = e^{(t_0+t_1 p)/\hbar} \, \mathrm{Euler}_T(\mathcal{V})\big(1 + o(\hbar^{-1})\big).$$

This follows from the same argument used to prove (11.45), which gave the asymptotic expansion $J_\mathcal{V} = e^{(t_0 + t_1 H)/\hbar} \operatorname{Euler}(\mathcal{V})(1 + o(\hbar^{-1}))$.

Now consider a change of variables of the form $t_0 \mapsto t_0 + f(q)\hbar + h(q)$, $t_1 \mapsto t_1 + g(q)$, and let J_T'' be the expression obtained from J_T by this substitution. Then the above asymptotic expansion implies

$$J_T'' = e^{(t_0 + t_1 p)/\hbar} \operatorname{Euler}_T(\mathcal{V})\big(e^{f(q)} + e^{f(q)}h(q)\hbar^{-1} + e^{f(q)}g(q)p\hbar^{-1} + o(\hbar^{-1})\big).$$

If we expand I_T in the same manner, we see that there is a unique choice for f, g, h such that $I_T \equiv J_T'' \bmod \hbar^{-2}$ (this is similar to what we did in (11.46)). Also note that f, g are homogeneous of degree 0 and h is homogeneous of degree 1 since $\tilde{S}_\mathcal{V}$ and $\tilde{S}_\mathcal{V}'$ are homogeneous of degree 0.

In order to prove the theorem, we must show that $I_T = J_T''$. With this goal in mind, first note that the change of variables $t_1 \mapsto t_1 + g(q)$ takes $q = e^{t_1}$ to $qe^{g(q)}$. It follows that applying the change of variables to

$$J_T = e^{(t_0 + t_1 p)/\hbar} \operatorname{Euler}_T(\mathcal{V})\, \tilde{S}_\mathcal{V}(q, \hbar)$$

gives

$$\begin{aligned}
J_T'' &= e^{(t_0 + f(q)\hbar + h(q) + (t_1 + g(q))p)/\hbar} \operatorname{Euler}_T(\mathcal{V})\, \tilde{S}_\mathcal{V}(qe^{g(q)}, \hbar) \\
&= e^{(t_0 + t_1 p)/\hbar} \operatorname{Euler}_T(\mathcal{V})\, e^{f(q) + h(q)/\hbar + g(q)p/\hbar}\, \tilde{S}_\mathcal{V}(qe^{g(q)}, \hbar) \\
&= e^{(t_0 + t_1 p)/\hbar} \operatorname{Euler}_T(\mathcal{V})\, \tilde{S}_\mathcal{V}''(q, \hbar),
\end{aligned}$$

where

$$\tilde{S}_\mathcal{V}''(q, \hbar) = e^{f(q) + h(q)/\hbar + g(q)p/\hbar}\, \tilde{S}_\mathcal{V}(qe^{g(q)}, \hbar).$$

If we then set

$$Z_{i,\mathcal{V}}'' = \int_{(\mathbb{P}^n)_T} \tilde{S}_\mathcal{V}'' \cup \phi_i = i_{q_i}^*(\tilde{S}_\mathcal{V}''),$$

then $I_T = J_T''$ is equivalent to $Z_{i,\mathcal{V}}' = Z_{i,\mathcal{V}}''$ for all i.

We next claim that the collection $\{Z_{i,\mathcal{V}}''\}_{i=0}^n$ is of class $\mathcal{P}$. To prove this, observe that the above formulas for $S_\mathcal{V}''$ and $Z_{i,\mathcal{V}}''$ imply

$$Z_{i,\mathcal{V}}''(q, \hbar) = e^{f(q) + h(q)/\hbar + g(q)\lambda_i/\hbar}\, Z_{i,\mathcal{V}}(qe^{g(q)}, \hbar).$$

since $i_{q_i}^*(p) = \lambda_i$. It follows from Lemma 11.2.15 that $\{Z_{i,\mathcal{V}}''\}_{i=0}^n$ is of class $\mathcal{P}$ since $\{Z_{i,\mathcal{V}}\}_{i=0}^n$ is.

Now the theorem follows easily. We know that $\{Z_{i,\mathcal{V}}\}$ and $\{Z_{i,\mathcal{V}}''\}$ are of class $\mathcal{P}$, and $I_T \equiv J_T'' \bmod \hbar^{-2}$ implies $Z_{i,\mathcal{V}}' \equiv Z_{i,\mathcal{V}}'' \bmod \hbar^{-2}$ for all i. Hence the hypotheses of the uniqueness lemma (Lemma 11.2.14) are satisfied, and we conclude that $Z_{i,\mathcal{V}}' = Z_{i,\mathcal{V}}''$ for all i. As noted above, this implies $I_T = J_T''$, and taking the nonequivariant limit as $\lambda_i \to 0$ show that $I_\mathcal{V}$ equals $J_\mathcal{V}$ after the required change of variables. This completes our proof of the Mirror Theorem. $\square$

We now make a few remarks on the relationship between the approaches to the Mirror Theorem discussed in this section and in Section 11.1. The recursion relation given in Lemma 11.2.11 plays a similar role to the gluing identity of Lemma 11.1.2 in Section 11.1.3. Combining the recursion relation with the polynomiality property of the double construction in Lemma 11.2.12 corresponds to the Eulerity property discussed in Section 11.1.3. Furthermore, the uniqueness property of Lemma 11.2.14 corresponds to Theorem 11.1.7 in Section 11.1.4. Note also that the -2 appearing in the bound $5d-2$ in the statement of Theorem 11.1.7 corresponds to the matching of 2 coefficients (those of $1, \hbar^{-1}$) in the proof of the uniqueness lemma given here. Finally, both approaches use mirror transformations to identify two different formal series. A more detailed comparison of these two approaches appears in an extensive footnote in [**Givental5**], which also gives a genus 1 version of the mirror theorem.

11.2.5. Toric Complete Intersections. We next turn to the extension of the Mirror Theorem to complete intersections in smooth toric varieties [**Givental4**]. The main result is reproduced as Theorem 11.2.16 below. We will then discuss how the Toric Mirror Theorem relates to one of the main themes of this book, namely mirror symmetry for Calabi-Yau manifolds as given by the Batyrev mirror construction. The section will conclude with some interesting examples.

Suppose that X_Σ is a smooth toric variety associated to a fan Σ, and assume that $\mathcal{L}_1, \dots, \mathcal{L}_\ell$ are line bundles on X_Σ generated by global sections. Then put $\mathcal{V} = \oplus_{i=1}^{\ell} \mathcal{L}_i$ and let $X \subset X_\Sigma$ be a smooth complete intersection defined by a generic global section of $\mathcal{V}$. In this situation, we want to define cohomology-valued functions $I_\mathcal{V}$ and $I_\mathcal{V}$ which generalize what we did in Section 11.2.1.

We first set up some notation. For each $\rho \in \Sigma(1)$, we abuse notation and let D_ρ also denote the cohomology class of the associated divisor D_ρ in $H^2(X_\Sigma)$. Following [**Givental4**], we put $\mathcal{L}_i(\beta) = \int_\beta c_1(\mathcal{L}_i)$ and $D_\rho(\beta) = \int_\beta D_\rho$. We also pick an integral basis $T_1, \dots, T_r$ of $H^2(X_\Sigma, \mathbb{Z})$ which lie in the closure of the Kähler cone. As usual, we set $\delta = \sum_{i=1}^{r} t_i T_i$.

We now define two cohomology-valued formal functions. We begin with $I_\mathcal{V}$, which is given by

$$I_\mathcal{V} = e^{(t_0+\delta)/\hbar} \, \mathrm{Euler}(\mathcal{V}) \times$$

$$(11.73) \qquad \sum_{\beta \in M(X_\Sigma)} q^\beta \, \frac{\prod_{i=1}^{\ell} \prod_{m=-\infty}^{\mathcal{L}_i(\beta)} (c_1(\mathcal{L}_i) + m\hbar) \prod_\rho \prod_{m=-\infty}^{0} (D_\rho + m\hbar)}{\prod_{i=1}^{\ell} \prod_{m=-\infty}^{0} (c_1(\mathcal{L}_i) + m\hbar) \prod_\rho \prod_{m=-\infty}^{D_\rho(\beta)} (D_\rho + m\hbar)}.$$

where $q_i = e^{t_i}$ and $q^\beta = \prod_{i=1}^{r} q_i^{\int_\beta T_i}$. Note that if Σ is the standard fan for $\mathbb{P}^n$, then we recover (11.38). Turning to $J_\mathcal{V}$, we define

$$J_\mathcal{V} = e^{(t_0+\delta)/\hbar} \, \mathrm{Euler}(\mathcal{V}) \times$$

$$(11.74) \qquad \left(1 + \sum_{\beta \neq 0} q^\beta PD^{-1} e_{1*} \left(\frac{\mathrm{Euler}(\mathcal{V}'_{\beta,2,1})}{\hbar - c_1(\mathcal{L}_1)} \cap [\overline{M}_{0,2}(X_\Sigma, \beta)]^{\mathrm{virt}} \right) \right),$$

where $[\overline{M}_{0,2}(X_\Sigma, \beta)]^{\mathrm{virt}}$ is the virtual fundamental class of $\overline{M}_{0,2}(X_\Sigma, \beta)$ and PD is Poincaré duality. Note that when X_Σ is convex, $[\overline{M}_{0,2}(X_\Sigma, \beta)]^{\mathrm{virt}}$ is just the usual fundamental class and the formula for $J_\mathcal{V}$ can be simplified. For example, when X_Σ is the convex variety $\mathbb{P}^n$, (11.74) reduces to (11.52).

In this situation, the variables q_i have degrees. As in Section 11.2.2, we define $\deg q_i$ by the equation

$$c_1(X_\Sigma) - c_1(\mathcal{V}) = \sum_{i=1}^{r} (\deg q_i) \, T_i.$$

We will assume that $X \subset X_\Sigma$ is a nef complete intersection in the sense of Section 5.5.3, which means that $-(K_{X_\Sigma} + \sum_{i=1}^{\ell} \mathcal{L}_i)$ is nef on X_Σ. When this occurs, we will assume that the basis $T_1, \dots, T_r$ of $H^2(X_\Sigma, \mathbb{Z})$ has been chosen so that

$-(K_{X_\Sigma} + \sum_{i=1}^{\ell} \mathcal{L}_i)$ lies in the cone generated by the T_i. This can always be arranged in the nef case. It follows that $\deg q_i \geq 0$ for all i.

We can now state Givental's version of the Toric Mirror Theorem.

THEOREM 11.2.16. *Let $X \subset X_\Sigma$ be a nef complete intersection, and let $I_\mathcal{V}$ and $J_\mathcal{V}$ be as in (11.73) and (11.74). Then $I_\mathcal{V}$ and $J_\mathcal{V}$ coincide after a triangular weighted homogeneous change of variables:*

$$t_0 \mapsto t_0 + f_0(q)\hbar + h(q), \quad t_i \mapsto t_i + f_i(q) \ \text{for} \ 1 \leq i \leq r,$$

where $f_0, f_1 \dots , f_k, h$ are weighted homogeneous power series and $\deg f_0 = \deg f_i = 0$, $\deg h = 1$.

PROOF. As one might expect, the strategy of the proof is to create equivariant versions of $I_\mathcal{V}$ and $J_\mathcal{V}$. This is why toric varieties are important, since the torus $T \subset X_\Sigma$ provides the needed group action. However, the formula (11.74) shows that we also need an equivariant version of the virtual fundamental class, and since localization plays such a central role in the proof, we see why a localization formula for the equivariant virtual fundamental class is essential. Hence we make contact with some of the issues discussed in Section 9.3.2.

We will not prove Theorem 11.2.16 here. The proof given in [**Givental4**] was incomplete at the time it was written. In particular, the proof was predicated on the existence of a good localization formula for the equivariant virtual fundamental class, as indicated above. Such a localization formula has since been established in [**GPa**]. Given the importance of Theorem 11.2.16, it would be desirable to have the remaining details of the proof written down carefully. We have full confidence in the conceptual framework created by Givental and believe that these details will be forthcoming, much as was done in [**BDPP, Pandharipande3**] for Proposition 11.2.1 and Theorem 11.2.2. $\qquad\square$

One can also ask how the formal function $J_\mathcal{V}$ defined in (11.74) relates to the Givental function J_X from Section 10.3.1, where $i : X \hookrightarrow X_\Sigma$ is the zero locus of a general section of $\mathcal{V}$. The basic idea is that $J_\mathcal{V}$ should compute those genus 0 gravitational correlators corresponding to cohomology classes which come from the ambient space. Thus we expect $J_\mathcal{V}$ to contain some but not all of the information present in J_X. In one case, the relation between $J_\mathcal{V}$ and J_X is easy: when $\mathcal{V} = 0$, we have $X = X_\Sigma$, and here one can show without difficulty that $J_X = J_\mathcal{V}$. When $\mathcal{V}$ is nonzero, the situation is more complicated, since $J_\mathcal{V}$ takes values in $H^*(X_\Sigma)$ and J_X takes values in $H^*(X)$. For $X_\Sigma = \mathbb{P}^n$, we showed in Proposition 11.2.3 that $J_\mathcal{V} = i_!(J_X)$. This is true more generally when X_Σ is convex and $\mathcal{V}$ is ample. However, we don't know if $J_\mathcal{V} = i_!(J_X)$ holds in the situation of Theorem 11.2.16. One problem is that $H^2(X_\Sigma) \to H^2(X)$ need not be surjective, so that J_X may involve more variables than $J_\mathcal{V}$. If we let $H^*(\mathcal{V})$ be the image of $i^* : H^*(X_\Sigma) \to H^*(X)$, then it may be that we first need to take an appropriate projection of J_X onto $H^*(\mathcal{V})$—this might be called the "toric" part of J_X—and then $J_\mathcal{V}$ should be $i_!$ applied to this toric part. In general, we feel that the exact relation between $J_\mathcal{V}$ and J_X is a question worthy of further study.

Finally, we should also make some comments about what Theorem 11.2.16 says in the case of a Calabi-Yau complete intersection. As in Chapter 4, we will suppose that Δ is a reflexive polytope and X_Σ is the toric variety obtained from a maximal projective subdivision. We will further assume that X_Σ is smooth.

For simplicity, we will consider the case when $V \subset X_\Sigma$ is an anticanonical divisor, which we can also assume to be smooth. Then the nef hypothesis of Theorem 11.2.16 is certainly satisfied, so that J_V and I_V are related by a coordinate change. As just explained, J_V tells us about the genus 0 gravitational correlators coming from the "toric" part of V. So presumably the formal function I_V should tell us something about the mirror V°.

We studied this in Section 5.5, and let's review what we discovered there. First recall from Section 5.5.3 that I_V satisfies the GKZ equations determined by a set $\mathcal{A}$ described in (5.42). Since $\Sigma(1) = \Delta^\circ \cap M - \{0\}$, we see that $\mathcal{A} = (\Delta^\circ \cap M) \times \{1\}$. Now let V° be the Batyrev mirror of V, which is determined by the reflexive polytope Δ°. Then recall from Section 5.5.2 that the GKZ equations for $\mathcal{A} = (\Delta^\circ \cap m) \times \{1\}$ are Picard-Fuchs equations for V°. This looks good, except for the following two complications. First, the GKZ equations don't contain all of the Picard-Fuchs equations, so that we haven't proved that I_V comes from periods of V° (though this is claimed in [**Givental4**] without proof). Second, Section 5.5.2 doesn't deal with the full moduli space of V°, but rather the smaller moduli space given by deformations of V° coming from the ambient toric variety.

It follows that we don't yet have a complete understanding of how Theorem 11.2.16 relates to mirror symmetry as discussed in earlier chapters of this book. In the case when the Calabi-Yau hypersurface V is a threefold, we are hopeful that the conjectures made in Section 8.6.4 should follow from Theorem 11.2.16. We will give an example of how this works in Example 11.2.5.1 below. In general, the relation between mirror symmetry and Theorem 11.2.16 is a question which should definitely be investigated further.

We will close this section with two examples which illustrate the power of Theorem 11.2.16. The first is an example that we have seen many times before in this book, namely Calabi-Yau hypersurfaces in the minimal desingularization of $\mathbb{P}(1,1,2,2,2)$.

Example 11.2.5.1. Let Σ be the fan obtained from the usual fan for $\mathbb{P}(1,1,2,2,2)$ by subdividing the cone spanned by $(-1,-2,-2,-2)$ and $(1,0,0,0)$, inserting the edge spanned by $(0,-1,-1,-1)$. For this fan, we have

$$
(11.75) \quad \begin{aligned} \Sigma(1) = \{ &(-1,-2,-2,-2), (1,0,0,0), (0,1,0,0), \\ &(0,0,1,0), (0,0,0,1), (0,-1,-1,-1) \}, \end{aligned}
$$

where we have abused notation slightly and identified an edge with its primitive generator.

This fan is the unique maximal projective subdivision of the standard fan of $\mathbb{P}(1,1,2,2,2)$. Note that the convex hull of the points listed in (11.75) is the polytope Δ° used in Batyrev's construction. In this context, we now get Calabi-Yau hypersurfaces $V \subset X_\Sigma$. In Example 6.2.4.3, we showed that the Kähler cone is spanned by the classes of the toric divisors D_1 and D_3, indexing toric divisors using the order in (11.75). We noted in that example the linear equivalences $D_1 \sim D_2$, $D_3 \sim D_4 \sim D_5$ and $D_6 \sim -2D_1 + D_3$, which allow us to express all divisor classes in terms of D_1 and D_3. We also observed that $V \in |4D_3|$. In the notation of this section, we therefore put $\mathcal{V} = \mathcal{O}_{X_\Sigma}(4D_3)$.

We order the Kähler cone generators as (D_3, D_1) to keep notation consistent with earlier chapters. Let $\{\beta_1, \beta_2\}$ be the basis of $M(X_\Sigma)$ dual to $\{D_3, D_1\}$, so that a general element $\beta \in M(X_\Sigma)$ is $\beta = d_1\beta_1 + d_2\beta_2$. The intersection numbers computed in Example 6.2.4.3 imply $\beta_1 = D_3^2 D_1$ and $\beta_2 = D_3^3 - 2D_3^2 D_1$ (here, we are thinking of β_i as an element of $H^6(X_\Sigma)$ via Poincaré duality). In Example 6.2.4.3 we also observed that $|D_1|$ defines a map $X_\Sigma \to \mathbb{P}^1$, so that $D_1^2 = 0$.

To see what the Toric Mirror Theorem says in this case, we begin by writing out $I_\mathcal{V}$. Adapting (11.73) to the present example shows that

$$I_\mathcal{V} = e^{(t_0 + t_1 D_3 + t_2 D_1)/\hbar}\, 4D_3 \sum_{d_1, d_2} \left(z_1^{d_1} z_2^{d_2} \times \right.$$

(11.76)
$$\left. \frac{\prod_{m=1}^{4d_1}(4D_3 + m\hbar) \prod_{m=-\infty}^{0}(D_3 - 2D_1 + m\hbar)}{\prod_{m=1}^{d_2}(D_1 + m\hbar)^2 \prod_{m=1}^{d_1}(D_3 + m\hbar)^3 \prod_{m=-\infty}^{d_1 - 2d_2}(D_3 - 2D_1 + m\hbar)} \right).$$

Notice we have put $z_i = e^{t_i}$ for $i = 1, 2$. We use z_i rather than q_i in order to be consistent with Chapters 5 and 6. The coordinates z_i were defined in Example 6.1.4.1 and are used in numerous examples in these two chapters.

The Toric Mirror Theorem says that $I_\mathcal{V}$ equals $J_\mathcal{V}$ after an appropriate change of variables. Furthermore, since

$$J_\mathcal{V} = e^{(t_0 + t_1 D_3 + t_2 D_1)/\hbar}\, 4D_3 \bigl(1 + o(\hbar^{-1})\bigr),$$

we can read off the change of variables by expanding $I_\mathcal{V}$. Using (11.76), we get

$$(11.77) \quad I_\mathcal{V} = e^{t_0/\hbar} 4D_3 \Bigl(y_0 + y_{1,1}\frac{D_3}{\hbar} + y_{1,2}\frac{D_1}{\hbar} + y_{2,1}\frac{D_1 D_3}{\hbar^2} + y_{2,2}\frac{D_3^2}{\hbar^2} + y_3\frac{D_1 D_3^2}{\hbar^3} \Bigr),$$

where

$$y_0 = \sum_{d_1 \geq 2d_2} z_1^{d_1} z_2^{d_2} \frac{(4d_1)!}{(d_2!)^2 (d_1!)^3 (d_1 - 2d_2)!}$$

$$y_{1,1} = t_1 y_0 + \sum_{d_1 \geq 2d_2} z_1^{d_1} z_2^{d_2} \frac{(4d_1)!}{(d_2!)^2 (d_1!)^3 (d_1 - 2d_2)!}\bigl(4\Psi(4d_1) - 3\Psi(d_1) - \Psi(d_1 - 2d_2)\bigr)$$

$$+ \sum_{d_1 < 2d_2} z_1^{d_1} z_2^{d_2} \frac{(-1)^{d_1+1}(2d_2 - d_1 - 1)!(4d_1)!}{(d_2!)^2 (d_1!)^3}$$

$$y_{1,2} = t_2 y_0 + \sum_{d_1 \geq 2d_2} z_1^{d_1} z_2^{d_2} \frac{(4d_1)!}{(d_2!)^2 (d_1!)^3 (d_1 - 2d_2)!}\bigl(2\Psi(d_1 - 2d_2) - 2\Psi(d_2)\bigr)$$

$$- 2 \sum_{d_1 < 2d_2} z_1^{d_1} z_2^{d_2} \frac{(-1)^{d_1+1}(2d_2 - d_1 - 1)!(4d_1)!}{(d_2!)^2 (d_1!)^3}.$$

Here, we have used the classical function

$$\Psi(n) = \sum_{i=1}^{n} \frac{1}{i}.$$

Notice that y_0, $y_{1,1}$ and $y_{1,2}$ are functions of t_1, t_2.

Using this, one easily sees that the functions f_0, f_1, f_2, h from Theorem 11.2.16 are given by

$$e^{f_0} = y_0, \quad f_1 = \frac{y_{1,1}}{y_0} - t_1, \quad f_2 = \frac{y_{1,2}}{y_0} - t_2, \quad h = 0,$$

so that the coordinate change is $t_0 \mapsto t_0 + f_0 \hbar$ and $t_i \mapsto t_i + f_i = y_{1,i}/y_0$ for $i = 1, 2$. Hence, for

$$(s_1, s_2) = \left(\frac{y_{1,1}(t_1, t_2)}{y_0(t_1, t_2)}, \frac{y_{1,2}(t_1, t_2)}{y_0(t_1, t_2)} \right),$$

the Toric Mirror Theorem implies that

$$(11.78) \qquad J_V(s_1, s_2) = \frac{I_V(t_1, t_2)}{y_0(t_1, t_2)}.$$

Setting $q_i = e^{s_i}$, the above change of variables becomes

$$(q_1, q_2) = \left(\exp\left(\frac{y_{1,1}}{y_0} \right), \exp\left(\frac{y_{1,2}}{y_0} \right) \right),$$

which is the mirror map in this case. The coordinate q_2 was computed by a different method in Example 6.3.4.2. The results are in complete agreement.

We will next explain how (11.78) relates to the "classical" notion of mirror symmetry, as formulated in terms of the A-model and B-model correlation functions. Our strategy will be to show that I_V/y_0 satisfies some differential equations which involve the B-model correlation functions. Hence J_V satisfies the same differential equations, which will lead to the desired equality of correlation functions.

We start with I_V. We know from Proposition 5.5.4 that I_V satisfies the GKZ equations coming from V°, but as noted above, these don't give all of the Picard-Fuchs equations of V°. However, we computed the missing Picard-Fuchs equations in Example 5.5.2.1—see (5.39) and (5.41). A straightforward computation shows that I_V satisfies these equations. (There should be a general theorem to this effect, but we don't know a proof.)

The Picard-Fuchs equations constructed in Example 5.5.2.1 used a particular choice for the holomorphic 3-form Ω. We also know that y_0 is the unique period (up to a constant) which is holomorphic at $z_1 = z_2 = 0$. It follows that $\widetilde{\Omega} = \Omega/y_0$ is the normalized holomorphic 3-form (up to a constant), so that I_V/y_0 satisfies the Picard-Fuchs equations for $\widetilde{\Omega}$. This remains true if we switch variables from $z_i = e^{t_i}$ to $q_i = e^{s_i}$.

In terms of the coordinates q_i, the normalized Yukawa couplings of V° lead to some nice Picard-Fuchs equations. In this situation, the normalized couplings are given by

$$Y_{ijk} = -\frac{1}{(2\pi i)^3} \int_{V^\circ} \widetilde{\Omega} \wedge \nabla_{\delta_i} \nabla_{\delta_j} \nabla_{\delta_k} \widetilde{\Omega},$$

where $\delta_j = \partial/\partial s_j$ and ∇ is the Gauss-Manin connection of the mirror family. This agrees with the definition given in Section 8.6.4, since there we wrote $q_j = e^{2\pi i u_j}$, while here we are using $q_j = e^{s_j}$.[2] Using the cohomology basis $T_1 = i^*(D_3)$, $T_2 = i^*(D_1)$ of $H^2(X)$, our goal is to prove

$$(11.79) \qquad \langle T_i, T_j, T_k \rangle = Y_{ijk} \quad \text{for all } i, j, k.$$

---

[2]This could a priori differ from the normalized couplings defined in Definition 5.6.3. There, we assumed the integrality conjecture and used the mirror map as defined in Section 6.3.1. Here, we don't assume the integrality conjecture and we use the toric version of the mirror map from Section 6.3.3. Using the detailed description of the monodromy given in [CdFKM], one can check that these agree in this case. But we prefer to show that one can still get good results without knowing everything about the monodromy.

Recall from Example 8.6.4.1 that we can compute Y_{ijk} explicitly. Hence (11.79) gives explicit formulas for all of the A-model correlation functions $\langle T_i, T_j, T_k \rangle$ of V.

To prove (11.79), we first note that we can describe the Gauss-Manin connection using the Y_{ijk}. More precisely, we claim that $H^3(V^\circ, \mathbb{C})$ has a basis $e_0, e_1, e_2, e^1, e^2, e^3$ such that $e_0 = \widetilde{\Omega}$ and

$$\nabla_{\delta_i}(e_0) = e_i, \quad \nabla_{\delta_i}(e_j) = \sum_k Y_{ijk} e^k, \quad \nabla_{\delta_i}(e^j) = \delta_{ij} e^0, \quad \nabla_{\delta_i}(e^0) = 0,$$

just as in (5.67). For the normalized Yukawa coupling defined in Section 5.6.4, this was proved in Proposition 5.6.1 and the discussion following the proof. Although the proof required the integrality conjecture, everything carries over to our situation provided we work over $\mathbb{C}$ rather than $\mathbb{Z}$.

Using the above basis, it is straightforward to compute that

$$\nabla_{\delta_1}\!\left(\frac{Y_{112}\nabla_{\delta_1}^2 - Y_{111}\nabla_{\delta_1}\nabla_{\delta_2}}{Y_{112}^2 - Y_{111}Y_{122}}\right)\widetilde{\Omega} = 0$$

since $\widetilde{\Omega} = e_0$. Hence this is a Picard-Fuchs equation, and since I_V/y_0 satisfies the Picard-Fuchs equations, we obtain

$$\frac{\partial}{\partial s_1}\left(\frac{Y_{112}\frac{\partial^2}{\partial s_1^2} - Y_{111}\frac{\partial^2}{\partial s_1 \partial s_2}}{Y_{112}^2 - Y_{111}Y_{122}}\right) I_V/y_0 = 0.$$

By the Toric Mirror Theorem (11.78), we conclude that

$$(11.80) \qquad \frac{\partial}{\partial s_1}\left(\frac{Y_{112}\frac{\partial^2}{\partial s_1^2} - Y_{111}\frac{\partial^2}{\partial s_1 \partial s_2}}{Y_{112}^2 - Y_{111}Y_{122}}\right) J_V(s_1, s_2) = 0,$$

where we are ignoring the dependence on t_0. We should mention that explicit calculations show that $Y_{112}^2 - Y_{111}Y_{122}$ is not identically zero.

Our next step is to relate this to the Givental function J_V of V. The inclusion $i : V \subset X_\Sigma$ induces an isomorphism $i^* : H^2(X_\Sigma) \to H^2(V)$, and we showed in Example 6.2.4.3 that the Kähler cone of V is spanned by $T_1 = i^*(D_3)$ and $T_2 = i^*(D_1)$. Thus J_V and J_V can be regarded as functions of s_1, s_2 (again ignoring t_0). However, we don't know that $i_!(J_V)$ and J_V are equal. When the ambient space is $\mathbb{P}^n$, we proved equality in Proposition 11.2.3. But our proof used the convexity of $\mathbb{P}^n$, which no longer is true for X_Σ. However, examining the proof of Proposition 11.2.3, one can check that $i_!(J_V) = J_V$ still follows, provided that

$$(11.81) \qquad j_*([\overline{M}_{0,2}(V, \beta)]^{\mathrm{virt}}) = \mathrm{Euler}(\mathcal{V}_{\beta,2}) \cap [\overline{M}_{0,2}(X_\Sigma, \beta)]^{\mathrm{virt}}.$$

We suspect this is true, but we don't yet have a complete proof. Hence, for the remainder of this example, we will *assume* that (11.81) holds. In particular, this allows us to assume that $i_!(J_V) = J_V$.

One can also check that $i_!$ is injective in our case. This, together with (11.80) and $i_!(J_V) = J_V$, implies that

$$(11.82) \qquad \hbar\frac{\partial}{\partial s_1}\left(\frac{Y_{112}\hbar^2\frac{\partial^2}{\partial s_1^2} - Y_{111}\hbar^2\frac{\partial^2}{\partial s_1 \partial s_2}}{Y_{112}^2 - Y_{111}Y_{122}}\right) J_V(s_1, s_2) = 0.$$

The factors of $\hbar$ were added in order to make (11.82) a quantum differential equation in the sense of Section 10.3. In order to apply Theorem 10.3.1, we expand (11.82) to obtain

$$\left(\frac{Y_{112}\hbar^3\frac{\partial^3}{\partial s_1^3} - Y_{111}\hbar^3\frac{\partial^3}{\partial s_1^2 \partial s_2}}{Y_{112}^2 - Y_{111}Y_{122}} + \hbar P(\hbar\partial/\partial s, q, \hbar)\right)J_V(s_1, s_2) = 0,$$

where $P(\hbar\partial/\partial s, q, \hbar)$ is some differential operator. Then Theorem 10.3.1 implies that the substitution $\hbar\partial/\partial s_i \mapsto T_i$, $\hbar \mapsto 0$ gives a polynomial relation in the small quantum cohomology of V. Thus

$$\frac{Y_{112}\, T_1 * T_1 * T_1 - Y_{111}\, T_1 * T_1 * T_2}{Y_{112}^2 - Y_{111}Y_{122}} = 0$$

in $H^*(V)$. Since

$$\int_V T_i * T_j * T_k = \langle T_i, T_j, T_k \rangle$$

by (8.7), the above equation implies

$$Y_{112}\, \langle T_1, T_1, T_1 \rangle - Y_{111}\, \langle T_1, T_1, T_2 \rangle = 0.$$

The argument just given works for general subscripts, so that

$$Y_{ijk}\, \langle T_a, T_b, T_c \rangle - Y_{abc}\, \langle T_i, T_j, T_k \rangle = 0$$

for arbitrary $1 \leq i, j, k, a, b, c \leq 2$. From here, it follows easily that there is a constant c such that

$$\langle T_i, T_j, T_k \rangle = cY_{ijk} \quad \text{for all } i, j, k.$$

Regarding $\langle T_i, T_j, T_k \rangle$ as a formal function of q_1, q_2, we have that $\langle T_1, T_1, T_1 \rangle(0, 0) = \int_V T_1^3 = 8$. Thus, in order to show $c = 1$, we must show that $Y_{111}(0, 0) = 8$. For this, let N_1 be the logarithm of the monodromy transformation about the boundary divisor $z_1 = 0$ in the complex moduli space of V°, keeping the notation of Example 5.6.2.1. The explicit calculations in [**CdFKM**] show that there are integral flat sections β^2, α_2 of $\mathcal{H}$ which are part of a $\mathbb{Z}$-basis and satisfy $\langle \alpha_2, \beta^2 \rangle = 1$ and $N^3(\beta^2) = -8\alpha_2$. Note that the monodromy in [**CdFKM**] was taken with the orientation opposite to the one taken here. An argument similar to that of Example 5.6.4.1 leads to $Y_{111}(0, 0) = 8$. Hence $c = 1$, and we have proven (11.79).

In Example 8.6.4.1, $\langle T_1, T_1, T_1 \rangle$ was denoted $\langle H, H, H \rangle$, and we have now proved the formula for this correlation function given in that example. We can also relate (11.79) to the Hodge-theoretic version of the Mirror Conjecture from Section 8.6.4. As explained in the discussion surrounding (8.75), the equality of the A-model and B-model correlation functions up to a constant implies that Conjecture 8.6.10 holds in this case. More precisely, we have proved that for the threefolds V and V° considered here, this conjecture follows from Theorem 11.2.16 and (11.81). Furthermore, using the previous paragraph, one can show that (V°, V) is a mathematical mirror pair in the sense of Definition 8.6.1.

However, there is more that can be said about this example, for we have yet to use the full power of the Toric Mirror Theorem. As we will see below, the equality (11.78) will give us a remarkable formula for the Gromov-Witten potential of V. We will use the following basis of the even cohomology of V:

$$T_0 = 1, \quad T_1 = i^*(D_3), \quad T_2 = i^*(D_1),$$

$$T^1 = \frac{i^*(D_1 D_3)}{4}, \quad T^2 = \frac{i^*(D_3^2 - 2D_1 D_3)}{4}, \quad T^0 = \frac{i^*(D_1 D_3^2)}{4}.$$

One can check that T^1, T^2 are dual to T_1, T_2 under cup product, and similarly for T^0 and T_0. In this situation, we have variables t_0, s_1, s_2, and according to Proposition 10.3.4, J_V is given by the formula

$$J_V = e^{t_0/\hbar}\left(1 + \hbar^{-1}(s_1 T_1 + s_2 T_2) + \hbar^{-2}\left(\frac{\partial \Phi}{\partial s_1} T^1 + \frac{\partial \Phi}{\partial s_2} T^2\right)\right.$$

(11.83)

$$\left. + \hbar^{-3}\left(s_1 \frac{\partial \Phi}{\partial s_1} + s_2 \frac{\partial \Phi}{\partial s_2} - 2\Phi\right)T^0\right),$$

where Φ is the Gromov-Witten potential of V.

We still are assuming (11.81), so that $J_V = i_!(J_{\mathcal{V}})$. Since $i_!(i^*(\alpha)) = \alpha \cup 4D_3$, applying $i_!$ to (11.83) gives

$$J_{\mathcal{V}}(s_1, s_2) = e^{t_0/\hbar}\, 4D_3\left(1 + s_1 \frac{D_3}{\hbar} + s_2 \frac{D_1}{\hbar} + \frac{\partial \Phi}{\partial s_1}\frac{D_1 D_3}{4\hbar^2} + \frac{\partial \Phi}{\partial s_2}\frac{D_3^2 - 2D_1 D_3}{4\hbar^2}\right.$$

$$\left. + \left(s_1 \frac{\partial \Phi}{\partial s_1} + s_2 \frac{\partial \Phi}{\partial s_2} - 2\Phi\right)\frac{D_1 D_3^2}{4\hbar^3}\right).$$

To compare this to I_V/y_0, note that (11.77) can be written in the form

$$\frac{I_V}{y_0}(t_1, t_2) = e^{t_0/\hbar} 4D_3\left(1 + \frac{y_{1,1}}{y_0}\frac{D_3}{\hbar} + \frac{y_{1,2}}{y_0}\frac{D_1}{\hbar} + \frac{y_{2,1} + 2y_{2,2}}{y_0}\frac{D_1 D_3}{\hbar^2}\right.$$

$$\left. + \frac{y_{2,2}}{y_0}\frac{D_3^2 - 2D_1 D_3}{\hbar^2} + \frac{y_3}{y_0}\frac{D_1 D_3^2}{\hbar^3}\right).$$

We will write the mirror map $(s_1, s_2) = (y_{1,1}/y_0, y_{1,2}/y_0)$ as $(s_1, s_2) = \Psi(t_1, t_2)$. As we saw in (11.78), the Toric Mirror Theorem implies that the above two expressions are equal via the change of variables $(s_1, s_2) = \Psi(t_1, t_2)$. Comparing coefficients, we see that the Gromov-Witten potential of V is given by the following explicit formula:

(11.84)
$$\Phi(\Psi(t_1, t_2)) = 2\left(\frac{y_{1,1}}{y_0}\frac{y_{2,1} + 2y_{2,2}}{y_0} + \frac{y_{1,2}}{y_0}\frac{y_{2,2}}{y_0} - \frac{y_3}{y_0}\right),$$

where $\Psi(t_1, t_2) = (y_{1,1}/y_0, y_{1,2}/y_0)$ is the mirror map. This is remarkably similar to the formula for the Gromov-Witten potential of the quintic threefold derived in Section 11.1.

A final observation is that comparing the above formulas for J_V and I_V/y_0 also gives some nice formulas for the A-model correlation functions. For example, we know that

$$\langle T_1, T_1, T_1\rangle = \frac{\partial^3 \Phi}{\partial s_1^3}.$$

Hence, if we differentiate $J_{\mathcal{V}}(s_1, s_2) = I_V/y_0(t_1, t_2)$ twice with respect to s_1 and compare the coefficients of $D_1 D_3$, we obtain

(11.85)
$$\langle T_1, T_1, T_1\rangle = \frac{\partial^2}{\partial s_1^2}\left(\frac{y_{2,1} + 2y_{2,2}}{y_0}\right).$$

This formula for $\langle T_1, T_1, T_1\rangle$ is *much* easier to derive than the formulas considered in Example 8.6.4.1. Another way to view (11.85) is that it gives an elegant formula for the normalized Yukawa coupling Y_{111}.

This example is very suggestive, for many of the computations should generalize to other toric Calabi-Yau threefolds. In particular:

- The formula (11.84) should generalize to any Calabi-Yau threefold V which is a nef complete intersection in a toric variety X_Σ. Since $H^2(X_\Sigma) \to H^2(V)$ need not be surjective, one would have to define the "toric" part of the Gromov-Witten potential of V. Then this toric part should be expressible explicitly in terms of I_V/y_0 by a formula similar to (11.84).
- The Gromov-Witten potential is a potential function for the A-model correlation functions $\langle T_i, T_j, T_k \rangle$. Hence (11.84) implies that the coefficients of I_V/y_0 give an explicit potential function (in the sense of Lemma 5.6.2) for the B-model correlation functions Y_{ijk} of V°. It should be possible to prove this directly, without using the Toric Mirror Theorem.
- In (11.85), we in essence gave a simple formula for Y_{111}. In general, there should be similar formulas for the normalized Yukawa couplings of V° in terms of I_V/y_0. As above, it should be possible to prove this directly,

We feel that there are many interesting open questions which concern 3-dimensional nef complete intersections in toric varieties.

For our second application of Theorem 11.2.16, we will study what happens when $V = 0$, so that X is the toric variety X_Σ. In this case, we will see that the Toric Mirror Theorem combines nicely with earlier results to describe the quantum cohomology of X_Σ.

Example 11.2.5.2. In this example, we will show that if $X = X_\Sigma$ is a smooth Fano toric variety, then the Batyrev quantum ring $H^*_\omega(X)$ from Example 8.1.2.2 agrees with the small quantum cohomology ring. We carry this out by combining Theorem 11.2.16 and Proposition 5.5.4. We will also see that when $-K_X$ is nef instead of Fano, $H^*_\omega(X)$ may differ from the quantum cohomology ring, though Theorem 11.2.16 still implies that there is a relation between the two rings. We will illustrate this for $X = \mathbb{F}_2$.

Since we are considering the toric variety $X = X_\Sigma$ itself, we let V be the trivial bundle in (11.73), and accordingly write I_V and J_V as I_X and J_X respectively. Note that when $X = \mathbb{P}^n$, this agrees with notation used in Proposition 11.2.1. We have

$$(11.86) \qquad I_X = e^{(t_0+\delta)/\hbar} \sum_{\beta \in M(X)} q^\beta \frac{\prod_\rho \prod_{m=-\infty}^{0}(D_\rho + m\hbar)}{\prod_\rho \prod_{m=-\infty}^{D_\rho(\beta)}(D_\rho + m\hbar)},$$

where $\delta = \sum_{i=1}^{r} t_i T_i$ and $D_\rho(\beta) = \int_\beta D_\rho$. We also have the formulas (11.39) for J_X from Section 11.2.1. Then we get the following generalization of Proposition 11.2.1.

PROPOSITION 11.2.17. *If $X = X_\Sigma$ is a smooth Fano toric variety, then we have $I_X = J_X$.*

PROOF. We know from (10.37) that $J_X = e^{(t_0+\delta)/\hbar}(1 + o(\hbar^{-1}))$. If we knew that the same were true for I_X when X is Fano, then the change of variables in Theorem 11.2.16 would be the identity, and then the proposition would follow immediately from the Toric Mirror Theorem. Hence it suffices to show that

$$(11.87) \qquad I_X = e^{(t_0+\delta)/\hbar}(1 + o(\hbar^{-1})).$$

To prove this, fix β in the Mori cone $M(X)$ and ρ in $\Sigma(1)$, and consider the fraction

$$\frac{\prod_{m=-\infty}^{0}(D_\rho + m\hbar)}{\prod_{m=-\infty}^{D_\rho(\beta)}(D_\rho + m\hbar)}$$

from (11.86). We will study the leading term when this is expanded as a power series in $\hbar^{-1}$. If $D_\rho(\beta) > 0$, then the leading term in $\hbar^{-1}$ is easily seen to be

$$\frac{1}{(D_\rho(\beta))!}\, h^{-D_\rho(\beta)}.$$

If $D_\rho(\beta) = 0$, then the fraction is just 1, and if $D_\rho(\beta) = -1$, then the fraction is just D_ρ. Finally, if $D_\rho(\beta) < -1$, then the leading term in $\hbar^{-1}$ is

$$(-1)^{-(D_\rho(\beta)-1)} D_\rho \left(-D_\rho(\beta) - 1\right)! \, \hbar^{-(D_\rho(\beta)+1)}.$$

In all cases, the leading power of $\hbar^{-1}$ is at least $D_\rho(\beta)$ and is strictly greater if and only if $D_\rho(\beta) < 0$.

Now, fixing β and multiplying over all ρ, we see that the leading power of $\hbar^{-1}$ in the coefficient of q^β in the expansion of (11.86) is at least $\sum_\rho D_\rho(\beta)$. However, we saw in Chapter 3 that $\sum_\rho D_\rho = -K_X$. It follows that in the coefficient of q^β, the leading power of $\hbar^{-1}$ is at least $-\int_\beta K_X$.

Since X is Fano, we see that $-\int_\beta K_X$ is strictly positive, so that the leading power of $\hbar^{-1}$ is at least one. However, in order to prove (11.87), we must show that it is at least two. The above discussion shows that the only way the leading power can be 1 is if $-\int_\beta K_X = 1$ and $D_\rho(\beta) \geq 0$ for all ρ. Since $-K_X = \sum_\rho D_\rho$, we see that there exists ρ_0 such that $D_{\rho_0}(\beta) = 1$ and $D_\rho(\beta) = 0$ for all $\rho \neq \rho_0$. We must show that this can't occur.

We will use Lemma 5.5.3 from Section 5.5.3 with $k = 0$. For $\beta \in H_2(X, \mathbb{Z})$, the lemma implies

$$(11.88) \qquad\qquad \sum_\rho D_\rho(\beta)\, v_\rho = 0,$$

where v_ρ is the primitive generator of $\rho \cap N$. This follows because $D_\rho(\beta) = (D_\rho \cdot \beta)$. In the situation of the preceding paragraph, (11.88) leads to $v_{\rho_0} = 0$, a contradiction. This completes the proof of (11.87), and the proposition now follows immediately from Theorem 11.2.16. $\qquad\square$

We will next use $I_X = J_X$ to study the small quantum cohomology ring of X when $X = X_\Sigma$ is Fano and smooth. The basic idea is that by Section 5.5.3, I_X satisfies a certain GKZ system. Hence J_X satisfies the same equations, and then the theory of quantum differential equations from Section 10.3 will imply the desired relations in quantum cohomology.

To carry out this strategy, we need to see what Proposition 5.5.4 looks like in this situation. We are dealing with the case when $\mathcal{V} = \oplus_{i=1}^{k}\mathcal{L}_i$ is the zero bundle, so that $k = 0$. Then (5.42) reduces to

$$\mathcal{A} = \{v_\rho : \rho \in \Sigma(1)\},$$

and the lattice Λ of Lemma 5.5.3 is precisely the lattice of integer relations among the v_ρ. Furthermore, the lemma identifies $H_2(X, \mathbb{Z})$ with Λ via the map which takes β to the relation (11.88).

But Λ is also the lattice of the corresponding $\mathcal{A}$-system, which means that for each $\beta \in \Lambda$, (5.47) gives the operator

$$\square_\beta = \prod_{D_\rho(\beta)>0} \partial_\rho^{D_\rho(\beta)} - \prod_{D_\rho(\beta)<0} \partial_\rho^{-D_\rho(\beta)}$$

since β corresponds to the relation (11.88). In this formula, $\partial_\rho = \partial/\partial\lambda_\rho$, and q^β is defined in terms of λ_ρ via

$$(11.89) \qquad q^\beta = \prod_\rho \lambda_\rho^{D_\rho(\beta)}$$

by (5.43). If we now let

$$\tilde{I} = e^\delta \sum_{\beta \in M(X)} q^\beta \frac{\prod_\rho \prod_{m=-\infty}^0 (D_\rho + m)}{\prod_\rho \prod_{m=-\infty}^{D_\rho(\beta)} (D_\rho + m)},$$

then Proposition 5.5.4 implies $\square_\beta \tilde{I} = 0$ for all β.

Note that $\tilde{I}$ is obtained from I_X by setting $t_0 = 0$ and $\hbar = 1$. We can simply multiply $\tilde{I}$ by $e^{t_0/\hbar}$, but we need to handle $\hbar$ more carefully: we put $\hbar$ back in using the operator $\square_{\beta,\hbar}$ obtained from $\square_\beta$ by replacing each ∂_ρ by $\hbar\partial_\rho$. Thus

$$\square_{\beta,\hbar} = \prod_{D_\rho(\beta)>0} (\hbar\partial_\rho)^{D_\rho(\beta)} - \prod_{D_\rho(\beta)<0} (\hbar\partial_\rho)^{-D_\rho(\beta)}.$$

We omit the easy proof that $\square_\beta \tilde{I} = 0$ implies $\square_{\beta,\hbar} I_X = 0$.

We next define

$$\square'_{\beta,\hbar} = \prod_{D_\rho(\beta)>0} \lambda_\rho^{D_\rho(\beta)} \, \square_{\beta,\hbar}$$

and observe that $\square'_{\beta,\hbar} I_X = 0$. If we set $\delta_\rho = \lambda_\rho \partial_\rho$, then adapting (5.51) to the present situation implies

$$
\begin{aligned}
(11.90) \qquad \square'_{\beta,\hbar} = {} & \prod_{D_\rho(\beta)>0} \hbar\delta_\rho(\hbar\delta_\rho - \hbar) \cdots (\hbar\delta_\rho - (D_\rho(\beta)-1)\hbar) \\
& - q^\beta \prod_{D_\rho(\beta)<0} \hbar\delta_\rho(\hbar\delta_\rho - \hbar) \cdots (\hbar\delta_\rho - (-D_\rho \cdot \beta - 1)\hbar).
\end{aligned}
$$

Since X is smooth and Fano, Proposition 11.2.17 implies that $I_X = J_X$, and we conclude that $\square'_{\beta,\hbar} J_X = 0$. But before we can apply the theory of Section 10.3, we need to write $\square'_{\beta,\hbar}$ as a differential operator in $\hbar\partial/\partial t_j$. Recall that the t_j are variables associated to the basis T_j of $H^2(X,\mathbb{Z})$. We can relate these to the λ_ρ as follows. If we let $\beta_j \in H_2(X,\mathbb{Z})$ be the dual basis to the T_j, then set

$$q_j = \prod_\rho \lambda_\rho^{D_\rho(\beta_j)}.$$

Then one can easily show that

$$\prod_j q_j^{\int_\beta T_j}$$

equals q^β as defined in (11.89). Setting $q_j = e^{t_j}$, we see that q^β has its usual meaning relative to the t_j.

In general, we can't express $\hbar\delta_\rho$ in (11.90) in terms of the $\hbar\partial/\partial t_j$. The key observation is that the dual torus $T^\circ = M \otimes \mathbb{C}^*$ acts on the λ_ρ via $m \otimes t \cdot \lambda_\rho = t^{\langle m, v_\rho \rangle} \lambda_\rho$ for $m \otimes t \in T^\circ$. The variables q_j are invariant under this action, and as we observed in (5.48), for T°-invariant functions, we have

$$(11.91) \qquad \hbar\delta_\rho = \sum_j D_\rho(\beta_j)\hbar\frac{\partial}{\partial t_j}.$$

This identity allows us to write (11.90) in terms of $\hbar\partial/\partial t_j$, which means that $\square'_{\beta,\hbar}$ is a quantum differential operator in the sense of Section 10.3. Then Theorem 10.3.1 implies that $\square'_{\beta,\hbar}$ gives a relation in quantum cohomology under the substitutions

$$\hbar\frac{\partial}{\partial t_j} \longmapsto T_j, \quad \hbar \longmapsto 0.$$

Applying this to (11.90), we obtain the relation

$$(11.92) \qquad \prod_{D_\rho(\beta)>0} D_\rho^{D_\rho(\beta)} = q^\beta \prod_{D_\rho(\beta)<0} D_\rho^{-D_\rho(\beta)}$$

in the small quantum cohomology of X.

We can use this to show that the Batyrev quantum ring $H^*_\omega(X)$ is isomorphic to the small quantum cohomology ring of X as follows. Let $H^*(X)$ have the small quantum product $*$, and let $\mathbb{C}[x_\rho]$ be a polynomial ring with a variable x_ρ for each $\rho \in \Sigma(1)$. Then $x_\rho \mapsto D_\rho$ defines a ring homomorphism $\mathbb{C}[x_\rho] \to H^*(X)$, and using (11.92), we see that the ideal

$$\left\langle \prod_{D_\rho(\beta)>0} x_\rho^{D_\rho(\beta)} - q^\beta \prod_{D_\rho(\beta)<0} x_\rho^{-D_\rho(\beta)} : \beta \in H_2(X,\mathbb{Z}) \right\rangle$$

is in the kernel of this map. But as β varies in $H_2(X,\mathbb{Z})$, the relations (11.88)

$$\sum_{D_\rho(\beta)>0} D_\rho(\beta)v_\rho = \sum_{D_\rho(\beta)<0} -D_\rho(\beta)v_\rho$$

span the lattice Λ. It follows that the above ideal coincides with the ideal (8.17), which is the form of the quantum Stanley-Reisner ideal $SR_\omega(\Sigma)$ given in [**Batyrev2**, Definition 5.1]. Thus $SR_\omega(\Sigma)$ lies in the kernel of our map $\mathbb{C}[x_\rho] \to H^*(X)$. This kernel also contains the ideal

$$P(\Sigma) = \left\langle \sum_{i=1}^s \langle m, v_i \rangle x_i : m \in M \right\rangle,$$

so that we get a well-defined ring homomorphism

$$\mathbb{C}[x_\rho]/(P(\Sigma) + SR_\omega(\Sigma)) \longrightarrow H^*(X).$$

Since Example 8.1.2.2 defined the Batyrev quantum ring $H^*_\omega(X)$ to be the above quotient ring, we obtain a ring homomorphism

$$\varphi : H^*_\omega(X) \longrightarrow H^*(X),$$

where we remind the reader that $H^*(X)$ is a ring under the small quantum product.

We now claim that φ is an isomorphism. To prove this, remember than we always assume that our basis T_j of $H^2(X)$ lies in the interior of the Kähler cone of X. This allows us to take the limit as $q_j \to 0$, in which case the above map φ becomes the isomorphism describing the usual cohomology ring of the toric variety X. Since we are dealing with finite dimensional vector spaces, it follows that φ is an isomorphism in a neighborhood of $q_j = 0$. Furthermore, if we think of the q^β as formal variables, it follows that φ is an isomorphism, so that when X is smooth and Fano, the Batyrev quantum ring coincides with the usual small quantum ring of X, as claimed.

If we weaken the hypothesis that X is Fano and instead assume that $-K_X$ is nef, then Theorem 11.2.16 still applies, but the result we get is more complicated because $I_X = J_X$ may fail to hold—there may be a nontrivial change of variables involved. The key point is that when $-K_X$ is only nef, we may have $I_X \neq e^{(t_0+\delta)/\hbar}\left(1 + o(\hbar^{-1})\right)$. Then Theorem 11.2.16 says that I_X and J_X are related by a nontrivial change of variables. In this situation, we still get interesting relations in the small quantum ring of X. Proposition 5.5.4 still applies, so that $\square'_{\beta,\hbar} I_X = 0$, where $\square'_{\beta,\hbar}$ is given by (11.90). But rather than $\square'_{\beta,\hbar} J_X = 0$, which leads to (11.92) by Theorem 10.3.1, we first have to change variables, so that we will get a "twisted" version of (11.92) in $H^*(X)$. This is the precise sense in which $H^*_\omega(X)$ is related to small quantum cohomology of X.

We will illustrate this for $X = \mathbb{F}_2$. We consider the standard fan Σ for $\mathbb{F}_2$ with

$$\Sigma(1) = \{(1,0),(0,1),(-1,2),(0,-1)\},$$

where we have identified the 1-dimensional cones of Σ with their primitive generators as usual. We denote the corresponding toric divisors by D_1, D_2, D_3, D_4, and we will abuse notation by letting D_i denote its divisor class in $H^2(\mathbb{F}_2)$. The Kähler cone of $\mathbb{F}_2$ is spanned by the class f of a fiber and the section H satisfying $H^2 = 2$. In terms of the toric divisors, we have $D_1 = D_3 = f$, $D_2 = H - 2f$, and $D_4 = H$. We have $-K_{\mathbb{F}_2} = \sum D_i = 2H$, which is nef. Note that D_2 is the -2 curve on $\mathbb{F}_2$. The Mori cone is generated by the -2 curve and the fiber, i.e., by $H - 2f$ and f.

We need to compare the functions $I_{\mathbb{F}_2}$ and $J_{\mathbb{F}_2}$, and since a change of coordinates will be required, we will follow the notation of the previous example and use variables $z_i = e^{t_i}$ for $I_{\mathbb{F}_2}$ and $q_i = e^{s_i}$ for $J_{\mathbb{F}_2}$. The cohomology basis of $H^2(\mathbb{F}_2)$ is $T_1 = f$ and $T_2 = H$.

Our first study $I_{\mathbb{F}_2}$, which is given by

$$I_{\mathbb{F}_2}(t_1,t_2) = e^{(t_0+t_1 f+t_2 H)/\hbar} \sum_{\beta \in M(\mathbb{F}_2)} z^\beta \times$$

$$\frac{\prod_{m=-\infty}^{0}(f+m\hbar)^2 \prod_{m=-\infty}^{0}(H-2f+m\hbar) \prod_{m=-\infty}^{0}(H+m\hbar)}{\prod_{m=-\infty}^{f\cdot\beta}(f+m\hbar)^2 \prod_{m=-\infty}^{(H-2f)\cdot\beta}(H-2f+m\hbar) \prod_{m=-\infty}^{H\cdot\beta}(H+m\hbar)}.$$

As in the proof of Proposition 11.2.17, we are especially concerned with the coefficient of $\hbar^{-1}$ in the expansion of $I_{\mathbb{F}_2}$. The argument of that proof shows that this coefficient is determined by those effective classes β such that $-K_{\mathbb{F}_2} \cdot \beta = 0$ and $D_i \cdot \beta < 0$ for precisely one i. It is easy to compute that the only such β are given by $\beta = kD_2$ for any $k > 0$. For this β, we have $D_1 \cdot \beta = D_3 \cdot \beta = k$, $D_2 \cdot \beta = -k$, and $D_4 \cdot \beta = 0$.

Now suppose that $\beta = kD_2 = k(H-2f)$ for some $k \geq 1$. Then one computes that $z^\beta = z_1^{f\cdot\beta} z_2^{H\cdot\beta} = z_1^k$, and the coefficient of z_1^k in the above formula for $I_{\mathbb{F}_2}$ is

$$\frac{\prod_{-2k+1}^{0}(H-2f+m\hbar)}{\prod_{1}^{k}(f+m\hbar)^2} = -(H-2f)\frac{(2k-1)!}{(k!)^2}\hbar^{-1} + o(\hbar^{-1}).$$

It follows that

$$I_{\mathbb{F}_2}(t_1,t_2) = e^{(t_0+t_1 f+t_2 H)/\hbar}\left(1 - (H-2f)F(z_1)\hbar^{-1} + o(\hbar^{-1})\right),$$

where

$$F(z_1) = \sum_{k=1}^{\infty} \frac{(2k-1)!}{(k!)^2} z_1^k.$$

Theorem 11.2.16 requires

$$f_0 = 0, \quad f_1 = 2F(z_1), \quad f_2 = -F(z_1), \quad h = 0,$$

and then the Toric Mirror Theorem implies

$$J_{\mathbb{F}_2}(s_1, s_2) = I_{\mathbb{F}_2}(t_1, t_2),$$

where

$$(s_1, s_2) = \big(t_1 + 2F(e^{t_1}), t_2 - F(e^{t_1})\big),$$

or, putting $z_i = e^{t_i}$ and $q_i = e^{s_i}$,

$$(11.93) \qquad (q_1, q_2) = \big(z_1 e^{2F(z_1)}, z_2 e^{-F(z_1)}\big).$$

Surprisingly, the change of variables $s_1 = t_1 + 2F(z_1)$ appeared earlier as (6.59) in Chapter 6. If we compare this with (6.60), we see that

$$s_1 = \log\left(\frac{1 - 2z_1 - \sqrt{1 - 4z_1}}{2z_1}\right), \quad \text{so} \quad q_1 = \frac{1 - 2z_1 - \sqrt{1 - 4z_1}}{2z_1}.$$

This implies $z_1 = q_1/(1 + q_1)^2$, and then one easily sees that the inverse of the above coordinate changes are given by

$$(11.94) \qquad (z_1, z_2) = \left(\frac{q_1}{(1 + q_1)^2}, q_2(1 + q_1)\right)$$

and

$$(11.95) \qquad (t_1, t_2) = \big(s_1 - 2\log(1 + e^{s_1}), s_2 + \log(1 + e^{s_1})\big).$$

These coordinate changes appear (with different variable names) in [**Givental4**] in an example which considers the projective bundles $\mathbb{P}(\mathcal{O}_{\mathbb{P}^1} \oplus \mathcal{O}_{\mathbb{P}^1}(-1) \oplus \mathcal{O}_{\mathbb{P}^1}(-1))$ and $\mathbb{P}(\mathcal{O}_{\mathbb{P}^1} \oplus \mathcal{O}_{\mathbb{P}^1} \oplus \mathcal{O}_{\mathbb{P}^1}(-2))$ on $\mathbb{P}^1$.

Now let's compute some relations in $H_\omega(\mathbb{F}_2)$ and in the small quantum cohomology of $\mathbb{F}_2$. Let $\beta \in H_2(\mathbb{F}_2, \mathbb{Z})$ be the Poincaré dual of f. We want to compute the operator $\square'_{\beta,\hbar}$. We let λ_i correspond to D_i and $\delta_i = \lambda_i \partial/\partial\lambda_i$. Using (11.90), one computes that

$$\square'_{\beta,\hbar} = \hbar\delta_1 \hbar\delta_2 - z_2$$

since $z^\beta = z_2$ (recall that we are now using z_i rather than q_i). By (11.91), we have

$$\hbar\delta_2 = \hbar\frac{\partial}{\partial t_2} - 2\hbar\frac{\partial}{\partial t_1}, \quad \hbar\delta_4 = \hbar\frac{\partial}{\partial t_2},$$

and then Proposition 5.5.4 implies

$$(11.96) \qquad \left(\hbar\frac{\partial}{\partial t_2}\left(\hbar\frac{\partial}{\partial t_2} - 2\hbar\frac{\partial}{\partial t_1}\right) - z_2\right) I_{\mathbb{F}_2}(t_1, t_2) = 0.$$

In the Batyrev quantum ring $H_\omega^*(\mathbb{F}_2)$, this corresponds to the relation

$$(11.97) \qquad H \cdot (H - 2f) = z_2.$$

Now we apply the Toric Mirror Theorem to see what this says about $J_{\mathbb{F}_2}$. The key point, of course, is that the change of variables will give a slightly different differential equation. If we apply the chain rule to (11.95), we get

$$\left(\hbar\frac{\partial}{\partial s_2} - 2\hbar\frac{\partial}{\partial s_1}\right) = \frac{1 - q_1}{1 + q_1}\left(\hbar\frac{\partial}{\partial t_2} - 2\hbar\frac{\partial}{\partial t_1}\right), \quad \hbar\frac{\partial}{\partial s_2} = \hbar\frac{\partial}{\partial t_2}.$$

Combining these and using the fact that q_1 is independent of t_2 by (11.93), we obtain

$$\frac{1-q_1}{1+q_1}\left(\hbar\frac{\partial}{\partial t_2}\left(\hbar\frac{\partial}{\partial t_2}-2\hbar\frac{\partial}{\partial t_1}\right)-z_2\right)=\frac{1-q_1}{1+q_1}\hbar\frac{\partial}{\partial t_2}\left(\hbar\frac{\partial}{\partial t_2}-2\hbar\frac{\partial}{\partial t_1}\right)-\frac{1-q_1}{1+q_1}z_2$$
$$=\hbar\frac{\partial}{\partial s_2}\left(\hbar\frac{\partial}{\partial s_2}-2\hbar\frac{\partial}{\partial s_1}\right)-q_2(1-q_1),$$

where the last line uses $z_2 = q_2(1+q_1)$ from (11.94).

In the above display, the operator on the left annihilates $I_{\mathbb{F}_2}(t_1,t_2)$, so that by Theorem 11.2.16, the operator on the right annihilates $J_{\mathbb{F}_2}(s_1,s_2)$. Hence

$$\hbar\frac{\partial}{\partial s_2}\left(\hbar\frac{\partial}{\partial s_2}-2\hbar\frac{\partial}{\partial s_1}\right)-q_2(1-q_1)$$

is a quantum differential operator, so that by Theorem 10.3.1, the substitution

$$\hbar\partial/\partial s_1 \longmapsto T_1 = f, \quad \hbar\partial/\partial s_2 \longmapsto T_2 = H,$$

gives the relation

$$(11.98) \qquad\qquad H*(H-2f) = q_2(1-q_1)$$

in the small quantum cohomology of $\mathbb{F}_2$. On the other hand, (11.97) shows that $H\cdot(H-2f)=z_2=q_2(1+q_1)$ in the Batyrev quantum ring of $\mathbb{F}_2$. We conclude that these rings are not the same.

We should also mention that there is a more elementary way to compute small quantum cohomology on $\mathbb{F}_2$, because $\mathbb{F}_2$ is a deformation of $\mathbb{F}_0 = \mathbb{P}^1 \times \mathbb{P}^1$. Let f again be the fiber of $\mathbb{F}_0$, and let s be the section coming from the fibration over the other $\mathbb{P}^1$. Since $\mathbb{F}_2$ and $\mathbb{F}_0$ are deformation equivalent, they have the same quantum cohomology by the Deformation Axiom. Identifying their cohomologies, we see that the fibers f correspond, and $H = s+f$. Since $\mathbb{P}^1 \times \mathbb{P}^1$ is Fano, we can compute its small quantum cohomology using the Batyrev quantum ring of $\mathbb{F}_0$. This gives the relations

$$s*s = q^f, \qquad f*f = q^s.$$

We then compute

$$H*(H-2f) = (s+f)*(s-f) = s*s - f*f = q^f - q^s = q^f - q^{H-f}.$$

Since $q^f = q_1^{f\cdot f}q_2^{H\cdot f} = q_2$ and $q^{H-f} = q_1^{f\cdot(H-f)}q_2^{H\cdot(H-f)} = q_1 q_2$, we see that the above calculation agrees with (11.98).

CHAPTER 12

Conclusion

We have now reached the end of the book. Along the way, we have seen some wonderful mathematics and hopefully have learned something about the phenomenon of mirror symmetry. In this final chapter, we will look back at what we've done and what remains to be done. We will also discuss briefly some of the many aspects of mirror symmetry not covered in earlier chapters.

12.1. Reflections and Open Problems

In the process of studying mirror symmetry, we've used an amazing amount of algebraic geometry: we've discussed toric varieties, variations of Hodge structure, mixed Hodge structures, differential equations with regular singular points, hypergeometric equations, the Frobenius method, Kähler cones, moduli spaces of many sorts, symplectic geometry, algebraic stacks, enumerative geometry, the Clemens conjecture, equivariant cohomology, and localization—just to mention a few. The way in which mirror symmetry draws together so many parts of algebraic geometry indicates that our common enterprise is more deeply connected than we realize.

But there is more to mirror symmetry than simply making intense use of known mathematics. In this book, we have also witnessed the introduction of new objects into the study of algebraic geometry, including Kähler moduli, stable maps, Gromov-Witten invariants, quantum cohomology, the A-variation of Hodge structure, quantum differential equations, Euler data, and cohomology-valued formal functions. When one combines these with some of the ideas from physics which are still mathematically ill-defined, we get some glimpses of what 21st century algebraic geometry may have in store for us.

We also want to emphasize that the subject of mirror symmetry is richer—much richer—than what is presented in this book. Besides a simply enormous physics literature which we have barely touched upon, there are also numerous mathematical aspects of mirror symmetry we haven't covered at all. We will give a partial list in Section 12.2. But even when we restrict ourselves to the vision of mirror symmetry presented in the previous chapters, there is much that remains to be done, as we will now explain.

12.1.1. What is Mirror Symmetry? The basic problem is that we have yet to formulate a fully satisfactory version of the Mirror Theorem or even the Mirror Conjecture. If we look back at what we've done so far, we can discern three distinct flavors of mirror symmetry:

- In Section 8.6, we explored the suggestion made in [**Morrison7**] that a mathematical version of mirror symmetry should be formulated as an isomorphism between the A-model variation of Hodge structure of a smooth

397

Calabi-Yau manifold and the geometric (or B-model) variation of Hodge structure of its mirror.

- In Section 11.1, we studied the work of [**LLY**], which formulated the Mirror Theorem as an equivalence of Euler data $\hat{P}$ and $\hat{Q}$ after a mirror transformation. This implies a relation between certain hypergeometric cohomology-valued functions, from which we could compute the Gromov-Witten potential of the quintic threefold.

- In Section 11.2, we discussed the theorems in [**Givental2, Givental4**], where the Mirror Theorem is formulated as the equality of the cohomology-valued functions I_V and J_V after a mirror transformation. The resulting equation gave an equivalent description of the Gromov-Witten potential of the quintic threefold.

These three approaches were previewed in Section 2.6 and intersect beautifully in the case of the quintic threefold in $\mathbb{P}^4$. But the above approaches to mirror symmetry also have some distinct differences, such as the following:

- The Hodge-theoretic approach of Section 8.6 only makes sense for Calabi-Yau manifolds.

- The techniques of [**LLY**] apply to a variety of situations, including the quintic threefold and the multiple cover formula studied in Section 9.2.2. But as of this writing, these results are limited to $\mathbb{P}^n$ as ambient space.[1]

- The theorems stated in [**Givental2, Givental4**] apply when we have a nef complete intersection X in a smooth toric variety X_Σ. There are nontrivial results even in the case when $X = X_\Sigma$, but these techniques do not apply to the multiple cover formula, for example.

The last two, of course, are exciting because one is finally able to actually prove some of the wonderful formulas of mirror symmetry. Their limitation becomes apparent when we remember the way equivariant cohomology and localization are used in the proofs of the Mirror Theorems in Chapter 11. These techniques require the presence of a group, which explains why toric varieties play such a prominent role. However, recall that these results also apply to more than just the Calabi-Yau case. It follows that the approaches of [**LLY**] and [**Givental2**] may be part of a broader "toric" version of mirror symmetry which includes Calabi-Yau toric complete intersections as a special case.

On the other hand, the Hodge-theoretic version presented in Section 8.6 has nothing to do with toric varieties. Any projective Calabi-Yau manifold V has an A-variation of Hodge structure, and one can ask if this is the same as the geometric variation of some other Calabi-Yau, which would then be the mirror V°. Furthermore, as indicated by the remarks following Example 11.2.5.1, the relation between the Hodge-theoretic version of mirror symmetry and Givental's version of the Mirror Theorem has yet to be worked out. Finally, as we will see in Section 12.2.6, there are versions of mirror symmetry which involve much more than what's been discussed in previous chapters. All of this suggests that we are still far from a complete understanding of the mathematics of mirror symmetry.

Finally, we remind the reader that according to physics, mirror symmetry is the equivalence of two physical theories which are at present ill-defined mathematically.

---

[1]We just learned (November 1998) that the authors of [**LLY**] will soon announce a toric version of the material covered in [**LLY**]. When their preprint appears, we should have a detailed proof of a Toric Mirror Theorem.

As these theories get put on firmer mathematical foundations, we will get a deeper understanding of the mathematics of mirror symmetry. This may fill in some of the gaps just mentioned but also may lead us in completely unanticipated directions.

12.1.2. Open Conjectures. As just explained, the major open problem in mirror symmetry is to formulate and prove a general Mirror Theorem. Related to this, there are also interesting questions about the existing versions of the Mirror Theorem. For instance, we should better understand the connection between Givental's version of the Toric Mirror Theorem from Section 11.2.5 and the Toric Mirror Conjectures from Section 8.6.4. In Example 11.2.5.1, we used this result together with (11.81) to derive formulas (11.84) and (11.85). It would be highly desirable to see how far these formulas can be generalized.

However, the problems presented by the Mirror Theorem aren't the full story. The reader should also remember that as we developed the tools needed to state the versions of mirror symmetry considered in this book, we encountered some interesting unsolved problems along the way. We will now review some of these conjectures, with the hope of encouraging people to work on them.

We begin with four conjectures concerning complex and Kähler moduli which we discussed in Chapters 5 and 6:

- The *Integrality Conjecture* from Section 5.2.2 explains how the integer structure of the weight filtration should interact with the monodromy at a maximally unipotent boundary point.
- Section 6.1.2 constructed some distinguished boundary points on the simplified moduli space of a Calabi-Yau toric hypersurface. Conjecture 6.1.4 asserts that these are maximally unipotent boundary points.
- The *Cone Conjecture* from Section 6.2.1 states that the Kähler cone of a Calabi-Yau manifold V of dimension > 2 should be rational polyhedral modulo the action of $\mathrm{Aut}(V)$.
- Conjecture 6.2.8 in Section 6.2.3 describes the toric part of the Kähler cone of a Calabi-Yau toric hypersurface.

All of these are interesting conjectures in their own right as well as being important components of mirror symmetry.

Besides these moduli conjectures, there are plenty of other open problems related to mirror symmetry for algebraic geometers to think about. Here are some that we find especially interesting:

- When we studied the GKZ equations in Section 5.5.2, we saw that these equations gave some but not all of the Picard-Fuchs equations. There should be a description of the missing Picard-Fuchs equations. Also, the cohomology-valued formal function I_V should satisfy the Picard-Fuchs equations of the mirror.
- Section 5.6.3 discussed the normalized Yukawa couplings and potential function of a Calabi-Yau threefold at a maximally unipotent boundary point. When the threefold is a toric complete intersection, it should be possible to compute the Yukawa couplings and potential function in terms of an appropriate I_V, as we did in Example 11.2.5.1.
- There are questions concerning the virtual fundamental class $[\overline{M}_{g,n}(X,\beta)]^{\mathrm{virt}}$ which deserve further study. Section 7.3.2 discussed equivalences between the various definition of Gromov-Witten invariants and virtual fundamental classes. There is also the question of how $[\overline{M}_{g,n}(X,\beta)]^{\mathrm{virt}}$ relates to

$[\overline{M}_{g,n}(Y,\beta)]^{\mathrm{virt}}$ when X is a complete intersection in Y. We saw in (11.81) that this is important to understanding the relation between Givental's functions J_X and J_Y.

- The definition of quantum cohomology involves an infinite sum, and in Section 8.1.1 we conjectured that this sum converges in a neighborhood of a large radius limit point. Note that this conjecture is needed in order for the A-variation of Hodge structure to make sense.

As above, we feel that these are interesting problems, independent of their connection to mirror symmetry.

Some of the most important unanswered questions in this book concern enumerative geometry. Initially, the predictions of [**CdGP**] for the quintic threefold appeared almost magical—somehow, physics seemed to give effortless solutions to enumerative problems which lay far beyond the reach of standard methods in algebraic geometry. Since then, mirror symmetry has emerged as a powerful tool for computing Gromov-Witten invariants, but understanding the enumerative meaning of a Gromov-Witten invariant remains a nontrivial problem. Even for the quintic threefold V, the instanton number n_{10} has a potentially complicated relation to degree 10 rational curves on V because the Clemens conjecture is not known in degree 10. Even if it were known, n_{10} would not be the number of such curves, as we explained in Section 7.4.4. Another good problem is Conjecture 7.4.5, which attempts to give a direct definition of instanton numbers, rather than defining them indirectly in terms of the Gromov-Witten invariants $\langle I_{0,0,\beta} \rangle$ (as we did in the text). But these questions about instanton numbers are only part of the picture. As shown by the examples and references given in Sections 7.4 and 8.3, mirror symmetry has led to an explosion of papers on enumerative geometry, many of which have nothing to do with Calabi-Yau manifolds or Gromov-Witten invariants.

As we said in the preface, this book should be regarded as a report on the work-in-progress known as mirror symmetry. While some parts of the field have solid foundations, others are still in a rapid state of flux. The definitive book on mirror symmetry has yet to be written. Nevertheless, we hope that the somewhat limited view of mirror symmetry presented in our book has conveyed some of the wonder and excitement of this amazing area of algebraic geometry.

12.2. Other Aspects of Mirror Symmetry

Having now completed our main goal, we note that there are many other aspects of mirror symmetry that we have not treated in this book. Mirror symmetry continues to be an active area of research as of this writing, and many new ideas are being developed. This has led to other new developments in mathematics and physics. We do not even attempt to give a complete list here, but instead give a few samples. Of the topics discussed below, some are purely mathematical in nature, while others are more closely related to the physics, though all have interesting things to say about the mathematics.

12.2.1. Rational Curves on K3 Surfaces. Fix a generic K3 surface containing a divisor class C with $C^2 = 2g - 2$. Then $|C|$ is a g-dimensional linear system of curves of arithmetic genus g. Let $n(g)$ be the number of rational curves in this linear system. Yau and Zaslow [**YZ**] conjectured the amazing formula

$$1 + \sum_{g=1}^{\infty} n(g)q^g = q/\Delta(q),$$

where $\Delta(q)$ is the usual modular form of weight 12. A mathematical exposition including proofs of intermediate steps was first written in [**Beauville2**]. The idea is that each rational curve occurring in this linear system is assigned a multiplicity equal to the Euler characteristic of its compactified Jacobian (the same Euler characteristic appears in [**YZ**] as well). This multiplicity can be related to stable maps [**FGvS**]. In addition, it is also shown in [**FGvS**] that the multiplicity is always positive, so that the Yau-Zaslow formula has a definite meaning as a count of rational curves with appropriate multiplicities.

The Yau-Zaslow formula has been generalized in [**Göttsche**] to a conjecture about curves with nodes on algebraic surfaces. A generating function is proposed to count the number of curves with n nodes in n-dimensional families of curves. In the case of a K3 or Abelian surface, the conjectured generating function can be expressed in terms of quasi-modular forms. This conjecture has been proven for K3 surfaces in [**BL**], and we supply a little more detail in this case. Consider a linear system of curves C with $C^2 = 2g + 2n - 2$. Then $|C|$ is a $(g+n)$-dimensional linear system of curves of arithmetic genus $g + n$. The set of curves in $|C|$ with n nodes can be expected to form a g-dimensional subfamily. Let $N_g(n)$ denote the number of curves of geometric genus g with n nodes passing through g generic points in this linear system. Then it is proven in [**BL**] that

$$\sum_{n=0}^{\infty} N_g(n)q^n = \left(\frac{d}{dq} G_2(q) \right)^g \frac{q}{\Delta(q)},$$

where $G_2(q)$ is the Eisenstein series

$$G_2(q) = -\frac{1}{24} + \sum_{k=1}^{\infty} \sigma(k)q^k,$$

$\sigma(k)$ being the arithmetic function $\sum_{d|k} d$. Note that G_2 and its derivatives are quasi-modular functions.

12.2.2. Mirror Symmetry for K3 Surfaces and Arnold's Strange Duality. In the late 1970's, various people showed how a certain duality for K3 surfaces could be used to explain Arnold's strange duality for exceptional unimodal critical points. In recent years, it was realized that this duality for K3 surfaces is part of the two-dimensional version of mirror symmetry [**Dolgachev3, Kobayashi, Ebeling, GN1, GN3**].

Mirror symmetry for a K3 surface has a different flavor. For a Calabi-Yau manifold V of dimension > 2, the Kähler cone is open in $H^2(V, \mathbb{C})$, but for a K3 surface S, this is no longer true, and the relation between the Kähler cone and the lattice $H^2(S, \mathbb{Z})$ is quite subtle. Hence it is no surprise that lattice theory plays an important role in the mirror symmetry of K3 surfaces. We saw some hints of this in the Voisin-Borcea construction given in Section 4.4.

One can also do string theory on a K3 surface S, as explained in [**AM3**]. There is still a SCFT moduli space, but because $h^{2,0}(S) \neq 0$, the deformations of the complex and Kähler structures no longer decouple, so that we no longer get the local product structure pictured in (1.3). This subject is too large to be summarized in a few paragraphs. We refer the reader to [**AM3, Aspinwall**] for further details and references to the literature.

12.2.3. The McKay Correspondence. If $G \subset \mathrm{SL}(n, \mathbb{C})$ is finite and $\mathbb{C}^n/G$ has a resolution Y with $K_Y = 0$, then the *McKay correspondence* [**Reid6**] asserts that irreducible representations of G should give a basis of $H^*(Y, \mathbb{Z})$ such that the character table of G corresponds to Poincaré duality on Y. This conjecture has consequences concerning the physicist's Euler number [**Vafa1, Roan2**] and the string theoretic Hodge numbers of Batyrev and Dais [**BD**].

12.2.4. Hyper-Kähler Manifolds. A compact complex manifold M is *holomorphically symplectic* if it has a closed nondegenerate holomorphic 2-form. By the Calabi-Yau theorem, M has a metric and a quaternionic action parallel with respect to the Riemannian connection, so that M is hyper-Kähler. In [**Verbitsky**], Verbitsky proves that a hyper-Kähler manifold is its own mirror.

12.2.5. D-branes and T-Duality. In this book, we have regarded mirror symmetry as the equality of two SCFTs associated to a mirror pair (V, ω) and (V°, ω°) (see the discussion in Chapter 1). The states in the SCFTs are called *perturbative states*, and these include the states corresponding to cohomology classes on V and V°. In recent years, it has been realized that the theory can be enlarged in a sense to include *non-perturbative states*, including those associated to *D-branes*. A D-brane is essentially the *worldvolume* swept out by a p-dimensional object as it propagates through spacetime. These objects are sometimes called Dp-branes for definiteness, and they are $(p+1)$-dimensional due to the extra time dimension. For example, a D1-brane looks exactly like a string worldsheet (as drawn in Chapter 1). The states themselves arise from open strings whose boundaries lie on the D-brane. See [**Polchinski1**] for a survey of D-branes. The physical theories associated to (V, ω) and (V°, ω°), *including non-perturbative states*, are presumed to be equal, making mirror symmetry an even stronger assertion.

In [**SYZ**], Strominger, Yau and Zaslow argue that mirror symmetry is what physicists call *T-duality*. We give a quick outline of the idea. Note that the moduli space of points on V° is isomorphic to V° itself. This is again naturally identified with a part of the moduli space of D0-branes on V°. To see this, recall from Chapter 1 that string theory is a 10-dimensional theory, of which V° contributes 6 dimensions. The time dimension is one of the remaining 4 dimensions. For any point $p \in V^\circ$, as p propagates in time, it sweeps out the worldline of a D0-brane. Mirror symmetry predicts that this should correspond to a part of the moduli space of D3-branes on V. Further analysis suggests that the mirror V° of a Calabi-Yau manifold V should be the moduli space (suitably complexified and compactified) of certain special Lagrangian tori on V. As explained in [**Morrison8**], T-duality may in this manner eventually lead to a purely mathematical construction of mirror symmetry. In this description, both V and V° admit maps to the same 3 (real) dimensional base, with generic fiber a (real) 3-torus. The fibers of V and V° in this fibration are presumed to be dual tori, hence the name "T-duality". This could lead to a topological construction of mirror manifolds. Aspects of this assertion have been checked for the Voisin-Borcea threefolds [**GW**] and K3 surfaces

[**GW, Morrison8**]. For elliptic K3 surfaces, T-duality is related to the Fourier-Mukai transform [**BBRP**]. Interesting consequences for topology are conjectured in [**Gross3**].

12.2.6. Homological or Categorical Mirror Symmetry. In 1994, Kontsevich [**Kontsevich3**] proposed a "homological" version of mirror symmetry where the Mirror Conjecture is formulated as an equivalence of categories. In this situation, V° being mirror to V means that the derived category of coherent sheaves on V° should be equivalent to the derived category of Lagrangian submanifolds of V with unitary local systems. The basic idea of creating a category out of Lagrangian submanifolds is due to [**Fukaya**]. In [**PZ**], it is suggested that one may want to restrict to special Lagrangian submanifolds of V. Note that Lagrangian submanifolds use the Kähler structure of V (the A-model), while coherent sheaves use the complex structure of V° (the B-model).

It is explained in [**Kontsevich3**] that certain deformations in the Lagrangian category associated to V should lead to a construction of the quantum cohomology of V as a deformation of the ordinary cup product. The deformations here take place naturally over all of $H^*(V)$, rather than just over $H^{1,1}(V)$ as in the definition of the A-model connection in Section 8.5. Since this should be related to the deformation theory of coherent sheaves on V°, this suggests a rich structure which goes beyond the usual deformations of complex structure. A hint of what this may entail is given by the formula

$$(12.1) \qquad HH^n(V^\circ) = \bigoplus_{p+q=n} H^q(V^\circ, \wedge^p \Theta_{V^\circ}),$$

where HH stands for Hochschild cohomology. This cohomology group is to be identified with the tangent space to a thickening of the complex moduli space of V°. Note that the summands in (12.1) are precisely the (p, q) eigenspaces of the operators $(Q, \bar{Q})$ from the $N = 2$ superconformal algebra written in the first line of (1.2). The idea of a thickening of the moduli space with tangent space (12.1) first appeared in [**Witten4**]. It is very nice that these spaces should arise naturally in mathematics. As explained in [**Kontsevich3**], $HH^2(V^\circ)$ consists of the usual first-order deformations of the complex structure of V° together with certain non-commutative deformations of the sheaf of algebras $\mathcal{O}_{V^\circ}$, together with some terms which are still mysterious from the point of view of deformations. Since $H^2(V^\circ, \mathcal{O}_{V^\circ})$ and $H^0(V^\circ, \wedge^2 \Theta_{V^\circ})$ vanish for Calabi-Yau threefolds of dimension at least 3, these last two terms are not a problem, but the meaning of $HH^n(V^\circ)$ for $n \neq 2$ is still something of a mystery. Also, [**Kontsevich3**] points out that the Lagrangian category of V has a conjectural A_∞ structure. (However, an A_∞-category is not a category in the usual sense—composition is no longer associative—so that one needs to be careful here.) This again suggests that on the mirror side, there should be richer structures in the category of coherent sheaves on V°. All of this is evidence that we have not yet discovered the full scope of mirror symmetry.

An example of this equivalence of categories has recently been worked out in [**PZ**] for the case of an elliptic curve. This article also points out that T-duality could follow from the homological version of mirror symmetry. In this context, given a point $p \in V^\circ$, the coherent sheaf $\mathcal{O}_{\{p\}}$ of the subvariety $\{p\} \subset V^\circ$ should correspond to a certain special Lagrangian submanifold of V. This is a version of the idea discussed in Section 12.2.5, where the mirror V° is the moduli space of certain (real) Lagrangian tori in V.

12.2.7. K3 and Elliptically Fibered Calabi-Yau Manifolds. Calabi-Yau manifolds with K3 fibrations or elliptic fibrations have been the subject of numerous papers in mathematics and physics. The interest of these fibrations is quite natural mathematically. In the study of linear systems on Calabi-Yau manifolds, K3 and elliptic fibrations play a special role because they arise naturally in the log-minimal model program for Calabi-Yau manifolds. In the ordinary minimal model program, one tries to understand varieties as fibrations where the fibers are minimal models. The analog for log-minimal models of Calabi-Yau varieties is for the fibers to be Calabi-Yau of lower dimension. A classification of these fibrations is begun in [**Oguiso**].

Mirror symmetry has also been studied for K3 fibrations. It is suspected that the mirror of the fiber should influence the mirror of the total space. This idea is discussed in [**GN2**], and examples of such spaces can be found in [**AKMS, HLY3**].

K3 and elliptic fibrations arise in physics as well, for different reasons. The main impetus for the study of K3 fibrations has been *duality*. More specifically, *type II* string theory associated to a Calabi-Yau threefold is believed to be *dual* to a heterotic string theory associated to the product of a K3 surface and an elliptic curve. This means that for certain parameter values of the respective theories, the theories become identical. Vector bundles on K3 surfaces are part of the data needed to define this particular type of heterotic string theory. Numerous mathematical checks of this have been performed.

Elliptically fibered Calabi-Yau threefolds arise in physics as spaces that can be used to construct models for *F-theory* [**Vafa3**]. Since bundles also play a key role in F-theory, this suggests further study of principal bundles over Calabi-Yau manifolds with elliptic fibrations [**FMW1**]. Mathematical studies of these bundles can be found in [**Donagi, FMW2**]. Closely related to this is a duality between F-theory and heterotic string theory. This duality arises in theories resulting from Calabi-Yau threefolds which admit both K3 fibrations and elliptic fibrations. Such dualities may be viewed as an extension of the analogous result in one dimension lower. Duality of F-theory associated to elliptically fibered K3 surfaces and heterotic string theory on an elliptic curve is supported mathematically by a beautiful identification of the moduli of bundles on an elliptic curve with the moduli of certain types of Weierstrass models of elliptically fibered K3 surfaces.

12.2.8. Conifold Transitions. A conifold transition occurs when a Calabi-Yau threefold acquires nodal singularities and has a resolution which is still Calabi-Yau. The SCFT arising from heterotic string becomes singular through this transition. However, it has been shown in [**GMS**] that *type II* string theory can avoid singularities at conifold transitions (roughly speaking) since electrically charged black holes associated to a vanishing cycle (thought of as a 3-brane[2]) on the original manifold become massless when the Calabi-Yau becomes singular, and this singularity compensates for the singularity in the theory that we mentioned before. These black holes then become elementary particles on the resolution. Assuming

---

[2]The D3-branes considered in Section 12.2.5 are special kinds of 3-branes. D-branes were not generally accepted until after [**GMS**] was written. It is now agreed that the 3-branes in [**GMS**] are in fact D3-branes.

a Kähler version of Reid's fantasy [**Reid5**], one could then connect the physical theories coming from any two Calabi-Yau threefolds. This has been verified for Calabi-Yau threefolds which are hypersurfaces in weighted projective spaces [**ACJM, CGGK**], though the singularities may be non-nodal. Some aspects of conifold transitions related to mirror symmetry are discussed in Section 6.2.4.

It is conjectured in [**Morrison9**] that the mirror of a conifold transition is a conifold transition (and more generally, that the mirror of an extremal transition is an extremal transition). This is checked in examples in [**BCFKvS1, BCFKvS2**].

12.2.9. Calabi-Yau Complete Intersections in Flag Manifolds. Here, the key observation, made in [**Sturmfels2**], is that the Grassmannian $G(k, n)$ can be deformed flatly to a projective toric variety $P(k, n)$. Furthermore, according to [**BCFKvS2**], $P(k, n)$ is Fano, i.e., it comes from a reflexive polytope. Using this, the authors of [**BCFKvS1**] give a mirror construction for a Calabi-Yau complete intersections $V \subset G(k, n)$ by combining the Batyrev-Borisov mirror construction from Chapter 4 with the extremal transitions discussed in Sections 6.2.4 and 12.2.8.

The basic ideas goes as follows. Suppose that $V \subset G(k, n)$ is a complete intersection of hypersurfaces of degree $d_1, \ldots, d_r$. As we degenerate $G(k, n)$ to $P(k, n)$, V degenerates to a toric complete intersection $\bar{V} \subset P(k, n)$. We resolve this using the smooth crepant resolution $\hat{P}(k, n) \to P(k, n)$, which then gives a resolution $V' \to \bar{V}$. If we view this as going from V' to $\bar{V}$ by contraction and then from $\bar{V}$ to V by smoothing, then we get one of the extremal contractions discussed in Section 6.2.4 (in this case, the transition is a conifold transition because the singularities of $\bar{V}$ are especially mild). As we discussed in that section, mirror symmetry suggests that there should be a "mirror" extremal transition, where the mirror V'° of V' degenerates to $\bar{V}^\circ$ and then resolves to the mirror V° of V.

In this situation, the mirror V'° of $V' \subset \hat{P}(k, n)$ is known by the Batyrev-Borisov mirror construction described in Section 4.3. By working out the corresponding degeneration and smoothing, [**BCFKvS1**] is able to describe an explicit mirror construction for Calabi-Yau complete intersections in a Grassmannian.

To computing Gromov-Witten invariants using this construction, [**BCFKvS1**] introduces the *factorial trick*, which the paper describes as a naive version of the Quantum Hyperplane Section Principle from Section 11.2.2. Another tool used in [**BCFKvS1**] is the relation between quantum cohomology and quantum differential equations, which we discussed in Section 10.3. Since the quantum cohomology of a Grassmannian is known, one obtains a lot of information about the quantum differential equations (called the *quantum $\mathcal{D}$-module* in [**BCFKvS1**]). Then, using the Toric Mirror Theorem from Section 11.2.5 and the generalization given in [**Kim2**], the instanton numbers of numerous 3-dimensional Calabi-Yau complete intersections in Grassmannians are computed.

All of this can be done in greater generality. Let $F(n_1, \ldots, n_\ell, n)$ denote the (partial) flag variety consisting of all flags $0 \subset W_1 \subset \cdots \subset W_\ell \subset \mathbb{C}^n$ where $\dim W_i = n_i$. In [**GL**], it is shown that $F(n_1, \ldots, n_\ell, n)$ degenerates to a toric variety $P(n_1, \ldots, n_\ell, n)$, generalizing the result of Sturmfels mentioned above. Since $P(n_1, \ldots, n_\ell, n)$ is Fano by [**BCFKvS2**], it follows that the above construction can be applied. We should also mention the paper [**Batyrev5**], which reviews the constructions just described and considers what happens in general when a Fano variety degenerates to a Fano toric variety.

12.2.10. Other Topics from Physics. There is a long list of recent developments in the physics of mirror symmetry which weren't discussed in earlier chapters. Some of these have already been mentioned in this section, but there are many others. For example, arguments from physics lead to the expectation that a supersymmetric field theory can be obtained as a limit of string theory as certain curves in a Calabi-Yau manifold are contracted. Thus Calabi-Yau geometry gives a tool for analyzing these supersymmetric theories, especially their gauge-theoretic aspects. These field theories have "periods" that can be solved using mirror symmetry, leading for example to a derivation of the Seiberg-Witten curve and differential without assuming any duality conjectures. See [**KKV, KMV**] for the use of mirror symmetry in this regard and [**KV, BJPSV, BV**] for other situations where quantum field theories are constructed using Calabi-Yau geometry. These theories are called *geometrically engineered* theories. An introduction to these theories can be found in [**Mayr**]. Mathematically, geometrically engineered theories give explicit predictions for the dimensions of smoothing components in the extremal transitions discussed in Section 6.2.4. The simplest example is the conifold transition, where the prediction from physics in [**GMS**] matches the mathematics. The mathematical calculation of the smoothing dimensions in this case follows from [**Clemens**]. As of this writing, there is no mathematical proof of the correctness of the prediction from physics. This prediction was checked in examples in [**KMP**].

APPENDIX A

Singular Varieties

Although complex manifolds are an important part of mirror symmetry, there are many situations where singular varieties occur naturally. For example, the Batyrev construction from Chapter 4 deals with potentially singular hypersurfaces in simplicial toric varieties. This appendix will review the types of singularities we will encounter.

All varieties considered in this book are *Cohen-Macaulay*, which means that all of the local rings are Cohen-Macaulay. The key feature of a Cohen-Macaulay variety X is that it has a dualizing sheaf, usually denoted ω_X. When X is smooth, ω_X is the usual sheaf Ω_X^d, where $d = \dim(X)$, and because of this, we will often write the dualizing sheaf of a general Cohen-Macaulay variety as $\widehat{\Omega}_X^d$. Cohen-Macaulay varieties are nice because they behave well with respect to duality theory—see [**Oda**] for a careful discussion. Another notation for $\widehat{\Omega}_X^d = \omega_X$ is $\mathcal{O}_X(K_X)$, where K_X is the canonical divisor. Note that in general, K_X is only a Weil divisor.

By standard commutative algebra, the dualizing sheaf ω_X of a Cohen-Macaulay variety X is a line bundle $\Leftrightarrow K_X$ is a Cartier divisor $\Leftrightarrow$ all of the local rings are Gorenstein. In this case, we say that X is *Gorenstein*.

A.1. Canonical and Terminal Singularities

A point of a normal variety X is a *canonical singularity* provided that rK_X is Cartier near the point for some positive integer r and there is a local resolution of singularities $f : Y \to X$ such that

$$(A.1) \qquad\qquad rK_Y = f^*(rK_X) + \sum_i a_i E_i,$$

where the sum is over all exceptional divisors E_i of f and $a_i \geq 0$ for all i. Furthermore, if $a_i > 0$ for all i, then the singularity is *terminal*. If X is Gorenstein, then we can take $r = 1$ in the above equation.

These types of singularities arise naturally when studying canonical models and minimal models of threefolds and higher dimensional varieties. More background on the singularities themselves can be found in [**Reid1**] and [**Reid4**]. For Calabi-Yau varieties, their relevance is obvious because the canonical class plays such a prominent role in the definition of Calabi-Yau. For example, a minimal Calabi-Yau variety V, as defined in Definition 1.4.1, has Gorenstein $\mathbb{Q}$-factorial terminal singularities. It follows that if V is not smooth, then any resolution $f : W \to V$ is no longer Calabi-Yau. This is because the singularities of V are terminal, so that $K_W = K_V + \sum_i a_i E_i$ is nontrivial since $K_V = 0$ and $a_i > 0$.

Another source of these singularities comes from toric varieties. In [**Reid1**], it is shown that Gorenstein toric varieties have at worst canonical singularities. Below, we will see that the same is true for Gorenstein orbifolds.

407

A.2. Orbifolds

We begin by recalling what it means for a variety to be an orbifold.

DEFINITION A.2.1. *A d-dimensional variety X is an orbifold if every $p \in X$ has a neighborhood analytically equivalent to $0 \in U/G$, where $G \subset \mathrm{GL}(d,\mathbb{C})$ is a finite subgroup with no complex reflections other than the identity and $U \subset \mathbb{C}^d$ is a G-stable neighborhood of the origin.*

In this definition, a *complex reflection* is an element of $\mathrm{GL}(d,\mathbb{C})$ of finite order such that $d - 1$ of its eigenvalues are equal to 1. The group G in Definition A.2.1 is called a *small* subgroup of $\mathrm{GL}(d,\mathbb{C})$, and $(U/G, 0)$ is called a *local chart* of X at p. Note that G is unique up to conjugacy by a theorem of [**Prill**].

The terms *orbifold* and *V-manifold* are interchangeable, and are the same as being *quasi-smooth*. For a toric variety, these are equivalent to being *simplicial*, and here, the group G is Abelian. Thus a simplicial toric variety has at worst finite Abelian quotient singularities.

A key intuition is that "over $\mathbb{Q}$", orbifolds behave like manifolds. For example, the singular cohomology of a compact orbifold satisfies Poincaré duality, but only with $\mathbb{Q}$ coefficients. Similarly, for a complete toric variety, the combinatorial description of $H^*(X,\mathbb{Z})$ when X is smooth works for $H^*(X,\mathbb{Q})$ when X is simplicial.

We next study Gorenstein orbifolds. It is well known that every Gorenstein orbifold has at worst canonical singularities [**Reid1**]. However, we also have the following folklore result, which says that in the terminal case, a Gorenstein orbifold has rather small singularities.

PROPOSITION A.2.2. *If a Gorenstein orbifold has at worst terminal singularities, then its singular locus has codimension ≥ 4.*

PROOF. We can reduce to $\mathbb{C}^d/G$, where G is a small subgroup. Given $g \in G$ of order r, let ζ be a primitive rth root of unity and write the eigenvalues of g as ζ^{a_i} where $0 \leq a_i < r$. By [**Reid1**], Gorenstein implies $\sum_{i=1}^{n} a_i \equiv 0 \bmod r$, and by [**Reid2**], terminal implies $\sum_{i=1}^{n} a_i > r$.

According to [**Prill**], the singular locus of $\mathbb{C}^n/G$ is the set

$$\big\{ x \in \mathbb{C}^n : g(x) = x \text{ for some } g \in G - \{1\} \big\}/G.$$

If the singular locus has codimension < 4, then we can find $g \neq 1$ in G whose eigenvalues (suitably reordered) are $\zeta^a, \zeta^b, \zeta^c, 1, \dots, 1$. Then the Gorenstein and terminal conditions imply $r|(a+b+c)$ and $a+b+c > r$. It follows that $a+b+c = 2r$. Then, applying the terminal criterion to g^{-1} easily yields a contradiction. We leave the details to the reader. $\qquad\square$

When the orbifold is also toric, Proposition A.2.2 is proved in [**Batyrev4**, Thm. 2.2.9]. In dimension three, the proposition has the following immediate corollary.

COROLLARY A.2.3. *A 3-dimensional Gorenstein orbifold with at worst terminal singularities is smooth.* $\qquad\square$

In terms of the definition of Calabi-Yau from Chapter 1, we see that an orbifold V of dimension d is Calabi-Yau if and only if its canonical class is trivial and $H^i(V, \mathcal{O}) = 0$ for $i = 1, \dots, d-1$. This is because trivial canonical class implies Gorenstein, which for an orbifold implies canonical. Furthermore, the above corollary shows that a 3-dimensional minimal Calabi-Yau orbifold is smooth, since minimal implies that the singularities are at worst terminal.

We next define a *suborbifold* of an orbifold.

DEFINITION A.2.4. *Given an orbifold X, a subvariety $Y \subset X$ is a suborbifold if for every $p \in W$ there is a local chart $(U/G, 0)$ of X at p such that the inverse image of Y in U is smooth at 0.*

It is easy to see that a suborbifold of an orbifold is again an orbifold. However, the converse is not true: a subvariety of X which is an orbifold need not be a suborbifold. This is because the singularities of a suborbifold are intimately related to the singularities of the ambient space.

A.3. Differential Forms on Orbifolds

There is a nice theory of differential forms for orbifolds. Given $p \in X$, consider a local chart $(U/G, 0)$. Then a C^∞ form on U/G is defined to be a G-invariant C^∞ form on U. There is a natural notion of patching of forms defined on different charts. This enables us to define the de Rham cohomology groups $H^*_{DR}(X, \mathbb{R})$, which are isomorphic to the usual ones (see [**Satake**]).

We can also do Hodge theory on orbifolds. Holomorphic p-forms on an orbifold are defined using local charts $(U/G, 0)$ and are called *Zariski p-forms* on X. They determine a sheaf we will denote $\widehat{\Omega}^p_X$. The Zariski sheaves have the following simple characterization.

PROPOSITION A.3.1. *If X is an orbifold and $j : X_0 \subset X$ is the inclusion of the smooth locus of X, then $\widehat{\Omega}^p_X = j_*(\Omega^p_{X_0})$, where $\Omega^p_{X_0}$ is the usual sheaf of holomorphic p-forms on the complex manifold X_0.*

An orbifold X is Cohen-Macaulay, and one can show that if $d = \dim(X)$, then $\widehat{\Omega}^d_X$ is the dualizing sheaf of X. Thus our notation is consistent with the previous section. The sheaves $\widehat{\Omega}^p_X$ have various other nice properties, including:

- There is a differential $d : \widehat{\Omega}^p_X \to \widehat{\Omega}^{p+1}_X$ such that $(\widehat{\Omega}^*_X, d)$ is a resolution of the constant sheaf $\mathbb{C}_X$.
- There is a natural product $\widehat{\Omega}^p_X \otimes \widehat{\Omega}^q_X \to \widehat{\Omega}^{p+q}_X$ such that the natural map $\widehat{\Omega}^p_X \to \underline{\mathrm{Hom}}_{\mathcal{O}_X}(\widehat{\Omega}^{d-p}_X, \widehat{\Omega}^d_X)$ is an isomorphism.

We can also define Hodge groups $H^{p,q}(X)$, and the Dolbeault theorem

$$H^q(X, \widehat{\Omega}^p_X) \simeq H^{p,q}(X)$$

applies in this situation (see [**Baily1**]). For an orbifold, the natural map to intersection cohomology

$$H^*(X, \mathbb{Q}) \longrightarrow IH^*(X, \mathbb{Q})$$

is an isomorphism. Since the intersection homology of a projective variety has a natural Hodge structure (see [**Saito**]), we see that $H^k(X, \mathbb{C}) = \oplus_{p+q=k} H^{p,q}(X)$ has a pure Hodge structure (but only over $\mathbb{Q}$) and satisfies the Hard Lefschetz theorem. This has interesting consequences for the combinatorics of simplicial polytopes (see [**Fulton3**, Section 5.6]). Other references for the Hodge theory of orbifolds are [**Steenbrink1, Steenbrink2**].

A Kähler form on an orbifold is a real, smooth $(1,1)$-form which is positive at every point (on a chart $(U/G, 0)$, this means its pullback to U is positive). For more details about Kähler forms on orbifolds, see [**AGM1, Baily2**].

A.4. The Tangent Sheaf of an Orbifold

Any algebraic variety X has a tangent sheaf Θ_X. When X is smooth, we also know that Θ_X is dual to $\widehat{\Omega}^1_X$, and that $H^1(X, \Theta_X)$ classifies infinitesimal deformations of X. The first of these facts continues to hold when X is an orbifold.

PROPOSITION A.4.1. *If X is an orbifold, then $\Theta_X \simeq \underline{\mathrm{Hom}}_{\mathcal{O}_X}(\widehat{\Omega}^1_X, \mathcal{O}_X)$.*

PROOF. Let Ω^1_X be the sheaf of Kähler 1-forms on X. The universal property of Ω^1_X implies $\Theta_X = \underline{\mathrm{Hom}}_{\mathcal{O}_X}(\Omega^1_X, \mathcal{O}_X)$. If $j : X_0 \to X$ is the inclusion of the smooth part of X, then the argument of page 128 of [**Oda**] shows that we can compute $\underline{\mathrm{Hom}}_{\mathcal{O}_X}(-, \mathcal{O}_X)$ by restricting to X_0. From here, the proposition follows easily. $\qquad\square$

As for $H^1(X, \Theta_X)$, recall that the infinitesimal deformations of a general variety X are classified by $\mathrm{Ext}^1(\Omega^1_X, \mathcal{O}_X)$. For an arbitrary orbifold, this may differ from $H^1(X, \Theta_X)$. But there is one case where they agree.

PROPOSITION A.4.2. *If X is a Gorenstein orbifold with at worst terminal singularities, then $H^1(X, \Theta_X)$ classifies infinitesimal deformations of X.*

PROOF. A result of [**Schlessinger**] implies that $\underline{\mathrm{Ext}}^1(\Omega^1_X, \mathcal{O}_X)$ vanishes for an orbifold if its singularities have codimension at least three. This is true by Proposition A.2.2, and then the result follows immediately from the local to global spectral sequence for Ext. $\qquad\square$

By [**Batyrev4**, Thm. 2.2.9], this proposition is also true for varieties, not necessarily orbifolds, which have at worst terminal toroidal singularities.

For an arbitrary orbifold X, one can also prove that $\Theta_X = j_*\Theta_{X_0}$ and that $\widehat{\Omega}^1_X$ is the double dual of Ω^1_X. When X is a Gorenstein orbifold, the isomorphism $\widehat{\Omega}^{d-1}_X \simeq \underline{\mathrm{Hom}}_{\mathcal{O}_X}(\widehat{\Omega}^1_X, \widehat{\Omega}^d_X)$ and Proposition A.4.1 imply that

$$(A.2) \qquad\qquad \widehat{\Omega}^{d-1}_X \simeq \Theta_X \otimes \widehat{\Omega}^d_X.$$

We can apply this to Calabi-Yau varieties as follows.

PROPOSITION A.4.3. *The infinitesimal deformations of a minimal Calabi-Yau orbifold V of dimension d are classified by*

$$H^1(V, \Theta_V) \simeq H^1(V, \widehat{\Omega}^{d-1}_V).$$

PROOF. A minimal Calabi-Yau orbifold has Gorenstein terminal singularities, so that $H^1(V, \Theta_V)$ classifies infinitesimal deformations by Proposition A.4.2. Then we are done by (A.2) since $\widehat{\Omega}^d_V$ is trivial. $\qquad\square$

A.5. Symplectic Orbifolds

Finally, one can define orbifolds in the C^∞ category, and there is also the notion of a *orbifold diffeomorphism*. Then a *symplectic orbifold* is a C^∞ orbifold with a closed, nondegenerate 2-form (this is defined using local charts). Furthermore, the process of symplectic reduction works naturally in this case. See [**Audin**, II-3.6] for the details.

APPENDIX B

Physical Theories

In this appendix, we summarize some of the key points of physical theories mentioned in this book. The first section gives background on some basic physical theories, leading up to quantum field theory by the progression

$$\begin{array}{ccc} \text{Classical} & & \text{Classical} & & \text{Quantum} \\ \text{Mechanics} & \longrightarrow & \text{Field Theory} & \longrightarrow & \text{Field Theory} \end{array}$$

Each of these theories has a *Lagrangian* and a *Hamiltonian* formulation. In quantum field theory, we will see that Lagrangians are better suited for path integrals while Hamiltonians lead to an algebra of self-adjoint operators on a Hilbert space. We can easily pass from a Lagrangian to a Hamiltonian formulation, so we will always start with a Lagrangian.

It will be helpful if the reader has some familiarity with the rudiments of classical mechanics (as in [**Arnold**]). We will discuss classical field theory, quantum field theory and some gauge theory. The remaining sections of the appendix will describe nonlinear sigma models, conformal field theories, Landau-Ginzburg theories, gauged linear sigma models, and finally topological quantum field theories.

We refer the interested reader to the lecture notes [**DEFJKMMW**] from the 1996–97 IAS Quantum Field Theory program for a more detailed and more mathematical treatment of some of the theories considered in this appendix.

B.1. General Field Theories

In this section, we explain what a field theory is. We do not attempt to be self-contained, but rather will give some simple examples of the kind of theories that are mentioned in the main text while omitting some essential background material. For more details, see [**Rabin**], whose treatment we follow closely but not exactly.

We will start with a classical theory of fields on a spacetime M. Typical examples of M can be 4-dimensional Minkowski space or the world sheet of a string (i.e., a Riemann surface). A *classical field* on M will be loosely defined as either a function, a differential form, a section of a bundle on M, or a connection on a bundle. It is not desirable at this point to limit the scope of our discussion by making a more precise definition. We will obscure this further by sometimes speaking of fields as if they were functions.

Physical dynamics is determined by the *Lagrangian density*, which is a functional of the fields and their derivatives.[1] The choice of Lagrangian is greatly constrained by symmetries, and is determined by the particular physics that the field theory is supposed to model. The basic principle of dynamics is that the fields

---

[1]This is imprecise, since we need to be able to integrate the Lagrangian density. The reader will note that we will always multiply the Lagrangian density by a suitable volume form before integrating.

must be minima of the *action*, which is an integral of the Lagrangian density over volumes $V \subset M$. The classical field equations can then be determined from the Euler-Lagrange equations for the action integral, assuming the values of the fields at different points are treated as independent dynamical variables. This will need to be modified when the Lagrangian has symmetries, as we will discuss later in this section.

As our first illustration, we take $M = M^4$ to be Minkowski 4-space. To set notation, we use coordinates $x = (x^0, x^1, x^2, x^3)$ on M, with x^0 denoting time. Sometimes, we will write instead $x = (t, \vec{x})$, with $\vec{x}$ denoting the spatial variables and t denoting time. We further let $\partial_i = \partial/\partial x^i$, and put $\vec{\partial} = (\partial_0, \partial_1, \partial_2, \partial_3)$. We also will use the Lorentzian inner product on M^4 of signature $(1, 3)$, so that for example, the wave operator $\partial_0^2 - \sum_{i=1}^{3} \partial_i^2$ can be denoted by $\square = \vec{\partial} \cdot \vec{\partial}$. The symbol $d^4 x$ will denote the Euclidean volume density on M.

We now consider a scalar field of mass m. Mathematically, this is a real valued function $\phi(x)$ on M^4. This field satisfies the Klein-Gordon equation

$$(B.1) \qquad (\square + m^2)\phi(x) = 0.$$

This equation can also be obtained from the *action* over $V \subset M$, which is defined to be

$$(B.2) \qquad S = S(\phi) = \frac{1}{2} \int_V d^4 x (\vec{\partial}\phi \cdot \vec{\partial}\phi - m^2 \phi^2) = \int_V d^4 x \mathcal{L}(\phi, \vec{\partial}\phi),$$

where

$$(B.3) \qquad \mathcal{L}(\phi, \vec{\partial}\phi) = \frac{1}{2}(\vec{\partial}\phi \cdot \vec{\partial}\phi - m^2 \phi^2)$$

is the *Lagrangian density*. As usual, ϕ is stationary for the action S provided that we have the Euler-Lagrange equation

$$(B.4) \qquad \frac{\delta \mathcal{L}}{\delta \phi} - \vec{\partial} \cdot \frac{\delta \mathcal{L}}{\delta(\vec{\partial}\phi)} = 0,$$

where $\delta \mathcal{L}/\delta \phi$ and $\delta \mathcal{L}/\delta(\vec{\partial}\phi)$ denote variational derivatives. Applying this to (B.3) yields (B.1).

We next recall the Hamiltonian formulation of classical mechanics. We start with a Lagrangian function $L = L(q, \dot{q}, t)$, where $q = q_i(t)$ are coordinates, t is time, and the dot denotes a time derivative. Intrinsically, L is a function on the tangent bundle T_M of M. The *conjugate momentum* of q_i is defined to be $p_i = \partial L/\partial \dot{q}_i$, which is natural terminology since for a free point particle in the presence of a potential, the conjugate momentum of the coordinate function q_i is just the corresponding component of the momentum. Note that (p_i, q_i) form a system of local coordinates on the cotangent bundle T_M^*. The manifold T_M^* has a canonical symplectic structure, which in local coordinates is described by the form

$$\omega = \sum_i dp_i \wedge dq_i.$$

The equation $p_i = \partial L/\partial \dot{q}_i$ locally defines a map $T_M \to T_M^*$ which we assume to be an isomorphism.

We define the Hamiltonian via

$$H = \sum_i p_i \dot{q}_i - L.$$

The Hamiltonian is a function on T_M^*, where we have used the above isomorphism to interpret L as being defined on T_M^*. Then Hamilton's equations are

$$\dot{q}_i = \frac{\partial H}{\partial p_i}, \quad \dot{p}_i = -\frac{\partial H}{\partial q_i},$$

though for our purposes, the equivalent equations

$$(B.5) \qquad \dot{A} = [A, H] = \sum_{i=1}^{3} \left(\frac{\partial A}{\partial q_i} \frac{\partial H}{\partial p_i} - \frac{\partial A}{\partial p_i} \frac{\partial H}{\partial q_i} \right),$$

where $A = q_i$ or p_i, are easier to generalize. The expression $[A, H]$ is an example of a *Poisson bracket*. Intrinsically, the symplectic form ω defines an isomorphism $I : T_M^* \to T_M$, and the Poisson bracket of two functions f, g may be rewritten as

$$[f, g] = \langle df, I(dg) \rangle.$$

Hamilton's equations are a consequence of the Euler-Lagrange equations for the action L. Notice also that we have the commutation relations (using the Poisson bracket defined above)

$$(B.6) \qquad \begin{aligned} [q_i, q_j] &= [p_i, p_j] = 0 \\ [q_i, p_j] &= \delta_{ij}. \end{aligned}$$

We now quantize this classical system. We get a Hilbert space $\mathcal{H}$ of states, such that functions f on T_M^* get replaced by self-adjoint operators $\mathbf{f}$ on $\mathcal{H}$. The classical values of f are reproduced as eigenvalues of the operators. For example, if there is a state $\psi \in \mathcal{H}$ for which $\mathbf{q}_i \cdot \psi = q_i^0 \psi$, then we think of ψ as a quantum state corresponding to a particle whose i^{th} coordinate has the value q_i^0. Similarly, the eigenvalues of the operators $\mathbf{p}_i$ are identified with classical values of the ith component of the momentum. The inner product on the Hilbert space $\mathcal{H}$ can be used to determine the probabilities of observing a quantum state to have a definite classical value.

Following the method of *canonical quantization*, we replace the Poisson brackets (B.6) with the commutators

$$\begin{aligned} [\mathbf{q}_i, \mathbf{q}_j] &= [\mathbf{p}_i, \mathbf{p}_j] = 0 \\ [\mathbf{q}_i, \mathbf{p}_j] &= i\delta_{ij}. \end{aligned}$$

The equations of motion (B.5) get replaced by the statement that (assuming no explicit time dependence in H) that the propagation of a state through a time interval t is given by the unitary operator

$$(B.7) \qquad U_t = e^{-iHt},$$

where we have chosen units in which Planck's constant $\hbar$ is 1. This is a lightning description of the *Schrödinger picture* of quantum mechanics, where the operators are constant and the states evolve in time. Quantum mechanics can equivalently be formulated using the *Heisenberg picture*, where the states are constant and the operators evolve in time. The Schrödinger and Heisenberg pictures can be compared by conjugation with U_t. In particular, an operator $\mathbf{A}$ in the Schrödinger picture gets replaced with the time-dependent operator $\mathbf{A}(t) = U_t^{-1} \mathbf{A} U_t$ in the Heisenberg picture. We then calculate

$$(B.8) \qquad \frac{d}{dt} \mathbf{A}(t) = -i[\mathbf{A}(t), \mathbf{H}],$$

which is the quantum-mechanical version of (B.5). In (B.8), we have used the usual definition of the commutator of operators $[\mathbf{A}, \mathbf{H}] = \mathbf{A}\mathbf{H} - \mathbf{H}\mathbf{A}$. In quantum field theory, we will use the Heisenberg picture without further comment.

Adapting this to field theory, we define the *conjugate momentum* $\pi(x)$ to be

$$\pi(x) = \frac{\delta \mathcal{L}}{\delta(\partial_0 \phi)}.$$

Assuming that we have a single field $\phi(x)$, the *Hamiltonian* (or energy) H is defined as the spatial integral (i.e., the integral over the variables not containing the time variable) of $\pi(x)\partial_0\phi(x) - \mathcal{L}$, where $\mathcal{L}$ is the Lagrangian density. The Hamiltonian is a functional of $\phi(x)$ and $\pi(x)$, so that the derivatives of $\phi(x)$ have been eliminated. In terms of H, the Euler-Lagrange equation (B.4) yields Hamilton's equations for the propagation in time of any observable O, which mathematically is just a functional of the fields $\phi(x)$ and $\pi(x)$. These equations, analogous to (B.5), are

$$\dot{O} = [O, H] = \int \left(\frac{\delta O}{\delta \phi(x)} \frac{\delta H}{\delta \pi(x)} - \frac{\delta O}{\delta \pi(x)} \frac{\delta H}{\delta \phi(x)} \right) dx^1 dx^2 dx^3,$$

and letting $O = \phi$ or π gives Hamilton's field equations

$$\dot{\phi} = [\phi, H], \quad \dot{\pi} = [\pi, H].$$

In the case of our scalar field $\phi(x)$, the conjugate momentum is $\pi(x) = \partial_0 \phi(x)$ and the Hamiltonian is given by

$$\text{(B.9)} \qquad H = \frac{1}{2} \int (\pi(x)^2 + |\vec{\nabla}\phi(x)|^2 + m^2\phi^2) dx^1 dx^2 dx^3.$$

The Poisson brackets (field analogs of (B.6)) are

$$\text{(B.10)} \qquad \begin{aligned} [\phi(\vec{x}, t), \phi(\vec{y}, t)] &= [\pi(\vec{x}, t), \pi(\vec{y}, t)] = 0 \\ [\phi(\vec{x}, t), \pi(\vec{y}, t)] &= \delta(\vec{x} - \vec{y}). \end{aligned}$$

The next step is to proceed to a quantum field. This step is not well-defined mathematically in the desired generality. The idea is that we seek a Hilbert space $\mathcal{H}$ of states, and that classical fields $\phi(x)$ get replaced by quantum fields, denoted $\Phi(x)$, which are operators on $\mathcal{H}$ depending on x and behave as distributions in the variable x. The classical values of the fields are reproduced as eigenvalues of the operators. For example, if there is a state $\psi \in \mathcal{H}$ for which $\Phi(x) \cdot \psi = f(x)\psi$ for some function $f(x)$, then we think of this quantum state as corresponding to the classical field with value $\phi(x) = f(x)$. The inner product is used to determine relative probabilities much as in ordinary quantum mechanics.

Using canonical quantization, we let $\Pi(\vec{x}, t)$ denote the operator-valued distribution corresponding to $\pi(\vec{x}, t)$ and replace the Poisson brackets (B.10) with

$$\begin{aligned} [\Phi(\vec{x}, t), \Phi(\vec{y}, t)] &= [\Pi(\vec{x}, t), \Pi(\vec{y}, t)] = 0 \\ [\Phi(\vec{x}, t), \Pi(\vec{y}, t)] &= i\delta(\vec{x} - \vec{y}). \end{aligned}$$

These equal-time commutators are identities of operator-valued distributions.

We now show how this formalism gives an interpretation of the quantum field $\Phi(x)$ as being comprised of quanta of particles. Namely, if we return to the Klein-Gordon equation (B.1) and apply standard Fourier transform methods for solving PDEs, we can write the classical field ϕ in the form

$$\text{(B.11)} \qquad \phi(x) = \int \left(a(\vec{k})e^{i(\vec{k}\cdot\vec{x} - E_{\vec{k}}t)} + a^\dagger(\vec{k})e^{-i(\vec{k}\cdot\vec{x} - E_{\vec{k}}t)} \right) \frac{dk^1\, dk^2\, dk^3}{\sqrt{(2\pi)^3 2E_{\vec{k}}}}$$

for certain complex conjugate functions a and $a^\dagger$ of $\vec{k} = (k^1, k^2, k^3)$. In (B.11), we have put $E_k = \sqrt{\vec{k} \cdot \vec{k} + m^2}$. The Klein-Gordon equation (B.1) is satisfied because of $E_k^2 = \vec{k} \cdot \vec{k} + m^2$, and on the *mass shell* defined by $E_k = \sqrt{\vec{k} \cdot \vec{k} + m^2}$, the expression $dk^1\, dk^2\, dk^3 / \sqrt{E_k}$ is invariant under Lorentz transformations of (E_k, k^1, k^2, k^3). The quantity E_k is the total energy of a relativistic particle of mass m and momentum $\vec{k}$. For the mathematician who has never seen this, it is worth pointing out that for $|\vec{k}| \ll m$ we have the approximation $E_{\vec{k}} \sim m + (\vec{k} \cdot \vec{k})/(2m)$. The first term is the rest energy of the particle (usually written $E = mc^2$, but we have chosen units where $c = 1$), and the second term is the classical kinetic energy of a particle of mass m and momentum $\vec{k}$.

Back in the quantum theory, we claim that $a(\vec{k}), a^\dagger(\vec{k})$ have natural particle interpretations. To begin with, $a(\vec{k}), a^\dagger(\vec{k})$ are operator-valued distributions, and one can show that they satisfy the commutation relations

(B.12)
$$[a(\vec{k}), a(\vec{k}')] = [a^\dagger(\vec{k}), a^\dagger(\vec{k}')] = 0$$
$$[a(\vec{k}), a^\dagger(\vec{k}')] = \delta(\vec{k} - \vec{k}').$$

Furthermore, one also shows that substitution back into the Hamiltonian (B.9) yields

$$H = \int \left(E_{\vec{k}} a^\dagger(\vec{k}) a(\vec{k}) + \frac{1}{2} E_{\vec{k}} \delta(\vec{0}) \right) dk^1\, dk^2\, dk^3,$$

where $\vec{0}$ is the 3-dimensional zero vector. The last term is mathematically meaningless, since it is an infinite scalar operator. The solution is to simply ignore this term, as the subtraction of a constant from the Hamiltonian has no effect on the dynamics. We finally arrive at

(B.13)
$$H = \int E_{\vec{k}} a^\dagger(\vec{k}) a(\vec{k})\, dk^1\, dk^2\, dk^3.$$

This has a nice intuitive description. The commutation relations (B.12) and the Hamiltonian (B.13) show that application of an operator $a(\vec{k})$ will decrease energy. Since the vacuum is to be the lowest energy state, we may define the vacuum state V as the state annihilated by all of the $a(\vec{k})$. Using (B.12) and $a(\vec{k})(V) = 0$, it follows that the state $a^\dagger(\vec{k}')(V)$ is an eigenstate with energy (= eigenvalue) $E_{\vec{k}'}$:

$$H(a^\dagger(\vec{k}')(V)) = \int E_{\vec{k}} a^\dagger(\vec{k}) a(\vec{k}) a^\dagger(\vec{k}')(V)\, dk^1\, dk^2\, dk^3$$
$$= \int E_{\vec{k}} a^\dagger(\vec{k}) \left(a^\dagger(\vec{k}') a(\vec{k}) + \delta(\vec{k} - \vec{k}') \right)(V)\, dk^1\, dk^2\, dk^3$$
$$= E_{\vec{k}'} a^\dagger(\vec{k}')(V).$$

This leads to the interpretation of $a^\dagger(\vec{k})$ as a *creation operator*: its application creates a scalar particle of momentum $\vec{k}$. Similarly, $a(\vec{k})$ is an *annihilation operator*, so that its application annihilates a particle of momentum $\vec{k}$.

Our conclusion is that in general, the formula (B.11) expresses the quantum field $\Phi(x)$ as a superposition of elementary particle states.

While we don't expect the above to be convincing to those who are seeing this for the first time, we want to explicitly make the point that by starting with a Lagrangian density and formal quantization procedures, we have arrived at the quantum structure of matter and a reformulation of dynamics in terms of these quanta. This is another way of saying that the dynamics of a quantum field are completely determined by the Lagrangian density. We should also mention that it is possible to put some (but not all) of the above on a firm mathematical foundation— see [**Rabin**, Lect. 8].

So far, we have concentrated on the Hamiltonian formulation of quantum field theory. As mentioned earlier, there is also a Lagrangian formulation, which involves the notion of a path integral. Suppose that at times $t < t'$, the classical fields have values $\phi_0(\vec{x}, t)$ and $\phi_1(\vec{x}, t')$. We will let $|\phi_0(\vec{x}, t)\rangle$ and $|\phi_1(\vec{x}, t')\rangle$ denote the quantum states with respective eigenvalues $\phi_0(\vec{x}, t)$, $\phi_1(\vec{x}, t')$ under the operator-valued distribution $\Phi(\vec{x}, t)$. In other words, we have

$$(\text{B.14}) \qquad \Phi(\vec{x}, t)|\phi_0(\vec{x}, t)\rangle = \phi_0(\vec{x}, t)\,|\phi_0(\vec{x}, t)\rangle,$$

with a similar expression for ϕ_1. The left hand side of (B.14) involves the action of an operator and the right hand side involves multiplication by a distribution.

The Hilbert inner product of these two states is $\langle\phi_1(\vec{x}, t') \mid \phi_0(\vec{x}, t)\rangle$. Its physical meaning is that its squared norm gives the probability density for the state $\phi_0(\vec{x}, t)$ to propagate into $\phi_1(\vec{x}, t')$. Formal calculations suggest that if $S(\phi)$ is the action generalizing (B.2), then

$$\langle\phi_1(\vec{x}, t') \mid \phi_0(\vec{x}, t)\rangle = N \int [D\phi]e^{iS(\phi)}$$

where on the right hand side, the *path integral* is over the space of all paths within the space of fields with initial point ϕ_0 and with terminal point ϕ_1. The factor N is a suitable normalization factor, and $[D\phi]$ is an appropriate measure on the space of paths. The path integral (or functional integral) is not well-defined mathematically at present.

The path integral also allows for the computation of certain physical *correlation functions*, sometimes called *n-point functions*. Pick n points $x_1, \ldots, x_n \in M$. The n-point function is formally defined to be

$$(\text{B.15}) \qquad \langle\Phi(x_1) \cdots \Phi(x_n)\rangle = N \int [D\phi]\phi(x_1) \cdots \phi(x_n)e^{iS(\phi)}$$

These are functions of the points $x_1, \ldots, x_n$ and of the types of fields associated to these points, but not the particular values of the fields, as these are integrated over. In our example of a scalar field, there is only one type of field, but for example the Yukawa couplings are a type of three-point function where different types of fields are associated to the different points. The Yukawa couplings are three-point functions in a *topological quantum field theory*, which greatly simplifies matters. We will discuss topological quantum field theories in Appendix B.6.

These quantities are of intrinsic physical interest. For example, the two-point functions are just the Green's functions of the theory. In a very real sense, one can argue that the n-point functions contain all of the physical predictions of the theory.

While path integrals such as (B.15) cannot be formulated rigorously, there are some accepted heuristics for their calculation that do have partial justifications. The most important one is the stationary phase method, whereby the path integral can be "localized" to an integral over the space of stationary points of S, a space which is often finite-dimensional. As a naive explanation of how this arises, note that if ϕ_0 is stationary for S, then the phase of $e^{iS(\phi)}$ will vary slowly near ϕ_0, allowing a nonzero contribution to the path integral locally near ϕ_0. If ϕ_0 is not stationary, then the more rapidly oscillating phase will tend to cause cancelation. Thus states near ϕ_0 are more likely to occur and contribute to physical processes. This point provides a link between a classical and a quantum field theory. The stationary phase method is rigorous for finite-dimensional parameter spaces; but in a path integral, the spaces are infinite-dimensional.

There is a subtlety here that may be worth mentioning. In the path integral (B.15), the terms $\phi(x_i)$ are distributions, hence commute. On the other hand, the *operators* $\Phi(x_i)$ do not commute in general, so there appears to be an inconsistency. The resolution is to define the *time-ordered product* $T(\Phi(x_1)\cdots\Phi(x_n))$ to be the operator obtained by applying the $\Phi(x_i)$ in chronological order. Then we have

$$\langle V \mid T(\Phi(x_1)\cdots\Phi(x_n)) \mid V\rangle = N \int [D\phi]\phi(x_1)\cdots\phi(x_n)e^{iS(\phi)}$$

where V again denotes the vacuum state.

We will also need some understanding of a *gauge theory*. Here, the action is invariant under a continuous group of local transformations on M. This group is usually infinite-dimensional, the typical example being Yang-Mills theory, where the gauge group is the space of maps to a finite-dimensional Lie group. Our very modest goal is to explain the role of what is called *gauge fixing* in the context of our discussion of quantum field theory, as well as the notion of a conserved quantity.

The simplest example is the classical theory of electricity and magnetism in the absence of charged matter. We make no pretense at being self-contained here, and are merely attempting to give a flavor. The interested reader is referred to [**BD**] for more details. We take $M = M^4$ to be Minkowski 4-space with the usual coordinates $(x^0,\dots,x^3) = (t,\vec{x})$. The field will be a real 1-form $\phi = A_i dx^i$,[2] which is the *electromagnetic potential*. We also let $\vec{A} = (A_1, A_2, A_3)$. The *field strength* of ϕ is defined to be $F = d\phi = F_{ij}dx^i \wedge dx^j$. The electric field $\vec{E}$ and magnetic field $\vec{B}$ can be extracted from the field strength by putting

$$F = F_{ij}dx^i \wedge dx^j = dt \wedge \left(\sum_{i=1}^{3} E_i dx^i\right) + \vec{B} \cdot (dx^1 \wedge dx^2 \wedge dx^3),$$

where $\cdot$ is contraction with $\vec{B}$, thought of as the vector field $B^i\partial/\partial x^i$. With these identifications, the electromagnetic field strength F is a closed 2-form, which is equivalent to two of the Maxwell equations

$$\vec{\nabla} \times \vec{E} = -\dot{\vec{B}}, \quad \vec{\nabla} \cdot \vec{B} = 0.$$

The other two equations arise from the Lagrangian density

$$(B.16) \qquad \mathcal{L} = -\frac{1}{4}F_{ij}F^{ij} = \frac{1}{2}(|\vec{E}|^2 - |\vec{B}|^2).$$

Here, the Euler-Lagrange equations take the form

$$\vec{\nabla} \cdot \vec{E} = \vec{0}, \quad \vec{\nabla} \times \vec{B} = \dot{\vec{E}},$$

---

[2]We follow the Einstein summation convention and sum over repeated upper and lower indices. Also, indices are raised and lowered without comment by contraction with the metric tensor $g_{ij} = \mathrm{diag}(1, -1, -1, -1)$.

and the Hamiltonian is computed to be

$$H = \frac{1}{2}\int (|\vec{E}|^2 + |\vec{B}|^2)dx^1 \wedge dx^2 \wedge dx^3.$$

The field strength $F = d\phi$ and hence the associated Lagrangian $\mathcal{L}$ are unchanged by the addition of an exact form to $\phi = A_i dx^i$, i.e, by a substitution which has the form

$$(B.17) \qquad\qquad A_i \mapsto A_i + \frac{\partial \Lambda(x)}{\partial x^i},$$

where $\Lambda(x)$ is an arbitrary real function on M. This substitution is associated to the *gauge group* of local transformations $e^{i\Lambda(x)}$; in fact, if we identify $\phi = \sqrt{-1}A_i dx^i$ with a connection form on a principal $U(1)$ bundle over M, then the gauge transformation $e^{i\Lambda(x)}$ on the $U(1)$ bundle induces (B.17). It is therefore sufficient to choose a slice of the parameter space of the $\phi = A_i dx^i$ which has the property that every possible ϕ is equivalent to one of the ϕ in the slice via a gauge transformation (B.17). Such a slice is called a *gauge choice*[3]. We will make the gauge choice in two steps as follows. First, we can assume that

$$(B.18) \qquad\qquad A_0 = 0.$$

Unfortunately, we can't just make this gauge choice and then quantize, since we also need to satisfy the Euler-Lagrange equation $\delta\mathcal{L}/\delta A_0 = 0$. This is Gauss' law $\vec{\nabla}\cdot\vec{E} = \vec{0}$, which becomes $\vec{\nabla}\cdot\partial\vec{A}/\partial t = 0$ since $A_0 = 0$. Hence $\vec{\nabla}\cdot\vec{A}$ is independent of time. Now solve Poisson's equation to find a time-independent function f with $\vec{\nabla}\cdot\vec{\nabla}f = \vec{\nabla}\cdot\vec{A}$. Replacing $\phi = A_i dx^i$ by $\phi - df$ does not affect A_0, and thus we can assume that our gauge choice also satisfies the equation

$$(B.19) \qquad\qquad \vec{\nabla}\cdot\vec{A} = \vec{\nabla}\cdot(A_1, A_2, A_3) = 0.$$

In general, when a gauge choice is made, we cannot simply ignore a field that has been "gauged away."

We can now follow a quantization procedure similar to that for the Klein-Gordon field by fixing the gauge choice (B.18) and (B.19). We make a Fourier expansion

$$(B.20) \qquad \vec{A}(x) = \int \sum_{j=1}^{2} \vec{\epsilon}_j(k)\left(a_j(k)e^{-ik\cdot x} + a_j^\dagger(k)e^{ik\cdot x}\right)\frac{dk^1 \wedge dk^2 \wedge dk^3}{\sqrt{2k_0(2\pi)^3}}$$

with the following notation. The $\vec{\epsilon}_j$ are 3-dimensional polarization vectors, which are orthogonal to $\vec{k}$ by the Fourier transformation of the second gauge condition (B.19). This is why there are two independent polarization vectors. Finally, $k = (k_0, \vec{k})$ satisfies $k_0 = |\vec{k}|$ since each component of $\vec{A}(x)$ is a solution of the 4-dimensional wave equation (this follows from Maxwell's equations).

We are led to the interpretation of the $a_j(\vec{k})$ and $a_j^\dagger(\vec{k})$ as operators which respectively annihilate and create a quantum of the electromagnetic field with momentum $\vec{k}$ and polarization $\vec{\epsilon}_j(k)$. This is consistent with our earlier discussion of $a_j(\vec{k})$ and $a_j^\dagger(\vec{k})$.

---

[3]From the path integral point of view, restricting to such a slice is clearly necessary, since otherwise the gauge symmetries would cause the same path to be counted more often than necessary, giving undesirable infinities.

It is a crucial fact (Noether's theorem) that continuous groups of symmetries give rise to conserved quantities as follows. Letting the fields transform infinitesimally as $\phi \mapsto \phi + \psi \varepsilon$ under an infinitesimal gauge transformation (where ε is an infinitesimal parameter), the quantity

$$(\text{B.21}) \qquad \frac{\delta \mathcal{L}}{\delta \partial_i \phi} \psi$$

has zero divergence, and is called a *conserved current*. As a consequence, the spatial (fixed time) integral of a conserved current is constant in time, and is called a conserved charge.

When we discuss conformal field theories in Section B.3, the gauge group will be the conformal group. We will see the need to make a similar gauge choice, and we will see that the gauge group leads to similar conserved quantities which play an important role in the theory.

The quantum field theory discussed so far applies best to *bosons*, which are particles of integer spin. Several classical (i.e., non-quantum) bosonic fields can be combined by multiplying the fields, and this multiplication is commutative. But in order to describe field theories for *fermions* (particles of half-integer spin), one starts with classical fields which multiply in an anticommutative fashion. There are several ways to formulate this; one way to axiomatize anticommuting variables is through the use of superspaces. This will be described briefly in Section B.4. The quantization procedure for fermionic fields requires a modification of the procedure sketched above for bosonic fields. We will see some examples of fermionic variables in the nonlinear sigma models described in the next section. A good introduction to the difference between bosonic systems and fermionic systems can be found in [**Alvarez**, Lect. 1 and 2].

B.2. Nonlinear Sigma Models

To define a *nonlinear sigma model* (or *sigma model* for short), we start with a Kähler manifold (X, g) and a closed 2-form B. The sigma model will be a theory built upon maps from Riemann surfaces Σ to X. We think of a map $f : \Sigma \to X$ as describing the propagation of a string through X, and Σ is the *world sheet* swept out by the string as it propagates. The Riemann surface Σ is not fixed in the theory, but can be arbitrary. This allows for arbitrarily complicated ways in which a string can split up into two or more strings or have several strings join up. In other words, allowing Riemann surfaces of arbitrary genus automatically includes interactions between strings in the theory. This is a general feature of all string theories and not just sigma models.

The nonlinear sigma model also has anticommuting fermionic fields, which are sections of certain bundles on Σ given in the following table.

	Field	Bundle
	ψ_+^i	$K^{\frac{1}{2}} \otimes f^*(T_X^{1,0})$
(B.22)	$\psi_+^{\bar{i}}$	$K^{\frac{1}{2}} \otimes f^*(T_X^{0,1})$
	ψ_-^i	$\overline{K}^{\frac{1}{2}} \otimes f^*(T_X^{1,0})$
	$\psi_-^{\bar{i}}$	$\overline{K}^{\frac{1}{2}} \otimes f^*(T_X^{0,1})$

Here, K is the canonical bundle of Σ, $\overline{K}$ is its complex conjugate (antiholomorphic) bundle, $K^{1/2}$ is a choice of a square root of K (i.e., a spin bundle on Σ), T_X

is the complexified tangent bundle of X, and we have the usual decomposition $T_X = T_X^{1,0} \oplus T_X^{0,1}$. The indices $i, \bar{i}$ range from 1 to $n =$ the complex dimension of X, with the understanding that i will be used for holomorphic indices, while $\bar{i}$ will be used for antiholomorphic indices. Notation has been chosen to allow us to abuse notation by identifying the objects introduced in the above table with certain of their coefficients; the meaning should be clear from the context. We do this as follows. Pick a local holomorphic coordinate z on Σ and local complex coordinates x^i on X. Then for example ψ_+^i can denote a section of $K^{1/2} \otimes f^*(T_X^{1,0})$ or its coefficient of $(dz)^{1/2} \otimes \partial/\partial x^i$, where $(dz)^{1/2}$ denotes a local generator of $K^{1/2}$. Notationally, $x^{\bar{i}}$ is synonymous with $\bar{x}^i$.

We now can define the action S. We will also let indices I, J run over all possible values of $i, \bar{i}$, i.e., over both holomorphic and antiholomorphic indices, so that x^I ranges over the $2n$ coordinates x^i, $x^{\bar{i}}$, where as above n is the complex dimension of X. We put

$$
\begin{aligned}
S = \int_\Sigma \Big(&\frac{1}{2}(g_{IJ} + iB_{IJ})\partial_z x^I \partial_{\bar{z}} x^J \\
&+ \frac{\sqrt{-1}}{2} g_{i\bar{i}} \psi_-^{\bar{i}} D_z \psi_-^i + \frac{\sqrt{-1}}{2} g_{i\bar{i}} \psi_+^{\bar{i}} D_{\bar{z}} \psi_+^i \\
&+ \frac{1}{4} R_{i\bar{i}j\bar{j}} \psi_+^i \psi_+^{\bar{i}} \psi_-^j \psi_-^{\bar{j}} \Big)(\sqrt{-1}) dz \wedge d\bar{z},
\end{aligned}
$$
(B.23)

where R is the curvature tensor of the metric g on X and D_z is deduced from the Levi-Civita connection of the metric $f^*(g)$ on Σ. Note that S depends on both the map $f : \Sigma \to M$ and the $4n$ fermionic fields $\psi_\pm^i$, $\psi_\pm^{\bar{i}}$.

This strange-looking action is quite natural. Its bosonic part reduces to the area of Σ after the equations of motion have been taken into consideration. The fermionic terms have been added in such a way to maintain *conformal invariance* (to be discussed later in this appendix) and so as to be *supersymmetric*; that is, there is an infinitesimal symmetry of the action which perturbs fermions in bosonic directions, and vice versa. Actually, there are two of them since a sigma model has $N = 2$ supersymmetry. The supersymmetries are rather complicated. They can be written as follows:

$$
\begin{aligned}
\delta x^i &= \sqrt{-1}\,\alpha_- \psi_+^i + \sqrt{-1}\,\alpha_+ \psi_-^i \\
\delta x^{\bar{i}} &= \sqrt{-1}\,\tilde{\alpha}_- \psi_+^{\bar{i}} + \sqrt{-1}\,\tilde{\alpha}_+ \psi_-^{\bar{i}} \\
\delta \psi_+^i &= -\tilde{\alpha}_- \partial_z x^i - \sqrt{-1}\,\alpha_+ \psi_-^j \Gamma_{jm}^i \psi_+^m \\
\delta \psi_+^{\bar{i}} &= -\alpha_- \partial_z x^{\bar{i}} - \sqrt{-1}\,\tilde{\alpha}_+ \psi_-^{\bar{j}} \Gamma_{\bar{j}\bar{m}}^{\bar{i}} \psi_+^{\bar{m}} \\
\delta \psi_-^i &= -\tilde{\alpha}_+ \partial_{\bar{z}} x^i - \sqrt{-1}\,\alpha_- \psi_+^j \Gamma_{jm}^i \psi_-^m \\
\delta \psi_-^{\bar{i}} &= -\alpha_+ \partial_{\bar{z}} x^{\bar{i}} - \sqrt{-1}\,\tilde{\alpha}_- \psi_+^{\bar{j}} \Gamma_{\bar{j}\bar{m}}^{\bar{i}} \psi_-^{\bar{m}},
\end{aligned}
$$
(B.24)

where α_- and $\tilde{\alpha}_-$ are *infinitesimal* holomorphic sections of $K^{-\frac{1}{2}}$ while α_+ and $\tilde{\alpha}_+$ are infinitesimal antiholomorphic sections of $\bar{K}^{-\frac{1}{2}}$. Applying the transformations (B.24) and using the Kähler condition, one can show that the integrand of (B.23) is altered by an exact form, hence the action S is invariant.

In the heterotic string theory, the Lagrangian must be modified somewhat so that some of the fermionic fields $\psi_\pm^I$ live in a 16-dimensional space containing the root lattice of $E_8 \times E_8$. This allows the resulting theory to contain an $E_8 \times E_8$ gauge group, and this plays a role in the definition of the Yukawa couplings. These are the correlation functions (or n-point functions) in our theory. We should also mention that in the heterotic theory, the equations of motion for the fermions decouple into left- and right-moving equations (which in this context can be thought of as holomorphic and antiholomorphic parts). Hence the sigma model gives a $(2,2)$ string theory, which means that there are 2 holomorphic and 2 antiholomorphic supersymmetries.

It is easier to compute the correlation functions of interest to us using the *twisted* theories, namely the A-model and the B-model. These theories differ from the sigma model, but some of their correlation functions (including the ones mentioned in Chapter 1) can be shown to coincide with corresponding correlation functions in the sigma model.

The action for these theories is identical in form to (B.23), but the fields $\psi_\pm^I$ have different meanings. For the A-model, we replace (B.22) with

Field	Bundle
ψ_+^i	$f^*(T_X^{1,0})$
$\psi_+^{\bar{i}}$	$K \otimes f^*(T_X^{0,1})$
ψ_-^i	$\bar{K} \otimes f^*(T_X^{1,0})$
$\psi_-^{\bar{i}}$	$f^*(T_X^{0,1})$

while for the B-model, we use instead

Field	Bundle
ψ_+^i	$K \otimes f^*(T_X^{1,0})$
$\psi_+^{\bar{i}}$	$f^*(T_X^{0,1})$
ψ_-^i	$\bar{K} \otimes f^*(T_X^{1,0})$
$\psi_-^{\bar{i}}$	$f^*(T_X^{0,1})$

In other words, the fields coincide locally with the corresponding fields in the sigma model, but globally are twisted by the bundles $K^{\frac{1}{2}}$, $\overline{K}^{\frac{1}{2}}$, or their inverses.

These models are still supersymmetric (but now have only $N = 1$ supersymmetry). Furthermore, if V and V° are a mirror pair, then the A-model derived from a Calabi-Yau manifold V is mirror symmetric to the B-model derived from its mirror manifold V° (for corresponding choices of the complex structure of V° and Kähler structure on V).

We start with the A-model. Here, we can view the fields ψ_+^i and $\psi_-^{\bar{i}}$ as the holomorphic and antiholomorphic parts of a section of $f^*(T_X)$. Also, the A-model associates to each k-form η on X the *local operator* $\mathcal{O}_\eta(x)$ defined as follows. We can write the k-form as $\eta = h_{I_1 \ldots I_k}(x) dx^{I_1} \wedge \cdots \wedge dx^{I_k}$, where each dx^I is either dx^i or $dx^{\bar{i}} = d\bar{x}^i$. Then the corresponding local operator is defined to be

$$(\text{B.25}) \qquad \mathcal{O}_\eta(x) = h_{I_1 \ldots I_k}(x) \chi^{I_1} \cdots \chi^{I_k}(x),$$

where the χ^I equal either ψ_+^i or $\psi_-^{\bar{i}}$ and should be regarded as anticommuting operators on a Hilbert space $\mathcal{H}$.

The first term of (B.23) simplifies to $\int_\Sigma f^*(\omega)$, where $\omega = B + iJ$ as usual, J here being the Kähler class of the metric g. There is some simplification if the equations of motion (the generalized Euler-Lagrange equations discussed above) are used, and the remaining terms can then be compactly written in terms of a certain fermionic operator, the *BRST operator Q*. This operator is the supersymmetry transformation that survives the reduction in supersymmetry from the original $N = 2$ theory to the twisted $N = 1$ theory. In fact, modulo the equations of motion, we have

$$S = \frac{i}{2} \int_\Sigma \{Q, V\} (\frac{i}{2} dz \wedge d\bar{z}) + \frac{1}{2} \int_\Sigma f^*(\omega),$$

where V is a certain field whose particulars will not concern us and $\{Q, V\}$ is the anticommutator. Without going into any detail, the BRST operator plays an important role in quantum field theories. We have $Q^2 = 0$, and the operator δ_Q defined by $\delta_Q(\mathcal{O}) = \{Q, \mathcal{O}\}$ essentially defines a complex whose cohomology is called the BRST cohomology of the quantum field theory. We have the following formula for the anticommutator.

$$\{Q, \mathcal{O}_\eta\} = -\mathcal{O}_{d\eta},$$

which shows that $\eta \mapsto \mathcal{O}_\eta$ gives a map from the de Rham cohomology $H^k(X)$ of X to the BRST cohomology.

We now let $\omega_1, \ldots, \omega_n \in H^2(X)$, and turn to the calculation of the n-point correlation function $\langle \mathcal{O}_{\omega_1} \cdots \mathcal{O}_{\omega_n} \rangle$. By (B.23), the Euler-Lagrange equation coming from x^i is $\partial x^i / \partial \bar{z} = 0$, so that f is holomorphic. The path integral accordingly reduces to an integral over the space of holomorphic maps. We next put $\gamma = f_*[\Sigma]$ and write the integral as a sum over γ, writing the term for $\gamma = 0$ separately. The result is

$$\langle \mathcal{O}_{\omega_1} \cdots \mathcal{O}_{\omega_n} \rangle = \int_X \omega_1 \wedge \cdots \wedge \omega_n +$$
$$\sum_{\gamma \neq 0} \int [D\phi][D\chi]D[\psi] e^{-\frac{1}{2}\{Q, \int V\}} \prod_i \mathcal{O}_{\omega_i} e^{\int_\gamma \omega}$$
$$= \int_X \omega_1 \wedge \cdots \wedge \omega_n + \sum_\gamma N_{\gamma, \omega_i} q^\gamma,$$

where the notation q^γ is as explained in Chapter 7, and where N_{γ, ω_i} is loosely defined as follows. Pick points $p_1, \ldots, p_n \in \mathbb{P}^1$ and choose representative cycles Z_i whose fundamental class is Poincaré dual to the classes ω_i. Then N_{γ, ω_i} denotes the number of holomorphic maps $f : \mathbb{P}^1 \to X$ such that $f(p_i) \in Z_i$ for each i. This is not quite a rigorous definition because the set of such maps may be infinite, but can be made precise through the notion of a Gromov-Witten invariant, which is discussed in Chapter 7. As explained in Chapter 8, these invariants are the crucial ingredient in the construction of quantum cohomology. As we will observe in Appendix B.6, the A-model is a topological quantum field theory, which implies that the three-point correlation functions define an associative ring structure. The second equality above results from carrying out the appropriate integral over the space of critical points of χ and ψ. This is an example of the stationary phase method mentioned in Appendix B.1. The nonconstant holomorphic maps $f : \Sigma \to X$ are called *worldsheet instantons*, and the above form of the n-point function is called an *instanton sum*.

For the B-model, the field $\eta^{\bar{i}} = \psi_+^{\bar{i}} + \psi_-^{\bar{i}}$ transforms as a section of $f^*(T_X^{0,1})$, and similarly, $\theta_i = g_{i\bar{i}}(\psi_+^{\bar{i}} - \psi_-^{\bar{i}})$ transforms as a $(1, 0)$-form. The B-model also has local operators, which come from $(0, q)$-forms with values in $\wedge^p T_X^{1,0}$. If we write such a form θ locally as

$$\theta = h^{j_1\cdots j_p}_{\bar{i}_1\cdots\bar{i}_q} d\bar{z}^{\bar{i}_1} \wedge \cdots \wedge d\bar{z}^{\bar{i}_q} \otimes \frac{\partial}{\partial z_{j_1}} \wedge \cdots \wedge \frac{\partial}{\partial z_{j_p}},$$

then the local operator $\mathcal{O}_\theta$ is defined by

$$\mathcal{O}_\theta = h^{j_1\cdots j_p}_{\bar{i}_1\cdots\bar{i}_q} \eta^{\bar{i}_1} \cdots \eta^{\bar{i}_q} \theta_{j_1} \cdots \theta_{j_p}.$$

As with the A-model, we get a simplified version of the action when the equations of motion are taken into consideration. The BRST operator of this theory, again denoted Q, satisfies $\{Q, \mathcal{O}_\theta\} = -\mathcal{O}_{\bar{\partial}\theta}$. Thus the association $\theta \mapsto \mathcal{O}_\theta$ maps the Dolbeault cohomology $H^q(X, \wedge^p(T_X^{1,0}))$ to the BRST cohomology of the B-model.

The B-model correlation function follows by computation from the simplified form of the action just mentioned. In principle, this correlation function could contain an instanton sum, but certain *non-renormalization theorems* say that this is not the case. Another way to argue this point is by mirror symmetry: the A-model correlation function does not depend on complex moduli, hence the B-model correlation function of the mirror manifold does not depend on the Kähler moduli. However, each term with $\gamma \neq 0$ in an instanton sum explicitly depends on the Kähler moduli due to the term q^γ. A different, heuristic argument for the absence of instanton corrections was given in Section 1.2.

Finally, we note that in order for our sigma model to avoid certain quantum anomalies, (X, g) must in fact be a Calabi-Yau threefold. The requirement that X have real dimension 6 will be explained in Appendix B.3, and the Kähler condition is essential for $N = 2$ supersymmetry. Finally, the Calabi-Yau condition, which was discussed briefly in Section 1.3, is more subtle and would require a digression into renormalization of quantum field theories that is beyond the scope of these appendices. See [**Hübsch**] for an exposition of more of the details. Also, a discussion of supersymmetric sigma models can be found in [**GO**].

B.3. Conformal Field Theories

Roughly speaking, a conformal theory is a theory of fields on Riemann surfaces which respects the group of conformal transformations of surfaces. We will attempt to make this more precise. A detailed introduction to the subject appears in [**Ginsparg**]. The connection to string theory is that the Riemann surfaces occur as the surface swept out by the string as it propagates in time. We will therefore refer to this Riemann surface as the world sheet.

A *conformal mapping* of a Riemann surface is a self-mapping preserving angles and orientation. Such mappings are holomorphic. We want to study conformal mappings locally by working in a two-dimensional disk Δ with complex coordinate $z = x+iy$. Unfortunately, there is no group of local conformal transformations, since neighborhoods will not be preserved and so mappings cannot be composed. On the other hand, there is a well-defined conformal algebra—an infinitesimal conformal transformation induces a real vector field on the underlying real surface which can be written in local holomorphic coordinates as $f(z)\partial/\partial z + \bar{f}(\bar{z})\partial/\partial\bar{z}$, where $f(z)$ is holomorphic. It is customary to separate holomorphic and antiholomorphic parts by choosing bases $\ell_n = -z^{n+1}\partial_z$ and $\bar{\ell}_n = -\bar{z}^{n+1}\partial_z$ which satisfy the Virasoro algebra relations

$$[\ell_n, \ell_m] = (n - m)\ell_{n+m}, \quad [\bar{\ell}_m, \bar{\ell}_n] = (n - m)\bar{\ell}_{m+n}.$$

The conformal algebra can be recovered from the Virasoro algebra as the subalgebra generated by the $\ell_n + \bar{\ell}_n$ and $i(\ell_n - \bar{\ell}_n)$. A conformal field theory includes a representation of the Virasoro algebra, though one needs to be careful because certain representations are subject to quantum anomalies.

We next define the fields of interest in a conformal field theory.

DEFINITION B.3.1. *A primary field of weight $(h, \bar{h})$ is a field $\Phi(z, \bar{z})$ which transforms as $\Phi(z, \bar{z}) \mapsto (\partial f/\partial z)^h (\partial f/\partial \bar{z})^{\bar{h}} \Phi(f(z), \bar{f}(z))$ under a conformal transformation $z \mapsto f(z)$.*

In other words, the expression $\Phi(z, \bar{z}) dz^h \otimes d\bar{z}^{\bar{h}}$ is invariant when $\Phi(z, \bar{z})$ is a primary field.

The gauge group of a conformal field theory is the local conformal group. As in Section B.1, this means that we need to make a gauge choice. Using the local coordinate $z = x + iy$, the gauge choice is determined by fixing the metric h on the world sheet to be $h = dx^2 + dy^2$. But as in Section B.1, we must also remember to impose the conditions $\delta S/\delta h^{ij} = 0$ from the Euler-Lagrange equations.

Unfortunately, the quantization is not quite as straightforward as in the case of electricity and magnetism. For instance, maintaining conformal invariance requires the introduction of certain auxiliary fields, called *ghosts*, in the process of quantization. Additional ghosts are needed for supersymmetry. In addition, although the Euler-Lagrange equation $\delta S/\delta h^{ij} = 0$ holds classically, it works a little differently at the quantum level. To explain what happens, we introduce the energy-momentum tensor with components

$$T_{ij} = \frac{1}{\sqrt{h}} \frac{\delta S}{\delta h^{ij}},$$

which is a symmetric tensor. Here, h as usual denotes the determinant of the matrix h_{ij} and h^{ij} denotes the inverse matrix. In holomorphic coordinates, the energy-momentum tensor has components $T(z) = T_{zz}$, $\bar{T}(\bar{z}) = T_{\bar{z}\bar{z}}$, and $T_{z\bar{z}}$. Similar to what we did in (B.21), the conservation law arising from conformal invariance under $h \mapsto \phi h$ for any function ϕ leads to $T_{z\bar{z}} = 0$. Energy-momentum conservation thus leads to the conclusion that $T(z)$ is holomorphic while $\bar{T}(\bar{z})$ is antiholomorphic (explaining the notation). Hence, the classical equations of motion (from the Euler-Lagrange equations) include the constraints $T(z) = \bar{T}(\bar{z}) = 0$.

In the quantum theory, these equations are modified using *radial quantization*. This means that in local coordinates, the equal time curves are chosen to be concentric circles, whose radii are the curves with fixed spatial coordinate (the world sheet of a string contains one spatial coordinate and one time coordinate locally, but the distinction between space and time is not invariantly defined). Formal manipulations and the fact that $T(z)$ is a conserved quantity yield that if $\Phi(z, \bar{z})$ is a primary field of weight $(h, \bar{h})$, then there is an identity of operators, the *operator product expansion*:

$$T(z)\Phi(w, \bar{w}) = \frac{h}{(z-w)^2}\Phi(w, \bar{w}) + \frac{\partial_w \Phi(w, \bar{w})}{z-w} + R(z, w, \bar{w}),$$

where $R(z, w, \bar{w})$ remains finite when $z = w$. There is a similar formula for $\bar{T}(\bar{z})\Phi(w, \bar{w})$. This transformation property of T suggests that T should be a primary field of weight $(2, 0)$. Unfortunately, if this were true, it would lead to a quantum anomaly. Hence, the actual formula requires another term in the operator product expansion:

$$(B.26) \qquad T(z)T(w) = \frac{c/2}{(z-w)^4} + \frac{2}{(z-w)^2}T(w) + \frac{\partial_w T(w)}{z-w},$$

where c is a constant known as the *central charge*. Expanding $T(z) = \sum_n L_n z^{-(n+1)}$ and substituting into (B.26), we get by contour integration

$$(B.27) \qquad [L_n, L_m] = (n-m)L_{n+m} + \frac{c}{12}(n^3 - n)\delta_{n+m,0}$$

where $\delta_{n+m,0}$ is the identity operator for $n + m = 0$ and is zero otherwise. In other words, quantum mechanically, we do not get a representation of the Virasoro algebra, but rather a representation of a central extension of the Virasoro algebra. We should mention that (B.27) appears in Section 10.1.4 in our discussion of the Virasoro conjecture.

For perturbative string theory to be consistent, we need an anomaly-free representation of the Virasoro algebra itself. By (B.27), we avoid getting a central extension provided the *overall* central charge c of the theory vanishes. In the conformal field theory arising from the sigma model on a manifold X of real dimension $2k$, each boson x^I contributes $+1$ to c while the $2k$ fermion pairs $\psi^i_\pm$, $\psi^{\bar{i}}_\pm$ each contribute $+1/2$, for a total contribution of $2k \cdot 1 + 2k \cdot \frac{1}{2} = 3k$. But as we have seen, quantization requires ghosts to remain invariant under reparametrization of the world sheet (the conformal group) and supersymmetry. The ghosts which compensate for world sheet reparametrization contribute -26 to c while those compensating for supersymmetry contribute $+11$. The equation $3k - 26 + 11 = 0$ is satisfied only for $2k = 10$. Four of these dimensions are taken to be the spacetime dimensions of our familiar world (for example, Minkowski four space is a common model), leaving 6 free real dimensions. This is the reason for restricting to Calabi-Yau manifolds of complex dimension 3.

A similar but simpler argument shows that a consistent bosonic string must propagate in a 26-dimensional space.

We close with a few comments about the representation of the $N = 2$ superconformal algebra given by a sigma model (X, g). This representation contains the operators $T(z), \bar{T}(\bar{z})$ as above. In addition, the holomorphic supersymmetries give a $U(1) = SO(2)$ symmetry acting naturally on the sigma model fermions

$$\begin{pmatrix} \psi^i_+ \\ \psi^i_- \end{pmatrix}$$

The conserved current resulting from this $U(1)$ symmetry is denoted by $J(z)$. The spatial integral of J is the operator Q which appeared in Section 1.1 as the generator of the right-moving (holomorphic) sector of the theory which infinitesimally rotates the two supersymmetries. Similarly, we obtain $\bar{J}$ from the antiholomorphic supersymmetries and the $U(1)$ action on

$$\begin{pmatrix} \psi^{\bar{i}}_+ \\ \psi^{\bar{i}}_- \end{pmatrix}$$

which leads to an operator $\bar{Q}$ for the left-moving (antiholomorphic) sector. We have seen above how $H^1(X, T_X)$ and $H^1(X, \Omega^1_X)$ become operators in the theory; applying these operators to the vacuum state, we obtain quantum states corresponding to $H^1(X, T_X)$ and $H^1(X, \Omega^1_X)$. These states are eigenvectors for the operators $Q, \bar{Q}$

as discussed in Section 1.1. One of the main motivations for mirror symmetry is the effect of changing the sign of the generator Q on the corresponding eigenvalues.

B.4. Landau-Ginzburg Models

The Landau-Ginzburg (LG) theories we describe here are also supersymmetric theories. For definiteness, we will restrict discussion to theories with $(2,2)$ supersymmetry. These can be described using the formalism of superspaces and superfields. We digress now to describe this formalism.

A *supermanifold* is a topological space M with local coordinates

$$(x^1, \dots, x^n, y^1, \dots, y^m)$$

where the variables x^i commute with each other and with the y^i, while the y^i anticommute. While a precise mathematical definition can be given (see [**Manin1**]), the above rough definition suffices for present purposes. We can think of M as a manifold with local coordinates $(x^1, \dots, x^n)$ together with extra anticommuting variables y^i. A *superfield* is just a function on superspace. A superfield can be written uniquely as a sum $\sum_I f_I(x)y^I$, where $f_I(x)$ is a function of $x^1, \dots, x^n$, $I = \{i_1, \dots, i_k\} \subset \{1, \dots, m\}$ is an index set, and $y^I = y^{i_1} \cdots y^{i_k}$ if I has been written with $i_1 < \dots < i_k$. In Section 8.2.1, we showed that $H^*(X, \mathbb{C})$ has a natural structure as a complex supermanifold.

In our case, the underlying manifold is the world sheet Σ, so we will take $n = 2$. To obtain $(2,2)$ supersymmetry, we will also require $m = 4$. The variables x^1 and x^2 will be replaced by a local complex coordinate z, while the y^i will be denoted by θ^i and $\bar{\theta}^i$, where $i = 1,\ 2$. The $\theta^i, \bar{\theta}^i$ are not merely local expressions, but they are *spinors*, i.e., they transform as spinors under one of the (1-dimensional) spin representations of so(2), with θ^1, θ^2 belonging to one of the representations and $\bar{\theta}^1, \bar{\theta}^2$ belonging to the other representation. For more details about spin geometry from a mathematical perspective, the reader is referred to [**LM**].

Thus, a superfield Φ can be locally expressed as

$$\Phi = f(z) + \phi_i(z)\theta^i + \bar{\phi}_i(z)\bar{\theta}^i + F_{ij}(z)\theta^i\theta^j + \cdots,$$

where some quadratic terms in the anticommuting variables, and all of the higher order terms in the anticommuting variables, have been omitted. There are supersymmetry transformations $Q^i, \bar{Q}^i$ that can be explicitly written down, and the anticommutators take on a simple form (see, for example, [**WB**] for the analogous formulas for 4-dimensional supersymmetry). This leads to the construction of the $N = 2$ *supersymmetry algebra*. Note that the supersymmetry transformations are not to be confused with the $U(1)$ charges $Q, \bar{Q}$ that we have discussed earlier.

The space of all superfields gives a representation of the $N = 2$ supersymmetry algebra. The subspace of chiral superfields is an invariant subspace. This can be constructed by explicitly writing down differential operators $\bar{D}_i$ in superspace which commute with the supersymmetry operators, and defining a *chiral superfield* to be a superfield Φ satisfying $\bar{D}_i\Phi = 0$ for $i = 1, 2$ (see e.g., [**WB**] for a definition of chiral superfields). The operators $\bar{D}_i$ involve differentiation with respect to the anticommuting variables. In fact, $\bar{D}_i$ includes a term $\partial/\partial\bar{\theta}^i$, which can be defined in the natural way. Thus chiral superfields behave much like holomorphic functions do in ordinary space.

Landau-Ginzburg theories are certain supersymmetric theories which can be simply described by a superpotential, which is a weighted homogeneous polynomial

$F(\Phi_1, \ldots, \Phi_k)$ of 2-dimensional chiral superfields. Here, weighted homogeneous refers to the assignment of a weight d_i to Φ_i. In the case of interest to us, we will choose F to be weighted homogeneous of degree $d = \sum_i d_i$.

In the action for the Landau-Ginzburg theory, the important term is

$$(B.28) \qquad \int d^2z\, d^2\theta\, F(\Phi_i),$$

where z is a local holomorphic coordinate on the world sheet, Φ_i are chiral superfields, and θ^j are fermionic superspace coordinates. Writing ϕ_i for the top (bosonic) components of the Φ_i, the classical equations of motion give $\partial F/\partial\phi_i = 0$. There is the obvious classical solution $\phi_i = 0$, and in a Landau-Ginzburg theory, this is the only solution.

Turning to the quantum theory, there is a unique vacuum state associated to the unique classical vacuum. Denoting the vacuum state by V, we can create new states as $P(\phi_i)(V)$ for any polynomial P. These states are clearly isomorphic to the Jacobian ring $\mathbb{C}[\phi_1, \ldots, \phi_k]/\langle\partial F/\partial\phi_i\rangle$, also called the chiral ring.

In the previous paragraph, notice how quickly we moved from something fairly exotic sounding (fermionic superspace coordinates, etc.) to something well known to algebraic geometers (the Jacobian ring of a weighted homogeneous polynomial). The condition that the origin is the only solution of $\partial F/\partial\phi_i = 0$ is the same as the condition for quasi-smoothness of the hypersurface $F = 0$ in the weighted projective space. This is typical of how standard mathematical objects can suddenly appear in the middle of these physical theories.

We are actually interested in related theories, namely Landau-Ginzburg orbifolds. In the case we are interested in, we let η be a primitive d^{th} root of unity and take an orbifold by the action of $(\eta^{d_1}, \ldots, \eta^{d_k})$. By this, we mean that we don't just consider well-defined superfields on the world sheet Σ, but rather those that transform by the ith power of η at the ends of a string. The resulting states are said to belong to the ith twisted sector. This will give rise to twisted vacua V_i. We can create more states by forming expressions of the form $P(\phi_j)(V_i)$. The Landau-Ginzburg orbifold will contain those states which are invariant under η.

As we will see below, there is a close connection between Calabi-Yau manifolds and Landau-Ginzburg orbifolds. For now, we illustrate with one example from a different viewpoint.

Example B.4.1. Suppose $\Phi_1, \ldots, \Phi_5$ have respective weights $1, 1, 2, 2, 2$, and let F be weighted homogeneous of weight 8 in the variables $\Phi_1, \ldots, \Phi_5$. Before working out some of the states in this theory, we note that this example is related to degree 8 hypersurfaces in $\mathbb{P}^4(1, 1, 2, 2, 2)$, which were studied in [**CdFKM, HKTY1**]. This weighted projective space has A_1 singularities along the $\mathbb{P}^2$ defined by $x_1 = x_2 = 0$. Blowing this up yields a ruled threefold over $\mathbb{P}^2$ as exceptional divisor. Restricting to the hypersurface, the exceptional divisor becomes a ruled surface over a quartic (genus 3) curve. The image of the Abel-Jacobi map for the family of fibers is thus a 3-dimensional subspace of $H^{2,1}$. It can be shown that these correspond to deformations of the Calabi-Yau threefold that cannot be realized as a hypersurface inside $\mathbb{P}^4(1, 1, 2, 2, 2)$. This 3-dimensional subspace is complementary to the codimension 3 subspace $H^{2,1}_{\text{poly}}(V^\circ)$ of $H^{2,1}(V^\circ)$ consisting of deformations of the hypersurface. A geometric description of these deformations is given in Example 6.2.4.3.

Let's now turn to the LG theory. We have the states $P(\phi_i)(V)$. It turns out that the states which correspond to the moduli parameters of the theory are those for which P has degree 8. We thus get the degree 8 component of the Jacobian ring, which has dimension 83. This construction is familiar from algebraic geometry.

So where are the 3 missing deformations? They appear in one of the twisted sectors—the fourth, to be precise. The twisted vacuum V_4 is not invariant, and to correspond to a deformation, it turns out that $P(\phi_i)(V_4)$ must be such that P has weight 2. But the variables ϕ_3, ϕ_4, ϕ_5 are invariant under the action of η^4, while ϕ_1 and ϕ_2 are not. So the states are just $P(\phi_3, \phi_4, \phi_5)(V_4)$, which is 3-dimensional, as expected.

The above discussion is based on calculations in [**Vafa1**], whose work also includes miraculous formulas for cohomology of Calabi-Yau manifolds derived from considerations of Landau-Ginzburg theory.

B.5. Gauged Linear Sigma Models

The gauged linear sigma model (GLSM) of Witten is a very useful model which in an appropriate sense "interpolates" between Calabi-Yau models and Landau-Ginzburg theories. As with the Landau-Ginzburg model, this theory can be formulated in terms of chiral superfields, which were outlined in Appendix B.4. For more details, see [**Witten5**]. A mathematical comparison between the GLSM and the nonlinear sigma model discussed in Appendix B.2 plays a crucial role in the mirror theorems discussed in Chapter 11.

The GLSM is a gauge theory, with gauge group $U(1)^s$ for some s. Associated to the gauge group are 2-dimensional *vector superfields* V_a, $a = 1, \ldots, s$. The theory also contains $k + r$ chiral superfields Φ_i, $i = 1, \ldots, k$ and P_j, $j = 1, \ldots, r$ of respective charges $Q_{i,a}$ and $q_{j,a}$ with $\sum_j q_{j,a} = -\sum_i Q_{i,a}$ for each a. The charges tell us how the superfields transform under the symmetry $U(1)_a$ ($=$ the ath factor of $U(1)^s$). The theory will also be determined by a superpotential W, which as we will see relates equations of Calabi-Yau hypersurfaces to Landau-Ginzburg potentials. The superpotential is a holomorphic function of the superfields which is invariant under the $U(1)^s$ action. We restrict our attention to superpotentials of the form $W = \sum_a P_a G_a(\Phi_1, \ldots, \Phi_n)$.

The action itself is a sum of four terms

$$\text{(B.29)} \qquad S = S_{\text{kin}} + S_W + S_{\text{gauge}} + S_{D,\theta},$$

whose precise form will not concern us. The only point we will make is that the action is a function of all the superfields. There are a few constants in the action which measure the strength of certain couplings in the theory: e_j is the gauge coupling for $U(1)_j$, and r_j determines the strength of a certain supersymmetric interaction. These constants enter explicitly in (B.30) and (B.31) below.

For the time being, we will make the simplifying assumption $r = 1$ and write P instead of P_1. Thus the superpotential is $W = PG$ for some polynomial G. (We will return to the case $r \geq 1$ later in the section.) When we group the terms in the action into commuting (bosonic) and anticommuting (fermionic) variables and take the equations of motion into consideration, the bosonic part of the action is

$$\text{(B.30)} \qquad U(\phi_i, p) = \sum_j \frac{1}{2e_j^2} D_j^2 + \sum_\alpha |F_\alpha^2|$$

with

$$(B.31) \qquad D_j = -e_j^2 \Big(\sum_i Q_{i,j} |\phi_i|^2 - q_j |p|^2 - r_j \Big)$$

and

$$(B.32) \qquad \sum_\alpha |F_\alpha^2| = |G|^2 + |p|^2 \sum_i \Big| \frac{\partial G}{\partial \phi_i} \Big|^2.$$

In (B.31) and (B.32), the ϕ_i and p are certain bosonic components (the top components) of the superfields Φ_i and P. In other words, the ϕ_i and p can be thought of as ordinary functions.

We will see that the solutions to the classical equations of motion look very different for different values of the r_j. Witten refers to the transitions as *phase transitions* in the theory [**Witten5**].

Example B.5.1. Suppose that $k = 5$, $s = 1$, and $Q_{i,1} = 1$ for each i. Choose a superpotential $W = PG(\Phi_1, \dots, \Phi_5)$, where G is a general quintic polynomial in the Φ_i. To minimize the potential, we see that each term (B.31) vanishes, and each term on the right hand side of (B.32) vanishes. Note that the vanishing of (B.31) says that $\sum_{i=1}^5 |\phi_i|^2 - 5|p|^2 - r_1 = 0$. There are two cases to consider:

- $r_1 > 0$. In this case, the vanishing of (B.31) shows that $(\phi_1, \dots, \phi_5) \neq 0$. Then the vanishing of $|p|^2 \sum_i |(\partial G/\partial \phi_i)|^2$ implies that $p = 0$, since $(\phi_1, \dots, \phi_5) \neq 0$ and G is general (i.e., smooth). The vanishing of (B.31) now says that $\sum |\phi_i|^2 = r_1$. We can now take the gauge symmetry into consideration and mod out by $U(1)$, which acts by scalar multiplication on $(\phi_1, \dots, \phi_5)$ by choice of the Q_i. This leave us with $\mathbb{P}^4$. Finally, from the right hand side of (B.32) we see that we now put $G = 0$; this leaves us with the space of classical solutions equal to the quintic threefold.

- $r_1 < 0$. In this case, the vanishing of (B.31) shows that $p \neq 0$. Then from the vanishing of the terms on the right hand side of (B.32) we see first that each $\partial G/\partial \phi_i$ must vanish. Since G is general, this implies that $(\phi_1, \dots, \phi_5) = (0, \dots, 0)$. The vanishing of (B.31) then gives $|p| = \sqrt{-r_1/5}$. The value of p itself is then well-defined up to the gauge transformation $p \mapsto u^5 p$ where $u \in U(1)$. Note that if u is a fifth root of unity, then p is preserved. We thus get a Landau-Ginzburg orbifold.

Thus the GLSM interpolates between Calabi-Yau theories and Landau-Ginzburg theories.

More generally, we can describe a GLSM in terms of toric geometry as follows. The $U(1)^s$ action on the Φ_i gives an action of $\mathbb{C}^k$ with coordinates $\phi_1, \dots, \phi_k$. Using the terminology of Section 3.3.3, this action is Hamiltonian and its moment map $\mu : \mathbb{C}^s \to \mathbb{R}^s$ is given by

$$\mu(\phi_1, \dots, \phi_k) = \frac{1}{2} \Big(\sum_i Q_{i,1} |\phi_i|^2, \dots, \sum_i Q_{i,s} |\phi_i|^2 \Big).$$

Then one can show that the Calabi-Yau phases of the GLSM correspond (roughly) to when $\vec{r} = (r_1, \dots, r_s)$ lies in a certain cone in $\mathbb{R}^s$ which is the Kähler cone of the toric variety

$$X_{\vec{r}} = \mu^{-1}(\vec{r})/U(1)^s.$$

Here, we are using the symplectic construction of toric varieties discussed in Section 3.3.4. This method of constructing toric varieties is a crucial ingredient of a GLSM. Furthermore, $G \in \mathbb{C}[\phi_1, \ldots, \phi_k]$ is a homogeneous polynomial in the sense of Section 3.2.3, so that the vanishing of $G = 0$ gives a Calabi-Yau hypersurface inside of $X_{\vec{r}}$. Hence, as we vary $\vec{r}$, the Calabi-Yau phases of the GLSM correspond to certain cones in the GKZ decomposition of Section 3.4. But we also have other phases of the theory, such as the Landau-Ginzburg ones, which come from cones outside the GKZ decomposition. These correspond to cones in the *enlarged GKZ decomposition* or *enlarged secondary fan* described in [**MP1**].

So far, we have been assuming $r = 1$ in our discussion. When $r > 1$, the superpotential is $W = \sum_a P_a G_a$, where G_a are again homogeneous polynomials (in the toric sense). A Calabi-Yau phase of such a GLSM would still give a toric variety, but now we would have the Calabi-Yau complete intersection $G_1 = \cdots = G_r = 0$ to consider.

There is an important difference between the linear sigma model and the nonlinear sigma model. For definiteness, consider the above example, supposing that $r_1 > 0$, so that we are in the Calabi-Yau phase. In the nonlinear sigma model, we have fields which are local coordinates on $\mathbb{P}^4$. In the linear model, we have fields which are coordinates on $\mathbb{C}^5$, then we mod out by scalars. Thus, for the nonlinear sigma model, we need to consider holomorphic instantons, which are holomorphic maps from genus 0 curves C to $\mathbb{P}^4$ of a certain degree d which lie on $G = 0$. For the GLSM, we use 5-tuples $(f_1(s,t), \ldots, f_5(s,t))$, where the f_i are homogeneous of a fixed degree d and which satisfy $G(f_1, \ldots, f_5) = 0$. In the above, (s,t) are thought of as homogeneous coordinates on $\mathbb{P}^1$. But this is not an exact correspondence: a 5-tuple $(f_1(s,t), \ldots, f_5(s,t))$ need not arise from a degree d holomorphic map, as the f_i can have a common factor and therefore attempting to interpret it as a map will result in lower degree after removal of base points. These 5-tuples or *naive instantons* are easier to deal with, and have been used to give strong support for mirror symmetry in the context of the GLSM (see [**MP1**]). The map $\varphi : M_d \to N_d$ defined in Section 11.1 can be described as comparing holomorphic instantons and naive instantons.

B.6. Topological Quantum Field Theories

In Appendix B.1, we gave a brief introduction to general quantum field theories. Such a theory is described by a Hilbert space $\mathcal{H}$ of states and certain operators on $\mathcal{H}$ which encode the propagation and interaction of particles. We now explain what it means to be a *topological quantum field theory* (TQFT). We will see that Frobenius algebras and quantum cohomology arise naturally in this context.

Let's start by discussing a point particle theory. Let $\mathcal{H}$ be the space of states. If $\mathbf{H}$ is the operator associated to the Hamiltonian function H by the quantization procedure, then as noted in (B.7), the propagation of states in time is given by the unitary operator $\exp(-i\mathbf{H}t)$. We can describe this propagation schematically by an interval of length t. We can associate to each endpoint of the interval the Hilbert space $\mathcal{H}$, and to the interval the operator

$$e^{-i\mathbf{H}t} : \mathcal{H} \to \mathcal{H}.$$

This is a $(0 + 1)$-dimensional quantum field theory. The 0 refers to the dimension of the endpoints, to which $\mathcal{H}$ is associated. The $+1$ refers to the additional time dimension which is needed to connect the endpoints.

To make the transition to TQFTs, we ask for which $(0+1)$-dimensional theories does the operator $\exp(-i\mathbf{H}t)$ depend *only* on the topology of the interval. This is easy to determine, for we are asking when $\exp(-i\mathbf{H}t)$ is independent of t. By differentiation at $t = 0$, we conclude that necessarily $\mathbf{H} = 0$. Thus $\exp(-i\mathbf{H}t)$ is the identity for any t. In other words, this system has no dynamics whatsoever, and is completely trivial and uninteresting as a physical theory.

Such theories are $(0+1)$-dimensional topological quantum field theories. As we will see, $(d+1)$-dimensional TQFTs are far from trivial when $d > 0$.

We now give an axiomatic definition of a topological quantum field theory. We use complex coefficients for definiteness. In order to incorporate fermions, we would need to use supervector spaces, which are $\mathbb{Z}/2\mathbb{Z}$-graded vector spaces satisfying natural sign rules [**Manin1**]. We leave the necessary revisions to the interested reader. See [**Quinn**] for a broader treatment.

DEFINITION B.6.1. *A $(d+1)$-dimensional topological quantum field theory consists of the following data:*

- *To each closed oriented d-dimensional manifold Y is associated a finite-dimensional complex vector space $Z(Y)$. This vector space behaves functorially under isomorphisms of Y.*
- *To each $(d+1)$-dimensional oriented manifold X whose boundary ∂X is a closed oriented d-dimensional manifold is associated an element $Z_X \in Z(\partial X)$. This element behaves functorially under isomorphisms of X.*

This data satisfies the following additional axioms:

Axiom 1: $Z(Y_1 \amalg Y_2) = Z(Y_1) \otimes Z(Y_2)$.

Axiom 2: *Considering the empty set as a closed d-dimensional oriented manifold, we have $Z(\emptyset) = \mathbb{C}$.*

Axiom 3: *Considering the empty set as a $(d+1)$-dimensional oriented manifold with empty boundary, we have $Z_\emptyset = 1 \in Z(\emptyset) = \mathbb{C}$.*

Axiom 4: *Let $\overline{Y}$ denote Y with the opposite orientation. Then $Z(\overline{Y}) \simeq Z(Y)^*$, this isomorphism behaving functorially under isomorphisms of Y. In particular, if $\partial X = (\amalg_{i=1}^k \overline{Y_i}) \amalg (\amalg_{j=1}^\ell Y_j')$, then we may view*

$$Z_X \in \mathrm{Hom}_{\mathbb{C}}\big(\otimes_{i=1}^k Z(Y_i), \otimes_{j=1}^\ell Z(Y_j')\big),$$

where we have used this axiom together with Axiom 1. We also assume that with the natural identifications, $Z_{\overline{X}}$ is the adjoint of Z_X.

Axiom 5: *Let I denote the oriented interval $[0,1]$. Let Y be any closed oriented d-manifold without boundary, and put $X = Y \times I$. Then $\partial X = \overline{Y} \amalg Y$. We require*

$$Z_X = 1_{Z(Y)} \in \mathrm{Hom}_{\mathbb{C}}(Z(Y), Z(Y)),$$

where we have used Axiom 4.

Axiom 6: *If $\partial X = \overline{Y_1} \amalg Y_2$ and $\partial X' = \overline{Y_2} \amalg Y_3$, then we can glue X and X' together along Y_2 to form a new manifold $X \cup_{Y_2} X'$ with boundary $\overline{Y_1} \amalg Y_3$. By Axiom 4, we write $Z(\partial X) = \mathrm{Hom}_{\mathbb{C}}(Z(Y_1), Z(Y_2))$, $Z(\partial X') = \mathrm{Hom}_{\mathbb{C}}(Z(Y_2), Z(Y_3))$, and $Z(\partial(X \cup_{Y_2} X')) = \mathrm{Hom}_{\mathbb{C}}(Z(Y_1), Z(Y_3))$. We require that $Z_{X \cup_{Y_2} X'} = Z_{X'} \circ Z_X$.*

These axioms are easy to understand when $d = 0$. We can put $\mathcal{H} = Z(\text{point})$, which is a finite-dimensional vector space over $\mathbb{C}$, and then Axiom 5 applied to

$Y =$ point says that $Z_I = 1_\mathcal{H}$. This was the conclusion of our discussion of $(0+1)$-dimensional TQFTs at the beginning of the section. Furthermore, since S^1 is an oriented 1-dimensional manifold with empty boundary, we have $Z(S^1) \in Z(\emptyset) = \mathbb{C}$. We can calculate this number as follows. We realize S^1 as a union of two copies of the unit interval I:

We now use Axiom 6 with:

$$X = I, \ Y_1 = \emptyset, \ Y_2 = \overline{\text{point}} \ \text{II point}, \ X' = \overline{I}, \ Y_3 = \emptyset.$$

Combining this with Axioms 2, 4, and 5, we calculate that $Z(S^1) = \dim \mathcal{H}$. Since an arbitrary 1-dimensional manifold is a disjoint union of S^1, $\overline{S^1}$, I and $\overline{I}$, the above axioms imply that a $(0+1)$-dimensional TQFT is uniquely determined by $\mathcal{H} = Z(\text{point})$. Furthermore, one can show that all finite-dimensional complex vector spaces $\mathcal{H}$ come from $(0+1)$-dimensional TQFTs in this way.

We now turn our attention to the case when $d = 1$. Our goal is to characterize $(1+1)$-dimensional TQFTs. For this purpose, we will need the following definition.

DEFINITION B.6.2. *A commutative Frobenius algebra is a commutative, associative algebra* $(\mathcal{A}, *)$ *with a unit* 1, *together with a nondegenerate inner product* $\langle \ , \ \rangle$ *on* $\mathcal{A}$, *satisfying*

(B.33) $$\langle a * b, c \rangle = \langle a, b * c \rangle$$

for all $a, b, c \in \mathcal{A}$.

In the literature, other equivalent definitions are sometimes given. For instance, a *trace map* $\epsilon : \mathcal{A} \to \mathbb{C}$ satisfying certain properties is given in some axiom systems. With our conventions, we simply put $\epsilon(a) = \langle a, 1 \rangle$.

Frobenius algebras have very rich structure. We can define a *three-point correlation function* $\langle \ , \ , \ \rangle : \mathcal{A}^{\otimes 3} \to \mathbb{C}$ by

$$\langle a, b, c \rangle = \langle a * b, c \rangle$$

for all $a, b, c \in \mathcal{A}$. Combining (B.33) and commutativity, we see that the three-point function is totally symmetric in its arguments. Similarly, for any n we can define the n-point correlation function by

(B.34) $$\langle a_1, \dots, a_n \rangle = \langle a_1 * \dots * a_{n-1}, a_n \rangle,$$

which can be shown to be totally symmetric in each of its arguments.

Now suppose that we have a $(1+1)$-dimensional topological quantum field theory. Fix once and for all a standard oriented closed disk $\overline{\Delta} = \{z : |z| \leq 1\}$, and let S^1 denote its boundary. We put $\mathcal{H} = Z(S^1)$, and denote by T_0 the element $Z_{\overline{\Delta}} \in Z(\partial\overline{\Delta}) = Z(S^1) = \mathcal{H}$. We will show that $\mathcal{H}$ has a natural structure as a Frobenius algebra with T_0 as identity element.

It is easy to construct an inner product on $\mathcal{H}$. The isomorphism $\overline{S^1} \simeq S^1$ given by complex conjugation induces by Axiom 4 an isomorphism $\mathcal{H} \simeq \mathcal{H}^*$. We let $\langle \ , \ \rangle$ denote the corresponding nondegenerate inner product on $\mathcal{H}$.

We next need to construct a product $* : \mathcal{H} \otimes \mathcal{H} \to \mathcal{H}$. For this, consider the "pair of pants" Σ given by

(B.35)

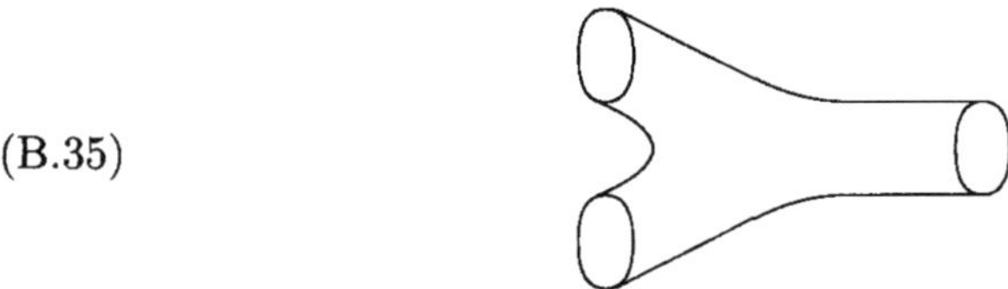

Choose oriented isomorphisms of the boundary components of Σ with $\overline{S^1} \amalg \overline{S^1} \amalg S^1$, the two copies of $\overline{S^1}$ corresponding to the left hand boundary circles of (B.35) and the S^1 corresponding to the right hand boundary circle. By Axiom 4, we get an element $Z_\Sigma \in \mathrm{Hom}_{\mathbb{C}}(\mathcal{H} \otimes \mathcal{H}, \mathcal{H})$. Let $*$ denote the product on $\mathcal{H}$ defined by Z_Σ.

PROPOSITION B.6.3. *The product $*$ is commutative and associative. The element T_0 is an identity for $*$, and for all $a, b, c \in \mathcal{H}$, we have the identity*

$$\langle a * b, c \rangle = \langle a, b * c \rangle.$$

Thus $\mathcal{H}$ is a Frobenius algebra under $*$.

PROOF. The chosen isomorphism $\partial\Sigma \simeq \overline{S^1} \amalg \overline{S^1} \amalg S^1$ can be replaced by another such isomorphism obtained by composing it with the isomorphism of $\overline{S^1} \amalg \overline{S^1} \amalg S^1$ which switches its first two components. Commutativity follows by functoriality. Now glue $\overline{\Delta}$ to a boundary $\overline{S^1}$ in (B.35), forming a space we call X. Using Axiom 6, we identify Z_X with the endomorphism $\phi \mapsto T_0 * \phi$ of $\mathcal{H}$. On the other hand, X has an orientation-preserving homeomorphism to the cylinder $S^1 \times I$. By Axiom 5, we see that $Z_X = 1_{\mathcal{H}}$. Comparing these two computations, we see that T_0 is the identity for $*$. Now choose an isomorphism of $\partial\Sigma$ with $\overline{S^1} \amalg \overline{S^1} \amalg \overline{S^1}$. Then $Z_\Sigma \in \mathrm{Hom}_{\mathbb{C}}(\mathcal{H} \otimes \mathcal{H} \otimes \mathcal{H}, \mathbb{C})$, where we are abusing notation by using the same symbol Z_Σ as before. If we compose this identification of the boundary of Σ with the isomorphism $\overline{S^1} \simeq S^1$ on the third boundary component, we compute $Z_\Sigma(a, b, c) = \langle a * b, c \rangle$. Making this identification on the first boundary component, we are led to $Z_\Sigma(a, b, c) = \langle a, b * c \rangle$. The equality (B.33) follows. Finally, to prove associativity, let Σ' be a 2-sphere with 4 disks removed and identify $\partial\Sigma'$ with $\overline{S^1} \amalg \overline{S^1} \amalg \overline{S^1} \amalg S^1$. This gives $Z_{\Sigma'} \in \mathrm{Hom}_{\mathbb{C}}(\mathcal{H} \otimes \mathcal{H} \otimes \mathcal{H}, \mathcal{H})$. Now decompose Σ' into 2 pairs of pants as follows:

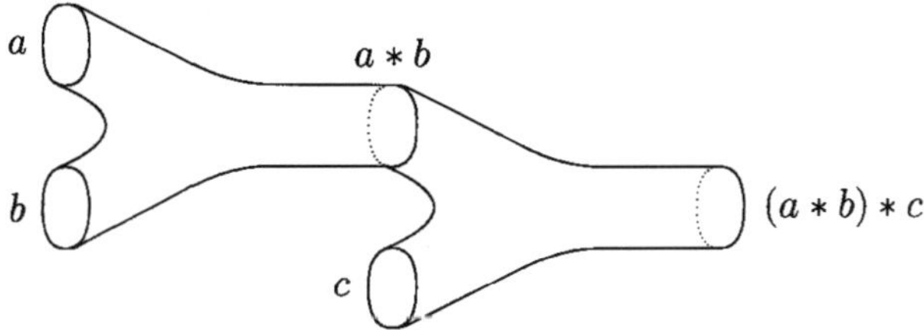

This shows that $Z_{\Sigma'}(a, b, c) = (a * b) * c$. However, since we're dealing with a topological quantum field theory, we can decompose Σ' into 2 pairs of pants in a different way, namely:

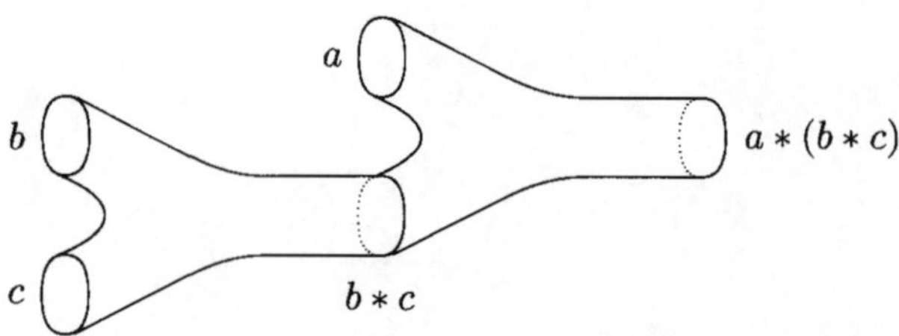

This second picture gives $Z_{\Sigma'}(a, b, c) = a * (b * c)$. Combining this with the previous computation of $Z_{\Sigma'}(a, b, c)$, we conclude that $(a * b) * c = a * (b * c)$, as desired. $\square$

It is not hard to see that a $(1+1)$-dimensional TQFT is uniquely determined by the finite-dimensional Frobenius algebra $\mathcal{H}$ described in Proposition B.6.3. Rather than give a proof, we explain how the axioms work in practice. For example, the 2-torus T^2 can be obtained by gluing together two cylinders. A straightforward modification of the calculation of $Z(S^1)$ in the $(0+1)$-dimensional case shows that $Z(T^2) = \dim \mathcal{H}$. Gluing together two disks to get S^2, we are led to $Z(S^2) = \langle T_0, T_0 \rangle$. Computing $Z(\Sigma)$ for other closed oriented 2-dimensional surfaces Σ is equally straightforward.

Once we have the Frobenius algebra $\mathcal{H}$ of a $(1+1)$-dimensional TQFT, one can also interpret the n-point functions (B.34) in terms of the TQFT. For example, when $n = 4$, let Σ' be a 2-sphere with 4 disks removed, and identify the boundary of Σ' with $\overline{S^1} \amalg \overline{S^1} \amalg \overline{S^1} \amalg \overline{S^1}$. Then, adapting the argument given in the proof of Proposition B.6.3, one can show that the 4-point function $\langle a, b, c, d \rangle$ is given by

$$\langle a, b, c, d \rangle = Z_{\Sigma'}(a, b, c, d)$$

for all $a, b, c, d \in \mathcal{H}$. More generally, the n-point function can be defined using a 2-sphere from which n disks have been removed.

As we have observed in Appendix B.2, the A-model is a $(1+1)$-dimensional TQFT depending on a choice of complexified Kähler class. So what is the underlying Frobenius algebra? We have already identified $\mathcal{H}$ with $H^*(X)$ in Appendix B.2. It turns out that the inner product is defined by

$$\langle T_1, T_2 \rangle = \int_X T_1 \cup T_2,$$

and the product $*$ is just the quantum product![4] We thus get an associative quantum cohomology ring. The trace map is $\epsilon(T) = \int_X T$. If X is general (i.e., not necessarily Calabi-Yau), this quantum product arises from the physical theory known as the *topological sigma model*. As we have seen, the quantum product is encoded by the pair of pants (B.35), which can be glued together to get Riemann surfaces of higher genus Σ_g with disks removed. Orienting the boundary circles appropriately, we get *genus g correlation functions* $Z_{\Sigma_g} : \mathcal{H}^{\otimes n} \to \mathbb{C}$. These generalize the n-point correlation functions discussed above, which are all genus 0 correlation functions. Identifying $\mathcal{H}$ with $H^*(X, \mathbb{C})$, these higher genus correlation functions are identified with some of the invariants of [**RT1**]. Note that in [**RT1**], the complex structure

---

[4]We actually have to work with supervector spaces here. This is clear, since quantum product is not commutative, but rather supercommutative (Theorem 8.1.4).

was fixed. Hence the invariants are purely topological, not depending on the complex structure.

The Frobenius algebra structure is how physicists knew of the existence of quantum cohomology before mathematicians could prove it. We have given an extensive mathematical treatment of quantum cohomology in Chapter 8. The physical explanation of quantum cohomology was observed in [**LVW**].

Finally, we should also mention that $(2+1)$-dimensional topological quantum field theories are related to the Jones polynomials. A nice exposition can be found in [**Atiyah**].

Bibliography

We use the following conventions for labeling references in the text and in the bibliography:

- References with one author are labeled by the author's last name.
- References with multiple authors are labeled using the first letter of the authors' last names in alphabetical order. When different groups of authors would share the same label, we distinguish between these by using the first two letters of one of the authors' last names.
- If there is more than one reference for the same author or group of authors, they are ordered by publication year.

In addition to the usual bibliographic information, we include references to the e-print archives math.AG (which includes alg-geom), math.DG (which includes dg-ga), and hep-th whenever possible. These archives are accessible on the WWW at the URL http://xxx.lanl.gov/.

[AKa] A. Albano and S. Katz, *Lines on the Fermat quintic threefold and the infinitesimal generalized Hodge conjecture*, Trans. AMS **324** (1991), 353–368.

[Alexeev] V. Alexeev, *Moduli spaces $M_{g,n}(W)$ for surfaces*, in *Higher-Dimensional Complex Varieties (Trento, 1994)*, de Gruyter, Berlin, 1996, 1–22, alg-geom/9410003.

[AN] V. Alexeev and V. Nikulin, *Classification of Del Pezzo surfaces with Log-terminal singularities of index ≤ 2, and involutions of K3 surfaces*, Soviet Math. Dokl. **39** (1989), 507–511.

[AKl] A. Altmann and S. Kleiman, *Foundations of the Theory of Fano Schemes*, Compositio Math. **34** (1977), 3–47.

[Aluffi] *Quantum Cohomology at the Mittag-Leffler Institute*, (P. Aluffi ed.), Appunti della Scuola Normale Superiore, Pisa, 1997.

[Alvarez] O. Alvarez, *Lectures on quantum mechanics and the index theorem*, in *Geometry and Quantum Field Theory* (D. Freed and K. Uhlenbeck, eds.), IAS/Park City Mathematics Series **1**, AMS, Providence, RI, 1995, 273–322.

[Arnold] V. Arnold, *Mathematical Methods of Classical Mechanics*, Springer-Verlag, New York, 1989.

[AMRT] A. Ash, D. Mumford, M. Rapoport and Y.-S. Tai, *Smooth Compactifications of Locally Symmetric Varieties*, Math Sci Press, Brookline, MA, 1975.

[Aspinwall] P. Aspinwall, *K3 Surfaces and String Duality*, in *Fields, Strings and Duality (TASI 96, Boulder, CO)* (C. Efthimiou and B. Greene, eds.), World Scientific, River Edge, NJ, 1997, 421–540, hep-th/9611137.

[AGM1] P. Aspinwall, B. Greene and D.R. Morrison, *The monomial-divisor mirror map*, Int. Math. Res. Notices (1993), 319–337, hep-th/9309007.

[AGM2] P. Aspinwall, B. Greene and D. Morrison, *Multiple mirror manifolds and topology change in string theory*, Phys. Lett. **B303** (1993), 249–259, hep-th/9301043.

[AGM3] P. Aspinwall, B. Greene and D. Morrison, *Measuring small distances in $N = 2$ sigma models*, Nuclear Physics **B420** (1994) 184, hep-th/9311042.

[AM1] P. Aspinwall and D. Morrison, *Topological field theory and rational curves*, Comm. in Math. Phys. **151** (1993), 245–262, hep-th/9110048.

[AM2] P. Aspinwall and D. Morrison, *Chiral rings do not suffice: $N = (2,2)$ theories with nonzero fundamental group*, Phys. Lett. **B344** (1994), 79–86, hep-th/9406032.

[AM3] P. Aspinwall and D. Morrison, *String theory of K3 surfaces*, Mirror Symmetry II (B. Greene and S.-T. Yau, eds.), AMS/IP Stud. Adv. Math. **1**, AMS, Providence, RI, 1997, 703–716, hep-th/9404151.

[AS] A. Astashkevich and V. Sadov, *Quantum cohomology of partial flag manifolds*, Comm. Math. Phys. **170** (1995), 503–528, hep-th/9401103.

[Atiyah] M. Atiyah, *The Geometry and Physics of Knots*, Cambridge University Press, Cambridge, 1990.

[AB] M. Atiyah and R. Bott, *The moment map and equivariant cohomology*, Topology **23** (1984), 1–28.

[Audin] M. Audin, *The Topology of Torus Actions on Symplectic Manifolds*, Progress in Math. **93**, Birkhäuser, Boston-Basel-Berlin, 1991.

[ACJM] A. Avram, P. Candelas, D. Jančić and M. Mandelberg, *On the connectedness of the moduli space of Calabi-Yau manifolds*, Nuclear Physics **B465** (1996), 458–472, hep-th/9511230.

[AKMS] A. Avram, M. Kreuzer, M. Mandelberg and H. Skarke, *Searching for K3 fibrations*, Nuclear Physics **B494** (1997), 567–589, hep-th/9610154.

[Baily1] W. Baily, Jr., *The decomposition theorem for V-manifolds*, Amer. J. Math. **78** (1956), 862–888.

[Baily2] W. Baily, Jr., *On the embedding of V-manifolds in projective space*, Amer. J. Math. **79** (1957), 403–430.

[BBRP] C. Bartocci, U. Bruzzo, D. Hernández Ruipérez and J. Muñoz Porras, *Mirror symmetry on elliptic K3 surfaces via Fourier-Mukai transform*, Comm. Math. Phys. **195** (1998), 79–93, alg-geom/9704023.

[Batyrev1] V. Batyrev, *On the classification of smooth projective toric varieties*, Tôhoku Math. J. **43** (1991), 569–585.

[Batyrev2] V. Batyrev, *Quantum cohomology rings of toric manifolds*, in Journées de Geómétrie Algébrique d'Orsay (Orsay, 1992), Astérisque **218**, Soc. Math. France, Paris, 1993, 9–34, alg-geom/9310004.

[Batyrev3] V. Batyrev, *Variations of the mixed Hodge structure of affine hypersurfaces in algebraic tori*, Duke Math. J. **69** (1993), 349–409.

[Batyrev4] V. Batyrev, *Dual polyhedra and mirror symmetry for Calabi-Yau hypersurfaces in toric varieties*, J. Algebraic Geometry **3** (1994), 493–535, alg-geom/9310003.

[Batyrev5] , V. Batyrev, *Toric degenerations of Fano varieties and constructing mirror manifolds*, preprint, 1997, math.AG/9712034.

[Batyrev6] , V. Batyrev, *On the classification of toric Fano 4-folds*, preprint, 1998, math.AG/9801107.

[BB1] V. Batyrev and L. Borisov, *On Calabi-Yau complete intersections in toric varieties*, in *Higher-Dimensional Complex Varieties (Trento, 1994)*, de Gruyter, Berlin, 1996, alg-geom/9412017.

[BB2] V. Batyrev and L. Borisov, *Mirror duality and string-theoretic Hodge numbers*, Invent. Math. **126** (1996), 183–203, alg-geom/9509009.

[BB3] V. Batyrev and L. Borisov, *Dual cones and mirror symmetry for generalized Calabi-Yau manifolds*, in *Mirror Symmetry II* (B. Greene and S.-T. Yau, eds.), AMS/IP Stud. Adv. Math. **1**, AMS, Providence, RI, 1997, 71–86, alg-geom/9402002.

[BCFKvS1] V. Batyrev, I. Ciocan-Fontanine, B. Kim, and D. van Straten, *Conifold transitions and mirror symmetry for complete intersections in Grassmannians*, Nuclear Physics **B514** (1998), 640–666, alg-geom/9710022.

[BCFKvS2] V. Batyrev, I. Ciocan-Fontanine, B. Kim, and D. van Straten, *Mirror symmetry and toric degenerations of partial flag manifolds*, preprint, 1998, math.AG/9803108.

[BC] V. Batyrev and D. Cox, *On the Hodge structure of projective hypersurfaces in toric varieties*, Duke Math. J. **75** (1994), 293–338, alg-geom/9306011.

[BD] V. Batyrev and D. Dais, *Strong McKay correspondence, string-theoretic Hodge numbers and mirror symmetry*, Topology **35** (1996), 901–929, alg-geom/9410001.

[BvS] V. Batyrev and D. Van Straten, *Generalized hypergeometric functions and rational curves on Calabi-Yau complete intersections in toric varieties*, Comm. Math. Phys. **168** (1995), 493–533, alg-geom/9307010.

[BS] D. Bayer and M. Stillman, *Macaulay: a computer algebra system for commutative algebra and algebraic geometry*, Available by anonymous ftp from ftp.math.harvard.edu.

[Beauville1] A. Beauville, *Quantum cohomology of complete intersections*, Mat. Fiz. Anal. Geom. **2** (1995), 384–398, alg-geom/9501008.

[Beauville2] A. Beauville, *Counting rational curves on K3 surfaces*, preprint, 1997, alg-geom/9701019.

[Behrend] K. Behrend, *Gromov-Witten invariants in algebraic geometry*, Invent. Math. **127** (1997), 601–617, alg-geom/9601011.

[BF] K. Behrend and B. Fantechi, *The intrinsic normal cone*, Invent. Math. **128** (1997), 45–88, alg-geom/9601010.

[BM] K. Behrend and Yu. Manin, *Stacks of stable maps and Gromov-Witten invariants*, Duke J. Math. **85** (1996), 1–60, alg-geom/9506023.

[BEFFGK] K. Behrend, D. Edidin, B. Fantechi, W. Fulton, L. Göttsche, and A. Kresch. *Introduction to Stacks*, in preparation.

[BK1] P. Berglund and S. Katz, *Mirror symmetry for hypersurfaces in weighted projective space and topological couplings*, Nuclear Physics **B416** (1994), 481–538, hep-th/9311014.

[BK2] P. Berglund and S. Katz, *Mirror symmetry constructions: a review*, in Mirror Symmetry II (B. Greene and S.-T. Yau, eds.), AMS/IP Stud. Adv. Math. **1**, AMS, Providence, RI, 1997, 87–113, hep-th/9406008.

[BKK] P. Berglund, S. Katz and A. Klemm, *Mirror symmetry and the moduli space for generic hypersurfaces in toric varieties*, Nuclear Physics **B456** (1995), 153–204, hep-th/9506091.

[BCOV1] M. Bershadsky, S. Cecotti, H. Ooguri and C. Vafa, *Holomorphic anomalies in topological field theories* (with an appendix by S. Katz), Nuclear Physics **B405** (1993), 279–304, reprinted in *Mirror Symmetry II* (B. Greene and S.-T. Yau, eds.), AMS/IP Stud. Adv. Math. **1**, AMS, Providence, RI, 1997, 655–682, hep-th/9302103.

[BCOV2] M. Bershadsky, S. Cecotti, H. Ooguri and C. Vafa, *Kodaira-Spencer theory of gravity and exact results for quantum string amplitudes*, Commun. Math. Phys. **165** (1994), 311–428, hep-th/9309140.

[BJPSV] M. Bershadsky, A. Johansen, T. Pantev, V. Sadov and C. Vafa, *F-theory, Geometric Engineering and $N = 1$ Dualities*, Nuclear Physics **B505** (1997), 153–164, hep-th/9612052.

[BV] M. Bershadsky and C. Vafa, *Global Anomalies and Geometric Engineering of Critical Theories in Six Dimensions*, preprint, 1997, hep-th/9703167

[Bertram] A. Bertram, *Quantum Schubert calculus*, Adv. in Math. **28** (1997), 289–305, alg-geom/9410024.

[BCF] A. Bertram, I. Ciocan-Fontanine and W. Fulton, *Quantum multiplication with Schur polynomials*, preprint, 1997, alg-geom/9705024.

[BDW] A. Bertram, G. Daskalopoulos, and R. Wentworth, *Gromov invariants for holomorphic maps from Riemann surfaces to Grassmannians*, J. AMS **9** (1996), 529–571, alg-geom/9306005.

[BDPP] G. Bini, C. De Concini, M. Polito, and C. Procesi, *Givental's work*, preprint, 1998, math.AG/9805097.

[BD] J. Bjorken and S. Drell, *Relativistic Quantum Fields*, International Series in Pure and Applied Physics, McGraw-Hill, New York 1965.

[Bogomolov] F. Bogomolov, *Hamiltonian Kähler manifolds*, Dokl. Akad. Nauk. SSSR **243** (1978), no. 5, 1101–1104.

[Borcea1] C. Borcea, *On desingularized Horrocks-Mumford quintics*, J. Reine Angew. Math. **421** (1991), 23–41.

[Borcea2] C. Borcea, *K3 surfaces with involution and mirror pairs of Calabi-Yau manifolds*, in *Mirror Symmetry II* (B. Greene and S.-T. Yau, eds.), AMS/IP Stud. Adv. Math. **1**, AMS, Providence, RI, 1997, 717–743.

[Borisov1] L. Borisov, *Towards the mirror symmetry for Calabi-Yau complete intersections in Gorenstein Fano toric varieties*, preprint, 1993, math.AG/ 9310001.

[Borisov2] L. Borisov, *On Betti numbers and Chern classes of varieties with trivial odd cohomology groups*, preprint, 1997, alg-geom/9703023.

[BL] J. Bryan and N. C. Leung, *The enumerative geometry of K3 surfaces and modular forms*, preprint, 1997, alg-geom/9711031.

[BG] R. Bryant and P. Griffiths, *Some observations on the infinitesimal period relations for regular threefolds with trivial canonical bundle*, in *Arithmetic and Geometry Volume II* (M. Artin and J. Tate, eds.), Progress in Math **39**, Birkhäuser, Boston-Basel-Berlin, 1983.

[Cd] P. Candelas and X. de la Ossa, *Moduli Space of Calabi-Yau manifolds*, Nuclear Physics **B355** (1991), 455–481.

[CdFKM] P. Candelas, X. de la Ossa, A. Font, S. Katz and D. Morrison, *Mirror symmetry for two parameter models – I*, Nuclear Physics **B416** (1994), 481–538, reprinted in *Mirror Symmetry II* (B. Greene and S.-T. Yau, eds.), AMS/IP Stud. Adv. Math. **1**, AMS, Providence, RI, 1997, 483–543, hep-th/9308083.

[CFKM] P. Candelas, A. Font, S. Katz and D. Morrison, *Mirror symmetry for two parameter models – II*, Nuclear Physics **B429** (1994), 626–674, hep-th/9403187.

[CdGP] P. Candelas, X. de la Ossa, P. Green and L. Parkes, *A pair of Calabi-Yau manifolds as an exactly soluble superconformal field theory*, Nuclear Physics **B359** (1991), 21–74 and in *Essays on Mirror Manifolds* (S.-T. Yau, ed.), International Press, Hong Kong 1992, 31–95.

[CdK] P. Candelas, X. de la Ossa and S. Katz, *Mirror symmetry for Calabi-Yau hypersurfaces in weighted* $\mathbf{P}^4$ *and extensions of Landau-Ginzburg theory*, Nuclear Physics **B450** (1995), 267–292, hep-th/9412117.

[CDP] P. Candelas, E. Derrick and L. Parkes, *Generalized Calabi-Yau manifolds and the mirror of a rigid manifold*, Nuclear Physics **B407** (1993), 115–154, hep-th/9304045.

[Caporaso] L. Caporaso, *Counting curves on surfaces: A guide to new techniques and results*, preprint, 1996, alg-geom/9611029.

[CH1] L. Caporaso and J. Harris, *Parameter spaces for curves on surfaces and enumeration of rational curves*, Compositio Math. **113** (1998), 155–208, alg-geom/9608023.

[CH2] L. Caporaso and J. Harris, *Enumerating rational curves: the rational fibration method*, Compositio Math. **113** (1998), 209–236, alg-geom/9608024.

[CH3] L. Caporaso and J. Harris, *Counting plane curves of any genus*, Invent. Math. **131** (1998), 345–392, alg-geom/9608025.

[CCD] E. Cattani, D. Cox, A. Dickenstein, *Residues in toric varieties*, Compositio Math. **108** (1997), 35–76, alg-geom/9506024.

[CaK] E. Cattani and A. Kaplan, *Degenerating variations of Hodge structures*, in *Actes du Colloque de Théorie de Hodge, Luminy 1987*, Astérisque **179–180**, Soc. Math. France, Paris, 1989, 67–96.

[CKS1] E. Cattani, A. Kaplan, and W. Schmid, *Degeneration of Hodge structures*, Annals of Math. **123** (1986), 457–535.

[CKS2] E. Cattani, A. Kaplan, and W. Schmid, L^2 *and intersection cohomologies for a polarizable variation of Hodge structure*, Inv. Math. **87** (1987), 217–252.

[CGGK] T.-M. Chiang, B. Greene, M. Gross and Y. Kanter, *Black hole condensation and the web of Calabi-Yau manifolds*, Nuclear Physics Proc. Supp. **46** (1996), 82–95, hep-th/9511204.

[CiocanF] I. Ciocan-Fontanine, *Quantum cohomology of flag varieties*, Internat. Math. Res. Notices (1995), 263–277, math.AG/ 9505002.

[Clemens] C. H. Clemens, *Double Solids*, Advances in Math. **47** (1983) 671–689.

[CG] C. Clemens and P. Griffiths, *The intermediate Jacobian of the cubic threefold*, Annals of Math. **95** (1972), 281–356.

[ClK] H. Clemens and H. Kley, *Counting Curves which Move with Threefolds*, preprint, 1998, math.AG/9807109.

[CKM] H. Clemens, J. Kollár, and S. Mori, *Higher dimensional complex geometry*, Astérisque **166** (1988), Soc. Math. France, Paris, 1989.

[CL] E. Coddington and N. Levinson, *Theory of Ordinary Differential Equations*, McGraw-Hill, New York-Toronto-London, 1955.

[CJ] A. Collino and M. Jinzenji, *On the structure of small quantum cohomology rings for projective hypersurfaces*, preprint, 1996, hep-th/9611053.

[Cox] D. Cox, *The homogeneous coordinate ring of a toric variety*, J. Algebraic Geometry **4** (1995), 17–50, alg-geom/9210008.

[CM] B. Crauder and R. Miranda, *Quantum cohomology of rational curves*, in *The Moduli Space of Curves (Texel Island, 1994)* (R. Dijkgraaf, C. Faber and G. van der Geer, eds.), Progress in Math. **129**, Birkhäuser, Boston-Basel-Berlin, 1995, 33–80.

[Danilov] V. Danilov, *The geometry of toric varieties*, Russian Math. Surveys **33** (1978), 97–154.

[DK] V. Danilov and A. Khovanskii, *Newton polyhedra and an algorithm for computing Hodge-Deligne numbers*, Math. USSR Izvestiya **29** (1987), 279–298.

[Deligne1] P. Deligne, *Equations Différentielles à Points Singuliers Réguliers*, Lecture Notes in Mathematics **163**, Springer-Verlag, New York 1970.

[Deligne2] P. Deligne, *Local behavior of Hodge structures at infinity*, in *Mirror Symmetry II* (B. Greene and S.-T. Yau, eds.), AMS/IP Stud. Adv. Math. **1**, AMS, Providence, RI, 1997, 683–699.

[DEFJKMMW] *Quantum Fields and Strings: A Course for Mathematicians*, (P. Deligne, P. Etingof, D.S. Freed, L. Jeffrey, D. Kazhdan, J. Morgan, D.R. Morrison and E. Witten, eds.), two volumes, American Mathematical Society, to appear. Lecture notes and problem sets from the 1996–97 IAS Quantum Field Theory program are at http://www.math.ias.edu/~drm/QFT/.

[DM] P. Deligne and D. Mumford, *The irreducibility of the space of curves of a given genus*, Publ. Math. IHES **36** (1969), 75–110.

[DHSS] J. De Loera, S. Hosten, F. Santos and B. Sturmfels, *The polytope of all triangulations of a point configuration*, Documenta Mathematica **1** (1996) 103–119, http://www.mathematik.uni-bielefeld.de/documenta

[Demazure] M. Demazure, *Sous-groupes algébriques de rang maximum du groupe de Cremona*, Ann. Sci. Ecole Norm. Sup. **3** (1970), 507–588.

[DI] P. Di Francesco and C. Itzykson, *Quantum intersection rings*, in *The Moduli Space of Curves (Texel Island, 1994)* (R. Dijkgraaf, C. Faber and G. van der Geer, eds.), Progress in Math. **129**, Birkhäuser, Boston-Basel-Berlin, 1995, 81–148.

[DVV1] R. Dijkgraaf, E. Verlinde, and H. Verlinde, *Topological Strings in $d < 1$* Nuclear Physics **B352** (1991), 59–86.

[DVV2] R. Dijkgraaf, E. Verlinde, and H. Verlinde, *Notes on Topological String Theory and 2D Gravity*, in *String Theory and Quantum Gravity*, Proceedings of the Trieste Spring School 1990, (M. Green et al., eds.), World Scientific, Singapore, 1991, 91–156.

[Distler] J. Distler, *Notes on $N = 2$ σ-models*, Lecture given at 1992 Trieste Spring School, 1992, hep-th/9212062.

[Dolgachev1] I. Dolgachev, *Weighted projective varieties*, in *Group Actions and Vector Fields* (J.B. Carrell ed.), Lecture Notes in Mathematics **956**, Springer-Verlag, New York 1982, 34–71.

[Dolgachev2] I. Dolgachev, *Integral quadratic forms: Applications to algebraic geometry (d'après Nikulin)*, Séminaire Bourbaki 1982/83, Astérisque **105–106**, Soc. Math. France, Paris, 1983, 251–278.

[Dolgachev3] I. Dolgachev, *Mirror symmetry for lattice polarized K3 surfaces*, Algebraic geometry, 4, J. Math. Sci. **81** (1996), 2599–2630, alg-geom/9502005.

[Donagi] R. Donagi, *Principal bundles on elliptic fibrations*, Asian J. Math. **1** (1997), 214–223, alg-geom/9702002.

[Dubrovin1] B. Dubrovin, *Integrable systems in topological field theory*, Nuclear Physics **B379** (1992), 627–689, hep-th/9403187.

[Dubrovin2] B. Dubrovin, *The geometry of 2D topological field theories*, in *Integrable systems and quantum groups*, Lecture Notes in Mathematics **1620**, Springer-Verlag, New York, 1996, 120–348.

[Ebeling] W. Ebeling, *Strange duality, mirror symmetry, and the Leech lattice*, preprint, 1996, alg-geom/9612010.

[EG] D. Edidin and W. Graham, *Equivariant Intersection Theory*, Inv. Math. (to appear).

[EHX] T. Eguchi, K. Hori and C.-S. Xiong, *Quantum cohomology and Virasoro algebra*, Phys. Lett. **B402** (1997), 71–80, hep-th/9703086.

[EJX] T. Eguchi, M. Jinzenji, and C.-S. Xiong, *Quantum Cohomology and Free Field Representation*, Nuclear Physics **B510** (1998), 608–622, hep-th/9709152.

[EX] T. Eguchi and C.-S. Xiong, *Quantum Cohomology at Higher Genus: Topological Recursion Relations and Virasoro Conditions*, Adv. Theor. Math. Phys. **2** (1998), 219–229, hep-th/9801010.

[EJS] T. Ekedahl, T. Johnsen and D. Sommervoll, *Isolated rational curves on K3-fibered Calabi-Yau threefolds*, preprint, 1997, alg-geom/9710010.

[ES1] G. Ellingsrud and S. Strømme, *The number of twisted cubics on the general quintic threefold*, Math. Scand. **76** (1995).

[ES2] G. Ellingsrud and S. Strømme, *Bott's formula and enumerative geometry*, J. AMS **9** (1996), 175–193, alg-geom/9411005.

[EK1] L. Ernström and G. Kennedy, *Recursive formulas for the characteristic numbers of rational plane curves*, J. Algebraic Geometry **7** (1998), 141–181, alg-geom/9604019.

[EK2] L. Ernström and G. Kennedy, *Contact cohomology of the projective plane*, preprint, 1997, alg-geom/9703013.

[FGvS] B. Fantechi, L. Göttsche, and D. van Straten, *Euler number of the compactified Jacobian and multiplicity of rational curves*, preprint, 1997, alg-geom/9708012.

[Font] A. Font, *Periods and duality symmetries in Calabi-Yau compactifications*, Nuclear Physics **B391** (1993), 358–388, hep-th/9203084.

[FMW1] R. Friedman, J. Morgan and E. Witten, *Vector bundles and F theory*, Comm. Math. Phys. **187**, (1997), 679–743, hep-th/9701162.

[FMW2] R. Friedman, J. Morgan and E. Witten, *Vector bundles over elliptic fibrations*, preprint, 1997, alg-geom/9709029.

[Fukaya] K. Fukaya, *Morse homotopy, A^∞-category, and Floer homologies*, in *Proceedings of GARC Workshop on Geometry and Topology '93 (Seoul, 1993)* (H. Kim, ed.), Lecture Notes Ser., **18**, Seoul Nat. Univ., Seoul, 1993, 1–102.

[FO] K. Fukaya and K. Ono, *Arnold conjecture and Gromov-Witten invariants*, preprint, 1996.

[Fulton1] W. Fulton, *Introduction to Intersection Theory in Algebraic Geometry*, CBMS Series, Number 54, AMS, Providence, RI, 1984.

[Fulton2] W. Fulton, *Intersection Theory*, Springer-Verlag, New York-Berlin-Heidelberg, 1984.

[Fulton3] W. Fulton, *Introduction to Toric Varieties*, Princeton University Press, Princeton, 1993.

[FP] W. Fulton and R. Pandharipande, *Notes on stable maps and quantum cohomology*, in *Algebraic Geometry—Santa Cruz 1995* (Kollár, Lazarsfeld, and Morrison, eds.),

Proceedings of the 1995 Summer Research Institute in Algebraic Geometry, Proceedings of Symposia in Pure Mathematics, Volume 62 Part 2, 45–96.

[Gathmann1] A. Gathmann, *Counting rational curves with multiple points and Gromov-Witten invariants of blow-ups*, preprint, 1996, alg-geom/9609010.

[Gathmann2] A. Gathmann, *Gromov-Witten invariants of blow-ups*, preprint, 1998, math.AG/9804043.

[GKZ1] I. Gelfand, M. Kapranov and A. Zelevinsky, *Generalized Euler integrals and A-hypergeometric functions*, Adv. in Math. **84** (1990), 255–271.

[GKZ2] I. Gelfand, M. Kapranov and A. Zelevinsky, *Discriminants, Resultants and Multidimensional Determinants*, Birkhäuser, Boston-Basel-Berlin, 1994.

[Getzler1] E. Getzler, *Intersection theory on $\overline{M}_{1,4}$ and elliptic Gromov-Witten invariants*, J. AMS **10** (1997), 973–998, alg-geom/9612004.

[Getzler2] E. Getzler, *Topological Recursion Relations in Genus 2*, preprint, 1998, math.AG/9801003.

[GeP] E. Getzler and R. Pandharipande, *The number of elliptic space curves*, in preparation.

[Ginsparg] P. Ginsparg, *Applied conformal field theory*, in *Fields, Strings, and Critical Phenomena: Proceedings of the Les Houches Summer School 1988*, Elsevier Science Publishers, 1989.

[Givental1] A. Givental, *Homological geometry I, projective hypersurfaces*, Selecta Math. **1** (1995), 325–345.

[Givental2] A. Givental, *Equivariant Gromov-Witten Invariants*, Internat. Math. Res. Notices (1996), 613–663, alg-geom/9603021.

[Givental3] A. Givental, *Stationary phase integrals, quantum Toda lattices, flag manifolds and the mirror conjecture*, preprint, 1996, alg-geom/9612001.

[Givental4] A. Givental, *A mirror theorem for toric complete intersections*, preprint, 1997, alg-geom/9701016.

[Givental5] A. Givental, *Elliptic Gromov-Witten invariants and the generalized mirror conjecture*, preprint, 1998, math.AG/9803053.

[Givental6] A. Givental, *The mirror formula for quintic threefolds*, math.AG/9807070

[GK] A. Givental and B. Kim, *Quantum cohomology of flag manifolds and Toda lattices*, Comm. Math. Phys. **168** (1995), 609–641.

[GL] N. Gonciulea and V. Lakshmibai, *Degenerations of flag and Schubert varieties to toric varieties*, Transform. Groups **1** (1996), 215–248.

[Göttsche] L. Göttsche, *A conjectural generating function for numbers of curves on surfaces*, preprint, 1997, math.AG/9711012.

[GöP] L. Göttsche and R. Pandharipande, *The quantum cohomology of blow-ups of $\mathbb{P}^2$ and enumerative geometry*, J. Diff. Geom. **48** (1998), 61–90, alg-geom/9611012.

[GPa] T. Graber and R. Pandharipande, *Localization of virtual classes*, preprint, 1997, alg-geom/9708001.

[GM] A. Grassi and D. Morrison, *Automorphisms and the Kähler cone of certain Calabi-Yau manifolds*, Duke Math. J. **71** (1993), 831–838.

[GSW] M. Green, J. Schwarz, and E. Witten, *Superstring theory*, vols. 1 and 2, Cambridge University Press, Cambridge, 1987.

[GMS] B. Greene, D. Morrison, and A. Strominger, *Black hole condensation and the unification of string vacua*, Nuclear Physics **B451** (1995) 109, hep-th/9504145.

[GO] B. Greene and H. Ooguri, *Geometry and quantum field theory: A brief introduction*, in *Mirror Symmetry II* (B. Greene and S.-T. Yau, eds.), AMS/IP Stud. Adv. Math. 1, AMS, Providence, RI, 1997, 3–27.

[GPl] B. Greene and M. Plesser, *Duality in Calabi-Yau moduli space*, Nuclear Physics **B338** (1990), 15–37.

[Griffiths1] P. Griffiths, *On the periods of certain rational integrals* I, II, Ann. of Math. **90** (1969), 460–495, 498–541.

[Griffiths2] P. Griffiths (ed.), *Topics in Transcendental Algebraic Geometry*, Ann. of Math. Stud. **106**, Princeton University Press, Princeton, NJ, 1984.

[GH] P. Griffiths and J. Harris, *Principles of Algebraic Geometry*, Wiley-Interscience, New York, 1978.

[GN1] V. Gritsenko and V. Nikulin, *K3 surfaces, Lorentzian Kac-Moody algebras and mirror symmetry*, Math. Res. Lett. **3** (1996), 211–229, alg-geom/9510008.

[GN2] V. Gritsenko and V. Nikulin, *The arithmetic mirror symmetry and Calabi-Yau manifolds*, preprint, 1996, alg-geom/9612002.

[GN3] V. Gritsenko and V. Nikulin, *K3 surfaces with interesting groups of automorphisms*, preprint, 1997, alg-geom/9701011.

[Gross1] M. Gross, *Deforming Calabi-Yau threefolds*, Math. Ann. **308** (1997), 187–220, alg-geom/9506022.

[Gross2] M. Gross, *Primitive Calabi-Yau threefolds*, J. Differential Geom. **45** (1997), 288–318, alg-geom/9512002.

[Gross3] M. Gross, *Special Lagrangian fibrations I: topology*, preprint, 1997, alg-geom/9710006.

[GW] M. Gross and P. Wilson, *Mirror symmetry via 3-tori for a class of Calabi-Yau threefolds*, Math. Ann. **309** (1997), 505–531, alg-geom/9608004.

[GD] A. Grothendieck and J. Dieudonné, *Éléments de Géométrie Algébrique, IV, Quatriéme Partie*, Publ. Math. I. H. E. S. **32** (1967).

[Guillemin] V. Guillemin, *Moment Maps and Combinatorial Invariants of Hamiltonian T^n-spaces*, Progress in Math. **122**, Birkhäuser, Boston-Basel-Berlin, 1994.

[HM] J. Harris and D. Mumford, *On the Kodaira dimension of the moduli space of curves. With an appendix by W. Fulton*, Invent. Math. **67** (1982), 23–88.

[Hartshorne] R. Hartshorne, *Algebraic Geometry*, Springer-Verlag, 1977.

[Hori] K. Hori, *Constraints for Topological Strings in $D \geq 1$*, Nuclear Physics **B439** (1995), 395, hep-th/9411135.

[HKTY1] S. Hosono, A. Klemm, S. Theisen and S.-T. Yau, *Mirror symmetry, mirror map, and application to Calabi-Yau hypersurfaces*, Comm. Math. Phys. **167** (1995), 301–350, hep-th/9308122.

[HKTY2] S. Hosono, A. Klemm, S. Theisen and S.-T. Yau, *Mirror symmetry, mirror map and applications to complete intersection Calabi-Yau spaces*, Nuclear Physics **B433** (1995), 501–554, reprinted in *Mirror Symmetry II* (B. Greene and S.-T. Yau, eds.), AMS/IP Stud. Adv. Math. **1**, AMS, Providence, RI, 1997, 545–606, hep-th/9406055.

[HLY1] S. Hosono, B. Lian and S.-T. Yau, *GKZ-Generalized hypergeometric systems in mirror symmetry of Calabi-Yau hypersurfaces*, Comm. Math. Phys. **182** (1996), 535–577, alg-geom/9511001.

[HLY2] S. Hosono, B. Lian and S.-T. Yau, *Maximal degeneracy points of GKZ systems*, J. Amer. Math. Soc. **10** (1997) 427–443, alg-geom/9603014.

[HLY3] S. Hosono, B. Lian and S.-T. Yau, *Calabi-Yau varieties and pencils of K3 surfaces*, preprint, 1996, alg-geom/9603020.

[HSS] S. Hosono, M.-H. Saito and J. Stienstra, *On mirror symmetry conjecture for Schoen's Calabi-Yau 3 folds*, preprint, 1997, alg-geom/9709027.

[Hübsch] T. Hübsch, *Calabi-Yau Manifolds: A Bestiary for Physicists*, World Scientific, Singapore, 1992.

[Illusie] L. Illusie, *Complexe Cotangent et Déformations I, II*, Lecture Notes in Mathematics **239, 283**, Springer-Verlag, New York-Berlin-Heidelberg, 1971, 1972.

[Ionel] E. Ionel, *Genus 1 enumerative invariants in $\mathbb{P}^n$ with fixed j invariant*, Duke Math. J. **94** (1998), 279–324, alg-geom/9608030.

[IP] E. Ionel and T. Parker, *The Gromov Invariants of Ruan-Tian and Taubes*, Math. Res. Lett. **4** (1997), 521–532, alg-geom/9702008.

[Iversen] B. Iversen, *A fixed point formula for action of tori on algebraic varieties*, Inv. Math. **16** (1972) 229–236.

[JK1] T. Johnsen and S. Kleiman, *Rational curves of degree at most 9 on a general quintic threefold*, Comm. Algebra **24** (1996), 2721–2753, alg-geom/9510015.

[JK2] T. Johnsen and S. Kleiman, *Toward Clemens' conjecture in degrees between 10 and 24*, preprint, 1996, alg-geom/9601024.

[KK] A. Kabanov and T. Kimura, *Intersection numbers and rank one cohomological field theories in genus one*, Comm. Math. Phys. **194** (1998), 651–674, alg-geom/9706003.

[KSZ1] M. Kapranov, B. Sturmfels, and A. Zelevinsky, *Quotients of toric varieties*, Math. Ann. **290** (1991) 643–655.

[KSZ2] M. Kapranov, B. Sturmfels, and A. Zelevinsky, *Chow polytopes and general resultants*, Duke Math. J. **67** (1992), 189–218.

[Katz1] S. Katz,*Degenerations of quintic threefolds and their lines*. Duke Math. J. **50** (1983) 1127–1135.

[Katz2] S. Katz, *On the finiteness of rational curves on quintic threefolds*, Compositio Math. **60** (1986), 151–162.

[Katz3] S. Katz, *Gromov-Witten Invariants via algebraic geometry*, Nuclear Physics Proc. Supp. **46** (1996), 108–115, hep-th/9510218.

[KKV] S. Katz, A. Klemm, and C. Vafa, *Geometric Engineering of Quantum Field Theories*, Nuc. Phys. **B497** (1997) 173–195, hep-th/9609239.

[KMV] S. Katz, P. Mayr, and C. Vafa, *Mirror symmetry and Exact Solution of 4D $N = 2$ Gauge Theories, I*, Adv. Theor. Math. Phys. **1** (1998), 53–114, hep-th/9706110.

[KMP] S. Katz, D. Morrison and M. Plesser, *Enhanced gauge symmetry in type II string theory*, Nuclear Physics **B477** (1996), 105–140, hep-th/9601108.

[KQR] S. Katz, Z. Qin and Y. Ruan, *Enumeration of nodal genus-2 plane curves with fixed complex structure*, J. Alg. Geom. **7** (1998), 569–587, alg-geom/960601.

[KS] S. Katz and S. Strømme, SCHUBERT, a Maple package for intersection theory, http://www.math.okstate.edu/~katz/schubert.html.

[KV] S. Katz and C. Vafa, *Geometric Engineering of $N=1$ Quantum Field Theories*, Nucl. Phys. **B497** (1997) 196–204, hep-th/9611090.

[Kaufmann] R. Kaufmann, *The intersection form in $H^*(\overline{M}_{0n})$ and the explicit Künneth formula in quantum cohomology*, Internat. Math. Res. Notices (1996), 929–952, alg-geom/9608026.

[Kawamata1] Y. Kawamata, *Crepant blowing-up of 3-dimensional canonical singularities and its application to degenerations of surfaces*, Ann. Math. **127** (1988) 93–163.

[Kawamata2] Y. Kawamata, *On the cone of divisors of Calabi-Yau fiber spaces*, Internat. J. Math. **8** (1997), 665–687, alg-geom/9701006.

[Keel] S. Keel, *Intersection theory on moduli spaces of stable n-pointed curves of genus zero*, Trans. AMS **330** (1992), 545–574.

[Kim1] B. Kim, *Quantum cohomology of flag manifolds G/B and quantum Toda lattices*, preprint, 1996, alg-geom/9607001.

[Kim2] B. Kim, *Quantum Hyperplane Section Theorem for Homogeneous Spaces*, preprint, 1998, alg-geom/9712008.

[KlM] A. Klemm and P. Mayr, *Strong coupling singularities and non-abelian gauge symmetries in $N = 2$ string theory*, Nuclear Physics **B469** (1996) 37, hep-th/9601014.

[KS] A. Klemm and R. Schimmrigk, *Landau-Ginzburg string vacua*, Nuclear Physics **B411** (1994) 559–583, hep-th/9204060.

[KT1] A. Klemm and S. Theisen, *Considerations of one modulus Calabi-Yau compactifications: Picard-Fuchs equations, Kähler potentials and mirror maps*, Nuclear Physics **B389** (1993), 153–180.

[KT2] A. Klemm and S. Theisen, *Mirror maps and instanton sums for complete intersections in weighted projective space*, Mod. Phys. Lett. **A9** (1994), 1807–1818, hep-th/9304034.

[Knudsen] F. Knudsen, *Projectivity of the moduli space of stable curves, II*, Math. Scand. **52** (1983), 1225–1265.

[Kobayashi] M. Kobayashi, *Duality of weights, mirror symmetry and Arnold's strange duality*, preprint, 1995, alg-geom/9502004.

[Kollár] J. Kollár, *Flops*, Nagoya Math. J. **113** (1989) 14–36.

[Kontsevich1] M. Kontsevich, *Intersection theory on the moduli space of curves and the matrix Airy function*, Comm. Math. Phys. **147** (1992), 1–23.

[Kontsevich2] M. Kontsevich, *Enumeration of rational curves via torus actions*, in *The Moduli Space of Curves (Texel Island, 1994)* (R. Dijkgraaf, C. Faber and G. van der Geer, eds.), Progress in Math. **129**, Birkhäuser, Boston-Basel-Berlin, 1995, 335–368, hep-th/9405035.

[Kontsevich3] M. Kontsevich, *Homological algebra of mirror symmetry*, in *Proceedings of the International Congress of Mathematicians, 1994, Zürich*, Birkhäuser, Boston-Basel-Berlin, 1995, 120–139, alg-geom/9411018.

[KoM1] M. Kontsevich and Yu. Manin, *Gromov-Witten classes, quantum cohomology, and enumerative geometry*, Comm. Math. Phys. **164** (1994) 525–562, reprinted in *Mirror Symmetry II* (B. Greene and S.-T. Yau, eds.), AMS/IP Stud. Adv. Math. **1**, AMS, Providence, RI, 1997, 607–653, hep-th/9402147.

[KoM2] M. Kontsevich and Yu. Manin, *Quantum cohomology of a product (with Appendix by R. Kaufmann)*, Invent. Math. **124** (1996), 313–339.

[KoM3] M. Kontsevich and Yu. Manin, *Relations between the correlators of the topological sigma-model coupled to gravity*, preprint, 1997, alg-geom/9708024.

[Kostant] B. Kostant, *Flag manifold quantum cohomology, the Toda lattice, and the representation of highest weight ρ*, Selecta Math. New Series **2** (1996), 43–91.

[Kresch] A. Kresch, *Associativity relations in quantum cohomology*, preprint, 1997, alg-geom/9703015.

[KS2] M. Kreuzer and H. Skarke, *No mirror symmetry in Landau-Ginzburg Spectra!*, Nuclear Physics **B388** (1993), 113–130, hep-th/9205004.

[KS3] M. Kreuzer and H. Skarke, *On the classification of reflexive polyhedra*, Comm. Math. Phys. **185** (1997), 495–508, hep-th/9512204.

[Kuranishi] M. Kuranishi, *On the locally complete families of complex analytic structures*, Ann. Math. **75** (1962), 536–577.

[Landman] A. Landman, *On the Picard-Lefschetz transformations*, Trans. Amer. Math. Soc. **181** (1973), 89–126.

[LM] H. Lawson, Jr. and M.-L. Michelsohn, *Spin Geometry*, Princeton University Press, Princeton, NJ, 1989.

[LVW] W. Lerche, C. Vafa, and N. Warner, Nuclear Physics **B324** (1989) 427.

[LTi1] J. Li and G. Tian, *The quantum cohomology of homogeneous varieties*, J. Alg. Geom. **6** (1997), 269–305, alg-geom/9504009.

[LTi2] J. Li and G. Tian, *Virtual moduli cycles and Gromov-Witten invariants of algebraic varieties*, J. AMS **11** (1998), 119–174, alg-geom/9602007.

[LTi3] J. Li and G. Tian, *Virtual moduli cycles and Gromov-Witten invariants of general symplectic manifolds*, preprint, 1996, alg-geom/9608032.

[LTi4] J. Li and G. Tian, *Comparison of the algebraic and symplectic definitions of GW invariants*, preprint, 1997, alg-geom/9712035.

[LLY] B. Lian, K. Liu and S.-T. Yau, *Mirror principle I*, Asian J. Math. **1** (1997), 729–763, alg-geom/9712011.

[LY1] B. Lian and S.-T. Yau, *Arithmetic properties of mirror maps and quantum couplings*, Comm. Math. Phys. **176** (1996), 163–191, hep-th/9411234.

[LY2] B. Lian and S.-T. Yau, *Mirror maps, modular relations and hypergeometric series I*, preprint, 1995, hep-th/9507151.

[LTe] A. Libgober and J. Teitelbaum, *Lines on Calabi-Yau complete intersections, mirror symmetry, and Picard-Fuchs equations*, Int. Math. Res. Not. **1** (1993) 29–39, alg-geom/9301001.

[LiuT] X. Liu and G. Tian, *Virasoro Constraints for Quantum Cohomology*, preprint, 1998, math.AG/9806028.

[Looijenga] E. Looijenga, *New compactifications of locally symmetric varieties*, in *Proceedings of the 1984 Vancouver Conference in Algebraic Geometry* (J. Carrell et al, eds.), CMS Conference Proceedings **6**, AMS, Providence, RI, 1986, 341–364.

[Manin1] Yu. Manin, *Gauge Field Theory and Complex Geometry*, Grund. Math. Wiss. **289**, Springer-Verlag, Berlin 1988.

[Manin2] Yu. Manin, *Generating functions in algebraic geometry and sums over trees*, in *The Moduli Space of Curves (Texel Island, 1994)* (R. Dijkgraaf, C. Faber and G. van der Geer, eds.), Progr. Math. **129**, Birkhäuser, Boston-Basel-Berlin, 1995.

[Manin3] Yu. Manin, *Frobenius manifolds, quantum cohomology, and moduli spaces (Chapters I, II, III)*, MPI preprint 96-113, Max-Planck-Institut für Mathematik, Bonn, 1996.

[Mayr] P. Mayr, *Geometric construction of $N = 2$ gauge theories*, preprint, 1998, hep-th/9807096.

[MS] D. McDuff and D. Salamon, *J-holomorphic Curves and Quantum Cohomology*, University Lecture Series **6**, AMS, Providence, RI, 1994.

[Meurer] P. Meurer, *The number of rational quartics on Calabi-Yau hypersurfaces in weighted projective space* $\mathbf{P}(2, 1^4)$, Math. Scand. **78** (1996), 63–83, alg-geom/9409001.

[MR] T. Mora and L. Robbiano, *The Gröbner fan of an ideal*, J. Symbolic Comput. **6** (1988), 183–208.

[Morrison1] D. Morrison, *Picard-Fuchs equations and mirror maps for hypersurfaces*, in *Essays on Mirror Manifolds* (S.-T. Yau, ed.), International Press, Hong Kong 1992, alg-geom/9202026.

[Morrison2] D. Morrison, *Mirror symmetry and rational curves on quintic threefolds: A guide for mathematicians*, Jour. AMS **6** (1993) 223–247, alg-geom/9202004.

[Morrison3] D. Morrison, *Compactifications of Moduli Spaces Inspired by Mirror Symmetry*, in *Journées de Geómétrie Algébrique d'Orsay (Orsay, 1992)*, Astérisque **218**, Soc. Math. France, Paris, 1993, 243–271, alg-geom/9304007.

[Morrison4] D. Morrison, *Where is the large radius limit?*, preprint, 1993, hep-th/9311049.

[Morrison5] D. Morrison, *Beyond the Kähler cone*, in *Proceedings of the Hirzebruch 65 Conference on Algebraic Geometry (Ramat Gan, 1993)*, Israel Math. Conf. Proc. **9**, Bar-Ilan Univ., Ramat Gan, 1996, 361–376, alg-geom/9407007.

[Morrison6] D. Morrison, *Making enumerative predictions by means of mirror symmetry*, in *Mirror Symmetry II* (B. Greene and S.-T. Yau, eds.), AMS/IP Stud. Adv. Math. **1**, AMS, Providence, RI, 1997, 457–481, alg-geom/9504013.

[Morrison7] D. Morrison, *Mathematical Aspects of Mirror Symmetry*, in *Complex algebraic geometry (Park City, UT, 1993)*, 265–327, IAS/Park City Math. Ser. 3, AMS, Providence, RI, 1997, alg-geom/9609021.

[Morrison8] D. Morrison, *The geometry underlying mirror symmetry*, preprint, 1996, alg-geom/9608006.

[Morrison9] D. Morrison, *Through the Looking Glass*, alg-geom/9705028.

[MP1] D. Morrison, M. Plesser, *Summing the instantons: quantum cohomology and mirror symmetry in toric varieties*, Nuclear Physics **B440** (1995), 279–354, hep-th/9412236.

[MP2] D. Morrison, M. Plesser. *Towards mirror symmetry as duality for two-dimensional abelian gauge theories*, Nuclear Physics Proc. Suppl. **46** (1996) 177–186, hep-th/9508107.

[Mumford1] D. Mumford, *Geometric Invariant Theory*, Springer-Verlag, Berlin-Heidelberg-New York, 1965.

[Mumford2] D. Mumford, *Towards an Enumerative Geometry of the Moduli Space of Curves*, in *Arithmetic and Geometry, Volume II* (M. Artin and J. Tate, eds.), Progress in Math. **36**, Birkhäuser, Boston-Basel-Berlin, 1983, 271–328.

[Nikulin] V. Nikulin, *Discrete Reflection groups in Lobachevsky spaces and algebraic surfaces*, in *Proceedings of the International Congress of Mathematicians, Berkeley 1986*, AMS, Providence, RI, 1987, 654–671.

[Oda] T. Oda, *Convex Bodies and Algebraic Geometry*, Springer-Verlag, Berlin-Heidelberg-New York, 1988.

[OdP] T. Oda and H. Park, *Linear Gale transforms and Gelfand-Kapranov-Zelevinskij decompositions*, Tôhoku Math. J. **43** (1991), 375–399.

[Oguiso] K. Oguiso, *On algebraic fiber space structures on a Calabi-Yau threefold*, Int. J. Math. **4** (1993), 439–465.

[OgP] K. Oguiso and T. Peternell, *Calabi-Yau threefolds with positive second Chern class*, Comm. Anal. Geom. **6** (1998), 153–172.

[Pandharipande1] R. Pandharipande, *Counting elliptic plane curves with fixed j-invariant*, Proc. AMS **125** (1997), 3471–3479, alg-geom/9505023.

[Pandharipande2] R. Pandharipande, *A geometric construction of Getzler's elliptic relation*, preprint, 1997, alg-geom/9705016.

[Pandharipande3] R. Pandharipande, *Rational curves on hypersurfaces (after A. Givental)*, preprint, 1998, to appear in Séminaire Bourbaki, math.AG/9806133.

[PS] C. Peters and J. Steenbrink, *Infinitesimal variations of Hodge structure and the generic Torelli problem for projective hypersurfaces (after Carlson, Donagi, Green, Griffiths, Harris)*, in *Classification of Algebraic and Analytic Manifolds* (K. Ueno, ed.), Progress in Math. **39**, Birkhäuser, Boston-Basel-Berlin, 1983, 399–463.

[Polchinski1] J. Polchinski, *Lectures on D-Branes*, in *Fields, Strings and Duality (TASI 96, Boulder, CO)* (C. Efthimiou and B. Greene, eds.), World Scientific, River Edge, NJ, 1997, 293–356, hep-th/9611050.

[Polchinski2] J. Polchinski, *String theory*, vols. 1 and 2, Cambridge University Press, Cambridge, 1998.

[PZ] A. Polishchuk and E. Zaslow, *Categorical mirror symmetry: the elliptic curve*, Adv. Theor. Math. Phys. **2** (1998), 443–470, math.AG/9801119.

[Prill] D. Prill, *Local classification of quotients of complex manifolds by discontinuous groups*, Duke Math. J. **34** (1967), 375–386.

[QR] Z. Qin and Y. Ruan, *Quantum cohomology of projective bundles over $\mathbb{P}^n$*, Trans. AMS **350** (1998), 3615–3638, alg-geom/9510010.

[Quinn] F. Quinn, *Lectures on axiomatic topological quantum field theory*, in *Geometry and Quantum Field Theory* (D. Freed and K. Uhlenbeck, eds.), IAS/Park City Mathematics Series **1**, AMS, 1995, 325–459.

[Rabin] J. Rabin, *Introduction to quantum field theory for mathematicians*, in *Geometry and Quantum Field Theory* (D. Freed and K. Uhlenbeck, eds.), IAS/Park City Mathematics Series **1**, AMS, 1995, 185–269.

[Ran1] Z. Ran, *Bend, break and count*, preprint, 1997, alg-geom/9704004.

[Ran2] Z. Ran, *Unobstructedness of Calabi-Yau orbi-Kleinfolds*, J. Math. Phys. **39** (1998), 625–629.

[Ran3] Z. Ran, *Bend, break and count II: elliptics, cuspidals, linear genera*, preprint, 1997, alg-geom/9708013.

[Reid1] M. Reid, *Canonical 3-folds*, in *Journées de géométrie algébrique d'Angers* (A. Beauville, ed.), Sijthoff & Noordhoff, Alphen aan den Rijn, 1980, 273–310.

[Reid2] M. Reid, *Minimal models of canonical threefolds*, in *Algebraic Varieties and Analytic Varieties*, Adv. Stud. Pure Math. **1**, North-Holland, Amsterdam-New York and Kinokuniya, Tokyo, 1983, 131–180.

[Reid3] M. Reid, *Decomposition of toric morphisms*, in *Arithmetic and Geometry, Volume II* (M. Artin and J. Tate, eds.), Progress in Math. **36**, Birkhäuser, Boston-Basel-Berlin, 1983, 395–418.

[Reid4] M. Reid, *Young person's guide to canonical singularities*, in *Algebraic Geometry, Bowdoin 1985, Part 1* (S. Bloch, ed.), Proc. Sym. Pure Math. **46**, AMS, 1987, 345–414.

[Reid5] M. Reid, *The moduli space of 3-folds with $K = 0$ may nevertheless by irreducible*, Math. Ann. **278** (1987), 329–334.

[Reid6] M. Reid, *McKay correspondence*, preprint, 1997, alg-geom/9702016.

[Roan1] S.-S. Roan, *The mirror of Calabi-Yau orbifolds*, Int. J. Math. **2** (1991) 439–455.

[Roan2] S.-S. Roan, *Orbifold Euler characteristic*, in Mirror Symmetry II (B. Greene and S.-T. Yau, eds.), AMS/IP Stud. Adv. Math. **1**, AMS, Providence, RI, 1997, 129–140.

[Ruan1] Y. Ruan, *Symplectic topology and extremal rays*, Geom. Funct. Anal. **3** (1993), 395–430.

[Ruan2] Y. Ruan, *Topological sigma model and Donaldson type invariants in Gromov theory*, Duke Math. J. **83** (1996), 461–500.

[Ruan3] Y. Ruan, *Virtual neighborhoods and pseudo-holomorphic curves*, preprint, 1996, alg-geom/9611021.

[RT1] Y. Ruan and G. Tian, *A mathematical theory of quantum cohomology*, J. Diff. Geom. **42** (1995), 259–367.

[RT2] Y. Ruan and G. Tian, *Higher genus symplectic invariants and sigma model coupled with gravity*, Inv. Math. **130** (1997), 455–516, alg-geom/9601005.

[Saito] M. Saito, *Mixed Hodge modules*, Publ. Res. Inst. Math. Sci. (Kyoto Univ.) **26** (1990), 221–333.

[Satake] I. Satake, *On a generalization of the notion of a manifold*, Proc. Nat. Acad. Sci. U.S.A. **42** (1956), 359–363.

[Schimmrigk] R. Schimmrigk, *Critical superstring vacua from noncritical manifolds: a novel framework for string compactifications and mirror symmetry*, Phys. Review Letters **70** (1993), 3688–3691, hep-th/9210062.

[Schlessinger] M. Schlessinger, *Rigidity of quotient singularities*, Invent. Math. **14** (1971), 17–26.

[Schmid] W. Schmid, *Variation of Hodge structure: the singularities of the period mapping*, Inv. Math. **22** (1973), 211–319.

[Siebert1] B. Siebert, *Gromov-Witten invariants for general symplectic manifolds*, preprint, 1996, dg-ga/9608005.

[Siebert2] B. Siebert, *Algebraic and symplectic Gromov-Witten invariants coincide*, preprint, 1998, math.AG/9804108.

[ST] B. Siebert and G. Tian, *On quantum cohomology rings of Fano manifolds and a formula of Vafa and Intriligator*, Asian J. Math. **1** (1997), 679–695, alg-geom/9403010.

[Skarke] H. Skarke, *Weight systems for toric Calabi-Yau varieties and reflexivity of Newton polyhedra*, Mod. Phys. Lett. **A11** (1996), 1637–1652, alg-geom/9603007.

[Steenbrink1] J. Steenbrink, *Intersection form for quasi-homogeneous singularities*, Compositio Math. **34** (1977), 211–223.

[Steenbrink2] J. Steenbrink, *Mixed Hodge structure on the vanishing cohomology*, in Real and Complex Singularities: Proceedings Nordic Summer School, Oslo, 1976, Sijthoff & Noordhoff, Alphen aan den Rijn, 1977, 525–563.

[SYZ] A. Strominger, S.-T. Yau and E. Zaslow, *Mirror symmetry is T-duality*, Nuclear Physics **B479** (1996) 243–259, hep-th/9606040.

[Sturmfels1] B. Sturmfels, *Gröbner bases of toric varieties*, Tôhoku Math. J. **43** (1991), 249–261.

[Sturmfels2] B. Sturmfels, *Gröbner Bases and Convex Polytopes*, AMS, Providence, 1996.

[Szendroi] B. Szendroi, *On the period map of Calabi-Yau Manifolds*, preprint, 1997, alg-geom/9708011.

[Taubes] C. Taubes, *Counting pseudo-holomorphic submanifolds in dimension 4*, J. Diff. Geom. **44** (1996), 818–893.

[Tian] G. Tian, *Smoothness of the universal deformation space of compact Calabi-Yau manifolds and its Peterson-Weil metric*, in Mathematical Aspects of String Theory (S.-T. Yau, ed.), World Scientific, Singapore, 1987, 629–646.

[TX] G. Tian and G. Xu, *On the semi-simplicity of the quantum cohomology algebras of complete intersections*, Math. Res. Lett. **4** (1997), 481–488, alg-geom/9611035.

[Tjøtta] E. Tjøtta, *Rational curves on the space of determinantal nets of conics*, preprint, 1998, math.AG/9802037.

[Todorov] A. Todorov, *The Weil-Petersson geometry of the moduli space of $SU(n \geq 3)$ (Calabi-Yau) manifolds. I*, Comm. Math. Phys. **126** (1989), 325–346.

[Vafa1] C. Vafa, *String vacua and orbifoldized LG models*, Modern Phys. Lett. **A4** (1989) 1169–1185.

[Vafa2] C. Vafa, *Topological Landau-Ginzburg models*, Modern Phys. Lett. **A6** (1991), 337–346.

[Vafa3] C. Vafa, *Evidence for F-theory*, Nuclear Physics **B469** (1996), 403–418, hep-th/9602022.

[Vainsencher] I. Vainsencher, *Enumeration of n-fold tangent hyperplanes to a surface*, J. Algebraic Geom. **4**, 1995, 503–526, alg-geom/9312012.

[Vakil1] R. Vakil, *Counting curves of any genus on rational ruled surfaces*, preprint, 1997, alg-geom/9709003.

[Vakil2] R. Vakil, *Genus g Gromov-Witten invariants of Del Pezzo surfaces: Counting plane curves with fixed multiple points*, preprint, 1997, math.AG/ 9709004.

[Vakil3] R. Vakil, *The enumerative geometry of rational and elliptic curves in projective space*, preprint, 1997, alg-geom/9709007.

[Verbitsky] M. Verbitsky, *Mirror symmetry for hyperkaehler manifolds*, preprint, 1995, hep-th/9512195.

[Viehweg] E. Viehweg, *Weak positivity and the stability of certain Hilbert points, III*, Inv. Math. **101** (1990), 521–543.

[Vistoli] A. Vistoli, *Intersection theory on algebraic stacks and their moduli spaces*, Invent. Math. **97**, 613–670, 1989.

[Voisin1] C. Voisin, *Miroirs et involutions sur les surfaces K3*, in Journées de Geómétrie Algébrique d'Orsay (Orsay, 1992), Astérisque **218**, Soc. Math. France, Paris, 1993, 273–323.

[Voisin2] C. Voisin, *A mathematical proof of a formula of Aspinwall and Morrison*, Compositio Math. **104** (1996), 135–151.

[Voisin3] C. Voisin, *Symétrie Miroir*, Panoramas et Synthèses **2**, Soc. Math. France, Paris, 1996.

[WB] J. Wess and J. Bagger, *Supersymmetry and Supergravity*, Princeton U. Press, Princeton, NJ, 1992.

[WW] E. Whittaker and G. Watson, *A Course of Modern Analysis*, Fourth Edition, Cambridge U. Press, Cambridge, 1963.

[Wilson1] P. Wilson, *Calabi-Yau manifolds with large Picard Number*, Inv. Math. **98** (1989) 139–155.

[Wilson2] P. Wilson, *The Kähler cone on Calabi-Yau threefolds*, Inv. Math. **107** (1992), 561–583.

[Wilson3] P. Wilson, *Symplectic Deformations of Calabi-Yau threefolds*, J. Diff. Geom. **45** (1997), 611–637, alg-geom/9707007.

[Wilson4] P. Wilson, *Flops, Type III contractions and Gromov-Witten invariants on Calabi-Yau threefolds*, preprint, 1997, alg-geom/9707008.

[Witten1] E. Witten, *Topological Sigma Models*, Comm. Math. Phys. **118** (1988), 411–449.

[Witten2] E. Witten, *On the structure of the topological phase of two-dimensional gravity*, Nuclear Physics **B340** (1990), 281–332.

[Witten3] E. Witten, *Two dimensional gravity and intersection theory on moduli space*, in Surveys in Differential Geometry, Proceedings of the Conference on Geometry and Topology held at Harvard University, Cambridge, MA, April 27–29, 1990, Supplement to the Journal of Differential Geometry, no. 1, Lehigh University, Bethlehem, PA; distributed by the AMS, Providence, RI, 1991, 243–310.

[Witten4] E. Witten, *Mirror Manifolds and Topological Field Theory*, in *Essays on Mirror Manifolds* (S.-T. Yau, ed.), International Press, Hong Kong 1992, 120–159.

[Witten5] E. Witten, *Phases of $N = 2$ theories in two dimensions*, Nuclear Physics **B403** (1993), 159–222, reprinted in *Mirror Symmetry II* (B. Greene and S.-T. Yau, eds.), AMS/IP Stud. Adv. Math. **1**, AMS, Providence, RI, 1997, 143–211, hep-th/9301042.

[Witten6] E. Witten, *The Verlinde algebra and the cohomology of the Grassmannian*, in *Geometry, Topology, & Physics*, Conf. Proc. Lecture Notes Geom. Topology, VI, Internat. Press, Cambridge, MA, 1995, 357–422, hep-th/9312104.

[YZ] S.-T. Yau and E. Zaslow, *BPS states, string duality, and nodal curves on K3*, Nuclear Physics **B471** (1996), 503–512, hep-th/9512121.

Index

Selected Titles in This Series

(*Continued from the front of this publication*)